W9-DIJ-617

NUMERICAL METHODS IN GEOTECHNICAL ENGINEERING

McGRAW-HILL SERIES IN MODERN STRUCTURES:
Systems and Management
THOMAS C. KAVANAGH *Consulting Editor*

BAKER, KOVALEVSKY, AND RISH *Structural Analysis of Shells*
COOMBS AND PALMER *Construction Accounting and Financial Management*
DESAI AND CHRISTIAN *Numerical Methods in Geotechnical Engineering*
FORSYTH *Unified Design of Reinforced Concrete Members*
FOSTER *Construction Estimates from Take-Off to Bid*
GILL *Systems Management Techniques for Builders and Contractors*
JOHNSON *Deterioration, Maintenance and Repair of Structures*
JOHNSON AND KAVANAGH *The Design of Foundations for Buildings*
KOMENDANT *Contemporary Concrete Structures*
KREIDER AND KREITH *Solar Heating and Cooling*
O'BRIEN *Value Analysis in Design and Construction*
OPPENHEIMER *Directing Construction for Profit*
PARKER *Planning and Estimating Dam Construction*
PARKER *Planning and Estimating Underground Construction*
PULVER *Constructions Estimates and Costs*
RAMASWAMY *Design and Construction of Concrete Shell Roofs*
SCHWARTZ *Civil Engineering for the Plant Engineer*
TSCHEBOTARIOFF *Foundations, Retaining and Earth Structures*
TSYTOVICH *The Mechanics of Frozen Ground*
WOODWARD, GARDNER, AND GREER *Drilled Pier Foundations*
YANG *Design of Functional Pavements*
YU *Cold-formed Steel Structures*

NUMERICAL METHODS IN GEOTECHNICAL ENGINEERING

Edited by

CHANDRAKANT S. DESAI

Professor of Civil Engineering
Virginia Polytechnic Institute and State University

JOHN T. CHRISTIAN

Consulting Engineer
Stone & Webster Engineering Corporation

McGRAW-HILL BOOK COMPANY
New York St. Louis San Francisco Auckland Bogotá Düsseldorf
Johannesburg London Madrid Mexico Montreal New Delhi Panama
Paris São Paulo Singapore Sydney Tokyo Toronto

NUMERICAL METHODS IN GEOTECHNICAL ENGINEERING

1 2 3 4 5 6 7 8 9 0 F G F G 7 8 3 2 1 0 9 8 7

This book was set in Times New Roman.
The editors were Albrecht von Hagen,
Jeremy Robinson, and Madelaine Eichberg;
the cover was designed by Pencils Portfolio, Inc.;
the production supervisor was Milton J. Heiberg.
Kingsport Press, Inc., was printer and binder.

Library of Congress Cataloging in Publication Data
Main entry under title:

Numerical methods in geotechnical engineering.

(McGraw-Hill series in modern structures)
Includes index.
1. Engineering geology—Mathematics.
I. Desai, Chandrakant S. II. Christian, John T.
TA705.N85 624'.151'01515 76-45761
ISBN 0-07-016542-4

To
Patricia
and Lynda

CONTENTS

LIST OF CONTRIBUTORS

Arya, Santosh K., Senior Development Engineer, Department of Applied Mechanics and Engineering Science, University of California at San Diego, La Jolla, California

Booker, J. R., Senior Lecturer in Civil Engineering, University of Sydney, Sydney, Australia

Cheung, Y. K., Professor of Civil Engineering, University of Adelaide, Adelaide, Australia

Christian, John T., Consulting Engineer, Stone and Webster Engineering Corporation, Boston, Massachusetts

Clough, G. W., Associate Professor of Civil Engineering, Stanford University, Stanford, California

Coyle, H. M., Professor of Civil Engineering, Texas A & M University, College Station, Texas

Davis, E. H., Professor of Civil Engineering (Soil Mechanics), University of Sydney, Sydney, Australia

Desai, Chandrakant S., Professor of Civil Engineering, Virginia Polytechnic Institute and State University, Blacksburg, Virginia

Goodman, Richard E., Professor of Civil Engineering, University of California at Berkeley, Berkeley, California

Hirsch, T. J., Professor of Civil Engineering, Texas A & M University, College Station, Texas

Humpheson, C., Research Associate, Department of Civil Engineering, University of Wales, Swansea, U.K. (currently with McClelland Engineers, London)

Kulhawy, Fred H., Associate Professor of Civil Engineering, Cornell University, Ithaca, New York

Lowery, L. L., Associate Professor of Civil Engineering, Texas A & M University, College Station, Texas

Lytton, Robert L., Professor of Civil Engineering, Texas A & M University, College Station, Texas

Poulos, Harry G., Reader in Civil Engineering, University of Sydney, Sydney, Australia

Reese, Lymon C., T. U. Taylor Professor of Civil Engineering, University of Texas at Austin, Austin, Texas

Roësset, José M., Professor of Civil Engineering, Massachusetts Institute of Technology, Cambridge, Massachusetts

Schiffman, Robert L., Professor of Civil Engineering and Associate Director of Computing Center, University of Colorado, Boulder, Colorado

Smith, I. M., Senior Lecturer, Simon Engineering Laboratory, University of Manchester, Manchester, U.K.

St. John, Christopher, Assistant Professor of Civil and Mineral Engineering, University of Minnesota, Minneapolis, Minnesota

Tsui, Y., Assistant Professor of Civil Engineering, Duke University, Durham, North Carolina

Wittke, Walter, Professor and Director, Institut für Grundbau, Bodenmechanik, Felsmechanik und Verkehrswasserbau, Technische Hochschule Aachen, Aachen, West Germany

Zienkiewicz, O. C., Professor of Civil Engineering, University of Wales, Swansea, U.K.

PREFACE

In the last two decades there has been a great expansion in the power and availability of numerical procedures, e.g., those based on finite elements, finite differences, integral equations, and the method of characteristics. This has opened a new era in analysis and design for a wide range of problems in geotechnical engineering, and it has permitted the incorporation into analysis and design of complex factors (nonlinear behavior, dynamic loads, discontinuities, and non-homogeneities) that cannot be readily handled by the closed-form and empirical procedures used previously. The large, high speed computer has been essential to the phenomenal growth of numerical methods because it has made them efficient and economical. Numerical procedures are now accepted and employed in practice and are included in the research and teaching programs at most colleges and universities. However, there is no unified treatment of the theory and applications available at this time. The basic aim of this book is to present such a unified treatment of numerical methods in the area of geotechnical engineering. It combines the experience and knowledge of a number of recognized workers from different backgrounds and countries.

The first four chapters contain fundamental theory and procedures for formulation of various numerical methods, constitutive laws for geologic materials, and analysis of discontinuous media. The remaining seventeen chapters describe applications to a wide range of specific topics in geotechnical engineering. Each chapter, written by a recognized worker or workers in that particular field, covers the theory and special features relevant to the particular topic, the applications from the viewpoint of design and analysis, and the numerical properties of the procedures as aids to the user.

Adequate knowledge of behavior of geologic materials is vital for realistic answers from any numerical technique. A comprehensive review of the usefulness and limitations of constitutive laws used in the past is presented. Also included are comments and ideas on improvements and new developments in the stress-strain laws. Use and interpretation of data from numerical procedures require care and judgment. Often, a numerical scheme may be valid for one set of problems and deficient for another set in the same class; such numerical characteristics as convergence and stability can provide good indications of the usefulness of a procedure for reliable results. Wherever available, these properties have been covered in sufficient detail to aid the user in selecting the most suitable procedure and in adopting the optimum spatial and temporal meshes.

One of the main aims of the book is to describe and to suggest the potential of numerical procedures for applications. Each chapter contains typical examples of practical problems and correlations between numerical solutions and field or laboratory observations. In some instances, guides for design use and charts have been included.

The book has been planned so that it can be used as a text, as a tool for the practitioner, and as a compendium for the research worker. By using the first four chapters and by choosing required material from the remaining chapters, the book can be used as a text for one or two courses. The student is assumed to have basic backgrounds in mechanics, strength of materials, mathematics, matrix algebra, and material behavior, as well as understanding of soil and rock mechanics. The practitioner can use the book as an aid in analysis and design of many practical problems and in understanding the ranges of applicability and the limitations of numerical techniques. The researcher will find an up to date source of information and references in the book.

The engineering community is now involved in major changes of systems of units. In the English speaking world the conversion from the traditional English units to the SI units is referred to as *metrication;* but there are also significant differences between the conventional metric units and those prescribed under the SI. Wherever possible SI units have been used in this book, but since much of the practical material was developed over years of research and practice, the results have often been reported in units other than those of the SI. It is both impractical and unrealistic to convert everything to the SI, and the original units have therefore been retained in many cases. No confusion should result because units are identified and because no system of units is used that would be unfamiliar to the reader of this book.

The editors are thankful to the contributors for their cooperation and support, and to McGraw-Hill Book Company for their assistance and patience in seeing the book through the press. Thanks are due to G. C. Appel, Jr., and H. G. Poulos for reading the manuscript of some of the chapters.

Chandrakant S. Desai
John T. Christian

NUMERICAL METHODS IN GEOTECHNICAL ENGINEERING

CHAPTER

ONE

INTRODUCTION, NUMERICAL METHODS, AND SPECIAL TOPICS

Chandrakant S. Desai and John T. Christian

1-1 INTRODUCTION

The past decade has witnessed a tremendous growth of numerical methods for solution of problems in engineering science. The popularity and versatility of these techniques have been greatly enhanced by the availability of the large high-speed digital computer.

In the past, soil and rock mechanics were considered essentially empirical disciplines. The enormous complexities encountered in natural states of geologic media can make analytical closed-form approaches very difficult. Pioneering work by Terzaghi[1] imparted scientific and mathematical bases to many aspects of these subjects; in these developments solutions were often obtained on the basis of differential equations that were assumed to govern the physical systems. Needless to say, a large number of simplifying assumptions were necessary to obtain the closed-form solutions. Although this approach has provided useful solutions for many practical situations, it cannot yield realistic solutions for problems involving such complexities as nonhomogeneous media, nonlinear material behavior, in situ stress conditions, spatial and temporal variations in material properties, arbitrary geometries, discontinuities, and other factors imposed by geologic characteristics.

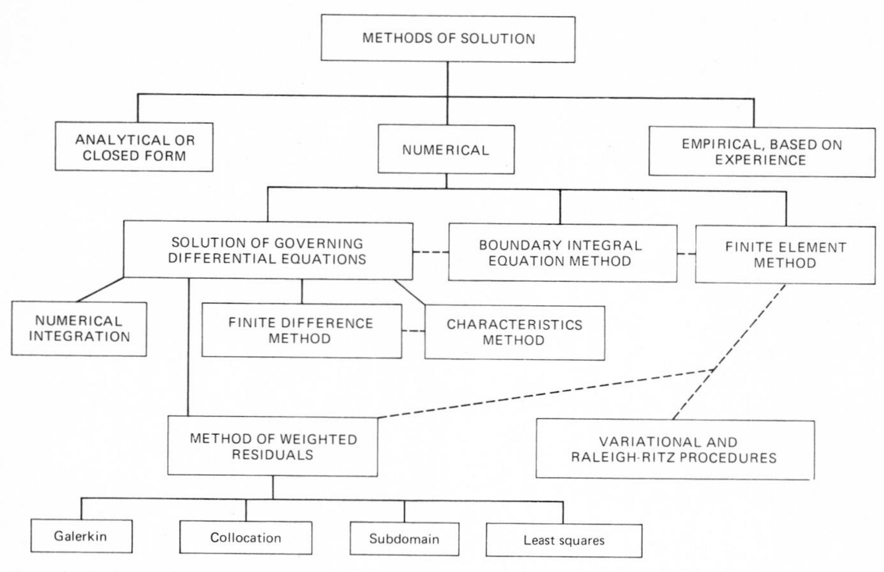

Figure 1-1 Numerical methods.[2]

Fortunately, the geotechnical engineer has been one of the first to recognize the usefulness of the newly developing numerical methods capable of accounting for a number of these factors. Consequently, in the last decade, applications of numerical techniques to soil and rock mechanics problems have grown at a fast pace.

Most Widely Used Methods

The finite element and the finite difference methods seem to be the most common procedures employed in geotechnical engineering. A number of other numerical schemes, e.g., numerical integration of governing equations, method of characteristics, the boundary-integral-equation method, and a combination of closed-form and numerical schemes, for solving governing equations have also been used. Some of the common procedures are shown schematically in Fig. 1-1.

Basic Principle

Most numerical techniques are based on the principle of *discretization*; in simple terms this can be defined as a procedure in which a complex problem of large extent is divided, or discretized, into smaller equivalent units, or components. We shall subsequently see that discretization can take different forms; in the finite difference method, for instance, we can consider that the basic governing equation is discretized, whereas in the finite element method the physical body or continuum that constitutes the system is discretized.

Categories of Problems

Problems in geotechnical engineering can be classified as steady (or equilibrium), transient (or propagation), and eigenvalue.[3,4] Table 1-1 shows some of the specific problems in these categories. As examples, we state below some elementary problems and their corresponding differential equations.

Problem	Category	Equation	
Steady flow or seepage in porous rigid media	Steady, or equilibrium	$\frac{\partial^2 u}{\partial x^2} + \frac{\partial^2 u}{\partial y^2} = 0$	(1-1)
One-dimensional consolidation	Transient	$\frac{\partial^2 u}{\partial x^2} = \frac{\partial u}{\partial t}$	(1-2)
One-dimensional wave propagation	Transient	$\frac{\partial^2 u}{\partial x^2} = \frac{\partial^2 u}{\partial t^2}$	(1-3)
Natural frequency of footing	Eigenvalue	$m \frac{\partial^2 u}{\partial t^2} + ku = 0$	(1-4)

Here u, the (unknown) dependent variable, can take various forms, e.g., fluid potential, pore water pressure, and displacement; m is the mass, and k is a constant.

Table 1-1 Categories of problems in geotechnical engineering

Steady, or equilibrium	Transient, or propagation	Eigenvalue
Static stress-deformation analyses for foundations, slopes, banks, tunnels, and other structures	Stress-deformation behavior of foundations slopes, banks, tunnels, and other structures under time-dependent forces	Natural frequencies of foundations and structures
Steady-state fluid flow	Viscoelastic analysis	
	Consolidation	
	Transient fluid flow and dispersion	
	Wave propagation	

In the mathematical literature, Eqs. (1-1), (1-2), and (1-3) are classified as elliptic, parabolic, and hyperbolic, respectively. These equations are special cases of the general equation[5,6]

$$A\frac{\partial^2 u}{\partial x^2} + B\frac{\partial^2 u}{\partial x\,\partial y} + C\frac{\partial^2 u}{\partial y^2} + D\frac{\partial u}{\partial x} + E\frac{\partial u}{\partial y} + Fu = G \tag{1-5}$$

in which the coefficients A to F are functions of x and y and G is a function of x, y, $\partial u/\partial x$, and $\partial u/\partial y$. The foregoing three categories can result from Eq. (1-5) if

$$B^2 - 4AC \begin{cases} < 0 & \text{elliptic} \\ = 0 & \text{parabolic} \\ > 0 & \text{hyperbolic} \end{cases} \tag{1-6}$$

1-2 FINITE DIFFERENCE METHOD

Before the era of finite element method, the finite difference method was perhaps the main numerical technique employed in geotechnical engineering. Although the finite element method possesses certain advantages over the finite difference method, the latter can be more suitable for certain classes of problems. The finite difference method has been covered extensively in many books.[3,5–8] Here, we shall present only introductory information.

Basic Concept

The discretization procedure is based on replacing continuous derivatives in equations governing the physical problem by the ratio of changes in the variable over a

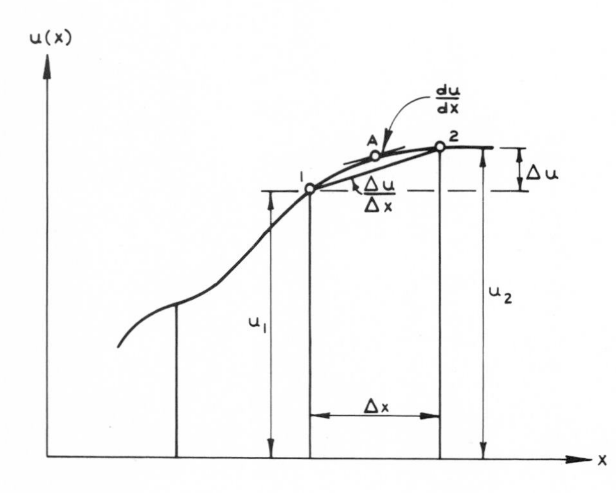

Figure 1-2 Finite difference approximation to first derivative.

small but finite increment. For example, the first derivative at point A in Fig. 1-2 is expressed as

$$\frac{du}{dx} = \lim_{\Delta x \to 0} \frac{\Delta u}{\Delta x} \approx \frac{\Delta u}{\Delta x} \tag{1-7}$$

As a result of these substitutions, a *differential* equation is transformed into a *difference* equation. The differential equations that we shall be concerned with will generally involve first, second, third, and fourth derivatives. We have therefore presented typical difference approximations to these four derivatives.

Difference Approximations

A number of procedures, e.g., Taylor series and interpolation polynomials, can be used for deriving approximations to various derivatives.[3,5] We shall use essentially the concept of a Taylor series expansion.

First derivative By expanding $u_{i-1,j}$ and $u_{i+1,j}$ (Fig. 1-3) into a Taylor series we obtain

$$u_{i+1,j} = u_{i,j} + \Delta x \frac{\partial u}{\partial x} + \frac{(\Delta x)^2}{2!}\frac{\partial^2 u}{\partial x^2} + \frac{(\Delta x)^3}{3!}\frac{\partial^3 u}{\partial x^3} + \frac{(\Delta x)^4}{4!}\frac{\partial^4 u}{\partial x^4} + \cdots \tag{1-8a}$$

$$u_{i-1,j} = u_{i,j} - \Delta x \frac{\partial u}{\partial x} + \frac{(\Delta x)^2}{2!}\frac{\partial^2 u}{\partial x^2} - \frac{(\Delta x)^3}{3!}\frac{\partial^3 u}{\partial x^3} + \frac{(\Delta x)^4}{4!}\frac{\partial^4 u}{\partial x^4} - \cdots \tag{1-8b}$$

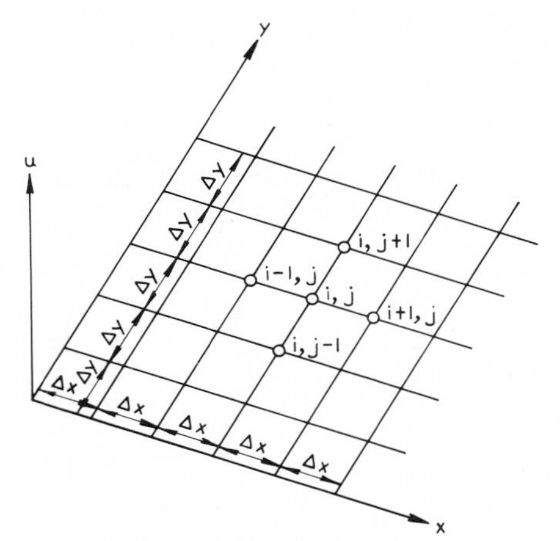

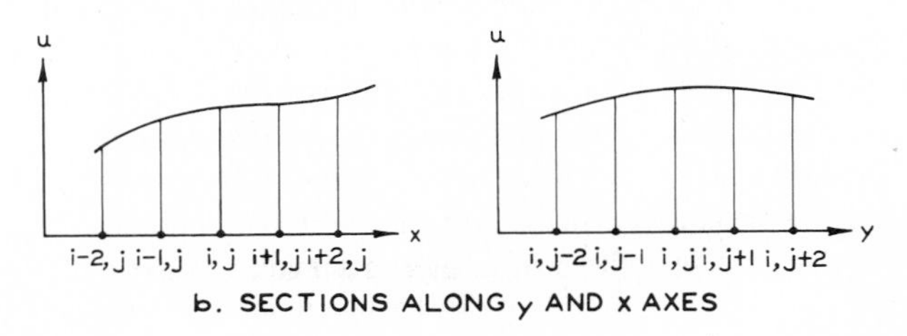

Figure 1-3 Finite difference approximations in one and two dimensions.

Simple manipulations of Eqs. (1-8*a*) and (1-8*b*) lead to

$$\frac{\partial u}{\partial x} = \frac{u_{i+1,j} - u_{i,j}}{\Delta x} + O(\Delta x) \tag{1-9a}$$

$$\frac{\partial u}{\partial x} = \frac{u_{i,j} - u_{i-1,j}}{\Delta x} + O(\Delta x) \tag{1-9b}$$

$$\frac{\partial u}{\partial x} = \frac{u_{i+1,j} - u_{i-1,j}}{2\,\Delta x} + O[(\Delta x)^2] \tag{1-9c}$$

which are called respectively the *forward-*, *backward-*, and *central-difference approximations* to the first derivative. The expression $O(\Delta x)$ denotes an error of order Δx and is the sum of the neglected terms in the series. It represents the error introduced in approximating the derivatives. $O[(\Delta x)^n]$ can be termed the *error of order n*. The error thus introduced can be called *discretization error*.

Higher derivatives By adding Eqs. (1-8*a*) and (1-8*b*) we obtain the finite difference analogs for the second derivative

$$\frac{\partial^2 u}{\partial x^2} = \frac{u_{i-1,j} - 2u_{i,j} + u_{i+1,j}}{(\Delta x)^2} + O[(\Delta x)^2] \tag{1-10}$$

Similarly, the expressions for the third and fourth derivatives are

$$\frac{\partial^3 u}{\partial x^3} = \frac{-u_{i-2,j} + 2u_{i-1,j} - 2u_{i+1,j} + u_{i+2,j}}{2(\Delta x)^3} + O[(\Delta x)^2] \tag{1-11}$$

and

$$\frac{\partial^4 u}{\partial x^4} = \frac{u_{i-2,j} - 4u_{i-1,j} + 6u_{i,j} - 4u_{i+1,j} + u_{i+2,j}}{(\Delta x)^4} + O[(\Delta x)^2] \tag{1-12}$$

We have included only typical difference forms for the four derivatives. A number of alternative difference forms can also be obtained.[3,5-8] Higher-order difference forms require use of additional neighboring points to define the derivatives.

Different Schemes and Subschemes

In the *explicit* scheme, we seek the approximate solution for u_i at time level $t + 1$ in terms of the known values of u_i at the previous time level t. Explicit schemes can often be formulated in which some values of u at time $t + 1$ are also known. The explicit procedures are relatively straightforward, permit step-by-step evaluation of u_i directly, and do not require solution of simultaneous equations. An *implicit* scheme, on the other hand, requires solution of a set of simultaneous equations at time level $t + 1$. The u_i at $t + 1$ occur as unknowns, and the right-hand side of the algebraic equations constitute the known values of u_i at time level t.

We shall now illustrate some common explicit and implicit schemes. For convenience, we shall use the parabolic equation governing one-dimensional consolidation as the basis for the illustrations:[1]

$$c_v \frac{\partial^2 u}{\partial x^2} = \frac{\partial u}{\partial t} \tag{1-13}$$

in which c_v is the coefficient of consolidation and u is the excess pore water pressure. A general form of the finite difference analog of Eq. (1-13) can be expressed as[5–7]

$$c_v \frac{\theta(\delta^2 u)_{i,t+1} + (1-\theta)\,\delta^2 u_{i,t}}{(\Delta x)^2} = \frac{u_{i,t+1} - u_{i,t}}{\Delta t} \tag{1-14}$$

where $\delta^2 u$ denotes

$$(\delta^2 u)_{i,t} = u_{i-1,t} - 2u_{i,t} + u_{i+1,t} \tag{1-15}$$

and θ can assume different values. The simplest explicit scheme, obtained by setting $\theta = 0$, is

$$u_{i,t+1} = u_{i,t} + \Delta T(u_{i-1,t} - 2u_{i,t} + u_{i+1,t}) \tag{1-16}$$

where $\Delta T = (c_v\,\Delta t)/(\Delta x)^2$.

An implicit form results if θ is adopted as unity. Hence

$$\Delta T(u_{i-1,t+1} - 2u_{i,t+1} + u_{i+1,t+1}) - u_{i,t+1} = -u_{i,t} \tag{1-17}$$

One commonly used implicit form is the Crank-Nicolson scheme, which results when $\theta = \frac{1}{2}$:

$$-\frac{\Delta T}{2}(u_{i-1,t+1} - 2u_{i,t+1} + u_{i+1,t+1}) + u_{i,t+1} = \frac{\Delta T}{2}(u_{i-1,t} - 2u_{i,t} + u_{i+1,t}) + u_{i,t} \tag{1-18}$$

The discretization error in the foregoing simple explicit and implicit schemes is of the order of $O[\Delta t + (\Delta x)^2]$, whereas that for the Crank-Nicholson scheme is of the order of $O[(\Delta t)^2 + (\Delta x)^2]$. At this stage, before discussing other schemes, we shall introduce the topics of accuracy, convergence, and stability of numerical schemes.

1-3 ACCURACY, CONVERGENCE, AND STABILITY

For a numerical scheme to yield realistic results, we must ascertain its reliability. Examination of such mathematical properties as errors, convergence, stability, and consistency[5–8] can give the idea of the scheme's dependability for general use. Often, the user of a numerical method judges its performance by solving a number of example problems; while this pragmatic approach is sometimes useful and necessary, we cannot establish the generality of the method simply on the basis of such quantitative analyses. It is essential to look at the mathematical bases of the numerical procedures.

The literature of mathematics of the finite difference method contains a number of procedures for analysis of convergence, stability, and consistency.[5–12] In this book we do not intend to go into the details except to point out the importance of these properties, to give a brief review, and to present elementary examples of derivation of stability and convergence criteria.

Accuracy of a numerical procedure can be studied in terms of convergence and stability. If the difference equations are consistent, stability implies convergence.[8] For a very large class of difference equations, the two properties convergence and stability are equivalent.

Stability and Convergence

Relevant to the concept of stability and convergence are the ideas about various kinds of errors that can enter into a numerical computation. To understand this, let $\bar{u}$ be the exact solution of the partial differential equation, say the linear equation (1-13); u^* be the exact solution of the partial difference equation, say Eq. (1-14); and u be the numerical solution of the partial difference equation.[9] The difference $\bar{u} - u^*$, called *discretization* or *truncation error*, arises because we use finite distances between mesh points and replace a continuous system by a finite system. The problem of *convergence* involves study of conditions under which u^* approaches $\bar{u}$. In other words, if during successive approximations u^* approaches $\bar{u}$ more and more closely, we say that the procedure is convergent or converges.

The difference $\bar{u} - u$ denotes numerical error. This error can be considered to be caused in several ways, e.g., faulty initial conditions, local truncation, and roundoff errors. We usually consider roundoff error, which is caused because the computer can store numbers up to only a finite number of decimal places. The study of conditions under which $\bar{u} - u$ will be small throughout the entire domain, spatial and temporal, constitutes the problem of *stability*. For instance, in a discretized region with Δx, Δy, Δz, and Δt as space and time increments, we examine what happens to the solution as $t \to \infty$.

1-4 SOME ADDITIONAL SCHEMES

Explicit Schemes

It is possible to obtain unconditionally stable explicit schemes for certain types of problems. Since these schemes may require less effort than the implicit schemes, we shall briefly describe a few of them.

In the *Dufort-Frankel procedure*[13] the finite difference form is expressed in terms of u at three time levels

$$\frac{1}{\Delta x}\left(\frac{u_{i-1,t} - u_{i,t-1}}{\Delta x} - \frac{u_{i,t+1} - u_{i+1,t}}{\Delta x}\right) = \frac{u_{i,t+1} - u_{i,t-1}}{2\,\Delta t} \tag{1-19}$$

In the *Saulev scheme*[6,14] the following two equations are used. For time level $t + 1$, the explicit scheme, Eq. (1-20), proceeds point by point in one direction, say in the positive x direction, and at the subsequent time level, Eq. (1-21) is used point by point in the negative x direction:

$$\frac{1}{\Delta x}\left(\frac{u_{i-1,t+1} - u_{i,t+1}}{\Delta x} - \frac{u_{i,t} - u_{i+1,t}}{\Delta x}\right) = \frac{u_{i,t+1} - u_{i,t}}{\Delta t} \tag{1-20}$$

$$\frac{1}{\Delta x}\left(\frac{u_{i-1,t+1} - u_{i,t+1}}{\Delta x} - \frac{u_{i,t+2} - u_{i+1,t+2}}{\Delta x}\right) = \frac{u_{i,t+2} - u_{i,t+1}}{\Delta t} \tag{1-21}$$

Note that the values of $u_{i-1,t+1}$ and $u_{i+1,t+2}$ must be known. For instance, in the case of consolidation, the values of pore water pressures at the top and bottom drainage surfaces are known at all time levels. Modified forms of the Saulev procedure have been proposed by Larkin[15] and Barkat and Clark.[16] Such *alternating-direction explicit* (ADE) schemes are found to be convergent and unconditionally stable[15,17] and have been used for solution of problems in geotechnical engineering.[18–23]

Alternating-Direction Implicit (ADI) Procedures

To illustrate the commonly used ADI scheme, we shall consider the parabolic equation expressed in terms of two space coordinates, x and y,

$$C\left(\frac{\partial^2 u}{\partial x^2} + \frac{\partial^2 u}{\partial y^2}\right) = \frac{\partial u}{\partial t} \tag{1-22}$$

where C denotes the material parameter. A simple implicit finite difference form for Eq. (1-22) is

$$C[(\delta_x^2 u)_{i,j,t+1} + (\delta_y^2 u)_{i,j,t+1}] = \frac{u_{i,j,t+1} - u_{i,j,t}}{\Delta t} \tag{1-23}$$

where $\quad (\delta_x^2 u)_{i,j,t+1} = u_{i-1,j,t+1} - 2u_{i,j,t+1} + u_{i+1,j,t+1}$

and so on. This procedure is unconditionally stable. For a generic point i (Fig. 1-4) the expanded version of Eq. (1-23) is

$$-\Delta T\, u_{i-1,j,t+1} - \Delta T\, u_{i,j-1,t+1} + (4\,\Delta T + 1)u_{i,j,t+1} - \Delta T\, u_{i,j+1,t+1} - \Delta T\, u_{i+1,j,t+1} = u_{i,j,t} \tag{1-24}$$

Equation (1-24) has five unknowns, and if it were written for all node points, we would obtain a set of equations whose coefficient matrix contains five diagonals

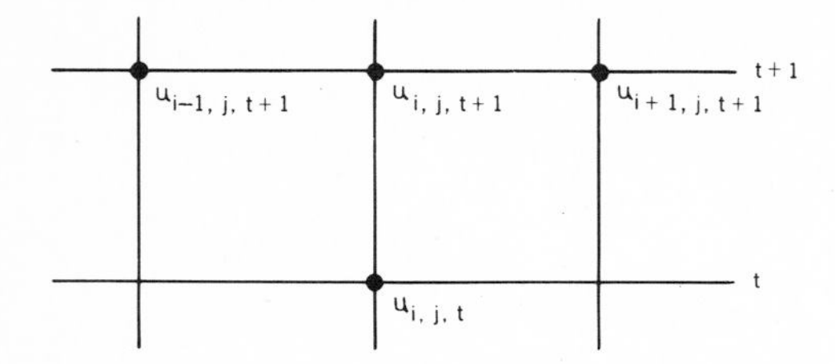

Figure 1-4 Implicit procedures.

with nonzero elements. As we shall discuss subsequently, schemes such as gaussian elimination and the Gauss-Seidel method can be used to solve these equations.

The ADI procedure is a modification of the implicit procedure and is detailed by Peaceman and Rachford,[24] Douglas,[25] and others.[5,6,26] This procedure yields a coefficient matrix that is tridiagonal instead of pentadiagonal, as given by Eq. (1-24), and hence renders the equations suitable for solution by more economical and efficient procedures. The procedure involves use of two finite difference equations

$$(\delta_x^2 \bar{u})_{i,j} + (\delta_y^2 u)_{i,j,t} = \frac{\bar{u}_{i,j} - u_{i,j,t}}{\Delta t/2} \tag{1-25}$$

and

$$(\delta_x^2 \bar{u})_{i,j} + (\delta_y^2 u)_{i,j,t+1} = \frac{u_{i,j,t+1} - \bar{u}_{i,j}}{\Delta t/2} \tag{1-26}$$

in which $\bar{u}$ denotes an intermediate or temporary value. These equations are applied over half the time step $\Delta t/2$. We can see that Eq. (1-25) is implicit in the x direction only whereas Eq. (1-26) is implicit in the y direction only. The ADI scheme has been used by various investigators.[21,27–30] It is unconditionally stable and involves a discretization error $O[(\Delta t)^2 + (\Delta x)^2]$, for $\Delta x = \Delta y$, and constant values of $\Delta x/\Delta y$. The ADI procedure can be extended for equations in three space coordinates.[31]

Alternating-Direction Explicit (ADE) Procedure

The alternating-direction explicit procedure, Eqs. (1-20) and (1-21), can be extended for two- and three-dimensional problems.[17] Accordingly, a finite difference equation for Eq. (1-22) (see Fig. 1-5) is

$$C\left[\left(\frac{u_{i-1,j,t+1} - u_{i,j,t+1}}{\Delta x} - \frac{u_{i,j,t} - u_{i+1,j,t}}{\Delta x}\right) + \left(\frac{u_{i,j+1,t+1} - u_{i,j,t+1}}{\Delta y} - \frac{u_{i,j,t} - u_{i,j-1,t}}{\Delta y}\right)\right] = \frac{u_{i,j,t+1} - u_{i,j,t}}{\Delta t} \tag{1-27}$$

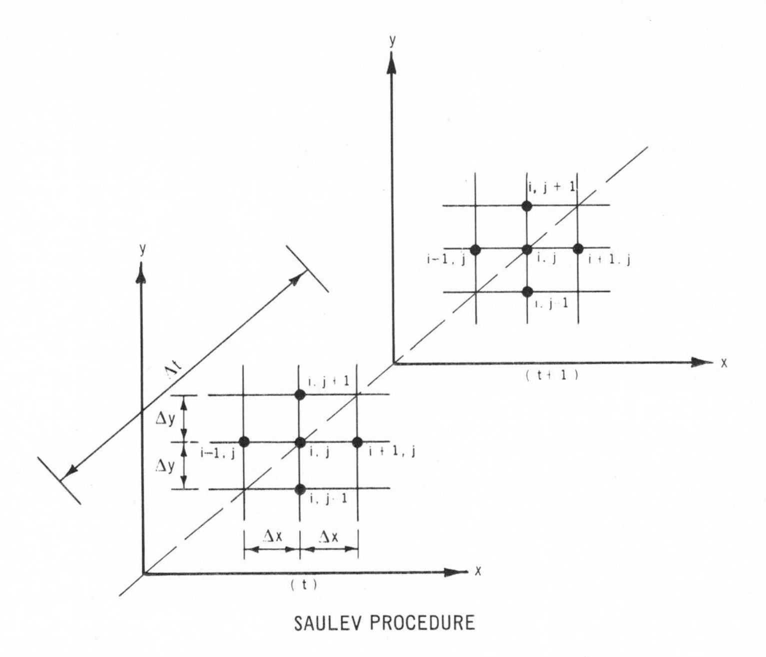

Figure 1-5 Explicit procedures.

A number of possible alternatives for the ADE method are described by Larkin,[15] where it is shown that the ADE schemes are unconditionally stable for positive time increments of any magnitude. Applications of the ADE procedure to three-dimensional problems are discussed by Allada and Quon.[17]

It has been observed[17] that there is no a priori reason to show that either the ADI or the ADE procedure would be better than the other from a physical or mathematical viewpoint. For an equation with one space coordinate [Eq. (1-13)], Desai and Johnson[22] found from a quantitative analysis that the behavior of an implicit procedure was somewhat better than the ADE procedures. However, the ADI schemes can take about five times as much computational time as the ADE schemes.[15,22] One of the requirements of ADE schemes is that the values of u at some boundary points such as $u_{i-1,t+1}$ [Eq. (1-27)] be known either from given boundary conditions or from computations. Notwithstanding some mathematical limitations of the ADE schemes, they are considered to be suitable for many practical situations.

1-5 BOUNDARY CONDITIONS

A *boundary condition* can be explained in simple terms as the value of the dependent variable (or its derivative) on the boundaries (or edges) of the region of interest. Since the basic differential equation, say Eq. (1-13), can be satisfied by

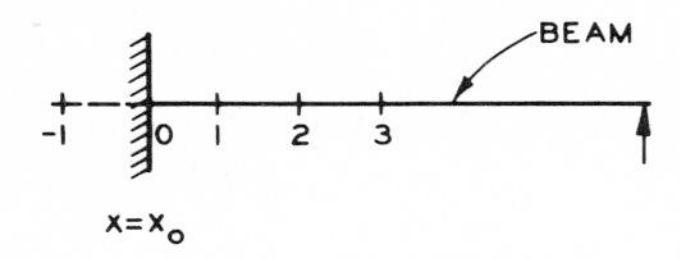

a. CANTILEVER BEAM

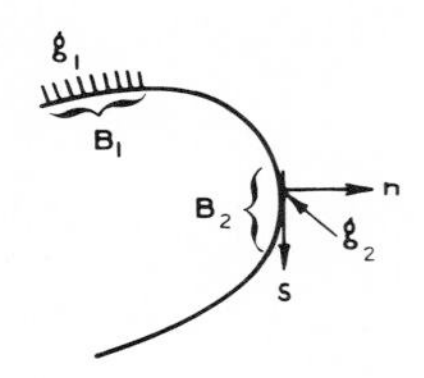

b. DIRICHLET AND NEUMANN TYPE BOUNDARY CONDITIONS

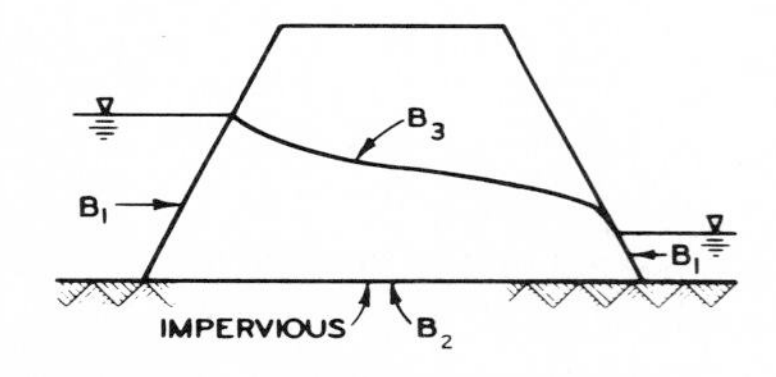

c. POTENTIAL, FLOW AND MIXED BOUNDARY CONDITIONS

Figure 1-6 Different boundary conditions.

many solutions, with the differential equation alone we cannot pinpoint a unique solution. In order to obtain a unique solution it is essential that additional information be provided. Such information is usually available in terms of boundary conditions. For example, in the case of a propped cantilever beam (Fig. 1-6*a*) fixed at one end, the value of the displacement is prescribed at the simply supported end, and the values of displacement and slope (gradient of displacement) are specified at the built-in end.

For an nth-order equation, we need n boundary conditions for a unique solution. If all n boundary conditions are specified at the same location, say at $x = x_0$, the problem can be termed an *initial-value problem.* On the other hand, if the boundary conditions are specified at more than one location, the problem is called a *boundary-value problem.* An example of the initial-value problem is the time-dependent problem of consolidation, and an example of a boundary-value problem is the static deformation of an axially loaded pile.

Categories

The boundary conditions are often classified as Dirichlet, Neumann and third, or mixed, boundary conditions (Fig. 1-6*b*). The Dirichlet condition can be expressed as

$$u = g_1 \tag{1-28a}$$

where u is the variable and g denotes its prescribed values on a part of the boundary B_1. For the consolidation problem, B_1 can correspond to the top of the layer and g_1 to the prescribed values of u.

The Neumann type of condition is given by

$$a\frac{\partial u}{\partial n} + b\frac{\partial u}{\partial s} = g_2 \tag{1-28b}$$

where a and b are parameters and n and s denote normal and tangent to the boundary B_2, respectively. An impervious layer where flow vanishes can correspond to Neumann conditions. In the mixed condition, both u and its gradient can be prescribed on B_3 as

$$a\frac{\partial u}{\partial n} + b\frac{\partial u}{\partial s} + cu = g_3 \tag{1-28c}$$

In the problem of seepage through a dam (Fig. 1-6*c*) the known heads on the upstream face of the dam constitute a Dirichlet condition, the condition of zero flow across the impervious base represents a Neumann condition, and the conditions of no flow across the free surface and a head equal to the elevation at the free surface represent a mixed condition.

In the finite difference method, the Dirichlet condition is introduced by setting the prescribed value of u at the given point. For instance, in the case of the cantilever with unyielding support (Fig. 1-6*a*) the displacement at the prop is set equal to zero.

For the specified slope at the fixed end (Fig. 1-6*a*)

$$\frac{du}{dx} = 0 \tag{1-29}$$

or since $(u_{-1} - u_1)/\Delta x \approx 0$, $u_{-1} = u_1$. The finite difference equations are modified to reflect the condition that $u_{-1} = u_1$. Similar finite difference forms can be introduced for other boundary conditions, e.g., prescribed second derivatives.

1-6 EXAMPLES OF FINITE DIFFERENCE METHOD

Figures 1-7 and 1-8 show some of the problems in geotechnical engineering for which the finite difference method has been used. Axially loaded piles are covered in Chaps. 7 and 10, beams and slabs in Chap. 5, consolidation in Chaps. 11 and

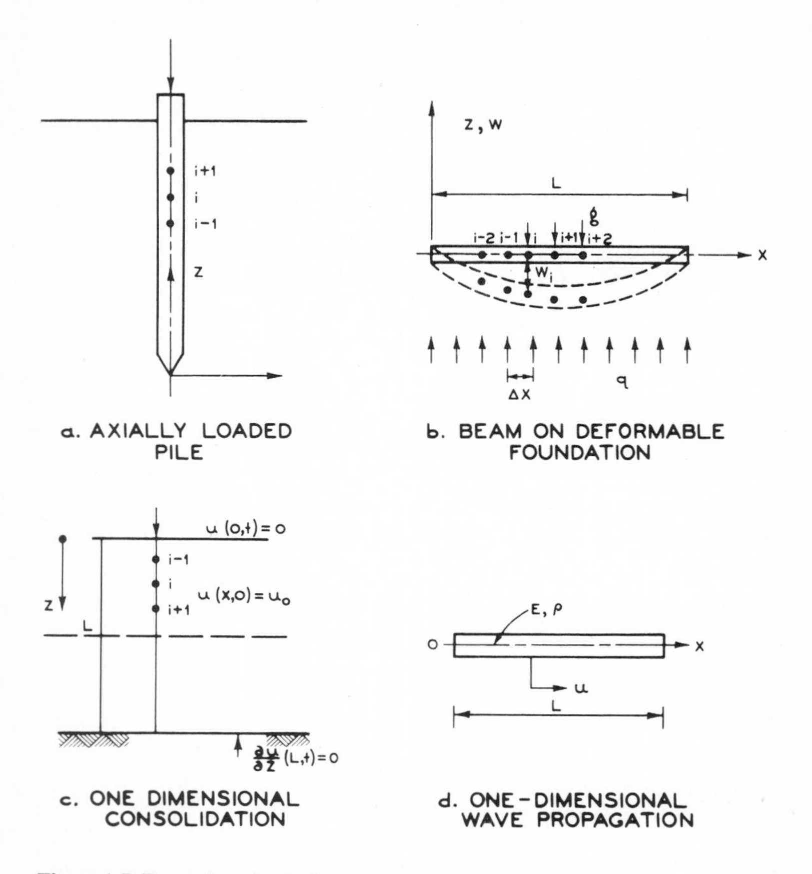

Figure 1-7 Examples of one-dimensional problems.

12, wave propagation in Chaps. 19 and 20, and seepage in Chap. 14. Here we describe a few of these problems for the sake of illustration and completeness.

Beams on Elastic Foundations and Laterally Loaded Piles

A form of the equation governing the bending behavior of a beam on an elastic foundation or a laterally loaded pile (Fig. 1-7*b*) can be written

$$\frac{d^2}{dx^2}\left(EI\frac{d^2w}{dx^2}\right) = -q(x) \tag{1-30}$$

where EI = flexural stiffness of beam (Fig. 1-7*b*)
w = deflection of beam or pile
$q(x)$ = intensity of applied load and/or soil reaction

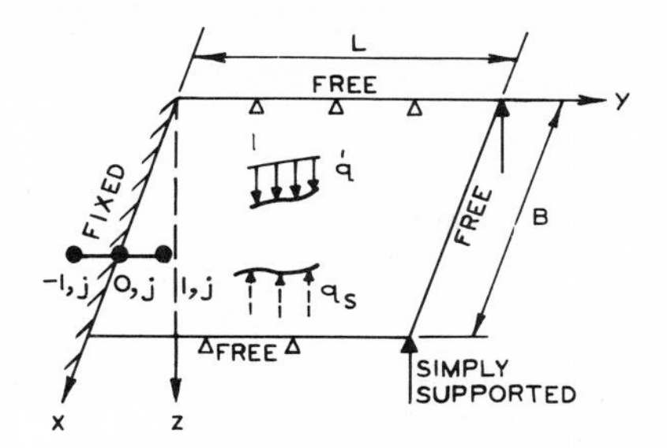

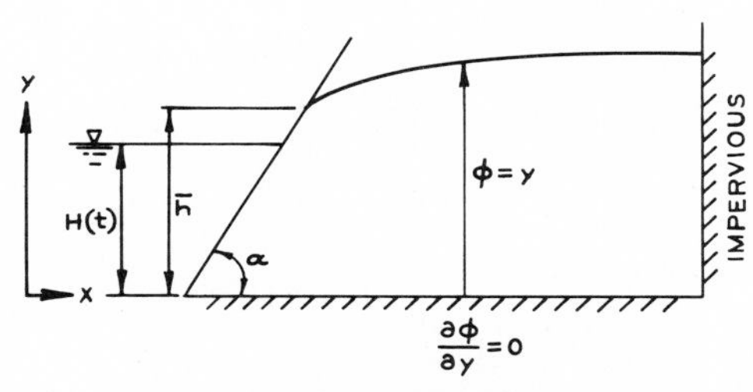

Figure 1-8 Examples of two-dimensional problems.

The last is often defined as[1,32,33]

$$q(x) = E_s w \tag{1-31}$$

where E_s is the modulus of soil reaction. A finite difference form for a beam or pile with uniform section corresponding to Eq. (1-30) is

$$(EI)_i \frac{w_{i-2} - 4w_{i-1} + 6w_i - 4w_{i+1} + w_{i+2}}{(\Delta x)^4} = -q_i \tag{1-32}$$

where Δx is the magnitude of increments into which the pile is subdivided; here Δx is assumed to be uniform.

If it is possible to assume that the beam is fixed at the left-hand side and free at the right-hand side, the boundary conditions corresponding to these physical conditions can be stated as

$$\frac{dw}{dx}(x = 0) = 0 \qquad \text{and} \qquad EI\frac{d^2w}{dx^2}(x = L) = 0 \tag{1-33}$$

One-dimensional consolidation The parabolic equation governing one-dimensional consolidation (Fig. 1-7c) is[1,22]

$$c_v \frac{\partial^2 u}{\partial z^2} = \frac{\partial u}{\partial t} \tag{1-34}$$

where c_v = coefficient of consolidation
u = pore water pressure
t = time

A number of explicit and implicit forms are possible, as stated in Eqs. (1-16) to (1-18).

The following typical initial and boundary conditions can occur in one-dimensional consolidation:

1. The pore water pressures are prescribed initially

$$u(x, 0) = u_0 \tag{1-35a}$$

2. The top of the layer allows free drainage

$$u(0, t) = 0 \tag{1-35b}$$

3. The bottom boundary is impervious

$$\frac{\partial u}{\partial z}(L, t) = 0 \tag{1-35c}$$

Two- and three-dimensional seepage Transient free-surface flow through two-dimensional porous soil media (Fig. 1-8*b*) is often assumed to be governed by the linearized equation[18–20]

$$\bar{h}(t)\left(k_x \frac{\partial^2 h}{\partial x^2} + k_y \frac{\partial^2 h}{\partial y^2}\right) = n \frac{\partial h}{\partial t} \tag{1-36}$$

where h = height of free surface
$\bar{h}(t)$ = mean external height
k_x = horizontal coefficient of permeability
k_y = vertical coefficient of permeability
t = time
n = effective porosity

The finite difference analog for Eq. (1-36) according to the ADE procedure is[18,20]

$$\begin{aligned} h_{i,j,t+1} = h_{i,j,t} &+ \bar{\beta}_x\left(\frac{h_{i-1,j,t+1} - h_{i,j,t+1}}{\Delta x1} - \frac{h_{i,j,t} - h_{i+1,j,t}}{\Delta x2}\right) \\ &+ \bar{\beta}_y\left(\frac{h_{i,j,t+1} - h_{i,j,t+1}}{\Delta y1} - \frac{h_{i,j,t} - h_{i,j-1,t}}{\Delta y2}\right) \end{aligned} \tag{1-37}$$

where $\bar{\beta}_x = \dfrac{k_x \bar{h}(t)}{0.5n(\Delta x1 + \Delta x2)}$ and $\bar{\beta}_y = \dfrac{k_y \bar{h}(t)}{0.5n(\Delta y1 + \Delta y2)}$

Since the head at the external boundary is prescribed as

$$h(x, x \tan \alpha, t) = H(t) \tag{1-38a}$$

the boundary values of $h_{i-1,j,t+1}$ are known at all times. Other initial and boundary conditions are

$$h(x, y, 0) = h_i \tag{1-38b}$$

and

$$\frac{\partial h}{\partial y} = 0 \quad \text{at impervious base} \tag{1-38c}$$

The condition in Eq. (1-38*c*) can be expressed in a finite difference form.

The foregoing ADE procedure can be extended for three-dimensional equations governing fluid flow, such as

$$k_x \frac{\partial^2 \phi}{\partial x^2} + k_y \frac{\partial^2 \phi}{\partial y^2} + k_z \frac{\partial^2 \phi}{\partial z^2} = n \frac{\partial \phi}{\partial t} \tag{1-39}$$

in which ϕ is the fluid potential given by

$$\phi = \frac{p}{\gamma} + z \tag{1-40}$$

where z = elevation head
p = pressure
γ = density

1-7 DISCONTINUITIES

Geologic media usually involve nonhomogeneous materials. In the case of arbitrary nonhomogeneities, it is relatively difficult to obtain finite difference equations to account for the variations of material properties. For such cases, the finite element method can provide an efficient and direct approach. For nonhomogeneities defined by distinct layers, it is possible to obtain modified finite difference analogs that will approximately represent the real system. We shall illustrate this procedure by using the problem of one-dimensional fluid flow and by considering a typical (vertical) boundary between two different media.

The physical condition that must be satisfied is that the flow across the boundary in Fig. 1-9 must be continuous. Let the coefficients of permeability of

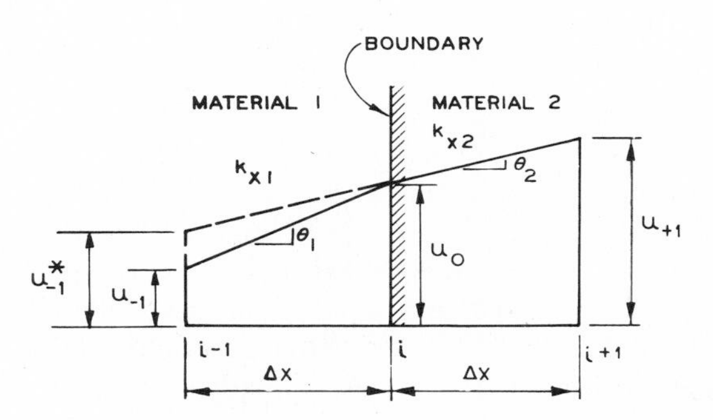

Figure 1-9 Nonhomogeneous media.

medium 1 and 2 be denoted respectively by k_{x1} and k_{x2}. Then the continuity-of-flow condition can be written

$$k_{x1}\theta_1 = k_{x2}\theta_2 \tag{1-41}$$

in which θ denotes the gradient of $u = \partial u/\partial x$. Assuming that the variation of u between the points $i - 1$, i, and $i + 1$ is linear, i.e.,

$$\theta_1 \,\Delta x + \theta_2 \,\Delta x = u_1 - u_{-1} \tag{1-42a}$$

substitution of the continuity condition in Eq. (1-42*a*) leads to

$$\theta_1 = (u_1 - u_{-1})\frac{1}{\Delta x(1 + 2k_{rx})} \tag{1-42b}$$

in which $k_{rx} = k_{x1}/k_{x2}$. Now

$$u^*_{-1} = u_{-1} + (\theta_1 - \theta_2)\,\Delta x \tag{1-42c}$$

Here the superscript * denotes a fictitious or temporary quantity. By substituting values of θ_1 and θ_2 in Eq. (1-42*c*) we obtain

$$u^*_{-1} = u_{-1} + (u_1 - u_{-1})\gamma_{x1} \qquad \text{where} \qquad \gamma_{x1} = \frac{1 - k_{\gamma x}}{1 + 2k_{\gamma x}} \tag{1-42d}$$

The finite difference equation, say Eq. (1-23), is written for point i, but u^*_{-1} is substituted for the value of u at $i - 1$. Similar expressions can be obtained for u_1^*, and for horizontal interface boundaries. Modifications for inclined interfaces can be generated by combining results of horizontal and vertical interfaces.[19,20]

Irregular Geometries

For simple configurations such as rectangular regions, we can adjust the mesh points to coincide with the boundaries. For an irregular boundary, however, the mesh points may not fall on the boundary, and modifications of the finite difference equation become necessary. Special procedures are necessary to account for irregular boundaries that introduce uneven meshes. A number of books have covered these procedures.[3,5] Here we mention that in the finite element method such special procedures are not required.

1-8 FINITE ELEMENT METHOD

In the last ten years, the finite element method has experienced tremendous growth, in both theoretical developments and applications. The engineer has been mainly responsible for the present use of the method. Often on the basis of intuitive considerations, with not too many mathematical proofs, he has exploited the power and potential of the method. Unlike the finite difference method, which came to the engineer with a considerable mathematical basis already existing, the

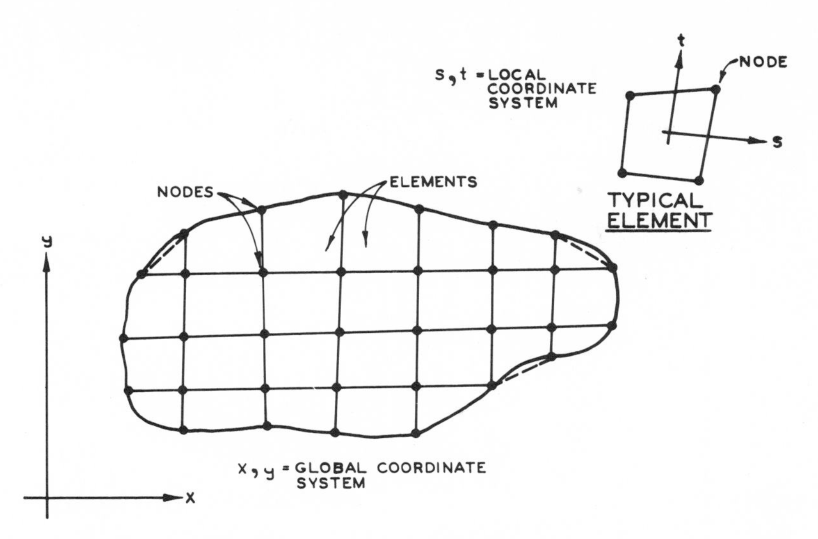

Figure 1-10 Subdivision into finite elements of arbitrary continuum.

mathematical characteristics of the finite element method have not been established fully. Significant activity has been recently concentrated toward laying down the mathematical bases of the method.

The geotechnical engineer was quick to recognize the usefulness of the finite element method for solution of complex problems that had defied conventional and closed-form solutions.

We shall present here a brief description of the finite element method. Wherever possible, we shall relate various steps to corresponding aspects of problems in geotechnical engineering. For details, the reader should refer to various papers and textbooks.[4,34,35]

Basic Steps

The finite element method can be considered to involve six basic steps.

Step 1: Discretization The discretization principle involves division of a continuum (Fig. 1-10) into an equivalent system of smaller continua; the smaller continua are called the *finite elements.* The intersections of the *nodal lines* separating the elements are called *nodal points.* The continuum can represent a physical body such as a pile-foundation system (Fig. 1-11), where we are interested in displacements (of the nodes). On the other hand, a rigid mass (soil) through which fluid flows can represent the continuum, and fluid potentials at the nodes are the unknowns. The quantities such as the displacement and the fluid potential are the *main,* or *primary, unknowns* of these problems. Often we need to compute *secondary* quantities such as stresses from displacements and quantity of flow from fluid potentials.

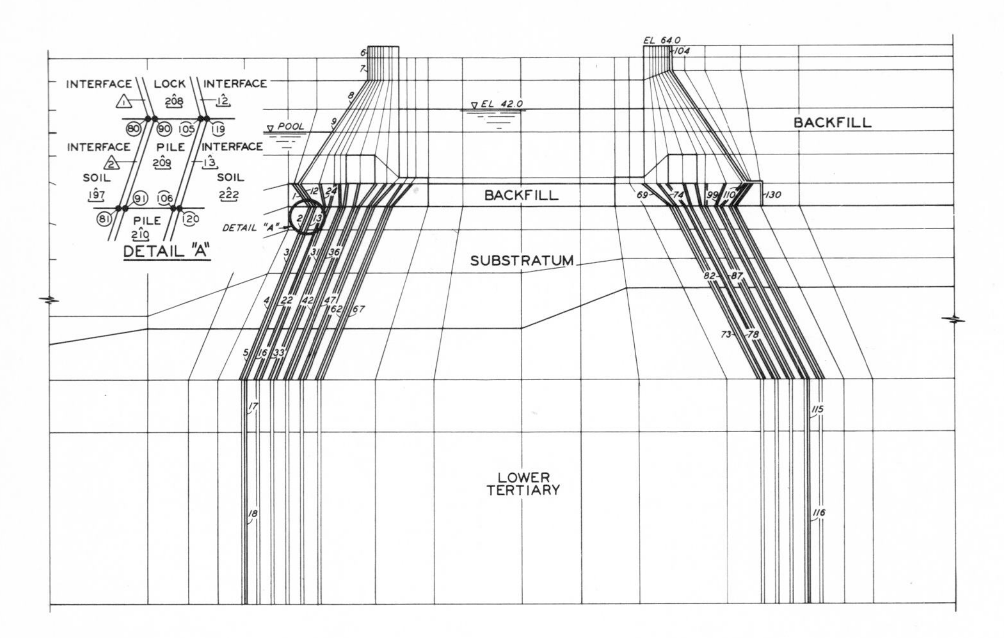

Figure 1-11 Finite element mesh for lock-pile-foundation system. (From Ref. 41, chap. 7).

A problem can be formulated alternatively in terms of stresses as primary unknowns or both displacements and stresses as primary unknowns for the stress-deformation problem. In the flow problem, formulation can be in terms of stream function or both fluid potential and stream function. For details of these equilibrium, hybrid, and mixed procedures, the reader should consult various publications.[4,36–41]

A basic characteristic of the finite element method is that the finite elements are analyzed and treated separately, one by one. Each element is assigned its physical or constitutive properties, and its *property* or *stiffness* equations are formulated. Subsequently, the elements are assembled to obtain equations for the total structure. The assembly procedure is essentially mechanical and involves putting together the element equations by observing certain conditions such as compatibility requirements.

Element types For a one-dimensional problem, we use line elements, linear or curved. Triangles and quadrilaterals are common shapes used for two-dimensional problems. In three-dimensional analyses, the common elements used are tetrahedra and hexahedra.

Step 2: Selection of approximation functions In this step, we assume a priori a pattern of solution for the unknown quantity such as the displacement over each element. The pattern is usually chosen in form of polynomials expressed in terms

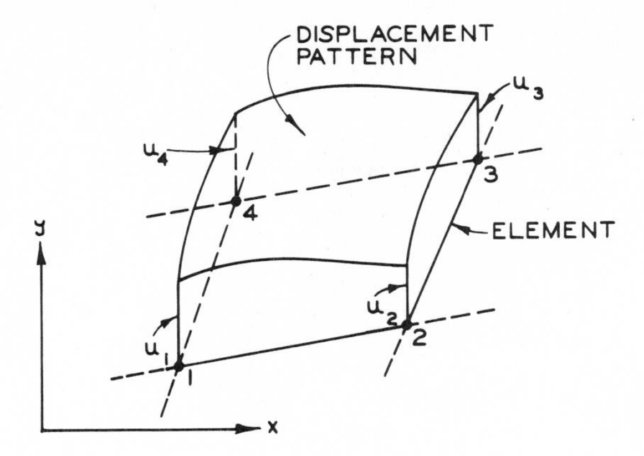

Figure 1-12 Schematic representation of displacement pattern.

of some generalized displacement or displacements of the nodes of the element. For instance, if u denotes a displacement, we can select displacement *functions*, or *models*, as polynomials of various degrees, such as

$$u = \alpha_1 + \alpha_2 x + \alpha_3 y + \alpha_4 x^2 + \alpha_5 xy + \alpha_6 y^2 \tag{1-43}$$

Here α_i are called *generalized coordinates* or *generalized displacement amplitudes*[4] and x, y are the cartesian coordinates. In matrix notation, Eq. (1-43) can be expressed as

$$u = \{\phi\}^T\{\alpha\} \tag{1-44}$$

where $\{\phi\}^T = [1 \quad x \quad y \quad \cdots]$ and $\{\alpha\}^T = [\alpha_1 \quad \alpha_2 \quad \cdots]$

By adopting polynomials of higher orders, we can usually approach the exact solution more and more closely. A schematic representation of assumed displacement pattern is shown in Fig. 1-12.

In three-dimensional problems the displacement functions for the other two components, v and w, can be expressed in a manner similar to that for component u. In fluid-flow problems, approximate models such as in Eq. (1-43) are chosen for the fluid potential ϕ. Both displacements and fluid pressure are approximated for coupled problems like consolidation and liquefaction.

A number of conditions must be satisfied for the chosen pattern to yield a satisfactory, consistent, and convergent solution. Details of the mathematics of these requirements and of various categories of functions such as conforming (or compatible) and nonconforming models can be found elsewhere.[4,35,42,43]

Finally, in the discretization procedure we seek solutions in terms of values of displacements at the nodes. This is done by evaluating $[\phi]$ and by performing the matrix computations

$$\{q\} = \begin{Bmatrix} u_1 \\ u_2 \\ \vdots \\ v_1 \\ v_2 \\ \vdots \end{Bmatrix} = \begin{bmatrix} [\phi_1] \\ [\phi_2] \\ [\phi_3] \\ [\phi_3] \end{bmatrix} \{\alpha\} = [A]\{\alpha\} \tag{1-45}$$

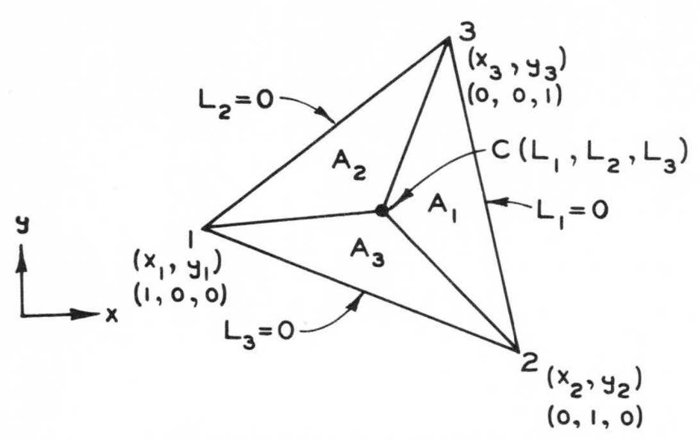

Figure 1-13 Local system for a triangular element.

Here $[A]$ contains values of ϕ_i in terms of the known coordinates x_i and y_i of the nodes. Finally,

$$\{u\} = [\phi][A^{-1}]\{q\} = [N]\{q\} \qquad \text{where} \quad \{u\}^T = [u \quad v] \tag{1-46}$$

Interpolation models We can avoid the operation of computing $[A^{-1}]$ if we choose the interpolation functions in matrix $[N]$ directly. In recent times, many developments of finite element procedures use interpolation models; hence, in what follows we shall adopt this concept. The matrix $[N]$ is composed of *interpolation functions*, or *shape functions*. An interpolation function assumes a value of unity at a particular node and a value of zero at all other nodes of the element.[4]

Interpolation functions are usually based on the use of local and natural coordinate systems. A *local coordinate system* pertains to a specific element, whereas the coordinate system for the entire body is called a *global system* (Fig. 1-10). The concept of *natural coordinates* allows definition of a point in the element in terms of dimensionless numbers. For instance, in the following we define natural coordinates L_i for a triangular element (Fig. 1-13):

$$L_1 = \frac{A_1}{A} \qquad L_2 = \frac{A_2}{A} \qquad L_3 = \frac{A_3}{A} \tag{1-47a}$$

Here $L_1 + L_2 + L_3 = 1$; A_1, A_2, and A_3 are areas of the component triangles; and A is the total area:

$$A = A_1 + A_2 + A_3 \tag{1-47b}$$

The coordinates L_i and (x_i, y_i) are related as follows:

$$\begin{Bmatrix} 1 \\ x \\ y \end{Bmatrix} = \begin{bmatrix} 1 & 1 & 1 \\ x_1 & x_2 & x_3 \\ y_1 & y_2 & y_3 \end{bmatrix} \begin{Bmatrix} L_1 \\ L_2 \\ L_3 \end{Bmatrix} \tag{1-48}$$

or

$$\begin{Bmatrix} L_1 \\ L_2 \\ L_3 \end{Bmatrix} = \frac{1}{2A} \begin{bmatrix} 2A_{23} & b_1 & a_1 \\ 2A_{31} & b_2 & a_2 \\ 2A_{12} & b_3 & a_3 \end{bmatrix} \begin{Bmatrix} 1 \\ x \\ y \end{Bmatrix} \tag{1-49}$$

where $\quad a_1 = x_3 - x_2 \qquad a_2 = x_1 - x_3 \qquad a_3 = x_2 - x_1$

$$b_1 = y_2 - y_3 \qquad b_2 = y_3 - y_1 \qquad b_3 = y_1 - y_2$$

$$A_{12} = \text{area of triangle } 12O$$

$$A_{23} = \text{area of triangle } 23O$$

$$A_{31} = \text{area of triangle } 31O$$

and $\quad 2A = a_3 b_2 - a_2 b_3 = a_1 b_3 - a_3 b_1 = a_2 b_1 - a_1 b_2$

Here O denotes the origin of the xy coordinate system.

Isoparametric Elements

The concept of isoparametric elements[4,44,45] has been used commonly for finite element formulations. It offers a number of advantages, e.g., efficient integrations and differentiations and handling of curved and arbitrary geometrical shapes. A detailed description of these elements is given in Ref. 4.

The term *isoparametric* implies common (*iso*-) parametric description of the unknown displacement and geometry of the element. The basic idea is to express both the displacement and the geometry of the element by using the same interpolation functions N_i.

We expressed displacements at a point in the element as

$$\{u\} = [N]\{q\} \tag{1-46}$$

In the isoparametric concept, we express the coordinates of a point in the element in terms of the same functions N_i

$$\{x\} = [N]\{x_n\} \tag{1-50}$$

where $\{x\}^T = [x \quad y]$ and $\{x_n\}^T = [x_1 \quad x_2 \quad \cdots]$ contains the coordinates of the nodal points.

As a simple example, let us consider the four-node quadrilateral isoparametric element (Fig. 1-14). For this element, the matrix $[N]$ in Eqs. (1-46) and (1-50) is composed of the following interpolation functions:

$$N_i = \frac{1}{4}(1 + ss_i)(1 + tt_i) \tag{1-51a}$$

or

$$N_1 = \frac{(1-s)(1-t)}{4} \qquad N_2 = \frac{(1+s)(1-t)}{4}$$
$$N_3 = \frac{(1+s)(1+t)}{4} \qquad N_4 = \frac{(1-s)(1+t)}{4} \tag{1-51b}$$

Step 3: Derivation of element equations A number of procedures are available for the derivation of equations defining properties of a finite element. Chief among them are the variational and residual methods. In recent times, such residual

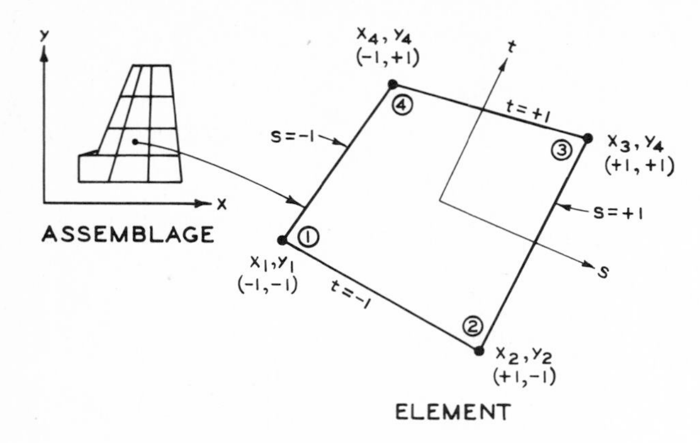

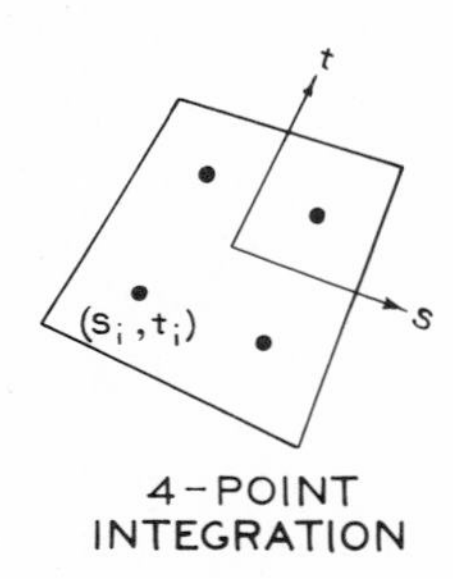

Figure 1-14 Quadrilateral isoparametric element.

methods as Galerkin's procedure are employed increasingly.[46-50] Residual methods are found to be more general and suitable for problems governed by both linear and nonlinear equations.

Use of any of the formulation procedures leads to element equations that can be expressed in matrix notation as

$$[k]\{q\} = \{Q\} \tag{1-52}$$

where $[k]$ is the element property matrix; e.g., it is the stiffness matrix in the displacement problem and the permeability matrix for the seepage problem. The vector $\{Q\}$ is the nodal forcing-parameter vector; in the displacement formulation it contains nodal forces, and for the seepage problem its terms can constitute applied fluid fluxes.

Displacement method The problems in solid mechanics are formulated by following one of the three procedures: displacement (or stiffness), stress (or equilibrium), and mixed.[4] In the displacement method, displacement in an element is assumed to be the unknown, and the element equations are derived by using the variational procedures based on the principle of *minimum potential energy*. In the equilibrium approach, stresses are assumed to be unknown, and the principle of *minimum complementary energy* is used. Both stresses and displacements are assumed to be unknowns in the mixed procedure, and special variational principles, e.g., those based on the Hellinger-Reissner functional, are used.

Most problems in geotechnical engineering have been formulated by using the displacement method. Some of the reasons for this choice are that the number and bandwidth of the final stiffness equations in the displacement method are smaller than those produced by other methods and it is relatively easier to establish approximation functions to satisfy compatibility requirements than it is to construct equilibrium or mixed models.[4] On the other hand, displacement formulation can be more sensitive to variations in such problem parameters as geometry, material properties, and stress-strain laws and may not provide a general procedure for such complex problems as simulation of excavation, interaction, and strain softening. In many problems where we are interested in precise distribution of stresses and in limit-equilibrium problems, the equilibrium and mixed procedures may be more suitable.[48] There are a few problems in which the mixed procedures have been used, but they are limited essentially to consolidation problems.

It is believed that in the future we will see increasing use of the mixed and equilibrium approaches because it is now recognized that the mixed procedures possess a number of advantages over the displacement method and can be more competitive with respect to generality, accuracy, and computational effort.[37] For our purposes we shall devote attention to the displacement method, and for details regarding the other methods the reader should refer to published literature.

In the early stages of the development of the finite element method for applications in geotechnical engineering, triangular (Fig. 1-15*a*) and quadrilateral elements (Fig. 1-15*b*) composed of four triangles were used. The triangle was of the constant-strain category, implying that the assumed displacement functions were linear. Details of elements are given in a number of publications.[4,51] In the recent past, the trend has been toward use of isoparametric elements. For two-dimensional problems, 4-node, 8-node, and 12-node isoparametric elements have been employed, and for three-dimensional cases the 8-node hexahedral element has been the common choice.

For interpolation functions, use of cubic and bicubic[52,53] spline functions is expected to increase for certain categories of problems. Many problems in geotechnical engineering may not require functions of higher order. Very often, from the viewpoint of tradeoff between accuracy and cost, lower-order elements may be suitable.[54,55]

As illustration, we shall describe the derivation of the element equation[4,56] for the four-node isoparametric quadrilateral element (Fig. 1-14). For the two-dimensional stress-deformation problem there are two degrees of freedom at each point in the element, displacement u in the x direction and displacement v in the y direction. In matrix notation

$$\{u\}^T = [u \quad v] \tag{1-53}$$

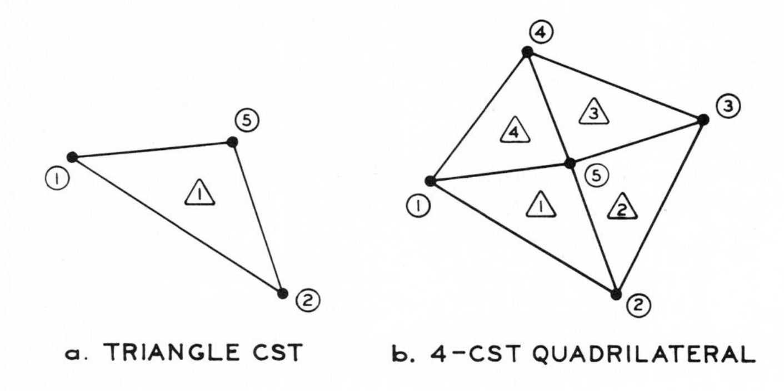

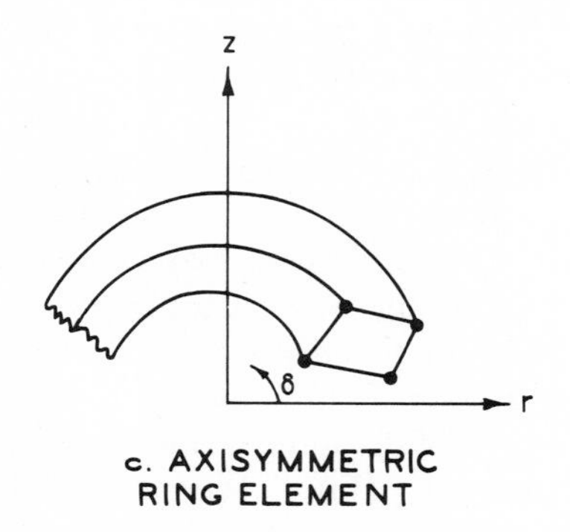

Figure 1-15 Some finite elements used in geotechnical engineering.

In terms of the nodal displacements we can write

$$\{u\} = \begin{bmatrix} N_1 & 0 & N_2 & 0 & N_3 & 0 & N_4 & 0 \\ 0 & N_1 & 0 & N_2 & 0 & N_3 & 0 & N_4 \end{bmatrix} \{q\} \tag{1-54}$$

where N_i are given in Eq. (1-51) and

$$\{q\}^T = [u_1 \quad v_1 \quad u_2 \quad v_2 \quad u_3 \quad v_3 \quad u_4 \quad v_4]$$

The geometry of the element is expressed by using the same $[N]$ as

$$\{x\} = \begin{Bmatrix} x \\ y \end{Bmatrix} = \begin{bmatrix} [N] & [0] \\ [0] & [N] \end{bmatrix} \{x_n\} \tag{1-55}$$

where $\qquad \{x_n\}^T = [x_1 \quad x_2 \quad x_3 \quad x_4 \quad y_1 \quad y_2 \quad y_3 \quad y_4]$

If we assume plane strain conditions, the strain-displacement relation for small strains is[4,57]

$$\{\epsilon\} = \begin{Bmatrix} \epsilon_x \\ \epsilon_y \\ \gamma_{xy} \end{Bmatrix} = \begin{Bmatrix} \dfrac{\partial u}{\partial x} \\ \dfrac{\partial v}{\partial y} \\ \dfrac{\partial u}{\partial y} + \dfrac{\partial v}{\partial x} \end{Bmatrix} = [B]\{q\} \tag{1-56a}$$

in which $[B]$ is obtained by taking appropriate derivatives of N_i

$$[B] = \begin{bmatrix} \dfrac{\partial N_1}{\partial x} & 0 & \dfrac{\partial N_2}{\partial x} & 0 & \dfrac{\partial N_3}{\partial x} & 0 & \dfrac{\partial N_4}{\partial x} & 0 \\ 0 & \dfrac{\partial N_1}{\partial y} & 0 & \dfrac{\partial N_2}{\partial y} & 0 & \dfrac{\partial N_3}{\partial y} & 0 & \dfrac{\partial N_4}{\partial y} \\ \dfrac{\partial N_1}{\partial y} & \dfrac{\partial N_1}{\partial x} & \dfrac{\partial N_2}{\partial y} & \dfrac{\partial N_2}{\partial x} & \dfrac{\partial N_3}{\partial y} & \dfrac{\partial N_3}{\partial x} & \dfrac{\partial N_4}{\partial y} & \dfrac{\partial N_4}{\partial x} \end{bmatrix} \tag{1-56b}$$

$$= [[B_1] \quad [B_2] \quad [B_3] \quad [B_4]] \tag{1-56c}$$

and

$$[B_i] = \begin{bmatrix} \dfrac{\partial N_i}{\partial x} & 0 \\ 0 & \dfrac{\partial N_i}{\partial y} \\ \dfrac{\partial N_i}{\partial y} & \dfrac{\partial N_i}{\partial x} \end{bmatrix} \tag{1-56d}$$

The global and local derivatives are related through the jacobian as

$$\begin{Bmatrix} \dfrac{\partial N_i}{\partial x} \\ \dfrac{\partial N_i}{\partial y} \end{Bmatrix} = [J]^{-1} \begin{Bmatrix} \dfrac{\partial N_i}{\partial s} \\ \dfrac{\partial N_i}{\partial t} \end{Bmatrix} \qquad \text{where} \quad [J]^{-1} = \frac{1}{|J|} \begin{bmatrix} \dfrac{\partial y}{\partial t} & -\dfrac{\partial y}{\partial s} \\ -\dfrac{\partial x}{\partial t} & \dfrac{\partial x}{\partial s} \end{bmatrix} \tag{1-57a}$$

and $|J|$ is the determinant of $[J]$,

$$|J| = \begin{vmatrix} \dfrac{\partial x}{\partial s} & \dfrac{\partial y}{\partial s} \\ \dfrac{\partial x}{\partial t} & \dfrac{\partial y}{\partial t} \end{vmatrix} = \sum_{i=1}^{n} \sum_{j=1}^{n} x_i \left(\frac{\partial N_i}{\partial s} \frac{\partial N_j}{\partial t} - \frac{\partial N_i}{\partial t} \frac{\partial N_j}{\partial s} \right) y_j \tag{1-57b}$$

where n is the number of nodes in the element.

The variational functional for the displacement method is given by the potential energy Π_p of the system, which can be expressed as[4]

$$\Pi_p = \iiint_V dU(u, v) - \iiint_V (\bar{X}u + \bar{Y}v)\, dV - \iint_{S_1} (\bar{T}_x u + \bar{T}_y v)\, dS_1 \qquad (1\text{-}58a)$$

where $dU(u, v)$ = strain energy per unit volume
$\bar{X}, \bar{Y}$ = components of prescribed body forces, e.g., weight of body
$\bar{T}_x, \bar{T}_y$ = prescribed surface tractions
V = volume of element
S_1 = surface on which tractions are specified

By assuming that the material behavior is linearly elastic, we can use the results from the theory of elasticity and express dU as

$$dU = \tfrac{1}{2}\{\epsilon\}^T\{\sigma\}\, dV \qquad (1\text{-}59)$$

where $\{\sigma\}^T = [\sigma_x \quad \sigma_y \quad \tau_{xy}]$ is the vector of components of stress.

Stresses and strains are connected through what we term *stress-strain laws* or *constitutive laws*. A simple constitutive law is Hooke's law defining linear elastic behavior. When we use a generalized Hooke's law, the stress-strain relation is[57]

$$\{\sigma\} = [C]\{\epsilon\} \qquad (1\text{-}60)$$

where $[C]$ is the stress-strain matrix. For a linear elastic and isotropic material, it is composed of pairs of parameters, Young's modulus E and Poisson's ratio ν; or bulk modulus K and shear modulus G; or Lamé's constants λ and μ. In Chaps. 2 and 3, we shall discuss various linear and nonlinear stress-strain relations used in geotechnical engineering. For the present, we shall consider the linear elastic isotropic behavior. Equation (1-58a) now becomes

$$\Pi_p = \frac{1}{2}\iiint_V (\{\epsilon\}^T[C]\{\epsilon\} - 2\{u\}^T\{\bar{X}\})\, dV - \iint_{S_1} \{u\}^T\{\bar{T}\}\, dS_1 \qquad (1\text{-}58b)$$

where $\{\bar{X}\}^T = [\bar{X} \quad \bar{Y} \quad \bar{Z}]$ and $\{\bar{T}\}^T = [\bar{T}_x \quad \bar{T}_y \quad \bar{T}_z]$

are the vectors of body forces and surface tractions, respectively. Substitution of Eqs. (1-56) into Eq. (1-58) leads to

$$\Pi_p = \frac{1}{2}\iiint_V (\{q\}^T[B]^T[C][B]\{q\} - 2\{q\}^T[N]\{\bar{X}\})\, dV - \iint_{S_1} \{q\}^T[N]^T\{\bar{T}\}\, dS_1 \qquad (1\text{-}58c)$$

By taking the first variation of Π_p with respect to (nodal) displacements and invoking the principle of stationary (minimum) potential energy

$$\delta\Pi_p = 0 \qquad (1\text{-}58d)$$

we obtain

$$[k]\{q\} = \{Q\} \tag{1-52}$$

where for the element in Fig. 1-14

$$[k] = h \int_{-1}^{1} \int_{-1}^{1} [B]^T[C][B]\,|J|\; ds\; dt \tag{1-59}$$

and
$$\{Q\} = h \int_{-1}^{1} \int_{-1}^{1} [N]^T\{\bar{X}\}\,|J|\; ds\; dt + h \int_{S_1} [N]^T\{\bar{T}\}\; dS_1 \tag{1-60}$$

in which h is the constant thickness of the element; for the plane strain case we adopt h as unity.

Axisymmetric Problems[51,57–59]

This idealization permits use of a two-dimensional simulation of a three-dimensional problem in which the stress-strain relation is given by

$$\begin{Bmatrix} \sigma_r \\ \sigma_z \\ \sigma_\theta \\ \tau_{yz} \end{Bmatrix} = \frac{E}{(1+\nu)(1-2\nu)} \begin{bmatrix} 1-\nu & \nu & \nu & 0 \\ & 1-\nu & \nu & 0 \\ & & 1-\nu & 0 \\ \text{sym} & & & \dfrac{1-2\nu}{2} \end{bmatrix} \begin{Bmatrix} \epsilon_r \\ \epsilon_z \\ \epsilon_\theta \\ \gamma_{rz} \end{Bmatrix} \tag{1-61}$$

Since the problem is symmetric with respect to both geometry and loading, only a unit angular sector (Fig. 1-15*c*) need be considered. Often a sector of 1 rad is used. The derivation of $[k]$ and $\{Q\}$ matrices will follow similar procedure except for some modifications in the final integrations. For example, the $[k]$ matrix for the element in Fig. 1-14 will now be

$$[k] = \int_{-1}^{1} \int_{-1}^{1} [B]^T[C][B]r\,|J|\; ds\; dt \tag{1-62}$$

Three-dimensional Problems

The common element used in geotechnical engineering is the hexahedral, or brick, element (Fig. 7-4*a*), with eight primary external nodes. With three degrees of freedom at each node (u, v, w), the stiffness matrix will be of order 24×24.

Using the isoparametric concept, we can express the displacements and coordinates in an element as

$$\{u\} = \begin{Bmatrix} u \\ v \\ w \end{Bmatrix} = \begin{bmatrix} N_1 & 0 & 0 & N_2 & 0 & 0 & \cdots & N_8 & 0 & 0 \\ 0 & N_1 & 0 & 0 & N_2 & 0 & \cdots & 0 & N_8 & 0 \\ 0 & 0 & N_1 & 0 & 0 & N_1 & \cdots & 0 & 0 & N_8 \end{bmatrix} \{q\} \tag{1-63}$$

and

$$\begin{Bmatrix} x \\ y \\ z \end{Bmatrix} = \begin{bmatrix} [N_1 \quad N_2 \quad \cdots \quad N_8] & [0] & [0] & \{x_n\} \\ [0] & [N_1 \quad N_2 \quad \cdots \quad N_8] & [0] & \{y_n\} \\ [0] & [0] & [N_1 \quad N_2 \quad \cdots \quad N_8] & \{z_n\} \end{bmatrix} \tag{1-64}$$

where $\{q\}^T = [u_1 \quad v_1 \quad w_1 \quad u_2 \quad v_2 \quad w_2 \quad \cdots \quad u_8 \quad v_8 \quad w_8]$

$\{x_n\}^n = [x_1 \quad x_2 \quad \cdots \quad x_8]$

. . . .

and $N_i = \frac{1}{8}(1 + rr_i)(1 + ss_i)(1 + tt_i) \qquad i = 1, 2, \ldots, 8$

Details of the derivation of stiffness matrix and load vector are similar to those for two-dimensional element and lead to general expressions of the same form as in Eq. (1-52). Stiffness matrices and load vectors are usually evaluated by numerical integration.[4,60,61]

Step 4: Assembling the element properties to form global equations Equations such as Eq. (1-52) are obtained for each element in the structure (Fig. 1-10). The next step is to combine those equations to obtain a stiffness relation for the entire system. This is done essentially by adding together the matrix equations for each element one by one. The addition, which is usually called the *direct stiffness method*, is performed to satisfy the basic physical condition that the structure should remain continuous. In other words, we satisfy the compatibility of displacements at nodal points across adjacent elements in the discretized assemblage (Fig. 1-11). Details of the assembly procedure are available in a large number of publications.

The stiffness relation for the entire body, often called the *global relation*, is expressed as

$$[K]\{r\} = \{R\} \tag{1-65}$$

where $[K]$ = global, or assemblage, stiffness matrix
$\{r\}$ = global nodal displacement vector
$\{R\}$ = global nodal force (forcing-parameter) vector

Often the element stiffness relations are formed with respect to a local coordinate system. It is necessary to transform such local relations into a global coordinate system before they are assembled to form Eqs. (1-65). Before the set of simultaneous equations (1-65) can be solved, the prescribed geometric boundary conditions are introduced into the system, a step that modifies the set of simultaneous equations. The natural boundary conditions, such as vanishing of slopes, are satisfied through the variational formulation and do not require special treatment.

Step 5: Computation of primary and secondary quantities In the displacement approach, nodal displacements are computed as primary quantities by solving Eqs. (1-65). Such quantities as stresses and strains are the secondary quantities computed from the nodal displacements. Table 1-2 lists primary unknowns and

Table 1-2 Primary unknowns and secondary quantities for various problems

Problem	Primary unknown	Secondary quantities
Stress deformation, static and dynamic foundations, dams and embankments, slopes, pavements, underground structures	Displacements†	Strains, stresses, accelerations, velocities
Seepage and flow	Fluid potentials‡	Velocities, quantity of flow
Coupled consolidation, liquefaction	Displacements and pore water pressure	Strains, stresses, quantity of flow

† Stresses or both stresses and displacements can be used.
‡ Stream function and/or velocity can be used.

secondary quantities commonly sought in some of the problems in geotechnical engineering. For the stress-deformation problem the (matrix) relations obtained during the formulation can be used. For instance, the strains and stresses can be found from

$$\{\epsilon\} = [B]\{q\} \tag{1-56a}$$

$$\{\sigma\} = [C]\{\epsilon\} = [C][B]\{q\} \tag{1-61a}$$

in which $\{q\}$ is the vector of nodal displacements, which is now known. The principal stresses and strains can then be evaluated by using appropriate formulas.[57]

Weighted Residual Methods; Galerkin's Method

As mentioned earlier, residual methods are becoming more popular for formulation of the element equations. In this approach, we operate directly on the governing differential equation and do not need a variational functional. Moreover, residual methods can be used for problems governed by nonlinear equations.

For details of the mathematics of residual methods, the reader should consult various publications.[3,4,12,35,46,49,50,62–66] Here we shall illustrate the Galerkin's (residual) method for a one-dimensional seepage problem governed by a linearized equation[49]

$$k_x \bar{h}(x, t)\frac{\partial^2 h}{\partial x^2} = n\frac{\partial h}{\partial t} \tag{1-66}$$

where h = head (height) of free surface (Fig. 1-16a)
n = effective porosity
$\bar{h}$ = mean head

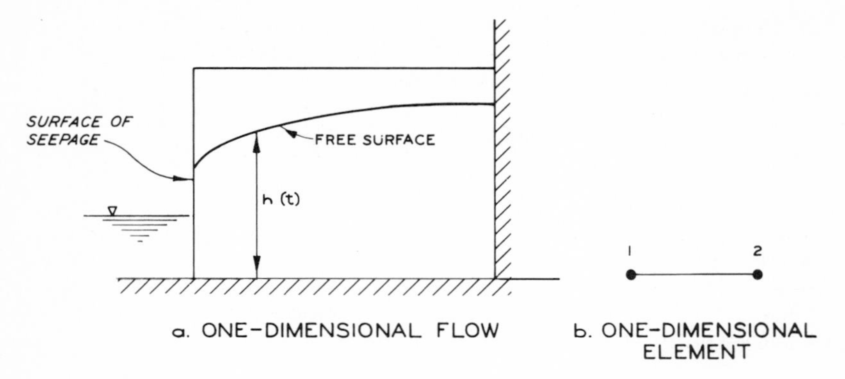

Figure 1-16 Flow idealized as one-dimensional.

We denote an approximate solution to h' in an element (Fig. 1-16*b*) as

$$h' = \sum N_i(x_i)h_i(t) \tag{1-67}$$

where N_i are the interpolation functions and h_i are the nodal heads. Substitution of h' in Eq. (1-66) will leave a residual

$$R = k_x \bar{h}(x, t)\frac{\partial^2 h'}{\partial x^2} - n\frac{\partial h'}{\partial t} \tag{1-68}$$

In residual methods, the weighted residual R is equated to zero. A number of different procedures such as collocation, subdomain, and Galerkin[3] methods can be used; they differ in the choice of the weighting functions.

Galerkin's method is the most commonly employed procedure in finite element applications. In this method, the interpolation functions N_i are used as weighting functions. Hence

$$\int_D RN_i \, dD = 0 \qquad i = 1, 2, \ldots, r \tag{1-69a}$$

where D denotes the flow domain and r the number of degrees of freedom. Now substitution of Eq. (1-68) in Eq. (1-69*a*) gives

$$\int_D \left[k_x \bar{h}(x, t)\frac{\partial^2}{\partial x^2} - n\frac{\partial}{\partial t} \right] \sum_{i=1}^{r} (N_i h_i) N_m \, dD = 0 \tag{1-69b}$$

which yields m simultaneous (linear) equations. By applying Green's theorem we obtain

$$\int_D \left[k_x \bar{h}(x, t)\frac{\partial N_m}{\partial x} \sum_1^r \frac{\partial N_i}{\partial x} \right] h_i \, dD + \int_D nN_m \sum_1^r N_i \dot{h}_i \, dD - \int_S N_m \left(k_x \sum_1^r \frac{\partial N_i}{\partial x} l_x \right) h_i \, dS = 0 \tag{1-69c}$$

in which l_x is the direction cosine of the normal to the boundary S and the dot denotes the derivative with respect to time. Equation (1-69c) leads to assemblage equations expressed in matrix form as

$$[K]\{r\} + [P]\{\dot{r}\} = \{R\} \tag{1-70}$$

where

$$K_{mi} = \sum_{1}^{M} \int_E \left[k_x \bar{h}(x, t) \frac{\partial N_m}{\partial x} \frac{\partial N_i}{\partial x} \right] dx$$

$$P_{mi} = \sum_{1}^{M} \int_E nNmN_i \, dx \qquad \text{and} \qquad R_m = -\sum_{1}^{M} \int_E N_m \bar{q} \, dS$$

where $[K]$ = permeability matrix
$[P]$ = porosity matrix
$\{R\}$ = forcing-parameter vector
E = element
M = number of elements
$\bar{q}$ = applied fluid flux

The summation is carried out for all elements in the domain.

1-9 ADDITIONAL EXAMPLES OF FINITE ELEMENT FORMULATIONS

In the previous section we illustrated derivations for typical stress-deformation and flow problems by using variational and Galerkin methods. In this section, we shall briefly state formulations for finite element equations for other relevant problems such as dynamics and earthquake analysis, seepage, and consolidation.

Dynamic Analysis†

Hamilton's variational principle,[4] used commonly in dynamic and earthquake analysis, is given by

$$\delta \int_{t_1}^{t_2} L \, dt = 0 \tag{1-71}$$

where

$$L = T - U - W_p$$

in which T = kinetic energy
U = strain energy
W_p = potential of external forces

For linearly elastic material,

$$L = \frac{1}{2} \iiint_V (\rho\{\dot{u}\}^T\{\dot{u}\} - \{\epsilon\}^T[C]\{\epsilon\} + 2\{u\}^T\{\bar{X}\}) \, dV + \iint_{S_1} \{u\}^T\{\bar{T}\} \, dS \tag{1-72}$$

† See also Chaps. 19 and 20.

Here the dot denotes the derivative with respect to time, and ρ is the density of the material.

The displacement components over an element can be expressed, as before, as

$$\{u\} = [N]\{q\} \tag{1-54}$$

Proper differentiations of u with respect to space (x, y, z) and time required in L and substitution in Eq. (1-72) lead to the element equations:

$$[m]\{\ddot{q}\} + [k]\{q\} = \{Q(t)\} \tag{1-73a}$$

where $[m]$ is the element mass matrix

$$[m] = \iiint_V \rho[N]^T[N]\, dV \tag{1-73b}$$

and $\{\ddot{q}\}$ denotes the vector of nodal accelerations.

If damping is included, the element equations are

$$[m]\{\ddot{q}\} + [c]\{\dot{q}\} + [k]\{q\} = \{Q(t)\} \tag{1-73c}$$

where $[c]$ is the element damping matrix. The damping and mass matrices can be assembled by following the procedure stated previously. The assemblage equations are then written

$$[M]\{\ddot{r}\} + [C]\{\dot{r}\} + [K]\{r\} = \{R(t)\} \tag{1-74}$$

The mass matrix can be formed either in lumped or consistent form. A number of procedures have been devised for defining the damping matrix. Details of the foregoing derivations and these aspects are given by Desai and Abel.[4]

Seepage†

The basic equation governing the fluid flow or seepage in porous rigid media can be expressed as

$$\frac{\partial}{\partial x}\left(k_x \frac{\partial \phi}{\partial x}\right) + \frac{\partial}{\partial y}\left(k_y \frac{\partial \phi}{\partial y}\right) + \frac{\partial}{\partial z}\left(k_z \frac{\partial \phi}{\partial t}\right) + \bar{Q} = n\frac{\partial \phi}{\partial t} \tag{1-75}$$

where k_x, k_y, k_z = coefficients of permeability (transmissivity, etc.) in x, y, z directions, respectively

$\phi = p/\gamma + z$ = fluid potential

p = pressure

γ = unit weight of water

z = elevation head

$\bar{Q}$ = specified fluid flux

n = effective porosity

† See also Chap. 14.

The boundary conditions (Sec. 1-5) that can occur in the flow problem are described in Chap. 14. The variational principle, used commonly, is[4,34,67]

$$A = \iiint\limits_V \left[\frac{1}{2} k_x \left(\frac{\partial \phi}{\partial x}\right)^2 + k_y \left(\frac{\partial \phi}{\partial y}\right)^2 + k_z \left(\frac{\partial \phi}{\partial z}\right)^2 - 2\left(\bar{Q} - n\frac{\partial \phi}{\partial t}\right)\phi\right] dV - \iint\limits_{S_2} \bar{q}\phi \, dS \tag{1-76}$$

We now express variation of fluid potential within an element as

$$\phi = \{N\}^T\{q\} \tag{1-77}$$

where $\{q\}$ is the vector of nodal fluid potentials. Then use of a variational principle (or a residual method[4,35,46,49,50]) leads to the element equations as

$$[k]\{q\} + [p]\{\dot{q}\} = \{Q(t)\} \tag{1-78a}$$

where

$$[k] = \iiint\limits_V [B]^T[R][B] \, dV$$

$$[p] = \iiint\limits_V [N]^T n[N] \, dV \qquad \text{and} \qquad \{Q\} = \iiint\limits_V [N]^T\{\bar{Q}\} \, dV$$

Here

$$[R] = \begin{bmatrix} k_x & 0 \\ 0 & k_y \end{bmatrix}$$

is the matrix of principal permeabilities, $[k]$ is the element (permeability) matrix, and $[p]$ is the element (porosity) matrix. Matrix $[B]$ is given by

$$\{g\} = [B]\{q\} \tag{1-78b}$$

where $\{g\}$ is the vector of gradients

$$\{g\}^T = \begin{bmatrix} \frac{\partial \phi}{\partial x} & \frac{\partial \phi}{\partial y} & \frac{\partial \phi}{\partial z} \end{bmatrix} \tag{1-78c}$$

For steady seepage, Eq. (1-78*a*) specializes to

$$[k]\{q\} = \{0\} \tag{1-78d}$$

We can use higher-order finite elements and follow similar procedures. The derivation for three-dimensional flow can be obtained in an analogous manner.[4,68]

Consolidation†

One-dimensional consolidation The basic equation governing one-dimensional consolidation is

$$c_v \frac{\partial^2 u_p}{\partial z^2} - \frac{\partial u_p}{\partial t} = \bar{Q}(t) \tag{1-79}$$

† See also Chaps. 11 and 12.

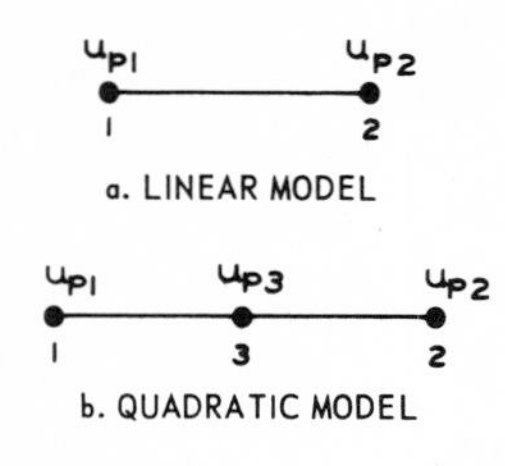

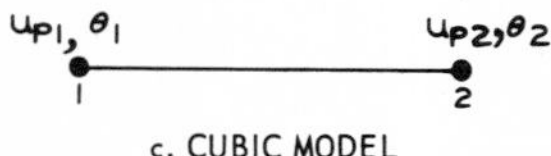

Figure 1-17 Approximation models for one-dimensional consolidation.

where c_v = coefficient of consolidation
u_p = excess pore water pressure
$\bar{Q}(t)$ = forcing parameter, e.g., externally applied fluid flux

Here, since we consider only one-dimensional deformations, it is necessary to define only one-dimensional line elements (Fig. 1-17). Various orders of approximation can be used for u_p. Some of the simple interpolation models are:[69]

Linear (Fig. 1-17*a*):

$$u_p = \left[\tfrac{1}{2}(1 - L) \quad \tfrac{1}{2}(1 + L)\right]\begin{Bmatrix} u_{p1} \\ u_{p2} \end{Bmatrix} \tag{1-80a}$$

Quadratic (Fig. 1-17*b*):

$$u_p = \left[\tfrac{1}{2}L(L - 1) \quad \tfrac{1}{2}L(L + 1) \quad (1 - L^2)\right]\begin{Bmatrix} u_{p1} \\ u_{p2} \\ u_{p3} \end{Bmatrix} \tag{1-80b}$$

Cubic (Fig. 1-17*c*):

$$u_p = \left[L_1^2(3 - 2L_1) \quad L_1^2 L_2 a \quad L_2^2(3 - 2L_2) \quad -L_1 L_2^2 a\right]\begin{Bmatrix} u_{p1} \\ \theta_1 \\ u_{p2} \\ \theta_2 \end{Bmatrix} \tag{1-80c}$$

Here θ is the gradient of $u_p = \partial u_p/\partial x$, a is the element length, and L, L_1, L_2 are local coordinates. Chapter 11 presents a number of other higher-order functions.

Following the procedure similar to that for the seepage problem gives the element equations

$$[k]\{q\} + [p]\{\dot{q}\} = \{Q(t)\} \tag{1-81}$$

where $[k]$ and $[p]$ are element property matrices and $\{q\}$ is the vector of element nodal pore water pressures (and gradients). The assembly procedure is essentially similar to that used for stress-deformation problems.

Consolidation is an example of a coupled phenomenon in which deformation of the soil skeleton and dissipation of pore water pressure take place simultaneously and both actions are interrelated. In one-dimensional consolidation under various assumptions,[1] is is permissible to consider the behavior of pore water pressure and the vertical deformation by separate equations. The former is governed by Eq. (1-79), and the latter is expressed through linear relation between strain (void ratio) and applied pressure

$$d\sigma = -\frac{1}{a_v}\,de \tag{1-82a}$$

where e = void ratio
$d\sigma$ = applied stress increment
a_v = coefficient of compressibility

The excess pore water pressure u_p, total applied stress σ, and effective stress σ', are connected[1] as follows:

$$\sigma = \sigma' + u \tag{1-82b}$$

Hence, once values of u_p are computed, the (vertical) deformation can be obtained from Eq. (1-82*a*).

Further details on one-dimensional consolidation are available in Chap. 11; two- and three-dimensional consolidation are covered in Chap. 12.

Other coupled phenomena Some of the coupled problems relevant in geotechnical engineering are *expansion* (or *swelling*), *electroosmosis*, and *liquefaction*. Some of these topics are covered in subsequent chapters.

1-10 TECHNIQUES FOR NONLINEAR ANALYSIS

Most problems in geotechnical engineering exhibit nonlinear behavior. The two main techniques used for nonlinear analyses are *incremental* and *iterative*. *Mixed* procedures are possible by a combination of the two techniques. We shall briefly state some salient aspects of these techniques.

In geotechnical engineering, the *material* nonlinearities, exhibited through variable material parameters, can be caused by a number of factors such as state of stress (strain), in situ stresses, joints and discontinuities, history of loading, temperature, and existence of fluids in the pores. The other kind of nonlinearity, called *geometric nonlinearity*, is caused by significant changes in the geometry of the loaded body; this is much less common in geotechnical engineering. Since the procedures for handling material nonlinearity can be extended for geometric nonlinearity, we shall cover only the techniques as they relate to material nonlinearity. The general question of both material and geometric nonlinearity is discussed by various workers.[4,66]

Incremental Procedure

Nonlinear behavior is approximated as piecewise linear, and the linear laws are used for each piece. For convenience, we have considered the stress-deformation problem, and the corresponding nonlinear behavior expressed through stress-strain ($\bar{\sigma}$-$\bar{\epsilon}$) curves (Chap. 2). The load-displacement relation is also shown in Fig. 1-18*a*. We shall consider the element equations (1-52) to illustrate the incremental procedure (Fig. 1-18*a*)

$$[k]\{q\} = \{Q\} \tag{1-83a}$$

Here $[k]$ is the stiffness matrix, which is no longer constant, as in the case of linear elastic behavior, but is a function of state of stress and loading; hence

$$[k] = [k(q, Q)] \tag{1-83b}$$

The matrix $[k]$ is function of $[C]$, which in turn is function of the state of stress $\{\sigma\}$ and strain $\{\epsilon\}$.

In the incremental procedure, the total load $\{Q\}$ is divided into "small" increments $\{\Delta Q\}$, which are applied one by one. The displacements and strains and stresses are computed for each load increment and are successively accumulated.

If the load is divided into m increments, and if $\{Q_0\}$ denotes the load vector corresponding to some initial loading, then

$$\{Q\} = \{Q_0\} + \sum_{i=1}^{m} \{\Delta Q_i\} \tag{1-84a}$$

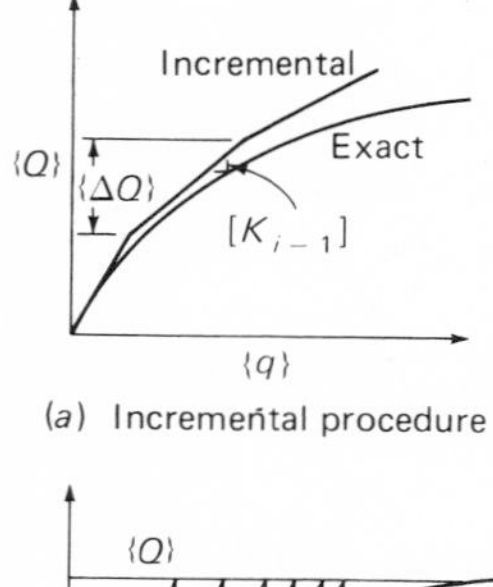

(*a*) Incremental procedure

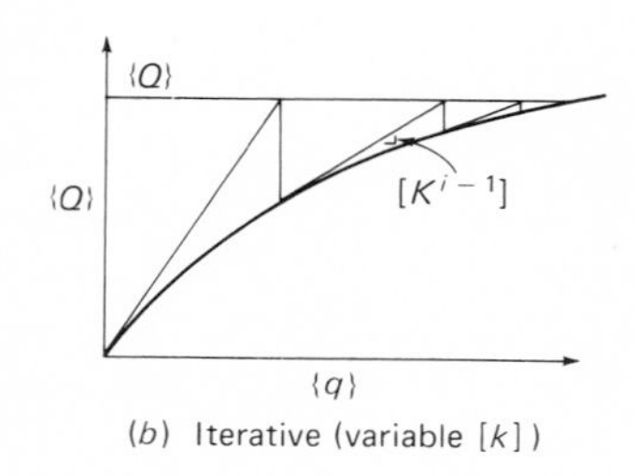

(*b*) Iterative (variable $[k]$)

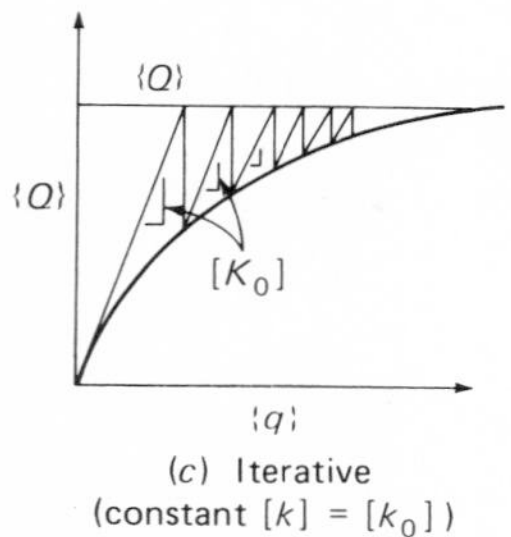

(*c*) Iterative (constant $[k] = [k_0]$)

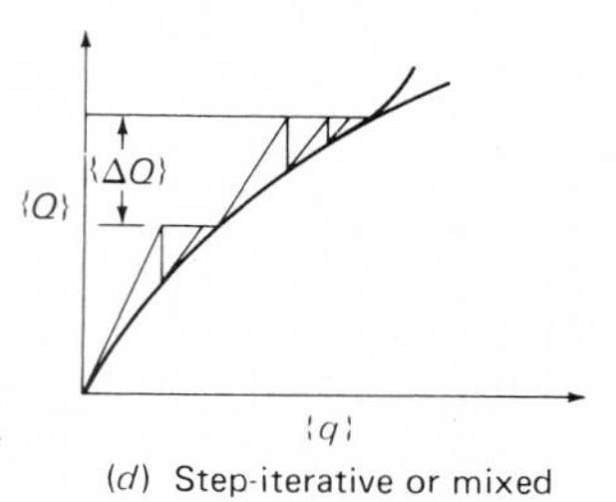

(*d*) Step-iterative or mixed

Figure 1-18 Techniques for nonlinear analysis. (From Ref. 4. Reprinted by permission of Van Nostrand Reinhold Company.)

At the nth stage of incremental loading

$$\{Q_n\} = \{Q_0\} + \sum_{i=1}^{n} \{\Delta Q_i\} \qquad (1\text{-}84b)$$

The displacements, strains, and stresses are

$$\{q_n\} = \{q_0\} + \sum_{i=1}^{n} \{\Delta q_i\} \qquad (1\text{-}85a)$$

$$\{\epsilon_n\} = \{\epsilon_0\} + \sum_{i=1}^{n} \{\Delta \epsilon_i\} \qquad (1\text{-}85b)$$

and

$$\{\sigma_n\} = \{\sigma_0\} + \sum_{i=1}^{n} \{\Delta \sigma_i\} \qquad (1\text{-}85c)$$

The incremental displacements $\{\Delta q_i\}$ are computed from

$$[k_t]\{\Delta q_n\} = \{\Delta Q_n\} \qquad n = 1, 2, 3, \ldots, m \qquad (1\text{-}86a)$$

The value of $[k_t]$ can be evaluated either as *tangent* stiffness or *secant* stiffness. In a simple scheme

$$[k_t] = [k_{n-1}] \qquad (1\text{-}86b)$$

i.e., the value of $[k_t]$ at the end of previous increment is used for the next increment. Alternative procedures, similar to Runge-Kutta schemes, can also be used to evaluate $[k_t]$.

Iterative, Mixed, and Other Procedures

In the *iterative* procedure (Fig. 1-18*b*, *c*) the total load $\{Q\}$ is applied at a time, and iterations are then performed to satisfy equilibrium on stress and strain. In the mixed procedure, both the incremental and iterative procedures are combined (Fig. 1-18*d*). Other procedures such as *initial-strain* and *initial-stress methods* are also used in geotechnical applications. Some of the subsequent chapters consider applications of these procedures. Comprehensive descriptions and comparisons between various procedures are given by Desai and Abel.[4]

1-11 TREATMENT OF TRANSIENT OR PROPAGATION PROBLEMS

In a time-dependent problem we first use the finite element method for the continuum in physical space. In the next step, we use a finite difference integration scheme for propagating the solution in the time domain. As typical examples relevant in geotechnical engineering we shall consider the problems of consolidation governed by the parabolic equation (1-81) and of dynamic and earthquake analysis, Eq. (1-74).

Consolidation

A number of time-integration schemes can be used. The simplest scheme would be the *forward-difference* type of procedure[5–7] expressed as

$$\{\dot{r}\} = \frac{\{r\}_{t+1} - \{r\}_t}{\Delta t} \tag{1-87a}$$

Substitution of Eq. (1-87*a*) in the assemblage equation corresponding to Eq. (1-70) leads to

$$\left([K] + \frac{1}{\Delta t}[P]\right)\{r\}_{t+1} = \{R\} + \frac{1}{\Delta t}[P]\{r\}_t \tag{1-87b}$$

Since $\{r\}_{t=0}$ are prescribed, the right-hand side in Eq. (1-87*b*) is known and the equations can be solved for $\{r\}_{0+\Delta t}$, and so on.

Another scheme, commonly employed in finite element applications,[22,46,54,70] first solves for values of unknowns at the time level at the middle of the time increment $t + \frac{1}{2}$. The equations to be solved at $t + \frac{1}{2}$ are given by

$$\left([K] + \frac{2}{\Delta t}[P]\right)\{r\}_{t+1/2} = \{R\} + \frac{2}{\Delta t}[P]\{r\}_t \tag{1-88a}$$

Then the values at $t + 1$ are evaluated from

$$\{r\}_{t+1/2} = \tfrac{1}{2}(\{r\}_{t+1} + \{r\}_t) \tag{1-88b}$$

It has been shown that both the foregoing schemes can be acceptable and stable when used with linear and cubic approximating models.[64,71] It was observed, however, that stability itself may not be the only governing factor in choosing a scheme; required accuracy also can play an important role.

Dynamic Analysis

The basic finite element equations relevant to this problem are Eqs. (1-74). We shall describe briefly a few of the common schemes.

According to the *Newmark's generalized acceleration method*,[72] the nodal displacements, velocities, and accelerations can be expressed as

$$\{r\}_{t+1} = \{r\}_t + \Delta t\{\dot{r}\}_t + (\tfrac{1}{2} - \beta)\,\Delta t^2\{\ddot{r}\}_t + \beta\,\Delta t^2\{\ddot{r}\}_{t+1} \tag{1-89a}$$

and

$$\{\dot{r}\}_{t+1} = \{\dot{r}\}_t + (1 - \gamma)\,\Delta t\{\ddot{r}\}_t + \gamma\,\Delta t\{\ddot{r}\}_{t+1} \tag{1-89b}$$

in which β and γ are dimensionless parameters.

Wilson and coworkers[73–75] have used integration schemes for the problems of structural dynamics and dynamic soil-structure interaction. In this approach[73] the acceleration at the nodes is assumed to vary linearly during the time interval between t and $t + 1$. Consequently, the nodal displacements are approximated by a cubic function. The relations between the nodal displacements, velocities, and acceleration are

$$\{\dot{r}\}_{t+1} = \frac{3}{\Delta t}\{r\}_{t+1} - \left(\frac{3}{\Delta t}\{r\}_t + 2\{\dot{r}\}_t + \frac{\Delta t}{2}\{\ddot{r}\}_t\right) \tag{1-90a}$$

and

$$\{\ddot{r}\}_{t+1} = \frac{6}{(\Delta t)^2}\{r\}_{t+1} - \left[\frac{6}{(\Delta t)^2}\{r\}_t + \frac{6}{\Delta t}\{\dot{r}\}_t + 2\{\ddot{r}\}_t\right] \tag{1-90b}$$

Note that if the values of $\beta = \frac{1}{6}$ and $\gamma = \frac{1}{2}$ are substituted in Eq. (1-89*a*), we obtain Eqs. (1-90). The scheme is conditionally stable and is widely used. A scheme based on Gurtin's method[77] and comparison of numerical characteristics such as stability of various schemes for dynamic analysis are given by Nickell.[76,78] Use of such other schemes as *explicit second central difference*, *Fox-Goodwin*, and *linear acceleration* for earthquake analysis are discussed by Ghaboussi and Wilson.[79]

We now return to the scheme that assumes linear variation of acceleration between t and $t + 1$. Substitution of the values of nodal velocities and accelerations [Eqs. (1-90*a*) and (1-90*b*)] in Eqs. (1-74) leads to the final assemblage equations

$$[\bar{K}]\{r\}_{t+1} = \{\bar{R}(t)\}_t \tag{1-91}$$

For example, with Eq. (1-90) we obtain

$$[\bar{K}] = [K] + \frac{3}{\Delta t}[C] + \frac{6}{(\Delta t)^2}[M] \tag{1-92a}$$

and

$$\{\bar{R}(t)\}_t = \{R(t)\}_t + [C]\left(\frac{3}{\Delta t}\{r\}_t + 2\{\dot{r}\}_t + \frac{\Delta t}{2}\{\ddot{r}\}_t\right) + [M]\left[\frac{6}{(\Delta t)^2}\{r\}_t + \frac{6}{\Delta t}\{\dot{r}\}_t + 2\{\ddot{r}\}_t\right] \tag{1-92b}$$

The *step-by-step* time-integration procedure discussed above is employed for problems in dynamic earth-structure interaction and earthquake analyses. Another procedure for dynamic analysis is *mode superposition*. For details of these procedures and comparisons see Chaps. 19 and 20 and other publications.[4,80,81]

1-12 SOLUTIONS OF EQUATIONS

A numerical procedure usually results in a set of simultaneous algebraic linear or nonlinear equations

$$[K]\{r\} = \{R\} \tag{1-65}$$

where it is required to compute unknowns such as nodal displacements and fluid potentials in $\{r\}$. A number of solution procedures are available.

Table 1-3 Methods of solution[52]

Direct methods	Iterative methods
Gauss elimination	Point-Jacobi
Conjugate gradient	Gauss-Seidel
Bandwidth elimination	Successive overrelaxation
Nested dissection	ADI
Frontal elimination	Symmetric successive over relaxation (SSOR)

Gaussian elimination is perhaps the commonest procedure employed for solution of linear equations generated in numerical techniques. It is a direct approach, as opposed to an *iterative* approach. In the direct elimination procedure, the matrix $[K]$ is reduced to a triangular form from which the unknowns $\{r\}$ are found directly. A number of modified versions of gaussian elimination, e.g., Gauss-Doolittle schemes, have also been successfully employed. Iterative procedures such as the Gauss-Seidel and successive overrelaxation methods for linear equations involve a series of approximations in which an initial estimate, or guess, is corrected or refined successively.

For nonlinear equations, repeated applications of the elimination and iterative schemes are required. Here, the well-known Newton-Raphson procedure has often been used.[82]

In most cases, the number of equations resulting from the use of the numerical methods [Eq. (1-65)] is very large. However, for most problems in geotechnical engineering, the matrix $[K]$ is *sparsely populated*, or banded, and is symmetric. These properties permit special schemes for handling the equations to obtain efficient computations. One of the schemes that has been commonly used in many finite element applications, the *bandwidth method*, takes advantage of the banded nature. An alternative scheme that improves efficient organization and ordering of elements and nodes is called the *wavefront-* or *frontal-solution scheme*.[83,84]

It is not possible to detail these methods within the scope of this book. A comprehensive review and evaluation of different methods for solutions of equations arising from finite element and finite difference methods is given by Traule[85] and Birkhoff and Fix.[52] Table 1-3 gives some of these methods.

1-13 METHOD OF CHARACTERISTICS

In the method of characteristics, a set of partial differential equations is converted into a set of ordinary differential equations; the latter is often solved by using the finite difference method.

In simple words, a characteristic can be explained as a path of transmission or propagation followed by a quantity or a disturbance such as a shock wave. In a sense, a characteristic is a line along which energy is propagated. The path followed by a vehicle along a road, the lines (or planes) along which an applied

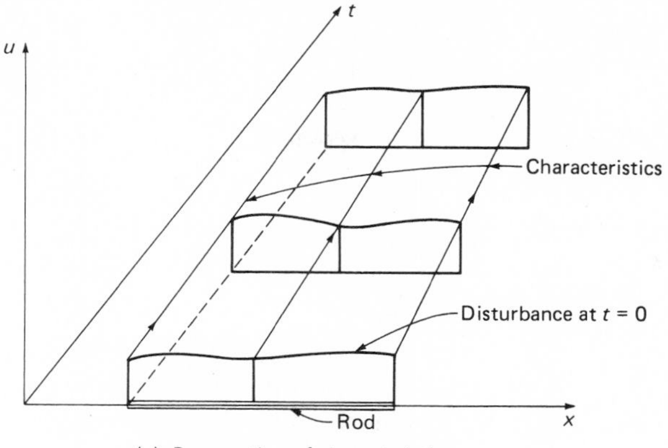

(a) Propagation of characteristics.

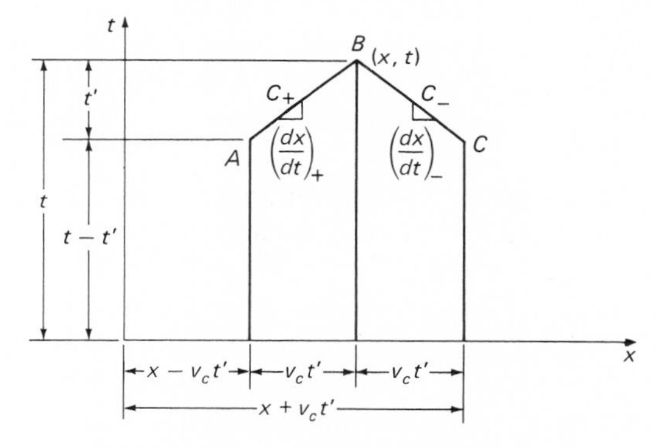

(b) Characteristics for linear wave equation.

Figure 1-19 Propagation of paths for characteristics in wave propagation.

load is transmitted in a geologic mass, and lines along which a disturbance caused by an impact is transmitted are some examples of characteristics.

For simple understanding, let us consider propagation of a wave and of displacement u with time t along a one-dimensional medium (rod) (Fig. 1-19). The initial displacements at $t = 0$ propagate as shown in Fig. 1-19. The paths of propagation of u in the xt plane are the characteristics. The propagation can take place in two directions, left to right and right to left. The paths in the left-to-right direction are denoted C_+ characteristics and those from right to left are denoted C_- characteristics. Solution of the problem involves finding values of u over the xt plane; i.e., if we can find the characteristics, we have the solution. The velocity of propagation $v = du/dt$ also propagates along the same characteristics as the displacement u.

A number of factors can influence the propagation of displacement. Chief among them are (1) initial conditions, (2) the laws governing the problem, often expressed by set(s) of differential equations, and (3) the physical properties of the medium.[86]

We now consider the example of propagation of longitudinal waves along a one-dimensional soil medium.[87] This problem can be assumed to be governed by the two differential equations

$$\frac{E}{\rho}\frac{\partial^2 u}{\partial x^2} = \frac{\partial^2 u}{\partial t^2} \qquad \text{or} \qquad \frac{E}{\rho}\frac{\partial^2 v}{\partial x^2} = \frac{\partial^2 v}{\partial t^2} \tag{1-93}$$

These equations are called *wave equations* and can be derived from the equations of motion (conservation of momentum) and the equations of continuity (conservation of mass).[86] They show that both u and velocity v propagate along the same paths. Here E is the modulus of elasticity, and ρ is the mass density of the medium.

A general solution to Eqs. (1-93) can be written as[88]

$$u = e^{v_c t D_x t} F(x) + e^{-v_c t D_x t} G(x) \tag{1-94}$$

where $v_c = \sqrt{E/\rho}$, $D_x \equiv \partial/\partial x$, and $F(x)$ and $G(x)$ are arbitrary functions. By using Taylor series we can write

$$F(x + v_c t) = e^{v_c t D_x t} F(x) \qquad \text{and} \qquad G(x - v_c t) = e^{-v_c t D_x t} G(x) \tag{1-95}$$

Then

$$u = F(x + v_c t) + G(x - v_c t) \tag{1-96}$$

Here F and G are determined by using initial and boundary conditions.

Note that G is invariant with respect to x and t. That is, its value is the same at point (x, t) and at any other point $(x - v_c t', t - t')$ (Fig. 1-19*b*), where t' is an increment of time. This is because

$$F(x - v_c t') = F[(x - v_c t') - v_c(t - t')] \tag{1-97}$$

The set of points $(x - v_c t', t - t')$ forms a line AB in the xt plane along which the propagation occurs with a velocity $v_c = \sqrt{E/\rho}$ (Fig. 1-19*b*). Similarly, the first term in Eq. (1-94) represents a wave traveling in the negative x direction with a velocity v_c. Physically this derivation illustrates that while the cross section of the medium at $x + \Delta x = x + v_c t'$ would experience a compressive stress, the cross section at x experiences a tensile stress.[89]

The characteristics for linear differential equations (1-93) form two sets of parallel lines. The region with such parallel and straight characteristics is called the *region of constant state*.[86] The characteristics for nonlinear equations are usually curved lines.

Example 1-1 The method of characteristics can be applied to a variety of problems in engineering. In fact, Eq. (1-93) governs a number of phenomena such as propagation of deflections in strings, beams, and struts, propagation of pressure and sound waves, and long-period waves. We consider here the problem of propagation of shear waves in a layered soil medium. The following description is adopted from Streeter, Wylie, and Richart.[87]

Figure 1-20*a* shows a model for transmission of a one-dimensional shear wave in a soil divided into five layers; the deformed shape of an element of soil in the xz plane under shear stresses is also shown in the figure. Figure 1-20*b* shows the characteristics in the zt plane. The partial differential equation of equilibrium is written

$$\frac{\partial \tau}{\partial z} - \rho \frac{\partial^2 u}{\partial t^2} = \frac{\partial \tau}{\partial z} - \rho \frac{\partial v}{\partial t} = 0 \tag{1-98}$$

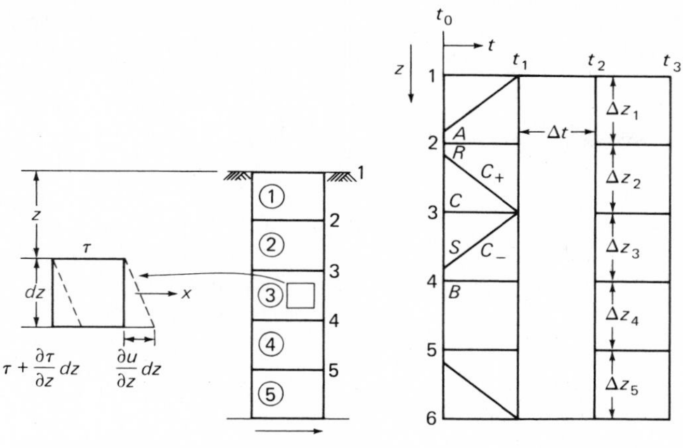

Figure 1-20 Example of one-dimensional wave propagation in layered mass.[87]

where τ = shear stress

u = displacement in x direction

v = velocity of particle = $\partial u/\partial t$

and displacement of a soil particle in the z direction is zero. If the material is assumed to be viscoelastic, the dynamic stress-strain relation is

$$\tau = G\gamma + \mu \frac{\partial v}{\partial t} = G \frac{\partial u}{\partial z} + \mu \frac{\partial^2 u}{\partial z\, \partial t} \tag{1-99}$$

where G = shear modulus = $\partial\tau/\partial\gamma$

μ = coefficient of viscosity

γ = shearing strain

We can write two additional equations

$$d\tau = \frac{\partial \tau}{\partial t} dt + \frac{\partial \tau}{\partial z} dz \qquad \text{and} \qquad dv = \frac{\partial v}{\partial t} dt + \frac{\partial v}{\partial z} dz \tag{1-100}$$

Combination of Eqs. (1-92) to (1-100) leads to

$$\begin{bmatrix} 1 & 0 & 0 & -\rho \\ 0 & 1 & -G + \dfrac{\mu}{\Delta t} & 0 \\ dz & dt & 0 & 0 \\ 0 & 0 & dz & dt \end{bmatrix} \begin{Bmatrix} \dfrac{\partial \tau}{\partial z} \\ \dfrac{\partial \tau}{\partial t} \\ \dfrac{\partial v}{\partial z} \\ \dfrac{\partial v}{\partial t} \end{Bmatrix} = \begin{Bmatrix} 0 \\ -\dfrac{\mu}{\Delta t} \dfrac{\partial v}{\partial z}\Big|_C \\ \partial \tau \\ \partial v \end{Bmatrix} \tag{1-101a}$$

or

$$[A]\{x\} = \{b\} \qquad \text{or} \qquad \{x\} = [A]^{-1}\{b\} \tag{1-101b}$$

Location of point C is shown in Fig. 1-20*b*. By noting that these equations are dependent and by expanding various determinants we obtain

$$\frac{dz}{dt} \pm \sqrt{\frac{G + \mu/\Delta t}{\rho}} = \pm v_s \tag{1-102}$$

where v_s is the shear velocity of the waves. Equations (1-102) indicate that there are two real values of the total derivative dz/dt and hence Eqs. (1-98) are hyperbolic partial differential equations. If the total derivative has only one real value, the equations are called *parabolic*, and if it has two complex values, the equations are *elliptic*.

As before, we can draw lines of propagation in the zt plane with slope given by Eq. (1-102); the lines with positive slope are the C_+ characteristics, and those with negative slope are the C_- characteristics (Fig. 1-20). The region of dependence is the area within which any disturbance can influence the values of u at point P.

Expansion of various determinants in Eq. (1-101*a*) leads to the following equations for the determination of the so-called *Riemann invariants*:[86]

$$\frac{d\tau}{dt}\frac{dz}{dt} - \left(G + \frac{\mu}{\Delta t}\right)\frac{dv}{dt} + \frac{\mu}{\Delta t}\frac{dv}{dz}\bigg|_C \frac{dz}{dt} = 0 \tag{1-103}$$

Substitution of dz/dt from Eq. (1-102) leads to the ordinary differential equations

C_+:

$$\frac{d\tau}{dt} - \rho v_s \frac{dv}{dt} + \frac{\mu}{\Delta t}\frac{dv}{dz}\bigg|_C = 0 \tag{1-104a}$$

$$\frac{dz}{dt} = v_s \tag{1-104b}$$

C_-:

$$\frac{d\tau}{dt} + \rho v_s \frac{dv}{dt} + \frac{\mu}{\Delta t}\frac{dv}{dz}\bigg|_C = 0 \tag{1-105a}$$

$$\frac{dz}{dt} = -v_s \tag{1-105b}$$

In order to solve for the velocity v and shear stress τ, we now use the finite difference method. The difference form of Eqs. (1-104) and (1-105) (Fig. 1-20) is

$$\tau_p - \tau_R - \rho v_{s2}(v_p - v_R) + \frac{\mu(v_B - v_A)}{\Delta z_2 + \Delta z_3} = 0 \tag{1-106a}$$

$$\tau_p - \tau_R + \rho v_{s3}(v_p - v_R) + \frac{\mu(v_B - v_A)}{\Delta z_2 + \Delta z_3} = 0 \tag{1-106b}$$

Equations (1-104*b*) and (1-105*b*) allow determination of increments Δz and Δt. From stability considerations

$$\Delta z \geqslant v_s \Delta t \tag{1-107}$$

Streeter et al.[87] have used the Ramberg-Osgood model (Chap. 2) to simulate nonlinear behavior of the soil:

$$\gamma = \frac{\tau}{G_0}\left(1 + \left|\frac{\tau}{\tau_y}\right|^{R-1}\right) \tag{1-108}$$

Use of Eq. (1-108) allows revision of the value of G with shear strain.

The values of τ and v are assumed to be known at $t = 0$. The following boundary conditions are then introduced. At the soil surface, $\tau_p = 0$; therefore, use of Eq. (1-106*b*) yields

$$v_p = v_s + \frac{1}{\rho v_{s1}}\left[\tau_s - \frac{\mu}{\Delta z_1}(v_B - v_C)\right] \tag{1-109a}$$

At the bottom, v_p is a known function of time, often given from a prescribed seismic motion. Then, using Eq. (1-106*a*), we obtain

$$\tau_p = \tau_R + \rho v_{sn}(v_p - v_R) - \frac{\mu}{\Delta z_{n-1}}(v_C - v_A) \tag{1-109b}$$

The final step is to use Eq. (1-106) to propagate the solution in time and compute the values of τ and v at $t_0 + \Delta t$, and so on.

Comments

The method of characteristics follows the physical process of propagation. Since we need solve only ordinary differential equations, the necessity for solving large systems of equations can be avoided. Thus the method can be found to be economical. At present it can be difficult to handle complexities such as nonhomogeneities and nonlinearities; however, the method shows good promise for further research and applications. Chapter 21 gives details of this method for plasticity solutions.

1-14 BOUNDARY-INTEGRAL-EQUATION METHOD

The boundary-integral-equation (BIE) method involves numerical solution of a set of integral equations that connect the boundary, or surface, tractions to boundary displacements. Unlike the finite element and finite difference methods, the BIE method is based on solution of integral rather than differential equations and consists in discretization of only the boundary or the surface of the body into a number of segments, or elements. The numerical solution is first obtained at the boundary, and then the solution at different points within the body is obtained from the solutions at the boundary. As a consequence, the number of physical dimensions to be considered in the BIE method is reduced by 1, resulting in savings in effort and computer time.

The method is commonly based on the *Fredholm equation*,[91] defined on the boundary as

$$\phi(x) - \lambda \int_a^b K(x, s)\phi(s)\, ds = f(x) \qquad a \leqslant x \leqslant b \tag{1-110}$$

where $\phi(x)$ = unknown function
$K(x, s)$, $\phi(s)$ = given functions
λ = a parameter
x = coordinate

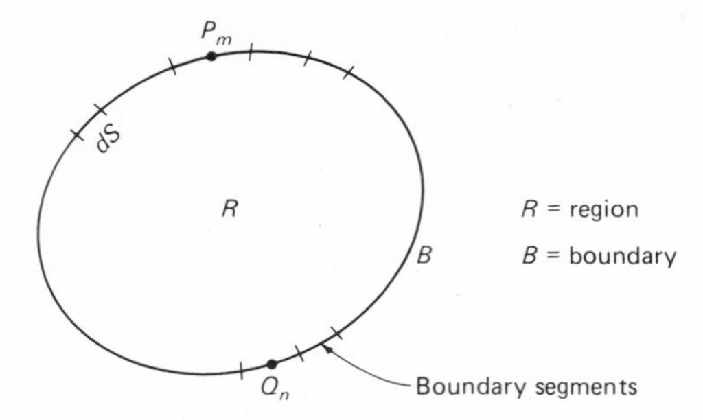

Figure 1-21 Boundary integral equation method.

$a \leqslant x \leqslant b$ denotes the domain of interest of the equation.

For the stress-deformation problems, Eq. (1-110) can be specialized to[92,93]

$$\frac{u_i(P)}{2} + \int_B u_j(Q)T_{ij}(Q, P)\, ds(Q) = \int_B t_j(Q)U_{ij}(Q, P)\, ds(Q) \qquad (1\text{-}111)$$

where P, Q = points on boundary of region R (Fig. 1-21)

T_{ij}, U_{ij} = second-order tensors corresponding to tractions and displacements for three orthogonal unit loads in elastic body and functions of distance between points P, Q and elastic moduli

u_i = boundary displacements

t_i = boundary tractions

Equation (1-111) is solved numerically by replacing the continuous boundary by N piecewise flat segments, or boundary elements, as

$$\frac{u_i(P_m)}{2} + \sum_{n=1}^{N} u_j(Q_n)\, \Delta T_{ij}(Q_n, P_m) = \sum_{n=1}^{N} t_i(Q_n)\, \Delta U_{ij}(Q_n, P_m) \qquad (1\text{-}112)$$

The values of ΔT_{ij} and ΔU_{ij} are obtained by using the geometrical properties of the segments and the elastic properties of the body. Finally, Eqs. (1-112) are arranged in the matrix form as[92,93]

$$[A]\{x\} = [B]\{y\} \qquad (1\text{-}113)$$

where $\{x\}$ contains $3N$ unknown nodal values of tractions or displacements and $\{y\}$ contains the product of various coefficients and the known boundary conditions.

Once the boundary displacements are computed by solving Eq. (1-113), stress at any point p in the interior of the body is obtained from

$$\sigma_{ij}(p) = \int_B t_k(Q)D_{kij}(Q, p)\, ds(Q) - \int_B u_k(Q)S_{kij}(Q, p)\, ds(Q) \qquad (1\text{-}114)$$

Details of the computation of tensors D_{kij} and S_{kij} and other terms are given by Cruse.[92,93]

The matrices in the set of algebraic equations (1-113) to be solved in the BIE method are full, and the equations can be solved easily by using standard elimination and iterative schemes. The BIE method has certain advantages over the finite element method and has recently been used for applications in geotechnical engineering.[94,95] It shows good promise for further research and applications; however, at this time it has been used essentially for linear problems.

1-15 THE FAST FOURIER TRANSFORM

The fast Fourier transform (FFT) method is one of the most powerful developments in numerical analysis in the past 10 years. It has greatly affected the way computers are used in signal precessing and dynamic analysis, and the frequency-domain methods described in Chaps. 19 and 20 depend on the use of the FFT. Most software centers have available a computer code for performing the FFT, but the algorithm is not intuitively obvious without some explanation. Details of this technique are rather complex, but in view of its importance a brief statement of FFT follows.

The complex Fourier transform of a continuous function $g(t)$ is

$$G(\omega) = \int_{-\infty}^{\infty} g(t)e^{-i\omega t}\, dt \tag{1-115}$$

and the inverse Fourier transform of $G(\omega)$ is

$$g(t) = -\frac{1}{2\pi}\int_{-\infty}^{\infty} G(\omega)e^{i\omega t}\, d\omega \tag{1-116}$$

Because many practical cases have functions $g(t)$ that vanish for $t < 0$ and $t > L$, Eq. (1-115) can be written

$$G(\omega) = \int_{0}^{L} g(t)e^{-i\omega t}\, dt \tag{1-117}$$

Now let $g(t)$ be known at N discrete points. The discrete Fourier transform can be defined at N frequencies ω_r

$$\omega_r = \frac{2\pi r}{L} \qquad r = 0, 1, 2, \ldots, N-1 \tag{1-118}$$

The times are

$$t_k = k\frac{L}{N} \qquad k = 0, 1, 2, \ldots, N-1 \tag{1-119}$$

By extension of Eq. (1-117) the discrete Fourier transform is

$$G_r = \sum_{k=0}^{N-1} g_k e^{-i2\pi rk/N} \qquad r = 0, 1, 2, \ldots, N-1 \tag{1-120}$$

In most cases the G_r are complex numbers.

The inverse discrete Fourier transform is

$$g_k = \frac{1}{N}\sum_{r=0}^{N-1} G_r e^{i2\pi rk/N} \qquad k = 0, 1, 2, \ldots, N-1 \tag{1-121}$$

Clearly, the inverse transform could be computed by a slight modification of the direct-transform computer program.

The problem is to compute all the terms G_r, given the g_k's. This would seem to require N^2 multiplications of the g_k by exponential terms. The importance of the FFT lies in the fact that it reduces the number of operations by a clever sequence of calculations.

Let it be true that N is even. Then it would be possible to divide the g_k into two sets, $a_j = g_{2j}$, containing all the even-numbered values, and $b_j = g_{2j+1}$, containing all the odd ones. Then the discrete Fourier transforms of the two new sets are

$$A_p = \sum_{j=0}^{(N/2)-1} a_j e^{-i4\pi pj/N} \qquad p = 0, 1, 2, \ldots, \frac{N}{2} - 1 \tag{1-122a}$$

and

$$B_p = \sum_{j=0}^{(N/2)-1} b_j e^{-i4\pi pj/N} \tag{1-122b}$$

The discrete Fourier transform of g_k is

$$G_r = \sum_{j=0}^{(N/2)-1} \left\{a_j \exp\left(-\frac{i4\pi rj}{N}\right) + b_j \exp\left[-\frac{i2\pi r}{N}(2j+1)\right]\right\}$$

$$r = 0, 1, 2, \ldots, N-1 \tag{1-123a}$$

or

$$G_r = \sum_{j=0}^{(N/2)-1} a_j e^{-i4\pi rj/N} + e^{-i2\pi r/N} \sum_{j=1}^{(N/2)-1} b_j e^{-i4\pi jr/N} \qquad r = 0, 1, 2, \ldots, N-1 \tag{1-123b}$$

For $0 \leqslant r < N/2$, Eqs. (1-122a) and (1-122b) give

$$G_r = A_r + B_r e^{-i2\pi r/N} \qquad r = 0, 1, 2, \ldots, \frac{N}{2} - 1 \tag{1-124}$$

Because the Fourier transform is periodic, A_r and B_r for $r > N/2$ must repeat the values for $r < N/2$, and therefore

$$G_{r+N/2} = A_r + B_r \exp\left[-\frac{i2\pi(r + N/2)}{N}\right]$$

$$= A_r - B_r \exp\left(-\frac{i2\pi r}{N}\right) \qquad r = 0, 1, 2, \ldots, \frac{N}{2} - 1 \tag{1-125}$$

Thus, if the Fourier transform of the two sets a_j and b_j is known, the transform of g_k is found by the combinations of Eqs. (1-124) and (1-125). The transforms of

a_j and b_j can each be computed in turn from the transforms of two subsets of $N/4$ terms. If N is a power of 2, the process can be continued until the summation signs in Eqs. (1-122*a*) and (1-122*b*) involve only one term each. In that case, $j = 0$, so that the single-point transforms are simply the discrete values g_k.

The FFT then proceeds in $\log_2 N$ steps. At each step m the terms are computed from the previous values by Eq. (1-125), with r ranging from 0 to $m - 1$. The values of $e^{-i2\pi r/N}$ can be computed and stored to save computations. It turns out that $N \log_2 N$ complex additions and $\frac{1}{2}N \log_2 N$ complex multiplications are required, and this is a clear saving over the N^2 multiplications required by the direct method.

The coding of the FFT is complicated by the fact that at each step the term r is combined with the term $r + N/2$. The resulting Fourier coefficients are in a scrambled order that can be decoded by writing the original location in binary form and reversing the order of the bits to find the location of the coefficients. For example, if there are 32 points in the data, location 18 is 10010; reversal of the bits gives 01001, and so G_{18} will be found as the ninth term. Alternatively, the data points can be rearranged, and the transform will come out unscrambled.

In addition, some codes are written with the assumption that the time is so scaled that a single cycle oscillates between 0 and 2π; others assume the cycle is between $-\pi$ and $+\pi$. Therefore, different computer codes will give different results, and the user should be aware of the many possible forms of FFT available.

There are many papers on the FFT. The original paper is by Cooley and Tukey;[96] hence the technique is also known as the Cooley-Tukey algorithm. This description has drawn heavily on the paper by Cochran et al.[97]

1-16 NUMERICAL CHARACTERISTICS OF FINITE ELEMENT PROCEDURES

The generality of a numerical procedure depends on its numerical characterics and can be studied by examining such properties as convergence, stability, and consistency. As mentioned previously, such properties for the finite difference method have been investigated extensively. Numerical properties of the finite element method, on the other hand, have not been established adequately, although many intuitive proofs and conclusions have been stated. Only recently has the analyst devoted time to studying numerical properties of the finite element methods.

The engineer has often studied numerical procedures quantitatively; here a given scheme is used for a number of problems and if it is found satisfactory, it is considered acceptable. This pragmatic approach is often necessary, but it may not yield a general scheme. For generality we should check numerical characteristics of the procedure. Stability and convergence criteria based on quantitative analyses will be given in the chapters on applications; for example, see Eqs. (14-15), (14-18), and 14-20). Knowledge of such criteria is highly useful to the user in selecting temporal and spatial mesh layouts to achieve optimum economy and accuracy.

A number of recent studies have used von Neumann, Lax-Richtmyer, perturbation, and other techniques for examining numerical accuracy and stability and for deriving stability criteria for finite element procedures for parabolic equations,[12,64,71,98,99] hyperbolic equations,[100–102] and dynamic applications.[66,76,78,102] The subject of the mathematical bases of finite element procedures is beyond the scope of this book, but a simple example for derivation of stability criteria for the one-dimensional consolidation equation (1-13) is given in Appendix 1A.

1-17 SPECIAL TOPICS

We consider here general details of some of the special topics relevent in geotechnical engineering. Specific details of these items can be found in subsequent chapters on applications.

Simulation of Embankment and Excavation

For a realistic evaluation of stresses and deformations in earth structures constructed sequentially, it is necessary to account for the path dependency and nonlinearity introduced by incremental construction. The conventional approach based on linear material behavior of computing displacements and stresses as if the structure were completed in a single lift does not account for these factors and may not yield realistic results. It is possible to include the effects of path dependency and nonlinearity in finite element formulations. One of the approaches used commonly in geotechnical applications was proposed by Goodman and Brown.[103] We describe briefly the main features of this approach.

Figure 1-22 shows a schematic representation of the processes of embankment and excavation. The in situ stresses under gravity or overburden loading are first introduced into the discretized mass. This can be done by performing a finite element cycle in which the in situ stresses in each element are computed. For the plane-strain idealization, for instance, the vector $\{\sigma_0\}^T = [\sigma_{x0} \quad \sigma_{y0} \quad \tau_{xy0}]$. If the value of the coefficient of the lateral earth pressure K_0 is known or assumed, the horizontal stress σ_{x0} can be replaced by

$$\sigma_{x0} = K_0 \sigma_{y0} \tag{1-126}$$

For soil or rock masses with a horizontal ground surface, the value of τ_{xy0} is assumed to be zero.

Embankment The in situ stresses are modified as the sequence of lifts (Fig. 1-22*a*) proceeds. For example, after the first lift, the state of stress and displacements is computed as

$$\{\sigma_1\} = \{\sigma_0\} + \{\Delta\sigma_1\} \qquad \{q_1\} = \{q_0\} + \{\Delta q_1\} \tag{1-127}$$

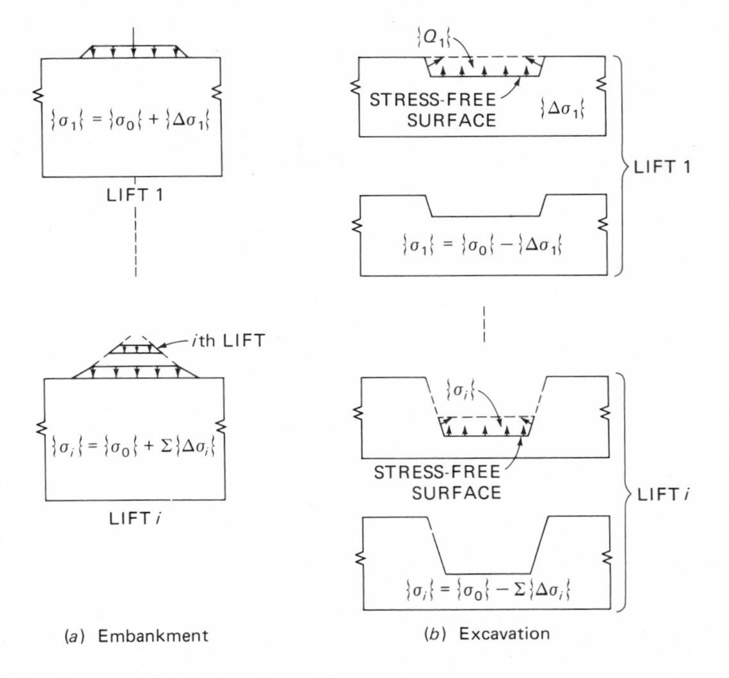

Figure 1-22 Analytic simulation of sequential embankment and excavation. (From Ref. 4. Reprinted by permission of Van Nostrand Reinhold Company.)

where $\{\Delta\sigma_1\}$ is the incremental stress due to lift 1. In general, after the ith lift

$$\{\sigma_i\} = \{\sigma_0\} + \sum_{j=1}^{i} \{\Delta\sigma_j\} \qquad \{q_i\} = \{q_0\} + \sum_{j=1}^{i} \{\Delta q_j\} \tag{1-128}$$

Excavation The excavated surface (Fig. 1-22*b*) is considered to be *stress*-free. Such a surface is assumed to be created by applying a set of equivalent forces $\{Q_i\}$ at the nodes on the surface in the direction opposite to the direction of stresses due to initial and subsequent loading conditions. The increment stresses and displacements are evaluated by using the recursive formulas

$$\{\sigma_i\} = \{\sigma_0\} - \sum_{j=1}^{i} \{\Delta\sigma_j\} \qquad \{q_i\} = \{q_0\} - \sum_{j=1}^{i} \{\Delta q_j\} \tag{1-129}$$

The foregoing procedure, as used in geotechnical problems, is discussed in Chaps. 15 and 16. However, the procedure may not be general. For instance, Christian and Wong[104] have recently reported that the results are influenced by such factors as number of sequential steps and shape of the elements. Desai[105,106] considered the problem with respect to simulation of excavation for analysis of a U-frame lock. It was suggested that the equivalent load for creating a stress-free surface can better be evaluated by computing the load vector directly

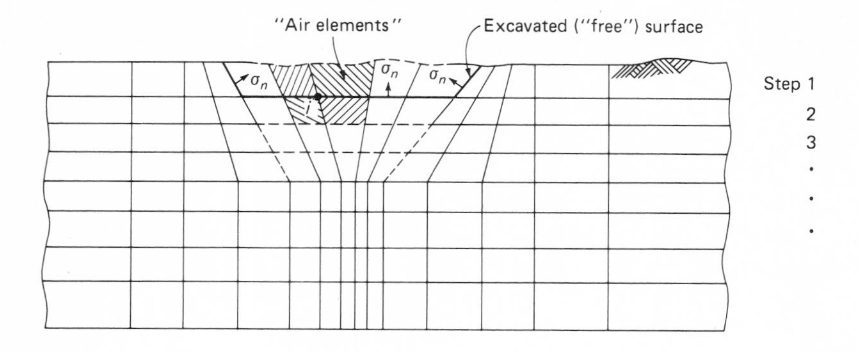

Figure 1-23 Simulation of excavation.[105]

from available displacements of element nodes near the free surface. Thus the incremental equivalent load to create a stress-free surface at node (Fig. 1-23) is

$$\{\Delta Q_0\}_i = \frac{1}{m} \sum_{j=1}^{m} \{\Delta Q_0\}^j \tag{1-130}$$

where i = node
j = elements surrounding node
m = number of such elements

and

$$\{\Delta Q_0\}^j = \iiint_V [B_j]^T[C_j][B_j]\{q\}_j \, dV \tag{1-131}$$

For a relatively simple geometry $\{\Delta Q_0\}_i$ may be computed simply from one element, that is, $m = 1$; for a complex system, however, it is necessary to use higher numbers of m. It may be noticed that the foregoing procedure is based on using incremental displacements, already computed, to find the equivalent loads. Chandrasekaran and King[107] used a similar procedure for simulating excavation.

For the procedure described in Chaps. 15 and 16, which can be considered a special case of Eq. (1-130), the equivalent load is computed by defining a separate interpolation function among a number of neighboring elements; this involves computation of additional terms and the inversion of a matrix. Moreover, the extraction of the equivalent load vector is limited by the order of the interpolation function chosen; hence, if one used a higher-order approximation in the original formulation, it may not be reflected in the equivalent load vector and may influence its accuracy.

Finally, realistic simulation of an excavation should include analysis of factors such as unloading, singularities like corners, existence of structures before excavation, artificial inclusion of "air" elements, and mathematical formulation.

Consideration of the latter may indicate that a mixed or hybrid procedure may be more suitable than the displacement method used in the past.[106]

Dewatering

The change in the state of stress caused by dewatering can be included in the finite element analysis as an equivalent load that modifies the in situ stresses in the soil. Such an equivalent load $\{Q_d\}$ can be computed as

$$\{Q_d\} = \iiint_V [B]^T\{\sigma_d\}\, dV \tag{1-132}$$

where $\{\sigma_d\}$ is the change in stress caused by reduction in pore water pressures caused by lowering the groundwater level. It is often possible to assume that the groundwater levels before and after pumping are horizontal. In that case, Eq. (1-132) can be applied directly. If dewatering is done in clayey soils, the groundwater level between wells will not be horizontal and the value of $\{\sigma_d\}$ should be computed to reflect the existence of a cone-of-depression in the groundwater table (Fig. 1-24*a*). This phenomenon can have significant influence on the subsequent behavior of the structure-foundation system. For example, Fig. 1-24*b* shows computed displacements from a finite element analysis[106] in the foundation of Arkansas Lock and Dam number 5 (LD5). It can be seen that the level of dewatering that yields the best correlation with observation lies somewhere between the bottom of the excavation and the final level in the wells. For problems of dewatering (or rise of the water table) it is important to choose proper values for material parameters for drained and undrained zones.

Discretization of "Infinite" Media

Applications of the finite element method to structures that have well-defined geometries and boundary conditions can be straightforward insofar as their subdivision and introduction of the boundary conditions are concerned. Most situations in geotechnical engineering, however, involve "infinite" media, and the boundary conditions are not exactly defined.

Besides determining the significant zones to be included in the mesh, it is also necessary to arrive at realistic boundary conditions. The former is possible since the influence of the perturbations such as applied loads and potential diminish with increasing distance from the points of their applications. The significant extents are usually determined by trial-and-error procedures in which the extents are varied and the resultant effects on the numerical solutions are examined. Criteria for determination of discretized boundaries for stress-deformation problems are covered by Desai and Abel[4] and in Chap. 16. Criteria for seepage problems are covered in Chap. 14. Dynamic problems represent a special case and are discussed in Chaps. 19 and 20.

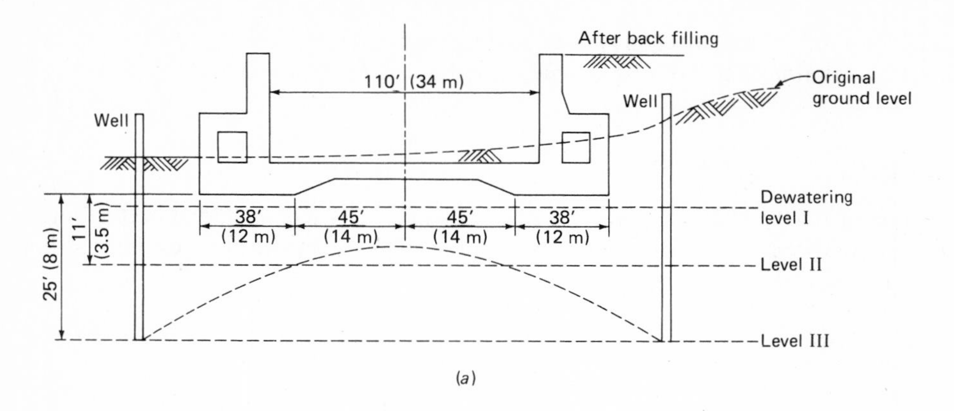

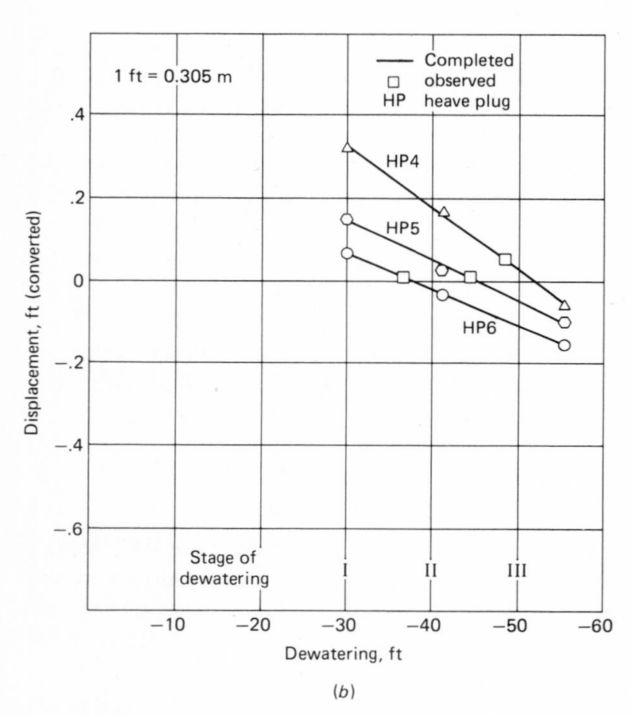

Figure 1-24 Simulation of dewatering.[105]

1-18 ASPECTS RELATED TO APPLICATIONS

We have presented introductory information on some common numerical methods and certain special aspects. It is hoped that the descriptions will help the student in his introduction to the basic ideas and will provide continuity toward applications in subsequent chapters. For detailed study of these topics, the reader may consult various textbooks and publications.

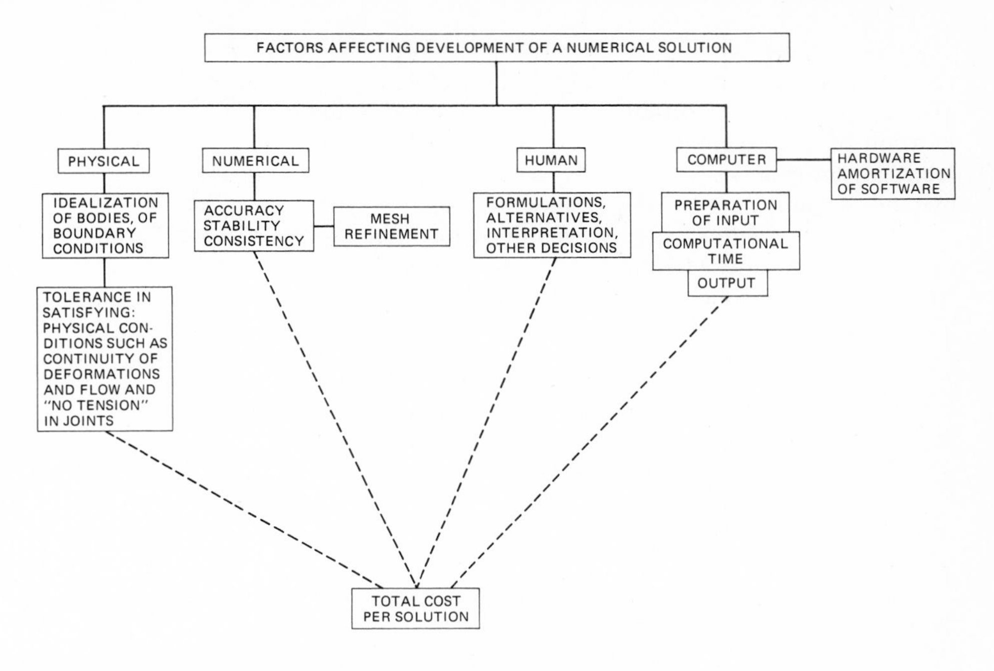

Figure 1-25 Factors in development of numerical method.[48]

The final aim of any solution (numerical) procedure will be optimum accuracy and economy. Tied in with this aim is the significant question of reliability of a procedure. Figure 1-25 shows a schematic representation of the factors that should be considered in arriving at the most suitable numerical scheme.[48] In geotechnical engineering, the physical factors can include idealization or discretization of field situations and precision in satisfaction of continuity of flow in a layered soil system. Adequate and sufficiently fine-mesh layouts become a factor for situations such as stress gradients in an excavation problem. Examination of such numerical characteristics as convergence and stability should be considered for a general procedure. Human factors are related to subjective considerations in formulating the problem and in choosing alternatives. Both software and hardware aspects of the computer used can also influence the numerical solution.

Numerical methods have reached a high level of development. We believe that the aspect of evaluation of available schemes leading to determination of optimum characteristics is a significant area that should be pursued vigorously now.

Applications

At this time, use of numerical methods as sole basis of design is limited to a very small class of problems. The methods have been found to be useful as tools of analysis; they can be used as alternative means of design and verification of conventional methods. A few studies have given attention to the aspect of generating design charts from numerical methods. Although this pursuit is difficult and requires considerable human and computer effort, it is vital from the viewpoint of the user-designer. Usefulness of numerical methods for such aspects as evaluation of existing structures, postfailure analysis, monitoring and control of construction, location of instrumentation, evaluation of laboratory tests, and computation of material parameters should also be recognized. Hardly any available conventional methods permit consideration of these factors.

Numerical methods have proved their potential as powerful tools for analysis of problems that defied solutions by conventional methods. Considerable work has already been done toward verification of the methods by comparing numerical predictions and observations. With increasing confidence, the methods will also provide viable tools as sole or alternative methods for analysis and design.

APPENDIX 1A

1A-1 CONVERGENCE AND STABILITY

We shall consider a few finite difference and finite element schemes and derive their convergence and stability criteria.

Simple Explicit Scheme

Consider the Taylor series expansion [Eq. (1-8)] and denote the discretization error $\epsilon = \bar{u} - u^*$. Representation of the parabolic equation (1-13) according to the Taylor series expansion is

$$c_v \frac{\bar{u}_{i+1,t} - 2\bar{u}_{i,t} + \bar{u}_{i-1,t}}{(\Delta x)^2} + O[(\Delta x)^2] = \frac{\bar{u}_{i,t+1} - \bar{u}_{i,t}}{\Delta t} + O(\Delta t) \tag{1-133}$$

The finite difference analog to Eq. (1-13) is

$$c_v \frac{u^*_{i+1,t} - 2u^*_{i,t} + u_{i-1,t}}{(\Delta x)^2} + O[(\Delta x)^2] = \frac{u^*_{i,t+1} - u^*_{i,t}}{\Delta t} + O(\Delta t) \tag{1-134}$$

Subtracting Eq. (1-134) from Eq. (1-133) leads to the following expression for ϵ:

$$c_v \frac{\epsilon_{i+1,t} - 2\epsilon_{i,t} + \epsilon_{i-1,t}}{(\Delta x)^2} + O[(\Delta t) + (\Delta x)^2] = \frac{\epsilon_{i,t+1} - \epsilon_{i,t}}{\Delta t} \tag{1-135}$$

Therefore

$$\epsilon_{i,t+1} = \epsilon_{i,t} + \Delta T(\epsilon_{i+1,t} - 2\epsilon_{i,t} + \epsilon_{i-1,t}) + O[(\Delta t)^2 + \Delta t(\Delta x)^2] \tag{1-136a}$$

or

$$\epsilon_{i,t+1} = \Delta T\epsilon_{i+1,t} + (1 - 2\,\Delta T)\epsilon_{i,t} + \Delta T\epsilon_{i-1,t} + O[(\Delta t)^2 + \Delta t(\Delta x)^2] \tag{1-136b}$$

Now assume that $0 < \Delta T \leqslant \frac{1}{2}$; then ΔT and $1 - 2\,\Delta T$ are nonnegative, which leads to the inequality

$$|\epsilon_{i,t+1}| \leqslant \Delta T|\epsilon_{i+1,t}| + (1 - 2\,\Delta T)|\epsilon_{i,t}| + \Delta T|\epsilon_{i-1,t}| + Q[(\Delta t)^2 + \Delta t(\Delta x)^2] \tag{1-137a}$$

or

$$|\epsilon_{i,t+1}| \leqslant |\epsilon_{i,t}| + Q[(\Delta t^2) + \Delta t(\Delta x)^2] \tag{1-137b}$$

Now let the upper bound of $|\epsilon_{i,t}|$ be expressed as

$$\|\epsilon_t\| = \max_i |\epsilon_{i,t}| \tag{1-138}$$

which needs to be studied for all nodes i to M and any time level t. With this notion, Eq. (1-138) leads to

$$\|\epsilon_{t+1}\| \leqslant \|\epsilon_t\| + Q[(\Delta t)^2 + \Delta t(\Delta x)^2] \tag{1-139}$$

By writing Eq. (1-139) successively for lower time levels we obtain

$$\begin{aligned}\|\epsilon_{t+1}\| &\leqslant \|\epsilon_t\| + Q[(\Delta t)^2 + \Delta t(\Delta x)^2] \\ &\leqslant \|\epsilon_{t-1}\| + 2Q[(\Delta t)^2 + \Delta t(\Delta x)^2] \\ &\leqslant \|\epsilon_0\| + nQ[(\Delta t)^2 + \Delta t(\Delta x)^2]\end{aligned} \tag{1-140}$$

By virtue of the initial and boundary conditions, $\epsilon_{i,0} = \epsilon_{0,t} = \epsilon_{m,t} = 0$. Hence $\|\epsilon_0\| = 0$ and

$$\|\epsilon_{t+1}\| \leqslant Qn\,\Delta t[\Delta t + (\Delta x)^2] \tag{1-141}$$

It can be seen from Eq. (1-141) that the simple explicit scheme is convergent because $\|\epsilon_{t+1}\|$ tends to zero as Δx and Δt tend to zero. This conclusion is dependent on the foregoing assumption that

$$\Delta T \leqslant \tfrac{1}{2} \tag{1-142}$$

Hence the simple explicit scheme is convergent only if $\Delta T \leqslant \frac{1}{2}$.

Let us now consider the question of stability. One of the common methods for stability is von Neumann's procedure.[6,9] According to this procedure, let the numerical error be expressed as

$$\epsilon = e^{\alpha t} e^{i\beta x} \tag{1-143}$$

where $\alpha = \alpha(\beta)$ is generally complex and β is a real number and member of the sequence of frequencies $\{\beta_n\}$.[9] For the original error (which can be introduced due to various reasons such as roundoff) not to grow with increasing t, it is necessary and sufficient that $|e^{\alpha\Delta t}| \leqslant 1$. Equation (1-16) leads to the finite difference expression for error

$$\Delta T(\epsilon_{i+1,t} - 2\epsilon_{i,t} + \epsilon_{i-1,t}) = \epsilon_{i,t+1} - \epsilon_{i,t} \tag{1-144}$$

Substituting Eq. (1-143) in Eq. (1-144) gives

$$\Delta T(e^{\alpha t} e^{i\beta(x+\Delta x)} - 2e^{\alpha t} e^{i\beta x} + e^{\alpha t} e^{i\beta(x-\Delta x)}) = e^{\alpha(t+\Delta t)} e^{i\beta x} - e^{\alpha t} e^{i\beta x} \tag{1-145}$$

Dividing both sides by $e^{\alpha t} e^{i\beta x}$ leads to

$$\Delta T(e^{i\beta\,\Delta x} - 2 + e^{-i\beta\,\Delta x}) = e^{\alpha\,\Delta t} - 1 \tag{1-146a}$$

or

$$2\,\Delta T(\cos\beta\,\Delta x - 1) = e^{\alpha\,\Delta t} - 1 \qquad \text{or} \qquad e^{\alpha\,\Delta t} - 1 = -4\,\Delta T \sin^2\frac{\beta\,\Delta x}{2} \tag{1-146b}$$

The condition that $|e^{\alpha\,\Delta t}| \leqslant 1$ leads to

$$-1 \leqslant 1 - 4\,\Delta T \sin^2\frac{\beta\,\Delta x}{2} \leqslant 1 \tag{1-147}$$

This inequality of the left is satisfied for all values of β if and only if

$$\Delta T \leqslant \tfrac{1}{2} \tag{1-148}$$

For the simple explicit scheme, the conditions for assuring convergence and stability are identical [Eqs. (1-142) and (1-148)].

$i-1$ i $i+1$ **Figure 1-26** Generic point for one-dimensional elements.

1A-2 STABILITY CRITERION FOR A FINITE ELEMENT SCHEME

The finite element equations corresponding to the one-dimensional consolidation problem are given in Eq. (1-88). Desai, Oden, and Johnson[64] and Desai and Lytton[71] have derived stability criteria for the consolidation problem by employing linear and cubic approximating models for pore water pressure and for the two time-integration schemes given in Eqs. (1-87*a*) and (1-88*b*). We shall illustrate herein a typical example for linear model [Eq. (1-80*a*)] with the integration scheme in Eq. (1-87*a*).

The element equations for linear model (Fig. 1-26) are

$$[k] = \frac{c_v}{a}\begin{bmatrix} 1 & -1 \\ -1 & 1 \end{bmatrix} \quad \text{and} \quad [p] = \frac{a}{6}\begin{bmatrix} 2 & 1 \\ 1 & 2 \end{bmatrix} \tag{1-149}$$

in which a is the length of an element. By considering two adjoining elements (Fig. 1-26) and by assembling their element equations the governing finite element–finite difference relation for the generic point i is

$$(-\Delta T + \tfrac{1}{3})u_{i-1,\,t+1/2} + (2\,\Delta T + \tfrac{4}{3})u_{i,\,t+1/2} + (-\lambda + \tfrac{1}{3})u_{i+1,\,t+1/2} = \tfrac{1}{3}u_{i-1,\,t} + \tfrac{4}{3}u_{i,\,t} + \tfrac{1}{3}u_{i+1,\,t} \tag{1-150}$$

The expression for error is obtained from Eq. (1-150):

$$(-\Delta T + \tfrac{1}{3})\epsilon_{i-1,\,t+1/2} + (2\,\Delta T + \tfrac{4}{3})\epsilon_{i,\,t+1/2} + (-\lambda + \tfrac{1}{3})\epsilon_{i+1,\,t+1/2} = \tfrac{1}{3}\epsilon_{i-1,\,t} + \tfrac{4}{3}\epsilon_{i,\,t} + \tfrac{1}{3}\epsilon_{i+1,\,t} \tag{1-151}$$

By substituting error [Eq. (1-143)] and following von Neumann's procedure, Eq. (1-151) reduces to

$$e^{(\alpha\,\Delta t)/2}[2\,\Delta T + \tfrac{4}{3} + (-\lambda + \tfrac{1}{3})(e^{-i\beta a} + e^{i\beta a})] = \tfrac{4}{3} + \tfrac{1}{3}(e^{-i\beta a} + e^{i\beta a}) \tag{1-152}$$

Therefore the stability condition is

$$\frac{4 + 2\cos\beta a}{4 + 6\,\Delta T + (2 - 6\,\Delta T)\cos\beta a} \leqslant 1 \tag{1-153}$$

in which $2\cos\beta a = e^{-i\beta a} + e^{i\beta a}$. This condition is satisfied for all values of $\Delta T \geqslant 0$. Hence, the scheme is unconditionally stable.

REFERENCES

1. Terzaghi, K.: "Theoretical Soil Mechanics," John Wiley & Sons, Inc., New York, 1943.
2. Desai, C. S.: Numerical Techniques for Design Analysis in Civil Engineering, *Proc., Symp. Mod. Trends Civ. Eng., Univ. Roorkee, India, November 1972.*
3. Crandall, S. H.: "Engineering Analysis," McGraw-Hill Book Company, New York, 1956.
4. Desai, C. S., and J. F. Abel: "Introduction to the Finite Element Method," Van Nostrand Reinhold Company, New York, 1972.
5. Carnahan, B., H. A. Luther, and J. O. Wilkes: "Applied Numerical Methods," John Wiley & Sons, Inc., New York, 1969.
6. Richtmyer, R. D., and K. W. Morton: "Difference Methods for Initial-Value Problems," Interscience Publishers, New York, 1967.
7. Forsythe, G. E., and W. R. Wasow: "Finite Difference Methods for Partial Differential Equations," John Wiley & Sons, Inc., New York, 1960.

8. Ralston, A., and H. S. Wilf: "Mathematical Methods for Digital Computers," 2 vols., John Wiley & Sons, Inc., New York, 1967.
9. O'Brien, G. G., M. A. Hyman, and S. Kaplan: A Study of the Numerical Solution of Partial Differential Equations, *J. Math. Phys.*, vol. 29, no. 3, pp. 223–251, January 1951.
10. Douglas, J.: On the Relation between Stability and Convergence in the Numerical Solution of Linear Parabolic and Hyperbolic Differential Equations, *J. Soc. Ind. Appl. Math.*, vol. 41, 1956.
11. Lax, P. D., and R. D. Richtmyer: Survey of the Stability of Linear Finite Difference Equations, *Commun. Pure Appl. Math.*, vol. 9, 1956.
12. Strang, G., and G. J. Fix: "An Analysis of the Finite Element Method," Prentice-Hall, Inc., Englewood Cliffs, N.J., 1973.
13. Dufort, F. C., and S. P. Frankel: Stability Conditions in the Numerical Treatment of Parabolic Differential Equations, *Math. Tables Aids Comput.*, vol. 7, pp. 135–152, 1953.
14. Saulev, V. K.: On a Method of Numerical Integration of the Equation of Diffusion, *Dokl. Akad. Nauk USSR*, vol. 115, p. 1077, 1957.
15. Larkin, B. K.: Some Stable Explicit Difference Approximations to the Diffusion Equation, *Math. Comput.*, vol. 18, pp. 196–202, 1964.
16. Barkat, H. Z., and J. A. Clark: On the Solution of the Diffusion Equation by Numerical Methods, *J. Heat Transfer Trans. ASME*, ser. C, vol. 88, pp. 421–427, 1966.
17. Allada, S. R., and D. Quon: A Stable, Explicit Numerical Solution of the Conduction Equation for Multi-dimensional Nonhomogeneous Media, *Heat Transfer Chem. Eng. Prog.*, ser. 64, vol. 62, pp. 151–156, 1966.
18. Desai, C. S.: Seepage in Mississippi River Banks: Analysis of Transient Seepage Using Viscous Flow Model and Numerical Methods, *U.S. Army Eng. Waterw. Expt. Stn., Vicksburg, Miss.* Misc. Pap. S-70-3, February 1970.
19. Desai, C. S., and W. C. Sherman: Unconfined Transient Seepage in Sloping Banks, *J. Soil Mech. Found. Div. ASCE*, vol. 97, no. SM2, February 1971.
20. Dvinoff, A. H., and M. E. Harr: Phreatic Surface Location after Drawdown, *J. Soil Mech. Found. Div. ASCE*, vol. 97, no. SM1, January 1971.
21. Koppula, S. D., and N. R. Morgenstern: Consolidation of Clay Layer in Two Dimensions, *J. Soil Mech. Found. Div. ASCE*, vol. 98, no. SM1, January 1972.
22. Desai, C. S., and L. D. Johnson: Evaluation of Some Numerical Schemes for Consolidation, *Int. J. Numer. Methods Eng.*, vol. 7, 1973.
23. Johnson, L. D., and C. S. Desai: A Numerical Procedure for Predicting Heave, *Proc. 2d Australia-New Zealand Conf. Geomech., Brisbane*, 1975.
24. Peaceman, D. W., and H. H. Rachford: The Numerical Solution of Parabolic and Elliptic Differential Equations, *J. Soc. Ind. Appl. Math.*, vol. 3, pp. 28–41, 1955.
25. Douglas, J.: On the Numerical Integration of Implicit Methods, *J. Soc. Ind. Appl. Math.*, vol. 3, pp. 42–65, 1955.
26. McCracken, D. D., and W. S. Dorn: "Numerical Methods and Fortran Programming," John Wiley & Sons, Inc., New York, 1964.
27. Remson, I., G. M. Hornberger, and F. J. Molz: "Numerical Methods in Subsurface Hydrology," Wiley-Interscience, New York, 1971.
28. Bredehoeft, J. D., and G. F. Pinder: Digital Analysis of Areal Flow in Multiaquifer Groundwater Systems: A Quasi Three-dimensional Model, *Water Resour. Res.*, vol. 6, no. 3, June 1970.
29. Freeze, R. A.: "The Mechanism of Natural Ground-water Recharge and Discharge," *Water Resour. Res.*, vol. 5, no. 1, February 1969.
30. Prickett, T. A., and C. G. Linquist: Selected Digital Computer Techniques for Groundwater Resource Evaluation, *Ill. State Water Supply Bull.* 55, Urbana, 1971.
31. Douglas, J.: Alternating Direction Method for Three Space Variables, *Numer. Math.*, vol. 4, pp. 41–63, 1962.
32. Matlock, H., and L. C. Reese: Generalized Solutions for Laterally Loaded Piles, *Trans. ASCE*, vol. 127, pt. I, proc. pap. 3770, pp. 1220–1249, 1962.
33. Timoshenko, S., and S. W. Krieger: "Theory of Plates and Shells," McGraw-Hill Book Company, New York, 1959.

34. Desai, C. S. (ed.): Applications of the Finite Element Method in Geotechnical Engineering. *Proc. WES Symp. Appl. Finite Element Method Geotech. Eng., U.S. Army Eng. Waterw. Expt. Stn., Vicksburg, Miss.*, 1972.
35. Zienkiewicz, O. C.: "The Finite Element Method in Engineering Science," McGraw-Hill Publishing Company, Ltd., London, 1971.
36. Pian, T. H. H., and P. Tong: Finite Element Methods in Continuum Mechanics, *Adv. Appl. Mech.*, vol. 12, 1972.
37. Noor, A. K., and C. M. Anderson: Mixed Isoparametric Elements for Saint-Venant Torsion, *Comput. Methods Appl. Mech. Eng.*, vol. 6, 1975.
38. Sandhu, R. S., and K. S. Pister: A Variational Principle for Linear Coupled Field Problems in Continuum Mechanics, *Int. J. Eng. Sci.*, vol. 8, no. 12, December 1970.
39. Christian, J. T., and B. J. Watt: Undrained Visco-elastic Analysis of Soil, Deformations, *Proc. Symp. Finite Element Methods Geotech. Eng., Vicksburg, September 1972.*
40. Meissner, U.: Generalized Variational Principles for Use in Flow Problems, *Proc. Int. Symp. Finite Element Methods Flow Problems*, University of Alabama Press, Huntsville, Ala., 1974.
41. Desai, C. S.: Analysis of Consolidation by Numerical Methods, *Proc. Symp. Soil Mech.: Recent Dev., Univ. New South Wales, Sydney, July 1975.*
42. Gallagher, R. H.: Analysis of Plate and Shell Structures, *Proc. Symp. Appl. Finite Element Methods Civil Eng., ASCE, Vanderbilt Univ., Nashville, Tenn., November 1969.*
43. Bogner, F. K., R. L. Fox, and L. A. Schmit: The Generation of Inter Element–Compatible Stiffness and Mass Matrices by the Use of Interpolation Formulas, *Proc. 1st Conf. Matrix Methods Struct. Mech., Wright-Patterson Air Force Base, Ohio, November 1966.*
44. Ergatoudis, B., B. M. Irons, and O. C. Zienkiewicz: Curved Isoparametric, "Quadrilateral" Elements for Finite Element Analysis, *Int. J. Solids Struct.*, vol. 4, no. 1, 1968.
45. Zienkiewicz, O. C., B. M. Irons, J. Ergatoudis, S. Ahmed, and I. C. Scott: Isoparametric and Associated Families for Two- and Three-dimensional Analysis," in I. Holland and K. Bell (eds.), "Finite Element Methods in Stress Analysis," TAPIR, Technical University of Norway, Trondheim, 1969.
46. Zienkiewicz, O. C., and C. J. Parekh: Transient Field Problems: Two-dimensional and Three-dimensional Analysis by Isoparametric Finite Elements, *Int. J. Numer. Methods Eng.*, vol. 2, no. 1, 1970.
47. Desai, C. S.: Finite Element Methods for Flow in Porous Media, *Int. Symp. Finite Element Methods Flow Probl., Univ. Wales, Swansea, 1974.*
48. Desai, C. S.: Overview, Trends, and Projections: Theory and Applications of the Finite Element Method in Geotechnical Engineering, *Proc., Symp. Appl. Finite Element Method Geotech. Eng., U.S. Army Eng. Waterw. Expt. Stn., Vicksburg, Miss., 1972.*
49. Desai, C. S.: An Approximate Solution for Unconfined Seepage, *J. Irrig. Drain. Div. ASCE*, vol. 99, no. IR1, March 1973.
50. Finlayson, B. A.: "The Methods of Weighted Residuals and Variational Principles," Academic Press, Inc., New York, 1972.
51. Wilson, E. L.: Structural Analysis of Axisymmetric Solids, *AIAA J.*, vol. 3, no. 12, December 1965.
52. Birkhoff, G., and G. J. Fix: Higher Order Linear Finite Element Methods, *Rep. to USAEC Off. Nav. Res.*, 1974.
53. Desai, C. S.: Nonlinear Analysis Using Spline Functions, *J. Soil Mech. Found. Div. ASCE*, vol. 97, no. SM2, February 1971; also see ibid., vol. 98, no. SM9, September 1972.
54. Desai, C. S., and L. D. Johnson: Evaluation of Two Finite Element Formulations for One-dimensional Consolidation, *Comput. Struct.*, vol. 2, no. 4, September 1972.
55. Desai, C. S.: Evaluation of Numerical Procedures for Fluid Flow, *ASCE Nat. Conv., Denver, Colo., November 1975*, prepr. 2609.
56. Desai, C. S.: Numerical Design-Analysis of Piles in Sands, *J. Geotech. Eng. Div. ASCE*, vol. 100, no. GT6, June 1974.
57. Timoshenko, S., and J. N. Goodier: "Theory of Elasticity," McGraw-Hill Book Company, New York, 1951.

58. Desai, C. S.: Solution of Stress-Deformation Problems in Soil and Rock Mechanics Using the FE Method," Ph.D. dissertation, University of Texas, Austin, 1968.
59. Desai, C. S., and L. C. Reese: Analysis of Circular Footings on Layered Soils, *J. Soil Mech. Found. Div. ASCE*, vol. 96, no. SM4, July 1970.
60. Hammer, P. C., and A. H. Stroud: Numerical Evaluation of Multiple Integrals, *Math. Tables Aids Comput.*, vol. 12, 1958.
61. Abramowitch, M., and I. A. Stegun (eds.): *Handbook of Mathematical Functions with Formulas, Graphs and Mathematical Tables, Natl. Bur. Stand. Appl. Math. Ser.* 55, Washington, 1964.
62. Szabo, B. A., and G. C. Lee: Derivation of Stiffness Matrices for Problems on Plane Elasticity by Galerkin Method, *Int. J. Numer. Methods Eng.*, vol. 1, no. 3, July 1969.
63. Hutton, S. G., and D. L. Anderson: Finite Element Method: A Galerkin Approach, *J. Eng. Mech. Div. ASCE*, vol. 97, no. EMS, March 1971.
64. Desai, C. S., J. T. Oden, and L. D. Johnson: Evaluation and Analyses of Some Finite Element and Finite Difference Procedures for Time-dependent Problems, *U.S. Army Eng. Waterw. Expt. Stn., Vicksburg, Miss., Misc. Pap.* S-75-7, April 1975.
65. Bruch, J. C.: Nonlinear Equation of Unsteady Ground-water Flow, *J. Hydrol. Div. ASCE*, vol. 99, no. HY3, March 1973.
66. Oden, J. T.: "Finite Elements of Nonlinear Continua," McGraw-Hill Book Company, New York, 1972.
67. Desai, C. S.: Finite Element Methods for Flow in Porous Media, chap. 8 in R. H. Gallagher et al. (eds.), "Finite Elements in Fluids," John Wiley & Sons, Ltd., Sussex, England, 1974–1975.
68. Desai, C. S.: A Finite Element Procedure for Three-dimensional Seepage, *Virginia Polytech. Inst. State Univ., Dept. Civ. Eng. Tech. Rep.*, Blacksburg, Va., June 1975.
69. Desai, C. S., and L. D. Johnson: Evaluation of Two Finite Element Formulations for One-dimensional Consolidation, *Comput. Struct.*, vol. 2, no. 4, September 1972.
70. Wilson, E. L., and R. E. Nickell: Applications of Finite Element Method to Heat Conduction Analysis, *Nucl. Eng. Des.*, 1966.
71. Desai, C. S., and R. L. Lytton: Stability Criteria for Two Finite Element Schemes for Parabolic Equation, *Int. J. Numer. Methods Eng.*, vol. 9, pp. 721–726, 1975.
72. Newmark, N. M.: A Method of Computation of Structural Dynamics, *Proc. ASCE*, vol. 85, no. EM 3, pp. 67–94, 1959.
73. Wilson, E. L., and R. W. Clough: Dynamic Response by Step-by-Step Matrix Analysis, *Symp. Use Comput. Civ. Eng.*, Laboratorio National de Engeniaria Civil, Lisbon, October 1962.
74. Wilson, E. L.: A Computer Program for the Dynamic Stress Analysis of Underground Structures, *U.S. Army Eng. Waterw. Expt. Stn. Contract Rep.* 1-175, Vicksburg, Miss., 1968.
75. Farhoomand, I., and E. L. Wilson: A Nonlinear Finite Element Code for Analyzing the Blast Response of Underground Structures, *U.S. Army Eng. Waterw. Expt. Stn. Contract Rep.* N-71-1, Vicksburg, Miss., January 1970.
76. Nickell, R. E.: On the Stability of Approximation Operators in Problems of Structural Mechanics, *Int. J. Solids Struct.*, vol. 7, no. 3, March 1971.
77. Gurtin, M. E.: Variational Principles for Linear Elastodynamics, *Arch. Ration. Mech. Anal.*, vol. 16, pp. 34–50, 1964.
78. Nickell, R. E.: Direct Integration Methods in Structural Mechanics, *J. Eng. Mech. Div. ASCE*, vol. 99, no. EM2, April 1973.
79. Ghaboussi, J., E. L. Wilson, and J. Isenberg: Finite Element for Rock Joints and Interfaces, *J. Soil Mech. Found. Div. ASCE*, vol. 99, no. SM10, October 1973.
80. Clough, R. W.: Analysis of Structural Vibrations and Dynamic Response, in R. H. Gallagher, Y. Yamada, and J. T. Oden (eds.), Recent Advances in Matrix Methods of Structural Analysis and Design, *Proc. U.S.-Jpn. Semin., 1969*, University of Alabama Press, Huntsville, Ala., 1971.
81. Idriss, I. M., J. Lysmer, R. Hwang, and H. B. Seed: A Computer Program for Evaluating the Seismic Response of Soil Structures by Variable Damping Finite Element Procedures, *Univ. Calif. Earthquake Eng. Res. Ctr. Rep.* EERC 73-16, Berkeley, July 1973.
82. Rabinowitz, P. (ed.): "Numerical Methods for Nonlinear Algebraic Equations," Gordon and Breach, Science Publishers, London, 1970.

83. Melosh, R. J., and R. M. Bamford: Efficient Solution of Load Deflection Equations, *J. Struct. Div. ASCE*, vol. 95, no. ST4, April 1969.
84. Irons, B. M.: A Frontal Solution Program for Finite Element Analysis, *Int. J. Numer. Methods Eng.*, vol. 2, no. 1, 1970.
85. Traule, J. L. (ed.): *Proc. Symp. Complexity Sequential Parallel Numer. Algorithms*, Academic Press, Inc., New York, 1973.
86. Abbott, M. B.: "An Introduction to the Method of Characteristics," American Elsevier Publishing Company, New York, 1966.
87. Streeter, V. L., B. E. Wylie, and F. E. Richart: Soil Motion Computations by Characteristics Method, *J. Geotech. Eng. Div. ASCE*, vol. 100, no. GT3, March 1974.
88. Pipes, L. A.: "Applied Mathematics for Engineers and Physicists," McGraw-Hill Book Company, New York, 1958.
89. Contractor, D. N.: Method of Characteristics, notes for short course on Numerical Methods in Geotechnical Engineering, Blacksburg, Virginia, June 1975.
90. Richart, F. E., J. R. Hall, and R. D. Woods: "Vibrations of Soils and Foundations," Prentice-Hall, Inc., Englewood Cliffs, N.J., 1970.
91. Mikhlin, S. G.: "Linear Integral Equations," Hindustan Publishing Corp., Delhi, 1960.
92. Cruse, T. A.: Numerical Solutions in Three-dimensional Elastostatics, *Int. J. Solids Struct.*, vol. 5, no. 12, pp. 1259–1273, December 1969.
93. Cruse, T.: An Improved Boundary-Integral Equation Method for Three-dimensional Elastic Stress Analysis, *Comput. Struct.*, vol. 4, no. 4, pp. 741–754, August 1974.
94. Banerjee, P. K., and R. Butterfiled: An Integral Equation Method for Analysis of Boundary Value Problems in Irregularly Stratified Media, *Int. Symp. Numer. Methods Soil Rock Mech., Univ. Karlsruhe, September 1975.*
95. Kobayashi, S., Y. Niwa, and F. Takuo: Applications of Integral Equation Method to Some Geotechnical Problems, *Proc. 2d Int. Conf. Numer. Methods Geomech., Blacksburg, Va., June 1976.*
96. Cooley, J. W., and J. W. Tukey: An Algorithm for the Machine Calculation of Complex Fourier Series, *Math. Comput.*, vol. 19, pp. 297–301, April 1965.
97. Cochran, W. T., J. W. Cooley, D. L. Farin, et al.: What Is the Fast Fourier Transform? *IEEE Trans. Audio Electroacoust.*, vol. AU-15, no. 2, pp. 45–55, July 1967.
98. Douglas, J., and T. Dupont: Galerkin Methods for Parabolic Equations, *SIAM J. Numer. Anal.*, vol. 7, no. 4, 1970.
99. Yalamanchili, R. V. S., and S. C. Chu: Stability and Oscillation Characteristics of Finite Element, Finite Difference and Weighted Residuals Methods for Transient Two-dimensional Heat Conduction in Solids, *J. Heat Transfer Trans. ASME*, May 1973, pp. 235–241.
100. Oden, J. T., and R. B. Fost: Convergence, Accuracy and Stability of Finite Element Approximations of a Class of Nonlinear Hyperbolic Equations, *Int. J. Numer. Methods Eng.*, vol. 6, no. 3, 1973.
101. Fujii, H.: Finite Element Schemes Stability and Convergence, *Proc. 2d U.S.-Jpn. Semin. Matrix Methods Struct. Mech.*, University of Alabama Press, Huntsville, Ala., August 1972.
102. Kreig, R. D., and S. W. Key: Comparison of Finite Element and Finite-Difference Methods, *Proc., ONR Symp. Numer. Methods Struct. Mech. Univ. Ill., Urbana, September 1971*, published as "Numerical and Computer Methods in Structural Mechanics," Academic Press, Inc., New York, 1973.
103. Goodman, L. E., and C. B. Brown: Dead Load Stresses and Instability of Slopes, *J. Soil Mech. Found. Div. ASCE*, vol. 89, no. SM3, May 1963.
104. Christian, J. T., and I. H. Wong: Errors in Simulating Excavation in Elastic Media by Finite Elements, *Soils Found.*, vol. 13, no. 1, March 1973.
105. Desai, C. S.: Finite Element Analysis of Locks and Some Problems Related to Sequential Construction, unpublished rep., 1972–1973.
106. Desai, C. S.: Soil-Structure Interaction and Simulation Problems, *Int. Symp. Numer. Methods Soil Rock Mech., Univ. Karlsruhe, September 1975.*
107. Chandrasekaran, and G. J. W. King: Simulation of Excavation Using Finite Elements, *Tech. Note J. Geotech. Eng. Div.*, vol. 100, no. GT9, September 1974.

CHAPTER

TWO

CONSTITUTIVE LAWS FOR GEOLOGIC MEDIA

John T. Christian and Chandrakant S. Desai

A central part of setting up a numerical treatment of a physical problem is the description of the relations between physical quantities such as stress, strain, and time. These are called *constitutive relations*, and this chapter sets out those which have been most widely and successfully used in numerical approaches to geotechnical problems. We begin by considering some fundamental concepts that apply to any material.

2-1 REVIEW

Review of Stress

Figure 2-1 shows an element of a larger continuous body. Forces are transmitted across each of the six faces of the element, and they can be described in terms of the stress tensor

$$\sigma_{ij} = \begin{bmatrix} \sigma_{xx} & \sigma_{xy} & \sigma_{xz} \\ \sigma_{xy} & \sigma_{yy} & \sigma_{yz} \\ \sigma_{xz} & \sigma_{yz} & \sigma_{zz} \end{bmatrix} \tag{2-1}$$

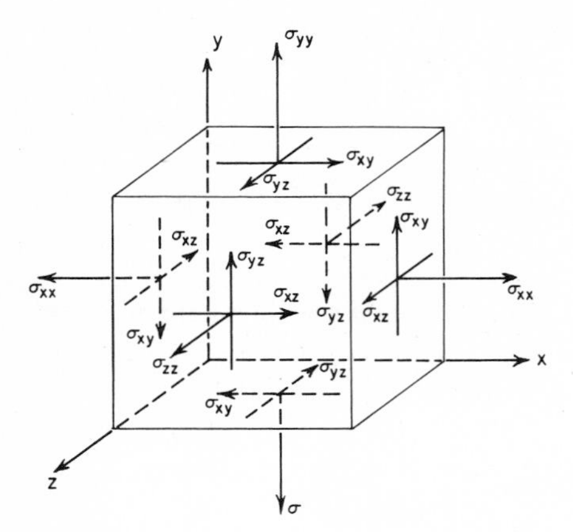

Figure 2-1 Sign convention for stress, tension positive.

Each component of the stress represents a force acting in a specific coordinate direction on a unit area oriented in a particular way. Thus σ_{xy} is the force in the positive x direction acting on a unit area whose outward normal is in the positive y direction. The terms σ_{xx}, σ_{yy}, and σ_{zz} are normal stresses, and the rest are shear stresses.

The sign convention of Fig. 2-1 is the usual one for continuum mechanics, in which stresses are positive when the senses of the force and the outward normal to the face on which the force acts are both positive or both negative. This makes tension positive. Because compression is more common than tension in geotechnical problems, the usual geotechnical convention is to make compression positive. If that convention is used, it makes sense to reverse the entire system, as shown in Fig. 2-2. In this book both conventions are used according to the demands of the problems treated, but the convention is identified where confusion might arise.

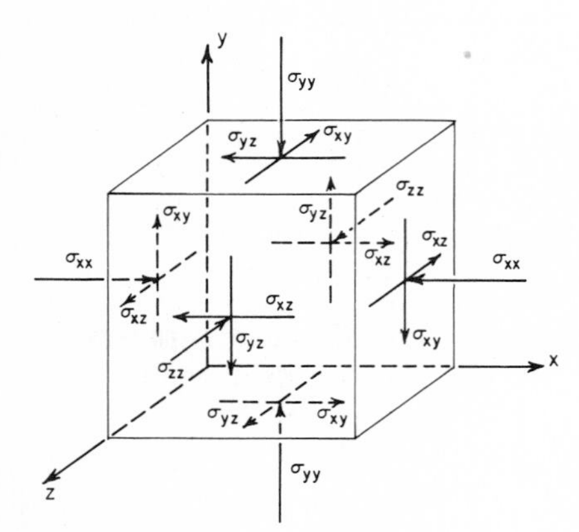

Figure 2-2 Sign convention for stress, compression positive.

The pressure in a fluid is a special case of the stress tensor. If p is the magnitude of the pressure, the corresponding stress tensor is

$$p_{ij} = \begin{bmatrix} p & 0 & 0 \\ 0 & p & 0 \\ 0 & 0 & p \end{bmatrix} \tag{2-2}$$

When pore pressures, total stresses, and effective stresses are being used, it is important to keep the sign convention clear. One approach is to use the same convention for all stresses; i.e., normal stresses and pore pressures are both positive in tension or both positive in compression. Some people have used a mixed convention with pore pressures positive in compression and normal stresses positive in tension, but this can lead to considerable confusion. The reader of the technical literature should be alert to the sign convention.

If the same sign convention is used for total stress and for pore pressure, the effective stress $[\bar{\sigma}]$ will then be

$$[\bar{\sigma}] = [\sigma] - [p] \tag{2-3}$$

or

$$\begin{bmatrix} \bar{\sigma}_{xx} & \bar{\sigma}_{xy} & \bar{\sigma}_{xz} \\ \bar{\sigma}_{xy} & \bar{\sigma}_{yy} & \bar{\sigma}_{yz} \\ \bar{\sigma}_{xz} & \bar{\sigma}_{yz} & \bar{\sigma}_{zz} \end{bmatrix} = \begin{bmatrix} \sigma_{xx} & \sigma_{xy} & \sigma_{xz} \\ \sigma_{xy} & \sigma_{yy} & \sigma_{yz} \\ \sigma_{xz} & \sigma_{yz} & \sigma_{zz} \end{bmatrix} - \begin{bmatrix} p & 0 & 0 \\ 0 & p & 0 \\ 0 & 0 & p \end{bmatrix} \tag{2-4}$$

Figure 2-3 shows two coordinate systems, x, y, z and a, b, c. The direction cosines are the cosines of the angles between pairs of coordinate directions. They can be denoted by the coordinate labels enclosed in parentheses; thus, (y, a) is the cosine between direction y and direction a. A point with coordinates (x, y, z) in one system will have coordinates (a, b, c) in the other system, defined by

$$\begin{Bmatrix} a \\ b \\ c \end{Bmatrix} = \begin{bmatrix} (x, a) & (y, a) & (z, a) \\ (x, b) & (y, b) & (z, b) \\ (x, c) & (y, c) & (z, c) \end{bmatrix} \begin{Bmatrix} x \\ y \\ z \end{Bmatrix} \tag{2-5}$$

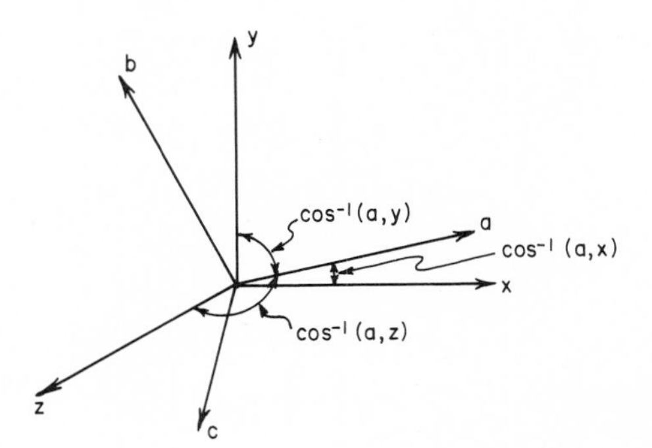

Figure 2-3 Rotation of coordinates and definition of direction cosines.

The 3×3 matrix in Eq. (2-5) is called the *rotation* or *transformation matrix* $[T]$. Any vector $\{v\}$ written in the x, y, z system can be transformed into $\{v^*\}$ in the a, b, c system by

$$\{v^*\} = [T]\{v\} \tag{2-6}$$

The stress $[\sigma]$ in the x, y, z system can be transformed into $[\sigma^*]$ in the a, b, c system by

$$[\sigma^*] = [T][\sigma][T]^T \tag{2-7}$$

From the theory of matrices one can find a set of rotations of coordinates in Eq. (2-7) that will give a stress $[\sigma^*]$ that has only diagonal or normal terms:

$$[\sigma^*] = \begin{bmatrix} \sigma_1 & 0 & 0 \\ 0 & \sigma_2 & 0 \\ 0 & 0 & \sigma_3 \end{bmatrix} \tag{2-8}$$

The diagonal terms are called the *principal stresses*. They are the solutions of the equation

$$\sigma^3 - I_1\sigma^2 + I_2\sigma - I_3 = 0 \tag{2-9}$$

where

$$I_1 = \sigma_{xx} + \sigma_{yy} + \sigma_{zz} \tag{2-10}$$

$$I_2 = \begin{vmatrix} \sigma_{xx} & \sigma_{xy} \\ \sigma_{xy} & \sigma_{yy} \end{vmatrix} + \begin{vmatrix} \sigma_{yy} & \sigma_{yz} \\ \sigma_{yz} & \sigma_{zz} \end{vmatrix} + \begin{vmatrix} \sigma_{xx} & \sigma_{xz} \\ \sigma_{xz} & \sigma_{zz} \end{vmatrix} \tag{2-11}$$

$$I_3 = \begin{bmatrix} \sigma_{xx} & \sigma_{xy} & \sigma_{xz} \\ \sigma_{xy} & \sigma_{yy} & \sigma_{yz} \\ \sigma_{xz} & \sigma_{yz} & \sigma_{zz} \end{bmatrix} \tag{2-12}$$

These are called, respectively, the first, second, and third invariants of the stress. Since the principal stresses are physical quantities independent of the choice of coordinate axes, the invariants must also be independent of the choice of axes and are therefore invariant under the rotations of coordinates like that described by Eq. (2-7).

The average normal stress is called the *octahedral normal stress* σ_{oct}, and it is also invariant:

$$\sigma_{\text{oct}} = \frac{I_1}{3} \tag{2-13}$$

$$[\sigma_{\text{oct}}] = \begin{bmatrix} \sigma_{\text{oct}} & 0 & 0 \\ 0 & \sigma_{\text{oct}} & 0 \\ 0 & 0 & \sigma_{\text{oct}} \end{bmatrix} \tag{2-14}$$

The deviator stress is then defined as

$$[s] = \begin{bmatrix} \sigma_{xx} - \sigma_{\text{oct}} & \sigma_{xy} & \sigma_{xz} \\ \sigma_{xy} & \sigma_{yy} - \sigma_{\text{oct}} & \sigma_{yz} \\ \sigma_{xz} & \sigma_{yz} & \sigma_{zz} - \sigma_{\text{oct}} \end{bmatrix} \tag{2-15}$$

The deviator stress also has principal values and invariants, which are identified by J_1, J_2, and J_3. By analogy to Eq. (2-9), the principal stresses are solutions to the equation

$$s^3 - J_2 s - J_3 = 0 \tag{2-16}$$

It should be noted that $J_1 = 0$ and J_2 has the sign opposite to that of I_2. The following useful relations can then be derived:

$$J_2 = -\begin{vmatrix} s_{xx} & s_{xy} \\ s_{xy} & s_{yy} \end{vmatrix} - \begin{vmatrix} s_{yy} & s_{yz} \\ s_{yz} & s_{zz} \end{vmatrix} - \begin{vmatrix} s_{xx} & s_{xz} \\ s_{xz} & s_{zz} \end{vmatrix} \tag{2-17a}$$

$$= \tfrac{1}{2}(s_{xx}^2 + s_{yy}^2 + s_{zz}^2) + s_{xy}^2 + s_{yz}^2 + s_{xz}^2 \tag{2-17b}$$

$$= \tfrac{1}{6}[(\sigma_{xx} - \sigma_{yy})^2 + (\sigma_{yy} - \sigma_{zz})^2 + (\sigma_{xx} - \sigma_{zz})^2] + \sigma_{xy}^2 + \sigma_{yz}^2 + \sigma_{xz}^2 \tag{2-17c}$$

$$= \tfrac{1}{6}[(\sigma_1 - \sigma_2)^2 + (\sigma_1 - \sigma_3)^2 + (\sigma_2 - \sigma_3)^2] \tag{2-17d}$$

If the coordinates are indicated by subscripts 1, 2, and 3 rather than x, y, and z, the invariants can be simply expressed by

$$J_2 = \frac{1}{2}\sum_{i=1}^{3}\sum_{j=1}^{3} s_{ij}s_{ij} = \frac{1}{2}s_{ij}s_{ij} \qquad \text{if } \textstyle\sum\text{'s omitted} \tag{2-18}$$

$$J_3 = \frac{1}{3}\sum_{i=1}^{3}\sum_{j=1}^{3}\sum_{k=1}^{3} s_{ij}s_{jk}s_{ki} = \frac{1}{3}s_{ij}s_{jk}s_{ki} \qquad \text{if } \textstyle\sum\text{'s omitted} \tag{2-19}$$

When the directions of the three principal stresses are taken as coordinates, it is easy to determine that planes which have outward normals forming equal angles with these three axes must have the octahedral normal stress σ_{oct} acting on them as the normal stress. There are eight such planes, as shown in Fig. 2-4; hence

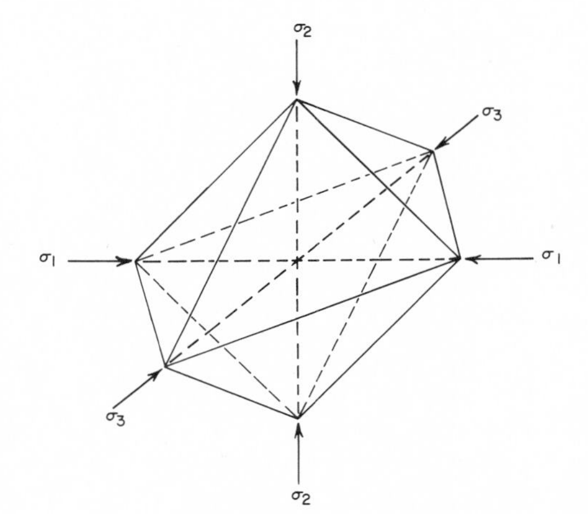

Figure 2-4 Octahedral planes.

the name octahedral stress. The resultant shearing stress acting on an octahedral plane is called the *octahedral shearing stress* τ_{oct}. Its magnitude is found from

$$\tau_{oct}^2 = \tfrac{1}{9}[(\sigma_1 - \sigma_2)^2 + (\sigma_1 - \sigma_3)^2 + (\sigma_2 - \sigma_3)^2] = \tfrac{2}{3}J_2 \tag{2-20}$$

Review of Strain

The deformation of a body is described by the strain. If the displacements are u, v, and w in the x, y, and z directions, respectively, the components of strain are

$$\begin{aligned} \epsilon_{xx} &= \frac{\partial u}{\partial x} \qquad \epsilon_{yy} = \frac{\partial v}{\partial y} \qquad \epsilon_{zz} = \frac{\partial w}{\partial z} \\ \gamma_{xy} &= \frac{\partial u}{\partial y} + \frac{\partial v}{\partial x} \qquad \gamma_{xz} = \frac{\partial u}{\partial z} + \frac{\partial v}{\partial x} \qquad \gamma_{yz} = \frac{\partial v}{\partial z} + \frac{\partial w}{\partial y} \end{aligned} \tag{2-21}$$

The same equations can be expressed more compactly by using a subscript notation in which u_1, u_2, and u_3 are the displacements in the x_1, x_2, and x_3 directions, respectively. The strain is then

$$\epsilon_{ij} = \frac{1}{2}\left(\frac{\partial u_i}{\partial x_j} + \frac{\partial u_j}{\partial x_i}\right) \qquad \gamma_{ij} = 2\epsilon_{ij} \tag{2-22}$$

The physical meaning of the components of strain is illustrated in Fig. 2-5. The first three are the normal strains, and the last three are shear strains.

It will be noted that there are two definitions of shear strain, the γ's and the ϵ's. The former are called the *engineering strains.* They are most useful for experimental work, but they are used in many numerical applications as well. The latter are called the *tensorial shear strains,* and they are more useful for theoretical derivations in which it is important to keep components clearly distinguished. From Fig. 2-5 it can be seen that the engineering definition of shear strain includes some rotation but the tensorial definition describes a pure deformation. No confusion need arise from the existence of two definitions of shear strain provided one is careful to identify which definition is being used.

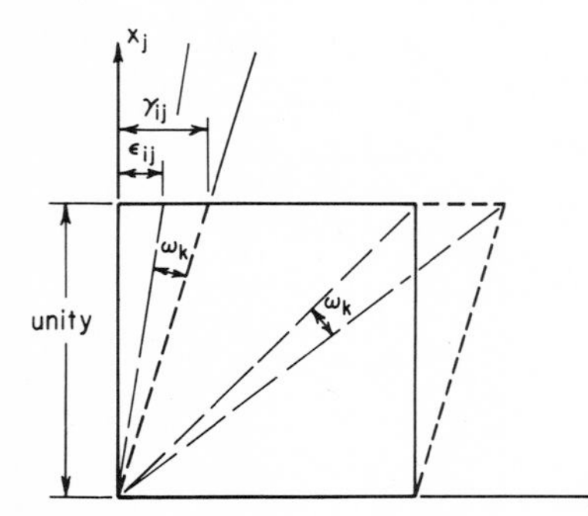

Figure 2-5 Engineering and tensorial shear strains.

The reason for defining a tensorial strain is that the complete set of strains then becomes a tensor:

$$\epsilon_{ij} = \begin{bmatrix} \epsilon_{xx} & \epsilon_{xy} & \epsilon_{xz} \\ \epsilon_{xy} & \epsilon_{yy} & \epsilon_{yz} \\ \epsilon_{xz} & \epsilon_{yz} & \epsilon_{zz} \end{bmatrix} \tag{2-23}$$

This tensor obeys all the transformation laws described in the previous section for stress. It has invariants, principal values, and so on.

The strain can be separated into an octahedral normal component

$$\epsilon_{\text{oct}} = \frac{\epsilon_{xx} + \epsilon_{yy} + \epsilon_{zz}}{3} \tag{2-24}$$

$$[\epsilon_{\text{oct}}] = \begin{bmatrix} \epsilon_{\text{oct}} & 0 & 0 \\ 0 & \epsilon_{\text{oct}} & 0 \\ 0 & 0 & \epsilon_{\text{oct}} \end{bmatrix} \tag{2-25}$$

and a deviatoric one

$$[e] = \begin{bmatrix} \epsilon_{xx} - \epsilon_{\text{oct}} & \epsilon_{xy} & \epsilon_{xz} \\ \epsilon_{xy} & \epsilon_{yy} - \epsilon_{\text{oct}} & \epsilon_{yz} \\ \epsilon_{xz} & \epsilon_{yz} & \epsilon_{zz} - \epsilon_{\text{oct}} \end{bmatrix} \tag{2-26}$$

If the volume of a small element is V, the volumetric strain ϵ_{vol} is defined by

$$\epsilon_{\text{vol}} = \frac{\Delta V}{V} = \frac{\partial u}{\partial x} + \frac{\partial v}{\partial y} + \frac{\partial w}{\partial z} = \epsilon_{xx} + \epsilon_{yy} + \epsilon_{zz} = 3\epsilon_{\text{oct}} \tag{2-27}$$

The volumetric strain is an important quantity, and it must be remembered that the volumetric strain is three times the octahedral strain.

A Comment on Large Strain and Displacement

The preceding sections are valid only when the displacements are such that the body undergoes infinitesimally small deformations and rotations. Since this is almost always the case in geotechnical problems, the infinitesimal theories are adequate. Should finite rotations or finite strains be present, one must consider them and the definitions of stress and strain become much more complicated. References 1 to 3 should be consulted for the necessary formulations.

Stress and Strain in a Plane

Many important problems require only the stresses and strains in a plane, say the xy plane. Thus, although σ_{zz} may exist, σ_{xz} and σ_{yz} are zero and can be ignored. The problem of finding principal stresses reduces to one involving three components, σ_{xx}, σ_{yy}, and σ_{xy}.

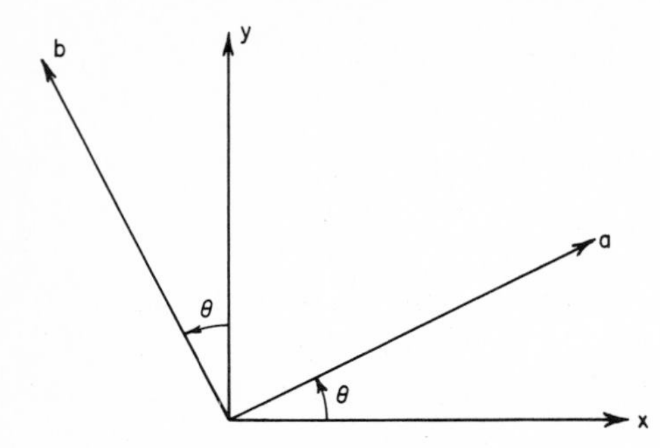

Figure 2-6 Rotation of coordinates in a plane.

Figure 2-6 shows the two sets of coordinate axes *xy* and *ab*. The rotation matrix [*T*] reduces to

$$[T] = \begin{bmatrix} \cos\theta & \sin\theta \\ -\sin\theta & \cos\theta \end{bmatrix} \tag{2-28}$$

The principal stresses become

$$\sigma_1 = \frac{\sigma_{xx} + \sigma_{yy}}{2} + \sqrt{\left(\frac{\sigma_{xx} - \sigma_{yy}}{2}\right)^2 + \sigma_{xy}^2} \tag{2-29a}$$

$$\sigma_2 = \sigma_{zz} \tag{2-29b}$$

$$\sigma_3 = \frac{\sigma_{xx} + \sigma_{yy}}{2} - \sqrt{\left(\frac{\sigma_{xx} - \sigma_{yy}}{2}\right)^2 + \sigma_{xy}^2} \tag{2-29c}$$

The maximum shear stress τ_{max} is

$$\tau_{max} = \frac{|\sigma_1 - \sigma_3|}{2} = \sqrt{\left(\frac{\sigma_{xx} - \sigma_{yy}}{2}\right)^2 + \sigma_{xy}^2} \tag{2-30}$$

The Mohr diagram is a convenient representation of the two-dimensional state of stress. Many versions exist; the one most common in geotechnical engineering is illustrated in Fig. 2-7. The assumption of positive compression is made.

It should be noted that the conventions for the signs of shear strain are different between the Mohr diagram and continuum mechanics. In the Mohr diagram positive shear stresses cause counterclockwise moments on the stressed body. Thus faces *a* and *b* have negative shear stresses and *c* and *d* positive ones in the Mohr circle. All these have positive shear stresses according to the convention of compression positive in continuum mechanics. This difference will not usually cause difficulty, but confusion can occur if one does not remember this fact.

Exactly the same two-dimensional analysis can be made for strain. The principal strains are

$$\epsilon_1 = \frac{\epsilon_{xx} + \epsilon_{yy}}{2} + \sqrt{\left(\frac{\epsilon_{xx} - \epsilon_{yy}}{2}\right)^2 + \epsilon_{xy}^2} \tag{2-31a}$$

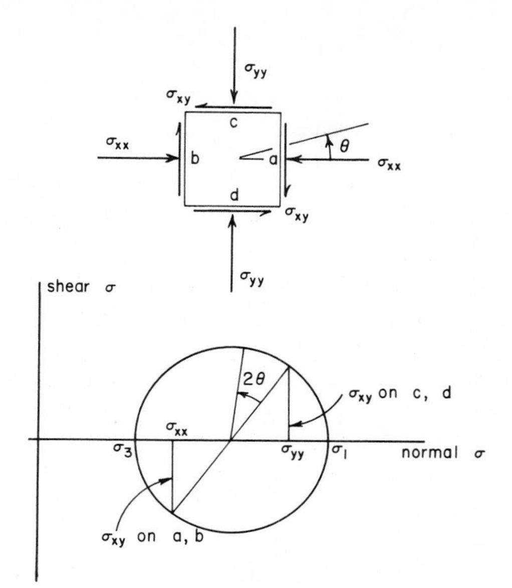

Figure 2-7 Mohr's circle.

$$\epsilon_2 = \epsilon_{zz} = 0 \tag{2-31b}$$

$$\epsilon_3 = \frac{\epsilon_{xx} + \epsilon_{yy}}{2} - \sqrt{\left(\frac{\epsilon_{xx} - \epsilon_{yy}}{2}\right)^2 + \epsilon_{xy}^2} \tag{2-31c}$$

and the maximum shear strain is

$$\gamma_{\mathbf{max}} = 2\epsilon_{\mathbf{max}} = \sqrt{(\epsilon_{xx} - \epsilon_{yy})^2 + \gamma_{xy}^2} \tag{2-32}$$

If no displacements occur out of the plane, the material is in a state of *plane strain.* If there are no forces out of the plane, there is *plane stress.*

2-2 STRESS-STRAIN BEHAVIOR

Types of Stress-Strain Behavior

Figure 2-8 illustrates several types of stress-strain curves that might result from a compression test on a cylindrical sample. Figure 2-8*a* shows a loading curve identical to the unloading curve. All strains are recovered when the load is removed. This is *elastic* behavior. If the relation is linear, as in Fig. 2-8*b*, the material is *linearly elastic.*

When some of the strains are not recovered on unloading (Fig. 2-8*c*), the unrecovered strains are called *plastic* strains and the material exhibits plastic behavior. Figure 2-8*d* shows *rigid-plastic* behavior, in which there are no elastic or recoverable strains.

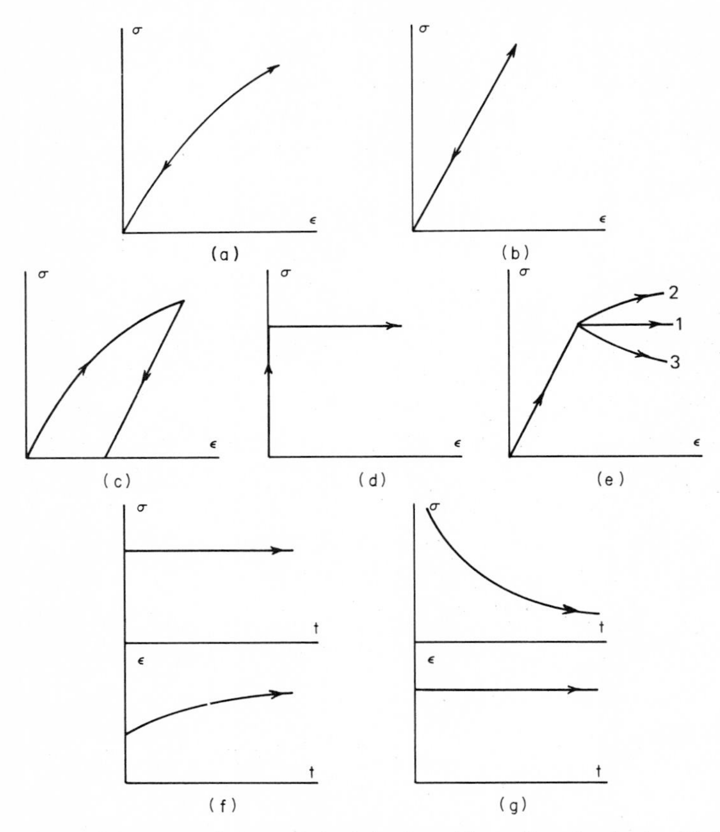

Figure 2-8 Types of stress-strain behavior: (*a*) nonlinearly elastic; (*b*) linearly elastic; (*c*) nonelastic, or plastic; (*d*) rigid, perfectly plastic; (*e*) elastoplastic: (1) perfectly plastic, (2) strain hardening, (3) strain softening: (*f*) viscoelastic creep at constant stress; (*g*) viscoelastic relaxation at constant strain.

Usually there are some recoverable strains and some unrecoverable ones (Fig. 2-8*e*), leading to *elastoplastic* behavior. In the example in Fig. 2-8*e* there are three types of linearly elastoplastic behavior, distinguished by the behavior after the stresses have reached the yield stress σ_y. Curve 1 is a *perfectly plastic* case; the yield stress is not affected by strains in the plastic range. Curve 2 is a case of *strain hardening*, and curve 3 is a case of *strain softening*.

All the above cases involve strains and stresses that occur simultaneously. *Viscous* and *viscoelastic* materials have strains that develop while the load is constant (Fig. 2-8*f*) or stresses that decrease while the strain stays constant (Fig. 2-8*g*). Linearly viscoelastic materials have a linear relation between stress and strain at a given time. In other words, if a load is applied for a time *t*, some strain develops, and if twice the load is applied for the same time *t*, twice the strain develops. There is almost never a linear relation between stress and strain on the one hand and time on the other.

Linear elasticity The basic relations between stress and strain, upon which many of the others are based, are those of linear elasticity. For most numerical purposes it is most useful to write the stresses and strains as vectors:

$$\{\sigma\}^T = [\sigma_{xx} \quad \sigma_{yy} \quad \sigma_{zz} \quad \sigma_{xy} \quad \sigma_{yz} \quad \sigma_{zx}] \tag{2-33}$$

(often, the normal stresses σ_{xx}, σ_{yy}, and σ_{zz} are denoted with only one subscript as σ_x, σ_y, σ_z, and the shear stresses σ_{xy}, σ_{yz}, σ_{zx} are denoted as τ_{xy}, τ_{yz}, and τ_{zx}, respectively)

and
$$\{\epsilon\}^T = [\epsilon_{xx} \quad \epsilon_{yy} \quad \epsilon_{zz} \quad \epsilon_{xy} \quad \epsilon_{yz} \quad \epsilon_{zx}] \tag{2-34}$$

(here the components of normal strain are often denoted by using only one subscript as ϵ_x, ϵ_y, and ϵ_z, and the shear strain components are denoted by γ_{xy}, γ_{yz}, γ_{zx}).

When plane conditions apply, these vectors can be reduced to

$$\{\sigma\}^T = [\sigma_{xx} \quad \sigma_{yy} \quad \sigma_{xy}] \tag{2-35}$$

and
$$\{\epsilon\}^T = [\epsilon_{xx} \quad \epsilon_{yy} \quad \epsilon_{xy}] \tag{2-36}$$

Since there are 6 independent components of stress and 6 of strain, 36 coefficients are needed to relate them linearly in the most general way. However, by considering the energy stored in a strained linearly elastic body one can show that the coefficients must form a symmetric array. The term relating, say, σ_{xx} to ϵ_{yy} must be the same as the one relating σ_{yy} to ϵ_{xx}. This symmetry of the stress-strain relation reduces the number of independent terms to 21. The relationship can be expressed in matrix form

$$\{\sigma\} = [C]\{\epsilon\} \tag{2-37}$$

where
$$[C] = \begin{bmatrix} C_{11} & C_{12} & C_{13} & C_{14} & C_{15} & C_{16} \\ & C_{22} & C_{23} & C_{24} & C_{25} & C_{26} \\ & & C_{33} & C_{34} & C_{35} & C_{36} \\ \text{sym} & & & C_{44} & C_{45} & C_{46} \\ & & & & C_{55} & C_{56} \\ & & & & & C_{66} \end{bmatrix} \tag{2-38}$$

The number of independent terms is reduced considerably by considering the planes or axes of isotropy in the material. In geotechnical engineering the two most commonly used types are complete isotropy and cross anisotropy. Cross anisotropy applies when there is some plane (usually horizontal) in which all stress-strain relations are isotropic; i.e., it makes no difference how the axes are chosen within the plane: the elastic constants are the same. The elastic constants for stresses and strains outside the plane are different. The resulting [C] matrix has five independent constants. If the z axis is normal to the plane of isotropy, the matrix becomes

$$[C] = \begin{bmatrix} C_{11} & C_{12} & C_{13} & 0 & 0 & 0 \\ C_{12} & C_{11} & C_{13} & 0 & 0 & 0 \\ C_{13} & C_{13} & C_{33} & 0 & 0 & 0 \\ 0 & 0 & 0 & C_{44} & 0 & 0 \\ 0 & 0 & 0 & 0 & C_{44} & 0 \\ 0 & 0 & 0 & 0 & 0 & C_{11} - C_{12} \end{bmatrix} \tag{2-39}$$

Total isotropy exists when the axes can be chosen arbitrarily with no effect on the elastic constants. Matrix $[C]$ then has only two independent terms and becomes

$$[C] = \begin{bmatrix} C_{11} & C_{12} & C_{12} & 0 & 0 & 0 \\ C_{12} & C_{11} & C_{12} & 0 & 0 & 0 \\ C_{12} & C_{12} & C_{11} & 0 & 0 & 0 \\ 0 & 0 & 0 & C_{11} - C_{12} & 0 & 0 \\ 0 & 0 & 0 & 0 & C_{11} - C_{12} & 0 \\ 0 & 0 & 0 & 0 & 0 & C_{11} - C_{12} \end{bmatrix} \tag{2-40}$$

The constants can be expressed in terms of several different parameters. The most common ones are as follows:

1. *Young's modulus E* This relates axial strain to axial stress in a simple tension or compression test

$$\sigma_{xx} = E\epsilon_{xx} \qquad \text{all other } \sigma\text{'s} = 0 \tag{2-41}$$

2. *Poisson's ratio v* This relates axial strain to transverse normal strain in a simple tension or compression test

$$\epsilon_{yy} = \epsilon_{zz} = -\nu\epsilon_{xx} \qquad \text{all other } \sigma\text{'s except } \sigma_{xx} = 0 \tag{2-42}$$

3. *Shear modulus G* This relates shear stress to shear strain

$$\sigma_{xy} = 2G\epsilon_{xy} = G\gamma_{xy} \tag{2-43}$$

4. *Bulk modulus K* This relates volumetric strain to average or octahedral stress

$$\sigma_{\text{oct}} = K\epsilon_{\text{vol}} \tag{2-44}$$

5. *Lamé's constants* λ, μ These relate the stresses and strains as follows:

$$\sigma_{xx} = \lambda\epsilon_{\text{vol}} + 2\mu\epsilon_{xx} \tag{2-45}$$

(with similar equations for σ_{yy} and σ_{zz}) and

$$\sigma_{xy} = 2\mu\epsilon_{xy} \tag{2-46}$$

(with similar equations for the other shears).

6. *Constrained modulus M* This relates axial strain to axial stress when the other two axial strains are held to zero:

$$\sigma_{xx} = M\epsilon_{xx} \qquad \text{all other } \epsilon\text{'s} = 0 \tag{2-47}$$

The following relations exist between these constants:

$$G = \mu = \frac{E}{2(1+v)} \qquad K = \frac{E}{3(1-2v)} = \lambda + \frac{2}{3}\mu$$
$$\lambda = \frac{vE}{(1+v)(1-2v)} \qquad M = \frac{E(1-v)}{(1+v)(1-2v)} \tag{2-48}$$

The stress-strain relation can be expressed in several ways, but the most useful are

$$[C] = \frac{E}{(1+v)(1-2v)}\begin{bmatrix} 1-v & v & v & 0 & 0 & 0 \\ v & 1-v & v & 0 & 0 & 0 \\ v & v & 1-v & 0 & 0 & 0 \\ 0 & 0 & 0 & 1-2v & 0 & 0 \\ 0 & 0 & 0 & 0 & 1-2v & 0 \\ 0 & 0 & 0 & 0 & 0 & 1-2v \end{bmatrix} \tag{2-49}$$

and

$$[C] = \begin{bmatrix} K+\frac{4}{3}G & K-\frac{2}{3}G & K-\frac{2}{3}G & 0 & 0 & 0 \\ K-\frac{2}{3}G & K+\frac{4}{3}G & K-\frac{2}{3}G & 0 & 0 & 0 \\ K-\frac{2}{3}G & K-\frac{2}{3}G & K+\frac{4}{3}G & 0 & 0 & 0 \\ 0 & 0 & 0 & 2G & 0 & 0 \\ 0 & 0 & 0 & 0 & 2G & 0 \\ 0 & 0 & 0 & 0 & 0 & 2G \end{bmatrix} \tag{2-50a}$$

and

$$[C] = \begin{bmatrix} M & M-2G & M-2G & 0 & 0 & 0 \\ M-2G & M & M-2G & 0 & 0 & 0 \\ M-2G & M-2G & M & 0 & 0 & 0 \\ 0 & 0 & 0 & 2G & 0 & 0 \\ 0 & 0 & 0 & 0 & 2G & 0 \\ 0 & 0 & 0 & 0 & 0 & 2G \end{bmatrix} \tag{2-50b}$$

For conditions of plane strain ϵ_{zz}, ϵ_{xz}, and ϵ_{yz} are all zero, so that the stress-strain relation becomes

$$\{\sigma\} = [C]\{\epsilon\} \tag{2-51a}$$

with

$$\{\sigma\}^T = [\sigma_{xx} \quad \sigma_{yy} \quad \sigma_{xy}] \qquad \{\epsilon\}^T = [\epsilon_{xx} \quad \epsilon_{yy} \quad \epsilon_{xy}] \tag{2-51b}$$

and

$$[C] = \begin{bmatrix} K + \frac{4}{3}G & K - \frac{2}{3}G & 0 \\ K - \frac{2}{3}G & K + \frac{4}{3}G & 0 \\ 0 & 0 & 2G \end{bmatrix} = \frac{E}{(1+\nu)(1-2\nu)} \begin{bmatrix} 1-\nu & \nu & 0 \\ \nu & 1-\nu & 0 \\ 0 & 0 & 1-2\nu \end{bmatrix}$$

$$= \begin{bmatrix} M & M - 2G & 0 \\ M - 2G & M & 0 \\ 0 & 0 & 2G \end{bmatrix} \tag{2-51c}$$

These relations can also be written in the inverse form:

$$\{\epsilon\} = [D]\{\sigma\} \tag{2-52a}$$

where

$$[D] = \frac{1}{E} \begin{bmatrix} 1 & -\nu & -\nu & 0 & 0 & 0 \\ -\nu & 1 & -\nu & 0 & 0 & 0 \\ -\nu & -\nu & 1 & 0 & 0 & 0 \\ 0 & 0 & 0 & 1+\nu & 0 & 0 \\ 0 & 0 & 0 & 0 & 1+\nu & 0 \\ 0 & 0 & 0 & 0 & 0 & 1+\nu \end{bmatrix} \tag{2-52b}$$

The cross-anisotropic relation of Eq. (2-39) can also be expressed in terms of a strain-stress matrix $[D]$. The usual situation calls for the xy plane of isotropy to be horizontal and the z axis of the anisotropic direction to be vertical. The terms of $[D]$ are then

$$D_{11} = D_{22} = \frac{1}{E_H} \qquad D_{33} = \frac{1}{E_V} \qquad D_{12} = D_{21} = -\frac{\nu_{HH}}{E_H}$$

$$D_{13} = D_{31} = D_{23} = D_{32} = -\frac{\nu_{VH}}{E_H} = -\frac{\nu_{HV}}{E_V} \tag{2-53}$$

$$D_{44} = D_{55} = \frac{1}{2G_{VH}} \qquad D_{66} = \frac{1 + \nu_{HH}}{E_H}$$

All other terms are zero. E_H and E_V can be interpreted as the Young's moduli for loading in the horizontal plane or along the vertical axis, respectively. The Poisson's ratio relating loading along one horizontal axis to strains along the other horizontal axis is ν_{HH}. The relation between extensional strains in the horizontal plane and vertical loadings or between vertical extensional strains and horizontal loadings is controlled by the other Poisson's ratio ν_{HV} or ν_{VH}. These are not independent of each other because of the symmetry of the $[D]$ and $[C]$ matrices. The fifth constant G_{VH} relates shear stresses out of the horizontal plane to shear strains out of the horizontal plane.

Drnevich[4] has suggested experimental procedures for obtaining the terms of the [C] matrix for cross-anisotropic materials. Using the constrained-modulus and shear-modulus notation of Eq. (2-50b), he proposes expressing the five independent terms with measured quantities

$$C_{11} = C_{22} = M_H \qquad C_{33} = M_V \qquad C_{44} = C_{55} = 2G_{VH}$$
$$C_{66} = 2G_{HH} \qquad C_{12} = C_{21} = M_H - 2G_{HH} \tag{2-54}$$
$$C_{13} = C_{23} = C_{31} = C_{32} \approx \frac{M_H + M_V}{2 - 2G_{VH}}$$

The last expression is only approximately correct because five independent measurements are required to obtain the five constants. A better measure might be $K_0 M_V$, where K_0 is the coefficient of lateral earth pressure at rest.

The terms in the anisotropic [C] matrix can also be found by inverting the [D] matrix to give

$$C_{11} = C_{22} = A\left(\frac{E_H}{E_V} - \nu_{VH}^2\right) = M_H \qquad C_{33} = A(1 - \nu_{HH}^2) = M_V$$
$$C_{12} = C_{21} = A\left(\frac{E_H}{E_V}\nu_{HH} + \nu_{VH}^2\right) = C_{11} - C_{66} \tag{2-55a}$$
$$C_{13} = C_{31} = C_{23} = C_{32} = A\nu_{VH}(1 + \nu_{HH})$$
$$C_{44} = C_{55} = 2G_{VH} \qquad C_{66} = \frac{E_H}{1 + \nu_{HH}} = 2G_{HH}$$

where
$$A = \frac{E_H}{(1 + \nu_{HH})[(E_H/E_V)(1 - \nu_{HH}) - 2\nu_{VH}^2]} \tag{2-55b}$$

$$\text{All other } C_{ij} = 0 \tag{2-55c}$$

2-3 NONLINEAR BEHAVIOR

Soil is not a linear material. The relations between stress and strain are much more complicated than the simple, linearly elastic ones described in the previous section. Therefore, in order to represent geotechnical problems realistically some form of nonlinear relation must be used, and the development and application of such relations have been important areas of research in recent years.

The various schemes for defining the constitutive behavior of soils and of rocks can be divided in three main groups:[5,6] (1) representation of given stress-strain curves by using curve-fitting methods, interpolation, or mathematical functions, (2) nonlinear elasticity theories, and (3) plasticity theories. We shall describe some of the main constitutive models used in geotechnical engineering.

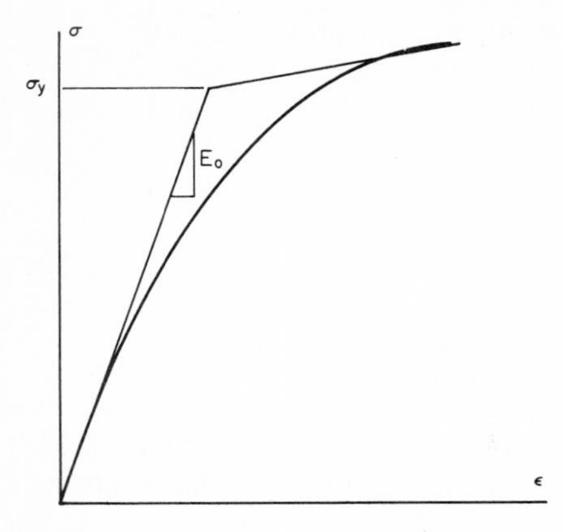

Figure 2-9 Bilinear model for nonlinear material.

Bilinear and Multilinear Models

The first group involves essentially a mechanical simulation of a set of given stress-strain curves. The simplest type of nonlinear relation is the bilinear one illustrated in Fig. 2-9. The material has initial moduli $[C_i]$ until the stresses reach a yield value σ_y, after which the moduli are changed to $[C_y]$. Before yield, therefore, the incremental stress-strain relation can be written as

$$\{\Delta\sigma\} = [C_i]\{\Delta\epsilon\} \tag{2-56a}$$

and after yield it is

$$\{\Delta\sigma\} = [C_y]\{\Delta\epsilon\} \tag{2-56b}$$

The usual way to develop bilinear relations is to change the Young's modulus from an initial value E_i to a yielded value E_y. This is especially attractive because Young's modulus can be factored out of the stiffness and the individual element stiffnesses can be computed only once and modified after yield by a simple multiplication by a constant. The drawback is that the bulk modulus is reduced as much as the shear modulus. The element becomes compressible just as it becomes highly deformable after yield, and often unreliable results can follow. It is much better to reduce the shear modulus and keep the bulk modulus constant.

An obvious problem in this kind of formulation is that the actual stress-strain relation is curved. Figure 2-10 shows two ways to fit a bilinear relation to a curved relation, but it is not clear immediately which is better.

In the next step, the nonlinear curve can be divided into a number of linear curves leading to the so-called *multilinear* or *piecewise linear models*. In the initial stages of the use of the finite element method in geotechnical engineering,[7–11] the piecewise linear approach involved the tabular or interpolation procedure,[5] in which the tangent moduli were computed by interpolation on the basis of a set of data points (σ_i, ϵ_i) (Fig. 2-11) on the given stress-strain curve(s). The tangent

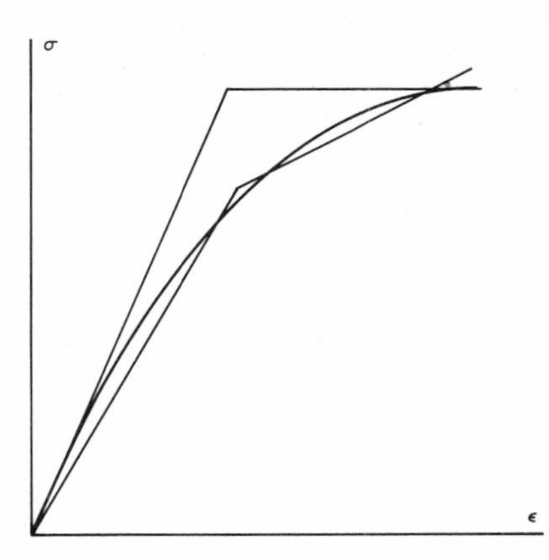

Figure 2-10 Options in choice of bilinear model.

modulus was defined as the slope of the chord between two computed points as

$$E_t = \frac{\sigma_i - \sigma_{i-1}}{\epsilon_i - \epsilon_{i-1}} \tag{2-57}$$

This method gives satisfactory and reliable answers as does the use of mathematical functions such as polynomials, hyperbolas, parabola, and splines. An advantage of the use of the mathematical functions is that, in contrast to the tabular form in which a number of data points are input, we need only a few parameters to describe the curves. On the other hand, use of mathematical functions can influence the accuracy in computing the tangent moduli. We note here that use of mathematical functions essentially constitutes the piecewise linear approach.

The Hyperbolic Relations

The widely used function for simulation of stress-strain curves in finite element analyses was formalized by Duncan and his colleagues,[12,13] using Kondner's[14] finding that the plot of stress vs. strain in a triaxial compression test is very nearly

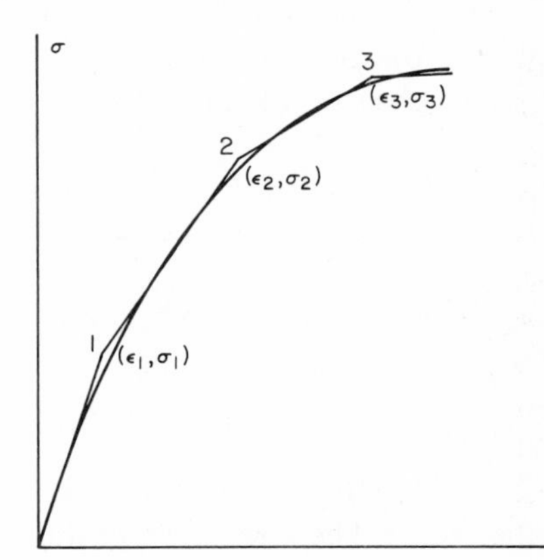

Figure 2-11 Multilinear model for nonlinear material.

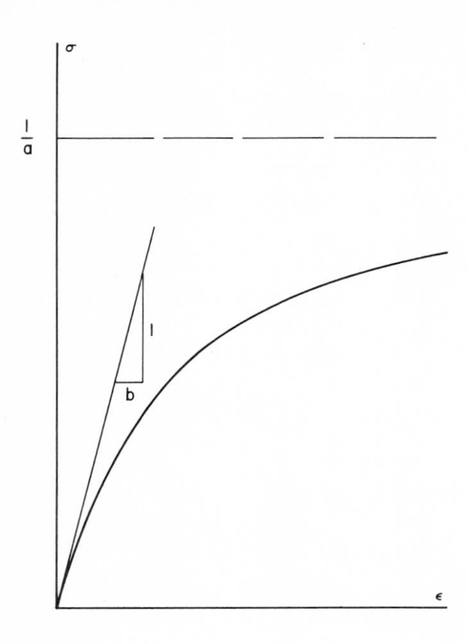

Figure 2-12 Hyperbolic model for nonlinear material.

a hyperbola. Figure 2-12 illustrates such a relation, which can be stated in equation form

$$\sigma = \frac{\epsilon}{b + a\epsilon} \tag{2-58a}$$

or

$$\frac{\epsilon}{\sigma} = b + a\epsilon \tag{2-58b}$$

In these equations the subscripts have been removed from the stresses and strains for clarity, so that σ and ϵ represent vertical stress and strain, respectively. The latter form of the equation plots as a straight line (Fig. 2-13), and, conversely, a plot with axes σ/ϵ and ϵ can be used to check whether the data from a test do fit a hyperbola or to find the parameters of the hyperbola from the test data.

To find the incremental stiffness it is first helpful to observe that at very small strains

$$\sigma = \frac{\epsilon}{b} \tag{2-58c}$$

so that $1/b$ is the initial Young's modulus E. At large strains the relation becomes

$$\sigma = \frac{1}{a} \tag{2-58d}$$

so that $1/a$ is the compressive strength; actually, $1/a$ is the asymptote. Since the compressive strength will be reached before the curve becomes asymptotic

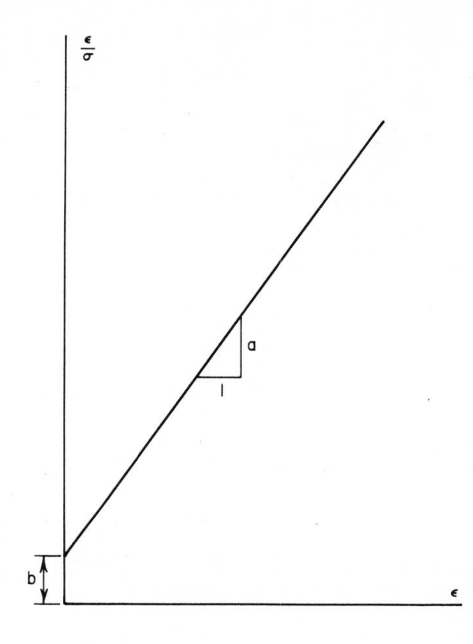

Figure 2-13 Hyperbolic model with transformed axes.

(Fig. 2-14), it is customary to require the compressive strength s to be R_f/a, where R_f is usually between 0.7 and 0.9. Thus

$$a = \frac{R_f}{s} \tag{2-58e}$$

Equation (2-58a) can also be solved for ϵ:

$$\epsilon = \frac{b\sigma}{1 - a\sigma} \tag{2-59}$$

The tangent modulus at any level of stress or strain is

$$E_t = \frac{\partial \sigma}{\partial \epsilon} = \frac{b}{(b + a\epsilon)^2} = \frac{1}{b}(1 - a\sigma)^2 = E_i\left(1 - \frac{R_f \sigma}{s}\right)^2 \tag{2-60}$$

For a Mohr-Coulomb material at failure

$$(\sigma_1 - \sigma_3)_f = \frac{2\sigma_3 + 2c \cos \phi}{1 - \sin \phi} \tag{2-61}$$

The term σ/s is the ratio between the existing $\sigma_1 - \sigma_3$ and s that would be available for the existing σ_3. The ratio is

$$\frac{\sigma}{s} = \frac{(\sigma_1 - \sigma_3)(1 - \sin \phi)}{2\sigma_3 \sin \phi + 2c \cos \phi} \tag{2-62}$$

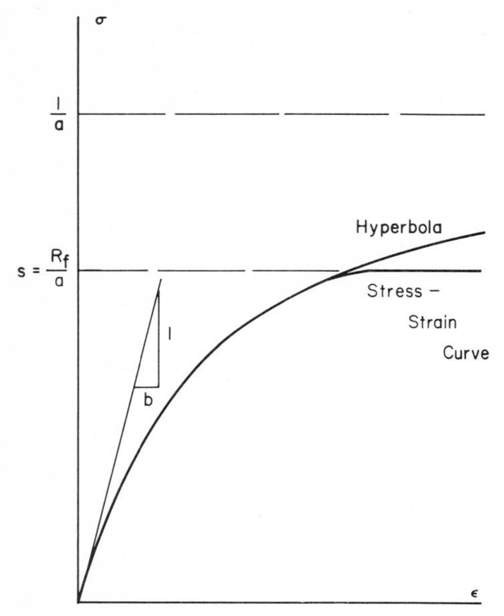

Figure 2-14 Hyperbolic model with strength cut off.

The tangent modulus now becomes

$$E_t = E_i \left[1 - \frac{R_f(\sigma_1 - \sigma_3)(1 - \sin \phi)}{2\sigma_3 \sin \phi + 2c \cos \phi}\right]^2 \tag{2-63a}$$

The initial modulus has been found to vary with the confining pressure according to[15]

$$E_i = Kp_a\left(\frac{\sigma_3}{p_a}\right)^n \tag{2-63b}$$

where p_a is the atmospheric pressure and K and n are constants to be determined.

The complete relation then becomes

$$E_t = Kp_a\left(\frac{\sigma_3}{p_a}\right)^n \left[1 - \frac{R_f(1 - \sin \phi)(\sigma_1 - \sigma_3)}{2c \cos \phi + 2\sigma_3 \sin \phi}\right]^2 \tag{2-63c}$$

For development of the full stiffness, a second elastic parameter is required. Kulhawy et al.[16] concluded from examination of many tests on sands and clays that the tangential value of Poisson's ratio satisfies

$$\nu_t = \frac{G - F \log (\sigma_3/p_a)}{\left\{1 - \dfrac{d(\sigma_1 - \sigma_3)}{Kp_a\left(\dfrac{\sigma_3}{p_a}\right)^n \left[1 - \dfrac{R_f(\sigma_1 - \sigma_3)(1 - \sin \phi)}{2c \cos \phi + 2\sigma_3 \sin \phi}\right]}\right\}^2} \tag{2-64}$$

where d, F, and G are constants to be determined experimentally. Values of these parameters for a number of soils are given in Ref. 16. The reader should remember that the G in Eq. (2-64) is not to be confused with the shear modulus.

The hyperbolic relations can also be applied to the direct determination of the shear modulus.[5] A derivation virtually identical to that leading to Eq. (2-63c) yields for the shear modulus

$$G_t = G_i \left[1 - \frac{R_f(\sigma_1 - \sigma_3)(1 - \sin \phi)}{2\sigma_3 \sin \phi + 2c \cos \phi} \right]^2 \tag{2-65}$$

where G is now and hereafter the shear modulus.

The initial value of the shear modulus G_i can be found in several ways. Hardin and Black[17,18] proposed the formula

$$G = 1230 \frac{(2.973 - e)^2}{1 + e} \bar{\sigma}_{\text{oct}}^{1/2} \text{OCR}^K \tag{2-66}$$

where all stresses and moduli are in pounds per square inch, e is the void ratio, OCR is the overconsolidation ratio, and K depends on the plasticity index of the soil according to Table 2-1. In situ shear-wave-velocity measurements can be used to find the shear modulus from the formula

$$G = \rho c_s^2 \tag{2-67}$$

where c_s is the measured shear-wave velocity and ρ is the mass density of the soil.

Laboratory tests like the resonant-column test can be used to find G_i. Recent research[19,20] has shown that secondary compression strongly influences the measurements of shear-wave velocity and modulus for clays. Therefore, laboratory values should be scaled up to account for sample disturbance. For Boston blue clay, Trudeau et al.[20] showed that the effect is 40 ft/s per log cycle of age of the clay, which gives about 250 ft/s for a recent glacial clay. The data also indicate that for clays the coefficient in Eq. (2-67) should be closer to 1630, as was also proposed by Hardin and Black.

Table 2-1 Value of K in Hardin-Black formula[18]

Plasticity index, %	K
0	0
20	0.18
40	0.30
60	0.41
80	0.48
⩾ 100	0.50

When the hyperbolic relation is used for the shear modulus, it is still necessary to find another elastic parameter. The most satisfactory arrangement is to describe the bulk modulus or constrained modulus either from laboratory tests or from in situ tests such as those involving compression-wave velocity measurements. Compression-wave velocities can be related to the constrained modulus by

$$M = \rho c_p^2 \tag{2-68}$$

where c_p is the compression-wave velocity. The hyperbolic relation using shear and compression waves has been described by Drnevich.[4]

In many cases it is satisfactory to consider a clay to be incompressible. This can be done by using a large value of bulk modulus as well as by using a special incompressible formulation, as described in Chap. 12 for consolidation problems. The stress-strain relation is best expressed in terms of the shear modulus, which becomes

$$G_t = G_i \left[1 - \frac{R_f(\sigma_1 - \sigma_3)}{2s_u} \right]^2 \tag{2-69}$$

Here the undrained shear strength s_u has been used because it is a reasonable description of the strength of a clay rapidly loaded; drained-strength parameters could also be used. Simon et al.[21] describe this type of relation and its application to several problems.

Many soils exhibit anisotropy, which is not accounted for in the above formulas. One way to introduce anisotropy was developed by Duncan and Dunlop.[22] They measured the nonlinear moduli in two orthogonal directions, as shown in Fig. 2-15. At any stage of the loading their procedure was to find the direction of the principal stress acting in each element and to compute the angle θ between that direction and the direction in which E_1 was measured. They then computed the apparent modulus from

$$E_\theta = E_1 - (E_1 - E_2) \sin^2 \theta \tag{2-70a}$$

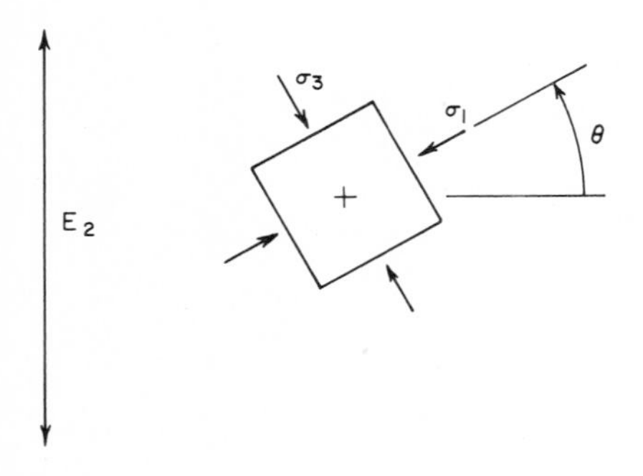

Figure 2-15 Definition of angle in anisotropic elastic model.

Comparison with laboratory tests revealed that a better fit was obtained if they used

$$E_\theta = E_1 - (E_1 - E_2)(\sin^2 \theta + 0.2 \sin 2\theta) \tag{2-70b}$$

If the material were generally anisotropic and one were to compute the apparent modulus for axial loading at an orientation θ from one of the principal axes of the material, the rotation of the elasticity matrix in Eq. (2-38) would lead to

$$E_\theta = E \cos^4 \theta + E_2 \sin^4 \theta = E_1 - (E_1 - E_2)(\sin^2 \theta - 0.25 \sin^2 2\theta) \tag{2-70c}$$

which is the opposite of the experimental observations.

It is often observed that soils are not very anisotropic at small strains but have different behavior near failure for different directions of loading. The undrained hyperbolic relation of Eq. (2-69) was modified by Simon et al.[21] by using $s_u(\theta)$ in the denominator instead of s_u; $s_u(\theta)$ is the shear strength that would exist for loading with principal stresses in the direction that now applies in the element. Two forms of anisotropic undrained strength often used are

$$s_u(\theta) = s_{uH} - (s_{uH} - s_{uV}) \cos^2 \theta \tag{2-71a}$$

and

$$\left(\frac{\sigma_{yy} - s_{uV}}{2} - \frac{\sigma_{xx} - s_{uH}}{2}\right)^2 + \sigma_{xy}^2 \frac{s_{u45}^2}{s_{uV}\, s_{uH}} = \left(\frac{s_{uV} + s_{uH}}{2}\right)^2 \tag{2-71b}$$

where s_{uV}, s_{uH}, and s_{u45} are the undrained strengths for compression loading vertically, horizontally, and at 45° to the vertical, respectively. Simon et al.[21] used Eq. (2-71b) in their work. Equation (2-71b) plots as an ellipse in Fig. 2-16, and the corresponding yield criterion gives simple expressions for bearing capacity.[23]

The hyperbolic simulation can be relatively easy to program for the computer, and the parameters have physical meaning to the geotechnical engineer. However, it is not the ultimate or consistent solution to the problems of constitutive laws for geologic media. It does not, in many respects, satisfy the definition of a constitutive law.[6] It is valid only for stress below the peak of the stress-strain curve,

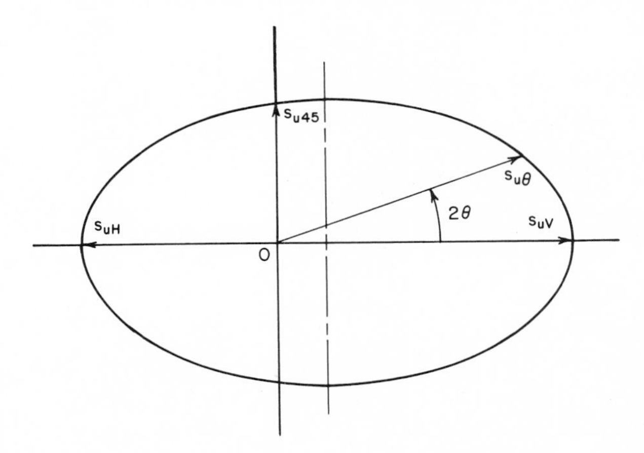

Figure 2-16 Anisotropic shear strength described by an ellipse.

usually for monotonically increasing loads. Although some applications have been made above that range, they involve a number of compromises and adjustments to avoid unreasonable answers. A second problem is that when anisotropy or other complications appear, the simplicity of the relation begins to disappear under correction factors. Dilatant materials cannot be treated with the hyperbolic relation, or if they are, they require Poisson's ratios larger than 0.5 or negative bulk moduli, which create potentially severe problems in the numerical solution of the equations.

The main limitation of the hyperbolic relations is the same as their strength. They are based directly on experimental observation with very little physical justification for the form of the relations or the magnitude of the parameters. Thus, they work well so long as the stresses and strains are similar to those under which the experimental observations were made. When different conditions exist, one cannot predict how well the relation will simulate reality.

Despite the above restrictions, these relations are often used in numerical work on geotechnical problems. Since a great deal of analysis is now being done on construction projects, there is reason to expect that a body of experience and evidence will be developed soon and that from this body it will be possible to improve the model and learn more about its limitations.

Ramberg-Osgood and Similar Models[24–26]

The Ramberg-Osgood function can provide an alternative and better simulation procedure than the hyperbolic function. This model can be expressed as

$$\epsilon = \frac{\sigma}{E_i} + \lambda\left(\frac{\sigma}{E_i}\right)^m \qquad \text{where} \qquad \lambda = \left(\frac{1}{\theta_2} - 1\right)\left(\frac{\sigma_2}{E_i}\right)^{1-m} \tag{2-72}$$

Here m is an exponent defining the shape of the curve and θ_2 is the ratio E_2/E_1 (Fig. 2-17).

An alternative and general mathematical function similar to the Ramberg-Osgood model can be obtained from the following function:[25,26]

$$\sigma = \frac{E_r \epsilon}{(1 + |E_r \epsilon/\sigma_v|^m)^{1/m}} + E_p \epsilon \tag{2-73}$$

where $E_r = E_i - E_p$
E_p = modulus in plastic zone
σ_y = plastic or yield stress

The tangent modulus E_t can be obtained by differentiation of Eq. (2-73):

$$E_t = \frac{d\sigma}{d\epsilon} = \frac{E_r}{(1 + |E_r \epsilon/\sigma_y|^m)^{(m+1)/m}} + E_p \tag{2-74}$$

For $m = 1$, the inverse form of Eq. (2-73) will reduce to the hyperbola [Eq. (2-58a)].

Procedures for incorporation of the effects of confining pressure or mean normal pressure p and stress paths in the model [Eq. (2-74)] have been proposed

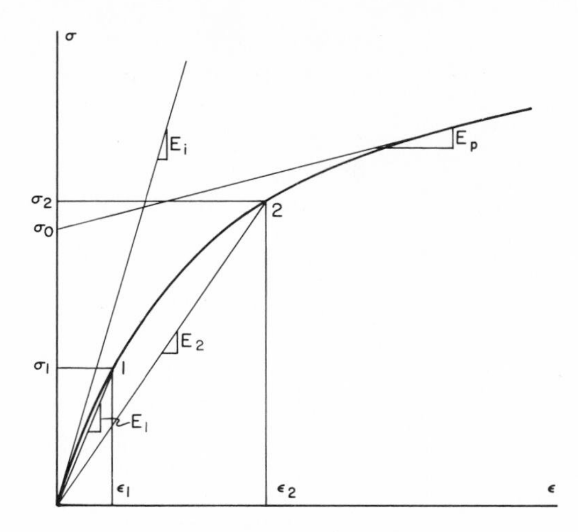

Figure 2-17 Ramberg-Osgood loading model.

by Desai and Wu.[26] For instance, use of regression analysis allows inclusion of σ_3 or p as

$$E_r = f_r(\sigma_3 \text{ or } p) \qquad m = f_m(\sigma_3 \text{ or } p)$$
$$\sigma_y = f_\sigma(\sigma_3 \text{ or } p) \qquad E_p = f_p(\sigma_3 \text{ or } p) \tag{2-75}$$

The relationships in Eqs. (2-75) for a sand were obtained as[26]

$$E_r = 146.68 + 198.75\sigma_3 + 79.5\sigma_3^2 - 5.07\sigma_3^3$$
$$m = 11.423 - 6.236\sigma_3 + 1.210\sigma_3^2 - 0.069\sigma_3^3 \tag{2-76}$$
$$\sigma_y = 0.743 + 2.536\sigma_3 + 0.066\sigma_3^2 \qquad E_p = 50 \text{ kg/cm}^2$$

Other Functions

Although the hyperbolic relation is the most popular among the empirical curve-fitting schemes now in use, other techniques are available. They permit a more general shape of curve to be modeled, and to a limited extent they may be able to represent some strain softening. A parabolic approach is used by Wong.[27]

Spline functions This approach, developed by Desai,[28,29] formalizes the use of cubic and bicubic splines for simulation of a set of stress-strain data (Fig. 2-18). The cubic spline function can be expressed as

$$s(\epsilon) = \frac{\epsilon_i - \epsilon}{\lambda_i}\sigma_{i-1} + \frac{\epsilon - \epsilon_{i-1}}{\lambda_i}\sigma_i + \frac{1}{6\lambda_i}[(\epsilon_i - \epsilon)^3 - \lambda_i^2(\epsilon_i - \epsilon)]\phi_{i-1}$$
$$+ \frac{1}{6\lambda_i}[(\epsilon - \epsilon_{i-1})^3 - \lambda_i^2(\epsilon - \epsilon_{i-1})]\phi_i \tag{2-77}$$

where s = spline function
ϵ_i, σ_i = data points
ϕ = second derivative of s
$\lambda_i = \epsilon_i - \epsilon_{i-1}$

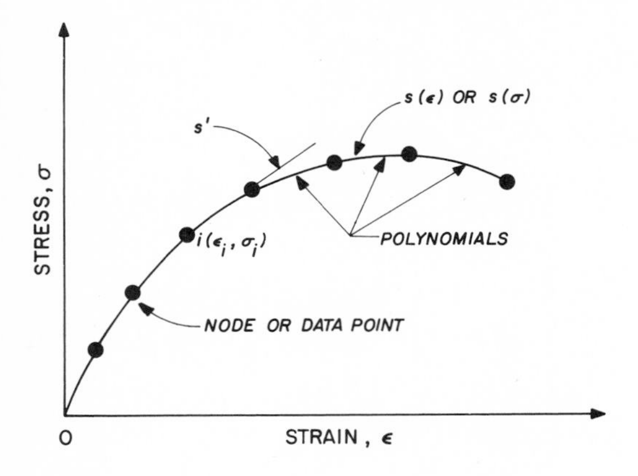

Figure 2-18 Spline-function model.

The spline function provides continuous first and second derivatives and hence is found to provide better simulation of curves compared with that given by the hyperbola, particularly in the initial ranges of the curve.[28] These functions are also powerful for formulation of finite element procedures. Sufficient experience has not been accumulated for evaluation of the general usefulness of this procedure, but it shows significant potential for successful use.

Polynomials If the stress-strain data are summarized by a set of $n + 1$ pairs of stress-strain data, it is easy to fit an nth-order polynomial through the points. This will in general be a very wiggly curve of almost no use for analytic purposes. However, a lower-order curve can also be fitted to the data by a least-squares fitting process. Many computer programs are available to do this, so that one need only appropriate a convenient subroutine to develop a relation such as

$$\tau = a_0 + a_1 \gamma + a_2 \gamma^2 + a_3 \gamma^3 + \cdots + a_n \gamma^n + \cdots \tag{2-78a}$$

Differentiation leads to the expression for the modulus

$$G = \frac{\partial \tau}{\partial \gamma} = a_1 + 2a_2 \gamma + 3a_3 \gamma^2 + \cdots + na_n \gamma^{n-1} + \cdots \tag{2-78b}$$

The same sort of generalizations are possible for the parabolic and hyperbolic relations. Wong[27] and Jaworski[30] used such a relation successfully to study the behavior of braced excavations, but further work is needed to identify the strengths and weaknesses of the approach.

Higher-order elasticity models The stress-strain laws based on the generalized Hooke's laws [Eqs. (2-37) to (2-50)] represent the lowest order of the higher-order elasticity models that have been used in geotechnical engineering. In most applications in the past, the incremental form of Eq. (2-37) has been used as a piecewise linear approximation to the nonlinear behavior of soils and rocks.

It is possible to employ higher-order elastic constitutive laws[6,31,32] allowing incorporation of a number of factors that cannot be accounted for by the piecewise linear approximation based on the incremental form of generalized Hooke's laws. Recently some activity has been directed toward use of higher-order elastic or hyperelastic and hypoelastic laws for description of the behavior of soils and rocks.

The hyperelastic models rely on finding constitutive laws by differentiation of a strain-energy function as

$$\sigma_{ij} = \frac{\partial U}{\partial I_1}\frac{\partial I_1}{\partial \epsilon_{ij}} + \frac{\partial U}{\partial I_2}\frac{\partial I_2}{\partial \epsilon_{ij}} + \frac{\partial U}{\partial I_3}\frac{\partial I_3}{\partial \epsilon_{ij}}$$

$$= \phi_1 \delta_{ij} + \phi_2 \epsilon_{ij} + \phi_3 \epsilon_{im}\epsilon_{mj} \tag{2-79a}$$

where

$$U = U(I_1, I_2, I_3) = a_0 + a_1 I_1 + a_2 I_2 + a_3 I_1^2 + a_4 I_1^3 + a_5 I_1 I_2 + a_6 I_3$$
$$+ a_7 I_1^4 + a_8 I_1^2 I_2 + a_9 I_1 I_3 + a_{10} I_2^2 \tag{2-79b}$$

and ϕ_i, $i = 1, 2, 3$, are response functions that satisfy the condition $\partial\phi_i/\partial I_j = \partial\phi_j/\partial I_i$. We can obtain different orders of hyperelastic models by retaining higher-order terms in Eq. (2-79*a*). For instance, if we keep terms up to the third power, we obtain a second-order hyperelastic law. Depending upon the order, the law can account for various factors; the third term in Eq. (2-79*b*) allows realistic inclusion of dilatancy, i.e., volume change under shear.

In the hyperelastic law, the stresses are expressed as functions of strains as

$$\sigma_{ij} = F_{ij}(\epsilon_{kl}) \tag{2-80a}$$

which indicates that the behavior is independent of loading path and that the material returns to the original unstrained state. For geologic materials, this may not be the case because materials do not return to the original states except at low levels and the behavior is path-dependent. Usually the material remembers its past history.[31] For such materials Truesdell[31] proposed a rate theory which is also known as the *hypoelastic formulation:*

$$\text{Rate of stress} = f(\text{rate of deformation}) \tag{2-80b}$$

Without time effects, we can write the general form of this law as[6]

$$d\sigma_{ij} = d\epsilon_{nn}\beta_0\delta_{ij} + \sigma_{mn}d\epsilon_{nn}\beta_1\delta_{ij} + \sigma_{mn}\sigma_{np}\beta_2\delta_{ij} + d\epsilon_{nn}\beta_3\sigma_{ij} + \sigma_{nn}d\epsilon_{mn}\beta_4\sigma_{ij}$$
$$+ \sigma_{mn}\sigma_{np}d\epsilon_{pm}\beta_5\sigma_{ij} + d\epsilon_{mn}\beta_6\sigma_{im}\sigma_{mj} + \sigma_{mn}d\epsilon_{mn}\beta_7\sigma_{is}\sigma_{sj}$$
$$+ \sigma_{mn}\sigma_{np}d\epsilon_{pm}\beta_8\sigma_{ij}\sigma_{sj} + \eta_5(\sigma_{im}d\epsilon_{mj} + d\epsilon_{im}\sigma_{mj})$$
$$+ \eta_6(\sigma_{im}\sigma_{mn}d\epsilon_{nj} + d\epsilon_{im}\sigma_{mn}\sigma_{nj}) + \eta_3 d\epsilon_{ij} \tag{2-80c}$$

By retaining different terms we get different orders of hypoelastic models. For example, the grade-zero hypoelastic model is the incremental Hooke's law:

$$d\sigma_{ij} = K\,d\epsilon_{nn}\delta_{ij} + 2G(d\epsilon_{ij} - \tfrac{1}{3}d\epsilon_{nn}\delta_{ij}) \tag{2-80d}$$

which requires determination of two parameters.

As the order increases, the number of material parameters required for the law increases. For the model in Eq. (2-80*c*) we require 12 parameters, and for the law of order we require 7 parameters. These parameters must be determined for representative laboratory tests. At this time, some studies are in progress toward the use of these models, but no general model has yet been evolved that can be adopted readily for practical use. In fact, it usually requires curve fitting and regression analysis to determine the parameters from a set of laboratory tests, and often the question of uniqueness may arise because it may be possible to fit more than one set of parameters to a set of laboratory data.

The hyperelastic and hypoelastic approaches show considerable promise for use in geotechnical engineering; however, further research will be needed before they can be reliably applied.

Simple Perfect Plasticity; Nonfrictional Models

All the constitutive relations described in the previous sections are called *deformation relations*. They relate the stress directly to the strain even though they may be expressed in the form of an instantaneous or tangent modulus. The relations arising from plasticity theory usually are incremental; i.e., the stresses and strains are related entirely by their incremental or differential behavior. Here it is not possible to relate total stress to total strain directly without knowledge of the loading path. We discuss here some of the simpler relations in some detail so that the procedures by which the more complicated ones are derived can be understood.

The yielding of a material obeying the theory of perfect plasticity is defined by the yield criterion f. This is a function of stress, strain, and other parameters such that when $f < 0$ the material is elastic and when $f = 0$ the material is in a plastic state. The function f is never greater than 0. More general formulations are possible, and some are presented here as well as in Chap. 3. The simple concept is best illustrated by an example.

Tresca's yield criterion states that plastic flow occurs when the maximum shear stress equals the allowable shear strength. In algebraic form, for plane strain

$$f(\sigma) = \left(\frac{\sigma_{xx} - \sigma_{yy}}{2}\right)^2 + \sigma_{xy}^2 - k^2 = 0 \tag{2-81}$$

where k is the shear strength.

If $f(\sigma) < 0$, the maximum shear stress is less than the shear strength and the material is elastic. If $f(\sigma) = 0$, the shear stress equals the shear strength and the material is plastic.

A central concept in plasticity theory is the theory of the plastic potential and the associated flow rule. This states that when a material is in the plastic state, the differential increments of strain are proportional to the outward normals to the yield criterion. Figure 2-19 illustrates this. The rule can also be expressed by

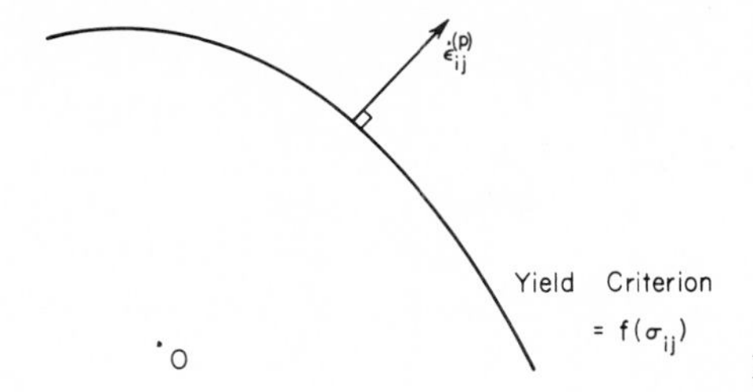

Figure 2-19 Associated flow rule for plasticity.

saying that the strain increments are proportional to the gradient of the yield criterion

$$\dot{\epsilon}_{ij}^{(p)} = \lambda \frac{\partial f}{\partial \sigma_{ij}} \tag{2-82}$$

In this equation λ is a factor of proportionality, and $\dot{\epsilon}_{ij}^{(p)}$ represents the plastic strain increment. Although the dot connotes a time derivative (or rate) and this term is often called the strain rate, it is not really a rate because no time derivative is involved. It is instead a differential increment.

For plane-strain conditions the components of the plastic strain increment become

$$\begin{gathered}\dot{\epsilon}_{xx}^{(p)} = \lambda \frac{\sigma_{xx} - \sigma_{yy}}{2} \qquad \dot{\epsilon}_{yy}^{(p)} = -\lambda \frac{\sigma_{xx} - \sigma_{yy}}{2} \\ \dot{\epsilon}_{xy}^{(p)} = \dot{\epsilon}_{yx}^{(p)} = \lambda\sigma_{xy} = \lambda\sigma_{yx}\end{gathered} \tag{2-83}$$

The last equation arises in this form because σ_{xy}^2 in Eq. (2-81) should properly be written $\sigma_{xy}\sigma_{yx}$. It is easy to see that the octahedral plastic strain increment is zero, and so the deviatoric components are

$$\begin{aligned}\dot{e}_{xx}^{(p)} &= \lambda \frac{\sigma_{xx} - \sigma_{yy}}{2} = \lambda \frac{s_{xx} - s_{yy}}{2} \\ \dot{e}_{yy}^{(p)} &= -\lambda \frac{\sigma_{xx} - \sigma_{yy}}{2} = -\lambda \frac{s_{xx} - s_{yy}}{2} \\ \dot{e}_{xy}^{(p)} &= \lambda\sigma_{xy} = \lambda s_{xy}\end{aligned} \tag{2-84a}$$

From Eqs. (2-52*b*) the elastic increments of deviatoric stress and strain are

$$\dot{e}_{xx}^{(e)} = \frac{1}{2G}\dot{s}_{xx} \qquad \dot{e}_{yy}^{(e)} = \frac{1}{2G}\dot{s}_{yy} \qquad \dot{e}_{xy}^{(e)} = \frac{1}{2G}\dot{s}_{xy} \tag{2-84b}$$

Then the total increments of deviatoric strain are

$$\begin{aligned}\dot{e}_{xx} &= \frac{1}{2G}\dot{s}_{xx} + \frac{\lambda(s_{xx} - s_{yy})}{2} \qquad & \dot{e}_{yy} &= \frac{1}{2G}\dot{s}_{yy} - \frac{\lambda(s_{xx} - s_{yy})}{2} \\ \dot{e}_{xy} &= \frac{1}{2G}\dot{s}_{xy} + \lambda s_{xy} \qquad & \dot{e}_{yx} &= \frac{1}{2G}\dot{s}_{yx} + \lambda s_{yx}\end{aligned} \tag{2-84c}$$

Now the first equation is multiplied by $(s_{xx} - s_{yy})/2$, the second by $(s_{yy} - s_{xx})/2$, the third by s_{yx}, and the fourth by s_{xy}. The results are then added to give

$$\frac{(\dot{e}_{xx} - \dot{e}_{yy})(s_{xx} - s_{yy})}{2} + \dot{e}_{xy}s_{yx} + \dot{e}_{yx}s_{xy}$$

$$= \frac{1}{2G}\left(\frac{s_{xx} - s_{yy}}{2}\dot{s}_{xx} - \frac{s_{xx} - s_{yy}}{2}\dot{s}_{yy} + s_{yx}\dot{s}_{xy} + s_{xy}\dot{s}_{yx}\right)$$

$$+ 2\lambda\left[\left(\frac{s_{xx} - s_{yy}}{2}\right)^2 + s_{xy}^2\right]$$

$$= \frac{1}{2G}\dot{f} + 2\lambda k^2 \tag{2-85}$$

The left-hand side of this equation is denoted by $\dot{W}$. Since $\dot{f} = 0$,

$$\lambda = \frac{\dot{W}}{2k^2} \tag{2-86a}$$

Further, $\dot{W}$ can be expressed in terms of total stresses and strains as

$$\dot{W} = \frac{\sigma_{xx} - \sigma_{yy}}{2}(\dot{\epsilon}_{xx} - \dot{\epsilon}_{yy}) + \sigma_{xy}\dot{\epsilon}_{xy} + \sigma_{yx}\dot{\epsilon}_{yx} \tag{2-86b}$$

Equations (2-86*a*) and (2-86*b*) can then be substituted into Eq. (2-84*c*) to obtain the incremental deviatoric stress-strain relation. Adding the octahedral components and collecting terms gives

$$\begin{Bmatrix} \dot{\sigma}_{xx} \\ \dot{\sigma}_{yy} \\ \dot{\sigma}_{xy} \end{Bmatrix} =$$

$$\begin{bmatrix} K + \frac{4}{3}G - \frac{G}{k^2}\left(\frac{\sigma_{xx} - \sigma_{yy}}{2}\right)^2 & K - \frac{2}{3}G + \frac{G}{k^2}\left(\frac{\sigma_{xx} - \sigma_{yy}}{2}\right)^2 & -\frac{G}{k^2}\sigma_{xy}\frac{\sigma_{xx} - \sigma_{yy}}{2} \\ & K + \frac{4}{3}G - \frac{G}{k^2}\left(\frac{\sigma_{xx} - \sigma_{yy}}{2}\right)^2 & \frac{G}{k^2}\sigma_{xy}\frac{\sigma_{xx} - \sigma_{yy}}{2} \\ \text{sym} & & G - \frac{G}{k^2}\sigma_{xy}^2 \end{bmatrix}$$

$$\cdot \begin{Bmatrix} \dot{\epsilon}_{xx} \\ \dot{\epsilon}_{yy} \\ \dot{\gamma}_{xy} \end{Bmatrix} \tag{2-87}$$

The differential relation of Eq. (2-87) can be converted into a finite form by replacing the dotted terms by finite increments. This is the form in which the equation is then used. If the increments are too large, improved accuracy can be obtained by successive iterations using updated values of the undotted parameters.

The second common nonfrictional yield criterion is the von Mises criterion

$$f(\sigma) = k^2 - J_2 \tag{2-88}$$

This incorporates the contribution of the intermediate principal stress and can be shown to relate yielding to the maximum strain energy. It is not clear whether this is more valid for soil, but it is considerably easier to deal with mathematically.

The same procedure is followed as for the Tresca criterion. The incremental differential plastic deviatoric strains are expressed as

$$\dot{e}_{ij}^{(p)} = \lambda s_{ij} \tag{2-89a}$$

The corresponding elastic deviatoric strains are

$$\dot{e}_{ij}^{(e)} = \frac{1}{2G}\dot{s}_{ij} \tag{2-89b}$$

and the combined elastic and plastic deviatoric strain increments are

$$\dot{e}_{ij} = \lambda s_{ij} + \frac{1}{2G}\dot{s}_{ij} \tag{2-89c}$$

Multiplying each of the equations implied by (2-89*c*) by the corresponding s_{ij} and adding all contributions gives

$$s_{ij}\dot{e}_{ij} = \lambda s_{ij}s_{ij} + \frac{1}{2G}s_{ij}\dot{s}_{ij} \qquad \sum\text{'s omitted} \tag{2-90a}$$

or

$$\dot{W} = 2\lambda J_2 + \frac{1}{2G}\dot{J}_2 = 2\lambda k^2 + 0 \tag{2-90b}$$

Hence

$$\lambda = \frac{\dot{W}}{2k^2} \tag{2-90c}$$

Substituting Eq. (2-90*c*) into Eq. (2-89*c*) and collecting terms gives

$$\dot{s}_{ij} = 2G\dot{e}_{ij} - \frac{G\dot{W}}{k^2}s_{ij} \tag{2-91}$$

The octahedral terms are now added in, and the entire set of equations can be written in expanded form

$$\{\dot{\sigma}\} = [C^e]\{\dot{\epsilon}\} - [C^p]\{\dot{\epsilon}\} \tag{2-92a}$$

where $[C^e]$ is the elasticity matrix of Eqs. (2-40) through (2-55) and $[C^p]$ is the plasticity matrix

$$[C^p] = \frac{G}{J_2}\begin{bmatrix} s_{xx}^2 & s_{xx}s_{yy} & s_{xx}s_{zz} & s_{xx}s_{yz} & s_{xx}s_{xz} & s_{xx}s_{xy} \\ & s_{yy}^2 & s_{yy}s_{zz} & s_{yy}s_{yz} & s_{yy}s_{xz} & s_{yy}s_{xy} \\ & & s_{zz}^2 & s_{zz}s_{yz} & s_{zz}s_{xz} & s_{zz}s_{xy} \\ \text{sym} & & & s_{yz}^2 & s_{yz}s_{xz} & s_{yz}s_{xy} \\ & & & & s_{xz}^2 & s_{xz}s_{xy} \\ & & & & & s_{xy}^2 \end{bmatrix} \tag{2-92b}$$

The last form of the relation, known as the *Prandtl-Reuss equations*, can be used directly in either plane strain or three-dimensional applications. It should be noted, however, that for plane strain cases there is a component of plastic strain normal to the plane and this is countered by an equal and opposite component of elastic strain. Eventually the stresses increase normal to the plane, and there ceases to be a significant normal plastic component, but for this reason one cannot ignore the out-of-plane plastic components in the formulation.

Hill[33] presents a thorough treatment of perfect plasticity as well as strain hardening. Direct applications to soil mechanics have been limited, but these relations have led to useful insights into the mechanics of bearing capacity for undrained soils.[34]

Perfect Plasticity, Frictional

The shear strength of soils is usually described by the Mohr-Coulomb law

$$\tau = c + \sigma_n \tan \phi \tag{2-93a}$$

where τ = available shear strength
σ_n = normal stress on possible failure plane (compression positive)
c = cohesion
ϕ = friction angle

The most fundamental version of the law relates shear strength to effective stresses, and the equation then becomes

$$\tau = \bar{c} + \bar{\sigma}_n \tan \bar{\phi} \tag{2-93b}$$

Here $\bar{c}$ and $\bar{\phi}$ are the effective stress parameters of shear strength.

Much of the classical work on plasticity theory that is the basis for bearing-capacity and slope-stability theories assumes that the Mohr-Coulomb relation is a yield criterion. Application of this idea to incremental analysis requires developing incremental relations between stress and strain, and Drucker and Prager[35] proposed that this could best be done by using a generalized form of the Mohr-Coulomb law that incorporates all the principal stresses:

$$J_2^{1/2} = k - \alpha I_1 \tag{2-94}$$

In this case tension is positive.

This has also been called the *extended von Mises yield criterion*. It is widely used as a three-dimensional description of the shear strength of soil and rock even though its development was motivated by mathematical convenience and there is evidence that the Mohr-Coulomb law fits the experimental data better.[36]

Reyes[37] developed incremental relations by writing

$$f = \alpha I_i + J_2^{1/2} - k = 0 \tag{2-95a}$$

and differentiating to obtain

$$\dot{\epsilon}_{ij}^{(p)} = \lambda \frac{\partial f}{\partial \sigma_{ij}} = \lambda \left(\alpha \, \delta_{ij} + \frac{s_{ij}}{2J_2^{1/2}} \right) \tag{2-95b}$$

It is now necessary to incorporate the equations for the elastic incremental behavior and to solve for λ. The equations can then be converted into forms similar to Eqs. (2-91) and (2-92*b*). The procedure is similar to that used for the von Mises criterion, but the algebra is considerably more complicated. References 5 and 37 to 39 describe the derivation in detail. The final form is

$$\frac{\dot{\sigma}_{ij}}{2G} = \dot{\epsilon}_{ij} - [A(\sigma_{kl}\,\delta_{ij} + \sigma_{ij}\,\delta_{kl}) + B\,\delta_{kl}\,\delta_{ij} + C\sigma_{ij}\sigma_{kl}]\dot{\epsilon}_{kl} \qquad \text{(2-96a)}$$

in which repeated subscripts are summed from 1 to 3 and

$$A = \frac{p-1}{6\alpha p J_2^{1/2}} = \frac{h}{pk}$$

$$B = \left(\alpha - \frac{I_1}{6J_2^{1/2}}\right)\frac{p-1}{3\alpha p} - \frac{3K\nu}{Ep} = \left(\alpha - \frac{I_1}{6J_2^{1/2}}\right)\frac{2h}{1 + 9\alpha^2 K/G} - \frac{3K\nu}{Ep}$$

$$C = \frac{1}{2kpJ_2^{1/2}} \qquad p = \frac{J_2^{1/2}}{k}\left(1 + \frac{9\alpha^2 K}{G}\right)$$

$$h = \alpha\left(1 + \frac{9K\nu}{E}\right) - \frac{I_1}{6J_2^{1/2}} = \frac{3K}{2G}\alpha - \frac{I_1}{6J_2^{1/2}}$$

For conditions of plane strain these become

$$\begin{aligned}
\frac{\dot{\sigma}_{xx}}{2G} &= (1 - 2A\sigma_{xx} - B - C\sigma_{xx}^2)\dot{\epsilon}_{xx} \\
&\quad + [(\sigma_{xx} + \sigma_{yy})A - B - C\sigma_{xx}\sigma_{yy}]\dot{\epsilon}_{yy} \\
&\quad + (-A\sigma_{xy} - C\sigma_{xx}\sigma_{xy})\dot{\gamma}_{xy} \\
\frac{\dot{\sigma}_{yy}}{2G} &= [(\sigma_{xx} + \sigma_{yy})A - B - C\sigma_{xx}\sigma_{yy}]\dot{\epsilon}_{xx} \\
&\quad + (1 - 2A\sigma_{yy} - B - C\sigma_{yy}^2)\dot{\epsilon}_{yy} \\
&\quad + (-A\sigma_{xy} - C\sigma_{yy}\sigma_{xy})\dot{\gamma}_{xy} \\
\frac{\dot{\sigma}_{xy}}{2G} &= (-A\sigma_{xy} - C\sigma_{xx}\sigma_{xy})\dot{\epsilon}_{xx} + (-A\sigma_{xy} - C\sigma_{yy}\sigma_{xy})\dot{\epsilon}_{yy} \\
&\quad + (\tfrac{1}{2} - C\sigma_{xy}^2)\dot{\gamma}_{xy}
\end{aligned} \qquad \text{(2-96b)}$$

Finite incremental forms can be easily written by replacing the dotted quantities with increments, that is, $\Delta\sigma_{xx}$ for $\dot{\sigma}_{xx}$.

The Mohr-Coulomb equation presents difficulties for three-dimensional analysis because it has "corners" where the yielding changes from, say, the $\sigma_1\,\sigma_2$ plane to the $\sigma_1\,\sigma_3$ plane. This problem also exists for the Tresca criterion. Where plane strain conditions apply, the Mohr-Coulomb equation can be used, for yielding will occur in the plane.

The incremental stress-strain relations can be derived from the Mohr-Coulomb equation in the form

$$f=\left[\left(\frac{\sigma_{xx}-\sigma_{yy}}{2}\right)^2+\tau_{xy}^2\right]^{1/2}-\frac{\sigma_{xx}+\sigma_{yy}}{2}\tan\alpha-d=0 \tag{2-97a}$$

where $d = c\cos\phi$ and $\sin\alpha = \tan\phi$ (2-97*b*)

The detailed derivation, which can be found in Hagmann,[40] gives for the incremental strain under conditions of plane strain

$$\dot{\epsilon}_{xx}^{(p)}=\frac{\lambda}{2q}(\sigma_{xx}-p+Aq)\qquad \dot{\epsilon}_{yy}^{(p)}=\frac{\lambda}{2q}(\sigma_{yy}-p+Aq)\qquad \dot{\epsilon}_{xy}^{(p)}=\frac{\lambda}{q}\sigma_{xy} \tag{2-98a}$$

where
$$q=\left|\frac{\sigma_1-\sigma_3}{2}\right|\qquad p=\frac{\sigma_1+\sigma_3}{2}\qquad A=\tan\alpha \tag{2-98b}$$

The incremental stress-strain relations, including the elastic components, become

$$[C^{ep}]=$$
$$\Gamma\begin{bmatrix}(1-\nu)(1-X)-\nu S-\bar{T} & \nu(1-X)-(1-\nu)S+\bar{T} & (1-2\nu)Y-U\\ & (1-\nu)(1-\bar{X})-\nu S-T & (1-2\nu)\bar{Y}-U\\ \text{sym} & & \frac{1}{2}[(1-2\nu)(1-2Z)-\bar{T}-T]\end{bmatrix} \tag{2-99a}$$

where

$$\begin{aligned}
&\Gamma=\frac{E}{(1+\nu)(1-2\nu-\bar{T}-T)} && X=N(p-\sigma_{xx}-Aq)^2\\
&\bar{X}=N(p-\sigma_{yy}-Aq)^2 && Y=N(p-\sigma_{xx}-Aq)\sigma_{xy}\\
&\bar{Y}=N(p-\sigma_{yy}-Aq)\sigma_{xy} && Z=N\sigma_{xy}^2\\
&S=N(p-\sigma_{xx}-Aq)(p-\sigma_{yy}-Aq)\\
&T=Nq(p-\sigma_{xx}-Aq)[\tan\alpha-A(1-2\nu)]\\
&\bar{T}=Nq(p-\sigma_{yy}-Aq)[\tan\alpha-A(1-2\nu)]\\
&U=Nq\sigma_{xy}[\tan\alpha-A(1-2\nu)] && N=\frac{1}{2q^2(1+A^2)}
\end{aligned} \tag{2-99b}$$

Substitution of the various terms of Eqs. (2-99*b*) into the stiffness Eq. (2-99*a*) will reveal that these relations are symmetric.

One objection to the Mohr-Coulomb or the Drucker-Prager relations used as yield criteria is that the associated flow rule implies very large extensional volumetric plastic strains. These are not observed experimentally, even for dense sands, for which some dilation is observed. Several researchers have suggested using nonassociated flow rules, and Davis[41] has developed extensive procedures for doing this in the context of classical plasticity theory. Hagmann et al.[40,42] developed the stiffness relations for an elastic, perfectly plastic material using a nonassociated flow rule based on the assumption that all the plastic dilatational strains predicted by the Mohr-Coulomb criterion with the associated flow rule are removed. Equation (2-98a) then becomes

$$\dot{\epsilon}_{xx}^{(p)} = \frac{\lambda}{2q}(\sigma_{xx} - p) \qquad \dot{\epsilon}_{yy}^{(p)} = \frac{\lambda}{2q}(\sigma_{yy} - p) \qquad \dot{\epsilon}_{xy}^{(p)} = \frac{\lambda}{q}\sigma_{xy} \tag{2-100}$$

The incremental stiffness corresponding to Eqs. (2-163) is

$$[C^{ep}] = \Gamma \begin{bmatrix} 1 - \nu^2 + \nu X(1 + 2\nu) - X + T & \nu(1 + \nu) - \nu X(1 + 2\nu) + X + T & (1 + \nu)(1 - 2\nu)Y \\ \nu(1 + \nu) - \nu X(1 + 2\nu) + X - T & 1 - \nu^2 + \nu X(1 + 2\nu) - X - T & -(1 + \nu)(1 - 2\nu)Y \\ (1 + \nu)(1 - 2\nu)Y - U & -(1 + \nu)(1 - 2\nu)Y - U & \dfrac{(1 - 2Z)(1 + \nu)(1 - 2\nu)}{2} \end{bmatrix} \tag{2-101a}$$

where

$$\Gamma = \frac{E}{(1 + \nu)^2(1 - 2\nu)} \qquad X = \frac{1}{2q^2}(p - \sigma_{xx})^2$$

$$Y = \frac{1}{2q^2}(p - \sigma_{xx})\sigma_{xy} \qquad Z = \frac{1}{2q^2}\sigma_{xy}^2 \tag{2-101b}$$

$$T = \frac{1 + \nu}{2q}(p - \sigma_{xx}) \tan \alpha \qquad U = \frac{1 + \nu}{2q}\sigma_{xy} \tan \alpha$$

Details of the derivation can be found in Refs. 40 and 42.

Lade and Duncan[43] have proposed a theory which incorporates special failure and yield criteria, a nonassociated flow rule, and an empirical work-hardening law for cohesionless soils. According to this theory, the failure criterion is

$$I_1^3 - k_1 I_3 = 0 \tag{2-102a}$$

or

$$f = \frac{I_1^3}{I_3} \tag{2-102b}$$

and

$$f = k_1 \qquad \text{at failure} \tag{2-102c}$$

where k_1 is a constant that depends on density of sand and the plastic potential function is

$$g = I_1^3 - k_2 I_3 \tag{2-103a}$$

where k_2 is a constant for a given value of f. The flow rule is given by

$$\Delta\epsilon_{ij}^p = \Delta\lambda \frac{\partial g}{\partial \sigma_{ij}} \tag{2-103b}$$

It is shown that the theory can simulate observed behavior from conventional triaxial and multiaxial tests and that the required parameters can be obtained from conventional triaxial tests.

The stiffness relations of Eqs. (2-101*a*) and (2-101*b*) are not symmetrical. This has important implications:

1. It implies that the material is unstable in the sense that it is possible to extract irrecoverable work from it in a closed cycle of loading. This definition of instability is credited to Drucker,[44] who also observed that assemblages of frictional blocks are, in this sense, unstable.[45] In other words, violation of the associated flow rule implies such instability, and it is to be expected that the incremental stress-strain relations will be unsymmetrical.
2. Incremental solutions must be carried out with care and precision, and large increments of loading must be avoided. The solutions are likely to be nonunique.
3. When finite element methods are used, the element stiffnesses and the global stiffness matrix will not be symmetric. The entire computer program, including the assembly of the equations and the solution routine, must be redesigned with this in mind. An example of this is found in Ref. 40.

Capped Yield Models

All the yield criteria discussed in the previous sections can be criticized on the grounds that they do not represent soil behavior adequately. The nonfrictional plastic models ignore the fact that soil and rock do have frictional components to

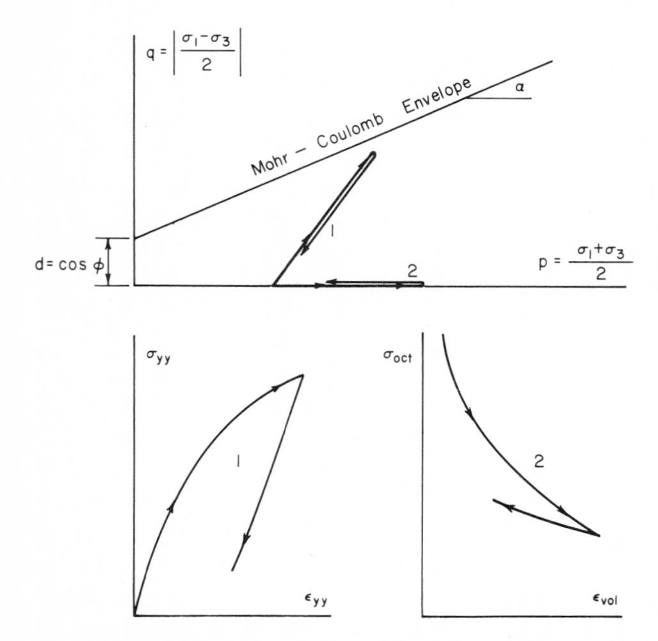

Figure 2-20 Stress-strain curves for drained loading of soil.

their shear strengths, although for undrained analysis of saturated soils (the $\phi = 0$ analysis) the nonfrictional criteria may be adequate first approaches.

The frictional criteria with associated flow rules predict unreasonably large dilatational strains, as has already been mentioned, but all the frictional criteria are subject to a more serious objection; i.e., yielding actually occurs well below the failure envelope of the Mohr-Coulomb equation, as illustrated in Fig. 2-20. If a sample were loaded along the stress path numbered 1, the elastic–perfectly plastic theory would predict a linear relation between stress and strain and a completely reversible one. The actually observed curve would show significant deviation from linearity and would usually not be reversible. Some permanent strain would remain after removal of the load. This is the plastic strain not accounted for in the theories that use the Mohr-Coulomb envelope as a yield criterion. The extreme example of yield below the Mohr-Coulomb envelope is the drained isotropic stress path numbered 2. This sequence of loading and unloading on a normally consolidated sample would give permanent changes of volume, which must be considered plastic strains even though there is no shear stress at all.

The difficulty of yielding at stresses below the failure envelope can be avoided if it is recognized that yielding may occur before the material reaches a failure envelope. Figure 2-21 shows a family of yield criteria forming "caps" to the open-ended Drucker-Prager envelope. As the stress changes from A to B, the yield criterion also moves as the result of strain hardening. The plastic-strain increments are always normal to the criterion corresponding to the current state of stress. If the stresses are reversed from B to A, the soil behaves elastically and the criterion remains fixed through point B. When the stress level is raised to a point on the current locus of the yield criterion, such as C, the material will again become plastic. Thus, the criterion is a reflection of past plastic strain.

The parameter governing the strain hardening is the plastic volumetric strain. As the soil is compressed along a virgin compression curve and develops some irrecoverable plastic volumetric strain, the yield criterion moves out along the p axis, becoming a larger cap. Another representation of the theory is shown in Fig. 2-22. Here a third axis has been added to represent the void ratio. Figure 2-21 can be considered a series of sections through Fig. 2-22. The plastic states are represented by the curved surface between the isotropic compression line and the critical-void-ratio line. The concept of capped models for granular soils and rocks

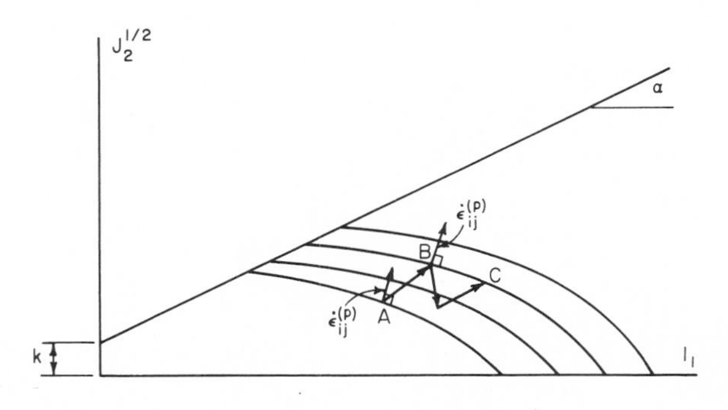

Figure 2-21 Strain hardening capped yield criterion.

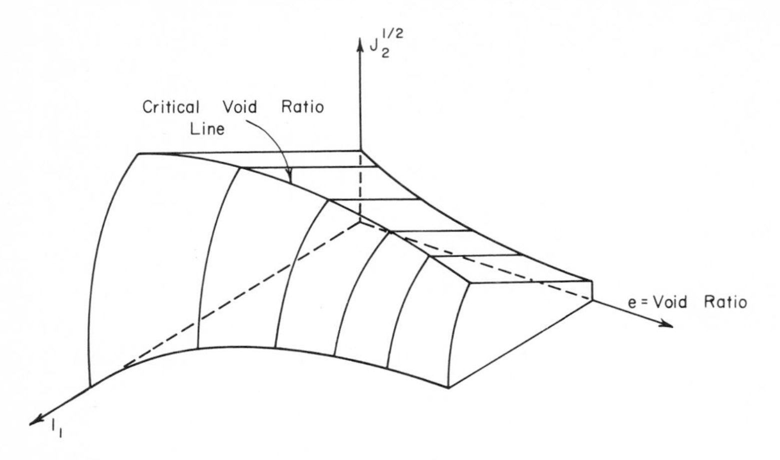

Figure 2-22 Yield surface and critical-void-ratio line.

under dynamic loads has been used by DiMaggio and Sandler[46] and Isenberg and Bagge.[47]

The theory has been the subject of extensive research and lively debate for two decades. Detailed expositions are found in Roscoe et al.,[48,49] Schofield and Wroth,[50] and in the Roscoe Memorial volume.[51] New papers and reports keep expanding and criticizing the theory, so that the details are continually changing.

Several questions have recurred in the development of the theory of capped yield criteria; some of the most significant are as follows:

1. What is the shape of the cap? A bullet shape was originally suggested, as shown in Fig. 2-21, but an ellipsoid may be preferable.[52] It has even been suggested that the cap does not cross the p axis.[53]
2. What happens when the soil is loaded on the dry side of the critical-void-ratio line (Fig. 2-23)? This is an open question. One suggestion is that the yield criterion continues beyond the critical-void-ratio line, as shown by the dotted line in Fig. 2-23, and that there is strain-softening behavior in this portion of

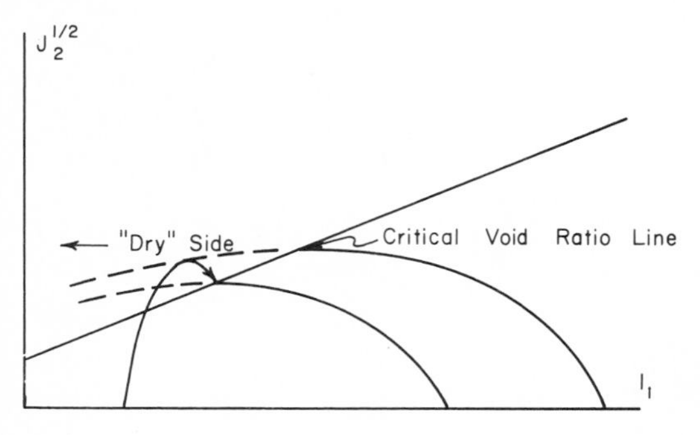

Figure 2-23 Loading path on dry side of critical.

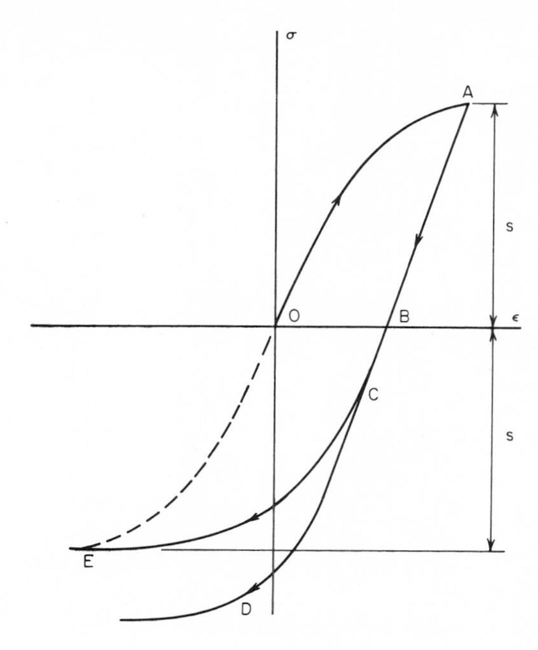

Figure 2-24 Anisotropic yielding or kinematic hardening.

the stress range. The criterion collapses back to a smaller size during plastic strain in the "dry" range. Since this would occur in a local zone of the soil where the stresses first reached the yield criterion, the shrinking of the yield criterion during strain softening might not be observable by exterior measurements on the soil sample. Another suggestion is that a Mohr-Coulomb criterion with a nonassociated flow rule might apply in the dry range.

3. Does the theory apply to tests other than the triaxial? Most of the experimental work was done with triaxial apparatus, but recent work in other devices seems to indicate that the general idea is valid.
4. How should one account for anisotropy of yielding? There is a great deal of evidence to indicate that soils which have been yielded in one direction of loading show a different yielding behavior when the loading is reversed.

Figure 2-24 illustrates the problem of anisotropic yielding for a simple case. A soil is loaded from *O* to *A* and unloaded from *A* to *B*. If it were reloaded to *A*, the theory would predict that the loading curve would go from *B* to *A* and then turn onto the yielded curve for further loading. Actually the reloading curve would not follow the rebound curve exactly, but the error involved in assuming that it does is relatively small. Now, if the reloading were in the other direction, the theory would predict the stress-strain path *BD*, with yielding at *D* at the same magnitude of stress as at point *A*. A more reasonable behavior would follow the line *BCE*. This means that the actual yielding behavior is anisotropic in that the yield criterion is not symmetric once yielding has occurred. The usual forms of the capped models do not account for this phenomenon. There is now active research into the nature of the yield criteria for soils. It can be expected that newer theories will incorporate the *kinematic hardening* of lines such as $OABCE$ rather than being limited to the present *isotropic hardening* of $OABD$. In the study of metals, isotropic hardening occurs when the yield criterion expands simultaneously in all directions, and kinematic hardening occurs when the criterion maintains its shape and size but translates during hardening. In a sense, the Roscoe model is isotropic with respect to the shear behavior and kinematic with respect to volumetric yield.

The exact form of the capped criterion varies with the needs of the analyst and his knowledge of the soil. The derivation of the incremental stress-strain relations starts from the algebraic expression of the yield function f. Then it can be shown that

$$\dot{\epsilon}_{ij}^{(p)} = \hat{G}\frac{\partial f}{\partial \sigma_{ij}}\frac{\partial f}{\partial \sigma_{k1}}\dot{\sigma}_{k1} \qquad \text{with} \qquad G = -\frac{1}{\left(\dfrac{\partial f}{\partial \epsilon_{mn}^{(p)}} + \dfrac{\partial f}{\partial \kappa}\dfrac{\partial \kappa}{\partial \epsilon_{mn}^{(p)}}\right)\dfrac{\partial f}{\partial \sigma_{mn}}} \tag{2-104}$$

In these equations f is a function of σ_{ij}, $\epsilon_{ij}^{(k)}$, and κ, which is a strain-hardening parameter.

Once a strain-hardening yield criterion has been chosen, the derivation of the incremental stress-strain relations is primarily a matter of algebra. A typical example is that provided by Hagmann's[40] derivation of the relations for a capped yield surface moving within the open end of the Mohr-Coulomb failure envelope. The strain hardening was set by the requirement that the plastic volumetric strains be logarithmically related to the volumetric stresses. The derivation is too long to include here. Similar results can be obtained for other strain-hardening criteria.

Strain Softening

The stress-strain curves of many soils exhibit peaks, like that illustrated in Fig. 2-25. It is difficult to represent such strain-softening behavior, and generally satisfactory procedures have not been developed although some promising starts

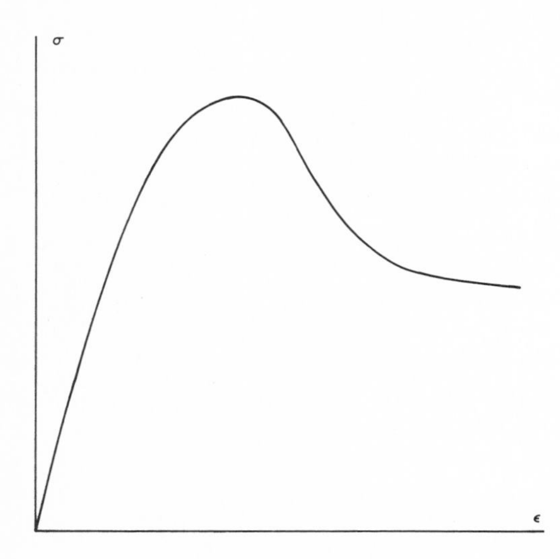

Figure 2-25 Strain softening.

have been made. A major part of the problem is that strain softening implies unstable behavior in many instances and can lead to nonunique results. This is a physical fact, not a result of analytical inexactitude.

Höeg[54] has applied the strain-softening theories of metal plasticity to soils by incorporating a strain-softening parameter in the von Mises yield criterion; his yield criterion is

$$f = k^2\left(1 + \frac{H}{3G}\right) - J_2 \tag{2-105}$$

where H is the slope of the relation between axial stress and axial plastic strain in a uniaxial loading test. If $H = 0$, the criterion is identical to the von Mises criterion. If $H > 0$, the material is strain-hardening, and if $H < 0$, it is strain-softening. The incremental stress-strain relations are identical to Eqs. (2-92*a*) and (2-92*b*) except that the coefficient of the matrix in Eq. (2-92*b*) is $(G/J_2)(1 + H/3G)$ instead of G/J_2.

Desai[6,55] has presented a hypothesis for defining the softening behavior after peak and a numerical procedure based on iterative relaxation scheme of the Newton-Raphson type. In this approach, it is suggested that the material parameters for forming stiffnesses beyond the peak stress be derived from a special function like

$$F_i = F(G_{i0}, \sigma, \delta\sigma, p, \sigma_p, \epsilon_p, \sigma_r, \epsilon_r, \phi_p, \phi_r, A, \beta_i) \tag{2-106}$$

where G_{i0} = initial modulus
$\delta\sigma$ = stress path
A = area below stress-strain curve(s)
β_i = parameters related to physical factors, e.g., density and overconsolidation

and subscripts p and r denote peak and residual conditions, respectively.

The distinguishing feature of this approach is that the strain-softening phenomenon is viewed as a process of modification (reduction) in strength of the material rather than as a change in the moduli computed as gradients of stress-strain curves, as is done in other studies. The moduli are derived from the function in Eq. (2-106) on the basis of laboratory tests and mathematical methods such as extremization and regression. Although some preliminary results have been obtained by using this approach,[55] further research will be required for establishing the validity of the approach.

Finally, the modulus or some component of the stiffness relation can be made negative. This flies in the face of convention and of the general belief that all stiffness matrices must be positive definite. The actual requirement is that the overall stiffness matrix for the system must be positive definite to obtain unique solutions. Some techniques, such as Höeg's, could very well have some elements with negative stiffness terms on the diagonal or negative terms dominating the stiffness. This does not necessarily cause a problem until the entire system becomes unstable. At that point physical collapse is imminent anyway.

The problem of the analysis of a soil system near collapse is similar to that of a structure at the point of buckling. The stiffness matrix becomes singular, which means that more than one solution can be found for the same loads. To follow such an analysis through the failure requires that displacements, not loads, be prescribed at the boundaries.

2-4 VISCOUS BEHAVIOR

Most numerical work has concentrated on elastic and plastic aspects of soil behavior, although most soils show some viscosity and many show a great deal. Hardin and Drnevich's[56,57] work as well as that of others has indicated that under repeated loadings sands and lean clays dissipate energy primarily by plastic or hysteretic processes rather than viscous ones, but some phenomena such as secondary compression are clearly viscous. The best way to treat viscous behavior in soils has not yet been found, but a few promising advances have been achieved. For a more detailed treatment of viscoplasticity the reader should refer to a text on continuum mechanics.

The simplest type of viscosity is that represented by a simple linear dashpot (Fig. 2-26*a*):

$$\tau = \eta \frac{\partial \gamma}{\partial t} \tag{2-107}$$

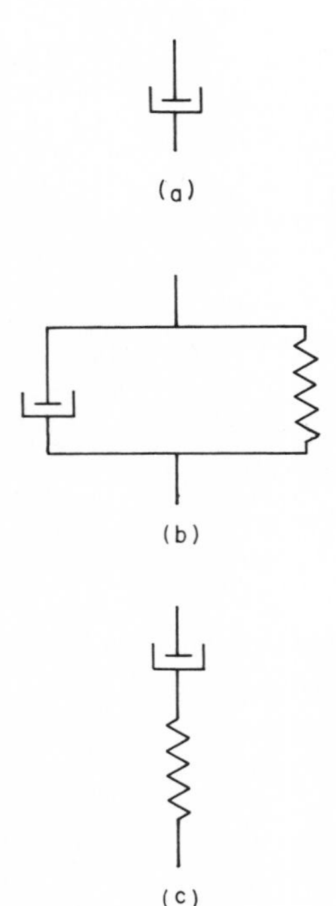

Figure 2-26 Viscoelastic conceptual models: (*a*) dashpot; (*b*) Kelvin-Voigt model; (*c*) Maxwell model.

It is possible to combine these with elastic springs in parallel (Fig. 2-26*b*) or in series (Fig. 2-26*c*). The parallel arrangement is called a *Kelvin-Voigt model*, and the series arrangement is a *Maxwell model*. The response of either is easily computed by combining the equations for the dashpot (2-107) with $\tau = G\gamma$ for the spring.

When constant stress is applied to the Kelvin-Voigt model, the strain describes a retarded elastic behavior:

$$\gamma = \frac{\tau}{G}\left[1 - \exp\left(-\frac{G}{\eta}t\right)\right] \tag{2-108}$$

in which $1/G$ is the equilibrium compliance constant. A constant stress in the Maxwell model gives the glassy response τ/G followed by steady creep:

$$\gamma = \frac{\tau}{G} + \frac{\tau t}{\eta} \tag{2-109}$$

A constant strain in the Maxwell model gives the relaxation behavior

$$\tau = G\gamma \exp\left(-\frac{Gt}{\eta}\right) \tag{2-110}$$

Several sets of springs and dashpots can be combined in series and parallel to fit an observed experimental creep or relaxation curve. These can then be described by higher-order differential equations. Solutions to the equations have been developed by such techniques as hereditary integrals and operational calculus. When numerical methods are used, a large number of states of stress and strain at earlier times must be retained in the computer. Since this rapidly becomes expensive and cumbersome, more efficient approaches must be tried.

One simplification was developed by Christian and Watt[58] for soils from earlier work by others in continuum mechanics. A set of linear viscoelastic Maxwell models in parallel can be described by the relaxation modulus, which establishes the decay of stress under constant strain γ_c:

$$\tau = \gamma_c\left[G_e + \sum_{i=1}^{N} G_i \exp\left(-\frac{t}{T_i}\right)\right] \tag{2-111}$$

where G_e is the elastic modulus of a spring in parallel with the Maxwell models. The terms in brackets are the relaxation modulus

$$G(t) = G_e + \sum_{i=1}^{N} G_i \exp\left(-\frac{t}{T_i}\right) \tag{2-112}$$

T_i $(= \eta_i/G_i)$ are the relaxation times for the N individual elements.

Similarly, the response to a constant stress τ of a set of Kelvin-Voigt models in series with a spring and a dashpot is expressed by

$$\gamma = \tau\left\{J_g + \frac{1}{\eta}t + \sum_{i=1}^{N} J_i\left[1 - \exp\left(-\frac{t}{T_i}\right)\right]\right\} \tag{2-113}$$

or by the creep compliance

$$J(t) = J_g + \frac{1}{\eta} t + \sum_{i=1}^{N} J_i \left[1 - \exp\left(-\frac{t}{T_i} \right) \right] \tag{2-114}$$

In these equations

$$J_i = \frac{1}{G_i} \tag{2-115}$$

and J_g $(= 1/G_g)$ and η are the constants for the spring and the dashpot in series. The term J_g represents the instantaneous or glassy response.

Either the creep compliance or the relaxation modulus can be computed from the observed curves of stress relaxation under constant strain or creep under constant stress. Watt[59] developed computer programs for evaluating the G_e, G_i, and T_i terms from an empirical relaxation or creep curve by least-squares fitting procedures. He also recognized that the relaxation modulus could be fitted equally well by a series

$$G(t) = \sum_{k=1}^{N} G_k \exp\left(-\frac{t}{T_k} \right) \tag{2-116}$$

Then the deviatoric stress at any time can be expressed in terms of the deviatoric strain history

$$s_{ij}(t) = 2 \sum_{k=1}^{N} G_k \int_0^t \exp\left(-\frac{t}{T_k} \right) \dot{e}_{ij}(\tau)\, d\tau \tag{2-117}$$

or in terms of new state variables

$$s_{ij}(t) = 2 \sum_{k=1}^{N} G_k q_{ijk}(t) \tag{2-118a}$$

$$q_{ijk}(t) = \int_0^t \exp\left[-\frac{(t-\tau)}{T_k} \right] \dot{e}_{ij}(\tau)\, d\tau \tag{2-118b}$$

These terms can be evaluated easily for each element of material by first writing

$$q_{ijk}(t) = \int_0^{t-h} \exp\left[-\frac{(t-\tau)}{T_k} \right] \dot{e}_{ij}(\tau)\, d\tau + \int_{t-h}^{t} \exp\left(-\frac{t-\tau}{T_k} \right) \dot{e}_{ij}(\tau)\, d\tau \tag{2-119}$$

If the time interval between $t - h$ and t is short, we can assume that the q's vary linearly over the interval, giving

$$\begin{aligned} q_{ijk}(t) &= \exp\left(-\frac{h}{T_k} \right) q_{ijk}(t-h) + \frac{e_{ij}(t) - e_{ij}(t-h)}{h} \int_{t-h}^{t} \exp\left(-\frac{t-\tau}{T_k} \right) d\tau \\ &= \beta_k q_{ijk}(t-h) + [e_{ij}(t) - e_{ij}(t-h)]\alpha_k \end{aligned} \tag{2-120a}$$

$$\alpha_k = \frac{T_k}{h}(1 - \beta_k) \qquad \beta_k = \exp\left(-\frac{h}{T_k} \right) \tag{2-120b}$$

It is then necessary only to keep in memory the q_{ijk} for each element in the solution, and the remainder of the formulation is similar to that for a regular finite element problem because the only unknown terms are the $e_{ij}(t)$. Watt further simplified the analysis by using an incompressible finite element formulation similar to those described in Chap. 12 for the initial solution of the consolidation problem. In that case the volumetric strains are zero, and the only elastic or viscous constant required in the analysis is $G(t)$.

An objection to this procedure is that many soils are not linearly viscoelastic. Singh and Mitchell[60] have developed from experiment and from rate-process theory the following relation for the strain rate of a soil:

$$\dot{\epsilon} = A \exp\left[\alpha D\left(\frac{t_1}{t}\right)^m\right] = A \exp\left[\bar{\alpha}\bar{D}\left(\frac{t_1}{t}\right)^m\right] \tag{2-121}$$

where $\dot{\epsilon}$ = strain rate (axial, shear, deviatoric, etc.)
A = strain rate at $t = t_1$ and $D = 0$ (extrapolated from data)
α = slope of linear portion of ln $\dot{\epsilon}$–vs.–D plot
D = deviator stress (soil-mechanics convention) $(= \sigma_1 - \sigma_3)$
D_{max} = D at failure in conventional strain rate
$\bar{D} = D/D_{max}$
$\bar{\alpha} = \alpha/D_{max}$
t = time after application of stress
t_1 = reference time
m = creep coefficient

It can be seen that the parameters A, α, and m must be evaluated from the best fit to observed data.

While this viscous model is well based in experiment, it is very difficult to use analytically. Therefore most applications to numerical work have not tried to follow the development of creep with time directly but have used equivalent elastic analyses. Edgers et al.,[61] for example, evaluated the state of stress by an elastic finite element program and then estimated the creep on the assumption that the stress did not change appreciably during creep, an assumption that is often nearly valid. In any case, the stress distribution can be improved by observing the strains and evaluating improved secant moduli. An alternative procedure is to use Eq. (2-121) to compute estimated stress-strain curves for different constant times. These can then be used as simple nonlinear elastic curves at the times of interest.

2-5 LOAD REVERSAL

In most of the preceeding sections we have assumed tacitly that the loadings were applied monotonically, i.e., that they were not applied and removed or applied in alternating senses. Two common problems occur in geotechnical engineering in which the assumption is not valid. First, there is the case in which soil or rock is excavated and then other soil or rock is placed. An unloading is followed by

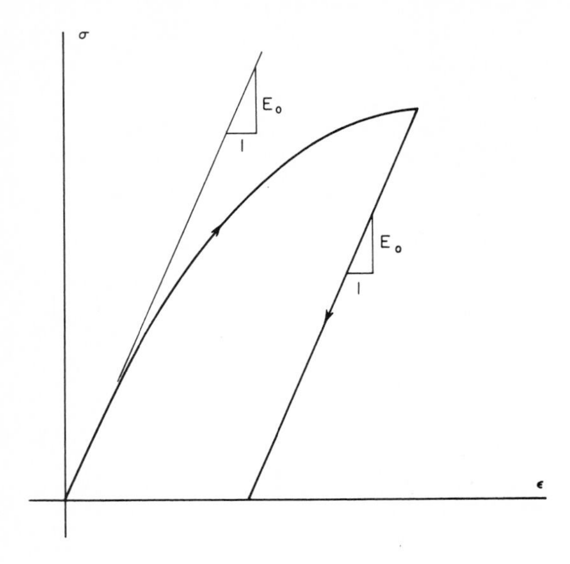

Figure 2-27 Simplest description of unloading modulus.

loading. A similar case of loading in different senses occurs when berms are built at the toe of a proposed embankment and then the remainder of the fill is placed. A second class of problems involves cyclic loading due to earthquakes or wind loads.

The simplest way to treat load reversal is to allow a material to unload along its virgin or elastic path, as in Fig. 2-27. This requires some bookkeeping in the computer program to keep track of what stress-strain path the material is on, but the programming is not too difficult.

In many cases the unloading is so extensive or so strong that it is more important than the original loading. Then it is necessary to have a better model for cyclic loading. One very useful one is the Ramberg-Osgood equation.[24] Figure 2-28 illustrates the shape of the stress-strain curve, which is defined by

$$\frac{\epsilon}{\epsilon_y} = \frac{\sigma}{\sigma_y} + \alpha \left| \frac{\sigma}{\sigma_y} \right|^r \quad \text{for virgin curve}$$

$$\frac{\epsilon - \epsilon_0}{2\epsilon_y} = \frac{\sigma - \sigma_0}{2\sigma_y} + \alpha \left| \frac{\sigma - \sigma_0}{2\sigma_y} \right|^r \quad \text{for reload} \tag{2-122}$$

The equation can be used for axial or deviatoric stresses and strains, depending on the wishes of the user. The parameters α and r must be found experimentally. Constantopoulos et al.[62] found that for shear modulus measurements the best agreement with the published data on behavior of soils under repeated loadings[56,57,63] was obtained with $\alpha = 0.05$ and r between 2.0 and 2.5. The great advantage of the Ramberg-Osgood model, in addition to its simplicity, is the fact

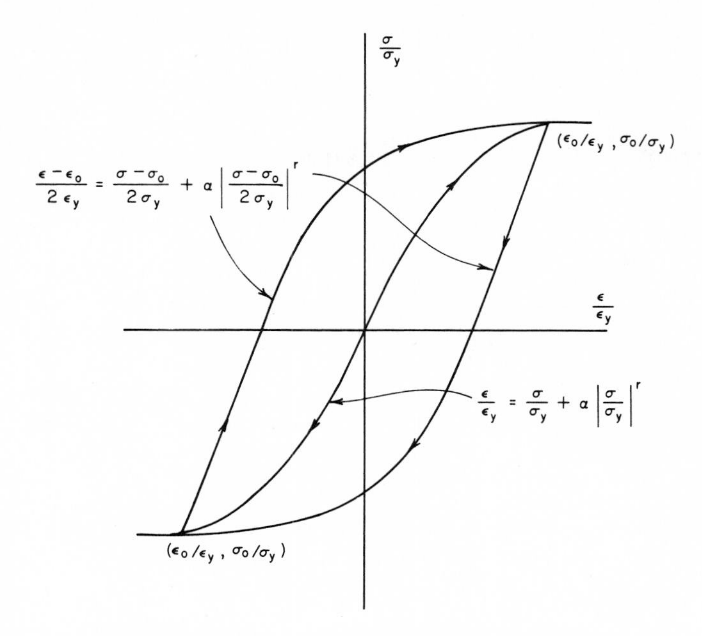

Figure 2-28 Ramberg-Osgood model for loading and unloading.

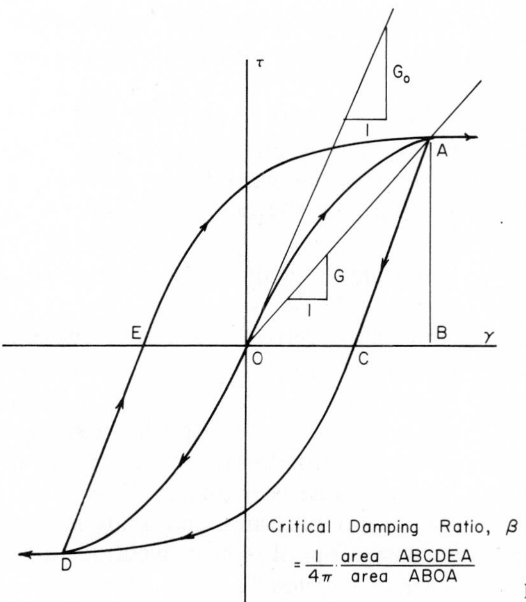

Figure 2-29 Calculation of equivalent linear hysteretic damping.

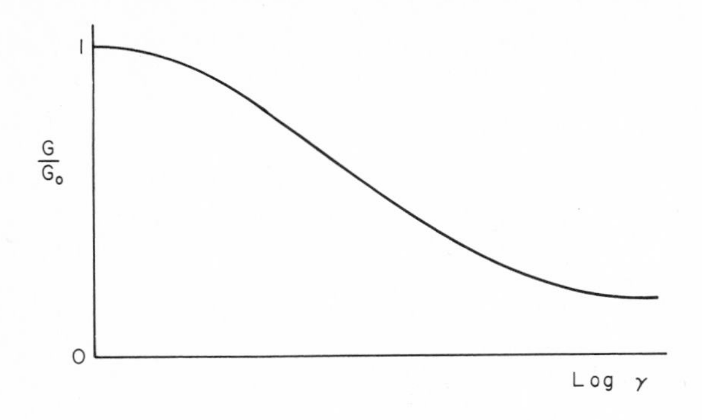

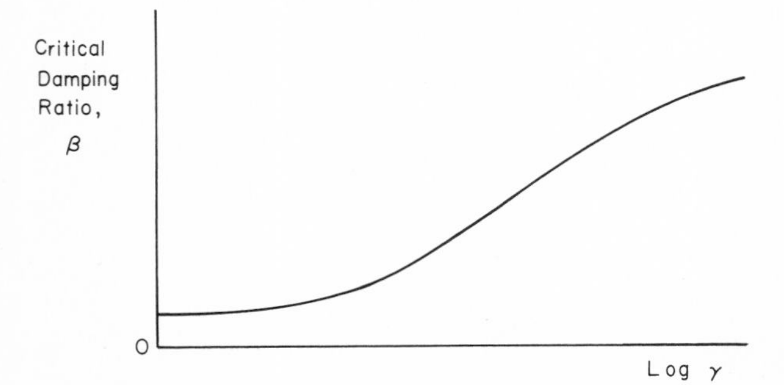

Figure 2-30 Secant modulus and damping as functions of strain.

that it carries within the equation for the reloading curve the parameters indicating where it will meet the virgin curve. Thus it is quite easy to ensure that the reloading curve will match the virgin curve on the other side of the strain axis if the reloading is large enough to cause yielding in the other direction.

As of this writing, there have been very few if any applications of the Ramberg-Osgood model to static soil-mechanics problems and a few to dynamic ones.[62] It is, however, an excellent and simple model that should see more use in the future.

Finally, the commonest way to treat problems of dynamic repeated loading is to evaluate equivalent shear moduli and damping. These are found by computing the secant modulus and the damping that would be needed to make a sample of soil loaded cyclically to a given peak strain and a linearly viscoelastic system respond at the same level (see Fig. 2-29). The results are plotted in the form of curves like those shown in Fig. 2-30. The analysis then proceeds by first assuming values of shear modulus and damping, then computing the response, finding an appropriate average value of the shear strain for each element, evaluating new moduli and damping for these strain levels, and repeating the analysis. The technique works well for engineering purposes, but it is not accurate if the analyses are aimed at finding exact responses of a system.[62]

REFERENCES

1. Biot, Maurice A.: "Mechanics of Incremental Deformations," John Wiley & Sons, Inc., New York, 1965.
2. Oden, J. T.: "Finite Elements of Nonlinear Continua," McGraw-Hill Book Company, New York, 1971.
3. Fung, H. C.: "Foundations of Solid Mechanics," Prentice-Hall, Inc., Englewood Cliffs, N.J., 1965.
4. Drnevich, Vincent P.: Constrained and Shear Moduli for Finite Elements, *Univ. Ky. Rep. Soil Mech.* 18, June 1974.
5. Desai, C. S., and J. F. Abel: "Introduction to the Finite Element Method," Van Nostrand Reinhold Company, New York, 1972.
6. Desai, C. S.: Overview, Trends and Projections: Theory and Applications of the FE Method in Geotechnical Engineering, *Proc. Symp. Appl. Finite Element Method Geotech. Eng. Vicksburg, Miss., September 1972.*
7. Clough, Ray W., and R. J. Woodward, III: Analysis of Embankment Stresses and Deformations, *J. Soil Mech. Found. Div. ASCE*, vol. 93, no. SM4, July 1967.
8. Vallabhan, C. V. G., and L. C. Reese: Finite Element Method Applied to Some Problems in Soil Mechanics, *J. Soil. Mech. Found. Div. ASCE*, vol. 94, no. SM2, March 1968.
9. Desai, C. S.: Solution of Stress-Deformation Problems in Soil and Rock Mechanics Using the Finite Element Method, Ph.D. dissertation, University of Texas, Austin, August 1968.
10. Desai, C. S., and L. C. Reese: Analysis of Circular Footings on Layered Soils, *J. Soil Mech. Found. Div. ASCE*, vol. 96, no. SM4, July 1970.
11. Desai, C. S., and L. C. Reese: Stress-Deformation and Stability Analysis of Deep Boreholes, *Proc. 2d Cong. Int. Soc. Rock Mech., Belgrade, 1970.*
12. Duncan, J. M., and C. Y. Chang: Non-linear Analysis of Stress and Strain in Soils, *J. Soil Mech. Found. Div. ASCE*, vol. 96, no. SM5, pp. 1629–1653, September 1970.
13. Chang, C. Y., and J. M. Duncan: Analysis of Soil Movement around a Deep Excavation, *J. Soil Mech. Found. Div. ASCE*, vol. 96, no. SM5, pp. 1655–1681, September 1970.
14. Kondner, R. L.: Hyperbolic Stress-Strain Response: Cohesive Soils, *J. Soil Mech. Found. Div. ASCE*, vol. 89, no. SM1, pp. 115–143, January 1963.
15. Janbu, Nilmar: Soil Compressibility as Determined by Oedometer and Triaxial Tests, *Proc. Eur. Conf. Soil Mech. Found. Eng., Wiesbaden, 1963*, vol. 1, pp. 19–25.
16. Kulhawy, F. H., J. M. Duncan, and H. B. Seed: Finite Element Analyses of Stresses and Movements in Embankments during Construction, *U.S. Army Eng. Waterw. Expt. Stn. Contract Rep.* 569-8, Vicksburg, Miss., 1969.
17. Hardin, B. O., and W. L. Black: Vibration Modulus of Normally Consolidated Clays, *J. Soil Mech. Found. Div., ASCE*, vol. 94, no. SM2, pp. 353–369, March 1968.
18. Hardin, B. O., and W. L. Black: Closure to Vibration Modulus of Normally Consolidated Clays, *J. Soil Mech. Found. Div. ASCE*, vol. 95, no. SM6, pp. 1531–1537, November 1969.
19. Marcuson, W. F., III, and H. E. Wahls: Time Effects on Dynamic Shear Modulus of Clays, *J. Soil Mech. Found. Div. ASCE*, vol. 98, no. SM12, pp. 1359–1373, December 1972.
20. Trudeau, P. J., R. V. Whitman, and J. T. Christian: Shear Wave Velocity and Modulus of a Marine Clay, *J. Boston Soc. Civ. Eng.*, vol. 61, no. 1, pp. 12–25, January 1974.
21. Simon, R. M., J. T. Christian, and C. C. Ladd: Analysis of Undrained Behavior of Loads on Clay, *Proc. Conf. Anal. Design Geotech. Eng. ASCE, Austin, Tex., June 1974*, pp. 51–84.
22. Duncan, J. M., and P. Dunlop: Behavior of Soils in Simple Shear Tests, *Proc. 7th Int. Conf. Soil Mech. Found. Eng., Mexico City, 1969*, vol. 1, pp. 101–109.
23. Davis, E. H., and J. T. Christian: Bearing Capacity of Anisotropic Cohesive Soils, *J. Soil Mech. Found. Div. ASCE*, vol. 97, no. SM5, pp. 753–769, May 1971.
24. Ramberg, W., and W. R. Osgood: Description of Stress-Strain Curves by Three Parameters, *Natl. Advis. Comm. Aeronaut.*, Tech. Note 902, Washington, D.C. 1943.
25. Richard, R. M., and B. J. Abbott: Versatile Elastic-Plastic Stress-Strain Formula, Tech. Note, *J. Eng. Mech. Div. ASCE*, vol. 101, no. EM4, August 1975.

26. Desai, C. S., and T. H. Wu: A General Function for Stress-Strain Curves, *Proc. 2d Int. Conf. Numer. Methods Geomech., Blacksburg, Va., June 1976.*
27. Wong, I. H.: Analysis of Braced Excavations, Sc.D. thesis, Department of Civil Engineering, Massachusetts Institute of Technology, Cambridge, Mass., 1971.
28. Desai, C. S.: Nonlinear Analysis Using Spline Functions, *J. Soil Mech. Found. Div. ASCE*, vol. 97, no. SM10, October 1971 and vol. 98, no. SM9, September 1972.
29. Ahlberg, J. H., E. N. Nilson, and J. L. Walsh: "The Theory of Splines and Their Applications," Academic Press, New York, 1967.
30. Jaworski, W. E.: An Evaluation of the Performance of a Braced Excavation, Sc.D. thesis, Department of Civil Engineering, Massachusetts Institute of Technology, Cambridge, Mass., 1973.
31. Eringen, A. C.: "Nonlinear Theory of Continuous Media," McGraw-Hill Book Company, New York, 1962.
32. Chang, T. Y., et al.: An Integrated Approach to the Stress Analysis of Granular Materials, *Calif. Inst. Tech. Rep. to Natl. Sci. Found.*, Pasadena, 1967.
33. Hill, R.: "The Mathematical Theory of Plasticity," Clarendon Press, Oxford, 1950.
34. Höeg, K., J. T. Christian, and R. V. Whitman: Settlement of Strip Load on Elastic-Plastic Soil, *J. Soil Mech. Found. Div. ASCE*, vol. 94, no. SM2, pp. 431–445, March 1968.
35. Drucker, D. C., and W. Prager: Soil Mechanics and Plastic Analysis or Limit Design, *Q. Appl. Math.*, vol. 10, no. 2, pp. 157–165, 1952.
36. Bishop, A. W.: The Strength of Soils as Engineering Materials, Sixth Rankine Lecture, *Geotechnique*, vol. 16, no. 2, pp. 91–128, June 1966.
37. Reyes, S. F.: Elastic-Plastic Analysis of Underground Openings by the Finite Element Method, Ph.D. thesis, University of Illinois, Urbana, 1966.
38. Christian, J. T.: Plane Strain Deformation Analysis of Soils, Ph.D. thesis, Department of Civil Engineering, Massachusetts Institute of Technology, Cambridge, Mass., 1966.
39. Shie, W. Y. J., and R. S. Sandhu: Application of Elasto-Plastic Analysis in Earth Structures, *Proc. Nat. Meet. Water Resour. Eng. ASCE, Memphis, Tenn., January 1970.*
40. Hagmann, A. J.: Prediction of Stress and Strain under Drained Loading Conditions, *MIT Dept. Civ. Eng., Rep.* R71-3, 1971.
41. Davis, E. H.: Theories of Plasticity and the Failure of Soil Masses, chap. 6, in I. K. Lee (ed.), "Soil Mechanics: Selected Topics," Butterworth & Co. (Publishers), London, 1968.
42. Hagmann, A. J., J. T. Christian, and D. J. D'Appolonia: Stress-Strain Models for Frictional Materials, *MIT Dept. Civ. Eng. Rep.* R70-18, 1970.
43. Lade, P. V., and J. M. Duncan: Elastoplastic Stress-Strain Theory for Cohesionless Soil, *J. Geotech. Eng. Div. ASCE*, vol. 101, no. GT10, October 1975.
44. Drucker, D. C.: Some Implications of Work Hardening and Ideal Plasticity, *Q. Appl. Math.*, vol. 7, no. 4, pp. 411–418, 1950.
45. Drucker, D. C.: Coulomb Friction, Plasticity, and Limit Loads, *J. Appl. Mech.*, vol. 21, pp. 71–74, 1954.
46. DiMaggio, F. L., and I. S. Sandler: Material Model for Granular Soils, *J. Eng. Mech. Div. ASCE*, vol. 97, no. EM3, June 1971.
47. Isenberg, J., and C. F. Bagge: Analysis of Steel-lined Penetration Shafts for Deeply Buried Structures, *Proc. Symp. Appl. Finite Element Method Geotech. Eng., Vicksburg, Miss., September 1972.*
48. Roscoe, K. H., A. N. Schofield, and C. P. Wroth: On the Yielding of Soils, *Geotechnique*, vol. 8, no. 1, pp. 25–53, March 1958.
49. Roscoe, K. H., A. N. Schofield, and A. Thurairajah: Yielding of Soils in States Wetter than Critical, *Geotechnique*, vol. 13, no. 3, pp. 211–240, September 1963.
50. Schofield, A. N., and C. P. Wroth: "Critical State Soil Mechanics," McGraw-Hill Publishing Company, Ltd., London, 1968.
51. Parry, R. H. G. (ed.): Stress-Strain Behaviour of Soils, *Proc. Roscoe Mem. Symp. Cambridge Univ.*, G. T. Foulis & Co., London, 1971.
52. Roscoe, K. H., and J. B. Burland: On the Generalized Stress-Strain Behavior of "Wet" Clay, pp. 535–609 in "Engineering Plasticity," Cambridge University Press, London, 1968.

53. Tatsuoka, F., and K. Ishihara: Drained Deformation of Sand under Cyclic Stress Reversing Direction, *Soils Found.*, vol. 14, no. 3, pp. 51–65, September 1974.
54. Höeg, K.: Finite Element Analysis of Strain-softening Clay, *J. Soil Mech. Founda. Div. ASCE*, vol. 98, no. SM1, pp. 43–58, January 1972.
55. Desai, C. S.: A Consistent Finite Element Technique for Work-softening Behavior, *Proc. Int. Conf. Comput. Methods Nonlinear Mech., Austin, Tex., 1974*, pp. 969–978.
56. Hardin, B. O., and V. P. Drnevich: Shear Modulus and Damping in Soils: Measurement and Parameter Effects, *J. Soil Mech. Found. Div. ASCE*, vol. 98, no. SM6, pp. 603–624, June 1972.
57. Hardin, B. O., and V. P. Drnevich: Shear Modulus and Damping in Soils: Equations and Curves, *J. Soil Mech. Found. Div. ASCE*, vol. 98, no. SM7, pp. 667–692, July 1972.
58. Christian, J. T., and B. J. Watt: Undrained Visco-elastic Analysis of Soil Deformations, *Proc. Symp. Appl. Finite Element Method Geotech. Eng., Vicksburg, Miss., 1972*, pp. 533–579.
59. Watt, B. J.: Analysis of Viscous Behavior in Undrained Soils, Sc.D. thesis, Department of Civil Engineering, Massachusetts Institute of Technology, Cambridge, Mass., 1969.
60. Singh, A., and J. K. Mitchell: General Stress-Strain-Time Function for Soils, *J. Soil Mech. Found. Div. ASCE*, vol. 94, no. SM1, pp. 31–46, January 1968.
61. Edgers, L., C. C. Ladd, and John T. Christian: Undrained Creep of Atchatalaya Layer Foundation Clays, *MIT Dept. Civ. Eng. Rep.* R73-16, February 1973.
62. Constantopoulos, I. V., J. M. Roesset, and J. T. Christian: A Comparison of Linear and Exact Non-linear Analyses of Soil Amplification, *5th World Conf. Earthquake Eng., Rome, 1973*, sess. 5B, pap. 225.
63. Seed, H. B., and I. M. Idriss: Soil Moduli and Damping Factors for Dynamic Response Analyses, *Univ. Calif. Earthquake Eng. Res. Center Rep.* 70-10, Berkeley, December 1970; also chap. 5 in Shannon & Wilson, Inc., and Agababian-Jacobsen Associates, "Soil Behavior under Earthquake Loading Conditions," state-of-the-art report to U.S. Army Eng. Corps, January 1972.

CHAPTER

THREE

VISCOPLASTICITY: A GENERALIZED MODEL FOR DESCRIPTION OF SOIL BEHAVIOR

O. C. Zienkiewicz and C. Humpheson

3-1 INTRODUCTION

The analysis of foundations, embankments, and other problems of soil mechanics has to date been largely based on either the prediction of collapse behavior using limit theorems of classical plasticity[1-3] or on stress-strain distribution studies following elasticity assumptions.[4,5] In many instances adequate results have been obtained despite the fact that (1) the soil does not obey classical plasticity assumptions on which limit theorems are based and (2) it shows at all stages of loading an irreversible strain violating elasticity assumptions. The reason for such relative success of the simplified assumptions lies in the relative insensitivity of the collapse load to the precise deformation law and of the stress distribution to the exact nature of the constitutive assumptions providing little deformation restraint is imposed on the masses (or in other words providing a situation close to that if statical determinacy exists).

If such conditions do not exist, serious errors can develop in the predictions of collapse loads[6] and deformations, and there is need for more precise analysis with a better definition of constitutive relations. The majority of such proposed relationships fall into the categories of (1) nonlinear elasticity in general form[7,8] or in piecewise linear form[9–11] and (2) strain-dependent[12,13] or ideal (strain-independent) plasticity; practically no attention has so far been given to more realistic material descriptions in which time effects like creep are considered. The neglect of time effects has usually been made as a simplification of already apparently complex laws and their involved numerical solution. In this chapter we shall introduce the concept of viscoplasticity and show that a general constitutive law based on this model is in fact capable of simply reproducing all kinds of plasticity as special cases with the additional possibility of considering real-time effects. Further it will be shown that this general model leads to computational procedures which in fact are simpler than those involved in more conventional plasticity computations.

Before proceeding with the introduction of this viscoplastic model we must restate the essential assumption of all soil mechanics behavior, i.e., the concept of effective stress.

3-2 ESSENTIAL ASSUMPTIONS OF POROUS TWO-PHASE MATERIAL ANALYSIS

In the analysis of the deformation behavior of a porous material such as soil it is important to realize that the total stress $\{\sigma\}$ is supported in part by the skeleton of the soil particles and in part by the interstitial fluid. The state of stress in the interstitial fluid can be described in the static state by a simple pressure p, and even if the fluid is in motion, the departure from this hydrostatic state is insignificant.

We now note an important experimentally observed fact, namely that a uniform increase of the hydrostatic pore pressure causes only insignificant strains in the soil skeleton. If this fact is true, it follows that any straining of the soil must be due to the difference of the total stress $\{\sigma\}$ and a hydrostatic stress $\{\sigma_h\}$ equal in magnitude to the pore pressure p.

In subsequent analysis we choose to define the stresses in a standard mechanics convention using tensions as positive, with the stress vector $\{\sigma\}$ defined by its components

$$\{\sigma\}^T = [\sigma_{xx} \quad \sigma_{yy} \quad \sigma_{zz} \quad \sigma_{xy} \quad \sigma_{yz} \quad \sigma_{zx}] \tag{3-1}$$

With p a hydrostatic, compressive pressure we thus write

$$\{\sigma_h\}^T = [-p \quad -p \quad -p \quad 0 \quad 0 \quad 0] = -\{m\}^T p \tag{3-2}$$

The stress responsible for soil deformation is the difference between total stress and pressure; we shall call this the *effective stress*, $\{\sigma_e\}$ (Fig. 3-1):

$$\{\sigma_e\} = \{\sigma\} - \{\sigma_h\} \tag{3-3}$$

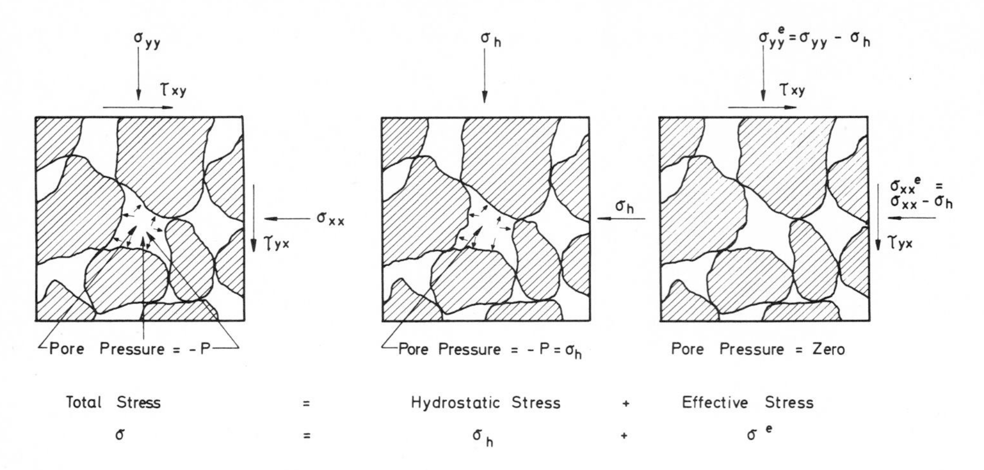

Figure 3-1 Total and effective stresses in a porous material.

Constitutive laws of soil mechanics are thus properly written in terms of the observable strain of the soil skeleton $\{\epsilon\}$,

$$\{\epsilon\}^T = [\epsilon_{xx} \quad \epsilon_{yy} \quad \epsilon_{zz} \quad 2\epsilon_{xy} \quad 2\epsilon_{yz} \quad 2\epsilon_{zx}] \tag{3-4}$$

and the effective stress $\{\sigma_e\}$.†

In the subsequent section we shall define such laws for a viscoplastic material.

At this stage it is convenient to consider the three classes of soil mechanics problems which occur in practice.

3-2.1 Long-Term Fully Drained Problems (Class I)

Here the pressure p is known a priori and can be determined by measurement or an uncoupled separate calculation in which the governing equation of seepage flow is solved; e.g., in steady state

$$\{\nabla\}^T[k]\{\nabla\}\{p\} = 0 \tag{3-5}$$

where the operator $\{\nabla\}$ is defined as

$$\{\nabla\}^T = \left[\frac{\partial}{\partial x} \quad \frac{\partial}{\partial y} \quad \frac{\partial}{\partial z}\right]$$

$[k]$ is the permeability tensor, and appropriate boundary conditions are applied. Once determined, the pressure p enters any calculation of deformation as a prescribed known function.

† The terms $2\epsilon_{xy}$, etc., represent the engineering shear strain, etc., in contrast to the conventional tensor notation. This is done simply to ensure that only six terms of a 3×3 tensor need be written in vector form.

3-2.2 Intermediate Time-coupled Settlement (Class II)

Here the pressure p is unknown and must be determined simultaneously with the deformation. The equation for solving the unknown deformation characterized by a displacement system $\{u\}$

$$\{u\}^T = [u \quad v \quad w] \tag{3-6}$$

will thus include p as the unknown, and in addition the seepage flow in its transient state will have to be solved; i.e.,

$$\{\nabla\}^T[k]\{\nabla\}\{p\} + \frac{\partial}{\partial t}\{\nabla\}^T\{u\} - \frac{\eta}{K_f}\frac{\partial p}{\partial t} = 0 \tag{3-7}$$

where

$$\{\nabla\}^T\{u\} = \frac{\partial u}{\partial x} + \frac{\partial v}{\partial y} + \frac{\partial w}{\partial z} = \epsilon_v \tag{3-8}$$

is the volumetric strain of the skeleton; in a unit time it represents the volume of fluid which must be supplied to a unit volume of space. K_f is the bulk modulus of the fluid phase, and η is the porosity. Thus the last term denotes the additional volume accommodated by the decrease of fluid volume per unit volume of space caused by change of pressure.

3-2.3 Rapid Loading, Undrained Behavior (Class III)

Here it is presumed that the loading is so fast that no seepage flow can occur. The fluid is thus trapped in the interstitial pores, and its *pressure changes* are related to volumetric changes of the skeleton by its elastic bulk modulus K_f. Thus we can write for changes of pressure Δp (assuming an incompressibility of the skeleton particles)

$$\Delta p = -\frac{K_f}{\eta}\epsilon_v = -\frac{K_p}{\eta}\{\nabla\}^T\{u\} = -\frac{K_f}{\eta}\{m\}^T\{\epsilon\} \tag{3-9}$$

This expression coupling the pressure changes to displacement and/or strains will be shown to lead to an analysis essentially similar to that applicable to class I.

3-3 ELASTOVISCOPLASTIC BEHAVIOR OF THE SOIL SKELETON

We shall now introduce the general ideas of elastoviscoplastic behavior linking the strains and effective stresses in the soil skeleton. It is convenient to introduce the concepts first by means of an uniaxial model and then to proceed with the generalization to a general stress state.

In Fig. 3-2 we summarize the behavior of elastic, ideally plastic, and viscous components; the figure shows the behavior of a purely elastoplastic and of an elastoviscoplastic assembly. In ideal plasticity the strains are zero where stresses

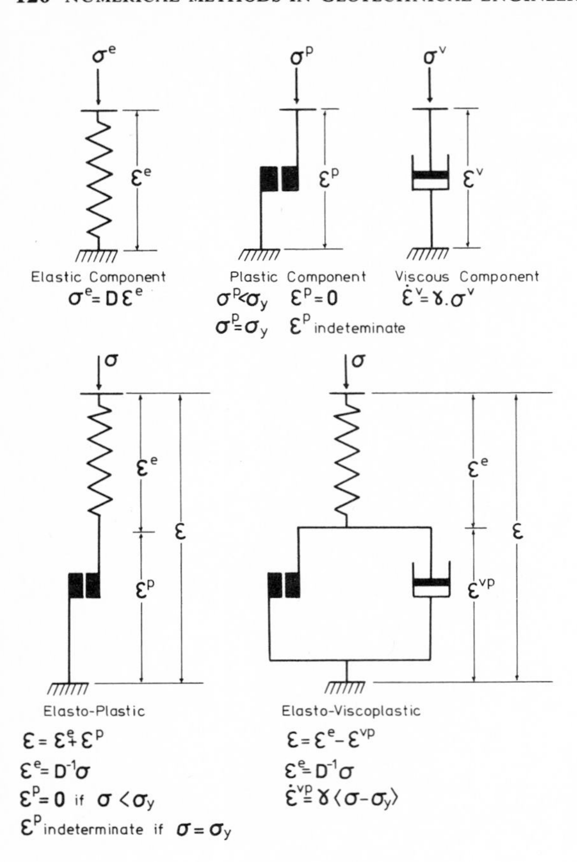

Figure 3-2 Uniaxial model of a simple elastoviscoplastic material.

are below a specified yield point, are indeterminate when this yield point is reached, and can in no circumstances exceed this yield point; this pinpoints one of the difficulties of specifying the stress-strain relationship here. On the contrary, for a viscoplastic model we can write explicitly that the total strain

$$\{\epsilon\} = \{\epsilon\}^e + \{\epsilon\}^{vp} \tag{3-10}$$

where the elastic strain $\{\epsilon\}^e$ is given by

$$\{\epsilon\}^e = [D^{-1}]^{\dagger}\{\sigma\} \tag{3-11}$$

and the *rate* of viscoplastic strain is uniquely determined by the stress, namely

$$\frac{\partial}{\partial t}\epsilon^{vp} = \dot{\epsilon}^{vp} = \gamma\langle\sigma - \sigma_y\rangle \tag{3-12}$$

† Here $[D]$ is the stress-strain matrix in the equation $\{\sigma_y\} = [D]\{\epsilon\}$. It is the same matrix $[C]$ used in other chapters; see Eq. (2-37). The reader may note that $[C]$ and $[D]$ are among a number of symbols used to define stress-strain relations.

where
$$\langle \sigma - \sigma_y \rangle = \begin{cases} 0 & \text{if } |\sigma| < |\sigma_y| \\ \sigma - \sigma_y & \text{if } |\sigma| \geq |\sigma_y| \end{cases}$$

In this model the stresses above the yield points are permissible and at any stress state at least the rate of strains is uniquely specified. Clearly the viscoplastic model is not only more positive in describing the behavior but also can be used to describe purely plastic phenomena when the rate of straining falls to zero.

Models of the viscoplastic kind can obviously be extended to deal with any complexities of behavior by introducing nonlinear dashpots in which some arbitrary function ψ specifies the viscoplastic rate, i.e.,

$$\dot{\epsilon}^{vp} = [\psi(\sigma - \sigma_y)] \tag{3-13}$$

by introducing the dependance of the yield stress σ_y on the total strain ϵ^{vp}

$$\sigma_y = \sigma_y(\epsilon^{vp}) \tag{3-14}$$

and finally by placing a number of viscoplastic models in series or parallel. For instance, Fig. 3-3 shows how standard creep and viscoplastic behavior can so be achieved.

To generalize the uniaxial to multiaxial stress-strain behavior it is necessary to introduce the concepts of a yield surface F† and a plastic potential Q,† well known in plasticity theory.

† Here F is used for yield surface; in Chap. 2, f is used; Q is a general symbol for plastic potential; in Chap. 2, we used function f for plastic potential for associated flow rule.

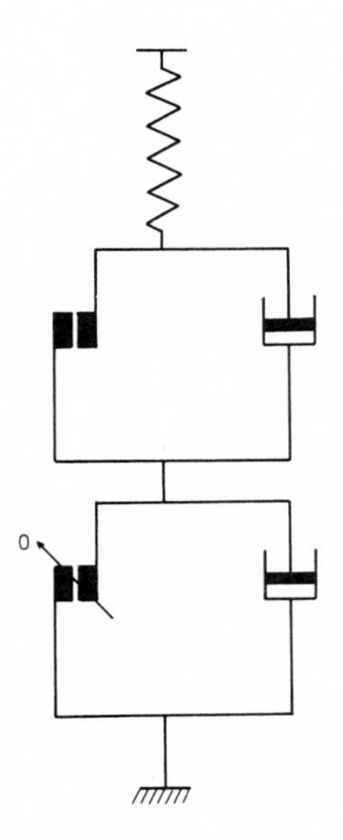

Figure 3-3 A series model of elastic and viscoplastic components.

Thus we state that (in plasticity) the limit between elastic recoverable behavior and plastic behavior is some combination of stresses and material constants which can give this limit when

$$F = F(\{\sigma_e\}, \{\epsilon\}^{vp}) = 0 \tag{3-15}$$

In the foregoing we have introduced the effective stress, as this controls the deformation of the skeleton. Immediately by analogy with the previous expressions we note that the viscoplastic strain rate will be proportional to the scalar quantity F, that is, when $F < 0$ the viscoplastic strain rate becomes zero (as $F = 0$ then); however, this does not define the relative magnitude of the components of the viscoplastic strain rate $\{\dot{\epsilon}\}^{vp}$

$$(\{\dot{\epsilon}\}^{vp})^T = [\dot{\epsilon}_{xx}^{vp} \quad \dot{\epsilon}_{yy}^{vp} \quad \dot{\epsilon}_{zz}^{vp} \quad \dot{\epsilon}_{xy}^{vp} \quad \dot{\epsilon}_{yz}^{vp} \quad \dot{\epsilon}_{zx}^{vp}] \tag{3-16}$$

and here we introduce the concept of a potential Q

$$Q = Q(\{\sigma\}, \{\epsilon\}^{vp}) \tag{3-17}$$

which defines the relative magnitudes by its gradient

$$\{\dot{\epsilon}\}^{vp} = \frac{\partial Q}{\partial\{\sigma\}} = \left[\frac{\partial Q}{\partial \sigma_{xx}} \quad \frac{\partial Q}{\partial \sigma_{yy}} \quad \cdots\right]^T \tag{3-18}$$

With the introduction of the potential concept we can now write a general multi-axial-statement analogy to the uniaxial form of laws[14,15]

$$\{\epsilon\} = \{\epsilon\}^{vp} + \{\epsilon\}^{e} \tag{3-19}$$

$$\{\epsilon\}^e = [D^{-1}](\{\sigma\}^e - \{\sigma_0\}^e) \tag{3-20}$$

$$\{\dot{\epsilon}\}^{vp} = \gamma F \frac{\partial Q}{\partial\{\sigma\}} = f(\sigma, \epsilon^{vp}) \tag{3-21}$$

where $[D]$ stands for the well-known matrix of elastic stresses and $\{\sigma_0\}^c$ was introduced as a reference state of effective stresses for which the strains changes are recorded.

We have not discussed in any detail the introduction of the concept of the plastic potential Q. This is clearly a completely arbitrary device to ensure the definition of strain rate components. There are, however, more profound reasons for its introduction into the theory of plasticity.[16] If we take

$$Q \equiv F$$

we have the possibility of proving certain limit theorems of classical plasticity—indeed these theorems presume this relationship on the basis of so-called stability postulates introduced by Drucker.[17,18] For soils these postulates are not generally valid, and the formulation used allows us to introduce arbitrary forms of Q. If $Q = F$ is assumed, we shall refer to *associative* viscoplasticity, and if $Q \neq F$, to *nonassociative*[19] behavior.

Again in the general context it is possible to introduce a series of multiaxial behavior models, and a very general discretization of viscoplastic behavior can always be written in a form[20]

$$\{\dot{\epsilon}\}^{vp} = \{f(\{\sigma\}, \alpha)\} \qquad \{\dot{\alpha}\} = \{g(\sigma, \alpha)\} \tag{3-22}$$

In practice we shall find the simpler description adequate.

3-4 SPECIFIC FORMS OF IDEAL AND STRAIN-DEPENDENT VISCOPLASTIC POTENTIALS USEFUL IN SOIL MECHANICS

3-4.1 Generalized Mohr-Coulomb Relationship

In the mechanics of porous, frictional media it has long been assumed that "failure" or yield is well defined by a unique relationship of the two extreme values of the principal stresses. As at this failure condition deformation proceeds at almost constant stress, it is natural to consider the failure and yield surfaces as coincident in the first approximation. Thus if σ_1^e, σ_2^e, and σ_3^e define the three principal stresses in the order

$$\sigma_1^e > \sigma_2^e > \sigma_3^e$$

the yield surface can be defined by a weighted relationship

$$\sigma_1^e = h_1(\sigma_3^e) \qquad \text{or} \qquad F = \sigma_1^e - h_1(\sigma_3^e) = 0 \tag{3-23a}$$

Alternatively we can write

$$\tau^e = \frac{\sigma_1^e - \sigma_3^e}{2} = h_2 \frac{\sigma_1^e + \sigma_3^e}{2} \qquad \text{or} \qquad F = \frac{\sigma_1^e - \sigma_3^e}{2} + h_2 \frac{\sigma_1^e + \sigma_3^e}{2} = 0 \tag{3-23b}$$

where the left-hand side will be recognized as the maximum shear stress. The above assumptions were originally introduced by Mohr[21] and Coulomb[22] and are found to hold well in actual tests.[23,24] The unique relationship is shown for a general case in Fig. 3-4*a* and in a linearized relationship in Fig. 3-4*b*. This linearized relationship can simply be derived from the usual assumption of a tangent envelope to the maximum principal-stress circles, and by inspection of Fig. 3-4*c* we can write this as

$$\frac{\sigma_1^e - \sigma_3^e}{2} = \left(\frac{\sigma_1^e + \sigma_3^e}{2} + \frac{c}{\tan \phi}\right) \sin \phi \tag{3-24}$$

in which c and ϕ denote the cohesion and internal friction angle, respectively.

The relationships written in terms of maximum and minimum principal stresses are inconvenient to transfer into a unique form defined in terms of the six

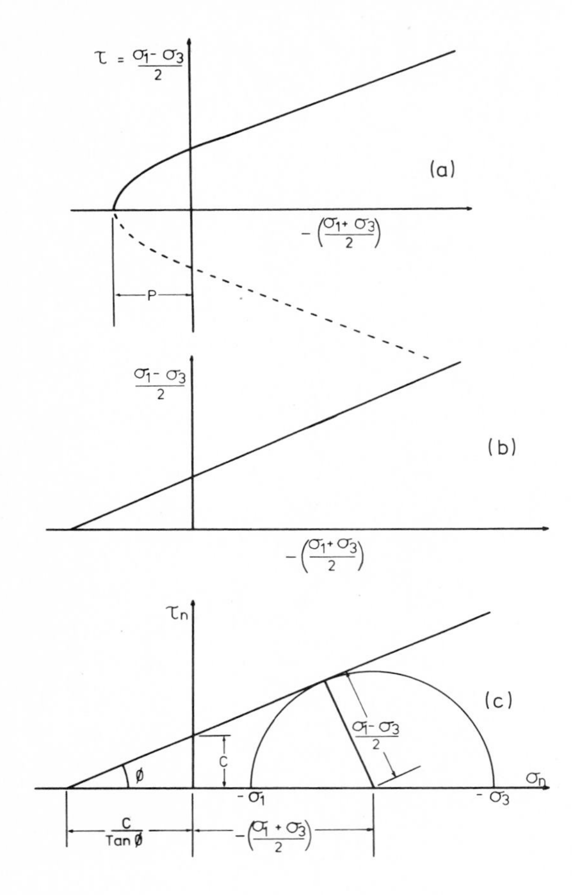

Figure 3-4 General and linearized Mohr-Coulomb relationships.

stress components. This is also difficult if these are defined in terms of the three standard-stress invariants

$$J_1 = \tfrac{1}{3}(\sigma^e_{xx} + \sigma^e_{yy} + \sigma^e_{zz}) = \sigma^e_m \tag{3-25}$$

$$J_2 = \bar{\sigma}^2 = \tfrac{1}{2}(S^2_x + S^2_y + S^2_z) + \sigma^{e2}_{xy}, \sigma^{e2}_{yz}, \sigma^{e2}_{zy} \tag{3-26}$$

$$J_3 = S_x S_y S_z + 2\sigma^e_{xy}\sigma^e_{yz}\sigma^e_{zx} - S_x\sigma^{e2}_{yz} - S_y\sigma^{e2}_{xz} - S_z\sigma^{e2}_{xy} \tag{3-27}$$

where $\quad S_x = \sigma^e_{xx} - \sigma^e_m \qquad S_y = \sigma^e_{yy} - \sigma^e_m \qquad S_z = \sigma^e_{zz} - \sigma^e_m \qquad$ (3-28)

A convenient alternative to the third invariant is defined by the Lode[25] angle θ_0 (see Appendix 3A):

$$-\frac{\pi}{6} \leqslant \theta_0 = \frac{1}{3}\sin^{-1}\left(-\frac{3\sqrt{3}}{2}\frac{J_3}{J_2^{3/2}}\right) \leqslant \frac{\pi}{6} \tag{3-29}$$

Hence the set

$$\sigma_m^e,\ \bar{\sigma}^e,\ -\frac{\pi}{6} \leqslant \theta_0 \leqslant \frac{\pi}{6} \tag{3-30}$$

is chosen to represent the stress invariants. This representation of the third invariant leads to the following expressions for the three principal stresses:

$$\begin{Bmatrix} \sigma_1 \\ \sigma_2 \\ \sigma_3 \end{Bmatrix} = \frac{2}{\sqrt{3}}\bar{\sigma}\begin{Bmatrix} \sin\,(\theta_0 + \frac{2}{3}\pi) \\ \sin\,\theta_0 \\ \sin\,(\theta_0 + \frac{4}{3}\pi) \end{Bmatrix} + \begin{Bmatrix} \sigma_m \\ \sigma_m \\ \sigma_m \end{Bmatrix} \tag{3-31}$$

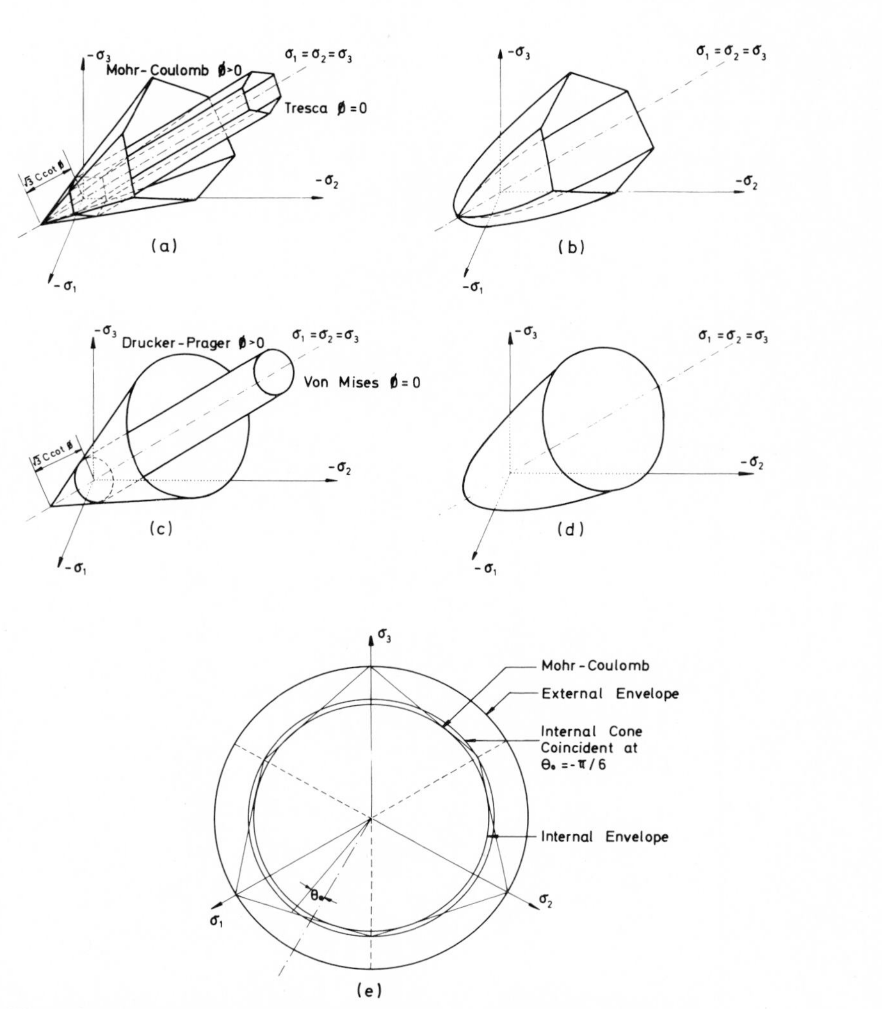

Figure 3-5 General and linearized yield surfaces in principal-stress space.

which immediately places them in the appropriate order of magnitude. The linearized description of the Mohr-Coulomb surface given in Eq. (3-24) can be written

$$\sigma_1^e - \sigma_3^e = 2c \cos \phi - (\sigma_1^e + \sigma_3^e) \sin \phi \tag{3-32}$$

which on substituting σ_1^e and σ_3^e from Eq. (3-32) gives

$$F = \sigma_m \sin \phi + \bar{\sigma} \cos \theta_0 - \frac{\bar{\sigma}}{\sqrt{3}} \sin \theta_0 \sin \phi - c \cos \phi = 0 \tag{3-33}$$

In Fig. 3-5*a* we show the form of the above yield surface in the space of the three principal stresses, and in Fig. 3-5*b* a more general case corresponding to the relationship sketched in Fig. 3-4*a* is represented.

It is of interest to note that when $\phi = 0$, the linearized Mohr-Coulomb relation becomes equivalent to the Tresca criterion of maximum shear stress and the expressions given above can still be used.

When a yield-surface equation of the above form has been obtained, it is a simple matter to adopt a *similar* surface to use for the nonassociated flow rules. In practice we shall use here for Q a surface obtained by the same expression as that for F but insert a different angle β in place of ϕ. To obtain the derivatives the expression defined for Q has to be appropriately differentiated. For completeness we derive the full derivative expressions in Appendix 3B.

3-4.2 Von Mises–Drucker-Prager Yield Surface

The yield surfaces defined in the preceding section have a serious drawback due to their angular nature in the principal (and hence also in the six-component) stress space. Whenever stresses are such that they fall on one of the "ridges" of the yield surfaces shown in Fig. 3-5*a*, the directions of derivatives are not determined and hence the viscoplastic strain rate cannot be easily determined.

In practical calculations it is common practice to "round" the corners and to assume that when the stress state *approaches* a corner (defined by the angle θ_0 approaching $\pm 30°$ within some tolerance), the strain direction is defined by a mean normal direction of the two intersecting surfaces. However, an alternative treatment which avoids this special differentiating is to approximate the angular yield surface by one which is smooth in the principal-stress space. Such smooth surfaces can be generalized by writing the yield criterion as a continuous relationship between the stress invariants.

To model the surfaces of Fig. 3-5*c* adequately it is sufficient to write down F as a linear relationship between the first two stress invariants σ_m and $\bar{\sigma}$. Such an approximation was proposed by Drucker and Prager[17] as

$$F = 3\alpha\sigma_m + \bar{\sigma} - k = 0 \tag{3-34}$$

where

$$\alpha = \frac{\sin \phi}{\sqrt{3}(3 + \sin^2 \phi)^{1/2}} \qquad k = \frac{\sqrt{3}\, c \cos \phi}{(3 + \sin^2 \phi)^{1/2}} \tag{3-35}$$

For $\alpha > 0$ this describes the surface of a right circular cone, but when $\alpha = 0$, it reduces to the well-known von Mises yield function, which is a cylinder in principal-stress space. Again a more general case corresponding to a surface of the type shown in Fig. 3-4*b* is shown in principal-stress space in Fig. 3-5*d*. A surface like this which is parabolic in shape has been used to represent the behavior of rock.[26] In terms of the parameters p and c shown on Fig. 3-4*c*, the expression of this yield surface is

$$F = (3\bar{\sigma}^2 + 3\sqrt{3}\,\beta k'\sigma_m)^{1/2} - k' = 0 \tag{3-36}$$

where $$\alpha' = \frac{c^2}{p} \qquad k'^2 = 3c^2 - \alpha'^2 \qquad \beta = \frac{\alpha'}{\sqrt{3}\,k'}$$

Referring to Fig. 3-5*a*, it is seen that on a particular π plane the Mohr-Coulomb yield surface is governed by the third invariant θ_0, which has a range $-\pi/6 \leqslant \theta_0 \leqslant \pi/6$, so that if the Mohr-Coulomb surface is to be represented by a circular cone, which one would represent the angular surface best? There are an infinite number of cones which could be chosen, each one having no more validity than the others. When $\theta = \pi/6$, the Mohr-Coulomb yield surface is given by

$$F = \frac{6 \sin \phi \sigma_m}{\sqrt{3}(3 - \sin \phi)} + \bar{\sigma} - \frac{6c \cos \phi}{\sqrt{3}(3 - \sin \phi)} = 0 \tag{3-37}$$

which represents a cone passing through the exterior corners of Mohr-Coulomb pyramid, whereas when $\theta = -\pi/6$, a cone intersecting the interior corners is obtained, with the corresponding surface given by

$$F = \frac{6 \sin \phi \sigma_m}{\sqrt{3}(3 + \sin \phi)} + \bar{\sigma} - \frac{6c \cos \phi}{\sqrt{3}(3 + \sin \phi)} = 0 \tag{3-38}$$

To obtain a lower bound on the Mohr-Coulomb surface for any π plane it is necessary to minimize Eq. (3-30) with respect to θ_0. On doing so it is found that the minimum value of F is reached when

$$\theta_0 = 2 \tan^{-1} \frac{\sqrt{3} - \sqrt{3 + \sin^2 \phi}}{\sin \phi} \tag{3-39}$$

which yields the following expression for F as substitution for θ_0:

$$F = \frac{\sqrt{3} \sin \phi}{\sqrt{3 + \sin^2 \phi}} \sigma_m + \bar{\sigma} - \frac{\sqrt{3}\, c \cos \phi}{\sqrt{3 + \sin^2 \phi}} = 0 \tag{3-40}$$

This expression is the same as that originally proposed by Drucker and Prager[17] in 1952. This particular form will always be a lower bound in the Mohr-Coulomb representation, whereas the surface represented by Eq. (3-37) will be an upper bound. Depending upon the stress combinations, the difference between these two bounds can give collapse loads differing by as much as 400 to 500 percent, so that, generally speaking, it is more realistic to use the full angular Mohr-Coulomb representation, the validity of which has been verified by experimental results.

3-4.3 Strain-dependent Viscoplasticity

The difficulty in accurately predicting the behavior of soil using either elastic or elastic–perfectly plastic constitutive laws has stimulated attempts to produce a model which gives a more realistic representation of observed soil behavior. Drucker, Gibson, and Henkel[12] suggested that the soil be treated as a work-hardening material which would eventually reach a perfectly plastic state. By so doing they reproduced some of the behavioral patterns of soil hitherto observed but as yet unaccounted for in analytical procedures. The proposed yield surface consisted of a Drucker-Prager failure envelope combined with a spherical end cap, the position and size of which depended upon the hydrostatic strain. This model demonstrated such phenomena as the increase in volume due to a decrease in hydrostatic stress and hence a decrease in yield strength and an increase or decrease in plastic volume due to yielding, the rate of which was a function of the state of yielding. The predictions produced from this model were in better agreement with experimental data than predictions from earlier models.

More recent work by Roscoe[13] has produced new possibilities for specific yield-function forms. Research has shown that elliptical yield surfaces fit experimental data more closely than previous solutions and that the phenomena of strain hardening and softening are automatically accounted in the problem formulation. Further details on this type of model can be found in Chap. 2.

A yield surface of this form which is only a function of the first two stress invariants is shown in Fig. 3-6*a*. This figure is drawn so that the *critical-state line*, or ultimate-failure surface, corresponds to the generalization of the Mohr-Coulomb representation given by Drucker and Prager. The normality principle assuming associated plastic behavior applies to the elliptical surface, thus defining the direction of plastic straining. Since at the intersection of the critical-state line and the ellipse the normal to the surface is vertical, there is at this point no component of plastic volumetric strain. Consequently any intersection to the right of this causes a plastic volumetric decrease and so strain hardening, whereas any intersection to the left results in a plastic volumetric increase and strain softening.

All the ellipses are assumed to pass through the same point a distance a from the origin. The size of the initial ellipse is governed by the preconsolidation pressure p_{c0} to which the soil has been subjected, so that if a soil has been overconsolidated at some stage, this could be quite large and the soil could sustain substantial loads before yielding occurred.

Following loading path 1-2, when the stress first exceeds the yield stress at 1, the plastic volumetric decrease causes the soil to harden and the ellipse to expand till eventually point 2 is reached. Here there can be further volumetric increase, and the soil is assumed to behave like an ideal material that flows as a frictional fluid at constant volume.

Similarly for the loading path 3-4; on reaching point 3 on the initial yield surface the volumetric increase causes strain softening, resulting in a decrease in the size of the ellipse. Eventually at point 4 the no-volume-change limit is reached, and again failure occurs at constant volume.

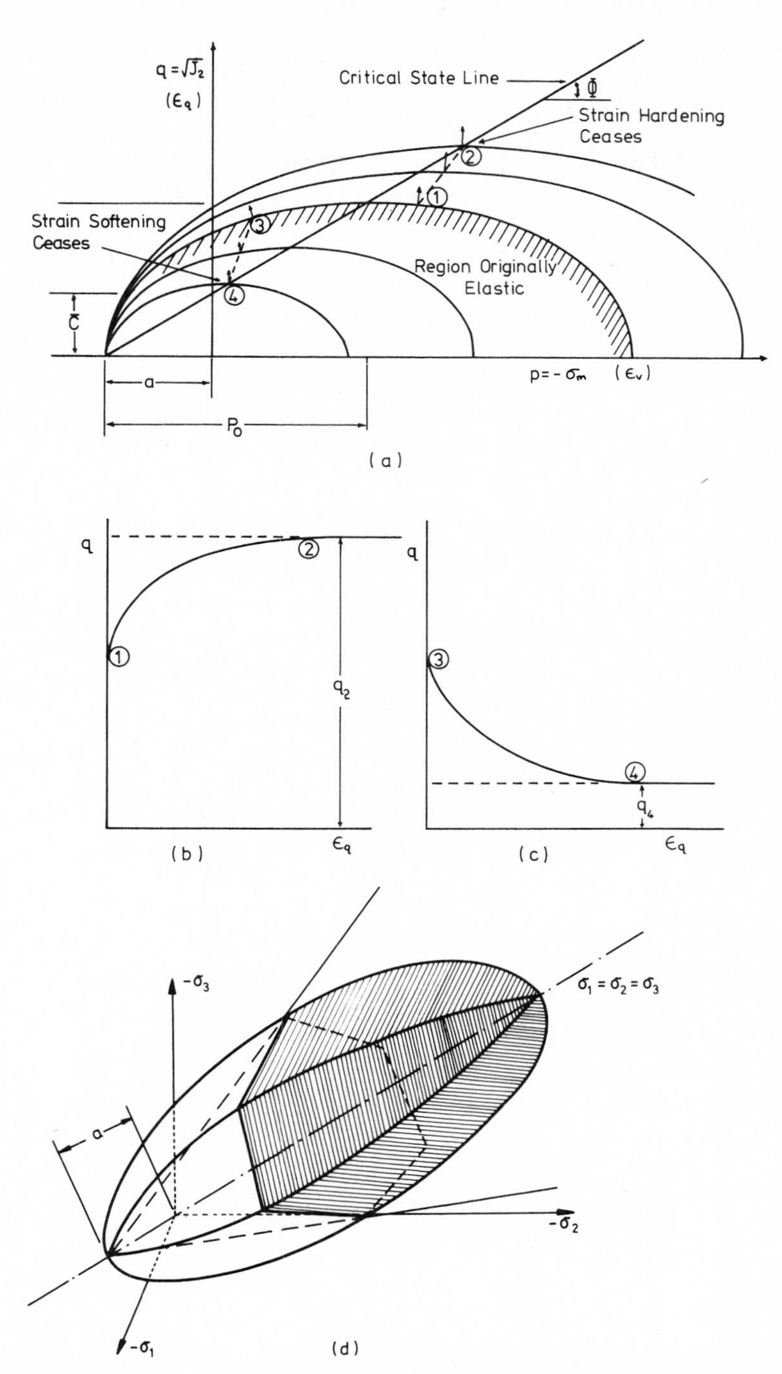

Figure 3-6 Critical-state strain-hardening or strain-softening yield surface: (*a*) graphical representation of yield surface in space of the two stress invariants p and q; (*b*) strain-hardening behavior; (*c*) strain-softening behavior; (*d*) generalization corresponding to the Mohr-Coulomb critical surface.

Since it has already been stated that the angular form of the Mohr-Coulomb surface is a better representation of the true behavior of soil than the Drucker-Prager generalization, it seems logical to combine a strain-dependent end cap of the critical-state type with a Mohr-Coulomb failure envelope. This combination yield function is shown in the space of the three-principal-stress space in Fig. 3-6*d*.

Identifying the critical-state line with the Mohr-Coulomb failure envelope leads to

$$a = \frac{\bar{c}}{\tan \Phi} = \frac{c}{\tan \phi}$$

$$\tan \Phi = \frac{3 \sin \phi}{\sqrt{3} \cos \theta_0 - \sin \theta_0 \sin \phi} = M_{cs} \tag{3-41}$$

$$\bar{c} = \frac{3c \cos \phi}{\sqrt{3} \cos \theta_0 - \sin \theta_0 \sin \phi}$$

The equation of the elliptical yield surface now becomes

$$F = \frac{3\bar{\sigma}^2}{(p_0 M_{cs})^2} + \frac{(-\sigma_m - p_0 + a)^2}{p_0^2} = 1 \tag{3-42}$$

This can be rewritten as

$$F = 3\bar{\sigma} + \frac{9 \sin^2 \phi[-\sigma_m(-\sigma_m - 2p_0 - 2c \cot \phi) + c \cot \phi(2p_0 - c \cot \phi)]}{(\sqrt{3} \cos \theta_0 - \sin \theta_0 \sin \phi)^2 \bar{\sigma}} = 0 \tag{3-43}$$

The size of the ellipse, that is, the parameter p_0, according to Roscoe, should be an exponential function of the plastic volumetric strain. Therefore the change in p_0 can be written

$$\Delta p_0 = f(\epsilon_v^p) = p_{c0} e^{\chi \epsilon_v{}^p} \tag{3-44}$$

where p_{c0} = initial value of p_0
χ = constant given by

$$\chi^\dagger = \frac{1 + e_0}{\lambda - k} \tag{3-45}$$

e_0 = initial-voids ratio
λ = compression index
k = swelling index

† See Roscoe and Burland[13] for the derivation of χ.

3-5 FINITE ELEMENT FORMULATION OF VISCOPLASTIC SOIL PROBLEMS

3-5.1 General Equilibrium Statement in Finite Element Displacement Discretization

Consider a general problem in which a total stress state $\{\sigma\}$ is to be determined within some volume Ω with boundaries Γ, on which we have either specified tractions $\{t\}$ (on Γ_t) or specified displacements $\{\bar{u}\}$ (on Γ_u). If the basic unknowns are the displacements $\{u\}$,

$$\{u\}^T = [u \quad v \quad w] \tag{3-46}$$

and we approximate these in the usual finite element manner by locally defined shape functions $\{N\} = N(x, y, z)$ and undetermined nodal parameters $\{a\}$ as

$$\{u\} = \{N\}^T\{a\} \tag{3-47}$$

we can determine the compatible strains $\{\epsilon\}$ as

$$\{\epsilon\} = [B]\{a\} \tag{3-48}$$

In the above expression the $[B]$ matrix depends only on the position coordinates x, y, z if displacements† are small. With body forces $\{b\}$ acting per unit volume we can write the approximate equilibrium equations from the principle of virtual work as[27]

$$\int_\Omega [B]^T\{\sigma\}\, d\Omega - \int_\Omega [N]^T\{b\}\, d\Omega - \int_{\Gamma_t} [N]^T\{t\}\, d\Gamma = 0 \tag{3-49}$$

in which the expression for $\{u\}$ is assumed to satisfy the prescribed displacement conditions on Γ_u. We note that all tractions are in terms of *total forces* on boundaries and that the above equations apply to total internal stress $\{\sigma\}$. As we are committed to laws relating effective stresses to the strains, it is convenient to rewrite the equilibrium statement using the definition

$$\{\sigma\} = \{\sigma_e\} + \{\sigma_h\} = \{\sigma_e\} - \{m\}\{p\} \tag{3-50}$$

Denoting the boundary and body-force terms in Eq. (3-54) as $\{F\}$, we have now

$$\int_\Omega [B]^T\{\sigma_e\}\, d\Omega - \int_\Omega [B]^T\{m\}^T\{p\}\, d\Omega - \{Q\} = 0 \tag{3-51}$$

If it is desired to represent the pore pressure p by its interpolation from nodal values $\{p\}$ as

$$\{p\} = \{\bar{N}\}^T\{p\}$$

† For treatment of large deformation see Ref. 27.

we can write

$$\int_{\Omega} [B]^T\{\sigma_e\}\, d\Omega - \{Q_p\} - \{Q\} = 0 \qquad \text{with } \{Q_p\} = \left(\int_{\Omega} [B]^T\{M\}\{\bar{N}\}\, d\Omega\right)\{p\} \tag{3-52}$$

The constitutive laws derived in Sec. 3-4.3 can be written

$$\{\sigma_e\} = [D](\{\epsilon\} - \{\epsilon\}^{vp}) = [D]([B]\{a\}^T - \{\epsilon\}^{vp}) + \{\sigma_0\} \tag{3-53}$$

$$\frac{d\{\epsilon\}^{vp}}{dt} = f(\{\sigma\}, \{\epsilon\}^{vp}) \tag{3-54}$$

These, together with the equilibrium statement provide a set of ordinary differential equations capable of being solved numerically in a variety of ways if the pressure $\{p\}$ is independently determined. The nature of the problem depends on how this is introduced, which we shall discuss in subsequent sections.

3-5.2 Solution Scheme to Long-Term (Drained) Conditions

Here the pressure p is known a priori, and the force $\{Q_p\}$ can be determined. We are thus faced with the simplest solution and can discuss the basic time-stepping numerical schemes in this context.

Clearly the most obvious scheme for solving the equation system (3-52) is to use a forward-difference scheme. Assuming thus that at some given time t_n we know $\{a\}$, $\{\epsilon\}^{vp}$ and hence $\{\sigma\}$, we can determine

$$\{\dot{\epsilon}_n\}^{vp} = \{f(\sigma_n, \epsilon_n^{vp})\} \equiv \{f_n\} \tag{3-55}$$

and hence the increment of viscoplastic strain can be approximated as

$$\Delta\{\epsilon_n\}^{vp} = \{f_n\}\, \Delta t \tag{3-56}$$

We now have for the total viscoplastic strain at t_{n+1}

$$\{\epsilon_{n+1}\}^{vp} = \{\epsilon_n\}^{vp} + \Delta\{\epsilon_n\}^{vp} \tag{3-57}$$

Combining the elastic constitutive law with the equilibrium statement, we can now compute $\{a_{n+1}\}$ from

$$[K]\{a_{n+1}\} = \int_{\Omega} [B]^T[D]\{\epsilon_{n+1}\}^{vp}\, d\Omega + \{Q_{p,n+1}\} + \{Q_{n+1}\} \tag{3-58}$$

or

$$\{a_{n+1}\} = [K^{-1}]\left(\int_{\Omega} [B]^T[D]\{\epsilon_{n+1}\}^{vp}\, d\Omega + \{Q_{p,n+1}\} + \{Q_{n+1}\}\right) \tag{3-59}$$

The new stress system $\{\sigma_{n+1}\}$ is found as

$$\{\sigma_{n+1}\} = [D]([B]\{a_{n+1}\} - \{\epsilon_{n+1}\}^{vp}) \tag{3-60}$$

This completes the time step, and the sequence can be repeated to any time desired.

Clearly an improvement of accuracy can be achieved if the new value of $\{\sigma_{n+1}\}$ is used iteratively, to give a better estimate of $\{\Delta\epsilon\}^{vp}$, and such corrections are often used. It is economical to do at least one such a revaluation of $\{\epsilon_{n+1}\}^{vp}$ and to use this improved value for the next increment calculation.[28]

The direct time-stepping scheme discussed can become unstable at large values of Δt. Cormeau[29] and Zienkiewicz[30] quote expressions for different forms of the viscoplastic relation which provide bounding values Δt for stability.

The algorithm described above is very simple in implementation and can be used for creep studies as well as for static plastic solutions in a limiting case. Alternative algorithms using improved predictor methods have been used and in some cases are more economical.[28]

3-5.3 Rapid Loading, Undrained Behavior

In such problems we shall assume that some known pore pressure p_0 preexists and that we are only concerned with determination of the excess pore pressure p. We thus rewrite the equilibrium equation as

$$\int_\Omega [B]^T\{\sigma_e\}\, d\Omega - \int_\Omega [B]^T\{m\}\{p\}\, d\Omega - \{Q\}^{po} - \{Q\} = 0 \tag{3-61}$$

with

$$p = -\frac{K_f}{\eta}\{m\}^T\{\epsilon\} = -\frac{K_f}{\eta}\{m\}^T[B]\{a\} \tag{3-62}$$

As the constitutive relations (3-53) remain unchanged, we can substitute these simultaneously and write

$$[\hat{K}]\{a\} - \int_\Omega [B]^T[D]\{\epsilon\}^{vp}\, d\Omega - \{Q\}^{po} - \{Q\} = 0 \tag{3-63}$$

with

$$[\hat{K}] = \int_\Omega [B]^T\left([D] + \{m\}\left[\frac{K_f}{\eta}\right]\{m\}^T\right)[B]\, d\Omega = \int_\Omega [B]^T[\hat{D}][B]\, d\Omega \tag{3-64}$$

Clearly the method of solution will proceed precisely as before with the elastic $[D]$ matrix augmented to $[\hat{D}]$ when stiffnesses are calculated. It is important to remember, however, that the ordinary $[D]$ matrix is still to be used in calculation of *effective stresses* [expression (3-53)] and that the pressure has to be separately determined by Eq. (3-62). *Program modification to deal with rapid loading is thus trivial but essential.*

3-5.4 Coupled Viscoplastic Consolidation Problem

Now the pressure p must be determined by solving the flow problem associated with changes of pore pressure. In Chap. 14 we show how the seepage equation can be discretized using finite element method. Starting from Eq. (3-51), we can write, using nodal values of p,

$$p = [\bar{N}]\{p\}$$

$$[H]\{p\} - [L]\frac{\partial\{a\}}{\partial t} - [S]\frac{\partial}{\partial t}\{p\} - \{R\} = \{0\}$$

where

$$[H_{ij}] = \int_\Omega \left(\frac{\partial\{\bar{N}_i\}^T[K]}{\partial x}\frac{\partial\{\bar{N}_j\}}{\partial x} + \frac{\partial[\bar{N}_i]^T[K]}{\partial y}\frac{\partial\{\bar{N}_j\}}{\partial y} + \frac{\partial\{\bar{N}_i\}^T[K]}{\partial z}\frac{\partial\{\bar{N}_j\}}{\partial z} \right) d\Omega \quad (3\text{-}67)$$

$$[L] = \int_\Omega \{\bar{N}\}^T[m][B]\, d\Omega \qquad [S] = \int_\Omega \{\bar{N}\}^T \frac{\eta}{K_f}[\bar{N}]\, d\Omega \quad (3\text{-}68)$$

and $\{R\}$ is a vector representing the boundary contributions.

If we note that the viscoplastic flow problem as stated in Secs. 3-5.1 and 3-5.2 given by Eq. (3-52)

$$\{F^p\} = [L]\{p\}^T \quad (3\text{-}69)$$

and therefore note the equilibrium equation as

$$[K]\{a\} - \int_\Omega [B]^T[D]\{\epsilon\}^{vp}\, d\Omega - [L]^T\{p\} - \{F\} = 0 \quad (3\text{-}70)$$

with $$\{\sigma_e\} = [D]([B]\{a\} - \{\epsilon\}^{vp}) \qquad \frac{d\{\epsilon\}^{vp}}{dt} = f(\sigma_e) \quad (3\text{-}71)$$

we have now an augmented system of ordinary differential equations in which $\{a\}$ and $\{p\}$ are the simultaneous unknowns.

Obviously once again there is a variety of time-marching procedures, ranging from forward time stepping to mixed schemes. In all, the variables $\{a\}$ and $\{p\}$ must be solved for simultaneously. For clarity it is convenient to rewrite the two equations (3-66) and (3-70) as

$$\begin{bmatrix} [K] & -[L]^T \\ [0] & [H] \end{bmatrix} \begin{Bmatrix} \{a\} \\ \{p\} \end{Bmatrix} + \begin{bmatrix} [0] & [0] \\ -[L] & [S] \end{bmatrix} \frac{d}{dt} \begin{Bmatrix} \{a\} \\ \{p\} \end{Bmatrix} - \begin{Bmatrix} \{\bar{Q}\} \\ \{R\} \end{Bmatrix} = 0 \quad (3\text{-}72)$$

in which

$$\{\bar{Q}\} = \{Q\} + \int_\Omega [B]^T[D]\{\epsilon\}^{vp}\, d\Omega = \{Q\} + \{Q\}^{vp} \quad (3\text{-}73)$$

and $$\frac{d\epsilon^{vp}}{dt} = f(\sigma_e) \qquad \{\sigma_e\} = [D]([B]\{a\} - \{\epsilon\}^{vp}) \quad (3\text{-}74)$$

Clearly $\{\bar{Q}\}$ is a nonlinear function of $\{a\}$ and t, and unless one of the nonlinear iterative schemes is adopted, it is simplest to adopt the forward time-stepping techniques discussed in the previous section and to compute $\{\dot{\epsilon}\}^{vp}$ from the starting values of the time interval. We shall thus take for any interval Δt_n $(t_{n+1} - t_n)$

$$\{\bar{Q}\}^{vp} \equiv \{Q_n\}^{vp} \quad (3\text{-}75)$$

given that assumption Eq. (3-72) can be solved using either forward, midstop (Crank-Nicholson), or backward time-stepping techniques with about equal computational effort providing the starting values of $\{a\}$, $\{p\}$, etc., are known.

We can thus write approximately

$$\begin{bmatrix} [K] & -[L]^T \\ [0] & [H] \end{bmatrix} \left[\begin{Bmatrix} \{a\} \\ \{p\} \end{Bmatrix}_n \theta + \begin{Bmatrix} \{a\} \\ \{p\} \end{Bmatrix}_{n+1} (1-\theta) \right] - \begin{bmatrix} [0] & [0] \\ -[L] & [S] \end{bmatrix}$$

$$\times \frac{1}{\Delta t_n} \left[\begin{Bmatrix} \{a\} \\ \{p\} \end{Bmatrix}_{n+1} - \begin{Bmatrix} \{a\} \\ \{p\} \end{Bmatrix}_n \right] - \begin{Bmatrix} \{\bar{Q}_{n+1/2}\} - \{Q_n\} \\ \{\bar{R}_{n+1}\} \end{Bmatrix} = \{0\} \tag{3-76}$$

with $\theta = 1, \frac{1}{2}, 0$ giving forward, midstop, or backward integration expressions, respectively.

Clearly for above the values of $\{a_{n+1}\}$ and $\{p_{n+1}\}$ can be calculated and the process continued. It is convenient to use either θ ranging from $\frac{1}{2}$ to zero as for these values stability problems do not arise providing Δt is still kept within the bounds discussed in Sec. 3-5.

It is important to remark that if at any time (including zero time) a sudden change of load occurs, an averaged value should be used (at zero time $\frac{1}{2}$ of a suddenly applied load, for instance). This serves to iron out the oscillation which often arises if this simple precaution is not taken.

It is of interest to write down the time-stepping solution explicitly for $\theta < 1$:

$$\begin{Bmatrix} \{a\} \\ \{p\} \end{Bmatrix}_{n+1} = \begin{bmatrix} +[K] & -[L]^T \\ -[L] & [H(1-\theta)\,\Delta t + S] \end{bmatrix}^{-1}$$

$$\times \begin{Bmatrix} \dfrac{\{Q\}_{n+1/2} + \{Q\}_n}{1-\theta} \\ \{R_{n+1/2}\} \end{Bmatrix} - \begin{bmatrix} \dfrac{[K]\theta}{1-\theta} & \dfrac{-[L]^T\theta}{1-\theta} \\ [L] & [H(\theta)\,\Delta t - S] \end{bmatrix} \begin{Bmatrix} \{a\} \\ \{p\} \end{Bmatrix} \tag{3-77}$$

If now we consider the case of a loading $\{Q\}$ suddenly applied to an equilibrium state with $a = 0$, $p = 0$, $\epsilon^{vp} = 0$, and take

$$\theta = \tfrac{1}{2} \qquad \Delta t \to 0$$

and as indicated previously

$$\{Q_{n+1/2}\} = \tfrac{1}{2}\{Q\}$$

we obtain

$$\begin{Bmatrix} \{a\} \\ \{p\} \end{Bmatrix} = \begin{bmatrix} [K] & -[L]^T \\ -[L] & [S] \end{bmatrix} \begin{Bmatrix} \{Q\} \\ \{0\} \end{Bmatrix} \tag{3-78}$$

It can be verified that this is precisely the solution obtained at the first (elastic) step of the rapid loading case with no drainage, as discussed in Sec. 3-5.3; however, the formulation is also valid for the case of complete incompressibility and is an alternative way of deriving the constrained solution for this.

3-6 SOME EXAMPLES OF APPLICATION OF THE VISCOPLASTIC FORMULATION

A few selected examples of application of the viscoplastic formulation are given here to illustrate the possibilities. In most the objective is to obtain an elasto-plastic state. Here clearly the exact nature of the law relating rates of viscoplastic strain to the stress state is of no importance, as we seek a stationary solution. For such a case the time is a purely artificial variable and the parameter γ need not be determined. The formulation shows a simple scheme for obtaining a solution for nonassociated laws, etc. In the first example, that of tunnel-lining stress computation, the knowledge of the time characteristics is of importance. This presents one further challenge to the engineer who seeks to correlate site observation and experiment with theory.

Example 3-1 Tunnel-lining design Here we consider the development of stresses in a tunnel excavated in deep solid rock. It is assumed that the changes of stress occur in the three stages illustrated in Fig. 3-7. The stresses existing at end of each stage are used as initial stresses for the start of the next.

When the tunnel excavation is completed, an immediate elastic unloading occurs, followed by some viscoplastic creep during time t_1. At that point an unstrained lining is inserted, and the only stresses which develop in it are due to continuing viscoplastic deformation.

In Fig. 3-8 we show the displacements which have developed in each stage and the final stresses in the lining. The time is defined in units of $1/\gamma$, and a Mohr-Coulomb type of behavior is assumed with associated flow and nonassociated incompressible flow. We see how sensitive the stresses and displacements are to the nature of the flow rule assumed.

The initial stress field assumed in this example did not take gravity into account so that a smaller region could be modeled utilizing two axes of symmetry.

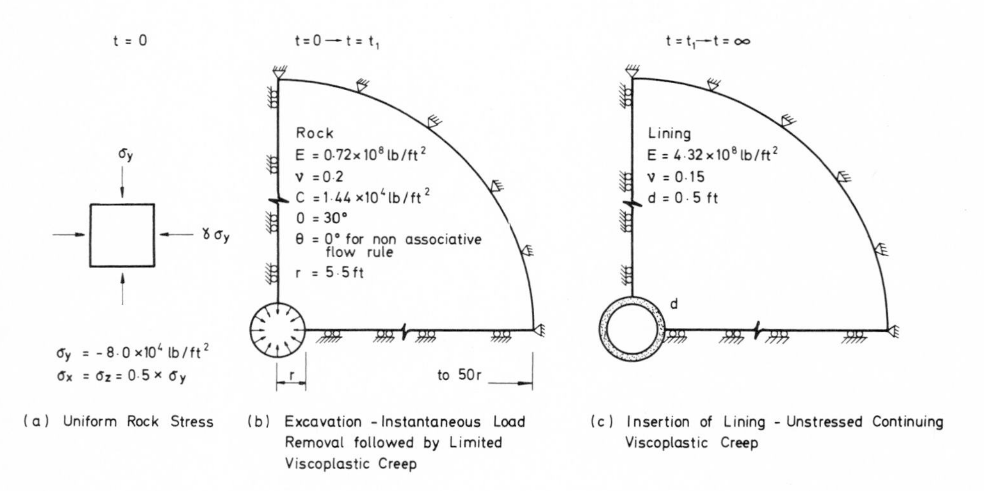

Figure 3-7 Stages of tunnel stress development. Each stage provides the initial stress for the next.

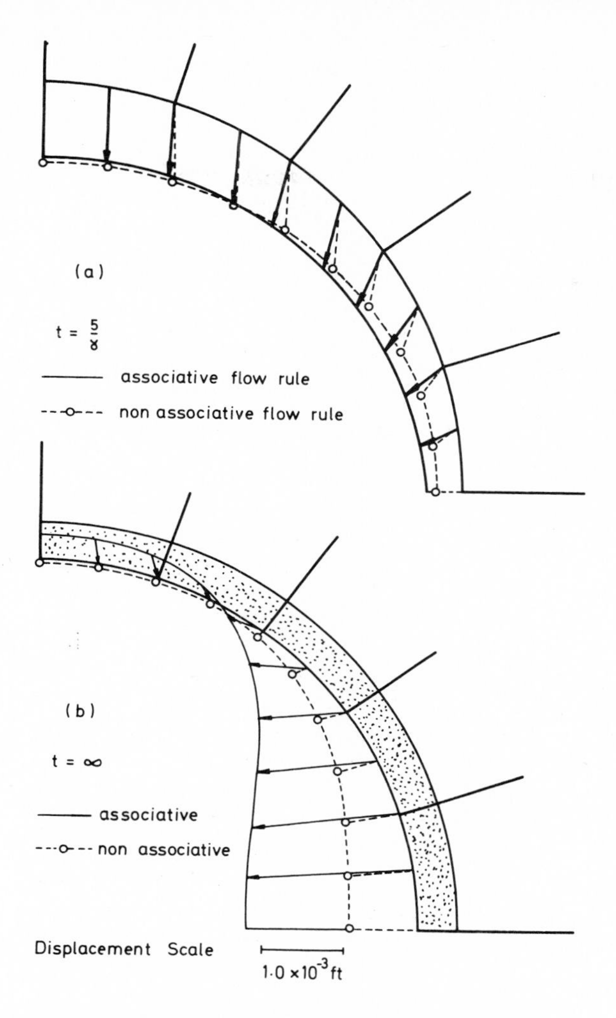

Figure 3-8 Displacements and stresses in tunnel lining.

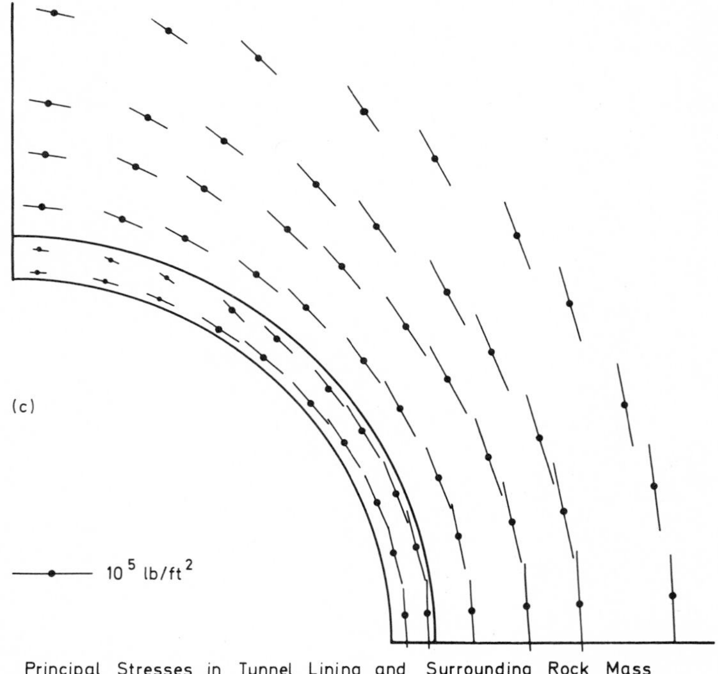

Principal Stresses in Tunnel Lining and Surrounding Rock Mass Using an Associative Flow Rule

Figure 3-8 (continued)

Example 3-2 Embankment stability In this problem the stability of an embankment is investigated using associated and nonassociated flow rules. The results of embankment studies are usually expressed in terms of a factor of safety with respect to collapse, and here we follow a similar procedure by determining the minimum value of cohesion for which a stable solution can be obtained.

The angle of friction ϕ was held constant at 20°, and the cohesion was reduced until failure occurred. The results of this analysis are presented in the form of a cohesion-vs.-displacement curve, shown in Fig. 3-9.

If we use an associated flow rule, we see that collapse occurred at a value of cohesion of $c = 3$ kN/m^2. With the values $c = 3$ kN/m^2 and $\phi = 20°$ a slip-circle analysis using the method of slices predicted a factor of safety of 1.01, showing agreement with the viscoplastic solution.

Figure 3-10 shows representations of velocity vectors and shear-strain-rate contours at collapse. Also shown on these diagrams is the slip circle, which provided the minimum factor of safety for the slip-circle analysis, and again there clearly is correlation between the plastic flow and the concentration of shear strains with this slip circle. For this problem there is very little difference between the predicted collapse cohesion using an associated or a nonassociated flow rule because the problem is relatively unconfined. Further discussion of this type of problem can be found in Ref. 15.

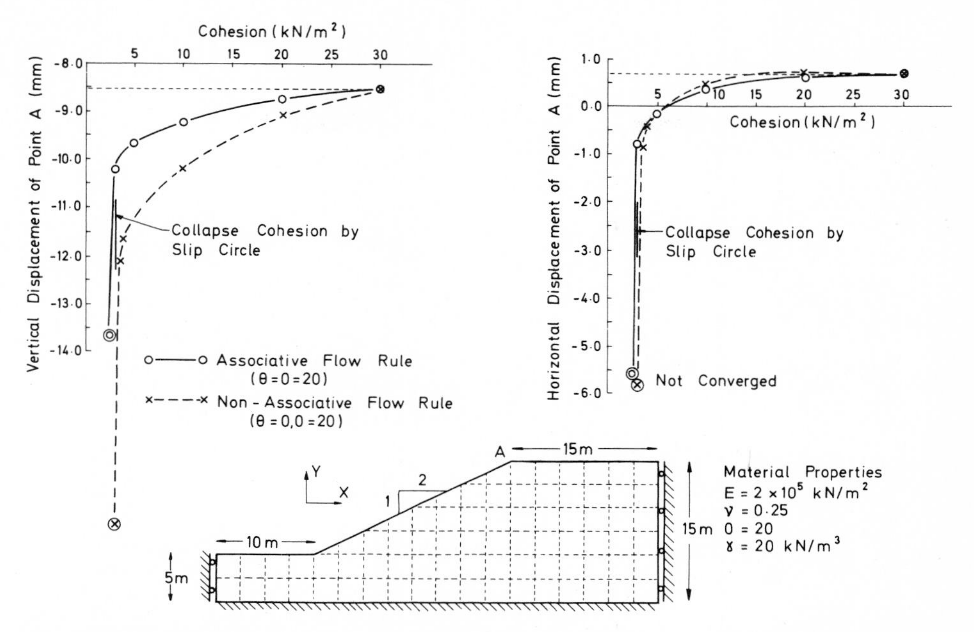

Figure 3-9 Embankment deformations resulting from reduction of cohesion; initial stress results from application of gravity.

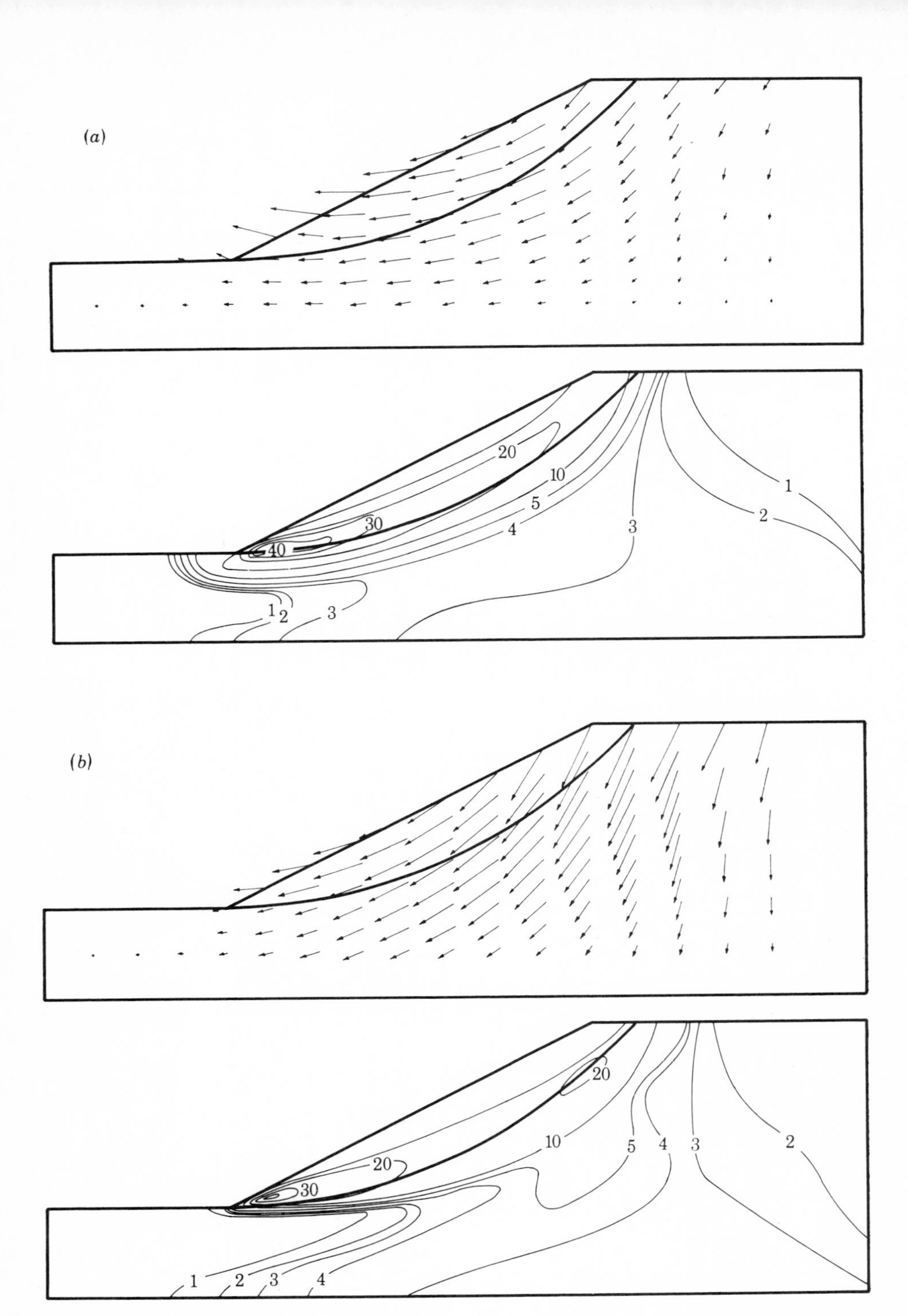

Figure 3-10 Plastic velocity distribution of collapse and contours at maximum shear strain. (a) Associated flow rule; (b) nonassociated flow rule ($\theta = 0$).

Example 3-3 Circular footing The final example investigates the stability of a 5-ft-radius circular footing resting on a weightless soil which is loaded until failure occurs. The geometry and material properties are shown in Fig. 3-11.

Where a Mohr-Coulomb failure surface with an associated flow rule was used, the collapse load was found to be given by $210 < 2f < 212$, whereas when a nonassociated flow rule was used, the collapse load was still the same although the displacements were significantly different. Figure 3-12 shows the load-settlement curves obtained for this problem, and, as would be expected, the nonassociated material has the larger settlements. Figure 3-13 shows the plastic flow patterns at collapse for both associated and nonassociated cases.

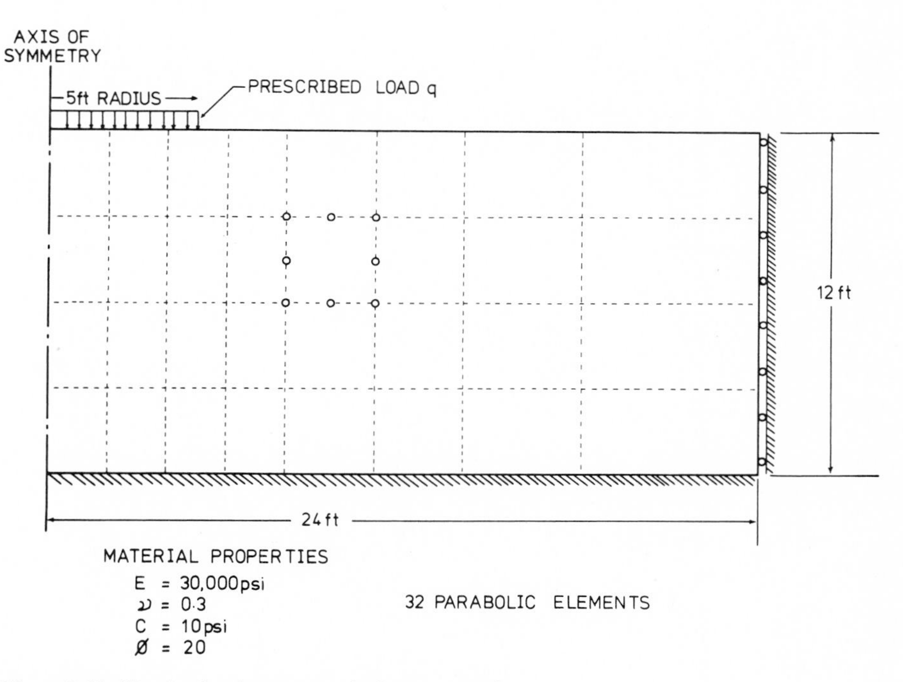

Figure 3-11 Circular footing, geometric data and mesh.

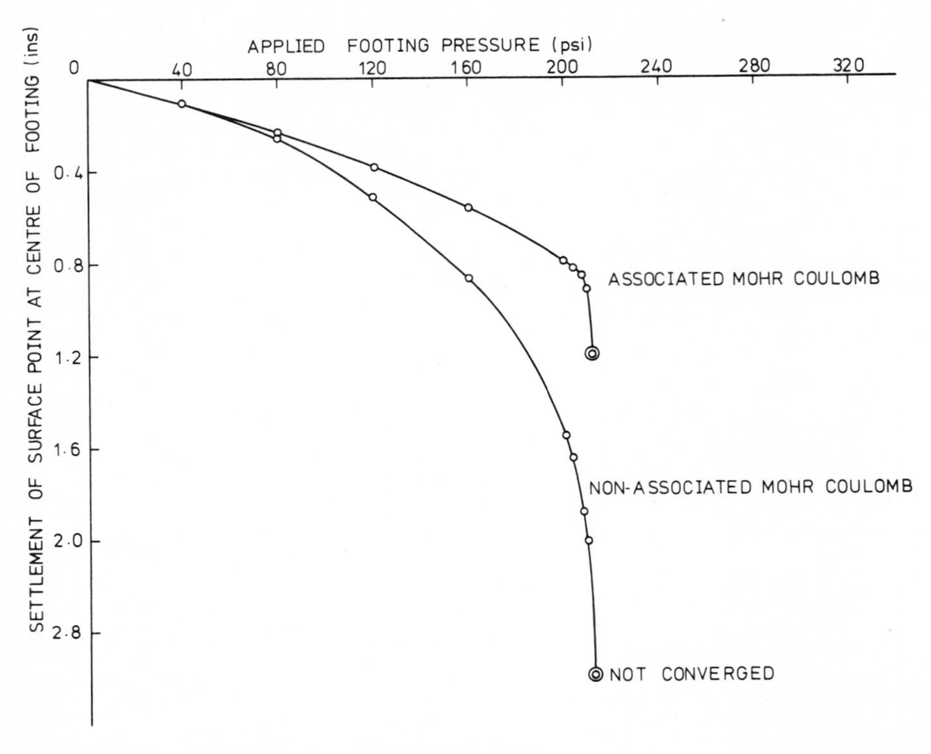

Figure 3-12 Load-displacement curves for circular footing.

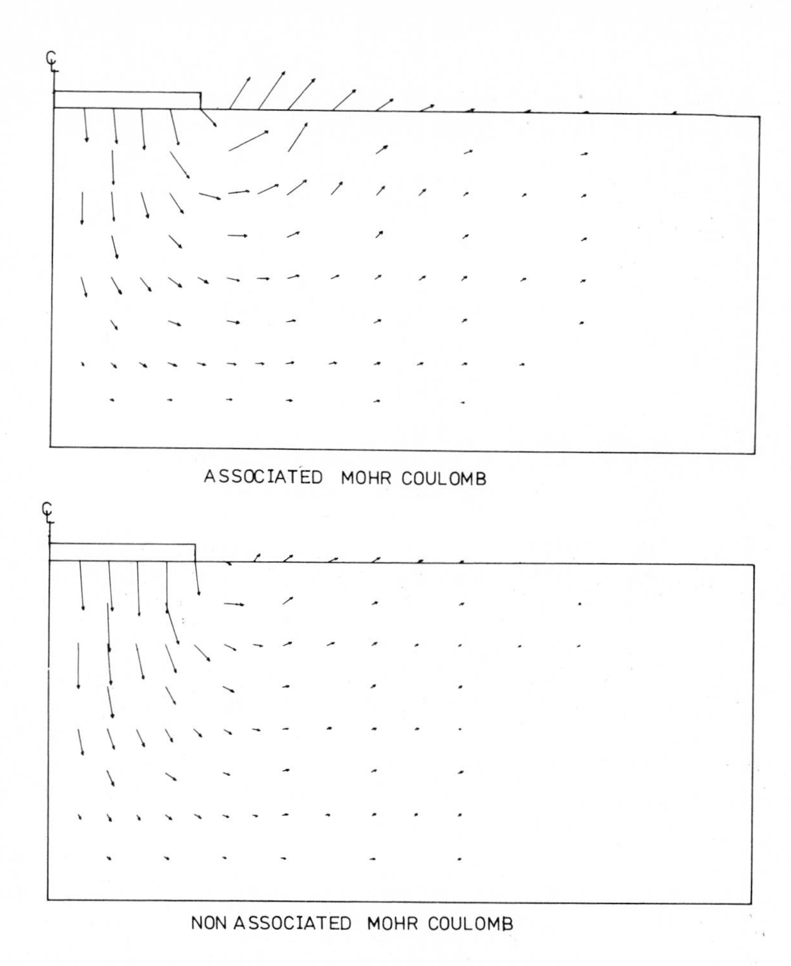

Figure 3-13 Circular footing, relative plastic velocities at collapse.

As stated earlier in this chapter the Mohr-Coulomb envelope can be generalized and expressed in terms of the first two stress invariants only. The validity of doing this can be questioned, as the exact size of the cone to be used is left to the discretion of the order. In order to illustrate this point the above problem was taken and solved using the two extreme cases, namely a cone outside the Mohr-Coulomb surface passing through the external corners and an internal cone which is tangential to the Mohr-Coulomb surface.

The results given by the external cone [Eq. (3-37)], the internal cone [Eq. (3-40)], and Mohr-Coulomb for both associated and nonassociated cases are summarized in Table 3-1.

The magnitude of the difference in the collapse loads can be clearly seen. The Mohr-Coulomb collapse load lies approximately midway between the other two,

Table 3-1

	Associated	Nonassociated
Mohr-Coulomb	212	212
Internal core [Eq. (3-40)]	142	132
External core [Eq. (3-37)]	290	260

Table 3-2

	Associated	Nonassociated
Mohr-Coulomb	152	147
Internal core [Eq. (3-40)]	148	138
External core [Eq. (3-37)]	344	284

and so if this is taken as being the correct collapse load, the others differ by about ± 30 percent.

The Mohr-Coulomb solution does not always lie midway between the two extremes, as can be illustrated by treating the footing as a uniform strip footing of width 10 ft and assuming plane strain conditions. Table 3-2 represents the relative collapse loads for this problem.

In this case the Mohr-Coulomb solution lies close to that given by the internal cone, while the external cone gives a collapse load which is more than 100 percent greater than the Mohr-Coulomb solution. The Mohr-Coulomb failure envelope is generally assumed to have more validity than the others, and the above problems illustrate the errors which can occur if these simplifications are used.

APPENDIX 3A

The stress invariants are expressed in terms of the hydrostatic and deviatoric components. The mean stress is taken as the first invariant and can be written as

$$\sigma_m = \tfrac{1}{3}(\sigma_{xx} + \sigma_{yy} + \sigma_{zz}) \tag{3-79}$$

The second and third invariants are associated with the deviatoric stresses

$$S_x = (\sigma_{xx} - \sigma_m) \qquad S_y = (\sigma_{yy} - \sigma_m) \qquad S_z = (\sigma_{zz} - \sigma_m) \tag{3-80}$$

and are usually written

$$J_2 = \bar{\sigma}^2 = \tfrac{1}{2}(S_x^2 + S_y^2 + S_z^2) + \tau_{xy}^2 + \tau_{yz}^2 + \tau_{zx}^2 \tag{3-81}$$

and

$$J_3 = S_x S_y S_z + 2\tau_{xy}\tau_{yz}\tau_{zx} - S_x\tau_{yz}^2 - S_y\tau_{xz}^2 - S_z\tau_{xy}^2 \tag{3-82}$$

The principal values of the deviatoric stresses are given as the roots of the cubic equation

$$S^3 - J_2 S - J_3 = 0 \tag{3-83}$$

The direct evaluation of the roots is not easy until we observe the similarity of Eq. (3-82) to the trigonometric identity

$$\sin^3 \theta - \tfrac{3}{4} \sin \theta + \tfrac{1}{4} \sin 3\theta = 0 \tag{3-84}$$

If we substitute $s = r \sin \theta$ into Eq. (3-83), we have

$$\sin^3 \theta - \frac{J_2}{r^2} \sin \theta - \frac{J_3}{r^3} = 0 \tag{3-85}$$

On equating with (3-84), we have

$$r = \frac{2}{\sqrt{3}} J_2^{1/2} = \frac{2}{\sqrt{3}} \bar{\sigma} \tag{3-86}$$

and

$$\sin 3\theta = -\frac{4J_3}{r^3} = -\frac{3\sqrt{3}}{2} \frac{J_3}{\bar{\sigma}^3} \tag{3-87}$$

The first root of Eq. (3-87) with θ determined for 3θ in the range $\pm \pi/2$ is a convenient alternative to the third stress invariant J_3. Thus the set

$$\sigma_m, \bar{\sigma}, \; -\frac{\pi}{6} \leqslant \theta = \frac{1}{3} \sin^{-1} \left(-\frac{3\sqrt{3}}{2} \frac{J_3}{\bar{\sigma}^3} \right) \leqslant \frac{\pi}{6} \tag{3-88}$$

is chosen to represent the stress invariants.

APPENDIX 3B

To calculate the viscoplastic strain rate the normal vector to the plastic potential is required. This entails the derivative of the plastic potential with respect to the stresses, as seen from

$$\{\dot{\epsilon}\}^{vp} = \gamma F \frac{\partial Q}{\partial \{\sigma\}} \tag{3-89}$$

Q being the expression for the plastic potential. For a Mohr-Coulomb material the plastic potential is given by

$$Q = \sigma_m \sin \beta + \bar{\sigma} \cos \theta_0 - \frac{\bar{\sigma}}{\sqrt{3}} \cos \theta_0 \cos \beta - c \cos \beta = 0 \tag{3-90}$$

β, the plastic potential angle, is analogous to ϕ in the expression for the Mohr-Coulomb yield surface. Since in all cases Q is a function of the stress invariants only, we can write

$$\begin{aligned} \frac{\partial Q}{\partial \{\sigma\}} &= \frac{\partial Q}{\partial \sigma_m} \frac{\partial \sigma_m}{\partial \{\sigma\}} + \frac{\partial Q}{\partial \bar{\sigma}} \frac{\partial \bar{\sigma}}{\partial \{\sigma\}} + \frac{\partial Q}{\partial J_3} \frac{\partial J_3}{\partial \{\sigma\}} \\ \frac{\partial Q}{\partial \{\sigma\}} &= C_1 \frac{\partial \sigma_m}{\partial \{\sigma\}} + C_2 \frac{\partial \bar{\sigma}}{\partial \{\sigma\}} + C_3 \frac{\partial J_3}{\partial \{\sigma\}} \end{aligned} \tag{3-91}$$

where
$$\frac{\partial \sigma_m}{\partial\{\sigma\}} = \frac{1}{3}[1 \quad 1 \quad 1 \quad 0 \quad 0 \quad 0] \tag{3-92}$$

$$\frac{\partial \bar{\sigma}}{\partial\{\sigma\}} = \frac{1}{2\bar{\sigma}} \begin{Bmatrix} S_x \\ S_y \\ S_z \\ 2\tau_{yz} \\ 2\tau_{xz} \\ 2\tau_{xy} \end{Bmatrix} \tag{3-93}$$

$$\frac{\partial J_3}{\partial\{\sigma\}} = \begin{Bmatrix} S_y S_z - \tau_{yz}^2 \\ S_x S_z - \tau_{xz}^2 \\ S_x S_y - \tau_{xy}^2 \\ 2(\tau_{yx}\tau_{yz} - S_x \tau_{yz}) \\ 2(\tau_{yz}\tau_{yx} - S_y \tau_{xz}) \\ 2(\tau_{zx}\tau_{zy} - S_z \tau_{xy}) \end{Bmatrix} + \frac{1}{3}\bar{\sigma}^2 \begin{Bmatrix} 1 \\ 1 \\ 1 \\ 0 \\ 0 \\ 0 \end{Bmatrix} \tag{3-94}$$

Therefore, for any yield surface it is only necessary to determine the constants C_1, C_2, and C_3 since the other terms are the same for all surfaces.

For the Mohr-Coulomb plastic potential by (a) it can easily be shown that

$$C_1 = \frac{\partial Q}{\partial \sigma_m} = \sin \beta \tag{3-95}$$

$$C_2 = \frac{\partial Q}{\partial \bar{\sigma}} = \cos \theta_0 \left[1 + \tan \theta_0 \tan 3\theta_0 + \frac{\sin 3\theta_0 (\tan 3\theta_0 - \tan \theta_0)}{\sqrt{3}} \right] \tag{3-96}$$

$$C_3 = \frac{\partial Q}{\partial J_3} = \frac{\sqrt{3}(\sin \theta_0 + 1/\sqrt{3} \cos \theta_0 \sin \beta)}{2\bar{\sigma}^2 \cos 3\theta_0} \tag{3-97}$$

Similar expressions for C_1, C_2, and C_3 can be written for any yield surface which is expressed in terms of the three stress invariants σ_m, $\bar{\sigma}$, and θ_0.

REFERENCES

1. Drucker, D. C.: Limit Analysis of Two and Three Dimensional Soil Mechanics Problems, *J. Mech. Phys. Solids*, vol. 1, pp. 217–226, 1953.
2. Rankine, W. J. M.: On the Stability of Loose Earth, *Phil. Trans. R. Soc. Lond.*, vol. 147, 1857.
3. Terzaghi, K.: "Theoretical Soil Mechanics," John Wiley & Sons, Inc., New York, 1943.
4. Boussinesq, J.: "Application des potentiels à l'étude de l'équilibre et du mouvement des solids élastique," Gauthier-Villars, Paris, 1883.
5. Mindlin, R. D.: Force at a Point in the Interior of Semi-infinite Solids, *Physics*, vol. 7, p. 195, 1936.
6. Davis, E. H., and J. R. Booker: The Effect of Increasing Strength with Depth on the Bearing Capacity of Clays, *Geotechnique*, vol. 23, no. 4, pp. 551–563, 1973.
7. Eringen, A. C.: "Nonlinear Theory of Continuous Media," McGraw-Hill Book Company, New York, 1962.
8. Desai, C. S.: Overview, Trends and Projections: Theory and Applications of the Finite Element Method in Geotechnical Engineering, *Proc. Symp. Appl. Finite Element Method Geotech. Eng., Vicksburg, Miss., September 1972.*
9. Desai, C. S., and L. C. Reese: Analysis of Circular Footings on Layered Soils, *J. Soil Mech. Found. Div. ASCE*, vol. 96, no. SM4, July 1970.

10. Duncan, J. M., and C. Y. Chang: Non-linear Analysis of Stress and Strain in Soils, *J. Soil Mech. Found. Div., ASCE*, vol. 96, no. SM5, pp. 1629–1653, 1970.
11. Desai, C. S.: Numerical Design Analysis of Piles in Sands, *J. Geotech. Eng. Div. ASCE*, vol. 100, no. GT6, June 1974.
12. Drucker, D. C., R. E. Gibson, and D. J. Henkel: Soil Mechanics and Work Hardening Theories of Plasticity, *Trans. ASCE*, vol. 122, pp. 338–346, 1957.
13. Roscoe, K. H., and J. B. Burland: On the Generalized Stress Strain Behavior of Wet Clay, pp. 535–609 in J. Heyman and F. A. Leckie (eds.), "Engineering Plasticity," Cambridge University Press, London, 1968.
14. Perzyna, P.: Fundamental Problems in Viscoplasticity, *Adv. Appl. Mech.*, vol. 9, pp. 243–377, 1966.
15. Zienkiewicz, O. C., C. Humpheson, and R. W. Lewis: Associated and Non-associated Visco Plasticity and Plasticity in Soil Mechanics, *Geotechnique*, vol. 25, pp. 671–689, 1975.
15a. Zienkiewicz, O. C., C. Humpheson, and R. W. Lewis: A Unified Approach to Soil Mechanics Including Plasticity and Viscoplasticity, *Proc. Int. Symp. Numer. Methods Soil Rock Mech., Univ. Karlsruhe*, September 1975.
16. Hill, R.: "The Mathematical Theory of Plasticity," Oxford University Press, Oxford, 1950.
17. Drucker, D. C., and W. Prager: Soil Mechanics and Plastic Analysis or Limit Design, *Q. Appl. Math.*, vol. 10, pp. 157–165, 1952.
18. Drucker, D. C.: Coulomb Friction, Plasticity and Limit Loads, *J. Appl. Mech.*, vol. 21, pp. 71–74, 1954.
19. Cox, A. D.: The Use of Non-associated Flow Rules in Soil Plasticity, *R. Am. Res. Dev. Establ. Rep.*, vol. (B)2163, 1963.
20. Zarka, J.: Constitutive Laws of Metals in Plasticity and Visco-plasticity, lecture presented at the Royal Institute of Technology, Stockholm, Sweden,
21. Mohr, O.: "Abhandlungen aus dem Gebiete der technischen Mechanik," 2d ed., W. Ernst, Berlin, 1914.
22. Coulomb, C. A.: Essai sur une application des règles des maximis et minimis à quelques problèmes de statique relatifs à l'architecture, *Mem. Acad. R. Pres. Div. Sav.*, vol. 5, no. 7, Paris, 1776.
23. Kirkpatrick, W. M.: The Condition of Failure in Sands, *Proc. 4th Int. Conf. Soil Mech. Found. Eng. London*, vol. 1, pp. 172–178, 1957.
24. Bishop, A. W.: The Strength of Soils as Engineering Materials, Sixth Rankine Lecture, *Geotechnique*, vol. 16, no. 2, pp. 91–130, 1966.
25. Lode, W.: Versuche über den Einfluss der mittbien Hauptspannug auf das Fliessen der Metalle Eisen, Kupfer und Nickel, *Z. Phys.*, vol. 36, pp. 913–939, 1926.
26. Meek, J. L.: Excavation in Rock: An appreciation of the Finite Element Method of Analysis, pp. 195–214, in Y. Yamada and R. H. Gallagher (eds.), "Theory and Practice in Finite Element Structural Analysis," University of Tokyo Press, Tokyo, 1973.
27. Zienkiewicz, O. C.: "The Finite Element Method in Engineering Science," McGraw-Hill Book Company, New York, 1971.
28. Zienkiewicz, O. C.: Visco-plasticity, Plasticity, Creep and Visco-plastic Flow, *Int. Conf. Comput. Methods Non-linear Mech., Univ. Texas, Austin, 1974.*
29. Cormeau, I. C.: Numerical Stability in Quasi-static Elasto-viscoplasticity, *Int. J. Numer. Methods Eng.*, vol. 9, 1974.
30. Zienkiewicz, O. C., and I. C. Cormeau: Visco-plasticity, Plasticity and Creep in Elastic Solids: A Unified Numerical Solution Approach, *Int. J. Numer. Methods Eng.*, vol. 8, pp. 821–845, 1974.

CHAPTER

FOUR

FINITE ELEMENT ANALYSIS FOR DISCONTINUOUS ROCKS

Richard E. Goodman and Christopher St. John

4-1 DISCONTINUOUS ROCKS

Rocks encountered in civil engineering construction usually break and deform along preexisting planes of weakness—joints, clay partings, minor faults, and other planar structures (Fig. 4-1). Unless the normal stress across these surfaces is greater than about half the unconfined compressive strength of the wall rock, they will present additional degrees of freedom to the rock mass, and calculations of rock behavior must consider their influence. This chapter concerns the methods by which the mechanical properties of a small number of weakness planes can be represented explicitly in a finite element analysis. Other means of ascertaining the mechanical influence of discontinuities are physical models, rigid block equilibrium calculations, and other numerical techniques such as explicit finite difference methods. At the present time, the physical-model method, as described by Krsmanovic,[1] is the most powerful approach for large three-dimensional problems including many discontinuities, particularly where progressive failure is of interest. On the other hand it is only a question of time before the numerical methods will be able to embrace these problems confidently and to realize fully their obvious advantages in relation to parameter studies.

Figure 4-1 A discontinuous rock mass.

Geological work at the surface, in underground exploratory tunnels, and in drill holes can provide reasonably accurate information on the orientation and absolute position of *important discontinuties*, so that it is reasonable to use deterministic methods with respect to them. *Joints*, on the other hand, can be known only statistically; i.e., their preferred orientations and average spacings can be expressed with a measure of confidence. If deterministic methods are used, it makes sense to discern first what the worst combinations of joint planes will be and then to analyze only these cases. There are usually so many joints that one cannot hope to input every one. Study of simple distorted models will thus help choose the conditions for the analysis. In addition to geological and structural information, the input will require description of the mechanical properties of each discontinuity. Field tests, laboratory tests on artificial or natural samples, and detailed geological description of the wall rock and filling material will make it possible for these properties to be assessed.[2] Since joint properties are highly stress-dependent, an estimate of the initial state of stress must supplement the characterization of the shear and normal deformability and strength.

4-2 MECHANICAL PROPERTIES OF JOINTS

We shall consider clean, initially open cracks, in which the mechanical properties depend upon the roughness and strength of the wall rock and the normal stress. Filled, cemented, and incipient discontinuities[3] are significantly different owing to the overriding influence of the filling or cement, but they can also be described with the approach offered here.

Compression

Consider a compression test specimen containing a rough, unfilled fracture parallel to the loaded ends (Fig. 4-2*a*). When a compressive load is applied, the total deformation is the sum of rock compression, which is mainly reversible, and joint closing, which is mainly irreversible. The compression curve (Fig. 4-2*a*) becomes asymptotic to the line representing compression of the rock material and is extremely close to it after a joint closing of V_{mc}. If the joint has an initial compression stress equal to σ_0, the compression curve can be obtained from the test result by translating the axes to positions Δv_0 and σ_0 with reduced maximum closure V_m, as shown in Fig. 4-2*b*. Then one can discuss joint opening as well as closing, as the

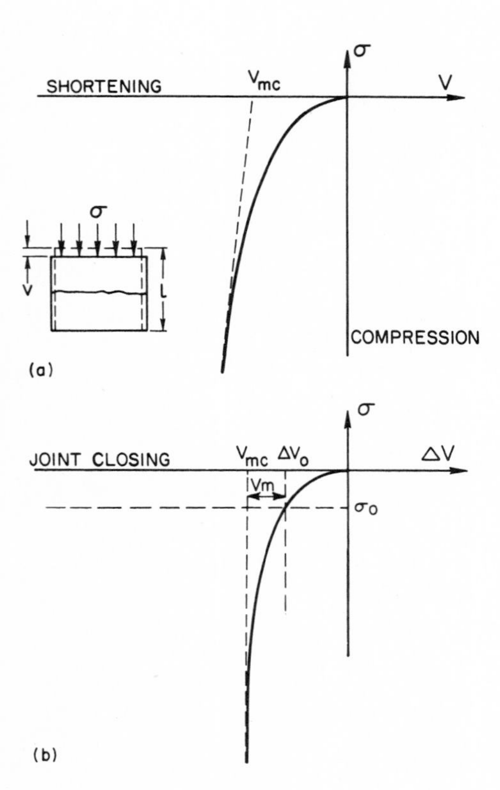

Figure 4-2 Normal deformation of a joint.

point is moved on the curve in decompression or further compression. When the original normal stress has been removed, there is no further resistance to motion. The joint normal stiffness, i.e., the slope of the joint-compression curve (Fig. 4-2*b*), varies continuously from 0 to ∞. To fit this relationship, we can use the hyperbolic expression

$$\frac{\sigma_n - \sigma_0}{\sigma_0} = C\left(\frac{\Delta v}{V_m - \Delta v}\right)^t \tag{4-1}$$

where $\Delta v < V_m$ and $\sigma_n < 0$ (tension positive). V_m depends upon σ_0 and will be calculated from the basic compression property V_{mc}, which will be measured in an experiment conducted with some initial seating load ξ. (V_{mc} depends upon the initial seating load of the specimen.) We do not yet know how the joint-compression curve is changed by shear stress.

Shear

There are three translational and three rotational degrees of freedom in conducting a shear test, and since no standard test has been specified, it must be understood that the shear-deformation constitutive relationship discussed herein may not be universally applicable. We shall presume that all rotations are restricted and that the normal deformation is free to occur continuously during shearing under constant normal stress. (It is convenient to refer to the normal force divided by the total joint area as *normal stress* even though the load may be transferred at as few as three points at low normal pressures.) Figure 4-3 shows the shear deformation $\Delta u(\tau)$ and normal deformation $\Delta v(\tau)$ from such a test. Important

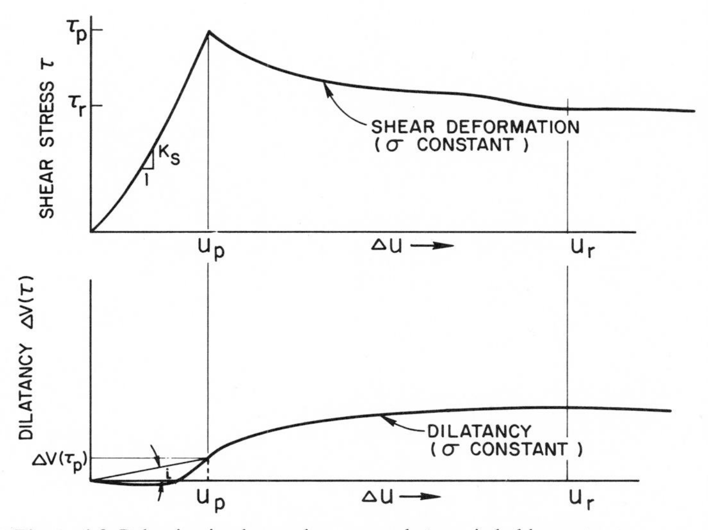

Figure 4-3 Behavior in shear when normal stress is held constant.

parameters are the shear stiffness k_s, the peak strength τ_p, the residual strength τ_r, the peak and residual displacements u_p and u_r, and the dilatancy rate tan i $[= \Delta V(\tau_p)/u_p]$. The dilatancy typically begins after the shear stress has reached a significant level and achieves its maximum rate (point of inflection on the dilatancy curve) when $\tau = \tau_p$; however, large dilatant displacements continue until the residual shear strength τ_r is reached. If dilatancy is prevented, the dilatancy tendency produces an enlarged normal stress and an increase in the peak shear strength (Fig. 4-4); this figure presumes that the shear stiffness k_s and peak displacement u_p increase somewhat with increased normal stress. Figure 4-5 shows such behavior in terms of k_s and σ for Glen Canyon sandstone. Dilatancy is principally dependent upon wall-rock roughness and, in particular, the angles offered by asperities blocking easy shear. Observing that each increment of shear deformation overcomes asperities of progressively longer base length and lower angle, Rengers[4] showed how measurements of roughness angles can be used to construct the constitutive curves. As the normal stress is increased, it becomes progressively more difficult to override asperities and an increasing number of them crush or shear to accommodate deformation, effectively reducing the dilatancy angle i. Barton[5] and Ladanyi and Archambault[6] found the dilatancy to be

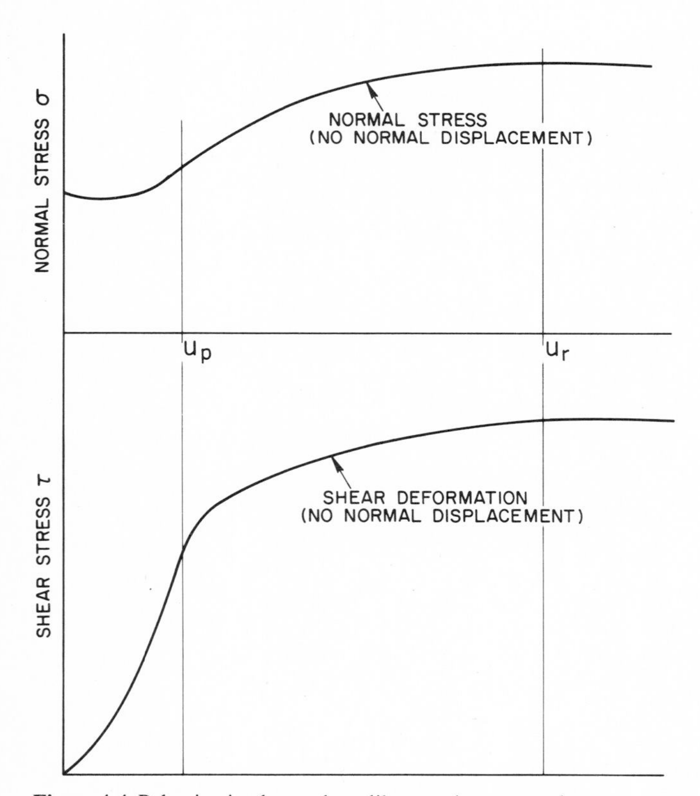

Figure 4-4 Behavior in shear when dilatancy is prevented.

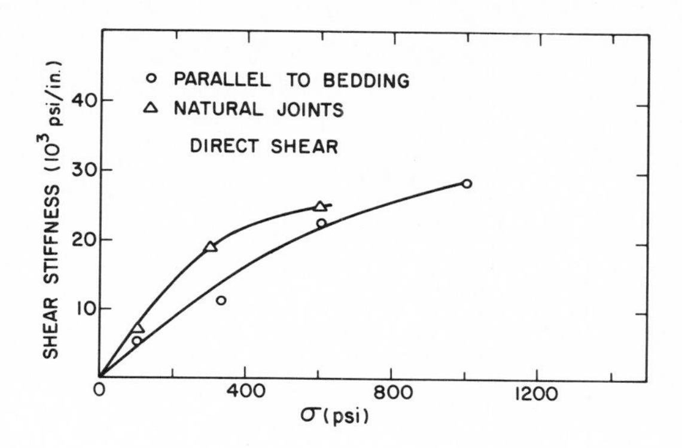

Figure 4-5 Shear stiffness k_s as a function of normal stress σ for Glen Canyon sandstone. (Courtesy of Dr. Howard Pratt, Terratek, Salt Lake City.)

entirely suppressed when the normal stress approached a value near the unconfined compressive strength q_u of the wall rock. Ladanyi and Archambault[6]† showed that dilatancy varies with normal stress according to

$$\tan i = \left(1 - \frac{\sigma_n}{q_u}\right)^4 \tan i_0 \qquad \sigma_n < q_u \tag{4-2a}$$

Similarly, the area of contact a_s varies with normal stress according to

$$a_s = \left(1 - \frac{\sigma_n}{q_u}\right)^{3/2} \qquad \sigma_n < q_u \tag{4-2b}$$

Ladanyi and Archambault's shear-strength equation is

$$\tau_p = \frac{\sigma_n(1 - a_s)(\tan i + \tan \phi_\mu) + a_s S_R}{1 - (1 - a_s) \tan i \tan \phi_\mu} \tag{4-3}$$

where tan i and a_s are given by Eqs. (4-2*a*) and (4-2*b*). The shear strength S_R of the rock composing the asperities varies with σ_n.[7] Barton[8] offered an empirical shear strength criterion accounting for the variation of dilatancy with normal stress and the shear strength of the asperities

$$\tau_p = \sigma_n \tan \left(R \log \frac{q_u}{\sigma_n} + \phi\right) \tag{4-4}$$

The factor R expresses the influence of roughness, varying linearly from 0 to 20 for the range from perfectly smooth to very rough. In Eqs. (4-3) and (4-4) the normal stress is the *effective stress* if the discontinuity contains a fluid under pressure p;

† We have substituted q_u as an estimate for Ladanyi's σ_t in both Eqs. (4-2*a*) and (4-2*b*).

that is, $\sigma_n = \sigma_{tot} - p$. The unconfined compressive strength q_u refers to the rock forming the asperities. Since weathering is often considerably more advanced along joints than through the body of rock, q_u may be considerably lower than values for unweathered rock and it should be obtained from results of Swedish hammer or scratch hardness tests on the wall rock. A comparison of Eqs. (4-3) and (4-4) for rough joints ($i_0 = 5°$, $R = 20$) showed Ladanyi's peak strength to be higher than Barton's in the range of stress: $0.5 < \sigma_n/q_u < 0.7$ and lower outside this range.

The other parameters required to complete the description of the mechanical behavior of the discontinuities are the ratio of residual to peak shear strength and the residual displacement u_r. Rough joints exhibit brittle behavior, with a residual strength as little as one-half or less the peak strength, attained after a displacement of the order of magnitude of 1 cm or less. In some joints, the ratio of residual to peak strength tends toward unity as the normal stress approaches q_u, but a general relationship has not been established.

Behavior of Jointed Masses

Rock masses including numerous joints contain many degrees of translational and rotational freedom. Since each discontinuity is nonlinear, computation of the mass behavior is path-dependent and sensitive to the initial stress state. Sliding on one perfectly parallel set of joints can produce imbrication of crossing joint sets, as sketched in Fig. 4-6*b*. Sliding on other sets then creates interblock cavities (Fig. 4-6*c*), while any rotation produces edge-to-face contacts and beam-type loadings (Fig. 4-6*d*). Under the latter conditions, splitting of some blocks as well as crushing of block corners will result at elevated loads.[9]

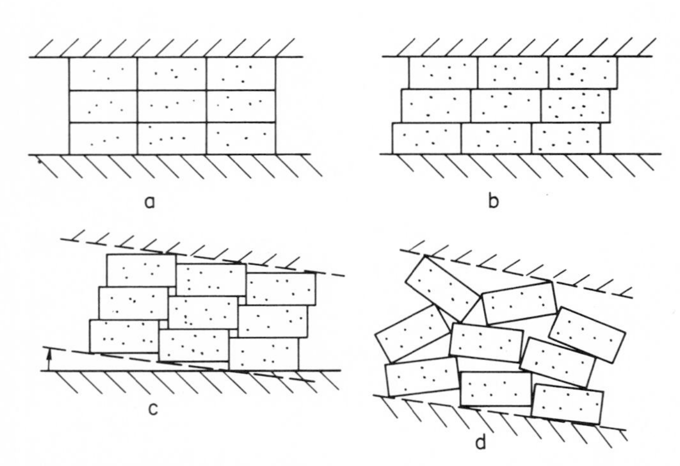

Figure 4-6 Deformation in jointed rock masses: (*a*) initial state; (*b*) shearing along one joint set; (*c*) shearing along two sets; (*d*) shearing and rotations.

4-3 FINITE ELEMENTS FOR JOINTS

Jointed rock masses can be modeled by solid elements linked by special *joint elements* consisting of two lines each with two nodal points (Fig. 4-7*a*). The *strain vector* for a joint element $\{\epsilon_J\}$ is defined by the relative displacements and rotations

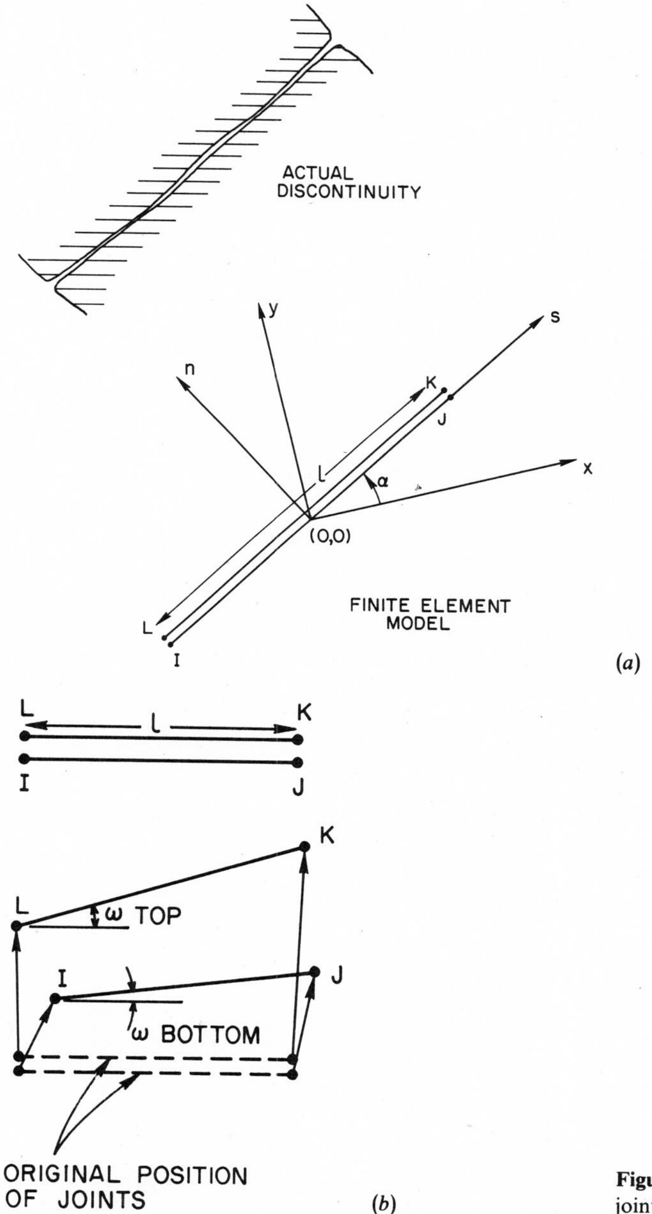

Figure 4-7 (*a*) Joint element; (*b*) joint-element rotation.

of the two walls as measured at the joint center

$$\{\epsilon_J\}^T = [\Delta u_0 \quad \Delta v_0 \quad \Delta\omega]$$

where u_0, v_0, and ω are shear, normal, and rotational "strains," respectively. The "strains" are related to nodal displacements by

$$\begin{Bmatrix} \Delta u_0 \\ \Delta v_0 \\ \\ \Delta\omega \end{Bmatrix} = \begin{bmatrix} -\frac{1}{2} & 0 & -\frac{1}{2} & 0 & \frac{1}{2} & 0 & \frac{1}{2} & 0 \\ 0 & -\frac{1}{2} & 0 & -\frac{1}{2} & 0 & \frac{1}{2} & 0 & \frac{1}{2} \\ \\ 0 & \frac{1}{l} & 0 & -\frac{1}{l} & 0 & \frac{1}{l} & 0 & -\frac{1}{l} \end{bmatrix} \begin{Bmatrix} u_I \\ v_I \\ u_J \\ v_J \\ u_K \\ v_K \\ u_L \\ v_L \end{Bmatrix} \tag{4-5}$$

The local joint stresses are expressed through the stress vector $\{\sigma\}_{sn}^T = [\tau_{sn} \quad \sigma_n \quad M_0]$ and are related to "strains" by

$$\begin{Bmatrix} \tau_{sn} \\ \sigma_n \\ M_0 \end{Bmatrix} = \begin{bmatrix} k_s & 0 & 0 \\ 0 & k_n & 0 \\ 0 & 0 & k_\omega \end{bmatrix} \begin{Bmatrix} \Delta u_0 \\ \Delta v_0 \\ \Delta\omega \end{Bmatrix} \tag{4-6}$$

where τ_{sn} = shear stress
σ_n = normal stress
M_0 = moment about center of joint
k_s = shear term
k_n = normal stiffness term

The value of k_ω can be evaluated by considering the moment and rotation when nodes I and J are fixed and all the force is applied to either node K or L (Fig. 4-8).

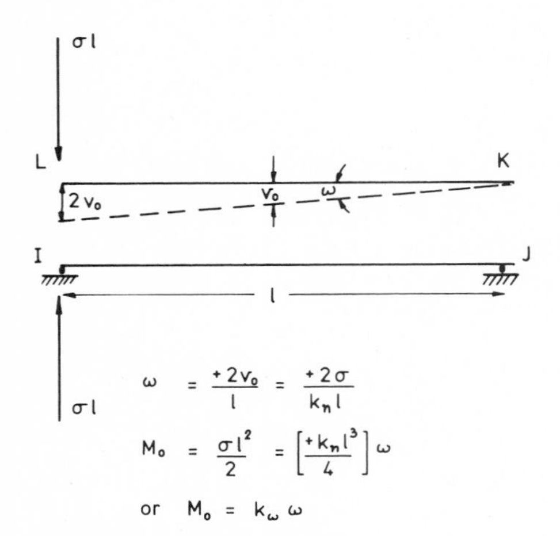

assume:

joint closure at each nodal point pair I, L and J, K is proportional to nodal point force there.

consider the case (*) when all the force is concentrated at nodal points I, L

then:

$$k_\omega = \frac{+k_n l^3}{4}$$

[(*) all other cases yield same result]

Figure 4-8 Rotational stiffness term.

This yields

$$k_\omega = \frac{l^3 k_n}{4} \tag{4-7}$$

If $F_L = -F_I$ and $F_K = -F_J$, nodal-point forces are related to $\{\sigma\}_{sn}$ by

$$\begin{Bmatrix} F_{sI} \\ F_{nI} \\ F_{sJ} \\ F_{nJ} \\ F_{sk} \\ F_{nk} \\ F_{sL} \\ F_{nL} \end{Bmatrix} = \begin{bmatrix} -\frac{l}{2} & 0 & 0 \\ 0 & -\frac{l}{2} & \frac{1}{l} \\ -\frac{l}{2} & 0 & 0 \\ 0 & -\frac{l}{2} & -\frac{1}{l} \\ \frac{l}{2} & 0 & 0 \\ 0 & \frac{l}{2} & \frac{1}{l} \\ \frac{l}{2} & 0 & 0 \\ 0 & \frac{l}{2} & -\frac{1}{l} \end{bmatrix} \begin{Bmatrix} \tau_{sn} \\ \sigma_n \\ M_0 \end{Bmatrix} \tag{4-8}$$

The local joint-element stiffness relates $\{F\}_{sn}$ to $\{u\}$ and is therefore

$$[K_{sn}] = \frac{l}{4} \begin{bmatrix} k_s & 0 & k_s & 0 & -k_s & 0 & -k_s & 0 \\ 0 & 2k_n & 0 & 0 & 0 & 0 & 0 & -2k_n \\ k_s & 0 & k_s & 0 & -k_s & 0 & -k_s & 0 \\ 0 & 0 & 0 & 2k_n & 0 & -2k_n & 0 & 0 \\ -k_s & 0 & -k_s & 0 & k_s & 0 & k_s & 0 \\ 0 & 0 & 0 & -2k_n & 0 & 2k_n & 0 & 0 \\ -k_s & 0 & -k_s & 0 & k_s & 0 & k_s & 0 \\ 0 & -2k_n & 0 & 0 & 0 & 0 & 0 & 2k_n \end{bmatrix} \tag{4-9}$$

Of course this local stiffness matrix must be rotated to find the term-by-term contributions to the structural stiffness matrix with respect to global x, y coordinates.

Another approach to deriving a joint-element stiffness is that proposed by Goodman, Taylor, and Brekke.[10] Rotation is not explicitly considered. A linear variation of displacement along the joint in the wall rock is assumed.

Along the top of the joint

$$\begin{Bmatrix} u_{\text{top}} \\ v_{\text{top}} \end{Bmatrix} = \frac{1}{2} \begin{bmatrix} 1 + \frac{2x}{l} & 0 & 1 - \frac{2x}{l} & 0 \\ 0 & 1 + \frac{2x}{l} & 0 & \frac{1 - 2x}{l} \end{bmatrix} \begin{Bmatrix} u_K \\ v_K \\ u_L \\ v_L \end{Bmatrix} \tag{4-10}$$

with a similar expression for displacements along the bottom of the joint. The shear and normal "strains" are

$$\begin{Bmatrix} \Delta u_0 \\ \Delta v_0 \end{Bmatrix} = \begin{bmatrix} u_{\text{top}} - u_{\text{bottom}} \\ v_{\text{top}} - v_{\text{bottom}} \end{bmatrix} \tag{4-11}$$

$$= \frac{1}{2} \begin{bmatrix} -A & 0 & -B & 0 & B & 0 & A & 0 \\ 0 & -A & 0 & -B & 0 & B & 0 & A \end{bmatrix} \begin{Bmatrix} u_I \\ v_I \\ u_J \\ v_J \\ u_K \\ v_K \\ u_L \\ v_L \end{Bmatrix} \tag{4-12}$$

where $A = 1 - 2x/l$ and $B = 1 + 2x/l$.

If we let k_s and k_n represent the shear and normal stiffness for a linear load step, then

$$\begin{Bmatrix} \tau \\ \sigma_n \end{Bmatrix} = \begin{bmatrix} k_s & 0 \\ 0 & k_n \end{bmatrix} \begin{Bmatrix} \Delta u_0 \\ \Delta v_0 \end{Bmatrix} \tag{4-13}$$

Considering the energy stored in the element then leads to the following local joint-element matrix†

$$k = \frac{l}{6} \begin{bmatrix} 2k_s & 0 & k_s & 0 & -k_s & 0 & -2k_s & 0 \\ & 2k_n & 0 & k_n & 0 & -k_n & 0 & -2k_n \\ & & 2k_s & 0 & -2k_s & 0 & -k_s & 0 \\ & & & 2k_n & 0 & -2k_n & 0 & -k_n \\ & \text{sym} & & & 2k_s & 0 & k_s & 0 \\ & & & & & 2k_n & 0 & k_n \\ & & & & & & 2k_s & 0 \\ & & & & & & & 2k_n \end{bmatrix} \tag{4-14}$$

This result can also be obtained as the limit, when the thickness e goes to zero, of a rectangular joint element with transversely isotropic filling material having Poisson's ratios equal to zero.[11]

† Equation (4-14) can also be derived in a manner similar to Eq. (4-9); we must assume that the distribution of forces along the joint element is proportional to that of displacement.

Since the displacement varies linearly in the 1968 element [Eq. (4-14)], the force will vary linearly across the element. For example, expanding $\{F\} = [K] \times \{U\}$ gives

$$F_{n_L} = k_n l(-\tfrac{1}{3}v_I - \tfrac{1}{6}v_J + \tfrac{1}{6}v_K + \tfrac{1}{3}v_L)$$

$$= \frac{k_n l}{3}(v_L - v_I) + \frac{k_n l}{6}(v_K - v_J) \tag{4-14a}$$

Similarly

$$F_{n_K} = \frac{k_n l}{3}(v_K - v_J) + \frac{k_n l}{6}(v_L - v_I)$$

The acceptability of a given deformation state can be examined separately in the left and right halves of the joint by defining joint "stresses" at the left $(-)$ and right $(+)$ quarter points, e.g.,

$$\sigma_{n(-)} = \frac{\tfrac{3}{4}F_{n_L} + \tfrac{1}{4}F_{n_K}}{l/2} = \frac{3F_{n_L} + F_{n_K}}{2l} \qquad \text{and} \qquad \sigma_{n(+)} = \frac{3F_{n_K} + F_{n_L}}{2l}$$

giving

$$\begin{aligned}\sigma_{n(-)} &= \tfrac{7}{12}k_n(v_L - v_I) + \tfrac{5}{12}k_n(v_K - v_J)\\ \sigma_{n(+)} &= \tfrac{7}{12}k_n(v_K - v_J) + \tfrac{5}{12}k_n(v_L - v_I)\end{aligned} \tag{4-15}$$

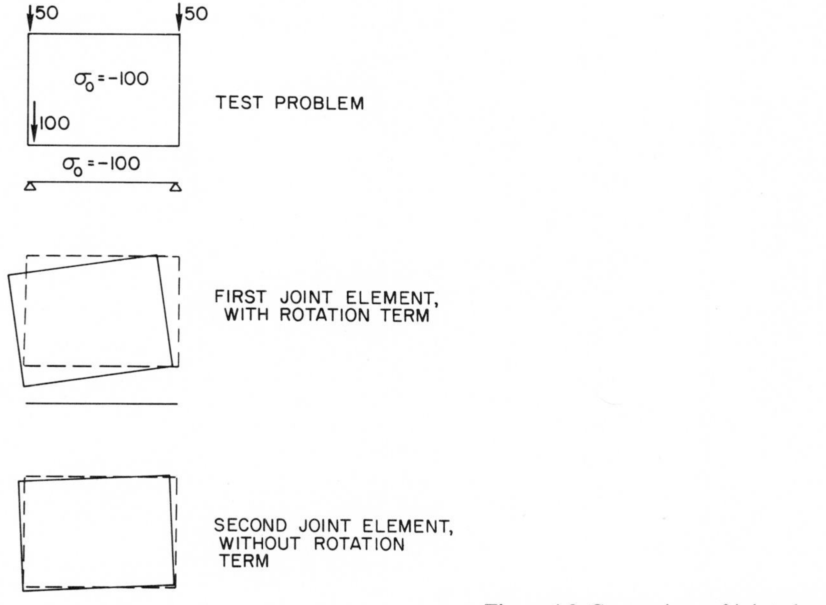

Figure 4-9 Comparison of joint elements.

Another approach is to work exclusively with nodal-point forces. For this purpose, modified elements can be defined (Fig. 4-11) extending from midpoint to midpoint of adjacent elements; this will be discussed later.

The response of the two joint elements, given by Eqs. (4-9) and (4-14), is compared in Fig. 4-9 for a simple test case involving an eccentric normal force. Since the derivation of the term k_ω assumed each nodal-point closure to be proportional to the nodal-point force there, the first joint element rotates with eccentric closure, pivoting about its end. The second joint pivots about an internal point, closing under the concentrated load but opening at the unloaded nodal point.

Joint elements for three-dimensional problems can be handled similarly, as discussed in Chap. 7. They may consist of two planar or curved surfaces with three or more nodal points defining each joint wall.

4-4 ITERATIVE SOLUTION TO SIMULATE REAL PROPERTIES OF JOINTS

The linear equations developed with the displacements of nodal points provide a first solution, but it will usually presume tension in some joints or excessive shear in others; and in any case, shear displacement will produce a dilatancy tendency which has not yet been accounted for. Through an iterative process, one hopes to compute deformations and stresses everywhere consistent with joint mechanics. A displacement output that meets this requirement is termed *acceptable*.

Various iterative processes are illustrated in Fig. 4-10, which considers a direct shear specimen with an initial shear load F_0 being forced by an initially

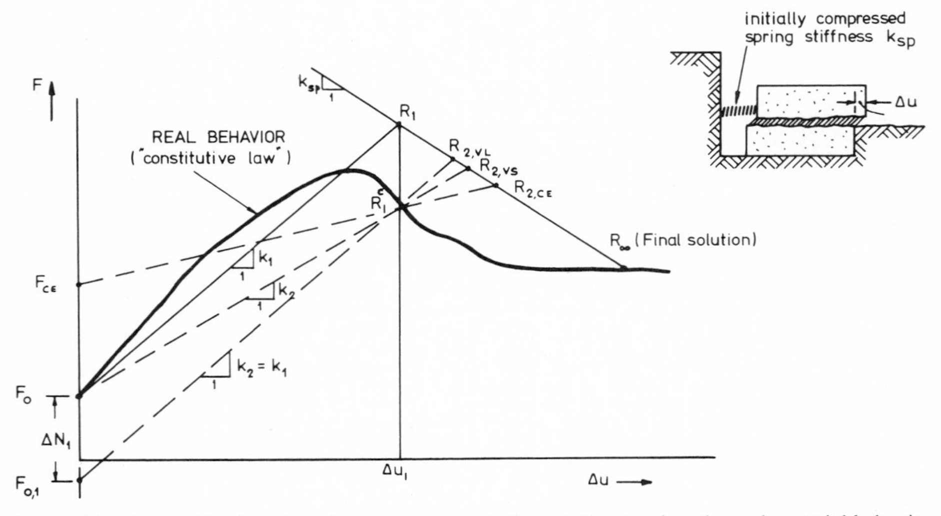

Figure 4-10 Alternative iterative schemes to constrain the solution to obey the real material behavior.

compressed spring. First choose the initial stiffness k_1, representative of the elastic portion of the load-deformation curve for the joint. Because of the initial load in the spring, the spring and joint come to equilibrium at point R_1, which is unacceptable. Restart the problem aiming the solution toward a point on the constitutive curve near R_1; for example, point R_1^C. For the second run, we can choose to redefine the stiffness to k_2 (*variable-stiffness method*) calculated to pass through R_1^C; if we do this, the new solution will be point $R_{2,VS}$, still unacceptable but closer to the correct answer R_∞.

To save recalculating the stiffness matrix, Zienkiewicz et al.[12] suggested restarting with the same stiffness ($k_2 = k_1$) but a new initial load $F_{0,1}$ (the *load-transfer method*); in this example $F_{0,1}$ is simply $F_0 + \Delta N_1$, where $\Delta N_1 = R_1^C - R_1$. This yields as a solution point $R_{2,VL}$. The load-transfer method generally requires more iterations than the variable-stiffness method to reach convergence but requires fewer calculations per iteration. It is the method used in the computer formulation discussed here. Other modification paths are possible and may be preferable in certain instances. For example,[13] if the stiffness matrix is to be altered at each run anyway, it will be only slightly more expensive to modify the load vector as well; then one can choose a path like that in Fig. 4-10 from F_{CE} with a slope such that the area under $F_{CE}R_1^C$ is the same as the area under the actual load-deformation curve up to R_1^C. Such an approach will usually speed convergence; however, it will not converge in all cases, as shown by Dubois.[14] All these methods can be viewed as modifications of the Newton-Raphson scheme for nonlinear functions (see Ref. 15) in which the stiffness is updated to the value of the slope of the constitutive surface in the neighborhood of the current solution.

The stiffness matrix, representing the coupling from one nodal point to another, was constructed from elements filling the space *between* the nodal points.

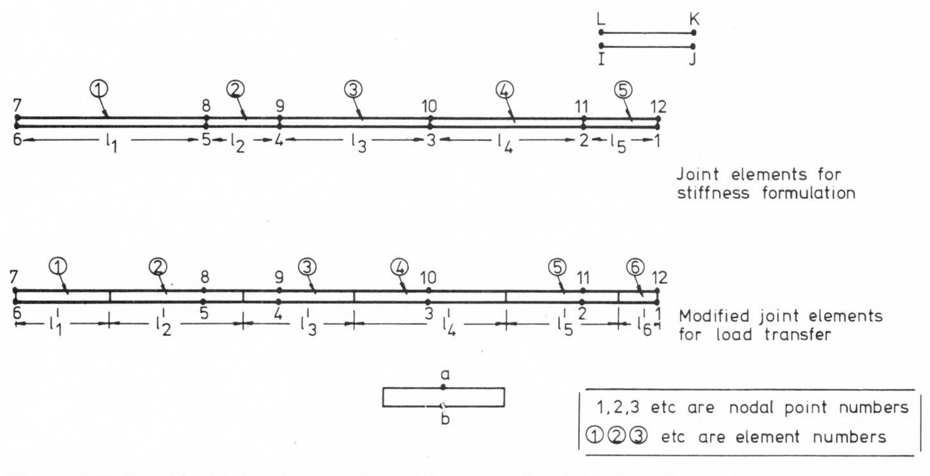

Figure 4-11 Interlaced joint elements formed by a set of nodal points along a joint.

The "solution" gives the forces and displacements *at* nodal points, and it is here that modifications to the load vector must be introduced. This can be done by separating each joint into left and right halves. Each nodal-point pair *a, b* of Fig. 4-11 has an associated length l' representing the sum of half-lengths of joint elements to the left and the right. Nodal-point forces at *a* and *b* are the sum of forces from these joint elements, and we can evaluate nodal-point "stresses" σ and τ at *a* and *b* by the quantities F_n/l' and F_s/l'. In other words, stiffness and stresses at a nodal point are the weighted averages of the stiffnesses and stresses in the joint elements which it occupies. While the stiffness formulation takes place between nodal points, the load transfer and load release occur at nodal points. More than one *release point* per joint can be formulated.

Joint Opening and Closing

Taking $C = 1$ and $t = 1$ in Eq. (4-1) gives

$$F_n = \left(\frac{\Delta v}{V_m - \Delta v} + 1\right) F_{n,0} \tag{4-16}$$

where $(F_n)_0$ is the initial internal force at a nodal point and Δv is the difference of normal displacements between a nodal-point pair caused by an increment of normal force $F_n - (F_n)_0$.

The fundamental joint property governing normal deformation is the maximum closure V_{mc} (defined as a positive quantity). It represents the maximum amount a joint can close starting from an initial seating load ξ. The maximum a joint can close if it has initial compressive stress σ_0 is (Fig. 4-12)

$$V_m = \frac{-V_{mc}}{\sigma_0} \xi \tag{4-17}$$

The computation is initiated with a unit normal stiffness (force/length^3)† derived by differentiating Eq. (4-16)

$$(k_n)_0 = \frac{\sigma_0}{V_m} = \frac{-\sigma_0^2}{\xi V_{mc}} \tag{4-18}$$

This value of normal stiffness produces a solution with displacements v_a and v_b at nodal points *a* and *b* along a modified joint element.‡ In general there will also be a contribution to displacements at *a* and *b* due to dilatancy $\Delta v(\tau)$. Dilatancy will be discussed later. From the initial stress σ_0, reference pressure ξ, and maximum closure V_{mc} of the two joints abutting nodal pair *a, b* we find the normal stiffness k_n'

† Assuming unit thickness ξ and σ_0 are negative, since they are compressive. V_m is a negative displacement. $(k_n)_0$ is always positive.

‡ In the modified element, *a* is *K* or *L* of a joint element while *b* is *I* or *J* of a joint element (Fig. 4-11).

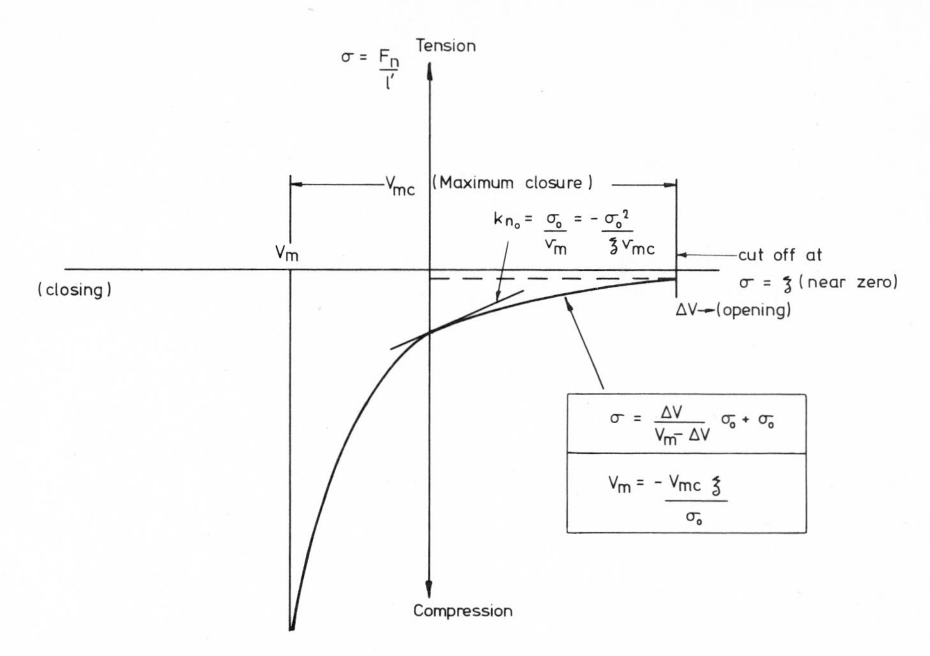

Figure 4-12 Relationship between normal force and normal displacement at nodal point.

(stress/length) using Eq. (4-18). If both joints have the same initial stress and maximum closure,

$$k_n' = \frac{\sigma_0}{V_m} = \frac{F_{n_i}}{V_m l'} \tag{4-19}$$

where l' is the length of the modified element and V_m is given by Eq. (4-17). Then the external force at node a of the pair a, b is

$$F_{n_1} = l'k_n' \,\Delta v_i + (F_{n,\,0})_i \tag{4-20}$$

Thus, as in Fig. 4-10, the first iteration produces point $R_1 = \{(F_n)_1 \quad \Delta v_1\}$. Only in a rare instance will R_1 be precisely on the constitutive curve [Eq. (4-16)].

Joint opening We first consider cases where Δv_1 is positive. In Fig. 4-13 Δv_1 is positive and R_1 is above the curve. A point on the constitutive curve with $\Delta v = \Delta v_1$ defines point R_1^C (compare with Fig. 4-10), and the distance $R_1^C - R_1$ determines the initial load for the second iteration $(F_n)_{0,\,2}$. For the $(i + 1)$st iteration

$$F_{n_i} = l'k_n' \,\Delta v_i + (F_{n,\,0})_i \qquad \text{and} \qquad (F_{n,\,0})_{i+1} = (F_{n,\,0})_i + \Delta N_i \tag{4-21}$$

where $$\Delta N_i = \left(\frac{\Delta v_i}{V_m - \Delta v_i}\right) F_{n,\,0} - F_{n_i} \qquad \text{and} \qquad \Delta v_i = v_a - v_b \tag{4-21a}$$

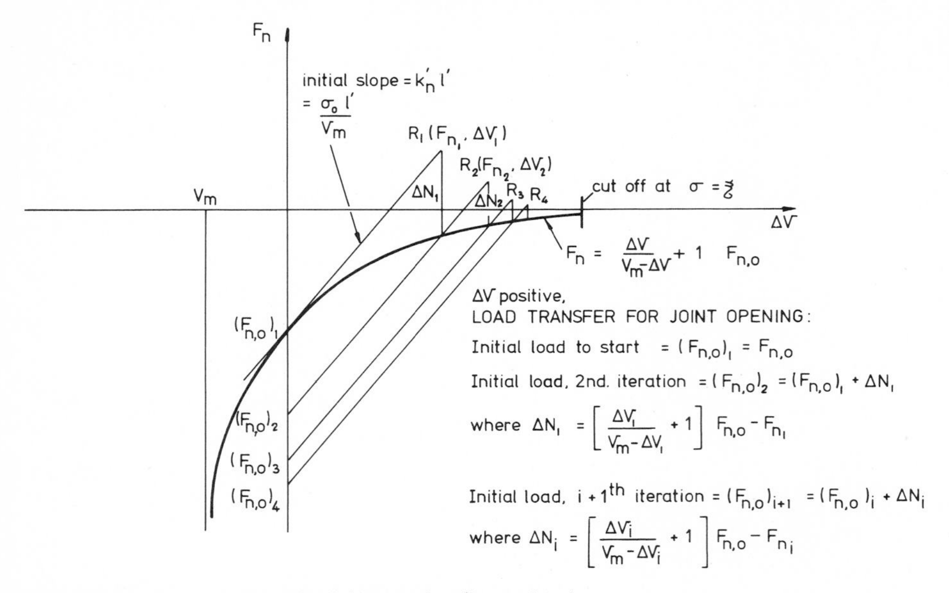

Figure 4-13 Iterative process for joint opening (Δv positive).

The process must be repeated until ΔN_i is smaller than a satisfactorily small number ϵ in all elements and other nonlinear constraints (to be discussed subsequently) have also been satisfied.

Combining Eqs. (4-21*a*) and introducing (4-17) gives a load-transfer formula for joint opening:

$$(F_{n,0})_{i+1} = \left[\frac{\Delta v_i V_m}{V_m - \Delta v_i} + (V_m - \Delta v_i)\right] k'_n l' \tag{4-22}$$

The iterations are the same whether R_i (Fig. 4-13) is above or below the constitutive curve; in the latter case, application of (4-22) will produce a positive value of N_i. Occasionally, iteration will yield an oscillating convergent series of $[\Delta N_i]$ values. Convergence can be speeded by applying numerical methods to the series, but this will not be considered here.

Joint closing This is handled similarly, except that the point R_1^C must be guided by the force rather than the displacement computed in the previous iteration (Fig. 4-14). For this purpose we rewrite the constitutive law[16] as

$$\Delta v = \frac{V_m(F_n - F_{n,0})}{F_n} \tag{4-23}$$

The results of the first iteration now show Δv_1 negative, with point $R_1 = \Delta v_1$, F_{n_1}. The initial load for the $(i + 1)$st iteration is

$$(F_{n,0})_{i+1} = k'_n l' \left[2\Delta v_i - V_m + \frac{V_m^2}{\Delta v_i + (F_{n,0})_i/(k'_n l')}\right] \tag{4-24}$$

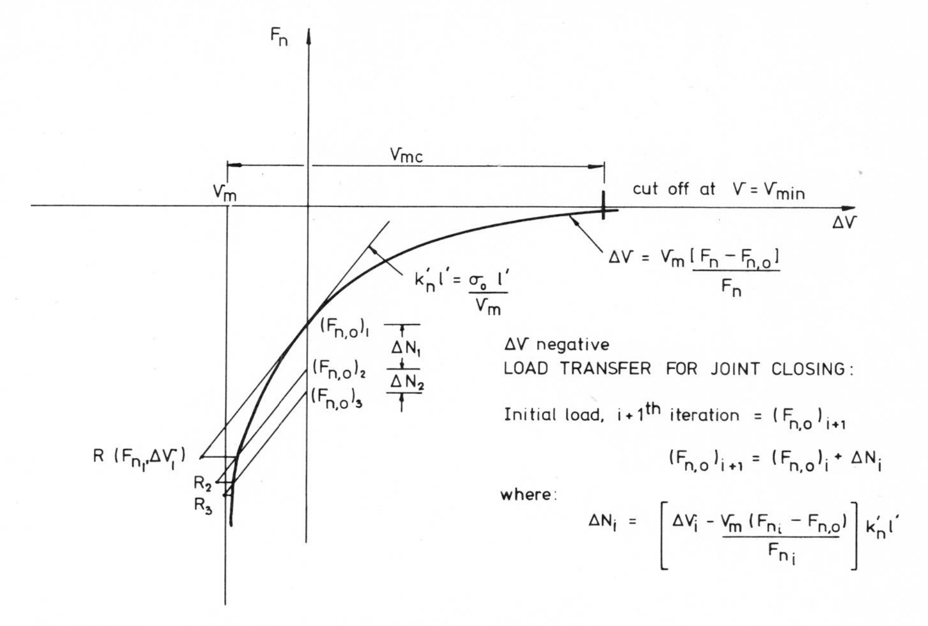

Figure 4-14 Iterative process for joint closing (Δv negative).

As in the case of joint opening, iterations are required whether R_i is to the left or the right of the constitutive curve; in the case depicted in Fig. 4-14, R_i is to the left resulting in ΔN_i negative, whereas if R_i is to the right, ΔN_i will be positive.

Joint shearing This can be treated as joint opening; the limiting shear-stress criterion to be imposed on the shear deformation behavior is analogous to the no tension criterion imposed on the normal stress-opening curve. Since we lack a universal model describing the shear deformation–shear stress behavior of joints, a simple constitutive law will be assumed (Fig. 4-15). In the initial region, the limiting stress τ_p depends upon σ according to a criterion of peak shear strength

$$\tau_p = f_1(\sigma) \tag{4-25}$$

If τ_p is exceeded, the strength falls, attaining a residual value τ_r when the displacement u_r has been attained

$$\tau_r = f_2(\sigma) \tag{4-26}$$

Unfortunately, we do not yet know enough about the variation of peak and residual displacements (u_p or u_r) with normal stress σ; therefore a simple linear model has been used

$$\begin{aligned} u_{p(+)} &= u_p = \frac{\tau_p - \tau_0}{k_s} \\ u_{r(+)} &= u_r = \frac{M\tau_p - \tau_0}{k_s} \qquad M > 1 \end{aligned} \tag{4-27}$$

A value of $M = 4$ has been inserted in our program.

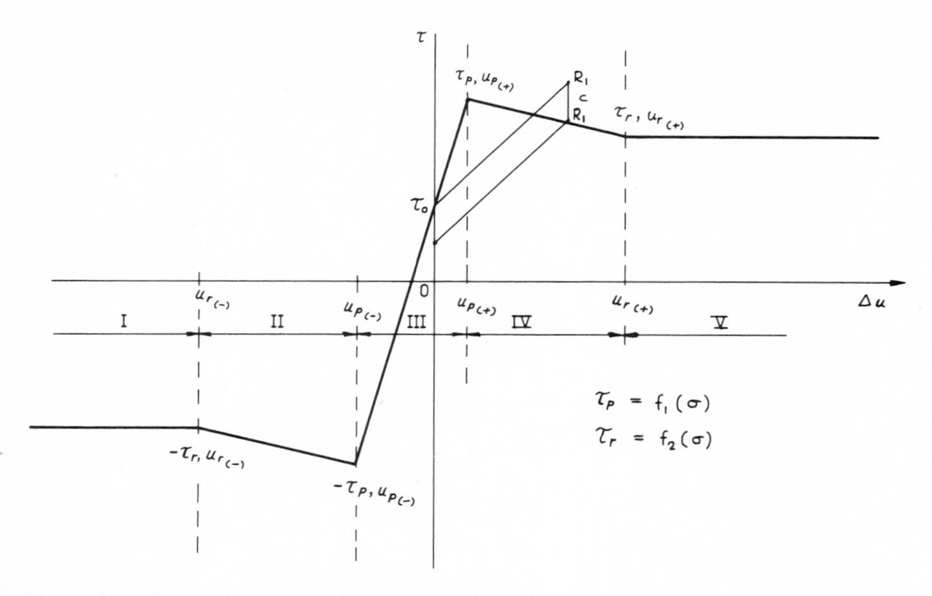

Figure 4-15 Constitutive law for shear-deformation behavior.

As before, the load transfer is in terms of forces in the modified joint elements. The computation begins with a stiffness k_s in each joint element, yielding $\Delta u_1 = u_a - u_b$ as the difference between the shear displacements of nodal points a and b along a modified joint element (Fig. 4-11). The corresponding shear force at nodal point a is

$$F_{s_1} = l'k_s' \,\Delta u_1 + (F_{s,0})_1 \tag{4-28}$$

where k_s' is the stiffness averaged from the joint elements sharing the nodal-point pair and $(F_{s,0})_1 = \tau_0 l'$ [compare with Eq. (4-19)]. Similarly, to begin the $(i + 1)$st iteration,

$$F_{s_1} = l'k_s' \,\Delta u_i + (F_{s,0})_{i+1} \qquad \text{and} \qquad (F_{s,0})_{i+1} = (F_{s,0})_i + \Delta S_i \tag{4-29}$$

with

$$\Delta S_i = \tau_i l' - F_{s_i}$$

The updated shear strength τ_i is obtained from formulation of Fig. 4-15. Combining Eqs. (4-29) gives

$$(F_{s,0})_{i+1} = \tau_i l' - k_s' l' \,\Delta u_i \tag{4-30}$$

where τ_i is given by (4-28), which in turn depends upon the choice of f_1 and f_2 [Eqs. (4-25) and (4-26)]. Any consistent specific experimental or empirical results can be used to define f_1 and f_2. We have input Ladanyi and Archambault's equation (4-3) and Barton's equation (4-4) as alternatives.

Unfortunately we know little about the variation of residual shear strength τ_r with σ_n. We can input f_2 in a consistent manner as follows. At high normal stresses, we suppose that the rock becomes plastic, i.e., exhibits a ratio $\tau_r/\tau_p = 1$.

We therefore presume that the ratio τ_r/τ_p increases from B_0 ($0 \leqslant B_0 \leqslant 1$) at $\sigma_n = 0$† to 1 at $\sigma_n = q_u$, the unconfined compressive strength

$$\begin{aligned} \tau_r &= \tau_p\left(B_0 + \frac{1 - B_0}{q_u}\sigma\right) \qquad && \sigma \leqslant q_u \\ \tau_r &= \tau_p && \sigma \geqslant q_u \end{aligned} \tag{4-31}$$

Typical values of B_0 are given by Krsmanovic[16] and Goodman.[3] B_0 is left as an input parameter of our computer program because it permits one to contrast the behavior of brittle and plastic joints.

Dilatancy Dilatancy must also be introduced into the analysis. Dilatancy describes the normal displacement (joint thickening) $\partial \Delta v(\tau)/\partial u(\tau)$ caused by shear. As shown in Fig. 4-3, it begins near peak load; when $\tau = \tau_p$, $\Delta v(\tau_p)/u_p = \dot{V} = \tan i$. That is, $\dot{V}$ is the secant dilatancy rate. Then as an approximation (Fig. 4-16) normal displacement caused by dilatancy at a shear displacement Δu is

$$\begin{aligned} \Delta v_i(\tau) &= -\tan i\left(|\Delta u| + \left|\frac{\tau_0}{k_s}\right|\right) \qquad && u_{r(-)} \leqslant \Delta u \leqslant u_{r(+)} \\ \Delta v_i(\tau) &= -\tan i\left(u_r + \left|\frac{\tau_0}{k_s}\right|\right) && \Delta u \geqslant u_{r(+)} \text{ or } \Delta u \leqslant u_{r(-)} \end{aligned} \tag{4-32}$$

The absolute value function and the minus sign ensure that the joint thickens regardless of the sign of the shear stress, as discussed by Goodman and Dubois;[13] this can be termed *doubly dilatant behavior*.

† This value will have to be obtained in general by extrapolating data to zero.

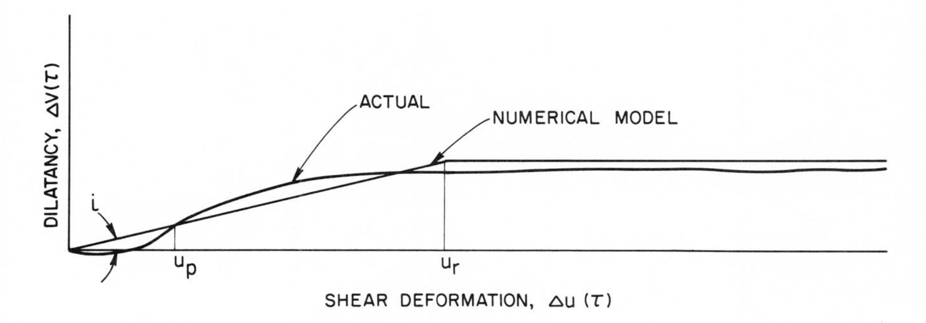

Figure 4-16 Simulation of dilatancy.

The variation of tan i with σ was given by Eq. (4-2a). Substituting Eq. (4-2a) in Eq. (4-32) with $\sigma_T = q_u$† and considering stresses in the modified joint element gives for the ith iteration

$$\Delta v_i(\tau) = \left[\frac{(F_n)_{i-1}}{l' q_u} - 1\right]^4 \tan i_0 \left[\Delta u_i + \frac{(F_s)_0}{K_s' l'}\right]$$

$$\text{for } u_{r(-)} \leqslant \Delta u \leqslant u_{r(+)} \tag{4-33}$$

and

$$\Delta v_i(\tau) = \left[\frac{(F_n)_{i-1}}{l' q_u} - 1\right]^4 \tan i_0 \left[u_r + \frac{(F_s)_0}{K_s' l'}\right]$$

$$\text{for } \Delta u \geqslant u_{r(+)} \qquad \text{or } \Delta u \leqslant u_{r(-)}$$

At the end of the ith iteration, we shall know the shear displacement Δu_i at each nodal-point pair. Inserting Δu_i in Eq. (4-33) for each modified joint element, we determine $\Delta v_i(\tau)$.

If all dilatancy is prevented by the adjacent elements, there must be external compressive forces applied to joint nodes a and b equal in magnitude to the dilatancy multiplied by the joint stiffness. When the joint stiffness is redefined as the slope of the compression curve [Eq. (4-16)] evaluated at the previous normal stress $(F_n)_{i-1}/l'$, the increment in initial normal stress in the joint due to dilatancy calculated for the ith iteration is

$$(\Delta\sigma_n)_{0,\,i} = \frac{\Delta v_i(\tau)}{\xi V_{mc}} \left[\frac{(F_n)_{i-1}}{l'}\right]^2 \tag{4-34}$$

To start the $(i + 1)$st iteration, the initial stress, incremented by $(\Delta\sigma_n)_{0,\,i}$, produces external forces on the neighboring elements, which may in fact deform, thereby automatically relaxing the initial assumption that dilatancy is prevented. For the first iteration we use $(F_n)_{i-1} = (F_n)_0$.

Updating the loads The final step in the load-transfer procedure is to rotate $(F_{n,\,0})_{i+1}$ and $(F_{s,\,0})_{i+1}$ to global coordinates and update the load vector. With the sign convention used in the modified joint elements, $(F_n)_i$ and $(F_s)_i$ are external forces at nodal point a (Fig. 4-11).

Table 4-1 summarizes the steps in the iteration scheme, and Table 4-2 summarizes the material-property descriptions that must be input. Table 4-3 reviews material-property relationships that have been assumed.

† As noted previously, q_u is a negative quantity here.

Table 4-1 Summary of steps in finite element analysis of jointed rock masses by the load-transfer method

$i = 1$

1. Read input: geometric and material properties; initial stresses; accelerations; pore pressures; and support loads
2. Form solid-element stiffness matrix for each type, orientation, and shape of solid element
3. Form joint-element stiffness matrix for each type, length, and orientation of joint element [Eq. (4-9) or (4-14)]
4. Assemble structural stiffness matrix $[K]$; actually this step is performed simultaneously with steps 2 and 3.
5. Assemble residual stress contributions to the load vector (also done simultaneously with steps 2 and 3); change sign and store in load vector
6. Add external forces from other sources: water pressures; weight; active and passive supports; total load vector = $\{F\}_i$
7. Invert the structural stiffness matrix $[K]$; for small computers this may be done outside the rest of the program as it need only be done once; store $[K]^{-1}$ (in large problems, banded solution schemes will be used)
8. Determine displacements by matrix multiplication:

$$\{u\} = [K]^{-1}\{F\}_i$$

9. Form modified joint-element stiffness and initial forces [Eqs. (4-19) and (4-20)] and relative displacements Δv_i and Δu_i [Eqs. (4-17) and (4-33)]
10. Determine normal force $(F_n)_i$ [Eq. (4-20)] and shear force $(F_s)_i$ [Eq. (4-30)] in each modified joint element
11. Find ΔN and ΔS in each element [Eqs. (4-21*a*) or (4-24) and (4-30)]; if $\sum \Delta N + \sum \Delta S > \epsilon$, where ϵ is some small number, update $(F_n)_0$ and $(F_s)_0$ [Eqs. (4-24) and (4-31)]; rotate to global coordinates and update $[F]$
12. $i = i + 1$; go to step 8

Table 4-2 Summary of material properties to be input

Rock†:

1. Deformability terms E_s, E_n, v_{sn}, G_{sn}, and orientation of bedding or banding, where E is Young's modulus, v is Poisson's ratio, and G is a shear modulus

Wall rock of discontinuities (Joints):

2. Compressive strength q_u (negative)
3. Ratio of tensile to compressive strength n if Ladanyi's formula is used

Discontinuities (Joints):

4. Shear stiffness k_s (may be a function of size)
5. Maximum amount V_{mc} that a joint can close from an initial seating stress ξ
6. Friction angle of a smooth joint ϕ_μ
7. Dilatancy angle i_0 giving correct peak dilatant displacement $v(\tau)$ in a test at zero confining pressure σ: $i_0 = \tan^{-1}[v(\tau_p)/u_p]$; here i_0 is a function of size and therefore must be input according to the scale of the mesh

† For constructing the load vector, other properties may be required, e.g., unit weight.

Table 4-3 Summary of material-property relationships assumed for joints†

Wall rock and discontinuities (Joints):

1. Peak shear strength:
   The asperities of joints obey Fairhurst's failure criterion consisting of a parabola fitted to the unconfined compression and tensile strengths[17]

and

Peak shear strength τ_p given by Ladanyi and Archambault's equation[6] with transition pressure $\sigma_T = q_u$ of wall rock

and

Dilatancy $\dot{V}$ and joint area of contact a_s vary with normal pressure, as given by Ladanyi and Archambault, with $\sigma_T = q_u$

or

Shear strength given by Barton's empirical equation

2. Residual shear strength $\tau_r = B(\sigma)\tau_p$, where $B(\sigma)$ decreases linearly from B_0 at $\sigma = 0$ to 1 at $\sigma = q_u$
3. The ratio of peak to residual displacements M, measured in a test beginning from $\tau = 0$, equals 4
4. Normal displacement obeys a parabolic law equation (4-16) with initial stiffness k_n determined by initial stress
5. Dilatancy stops when the shear deformation reaches the residual value u_r [Eqs. (4-32) and (4-33)]

† Most of these assumptions can readily be replaced by alternatives without major alterations.

4-5 ALTERNATIVE METHODS FOR INTRODUCING JOINTS IN ROCK COMPUTATIONS

Other approaches have been used to introduce joint mechanics into computations of rock-mass behavior. The methods outlined here may be termed explicit since the joint properties are actually input as distinct numbers in the structural stiffness matrix. Alternatively, the joints can be taken into account implicitly, e.g., by recognizing changed properties for the rock stiffness without actually drawing joint elements, or joints may be considered implicitly in the criteria for acceptability of the output at any stage, e.g., as done in the no tension analysis of Zienkiewicz, Valliappan, and King[12] or the ubiquitous joint described by Goodman and Duncan.[18] The trouble with these methods is that the nonlinear behavior of each individual joint and the kinematic constraints on the blocks imposed by the system of discontinuities cannot readily be duplicated. The latter is a severe limitation. Another method for modeling jointed rock masses, used by Cundall,[19] Burman,[20] and Byrne,[11] is to reduce the degrees of freedom of the structural assemblage by considering the rock blocks to be rigid bodies. The force displacement equations referred to nodal points are then transformed to refer to the centroids of rigid blocks. Cundall integrated the progressive motion of the block centroids through a dynamic relaxation procedure. Time is implicit when the nonlinear constraints of the joints are approximated by a load-transfer or variable-stiffness solution procedure.

Relative Displacement Coding

Ghaboussi, Wilson, and Isenberg[21] introduced a coding technique where only one of the walls of a joint is recognized explicitly as a line of nodal points in the mesh; the opposite joint wall is considered to belong to one of the contiguous blocks as an internal feature. Its effect is incorporated in the block stiffness through an expansion of the block's degrees of freedom.

To do this, first expand the 8 × 8 stiffness matrix $[K]$ of the quadrilateral rock block $IJKL$ to a 12 × 12 matrix $[K']$, as follows:

$$[K'] = [a]^T[K][a] \qquad \text{where} \qquad [a] = \begin{bmatrix} I & 0 & I \\ 0 & I & 0 \end{bmatrix}$$

in which $[I]$ is a rank 4 identity matrix. If side IJ of the block contacts side AB of a contiguous block across joint element $ABJI$, the stiffness terms of the upper left quarter of the matrix given in Eq. (4-9) or (4-14) must be added to rows and columns 9 through 12 of $[K']$. Since $u_I = u_A + \Delta u_A$, $v_I = v_A + \Delta v_A$, $u_J = u_B + \Delta u_B$, and $v_J = v_B + \Delta v_B$, the vector of nodal-point displacements for the block $IJKL$ can be written $[u] = [a][u']$, where

$$[u'] = [u_A \quad v_A \quad u_B \quad v_B \quad u_K \quad v_K \quad u_L \quad v_L \quad \Delta u_A \quad \Delta v_A \quad \Delta u_B \quad \Delta v_B]^T$$

The force vector for the computation is now changed to $[F'] = [a]^T[F]$. Solution of the structural equations then yield directly the displacements of nodal points A, B, K, and L and the relative displacements Δu_A, Δv_A, Δu_B, and Δv_B due to the joint. If there is a very stiff joint, the latter terms will simply be zero.

According to Ghaboussi et al.,[21] this technique allows one to compute successfully with high joint stiffness. Ordinarily, a very high stiffness term on the main diagonal tends to "fix" a nodal point. In the relative displacement scheme above, a high joint stiffness will zero only the relative displacement of the joint; the results should then be identical whether a mesh without a joint or one with a very stiff joint is used.

Example 4-1 Joint closing A block under an initial stress of 5 MN/m^2 compression is next to a joint with initial compression 1 MN/m^2, held by a constraint (not shown). When the constraint is removed, the momentary disequilibrium distresses the block and compresses the joint to restore equilibrium. The speed of convergence depends upon the initial stress in the joint.

Combining Eqs. (4-17) and (4-16) and substituting $(F_n)_0 = -\xi$ gives

$$F_n = \frac{\Delta v}{-V_{mc} - \Delta v}\xi + \xi$$

The normal deformation curve corresponding to $V_{mc} = 0.05$ and $\xi = -0.1$ passes through the points $(F_n, \Delta v)$ in Table 4-4.

Table 4-4 For $V_{mc} = 0.05$ and $\xi = -0.1$

Δv	$\sigma_x = \sigma_n = F_n$	Δv	$\sigma_x = \sigma_n = F_n$
0	−0.100	−0.042	−0.025
−0.01	−0.125	−0.045	−1.0
−0.02	−0.167	−0.046	−1.25
−0.03	−0.25	−0.048	−2.5
−0.035	−0.333	−0.049	−5.0
−0.04	−0.500		

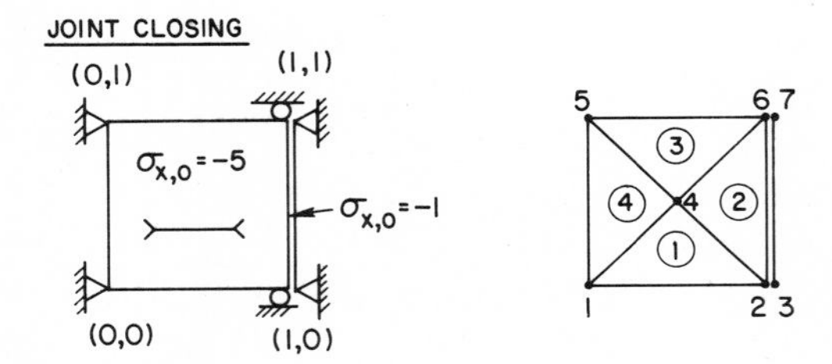

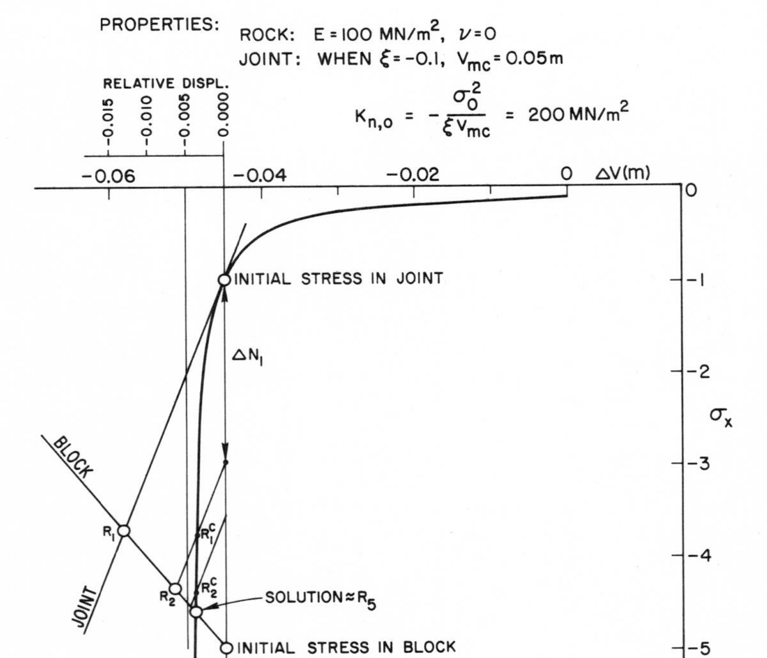

Figure 4-17 Joint closing.

As shown in the graphical solution (Fig. 4-17), about 5 iterations are required for convergence for the given data, with $\sigma_0 = -1\ \text{MN/m}^2$ in the joint, whereas when σ_0 is -0.5 and $-0.2\ \text{MN/m}^2$, convergence requires 12 and 35 iterations, respectively. In the latter case, it is likely that the requested number of iterations would fall short. On restarting the computations, several ways of accelerating the convergence are possible, as in Fig. 4-18. Say the first computation with initial joint stress A terminated after 2 iterations, yielding point R_2, and associated points C ($R = R_2^C$) and E [$= (F_o)_3$]. Three alternative accelerating schemes are shown.

1. A Newton-Raphson correction restarts from C with stiffness redefined to the tangent to the compression curve at C. The initial stress in the rock must then be changed, for restart, from B to D or the "solution" will shift. The displacements will have to be stored from the first run and added cumulatively on each restart. Convergence will be achieved in about six more iterations.
2. Another approach, the dashed lines, restarts at E with the stiffness redefined to the slope of the compression curve at C. This produces an overcorrection, but convergence is reached in about seven more iterations.
3. A third approach, the dotted lines, is readily programmed. The restart point is the initial starting point A, and the stiffness is redefined to that corresponding to line AC. Convergence is achieved in an additional 16 iterations.

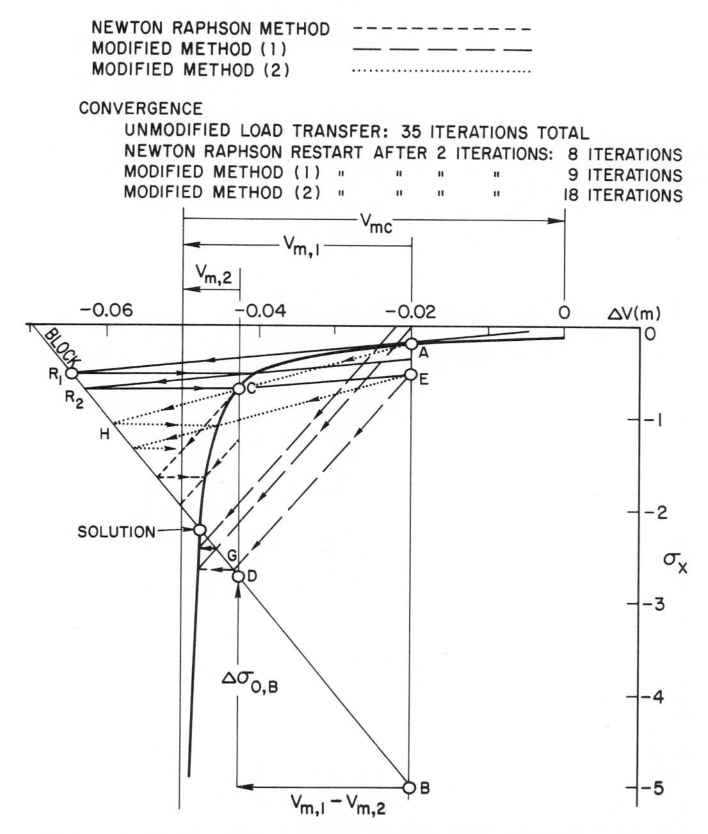

Figure 4-18 Joint closing; previous example with initial joint stress = 0.2 MN/m². Accelerated restart after two iterations.

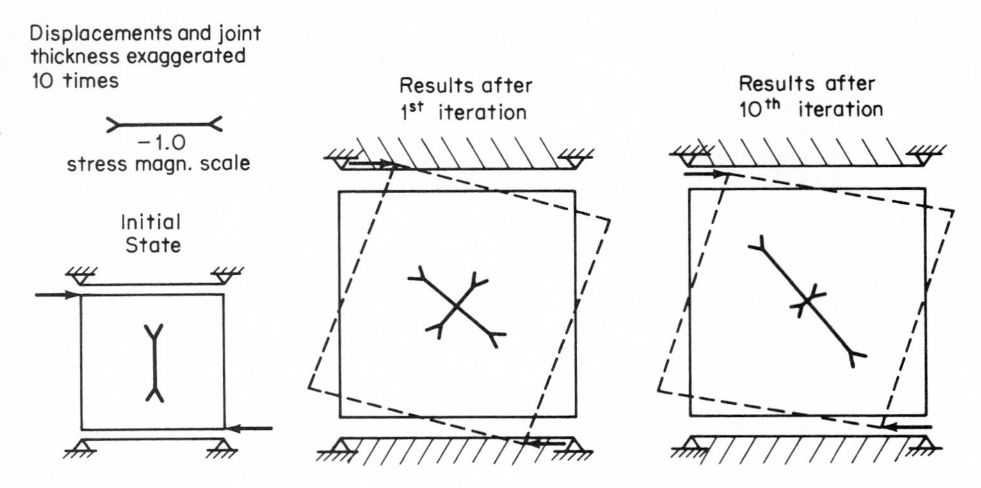

Figure 4-19 Effect of joint closing on stresses in a rotating block.

Example 4-2 Joint rotation Consider a block between two joints loaded by a shear couple as shown in Fig. 4-19. Rotation first closes the joint beyond the permissible maximum closure V_{mc}. Subsequent iterations produce a convergent final state of displacement and stress, as shown in Fig. 4-19.

Additional examples of computations with joints will be found elsewhere in this volume and in Goodman,[22] which contains a listing and description of a simple computer program utilizing the theory presented here.

References

1. Krsmanovic, D.: On the Results of Measurement of Stresses and Strains in the Rocky Mass of a Geostatical Model of the Grancarevo Dam, *Inst. Geotech. Found. Eng. Sarajevo Publ.* 3, 1971.
2. Goodman, R. E.: The Mechanical Properties of Joints, *Proc. 3d Congr. Int. Soc. Rock Mech., 1974*, vol. 1, pt. 2, suppl. rep. vol.
3. Goodman, R. E.: The Deformability of Joints, in Determination of the In-situ Modulus of Deformation of Rock, *ASIM Spec. Tech. Note* 477, pp. 174–196, 1970.
4. Rengers, N.: Influence of Surface Roughness on the Friction Properties of Rock Planes, *Proc. 2d Congr. Int. Soc. Rock Mech., Belgrade, 1970*, vol. 1, pap. 1–31.
5. Barton, N. R.: A Relationship between Joint Roughness and Joint Shear Strength, *Proc. Int. Symp. Rock Fracture, Int. Soc. Rock Mech., Nancy, 1971.*
6. Ladanyi, B., and G. Archambault: Simulation of Shear Behavior of a Jointed Rock Mass, *Proc. 11th Symp. Rock Mech. AIME, 1970*, pp. 105–125.
7. Jaeger, J. C., and N. G. W. Cook: Fundamentals of Rock Mechanics, Methuen & Co., Ltd., London, 1969.
8. Barton, N. R.: Estimating the Shear Strength of Rock Joints, *Proc. 3d Congr. Int. Soc. Rock Mech., 1974*, vol. 2.
9. De Rouvray, A., and R. E. Goodman: Finite Element Analysis of Crack Initiation in a Block Model Experiment, *Rock Mech.*, vol. 4, pp. 203–223, 1972.
10. Goodman, R. E., R. L. Taylor, and T. L. Brekke; A Model for the Mechanics of Jointed Rock, *J. Soil Mech. Found. Div., ASCE*, vol. 94, no. SM3, pp. 637–659, 1968.

11. Byrne, R. J.: Physical and Numerical Models in Rock and Soil Slope Stability, Ph.D. thesis, James Cook University of North Queensland, 1974.
12. Zienkiewicz, O. C., S. Valliappan, and I. P. King: Stress Analyses of Rock as a "No Tension" Material, *Geotechnique*, vol. 18, no. 1, 1968.
13. Goodman, R. E., and J. Dubois; Duplication of Dilatancy in Analysis of Jointed Rocks, *J. Soil Mech. Found. Div. ASCE*, vol. 98, no. SM4, pp. 399–422, 1972.
14. Dubois, J. J.: The Hyperplane Perturbation Method for the Analysis of Nonlinearities, Ph.D. thesis, University of California, Berkely, 1972.
15. Carnahan, B., H. Luther, and J. Wilkes: Applied Numerical Methods, John Wiley & Sons, Inc., New York, 1969.
16. Krsmanovic, D.: Initial and Residual Shear Strength of Hard Rocks, *Geotechnique*, vol. 17, no. 2, pp. 145–160, 1967.
17. Fairhurst, C.: On the Validity of the Brazilian Test for Brittle Materials, *Int. J. Rock Mech. Min. Sci.*, vol. 1, pp. 535–546, 1964.
18. Goodman, R. E., and J. M. Duncan: The Role of Structure and Solid Mechanics in the Design of Surface and Underground Excavations in Rock, *Proc. Conf. Struct. Solid Mech. Eng. Des., 1971*, pt. 2, pap. 105, pp. 1379–1404.
19. Cundall, P.: A Computer Model for Simulating Progressive, Large Scale Movements in Blocky Rock Systems, *Proc. Int. Symp. Rock Fracture, Int. Soc. Rock Mech.*, Nancy, 1971, pap. II-8.
20. Burman, B. C.: A Numerical Approach to the Mechanics of Discontinua, Ph.D. thesis, James Cook University of North Queensland, 1972.
21. Ghaboussi, J., E. L. Wilson, and J. Isenberg: Finite Element for Rock Joints and Interfaces, *J. Soil Mech. Found. Div. ASCE.* vol. 99, no. SM 1, 1973.
22. Goodman, R. E.: Methods of Geological Engineering in Discontinuous Rocks, West Publishing Company, St. Paul, Minn., 1975.

CHAPTER

FIVE

BEAMS, SLABS, AND PAVEMENTS

Y. K. Cheung

5-1 INTRODUCTION

All civil engineering structures are built on soil or rock foundations, and the live and dead loads acting on such structures are transmitted to the foundation through individual footings or foundation beams and slabs. The latter type of construction is especially suitable for structures under comparatively heavy concentrated loadings or for structures resting on relatively weak foundations. It is also used in special structures such as docks or swimming pools.

The primary difficulty in the analysis of beams and slabs on an elastic foundation lies in the determination of contact pressures between the structure and the foundation. In the past, this problem was often overcome by adopting some arbitrary simplification, e.g., assuming that the contact pressure would be linear (Fig. 5-1).

While such grossly simplified assumptions can be considered satisfactory for preliminary studies or for unimportant foundation beams, they should not be used for the analysis of important structures since no account whatsoever has been taken of the compatibility of deformation between the structure and foundation.

A more advanced hypothesis was proposed by Winkler,[1] who suggested that the contact pressure p at a point of the foundation can be considered proportional to the vertical displacement w of that point, or simply $p = kw$, in which k is known

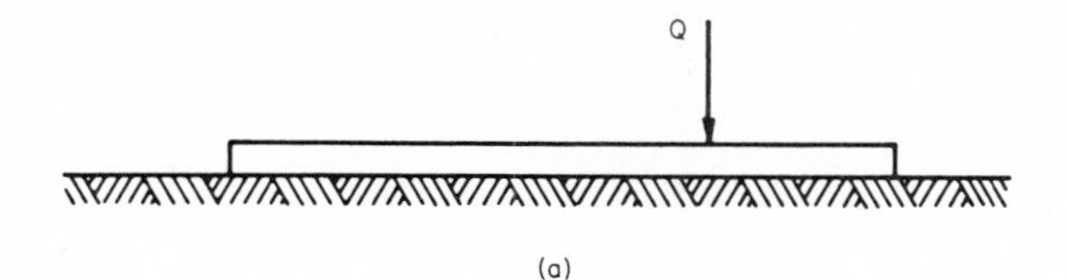

(a)

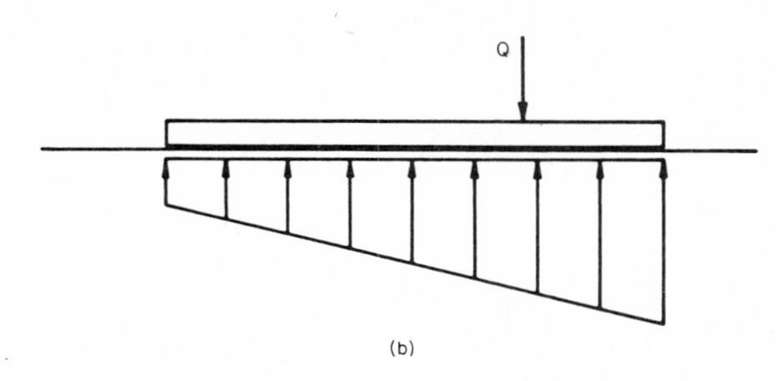

(b)

Figure 5-1 (*a*) Beam on elastic foundation; (*b*) linear contact pressure assumption.

as the modulus of subgrade reaction. Such a property is demonstrated by foundations consisting of a heavy liquid or independent springs. The behavior of an elastic beam on a spring foundation has been treated extensively in a text by Hetenyi.[2] A number of analytical solutions also exist for plates on spring foundations.

In practice, the behavior of both rock and soil foundations differs considerably from Winkler's hypothesis. According to Winkler, a load acting at a point will cause settlement only at that point, whereas the foundation is in reality a continuous body and a continuous settlement curve should result (Fig. 5-2*a* and *b*). Furthermore, it is a difficult task to establish a realistic value for the subgrade modulus of a soil since for the same intensity of loading the test results depend on the size of the loading area. In general, the hypothesis can only be applied to cases in which the compressible layer is extensive but relatively shallow (Fig. 5-2*c*).

In recent years, a new method of analysis[3] has been proposed in which the foundation is treated as a linearly elastic half-plane or half-space and the contact pressures are determined through equilibrium and compatibility conditions for the structure and foundation systems. Recently, the analysis of a loaded half-space made up of anisotropic elastic layers has been developed by Gerrard et al.,[4] using an integral transform technique.

Obviously, such an assumption of linear elasticity is suitable for a rock foundation but only approximately valid for a soil foundation, since soil is not ideally elastic and some permanent deformation normally remains after the removal of loading. This problem is alleviated to some extent by using a compressibility modulus of soil in place of the elastic modulus. Another important factor is that plastic flow often occurs near areas of stress concentration, although in practice this is normally restricted to a fairly small region because of the usually fairly low design bearing stress values.

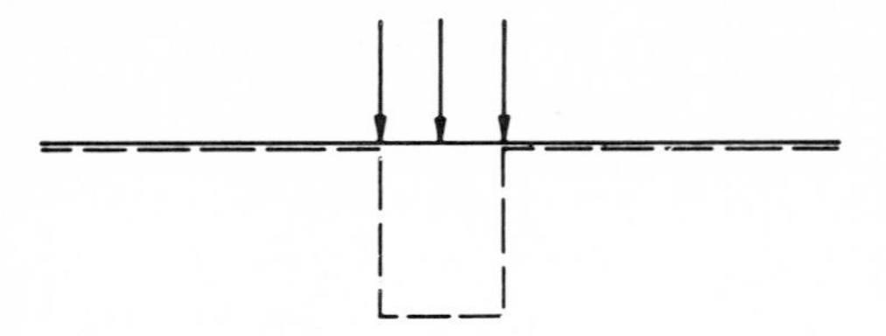

(a) WINKLER'S MODEL

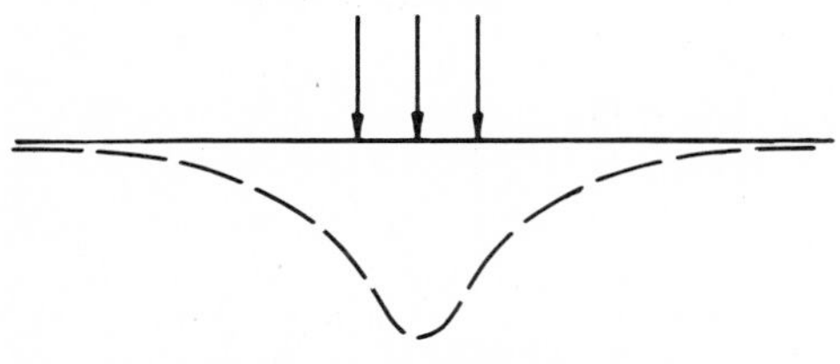

(b) TYPICAL SETTLEMENT CURVE ON SOIL FOUNDATION

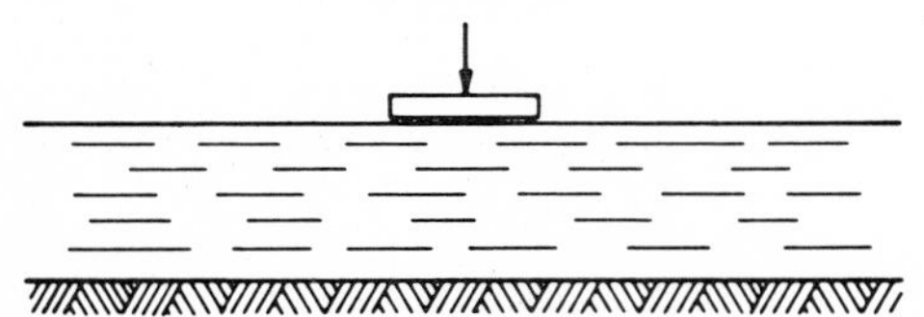

(c) BEAM ON SHALLOW COMPRESSIBLE LAYER

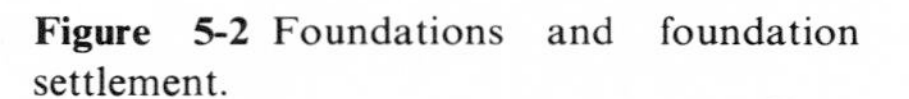

Figure 5-2 Foundations and foundation settlement.

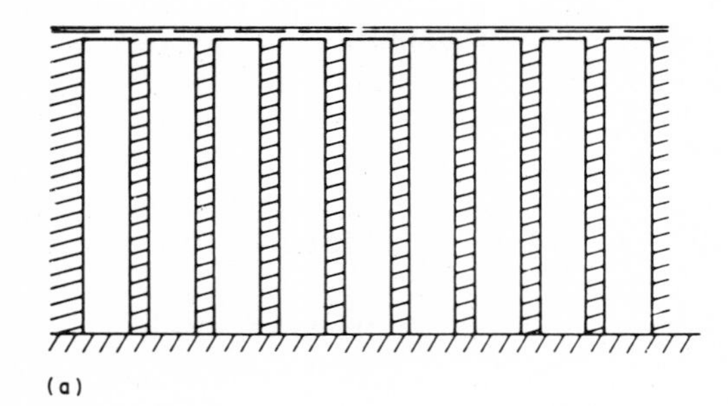

(a)

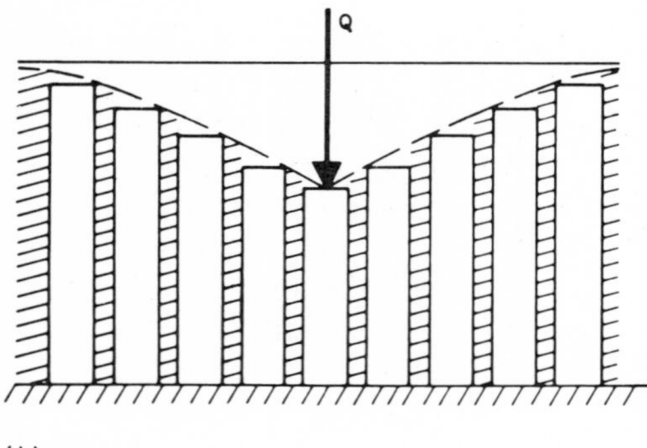

(b)

Figure 5-3 (*a*) Vertical columns representing foundation model; (*b*) deflected shape under load.

Vlasov and Leont'ev[5] proposed a two-parameter foundation model which is intermediate between the Winkler springs and the semi-infinite elastic bodies. In their method of analysis, only vertical displacements exist in the foundation, as the horizontal displacements are assumed to be zero everywhere. Thus the model consists of a series of vertical columns (Fig. 5-3), in which shear due to differential vertical displacements can be transmitted between the adjacent members through friction and bond forces.

In this chapter, nearly all computations will be based on the isotropic, linearly elastic half-plane or half-space model, and the problem is always discretized so that foundation and structure are connected only at a specified number of points. For convenience, all connecting points should lie on a straight line or on a rectangular grid. Problems dealing with Winkler foundations will be included only for completeness.

5-2 GEOTECHNICAL ASPECTS

Foundation Displacement Formulas

In Fig. 5-4 it is seen that the foundation reaction P_i represents the resultant of the contact pressure acting on a rectangle $a \times b$ with contact point i as the center points by integrating the Flamant, Boussinesq, or Cerruti[6,7] formulas for a point possible to relate this pressure P_i/ab to the surface displacements at all connecting points by integrating the Flamant, Boussinesq, or Cerruti[6,7] formulas for a point load over the area of the rectangle (Fig. 5-5).

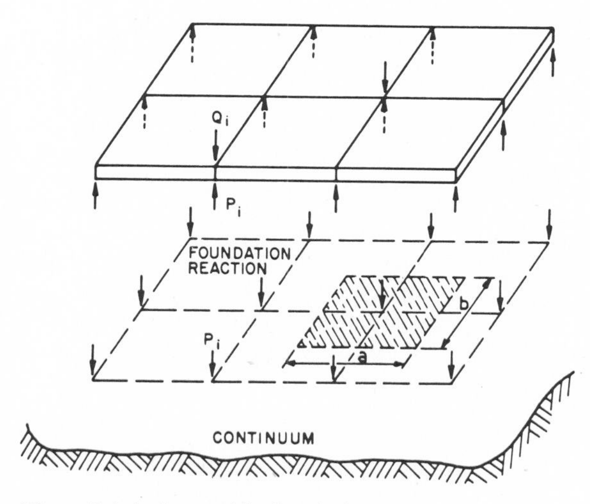

Figure 5-4 A plate and its foundation.

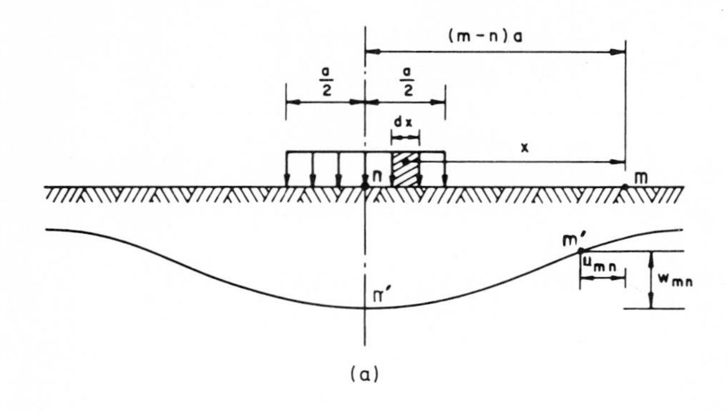

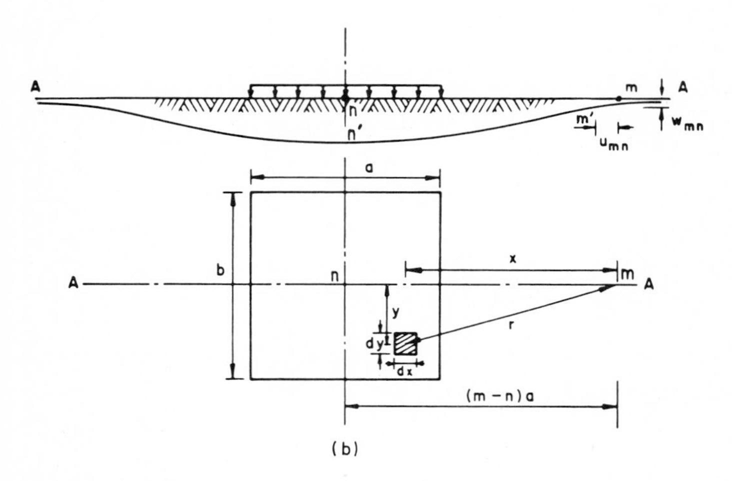

Figure 5-5 Vertical and horizontal relative displacements due to (*a*) uniformly loaded strip on isotropic half-plane and (*b*) uniformly loaded rectangular area on isotropic half-space.

Isotropic Half-Plane

The half-plane case is used for problems such as a beam on top of a deep brick wall (plane stress) or the central part of elongated structures such as retaining walls or docks (plane strain). For a plane stress situation the vertical deflection at any point i on the surface due to a unit vertical point load at point j on the same surface of an isotropic elastic half-plane is given by the Flamant[7] equation. Unit thickness has been assumed here for convenience:

$$w_{ij} = \frac{2}{\pi E_0} \ln \frac{d}{x} \tag{5-1}$$

Similarly the horizontal displacement u at point i due to load at point j is

$$u_{ij} = \begin{cases} \pm \dfrac{1 - \nu_0}{2E_0} & i \neq j \\ 0 & i = j \end{cases} \tag{5-2}$$

where d = constant to be determined
x = distance between point i and point j
E_0 = Young's modulus of foundation
ν_0 = Poisson's ratio of foundation

The corresponding equations for points m and n in the case of a vertically loaded strip of width a and magnitude V_n/a (vertical contact pressure), where n is taken as the center point of the strip (see Fig. 5-5a), can be obtained by integrating Eqs. (5-1) and (5-2) over the area involved:

$$w_{mn} = \frac{2V_n}{\pi E_0 a} \int_{(m-n-0.5)a}^{(m-n+0.5)a} \ln \frac{d}{x}\, dx \tag{5-3}$$

The constant d can be determined by choosing a specified point as the reference point for the deflection curve. Thus if it is assumed that $w_{nn} = 0$, we have

$$d = \frac{a}{2e} \tag{5-4}$$

where e is the base for the natural logarithm. Equation (5-3) is now rewritten as

$$w_{mn} = \frac{2V_n}{\pi E_0 a} \int_{(m-n-0.5)a}^{(m-n+0.5)a} \ln \left(\frac{a}{2ex}\right) dx = \frac{V_n}{\pi E_0} F_{mn} \tag{5-5a}$$

where $F_{mn} = A \ln A - B \ln B$, $A = 2(m - n) - 1$, and $B = 2(m - n) + 1$.

The coefficient F_{mn} varies only with the ratio x/a and is given in Table 5-1. When Eq. (5-2) is used, the horizontal displacement is given by

$$u_{mn} = \pm \frac{1 - \nu_0}{2E_0} V_n \tag{5-5b}$$

A similar set of formulas with identical coefficients can be derived for the horizontal contact pressure H_n/a (shear stresses) as

$$w_{MN} = \pm \frac{1 - \nu_0}{2E_0} H_n \qquad u_{MN} = \frac{H_n}{\pi E_0} F_{MN} \qquad F_{MN} = F_{mn} \tag{5-6}$$

where the subscripts M and N are used to indicate that the displacements are due to horizontal, and not vertical, contact pressures.

For plane strain it is only necessary to replace E_0 by $E_0/(1 - \nu_0^2)$ and ν_0 by $\nu_0/(1 - \nu_0)$.

Table 5-1 The coefficient F_{mn} for half-plane problems

$\frac{x}{a}$	F_{mn}	$\frac{x}{a}$	F_{mn}
0	0	11	−8.181
1	−3.296	12	−8.356
2	−4.751	13	−8.516
3	−5.574	14	−8.664
4	−6.154	15	−8.802
5	−6.602	16	−8.931
6	−6.967	17	−9.052
7	−7.276	18	−9.167
8	−7.544	19	−9.275
9	−7.780	20	−9.378
10	−7.991		

Isotropic Half-Space (Fig. 5-5*b*)

Displacement formulas corresponding to Eq. (5-1) have been derived by Boussinesq and Cerruti[6] for unit vertical and horizontal point loads, respectively:

$$w_{mn} = \frac{1 - \nu_0^2}{\pi E_0 r} \qquad u_{mn} = \frac{1 - \nu_0 - 2\nu_0^2}{2\pi E_0} \frac{x}{r^2}$$

$$w_{MN} = \frac{1 - \nu_0 - 2\nu_0^2}{2\pi E_0} \frac{x}{r^2} \qquad u_{MN} = \frac{1 - \nu_0^2}{\pi E_0} \frac{1}{r^2} + \frac{\nu_0(1 + \nu_0)}{\pi E_0} \frac{x^2}{r^3} \tag{5-7}$$

The vertical deflection w_{mn} at any point m along the x axis due to uniformly distributed vertical contact pressure V_n/ab over a rectangular area $a \times b$ with point n as the center can be obtained as before through the integration of Eq. (5-7). Thus

$$w_{mn} = \frac{2(1 - \nu_0^2)V_n}{\pi E_0 ab} \int_{(m-n-0.5)a}^{(m-n+0.5)a} \int_0^{b/2} \frac{dx\, dy}{\sqrt{x^2 + y^2}}$$

$$= \frac{(1 - \nu_0^2)V_n}{\pi E_0 ab} \left(B \sinh^{-1} \frac{1}{B} + \sinh^{-1} B - C \sinh^{-1} \frac{1}{C} - \sinh^{-1} C \right) \tag{5-8a}$$

where $$B = 2(m - n + 0.5)\frac{a}{b} \qquad C = 2(m - n - 0.5)\frac{a}{b} \tag{5-8b}$$

The deflection at the center of the rectangular area is

$$w_{nn} = 2\left(B \sinh^{-1} \frac{1}{B} + \sinh^{-1} B \right) \tag{5-8c}$$

Similar expressions can be obtained for the other displacement components:

$$u_{mn} = -\frac{(1-\nu_0-2\nu_0^2)}{2\pi E_0}\frac{V_n}{ab}\int_{(m-n-0.5)a}^{(m-n+0.5)a}\int_{-b/2}^{b/2}\frac{x\,dx\,dy}{x^2+y^2}$$

$$= \begin{cases} -\dfrac{(1-\nu_0-2\nu_0^2)V_n}{2\pi E_0 a}\left(B\tan\dfrac{1}{B}-\ln\dfrac{1}{1+B^2}-C\tan^{-1}\dfrac{1}{C}+\ln\dfrac{1}{1+C^2}\right) & m\neq n \\ 0 & m=n \end{cases}$$

$$w_{MN} = u_{mn}$$

$$u_{MN} = \frac{(1-\nu_0^2)H_n}{\pi E_0 ab}\int_{(m-n-0.5)a}^{(m-n+0.5)a}\int_{-b/2}^{b/2}\frac{dx\,dy}{x^2+y^2}+\frac{\nu_0(1+\nu_0)H_n}{\pi E_0 ab}\int_{(m-n-0.5)a}^{(m-n+0.5)a}\int_{-b/2}^{b/2}\frac{x^2\,dx\,dy}{(x^2+y^2)^{3/2}} \quad (5\text{-}9)$$

$$= \begin{cases} \dfrac{(1-\nu_0^2)H_n}{\pi E_0 a}\left(B\sinh^{-1}\dfrac{1}{B}+\sinh^{-1}B-C\sinh^{-1}\dfrac{1}{C}-\sinh^{-1}C\right) \\ \qquad +\dfrac{\nu_0(1+\nu_0)H_n}{\pi E_0 a}(\sinh^{-1}B-\sinh^{-1}C) & m\neq n \\ \dfrac{2(1-\nu_0^2)H_n}{\pi E_0 a}\left(B\sinh^{-1}\dfrac{1}{B}+\sinh^{-1}B\right)+\dfrac{2\nu_0(1+\nu_0)}{\pi E_0 a}(\sinh^{-1}B) & m=n \end{cases}$$

Table 5-2 Values of $\bar{F}_{mn}$

$m-n$	$\frac{1}{m-n}$	$\frac{b}{a}=\frac{2}{3}$	$\frac{b}{a}=1$	$\frac{b}{a}=2$	$\frac{b}{a}=3$	$\frac{b}{a}=4$	$\frac{b}{a}=5$
0	∞	4.265	3.525	2.406	1.867	1.542	1.322
1	1	1.069	1.038	0.929	0.829	0.746	0.678
2	0.500	0.508	0.505	0.490	0.469	0.446	0.424
3	0.333	0.336	0.335	0.330	0.323	0.315	0.305
4	0.250	0.251	0.251	0.249	0.246	0.242	0.237
5	0.200	0.200	0.200	0.199	0.197	0.196	0.193
6	0.167	0.167	0.167	0.166	0.165	0.164	0.163
7	0.143	0.143	0.143	0.143	0.142	0.141	0.140
8	0.125	0.125	0.125	0.125	0.124	0.124	0.123
9	0.111	0.111	0.111	0.111	0.111	0.111	0.110
10	0.100	0.100	0.100	0.100	0.100	0.100	0.099
11	0.091	...	...	0.091			
12	0.083	...	...	0.083			
13	0.077	...	...	0.077			
14	0.071	...	...	0.071			
15	0.067	...	...	0.067			
16	0.063	...	...	0.063			
17	0.059	...	...	0.059			
18	0.056	...	...	0.056			
19	0.051	...	...	0.051			
20	0.050	...	...	0.050			

In practice, w_{mn} is the most commonly used displacement component since horizontal contact pressures are often neglected in the formulations. For this reason it is convenient to write

$$w_{mn} = \frac{(1 - \nu_0^2)V_n}{\pi E_0 a} \bar{F}_{mn} \tag{5-10}$$

The coefficients of $\bar{F}_{mn}$ are tabulated in Table 5-2.

5-3 BEAMS ON ELASTIC CONTINUUM

Rigid Bar Method by Zemochkin[3]

Zemochkin proposed that the beam should be connected with the foundation at discrete regular intervals through a number of rigid bars. Using equilibrium conditions between the external loads and the rigid bar forces and compatability conditions of equal deflections for the beam and foundation at all the rigid bar connections, it is possible to compute the forces in the bars. Each bar force represents the resultant of the contact pressure acting over a rectangular area, so that the overall contact pressure distribution can either be represented by a stepped diagram or by smooth curves drawn through the center points of all the steps. Obviously, the larger the number of rigid bars used in the analysis, the more accurate the results but also the greater the computation effort.

Zemochkin used a mixed method in which the rigid bar forces and also some displacements are treated as unknowns. The beam system in Fig. 5-6*a* has been idealized into a clamped-end beam supported on rigid bars (Fig. 5-6*b*), which are numbered consecutively from 1 to n. If the rigid bars are now replaced by unknown forces X_1 to X_n, we obtain the primary structure given in Fig. 5-6*c*. Unlike ordinary beams, however, the clamped end is capable of both displacement y_0 and rotation ϕ_0, and thus the whole beam will undergo rigid-body displacements, as shown in Fig. 5-6*d*. In addition to the above-mentioned rigid-body displacements, there exist bending deflections of the beam due to external loads p and the unknown bar forces X_m and also foundation settlements due to the bar forces (Fig. 5-6*e*). From compatibility considerations, it can be established that the relative displacement at each bar location should be equal to zero, i.e.,

$$X_1\delta_{m1} + X_2\delta_{m2} + \cdots + X_m\delta_{mm} + \cdots + X_n\delta_{mn} - y_0 - a_m\phi_0 + \Delta_{mp} = 0 \qquad \text{where } \delta_{mi} = w_{mi} + y_{mi} \tag{5-11}$$

where w_{mi} = vertical displacement of foundation at m due to unit vertical force at i
y_{mi} = beam flexibility coefficient for point m due to unit vertical force at i
Δ_{mp} = beam deflection at point m due to external loads

In this way n simultaneous equations can be written. In addition, since the clamped end of the primary structure corresponds to the free end of the original

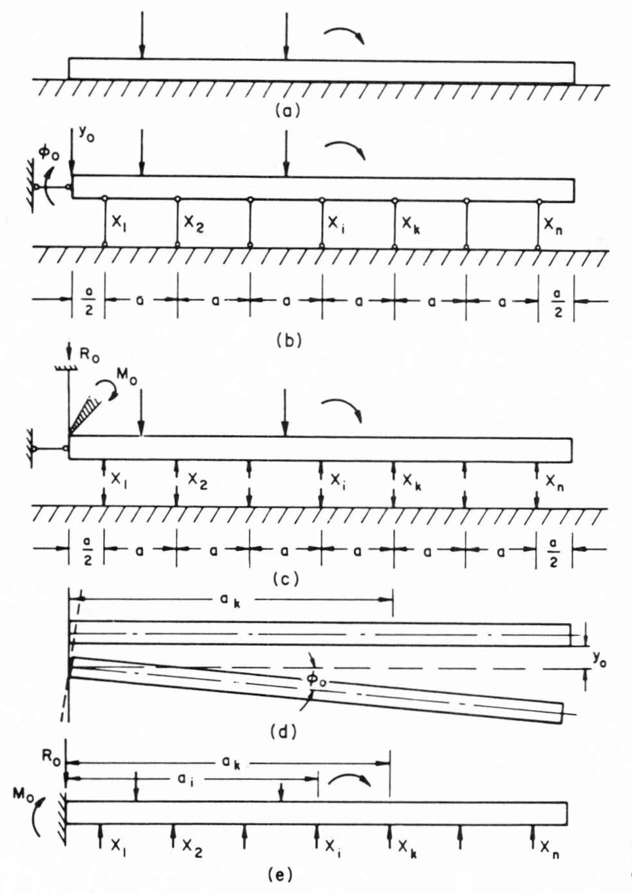

Figure 5-6 Zemochkin's proposed model of beam on elastic continuum.

beam, it can be concluded that the shear force and bending moment at the clamped end should also be zero, i.e.,

$$\begin{aligned} -X_1 - X_2 - \cdots - X_m - \cdots + \sum P &= 0 \\ -X_1 a_1 - X_2 a_2 - \cdots - X_m a_m - \cdots + \sum M_P &= 0 \end{aligned} \tag{5-12}$$

where $\sum P$ represents the sum of all external loads and $\sum M_P$ represents the sum of all moments with respect to the left end of the beam. Thus $n + 2$ equations can be established to solve the $n + 2$ unknowns $X_1, \ldots, X_n, y_0$, and ϕ_0.

The influence coefficient y_{mi} can be computed by such procedures as virtual-work and unit-load methods. It is found that for a constant-thickness beam (Fig. 5-7)

$$y_{mi} = \int_0^{a_i} \frac{(a_m - a_i + x)(x)\,dx}{EI} = \frac{a^3}{6EI}\left(\frac{a_i}{a}\right)^2\left(\frac{3a_m}{a} - \frac{a_i}{a}\right) \tag{5-13}$$

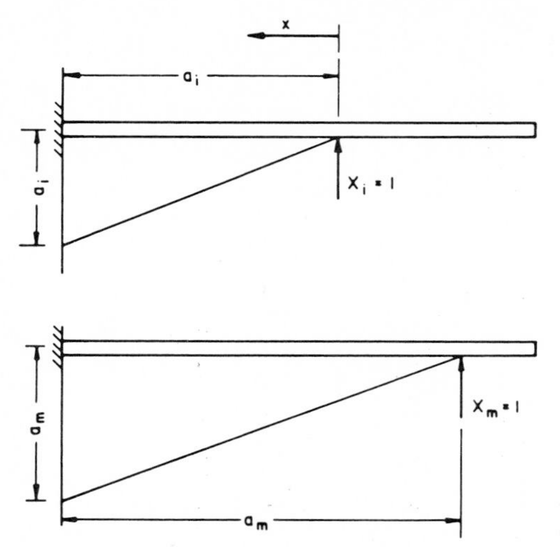

Figure 5-7 Bending-moment diagrams for computation of beam flexibility coefficients.

Example 5-1 A reinforced-concrete beam resting on top of a very deep brick wall is acted upon by three concentrated loads, $P_1 = 40$ t (tons), $P_2 = 100$ t, $P_3 = 20$ t (Fig. 5-8). For the beam $E = 2.1 \times 10^6$ t/m^2, and for the wall $E_0 = 3 \times 10^5$ t/m^2.

The beam can be conveniently divided into 10 segments (from experience it has been found that 6 to 10 segments will usually produce solutions of sufficient accuracy); thus the length of each segment a is equal to 0.5 m. The clamped end of the beam is placed at B. This is a plane stress problem, and thus Eq. (5-5a) can be used directly.

In all, ten compatibility and two equilibrium equations are established, and they are listed in Table 5-3. The equations can be solved for the 12 unknowns by using any standard library computer subroutines. The final contact pressure, bending-moment, and settlement-curve diagrams[8] are presented in Fig. 5-9.

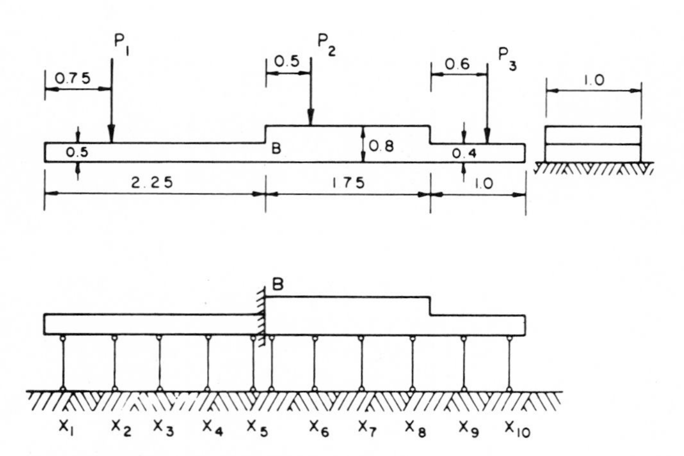

Figure 5-8 Beam and computation model.

Table 5-3 Final algebraic equations for Example 5-1[8]

X_1	X_2	X_3	X_4	X_5	X_6	X_7	X_8	X_9	X_{10}	ψ_0	y_0	Right-hand side
114.893	69.410	31.153	4.300	−6.154	−6.602	−6.967	−7.276	−7.544	−7.780	2	−1	−2907.58
69.410	48.476	21.837	2.430	−5.574	−6.154	−6.602	−6.967	−7.276	−7.544	1.5	−1	−1938.38
31.153	21.837	14.362	1.192	−4.751	−5.574	−6.154	−6.602	−6.967	−7.276	1	−1	−1005.09
4.300	2.430	1.192	1.795	−3.296	−4.751	−5.574	−6.154	−6.602	−6.967	0.5	−1	−287.17
−6.154	−5.574	−4.751	−3.296	0	−3.296	−4.751	−5.574	−6.154	−6.602	0	−1	0
−6.602	−6.154	−5.574	−4.751	−3.296	0.438	−2.200	−2.998	−3.164	−3.087	−0.5	−1	−98.60
−6.967	−6.602	−6.154	−5.574	−4.751	−2.200	3.506	2.839	4.013	5.819	−1	−1	−311.12
−7.276	−6.967	−6.602	−6.154	−5.574	−2.998	2.839	11.832	14.451	18.912	−1.5	−1	−589.38
−7.544	−7.276	−6.967	−6.602	−6.154	−3.164	4.013	14.451	28.431	36.799	−2	−1	−926.32
−7.780	−7.544	−7.276	−6.967	−6.602	−3.087	5.819	18.912	36.799	65.135	−2.5	−1	−1345.94
+2	+1.5	+1	+0.5	0	−0.5	−1	−1.5	−2	−2.5	0	0	35.00
−1	−1	−1	−1	−1	−1	−1	−1	−1	−1	0	0	160.00

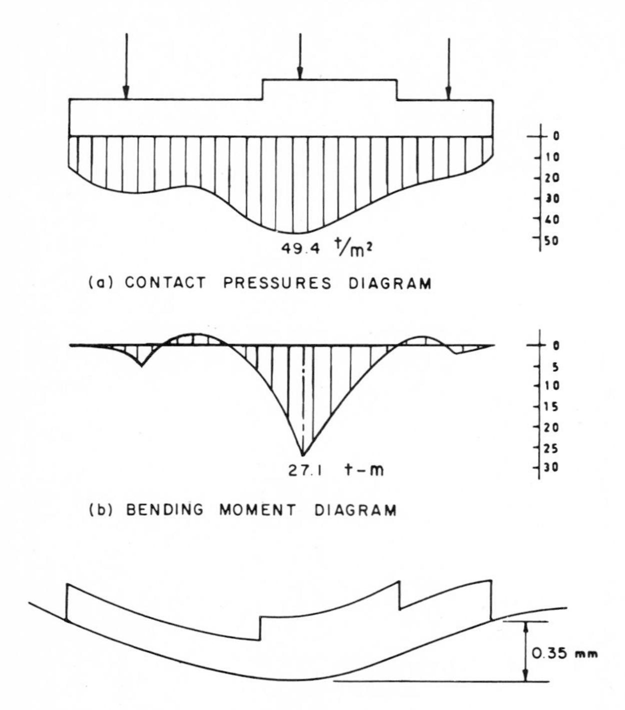

Figure 5-9 Results for Example 5-1.

Finite Difference Method[9]

In this method the beam is again divided into n equal segments of length a, and the contact pressure p acting on each segment is again assumed to be uniform over the length of each segment. The vertical deflection w_i at the center point of segment i is of course affected by the contact pressures at all segments and can be written as

$$w_i = w_i(p_1, p_2, \ldots, p_n) = a_{i1}p_1 + a_{i2}p_2 + \cdots + a_{in}p_n$$

The differential equation for a beam is

$$\frac{d^2w}{dx^2} = \frac{-M}{EI} \qquad (5\text{-}14a)$$

Expressed in finite difference form, this becomes

$$\frac{w_{i+1} - 2w_i + w_{i-1}}{a^2} = \frac{-M_i}{EI} \qquad (5\text{-}14b)$$

for the ith point, and M_i is the bending moment at the center of segment i caused by external load and contact pressure p.

In Eq. (5-14*b*), w_{i+1}, w_i, w_{i-1}, and M_i are all functions of $p_1, p_2, \ldots, p_n$; therefore $n-2$ equations can be established for $i = 2, 3, \ldots, n-1$. The end points are not used because they are free from any bending moment. The two other equations necessary for solving the n unknowns come from the two equilibrium equations $\sum M_0 = 0$ and $\sum P + \sum pa = 0$.

The foundation displacements due to the contact pressures are primarily given by Eqs. (5-5*a*) and (5-10), with some modifications.

Half-plane problem All vertical displacements are relative displacements in which the left end of the beam (point *0*) is regarded as the reference point. Thus for the displacement at the center of segment m caused by contact pressure acting at segment i we have

$$w_{mi} = \frac{p_i a}{\pi E_0}(F_{mi} - F_{0i}) \qquad (5\text{-}15a)$$

in which F_{mi} are the same coefficients as those given in Table 5-1, although F_{0i} must now be given in terms of half increments of a and are listed in Table 5-4.

It can be shown that Eq. (5-15*a*) is not valid for $i = 1$, and in order to compute the relative displacements w_{m1} the following equations are derived separately:

$$w_{11} = 1.386 \frac{p_i a}{E_0}$$

$$w_{m1} = \frac{2p_1}{E_0}[a \ln a - a_m \ln a_m + (a_m - a) \ln (a_m - a)] \qquad (5\text{-}15b)$$

in which a_m is the distance between point m and the left end of the beam.

In order to facilitate computation, Chai worked out Tables 5-5 and 5-6, assuming that a constant-section beam is divided into 10 equal segments. Table 5-5 gives the coefficient for obtaining the relative displacements, while in Table

Table 5-4 Coefficients F_{mi} or F_{0i}

x/a	F_{mi} or F_{0i}	x/a	F_{mi} or F_{0i}	x/a	F_{mi} or F_{0i}	x/a	F_{mi} or F_{0i}
0	0	3.0	−5.574	5.5	−6.793	8.0	−7.544
1.0	−3.296	3.5	−5.885	6.0	−6.967	8.5	−7.665
1.5	−4.159	4.0	−6.155	6.5	−7.128	9.0	−7.780
2.0	−4.751	4.5	−6.390	7.0	−7.276	9.5	−7.888
2.5	−5.205	5.0	−6.602	7.5	−7.415		

Table 5-5 Relative settlement in terms of contact pressures (half-plane case)

w_1	w_2	w_3	w_4	w_5	w_6	w_7	w_8	w_9	w_{10}
$1.386p_1$	$-1.910p_1$	$-3.365p_1$	$-4.188p_1$	$-4.769p_1$	$-5.216p_1$	$-5.581p_1$	$-5.890p_1$	$-6.158p_1$	$-6.394p_1$
$0.863p_2$	$4.159p_2$	$0.863p_2$	$-0.592p_2$	$-1.415p_2$	$-1.996p_2$	$-2.443p_2$	$-2.808p_2$	$-3.117p_2$	$-3.385p_2$
$0.454p_3$	$1.909p_3$	$5.205p_3$	$1.909p_3$	$0.454p_3$	$-0.369p_3$	$-0.950p_3$	$-1.397p_3$	$-1.762p_3$	$-2.071p_3$
$0.311p_4$	$1.134p_4$	$2.589p_4$	$5.885p_4$	$2.589p_4$	$1.134p_4$	$0.311p_4$	$-0.270p_4$	$-0.717p_4$	$-1.082p_4$
$0.235p_5$	$0.816p_5$	$1.639p_5$	$3.094p_5$	$6.390p_5$	$3.094p_5$	$1.639p_5$	$0.816p_5$	$0.235p_5$	$-0.212p_5$
$0.191p_6$	$0.638p_6$	$1.219p_6$	$2.042p_6$	$3.497p_6$	$6.793p_6$	$3.497p_6$	$2.042p_6$	$1.219p_6$	$0.638p_6$
$0.161p_7$	$0.526p_7$	$0.973p_7$	$1.554p_7$	$2.377p_7$	$3.832p_7$	$7.128p_7$	$3.832p_7$	$2.377p_7$	$1.554p_7$
$0.139p_8$	$0.448p_8$	$0.813p_8$	$1.260p_8$	$1.841p_8$	$2.664p_8$	$4.119p_8$	$7.415p_8$	$4.119p_8$	$2.664p_8$
$0.121p_9$	$0.389p_9$	$0.698p_9$	$1.063p_9$	$1.510p_9$	$2.091p_9$	$2.914p_9$	$4.369p_9$	$7.665p_9$	$4.369p_9$
$0.108p_{10}$	$0.344p_{10}$	$0.612p_{10}$	$0.921p_{10}$	$1.286p_{10}$	$1.733p_{10}$	$2.314p_{10}$	$3.137p_{10}$	$4.592p_{10}$	$7.888p_{10}$

Table 5-6 Sum of relative settlements in terms of contact pressures (half-plane case)

$w_3 - 2w_2 + w_1$	$w_4 - 2w_3 + w_2$	$w_5 - 2w_4 + w_3$	$w_6 - 2w_5 + w_4$	$w_7 - 2w_6 + w_5$	$w_8 - 2w_7 + w_6$	$w_9 - 2w_8 + w_7$	$w_{10} - 2w_9 + w_8$
$1.841p_1$	$0.632p_1$	$0.242p_1$	$0.134p_1$	$0.082p_1$	$0.056p_1$	$0.041p_1$	$0.032p_1$
$-6.592p_2$	$1.841p_2$	$0.632p_2$	$0.242p_2$	$0.134p_2$	$0.082p_2$	$0.056p_2$	$0.041p_2$
$1.841p_3$	$-6.592p_3$	$1.841p_3$	$0.632p_3$	$0.242p_3$	$0.134p_3$	$0.082p_3$	$0.056p_3$
$0.632p_4$	$1.841p_4$	$-6.592p_4$	$1.841p_4$	$0.632p_4$	$0.242p_4$	$0.134p_4$	$0.082p_4$
$0.242p_5$	$0.632p_5$	$1.841p_5$	$-6.592p_5$	$1.841p_5$	$0.632p_5$	$0.242p_5$	$0.134p_5$
$0.134p_6$	$0.242p_6$	$0.632p_6$	$1.841p_6$	$-6.592p_6$	$1.841p_6$	$0.632p_6$	$0.242p_6$
$0.082p_7$	$0.134p_7$	$0.242p_7$	$0.632p_7$	$1.841p_7$	$-6.592p_7$	$1.841p_7$	$0.632p_7$
$0.056p_8$	$0.082p_8$	$0.134p_8$	$0.242p_8$	$0.632p_8$	$1.841p_8$	$-6.592p_8$	$1.841p_8$
$0.041p_9$	$0.056p_9$	$0.082p_9$	$0.134p_9$	$0.242p_9$	$0.632p_9$	$1.841p_9$	$-6.592p_9$
$0.032p_{10}$	$0.041p_{10}$	$0.056p_{10}$	$0.082p_{10}$	$0.134p_{10}$	$0.242p_{10}$	$0.632p_{10}$	$1.841p_{10}$

Table 5-7 Settlement in terms of contact pressures for $b/a = 2$ (half-space case)

Multiplier $\dfrac{1 - \nu_0^2}{\pi E_0}$

w_1	w_2	w_3	w_4	w_5	w_6	w_7	w_8	w_9	w_{10}
$2.406p_1$	$0.929p_1$	$0.490p_1$	$0.330p_1$	$0.249p_1$	$0.199p_1$	$0.166p_1$	$0.143p_1$	$0.125p_1$	$0.111p_1$
$0.929p_2$	$2.406p_2$	$0.929p_2$	$0.490p_2$	$0.330p_2$	$0.249p_2$	$0.199p_2$	$0.166p_2$	$0.143p_2$	$0.125p_2$
$0.490p_3$	$0.929p_3$	$2.406p_3$	$0.929p_3$	$0.490p_3$	$0.330p_3$	$0.249p_3$	$0.199p_3$	$0.166p_3$	$0.143p_3$
$0.330p_4$	$0.490p_4$	$0.929p_4$	$2.406p_4$	$0.929p_4$	$0.490p_4$	$0.330p_4$	$0.249p_4$	$0.199p_4$	$0.166p_4$
$0.249p_5$	$0.330p_5$	$0.490p_5$	$0.929p_5$	$2.406p_5$	$0.929p_5$	$0.490p_5$	$0.330p_5$	$0.249p_5$	$0.199p_5$
$0.199p_6$	$0.249p_6$	$0.330p_6$	$0.490p_6$	$0.929p_6$	$2.406p_6$	$0.929p_6$	$0.490p_6$	$0.330p_6$	$0.249p_6$
$0.166p_7$	$0.199p_7$	$0.249p_7$	$0.330p_7$	$0.490p_7$	$0.929p_7$	$2.406p_7$	$0.929p_7$	$0.490p_7$	$0.330p_7$
$0.143p_8$	$0.166p_8$	$0.199p_8$	$0.249p_8$	$0.330p_8$	$0.490p_8$	$0.929p_8$	$2.406p_8$	$0.929p_8$	$0.490p_8$
$0.125p_9$	$0.143p_9$	$0.166p_9$	$0.199p_9$	$0.249p_9$	$0.330p_9$	$0.490p_9$	$0.929p_9$	$2.406p_9$	$0.929p_9$
$0.111p_{10}$	$0.125p_{10}$	$0.143p_{10}$	$0.166p_{10}$	$0.199p_{10}$	$0.249p_{10}$	$0.330p_{10}$	$0.490p_{10}$	$0.929p_{10}$	$2.406p_{10}$

5-6, each coefficient represents the sum of relative displacements $w_{i-1} - 2w_i + w_{i+1}$ due to each contact pressure. These coefficients are now used directly for setting up the $n - 2$ equations [see Eq. (5-14*b*)].

Half-space problem Equation (5-10) will be applicable here since the same rectangular area under contact pressures is used in deriving the expressions for displacements. Of course, the concentrated force V_n will now have to be substituted by $p_n ab$. The resulting modified equation will take the form

$$w_{mn} = \frac{1 - \nu_0^2}{\pi E_0} p_n b \bar{F}_{mn} \tag{5-16}$$

in which $\bar{F}_{mn}$ is given in Table 5-2.

As in the half-plane problem, it is possible to prepare tables which give foundation settlements and sum of displacement in terms of contact pressures for a given number of beam segments. Here some additional work is necessary because, unlike the half-plane case, the foundation settlement is dependent on the aspect ratio b/a of each contact-pressure-segment area a/b. Therefore for a beam divided into ten equal segments, with $b/a = 2$, Table 5-7 gives the foundation settlement coefficients while Table 5-8 gives the sum of settlements $w_{i-1} + 2w_i + w_{i+1}$ in terms of the contact pressures.

Example 5-2 Figure 5-10 shows a beam on half-space 4.5 m long and with $EI =$ 50,000 t·m², $b = 1.0$ m; $E_0 = 3000$ t/m², and $\nu_0 = 0.3$. Determine the contact pressure if the beam is acted upon by two symmetric loads of 50 t each at a distance of 1.0 m from the center of the beam.

If the beam is divided into nine segments, $a = 0.5$ m and $b/a = 2$. It is now possible to establish the following equation at points 2 to 5 (the other points are taken care of by symmetry,

Table 5-8 Sum of settlements in terms of contact pressures for $b/a = 2$ (half-space case)

Multiplier $\dfrac{1 - \nu_0^2}{\pi E_0}$

$w_3 - 2w_2 + w_1$	$w_4 - 2w_3 + w_2$	$w_5 - 2w_4 + w_3$	$w_6 - 2w_5 + w_4$	$w_7 - 2w_6 + w_5$	$w_8 - 2w_7 + w_6$	$w_9 - 2w_8 + w_7$	$w_{10} - 2w_9 + w_8$
$1.038p_1$	$0.279p_1$	$0.079p_1$	$0.031p_1$	$0.017p_1$	$0.010p_1$	$0.005p_1$	$0.004p_1$
$-2.954p_2$	$1.038p_2$	$0.279p_2$	$0.079p_2$	$0.031p_2$	$0.017p_2$	$0.010p_2$	$0.005p_2$
$1.038p_3$	$-2.954p_3$	$1.038p_3$	$0.279p_3$	$0.079p_3$	$0.031p_3$	$0.017p_3$	$0.010p_3$
$0.279p_4$	$1.038p_4$	$-2.954p_4$	$1.038p_4$	$0.279p_4$	$0.079p_4$	$0.031p_4$	$0.017p_4$
$0.079p_5$	$0.279p_5$	$1.038p_5$	$-2.954p_5$	$1.038p_5$	$0.279p_5$	$0.079p_5$	$0.031p_5$
$0.031p_6$	$0.079p_6$	$0.279p_6$	$1.038p_6$	$-2.954p_6$	$1.038p_6$	$0.279p_6$	$0.079p_6$
$0.017p_7$	$0.031p_7$	$0.079p_7$	$0.279p_7$	$1.038p_7$	$-2.954p_7$	$1.038p_7$	$0.279p_7$
$0.010p_8$	$0.017p_8$	$0.031p_8$	$0.079p_8$	$0.279p_8$	$1.038p_8$	$-2.954p_8$	$1.038p_8$
$0.005p_9$	$0.010p_9$	$0.017p_9$	$0.031p_9$	$0.079p_9$	$0.279p_9$	$1.038p_9$	$-2.954p_9$
$0.004p_{10}$	$0.005p_{10}$	$0.010p_{10}$	$0.017p_{10}$	$0.031p_{10}$	$0.079p_{10}$	$0.279p_{10}$	$1.038p_{10}$

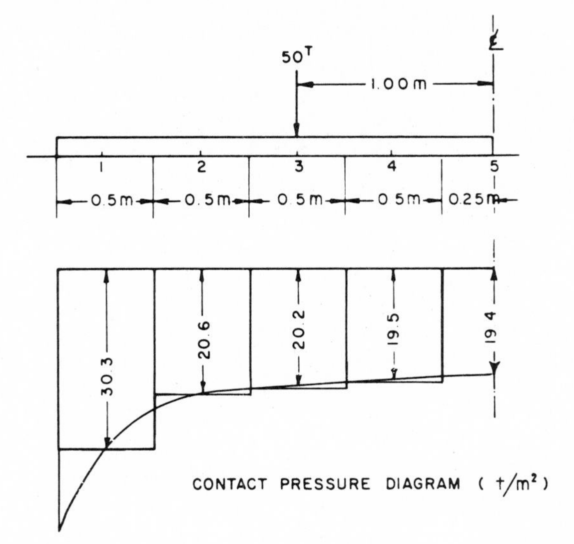

Figure 5-10 Example 5-2.

that is, $p_9 = p_1$, $p_8 = p_2$ and so on):

$$\frac{-M_2}{EI} = \frac{1}{a^2}(w_1 - 2w_2 + w_3)$$

$$= \frac{(1 - v_0^2)b}{\pi E_0 a^2}[(w_{11}p_1 + w_{12}p_2 + w_{13}p_3 + w_{14}p_4 + w_{15}p_5 + w_{16}p_4 + w_{17}p_3 + w_{18}p_2 + w_{19}p_1)$$

$$- 2(w_{21}p_1 + w_{22}p_2 + w_{23}p_3 + w_{24}p_4 + w_{25}p_5 + w_{26}p_4 + w_{27}p_3 + w_{28}p_2 + w_{29}p_1)$$

$$+ (w_{31}p_1 + w_{32}p_2 + w_{33}p_3 + w_{34}p_4 + w_{35}p_5 + w_{36}p_4 + w_{37}p_3 + w_{38}p_2 + w_{39}p_1)]$$

$$= \frac{(1 - v_0^2)b}{\pi E_0 a^2}(1.043p_1 - 2.944p_2 + 1.055p_3 + 0.310p_4 + 0.079p_5)$$

Similar equations can be derived for the other points:

$$\frac{-M_3}{EI} = \frac{(1 - v_0^2)b}{\pi E_0 a^2}(0.289p_1 + 1.055p_2 - 2.923p_3 + 1.117p_4 + 0.279p_5)$$

$$\frac{-M_4}{EI} = \frac{(1 - v_0^2)b}{\pi E_0 a^2}(0.096p_1 + 0.310p_2 + 1.117p_3 - 2.675p_4 + 1.638p_5)$$

$$\frac{-M_5}{EI} = \frac{(1 - v_0^2)b}{\pi E_0 a^2}(0.062p_1 + 0.158p_2 + 0.558p_3 + 2.076p_4 - 2.954p_5)$$

From statics, the moments at the center of segments 2 to 5 are

$$-M_2 = -p_1(0.5^2) - p_2\frac{0.5^2}{8} = -0.25p_1 - 0.031p_2$$

$$-M_3 = -0.5p_1 - 0.25p_2 - 0.031p_3$$

$$-M_4 = 50(0.5) - 0.75p_1 - 0.5p_2 - 0.25p_3 - 0.031p_4$$

$$-M_5 = 50 - p_1 - 0.75p_2 - 0.5p_3 - 0.25p_4 - 0.031p_5$$

Substituting the moment equations into the basic finite difference equations gives

$$1.056p_1 - 2.942p_2 + 1.055p_3 + 0.283p_4 + 0.079p_5 = 0$$

$$0.315p_1 + 1.068p_2 - 2.921p_3 + 1.117p_4 + 0.279p_5 = 0$$

$$0.135p_1 + 0.336p_2 + 1.130p_3 - 2.673p_4 + 1.038p_5 = 1.29$$

$$0.114p_1 + 0.197p_2 + 0.584p_3 + 2.089p_4 - 2.952p_5 = 2.58$$

A further equation is provided through equilibrium conditions:

$$0.5\left(p_1 + p_2 + p_3 + p_4 + \frac{p_5}{2}\right) = 50$$

The solution of the above five equations yields

$$p_1 = 30 \qquad p_2 = 20.6 \qquad p_3 = 20.2 \qquad p_4 = 19.5 \qquad p_5 = 19.4 \text{ t/m}^2$$

Stiffness Method of Analysis by Cheung[10,11]

In this method of analysis, the beam is divided into a number of elements. The stiffness matrices of all the beam elements are assembled together in the standard manner (see texts on matrix analysis of structures), while a foundation stiffness can be established by inverting a flexibility matrix computed from Eqs. (5-5), (5-8), and (5-9). The two stiffness matrices can be combined to give the overall matrix for the foundation-structure system. For any given external loading, the combined matrix is solved to yield the displacements, which in turn can be used for computing the beam forces and the contact pressures. With this approach, problems involving vertical as well as horizontal contact pressures are solved without any difficulty.

Beam stiffness matrix Consider a beam element with ends 1 and 2, as shown in Fig. 5-11. The stiffness matrix which connects the end forces P_1, P_2 and the end displacements D_1, D_2 can be written

$$\begin{Bmatrix} P_1 \\ P_2 \end{Bmatrix} = \begin{bmatrix} S_{11} & S_{12} \\ S_{21} & S_{22} \end{bmatrix} \begin{Bmatrix} D_1 \\ D_2 \end{Bmatrix} \tag{5-17a}$$

Note that all displacements and forces are referred to the neutral axis of the beam.

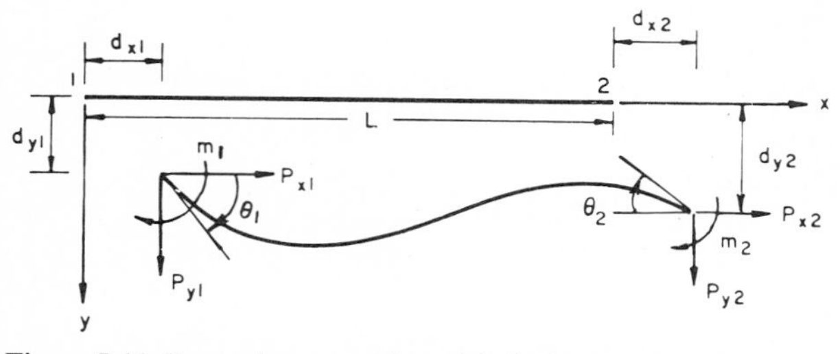

Figure 5-11 Beam element with nodal displacements and forces.

For a straight beam with constant cross section, the full form of Eq. (5-17*a*) is

$$\begin{Bmatrix} p_{x1} \\ p_{y1} \\ m_1 \\ p_{x2} \\ p_{y2} \\ m_2 \end{Bmatrix} = \begin{bmatrix} \frac{EA}{L} & & & & & \\ 0 & \frac{12EI}{L^3} & & & \text{sym} & \\ 0 & \frac{6EI}{L^2} & \frac{4EI}{L} & & & \\ -\frac{EA}{L} & 0 & 0 & \frac{EA}{L} & & \\ 0 & -\frac{12EI}{L^3} & -\frac{6EI}{L^2} & 0 & \frac{12EI}{L^3} & \\ 0 & \frac{6EI}{L^2} & \frac{2EI}{L} & 0 & -\frac{6EI}{L^2} & \frac{4EI}{L} \end{bmatrix} \begin{Bmatrix} d_x \\ d_{y1} \\ \theta_1 \\ d_{x2} \\ d_{y2} \\ \theta_2 \end{Bmatrix} \qquad (5\text{-}17b)$$

where p_{x1}, p_{y1} are the horizontal and vertical components of forces at node 1, respectively, and so on, and m_1 is the moment at node 1, and so on.

For soil foundations the horizontal contact stresses are usually ignored. In such cases, the columns and rows in Eq. (5-17*b*) corresponding to p_{x1}, p_{x2} and d_{x1}, d_{x2} should be deleted and a 4 × 4 matrix should be used.

The assembled beam stiffness equations are now written

$$\{P\} = [S]\{D\} \qquad (5\text{-}18)$$

in which $\{P\}$ and $\{D\}$ are, respectively, the nodal forces and nodal displacements, with reference to the beam axis.

Foundation stiffness matrix Figure 5-12 shows a beam resting on an elastic foundation. Because there are three degrees of freedom at each node (d_{xi}, d_{yi}, θ_i) along

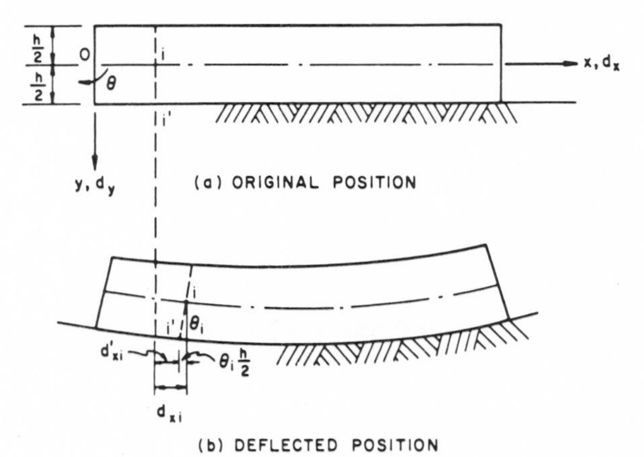

Figure 5-12 Beam and foundation displacements.

the beam axis but only two degrees of freedom (d_{xi}, d_{yi}) at each foundation node along the bottom face of the beam, it is evident that the force and displacement components for the two systems are not identical. Therefore before the stiffness matrices of the two systems can be combined, a transformation must be carried out; the process will be described in the following paragraphs.

For any set of reactive forces P' acting on the foundation, it is possible to compute the displacement at the same nodes by virtue of Eqs. (5-5), (5-6), (5-8), and (5-9). Thus we have

$$\{D'\} = [f'_f]\{P'\} \tag{5-19a}$$

where
$$\{D'\} = \begin{Bmatrix} D'_1 \\ D'_2 \\ \vdots \\ D'_n \end{Bmatrix} \qquad \{D'_i\} = \begin{Bmatrix} d'_{xi} \\ d'_{yi} \end{Bmatrix} \qquad \{P'\} = \begin{Bmatrix} P'_1 \\ P'_2 \\ \vdots \\ P'_n \end{Bmatrix} \qquad \{P'_i\} = \begin{Bmatrix} p'_{xi} \\ p'_{yi} \end{Bmatrix} \tag{5-19b}$$

and $[f'_f]$ is the foundation flexibility matrix.

In Eqs. (5-19b), $\{D'_i\} = \{d'_{yi}\}$ and $\{P'_i\} = \{p'_{yi}\}$ when the horizontal contact pressure is ignored.

Inversion of Eq. (5-19a) yields

$$\{P'\} = [S'_f]\{D'\} \tag{5-20}$$

in which $[S'_f]$ is the stiffness matrix of the foundation.

The complete stiffness formulation The two stiffness matrices given in Eqs. (5-18) and (5-20) should now be combined to form an overall stiffness matrix for a beam on an elastic continuum. As indicated previously, $\{D_i\}$ and $\{D'_i\}$ are not entirely compatible with each other, and the stiffness matrices will have to be modified before assembly. Two separate cases are discussed below:

1. Vertical contact pressure only. In this case

$$\{D_i\} = \begin{Bmatrix} d_{yi} \\ \theta_i \end{Bmatrix} \qquad \text{and} \qquad \{d_{yi}\} = \{d'_{yi}\}$$

   The customary procedure is to eliminate the moment-rotation terms in the beam stiffness matrix through a static condensation technique. The reduced beam stiffness can now be added onto the foundation stiffness matrix.
2. Vertical and horizontal contact pressures.

From Fig. 5-12 it can be seen that the following relationship can be established at any node i:

$$d'_{xi} = d_{xi} - \theta_i \frac{h}{2} \qquad d'_{yi} = d_{yi} \qquad p_{xi} = p'_{xi} \qquad p_{yi} = p'_{yi} \qquad M_i = p'_{xi} \frac{h}{2} \tag{5-21}$$

In matrix notation

$$\begin{Bmatrix} p_{xi} \\ p_{yi} \\ M_i \end{Bmatrix} = \begin{bmatrix} 1 & 0 \\ 0 & 1 \\ -\dfrac{h}{2} & 0 \end{bmatrix} \begin{Bmatrix} p'_{xi} \\ p'_{yi} \end{Bmatrix}$$

or

$$\{p_i\} = [r]\{p'_i\} \tag{5-22}$$

Similarly, we have

$$\begin{Bmatrix} d'_{xi} \\ d'_{yi} \end{Bmatrix} = \begin{bmatrix} 1 & 0 & -\dfrac{h}{2} \\ 0 & 1 & 0 \end{bmatrix} \begin{Bmatrix} d_{xi} \\ d_{yi} \\ \theta_i \end{Bmatrix} \tag{5-23}$$

or

$$\{D'_i\} = [r]^T\{D_i\}$$

For all nodes of the beam-foundation system, we have

$$\begin{Bmatrix} P_1 \\ P_2 \\ \cdot \\ P_n \end{Bmatrix} = \begin{bmatrix} [r] & & [0] \\ & [r] & \\ & \cdot & \\ [0] & & [r] \end{bmatrix} \begin{Bmatrix} P'_1 \\ P'_2 \\ \cdot \\ P'_n \end{Bmatrix} \tag{5-24}$$

or

$$\{P\} = [R]\{P'\}$$

and

$$\begin{Bmatrix} D'_1 \\ D'_2 \\ \cdot \\ D'_n \end{Bmatrix} = \begin{bmatrix} [r]^T & & [0] \\ & [r]^T & \\ & \cdot & \\ [0] & & [r]^T \end{bmatrix} \begin{Bmatrix} D_1 \\ D_2 \\ \cdot \\ D_n \end{Bmatrix} \tag{5-25}$$

or

$$\{D'\} = [R]^T\{D\}$$

The stage is now set for the transformation of the foundation stiffness matrix. From Eqs. (5-24), (5-20), and (5-25)

$$\{P\} = [R]\{P'\} = [R][S'_f]\{D'\} = [R][S'_f][R]^T\{D\} = [S_f]\{D\} \tag{5-26}$$

$[S_f]$ has been transformed to the beam nodal coordinates and therefore can be added onto the beam stiffness matrix. Thus in either case, any external load Q will be resisted jointly by both the beam and the foundation:

$$\{Q\} = [S]\{D\} + [S_f]\{D\} = [[S] + [S_f]]\{D\} \tag{5-27}$$

The beam displacements $\{D\}$ are now obtained by solving Eq. (5-27). The foundation displacements and contact pressures are obtained through Eqs. (5-25) and (5-20).

Slabs on Elastic Foundation

Introduction Earlier attempts to solve the problem of a slab on an elastic foundation were mainly concerned with the study of plates on a Winkler foundation using the finite difference method. Allen and Severn[12] studied the stresses in foundation rafts by solving two second-order differential equations. Later the finite difference method was also applied to the analysis of rectangular plates on an elastic half-space.[13,14] The principal drawback of the finite difference method is that special difference operators have to be used at and near the free edge and at a corner of a slab. Special formulation is needed for a reentrant corner.[12]

The finite element analysis of slabs on Winkler foundation and on half-space was introduced by Cheung.[15] The problem of a slab on a Winkler foundation was later also attempted by Severn[16] and Haddadin,[17] who investigated the effect of structural rigidity of frames on the response of the system. Yang[18] has analyzed a raft on the two-parameter foundation model,[5] and the treatment for half-space problems has been adapted and extended by Wardle et al.[19] to layered foundations. The 12 DOF noncompatible rectangular bending element[20] was used in all the above-mentioned publications. The use of the 16 DOF compatible rectangular element was reported by Yang,[21] and a 10-node triangular plate bending element was developed by Svec and Gladwell[22] specifically for contact problems. In the following paragraphs, the method developed by Cheung will be described in detail.

Basically, the analysis of slabs on an elastic continuum follows the procedure described previously. The primary differences are that (1) a plate bending element must be used in place of a beam element and (2) the nodal points for the plate structure form a plane grid and therefore no longer lie on one straight line as in the case of a beam. It is thus obvious that the displacement equations given in Eqs. (5-5), (5-6), and (5-8) to (5-10) are not directly applicable. A detailed discussion of the above two points is given below.

Although a host of plate bending elements, such as triangular, rectangular, and sector elements, are available, the element most suitable for the present purpose is a lower-order rectangular element like the one proposed by Zienkiewicz and Cheung.[20] The reasons for this choice are twofold: (1) the foundation model has been set up for contact pressure loads over a rectangular area, with the result that the nodes must lie on a rectangular grid, thus making a rectangular element an obvious choice over a triangular element (of the same complexity) because of its higher accuracy; (2) the rectangular patch pressure loads will represent the contact pressure variation by a stepped surface, which is a fairly crude approximation and incompatible with the use of refined higher-order elements for the plate model, in which curvature compatibilities are also required at the nodes.

Equation (5-16) should now be written

$$\begin{Bmatrix} P_1 \\ P_2 \\ P_3 \\ P_4 \end{Bmatrix} = \begin{bmatrix} S_{11} & S_{12} & S_{13} & S_{14} \\ S_{21} & S_{22} & S_{23} & S_{24} \\ S_{31} & S_{32} & S_{33} & S_{34} \\ S_{41} & S_{42} & S_{43} & S_{44} \end{bmatrix} \begin{Bmatrix} D_1 \\ D_2 \\ D_3 \\ D_4 \end{Bmatrix} \tag{5-28a}$$

Table 5-9 Approximate flexibility coefficients for plates

x/a	Exact	Approx.	x/a	Exact	Approx.
1	1.038	1.000	6	0.167	0.167
2	0.505	0.500	7	0.143	0.143
3	0.333	0.333	8	0.125	0.125
4	0.251	0.250	9	0.111	0.111
5	0.200	0.200	10	0.100	0.100

in which

$$\{P_i\} = \begin{bmatrix} P_{yi} \\ M_{xi} \\ M_{yi} \end{bmatrix} \qquad \{D_i\} = \begin{bmatrix} d_{yi} \\ \theta_{xi} \\ \theta_{yi} \end{bmatrix} \tag{5-28b}$$

The complete stiffness matrix $[S]$ can be found in Zienkiewicz and Cheung.[23]

As stated previously, the flexibility coefficients w_{mn} are derived for two points m and n such that n is the center of a rectangular patch load and m lies on one of the two axes of symmetry bisecting the rectangle.

Obviously, for a rectangular grid, the equations for computing the displacement coefficients are not applicable for the majority of the points, and a choice has to be made between deriving an additional set of displacement equations and using some approximate expressions. It was found that Eq. (5-7a) will yield sufficiently accurate flexibility coefficients for most of the points and can be used safely provided that w_{nn} is computed by using Eq. (5-8). In Table 5-9, the approximate coefficients are compared with the accurate values along the x axis, and it can be seen that the maximum error involved is only of the order of 3.8 percent for $x = a$ and the difference decreases rapidly as x increases. This is to be expected by virtue of St. Venant's principle.

For a thin plate, the horizontal contact pressures developed are usually quite small and can be neglected in the analysis.

Nonlinear behavior due to loss of contact[11] For relatively flexible beams and plates on comparatively stiff foundations and under the action of one or more concentrated loads, tensile contact pressures will often develop at or near the corners of a plate if the plate foundations are to remain in contact with each other at all nodal points. Such a situation is generally unnatural because no bond exists between the plate and foundation. Consequently, it should be assumed that separation will occur and that no contact pressures will exist at those nodal points. The procedure for simulating such no-contact-pressure situation can be outlined as follows:

1. Perform the analysis as given in the previous sections. A linear elastic solution can be obtained.
2. If all contact pressures are compressive, the problem is terminated; otherwise, proceed to next step.

3. Find which nodes are associated with tensile or zero contact pressures and eliminate the corresponding rows and columns of the foundation stiffness matrix $[S'_f]$. This corresponds to the situation in which no reactive forces are acting at those nodes.
4. Invert the new flexibility matrix. Before inversion it is advisable to replace all zero diagonal coefficients of the matrix with unit values. Of course, the zeros should go back to the corresponding positions after inversion.
5. Repeat step 1.

Temperature and moisture effects When a thermal gradient exists through the thickness of the slab, in general the slab will bend into concave or convex shapes, with possibly some loss of contact between the slab and the foundation. Such thermal effects can be incorporated into the finite element analysis without difficulty since an equivalent static loading due to the temperature gradient can be assumed to act at the nodes and the computation is carried on in the usual way.

The problem of moisture changes in the soil, in particular expansive clays, is a more complex one. Fraser and Wardle[24] recently presented an upper-bound solution in which a heave profile proposed by Aitchison[25] was adopted as the most severe deformation case. The heave profile is generally a mound shape with a flat central portion, the maximum heave Y_m being about 0.05 m (Fig. 5-13*a*). A second quantity, the overhang distance e (Fig. 5-13*a*) for the slab must be chosen so that it is consistent with the moisture change underneath the slab, and a value of $e/L = 0.2$ has been adopted in Ref. 24. The application of any external load will of course tend to reduce the magnitude of Y_m. In fact, such external loads might cause nodes which are originally in the overhang to come into contact with the somewhat flattened mound (Fig. 5-13*b*), and an iterative process which is simply a reversal of the loss-of-contact problem mentioned previously would have to be carried out.

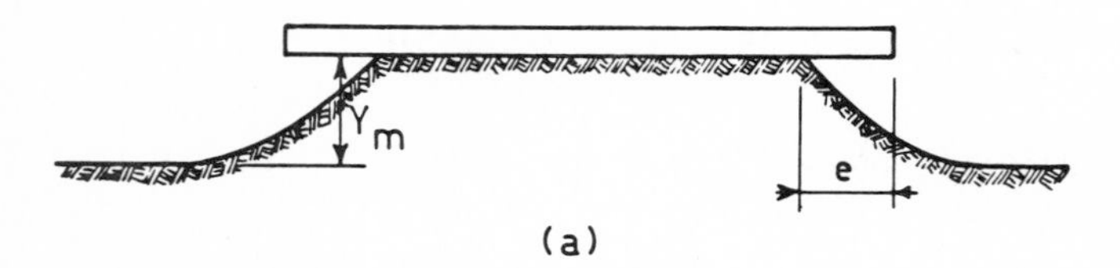

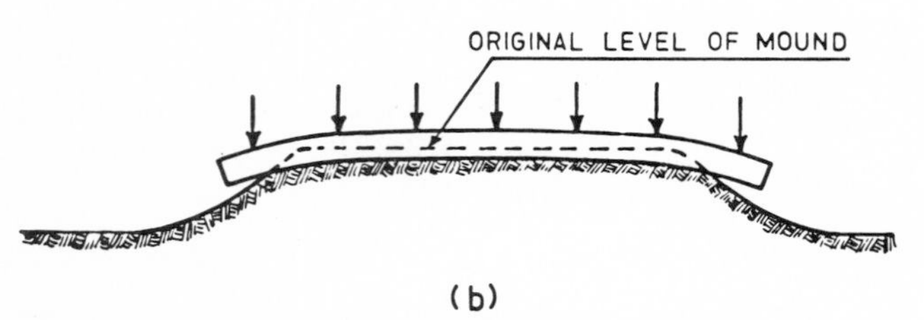

Figure 5-13 (*a*) Unloaded raft on mound; (*b*) profile of raft under load.

5-4 NUMERICAL EXAMPLES

Example 5-3 The beam example shown in Fig. 5-8 will be reanalyzed here by the stiffness method. The beam was divided into 20 elements of equal length, and a plane stress situation involving only vertical contact pressures was adopted for the analysis. The results are compared with the previous example given by Zemochkin.[3] The agreement is excellent, and from Fig. 5-14 it is seen that the discrepancies for both the maximum deflection and maximum contact pressure are below 1 percent.

Example 5-4 A 3.048-m-long concrete beam of uniform cross section (Fig. 5-15) resting on an elastic half-space is subjected to an external concentrated load of 1.814 t at the center of the beam and also to its own weight. Two studies were made. In the first only vertical contact pressures were considered, while in the second both vertical and horizontal contact pressures were taken into account. The results are tabulated in Table 5-10 and Fig. 5-16. Note that while

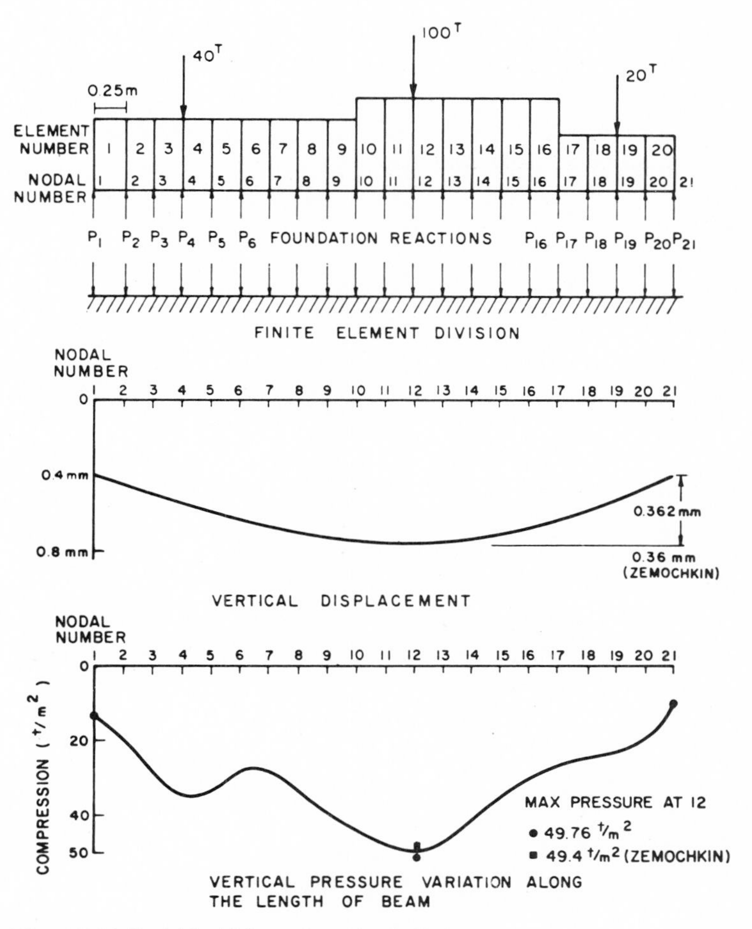

Figure 5-14 Variable-thickness beam on half-plane.

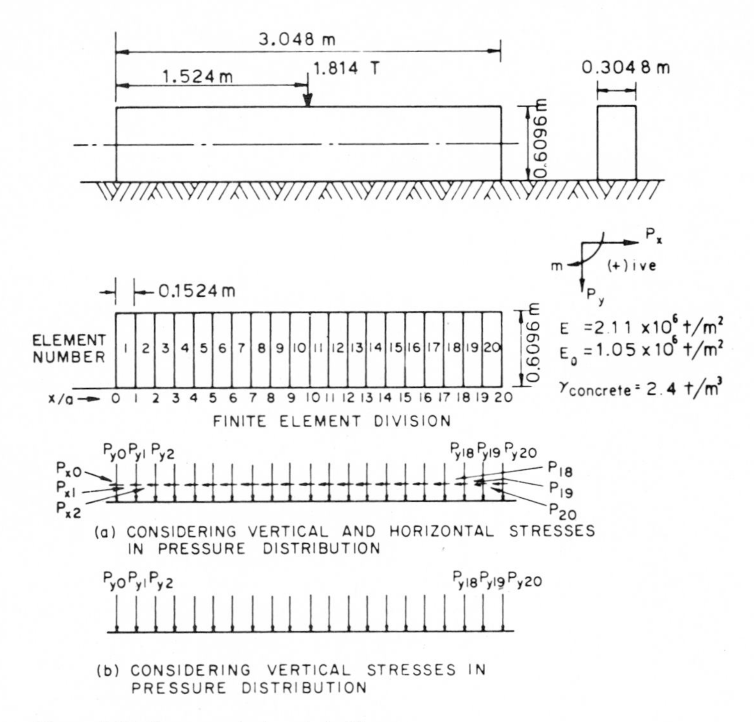

Figure 5-15 Beam on isotropic half-space.

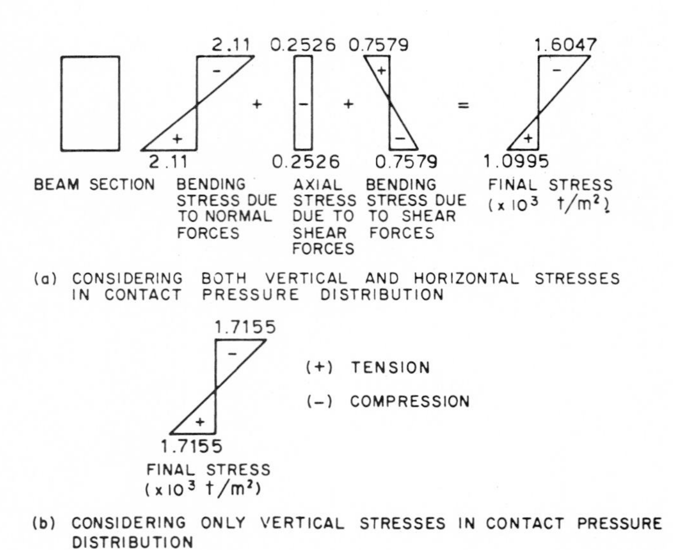

Figure 5-16 Stress distribution at the midsection of beam on isotropic half-space.

Table 5-10 Beam on elastic half-space (4000-lb central load and 300-lb/ft uniformly distributed load)

Point along length of beam	All force components considered		Vertical component only
	Horizontal contact pressure, 10^3 lb/ft^2	Vertical contact pressure, 10^3 lb/ft^2	Vertical contact pressure, 10^3 lb/ft^2
0	−0.064	0.204	0.1396
1	−0.0214	0.198	0.1656
2	−0.02476	0.277	0.2394
3	−0.0432	0.3593	0.3178
4	−0.079	0.458	0.4154
5	−0.1314	0.5816	0.5492
6	−0.1984	0.7424	0.7282
7	−0.2697	0.9418	0.9596
8	−0.32368	1.178	1.2294
9	−0.31358	1.4336	1.4998
10	0	1.5878	1.6496
11	0.31358	1.4336	1.4998
12	0.32368	1.178	1.2294
13	0.2697	0.9418	0.9596
14	0.1984	0.7424	0.7282
15	0.1314	0.5816	0.5492
16	0.079	0.458	0.4154
17	0.0432	0.3593	0.3178
18	0.02476	0.277	0.2394
19	0.0214	0.198	0.1656
20	0.064	0.204	0.1396

the difference in the magnitudes of vertical contact pressures is not really significant, the tensile stress in the beam has actually been reduced by 36 percent with the introduction of horizontal contact pressures. This reduction is primarily due to the compression and negative bending moment produced by the eccentric horizontal contact forces.

Example 5-5 Square plate on isotropic half-space with a concentrated load at center (Fig. 5-17) A series of computations was carried out to study the relationship between contact pressures and the relative rigidity of the plate-foundation system. In Figs. 5-17 and 5-18, the nondimensional coefficient pL^2/P at several points is plotted against a relative rigidity

$$\gamma = 180\pi \frac{E_0}{E_p}\left(\frac{a}{t}\right)^3$$

where a is mesh length and t is plate thickness and the results are as expected. For a very stiff plate, the contact pressure approaches a very high value (infinite for a rigid footing) at the corner, while for a very flexible plate, the contact pressure at the center reaches a very high value, as the plate offers little help in spreading out the load. Note that the contact pressure has become tensile at the edge of the plate (Fig. 5-17).

Example 5-6 Nonlinear behavior of beam on elastic foundation A beam under the combined action of constant uniformly distributed loading and an increasing central upward concentrated force is analyzed. The results presented in Fig. 5-19 show that before any tensile contact pressure

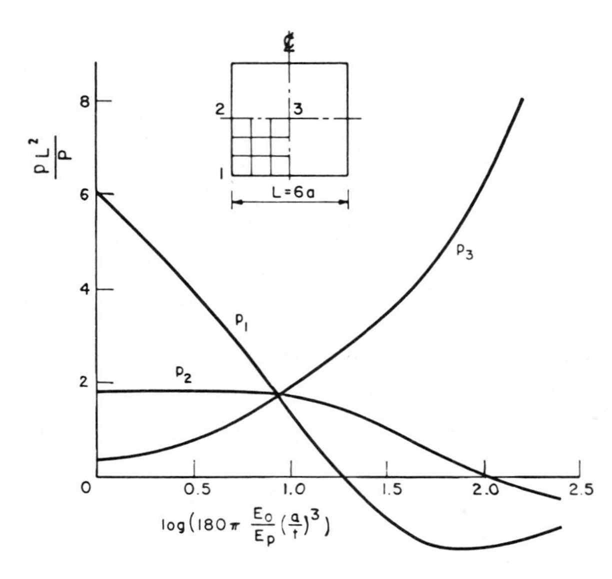

Figure 5-17 Contact pressures of square plate on isotropic half-space with concentrated load at center P at points 1, 2, and 3 for various values of γ.

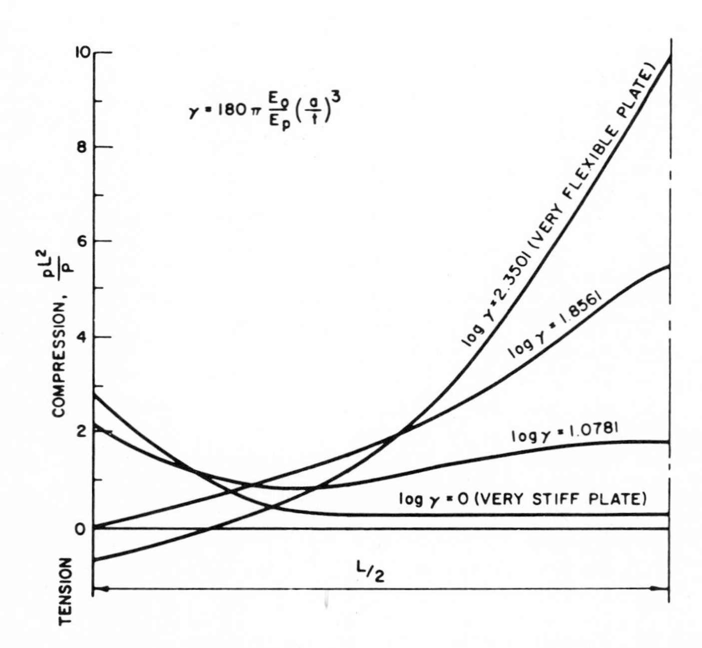

Figure 5-18 Contact pressure variation along center line of plate on isotropic half-space under concentrated load P.

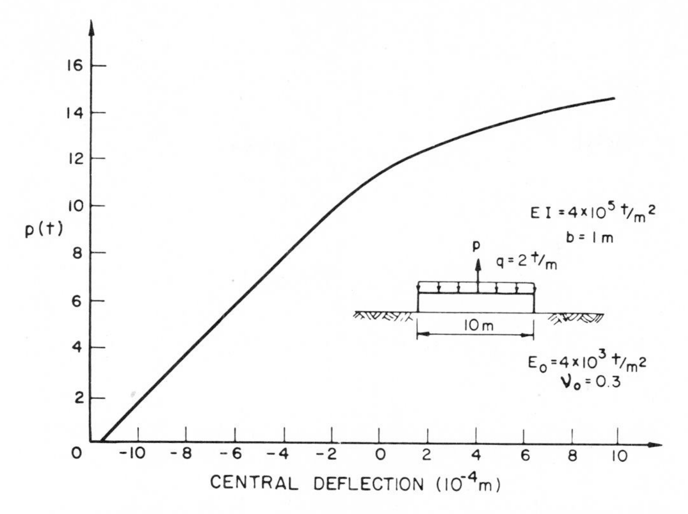

Figure 5-19 Nonlinear behavior of beam on half-space due to separation of contact surfaces.

is developed, the load-displacement curve is linear. Once tensile contact pressure appears, separation of the contact surface takes place, and as the loss of contact progresses, much larger deflections will result. The load-displacement curve thus becomes highly nonlinear.

Example 5-7 Nonlinear analysis of plate on half-space Figure 5-20 and Table 5-11 summarize the results. The phenomenon of lifting of corners is clearly demonstrated. Convergence is reached after only three iterations. At the end of each iteration the displacements computed are in fact the slab displacements. In order to obtain the foundation displacements, it is necessary to use Eq. (5-18) to compute the reactive forces first. These reactive forces, when operated on the modified foundation flexibility matrix, will yield the foundation displacements.

Table 5-11 Nonlinear response of plate on half-space

Nodal number	Slab displacement iterations			Foundation displacement iterations			Contact pressure iterations		
	(i)	(ii)	(iii)	(i)	(ii)	(iii)	(i)	(ii)	(iii)
1	70.1	62.7	62.4	70.1	77.9	78.0	−40.3	0	0
2	98.0	94.4	94.2	98.0	94.4	94.6	9	−0.4	0
3	118.6	117.2	117.1	118.6	117.2	117.1	23.3	22.1	21.8
4	126.3	125.5	125.4	126.3	125.5	125.4	28.4	28.0	28.0
5	133.4	131.6	131.5	133.4	131.6	131.5	16.6	15.7	15.6
6	162.9	162.4	162.4	162.9	162.4	162.4	25.4	25.6	25.6
7	174.9	174.8	174.7	174.9	174.8	174.7	29.1	29.4	29.4
8	206.4	206.6	206.6	206.4	206.6	206.6	42.6	43.0	43.0
9	228.0	228.4	228.4	228.0	228.4	228.4	53.4	53.9	53.9
10	265.3	266.0	266.0	265.3	266.0	266.0	84.3	84.8	84.9

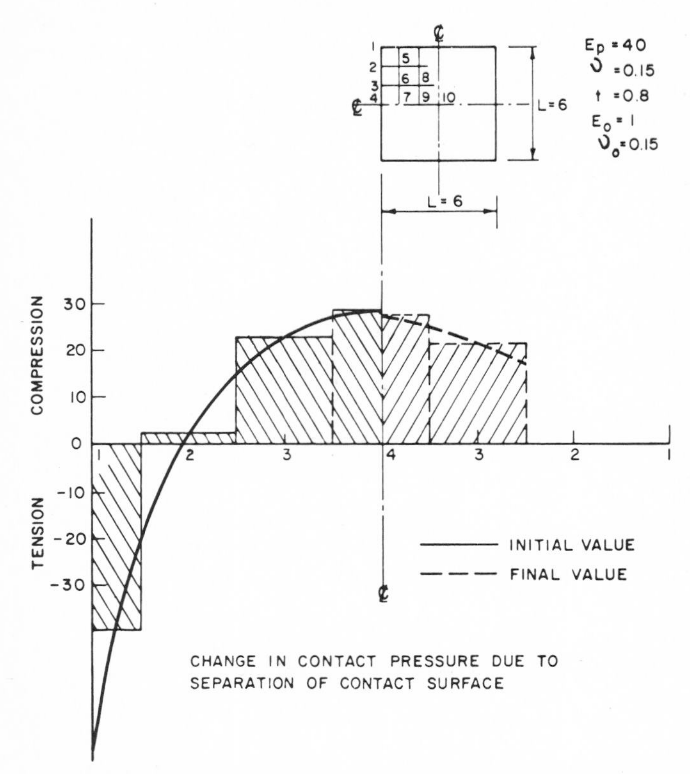

Figure 5-20 Plate on isotropic half-space with concentrated load ($P = 1000$) at center.

5-5 FINITE ELEMENT ANALYSIS OF PAVEMENTS

Two types of pavements are used. The flexible pavement is a semi-infinite multi-layer system with different material properties in different layers, and the load is distributed gradually to the layers underneath. A rigid pavement acts as a finite or infinite plate resting on an elastic foundation, and the Young's modulus of the plate material is usually very much higher than that of the foundation.

The flexible pavement has been analyzed by Duncan et al.,[26] using an axisymmetric triangular element.[27] It was concluded that for a three-layer system subjected to a uniform circular load displacements and stresses computed by the finite element method compare favorably with a layered system analysis provided that the nodal points are fixed at a depth of about 50 radii and allowed to slide vertically at a radial distance of about 12 radii from the axis of symmetry. In the same paper, nonlinear analyses were made of the resilient deflections of the Gonzales Bypass pavement for summer and winter conditions.

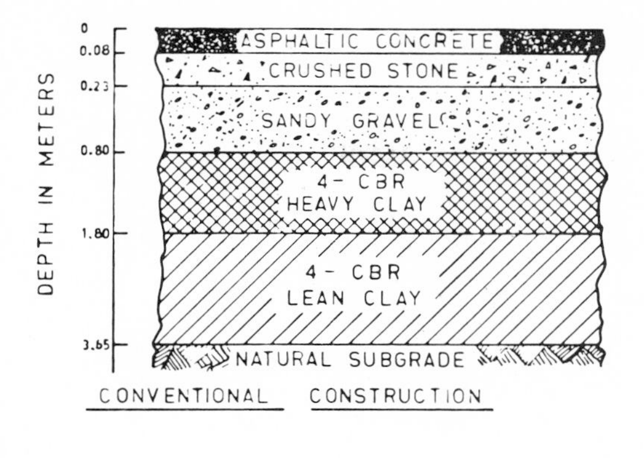

(a)

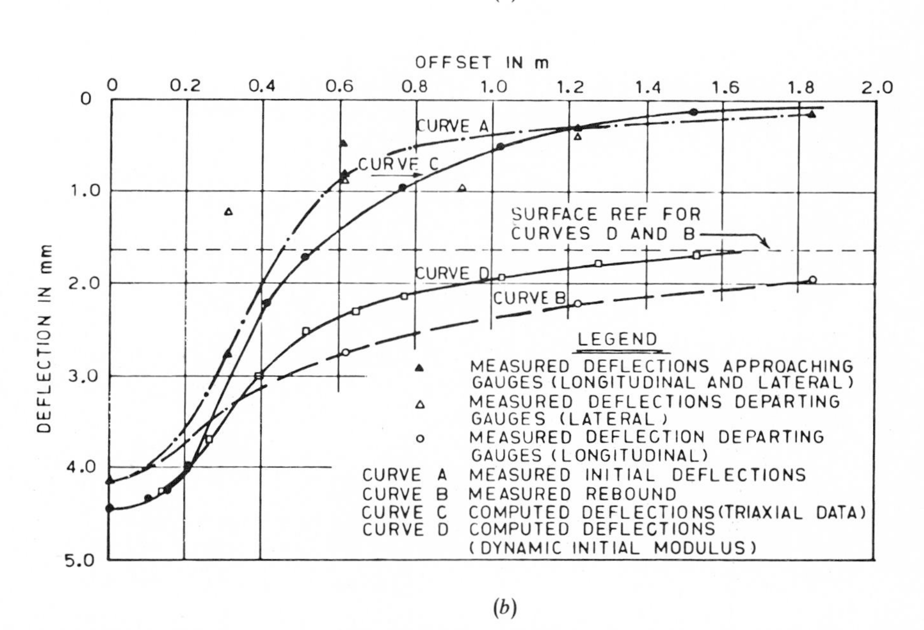

(b)

Figure 5-21 (a) Profile of test items used for analysis; (b) measured and computed deflection basins and measured rebound, conventional construction item.[28]

A nonlinear finite element analysis for heavily loaded airfield pavements was presented by Barker.[28] The computed deflections compared favorably with field measurements (Fig. 5-21) carried out at the Waterways Experiment Station at Vicksburg, Miss., and it was concluded that the method may be used as a basis of design for such pavements. For details of the test and the computation, the readers should refer to the original paper.[28]

The finite element analysis of rigid pavements has been attempted by Wang et al.,[29] using the procedure of plate on elastic foundations developed in Sec. 5-3.

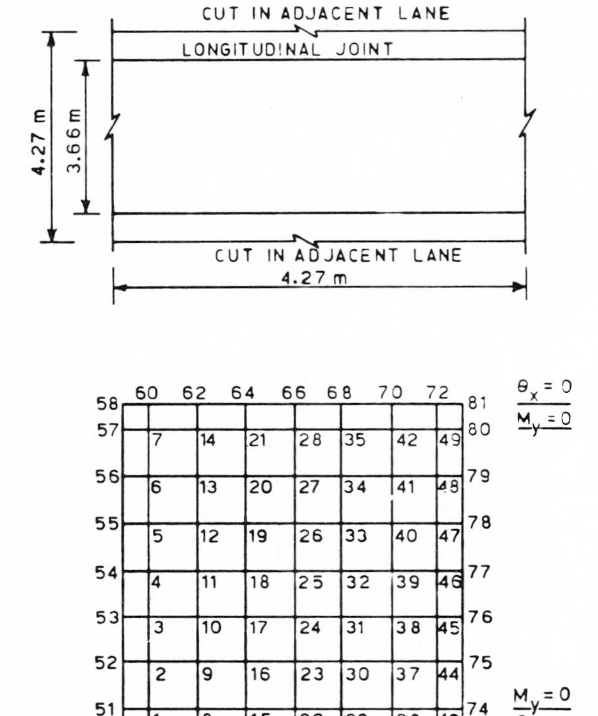

Figure 5-22 Typical airport runway and finite element mesh.

A highway or the runway of an airport consists generally of several lanes with longitudinal joints which can transfer shear forces but not bending moments, and such a special feature has been incorporated in the finite element analysis (Fig. 5-22). As is common for all numerical analysis, the plate must be cut at a distance sufficiently far from the location of load application to avoid excessive computation. The normal rotations are assumed to be zero at these cuts.

The effect of openings, e.g., openings for lighting pits in airport runways, on stresses in rigid pavements has also been investigated by the same writers.[30] Here a stress-concentration problem arises, and there are too many unknowns if the standard procedure is adopted. It was argued that since the stress concentration would occur primarily in the slab, it would be desirable to insert additional nodes in the slab which are assumed not to be in contact with the foundation. Such noncontact nodes (node 30 to node 89 of Fig. 5-23) can be eliminated by static condensation at the beginning of the analysis.

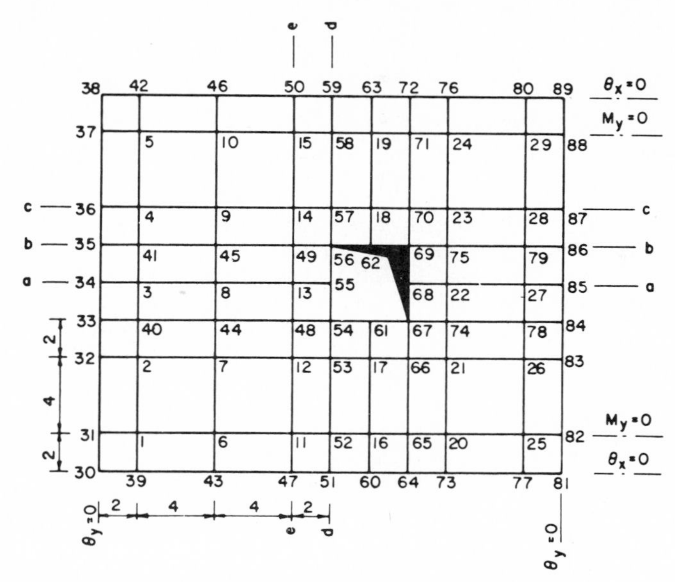

Figure 5-23 Finite element mesh for runway with opening.

REFERENCES

1. Winkler, E.: "Die Lehre von der Elastizität und Festigkeit," Prague, 1867.
2. Hetenyi, M.: "Beams on Elastic Foundations," University of Michigan Press, Ann Arbor, 1949.
3. Zemochkin, B. N., and A. P. Sinitsin: "Practical Methods for Calculating Beam and Plate on Elastic Foundation," 2d ed., Moskow, 1962 (in Russian).
4. Gerrard, C. M., and W. Jill Harrison: The Analysis of a Loaded Half Space Comprised of Anisotropic Layers, *CSIRO Div. Appl. Mech. Tech. Pap.* 10, Melbourne,
5. Vlasov, V. Z., and N. N. Leont'ev: "Beams, Plates and Shells on Elastic Foundations," Moscow, 1960 (in Russian); also *Israel Prog. Sci. Trans.*, 1966 (in English).
6. Love, A. E. H.: "A Treatise on the Mathematical Theory of Elasticity," Cambridge University Press, London, 1966.
7. Timoshenko, S., and J. N. Goodier: "Theory of Elasticity," 2d ed., McGraw-Hill Book Company, New York, 1951.
8. Pan, K. C.: "Beams and Frames on Elastic Foundations," Shanghai, 1960 (in Chinese).
9. Chai, S. W.: "Analysis of Beams on Elastic Foundations," Shanghai, 1962 (in Chinese).
10. Cheung, Y. K., and O. C. Zienkiewicz: "Plates and Tanks on Elastic Foundation: An Application of Finite Element Method," *Int. J. Solids Struct.*, vol. 1, pp. 451–461, 1965.
11. Cheung, Y. K., and D. K. Nag: Plates and Beams on Elastic Foundations: Linear and Nonlinear Behaviour, *Geotechnique*, June 1968.
12. Allen, D. N. de G., and R. T. Severn: The Stresses in Foundation Rafts. *Proc. Inst. Civ. Eng.*, vol. 15, January 1960; vol. 20, October 1961.
13. Pickett, G., M. E. Raville, W. C. Janes, and F. J. McCormick: Deflection, Moments and Reactive Pressures for Concrete Pavements, *Kans. State Coll. Eng. Exp. Stn. Bull.* 65, 1951.
14. Janes, W. C.: Digital Computer Solution for Pavement Deflections, *J. Portland Cement Ass.*, Apr. 30, 1962.

15. Cheung, Y. K.: The Finite Element Method in the Study of Plane Stresses, Plates and Shells with Special Reference to Arch Dams, Ph.D. thesis, University of Wales, 1964.
16. Severn, R. T.: The Solution of Foundation Mat Problems by Finite Element Method, *Struct. Eng.*, vol. 44, no. 6, June 1966.
17. Haddadin, M. J.: Mats and Combined Footings: Analysis by the Finite Element Method, *ACIJ.*, pp. 954–969, 1971.
18. Yang, T. Y.: A Finite Element Analysis of Plates on a Two Parameter Foundation Model, *Comput. Struct.*, vol. 2, pp. 593–614, 1972.
19. Wardle, L. J., and R. A. Fraser: Finite Element Analysis of a Plate on a Layered Cross-anisotropic Foundation, *Proc. Finite Element Methods Eng., Univ. New South Wales, August 1974.*
20. Zienkiewicz, O. C., and Y. K. Cheung: The Finite Element Method for the Analysis of Elastic Isotropic Slabs, *Proc. Inst. Civ. Eng., Aug. 28, 1964.*
21. Yang, T. Y.: Flexible Plate Finite Element on Elastic Foundation, *J. Struct. Div. Proc. ASCE,* vol. 96, no. ST10, October 1970.
22. Svec, O. J., and G. M. L. Gladwell: A Triangular Plate Bending Element for Contact Problems, *Int. J. Solids Struct.*, vol. 9, pp. 435–446, 1973.
23. Zienkiewicz, O. C., and Y. K. Cheung: "Finite Element Method in Structural and Continuum Mechanics," McGraw-Hill Book Company, New York, 1967.
24. Fraser, R. A., and L. J. Wardle: The Analysis of Stiffened Raft Foundations on Expansive Soil, *Symp. Recent Dev. Anal. Soil Behaviour Appl. Geotech. Struct., Univ. New South Wales, July 1975.*
25. Aitchison, G. D.: A Quantitative Description of the Stress-Deformation Behaviour of Expansive Soils, *Proc. 3d Int. Conf. Expansive Soils, Haifa, 1973*, vol. 2, pp. 79–82.
26. Duncan, J. M., C. L. Monismith, and E. L. Wilson: Finite Element Analyses of Pavements, *Highw. Res. Rec.* 228, 1968.
27. Clough, R. W., and Y. Rashid: Finite Element Analysis of Axisymmetric Solids, *J. Eng. Mech. Div. Proc. ASCE, February 1965.*
28. Barker, W. R.: Nonlinear Finite Element Analysis of Heavily Loaded Airfield Pavement Systems, *Proc. Symp. Appl. Finite Element Method Geotech. Eng., Vicksburg, Miss., May 1972.*
29. Wang, S. K., M. Sargious, and Y. K. Cheung: Advanced Analysis of Rigid Pavements, *Proc. ASCE,* vol. 98, no. TE1, February 1972.
30. Wang, S. K., M. Sargious, and Y. K. Cheung: Effect of Openings on Stresses in Rigid Pavements, *Proc. ASCE*, vol. 99, no. TE2, May 1973.

CHAPTER

SIX

SHALLOW FOUNDATIONS

John T. Christian

The analysis of shallow foundations calls for use of several techniques discussed in this book. Specific additional topics relevant to shallow foundations include mats (Chap. 5), consolidation (Chaps. 11 and 12), expansive soils (Chap. 13), excavations (Chap. 17), and dynamics (Chap. 20). This chapter is concerned with the analysis of the behavior of soils subjected to loads at or near the surface.

6-1 MODELING

Figure 6-1 shows a typical situation that often arises when a structure is supported by a shallow foundation on soil, and it also shows how this situation might be modeled analytically. Here the soil is represented by a number of finite elements. The structure could be represented by another set of finite elements, but for many practical applications it is adequate to consider only the loads imposed by the structure. In that case, the effect of the structure is taken care of by a set of forces applied at the surface.

Analysis of such a problem requires a finite element or finite difference program that can accept fairly general boundary conditions and loading patterns. Most problems of practical interest can be represented adequately by horizontally layered soil. The geometry will usually permit conditions of plane strain or axial

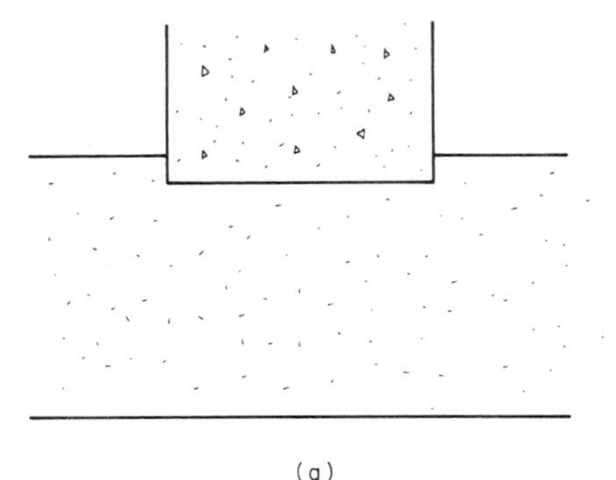

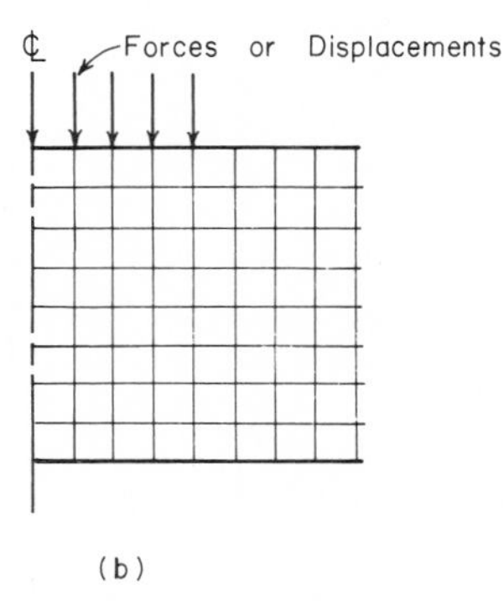

Figure 6-1 Shallow foundations: (*a*) actual problem; (*b*) finite element idealization.

symmetry. This greatly simplifies the analysis because it allows a two-dimensional program to be used. Of course, for some problems a fully three-dimensional analysis may be necessary.

The most significant things to be learned from such an analysis involve the effects of nonlinearity in soil behavior, and one must be able to represent the nonlinear stress-strain characteristics of the soil in the finite element or finite difference program. These nonlinear properties, some of which are described in Chap. 2, must be employed in combination with an incremental analysis; i.e., the program must permit the load to be applied in small increments and must permit the deformations to develop in small increments so that the evolving pattern of stress and strain can be followed.

Many programs have been written that are capable of performing such analyses,[1–6] and a great many of them are available to the engineering user. The programs differ in many particulars, e.g., the types of elements used, the stress-strain relations used, and the details of the incremental loading procedures. Nevertheless, all these programs perform essentially the same task. It is often possible to modify one of these programs to adapt to some particular requirement.

Far fewer cases have been reported in which the interactive problem including the stiffness of both the soil and the structure was modeled. Although it is today technically feasible to represent the soil and the structure in one finite element model and this is increasingly done for dynamic studies, most practical results have been obtained by somewhat simpler procedures. Focht et al.[7] reported on the analysis of a tall concrete building. In this case, the computer model represented the building. The soil was introduced by evaluating the compressibility of the underlying soils both to initial loads and as the result of consolidation and using the results to develop reaction compliances under various sections of the mat foundations. These were computed from elastic theory. A load applied at some point would also cause settlement at all the other points, and the compliances of the different points were coupled to take this into account. The results presented by Focht et al. indicate that the procedure worked very well. The most novel aspect of the analysis was the care taken by the authors to evaluate the properties of the soil in such a way that meaningful values of the compliances could be used. The analytical technique was in fact reasonably simple; the sophistication lay in the soil parameters.

One procedure is to use a standard structural analysis package together with foundation springs to analyze the behavior of buildings with shallow foundations. Another way of decoupling the problem is that described by Palmer et al.,[8] who analyzed the behavior of braced excavations by using one finite element model for the soil outside the excavation, one model for the soil inside the area of the excavation, and a third model for the sheeting. The deformations of each of these were made compatible by a manual iteration procedure. Although the technique is crude and somewhat cumbersome, it yielded useful results and was able to provide insight into the behavior of actual foundations.

Once a computer program has been written that permits nonlinear stress-strain relations and incremental loadings to be used, the next major difficulty is determining how the incremental loading will be carried out. In particular, should the load be divided into a large number of very small increments or should the increments be larger, and should iterations be performed within each increment to ensure that the stresses and strains stay on the stress-strain curve? Should a large number or a small number of elements be used? It is fairly obvious that the best results are most likely to be obtained with a large number of elements, with a load divided into many small increments, and with many iterations performed within each increment to ensure that the stresses and strains are corrected to approach the stress-strain curve. Unfortunately, these procedures take time and increase the cost of the analysis. Thus, a tradeoff must be evaluated. While relatively little has been reported on this problem, Hagmann[9] did describe one study in which he set up several standard problems for surface strip loads and then changed each of these parameters to see what improvements he obtained in the analytical results. His primary conclusion was that the most benefit was obtained by reducing the size of the increments of load, i.e., by dividing the load into a large number of small increments. Similar results for effects of size of load increment are reported by Desai.[10]

Another difficulty arises from the choice of elements. It is now popular to use isoparametric quadrilaterals in finite element analysis. While quadrilateral elements give good results for nonfrictional materials[11] in the sense that they indicate failure at approximately the load that would be predicted from bearing-capacity theory, they may sometimes yield unsatisfactory results for frictional materials.[11–13] They give estimates of ultimate load that are greatly in excess of those predicted by bearing-capacity theory, at times indicating failure loads two or three times the theoretical ones.

6-2 RESULTS OF SOME NUMERICAL ANALYSES

Figures 6-2 and 6-3 show typical results of a finite difference analysis for the effects of a strip surface loading on linearly elastoplastic soils. The initial state of stress was assumed not to influence the ultimate bearing capacity (Fig. 6-4), which was affected only by the shear strength. What the initial state of stress does affect is the level of load at which yielding begins to occur and the shape of the load-deformation curve. The ordinates of the load-deformation curve are inversely proportional to the initial modulus.

Desai[2,4,14] performed comprehensive numerical and experimental analyses for footings at ground surface and at shallow depths. These analyses included such factors as nonlinear soil behavior, layering, incremental load analysis, simulation of footings by uniform pressure and rigid displacements, use of various stress-strain models (tabular, hyperbolic, and spline; see Chap. 2), and use of numerical

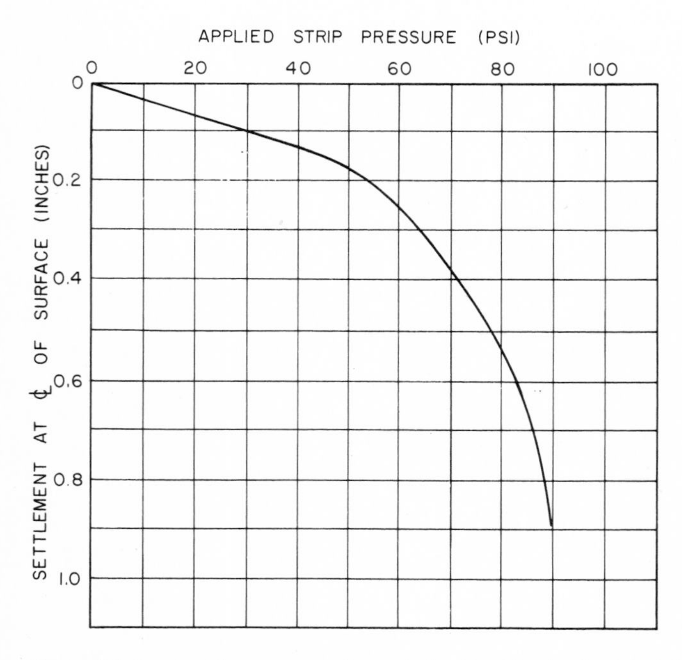

Figure 6-2 Load-settlement curve for strip load on elastic-plastic soil.[1]

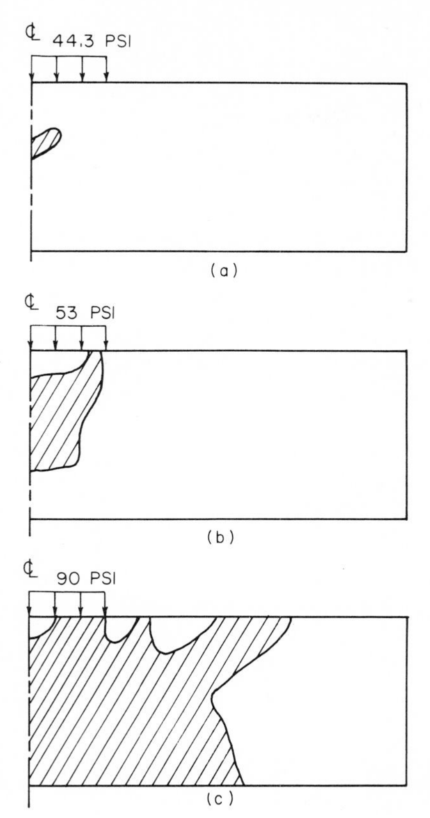

Figure 6-3 Spread of plastic zone for strip load on elastic-plastic soil.[1]

methods for design of foundations. Figure 6-5*a* and *b* shows comparisons between numerical predictions and laboratory observations for a footing on single layer of clay; the footing load was simulated by using two schemes: uniform pressure and rigid displacements. Similar results for a two-layered soil system are shown in Fig. 6-6*a* and *b*. The influence of surcharge on load-deformation behavior of a footing at shallow depth is shown in Fig. 6-7; here the footing rested at a depth of 1.2 in from the ground surface. The stress-stress behavior of the clays was derived from a series of triaxial tests conducted on samples cut from the layers used in the laboratory plate-loading tests.

The study showed that the finite element method can give satisfactory results for predictions of stresses and deformations in foundations at ground level and at shallow depths. Some details related to the use of numerical predictions for design analysis are discussed subsequently.

Figure 6-8 shows results developed by D'Appolonia.[15] In this case, the analysis was carried out by finite elements and the stress-strain model was a bilinearly elastic one rather than a truly elastic–perfectly plastic one. However, the effects of anisotropy in the shear strength are included. It can be seen that the results are similar to those of Fig. 6-4.

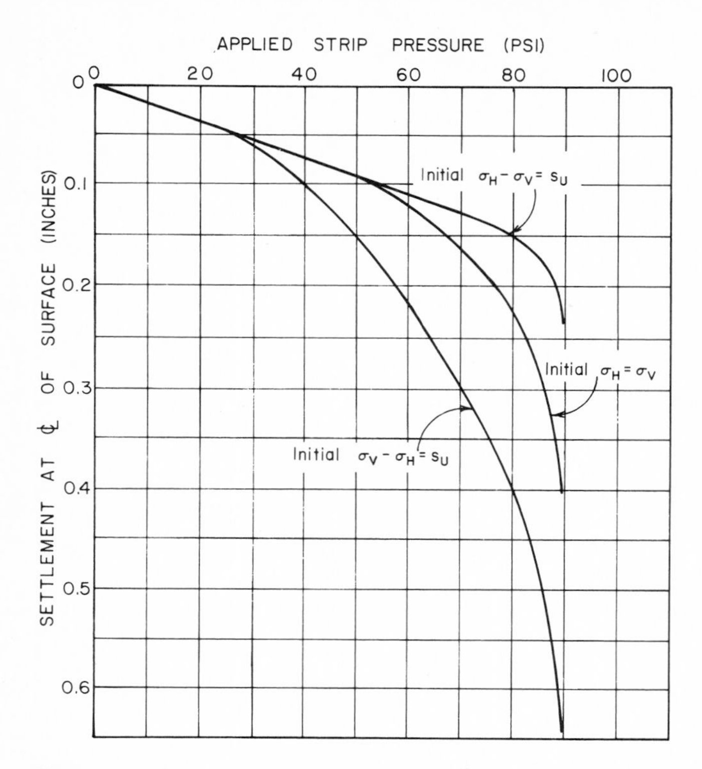

Figure 6-4 Effects of in situ stress conditions for strip load on elastic-plastic soil.[1]

The distribution of stresses at the point of first yield and at the end of the loading pattern are illustrated in Figs. 6-9 and 6-10, where the stresses have been normalized by dividing them by the applied surface traction. The normalized vertical stresses are remarkably similar in these two figures. The normalized horizontal stresses and the normalized shear stresses on the horizontal plane are not similar. This indicates that the effect of plastic yielding is to change the pattern not of vertical stress distribution but of the other stresses. A similar result for the analysis of surface loading on a linearly viscoelastic material[16] is shown in Fig. 6-11. Here, too, the effect of the nonelastic behavior is to change not the normalized vertical stresses but the normalized values of the other stresses.

Since the nonlinearity of the stress-strain relations appears in the horizontal stresses more strongly than it does in the vertical stresses, a useful simplification can be made. Suppose it is desired to find the stress distribution beneath the centerline of some surface load. Numerous elastic solutions have been published, which can be used to evaluate the vertical stress since this is not affected by the nonlinear behavior. The horizontal stress would then be computed from the same elastic solution. If the difference between the two exceeds twice the yield shear stress, a revised horizontal stress from the yielded material can be estimated by

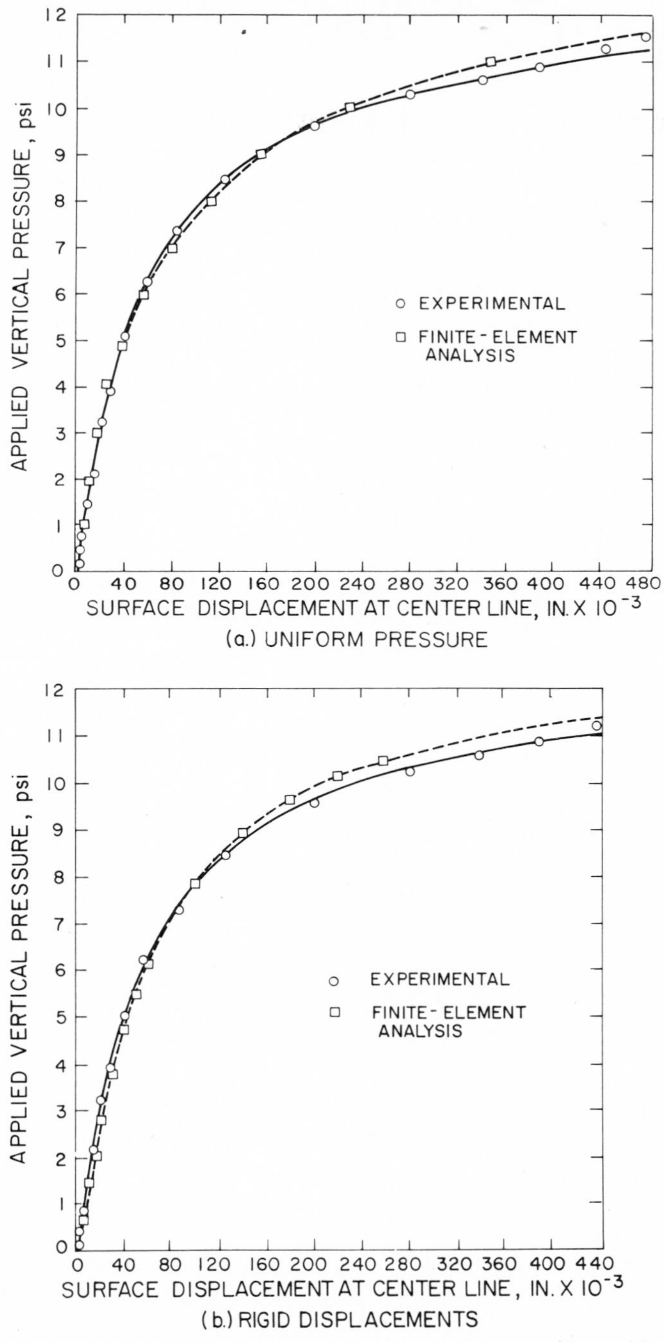

Figure 6-5 Load-displacement curves for single-layer system: (*a*) uniform pressure; (*b*) rigid displacement.[2,4]

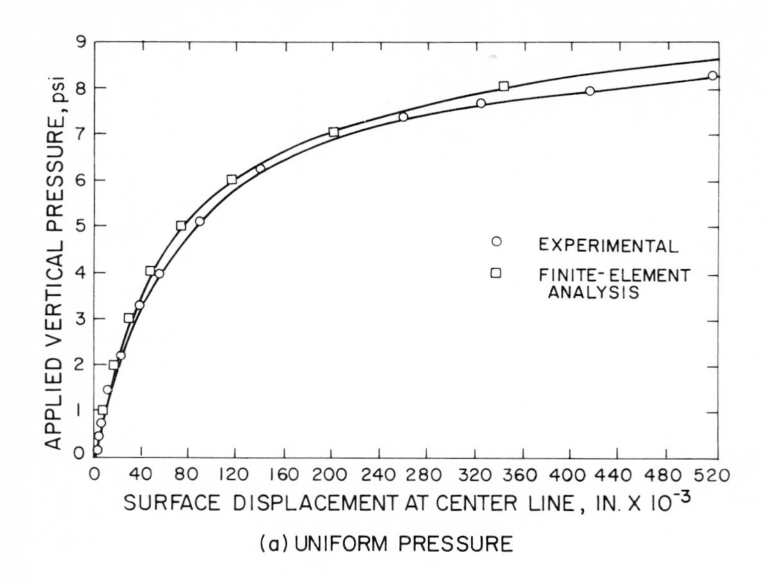

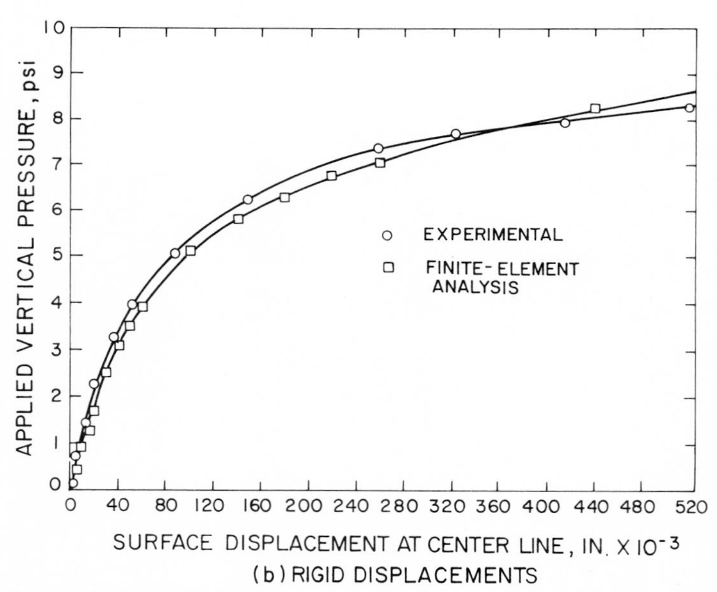

Figure 6-6 Load-displacement curves for two-layer system: (*a*) uniform pressure; (*b*) rigid displacement.[2,4]

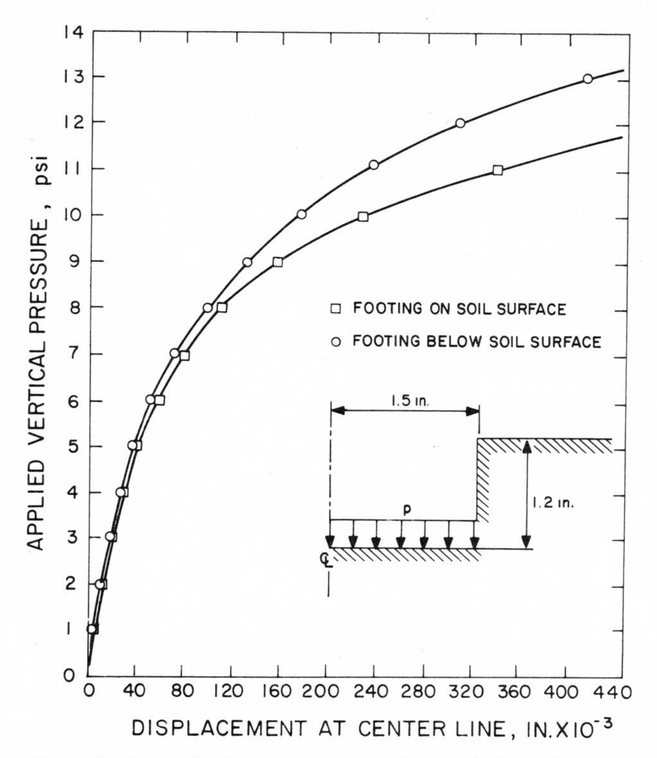

Figure 6-7 Load-displacement curve for footing below soil surface for single-layer system.[2,4]

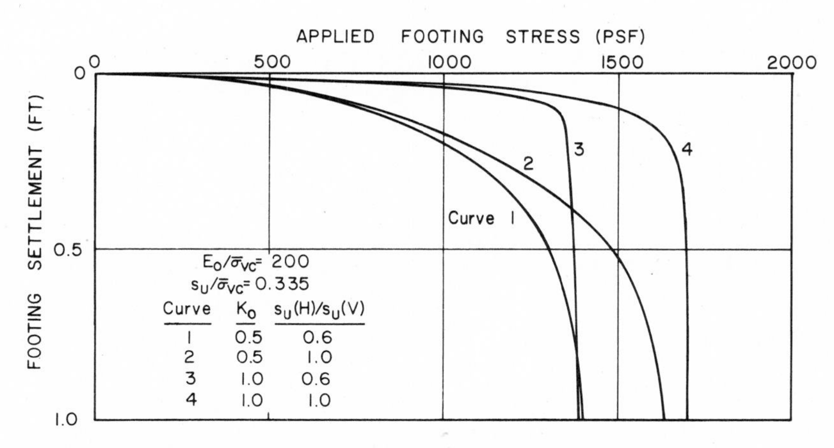

Curve	K_0	$s_u(H)/s_u(V)$
1	0.5	0.6
2	0.5	1.0
3	1.0	0.6
4	1.0	1.0

Figure 6-8 Effect of anisotropy on load-settlement behavior.[15]

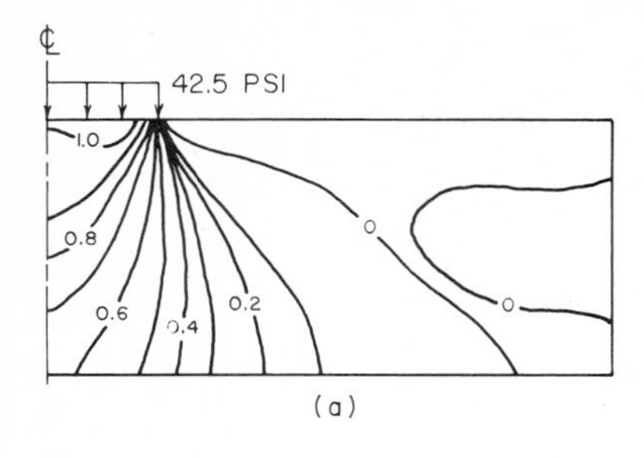

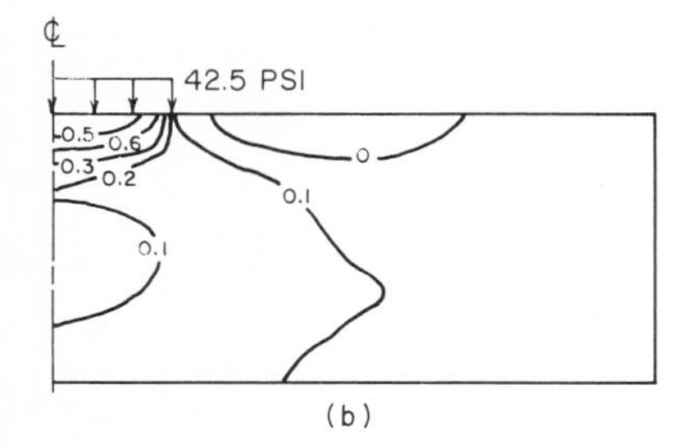

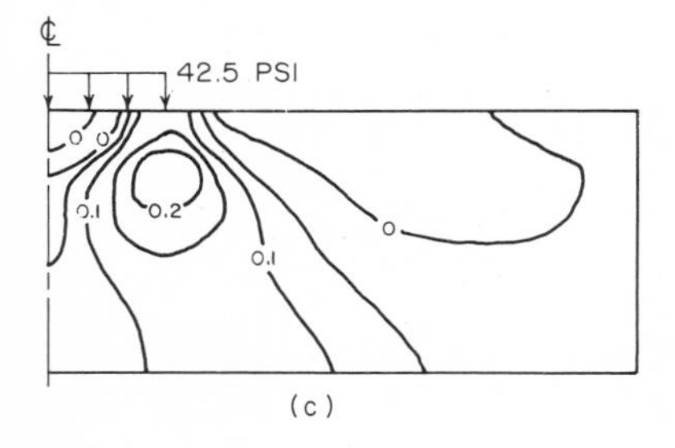

Figure 6-9 Normalized stress distribution at first yield for strip load on elastic-plastic soil: (*a*) vertical; (*b*) horizontal; (*c*) shear stress on horizontal plane.[1]

subtracting twice the shear strength from the vertical stress. Another way of looking at this is to plot the development of horizontal stress for some point under the centerline of the load. Initially the horizontal stress increases at a rate somewhat lower than that of the vertical stress. When yielding occurs, no more shear stress can be carried by that element of soil. Since the vertical stress will continue to increase at the same rate as before, from then on the horizontal stress must increase at the same rate as the vertical stress. There must be a break in the curve of horizontal stress vs. load. This is illustrated in Fig. 6-12, which is a plot of the horizontal stresses computed in the analysis of Figs. 6-2 and 6-3. Another consequence of this is that the pore pressures start to increase very rapidly because the octahedral stress is increasing more rapidly. One would expect a piezometer at this location to indicate a steady rise in pore pressure during loading until yielding occurred, at which point the rise in pore pressure would become much steeper. This has been observed in the field and has been used to interpret the results of field loading tests by using the break in the pore pressure curve to identify the point at which yielding first occurred.[17]

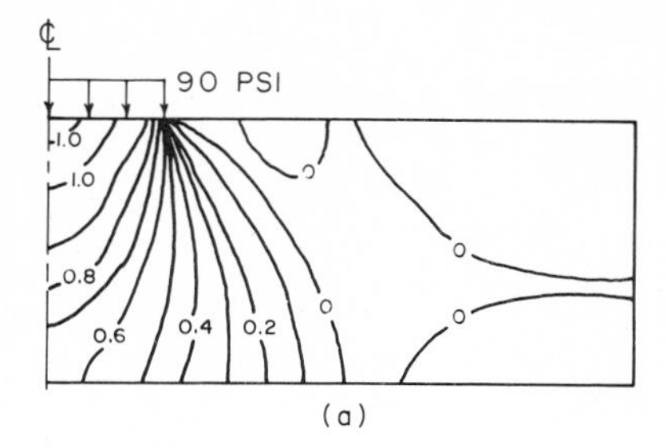

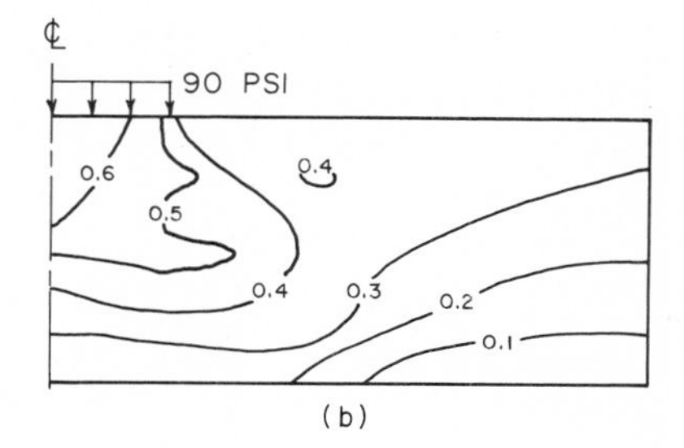

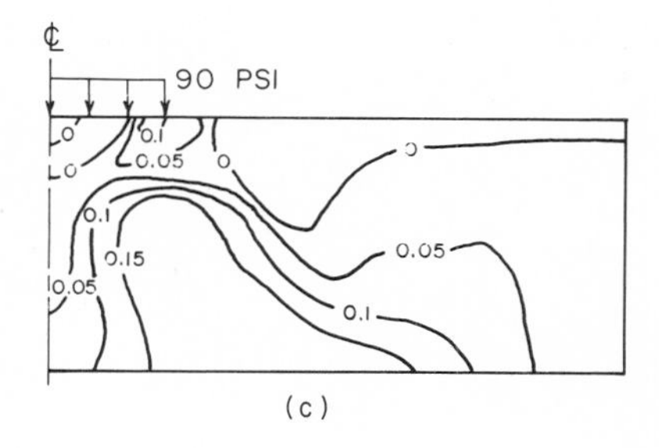

Figure 6-10 Normalized stress distribution at failure for strip load on elastic-plastic soil: (*a*) vertical; (*b*) horizontal; (*c*) shear stress on horizontal plane.[1]

D'Appolonia and others[18] have summarized many of the analytical results for strip loadings at the surface of layers of uniform soil. The soil is assumed to be linearly elastic, perfectly plastic, or bilinearly elastic with a very small modulus when the shear stress exceeds the yield stress. The magnitude of the initial shear stress before load is applied is expressed through a dimensionless parameter f

$$f = \frac{\bar{\sigma}_{V0} - \bar{\sigma}_{H0}}{2S_u(V)} = \frac{1 - K_0}{2S_u(V)/\bar{\sigma}_{V0}} \tag{6-1}$$

where $\bar{\sigma}_{V0}$ and $\bar{\sigma}_{H0}$ are the initial effective stresses in the vertical and horizontal directions, respectively, and $S_u(V)$ is the undrained shear strength for loading with the principal stress in the vertical direction. Clearly, f is merely the ratio between the maximum shear stress existing in the ground before loading and the undrained shear strength of the material. The second parameter used by these authors is the ratio between the settlement that would have occurred on a linearly elastic, isotropic material and the initial settlement that is computed for the yielding material.

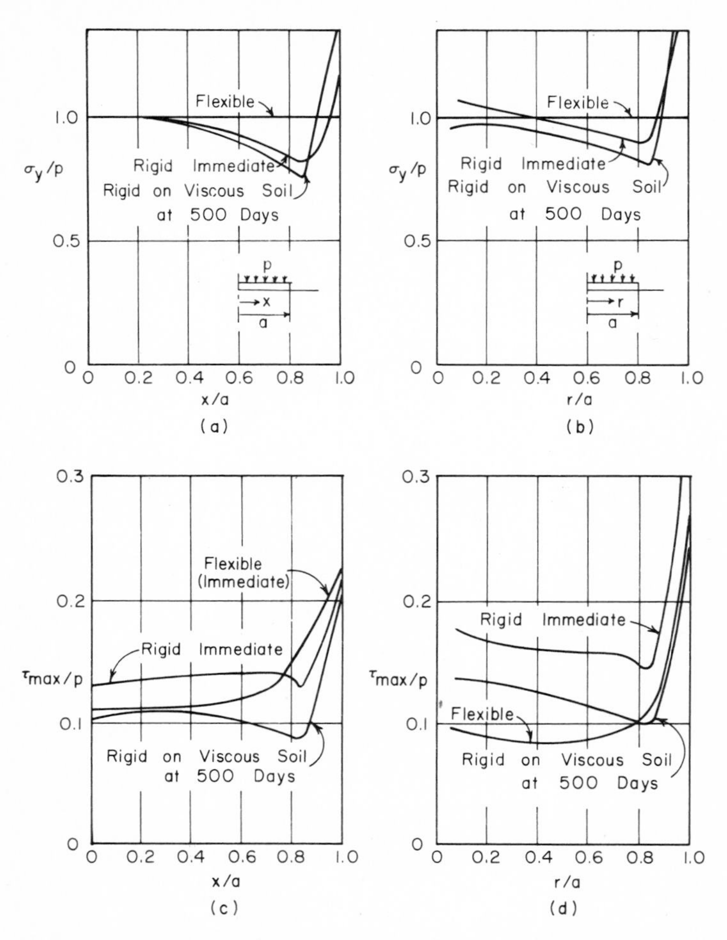

Figure 6-11 Influence of creep compliance on stress beneath footings: (*a*) strip footing, vertical stress; (*b*) circular footing, vertical stress; (*c*) strip footing, maximum shear stress; (*d*) circular footing, maximum shear stress.

Both materials have the same initial elastic modulus. This ratio, called S_R, is defined by

$$S_R = \frac{\rho_e}{\rho_i} \tag{6-2}$$

where ρ_e is the initial settlement for the elastic material and ρ_i is the initial settlement for the yielding material. The applied load on the strip can be made dimensionless by dividing it by the bearing capacity of the strip loading q_u.

Figure 6-13 shows the results of several analyses plotted with these dimensionless parameters. Before yield occurs, the ratio S_R is equal to 1. This is the top line of each of the figures. At failure, the applied stress ratio is equal to unity and

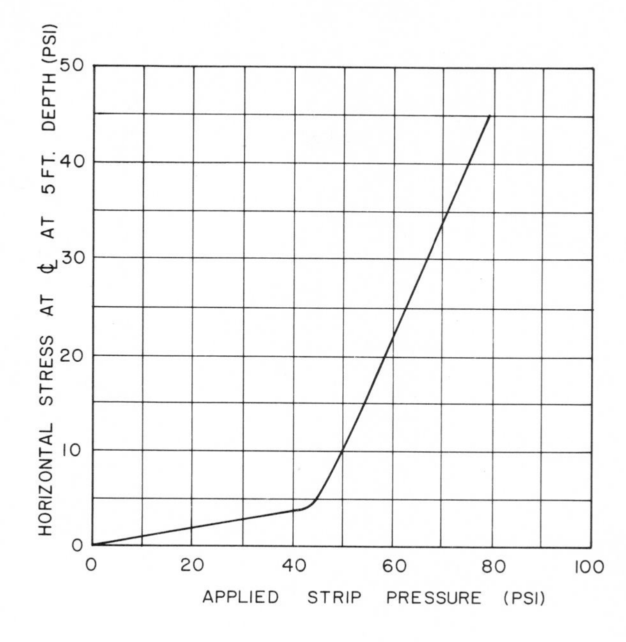

Figure 6-12 Horizontal stress vs. applied pressure for strip load on elastic-plastic soil.[1]

S_R is equal to 0. At intermediate points the settlement is determined by the parameter f. It can be seen that the larger f is, the larger are the settlements for the yielding material.

No similar charts have been prepared for materials whose shear strength includes a frictional component or for the effects of consolidation on these results. The charts exist only for undrained materials with purely cohesive shear strength. Nevertheless, charts of this sort are extremely useful in estimating the effects of nonlinear behavior on the settlement of structures and embankments.

The results presented so far can be summarized as follows:

1. The initial slope of the load-deformation curve is determined by the initial elastic properties of the soil or the initial deformation properties of the soil. This implies that the initial properties such as the initial shear modulus are very important parameters in this type of analysis.
2. The bearing capacity for nonfrictional materials is not affected by the initial stress distribution in the soil, at least for materials that are not strain-softening.
3. The normalized vertical stress, i.e., the vertical stress divided by the surface traction, is not significantly affected by plastic or viscous behavior. Consequently, the effects on the stress distribution for nonlinear materials will be apparent primarily in the normalized horizontal stresses and the normalized shear stresses.
4. A plot of horizontal stress vs. applied load under the centerline or a plot of pore pressure vs. applied load can be expected to show a sharp break when yielding occurs, and this phenomenon has been observed in the field.

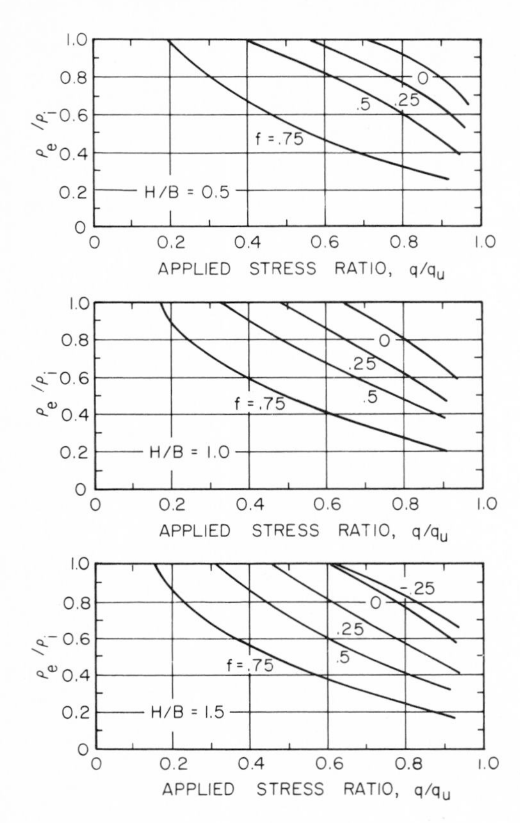

Figure 6-13 Relation between settlement ratio and applied stress ratio for strip foundation on homogeneous isotropic layer.[18]

6-3 SELECTION OF SOIL PARAMETERS

It is implicit in the preceding section that the engineer has some way of obtaining the appropriate values of the parameters describing the soil behavior. This is true whether he is doing a finite element analysis or using published results. By now there is a great deal of information available on the determination of both drained and undrained strength parameters for soil, although it must be admitted that there is still a great deal of confusion over how best to evaluate shear strength.

Unfortunately, there has not been nearly so much research into the determination of the other parameters required for the analysis. The initial modulus and the initial coefficient of lateral earth pressure at rest K_0 are not easy to measure.

The initial values of shear modulus or Young's modulus are hard to determine because it is very difficult to obtain a truly undisturbed sample of a soil. There is a great deal of evidence to suggest that one of the principal effects of minor sample disturbance is to reduce the initial stiffness of the soil (see also Chap. 7). The effect is much more pronounced on the initial modulus than it is on the ultimate shear strength of a soil sample. Therefore one must view with considerable suspicion the results of laboratory tests on "undisturbed" samples, but tests on compacted samples of material to be placed may provide very useful information on the initial moduli of the material. Two such tests are of immediate relevance.

First, the result of a conventional triaxial test or a direct shear test can be simulated by using mathematical functions (Chap. 2). Often, when a hyperbola is used, it is difficult to obtain good simulation for the initial portion of the stress-strain curve.[14] However, digression in the initial portions may not affect the prediction at larger strains.

A second type of test is the torsional resonant column test, described by a number of authors and used extensively by Hardin and Drnevich.[19,20] In this test a column of soil acts as a torsional spring with a large mass above it. The mass is caused to oscillate torsionally, and the frequency at which resonance occurs is measured. From the value of the resonant frequency it is possible to compute the stiffness in the soil spring and hence the shear modulus of the soil. Obviously, this test is subject to errors caused by initial disturbance of the samples. Marcuson and Wahls[21] and Trudeau et al.[22] have shown that one of the effects of sample disturbance is to reduce the effects of previous secondary pressure. When a sample is placed in a Hardin oscillator, consolidated to some level of stress, and subjected to a resonant column test, the value of the shear modulus is too low. If now the sample is allowed to undergo secondary compression without any change in the confining stress, the shear modulus will increase as secondary compression proceeds. A convenient way to express this is to use the shear wave velocity c_S, which is defined by

$$c_S = \sqrt{\frac{G}{\rho}} \tag{6-3}$$

where G is the shear modulus and ρ is the mass density of the material.

Trudeau et al.[22] found that for Boston blue clay the effect of secondary compression is to increase the shear wave velocity by approximately 40 ft/s per log cycle of time for secondary compression. They also found that the increase of the shear wave velocity is approximately linear with the logarithm of time, as illustrated in Fig. 6-14.

One widely used result of Hardin's work[23] is the Hardin and Black formula

$$G = 1230 \frac{(2.973 - e)^2}{1 + e} \mathrm{OCR}^K \sqrt{\bar{\sigma}_{\mathrm{oct}}} \tag{6-4}$$

where e = void ratio
OCR = overconsolidation ratio
K = exponent that depends on plasticity index (Fig. 6-15)
$\bar{\sigma}_{\mathrm{oct}}$ = octahedral effective stress

Both stress and shear modulus must be expressed in pounds per square inch for this formula to be valid. In a closure to their original paper, Hardin and Black[24] suggested that for a very small strain levels for clays the coefficient 1230 should be replaced by 1630. The results of Trudeau's work[22] support this contention for Boston blue clay. Therefore, when one is dealing with a clay deposit, it is advisable to use somewhat larger values of the coefficient in the Hardin and Black equation.

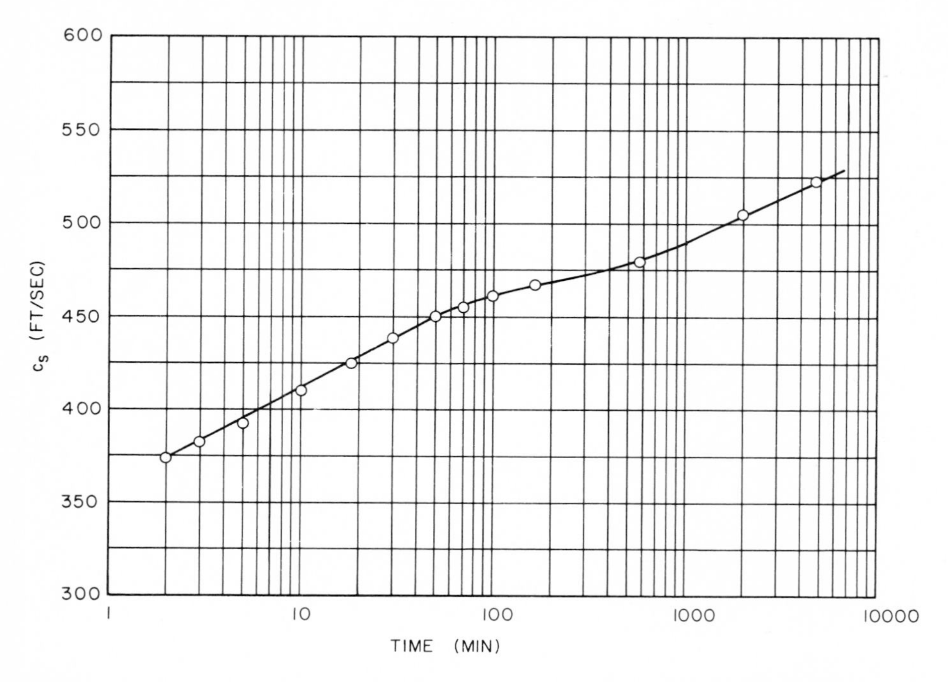

Figure 6-14 Shear wave velocity vs. time for secondary compression of Boston blue clay.[22]

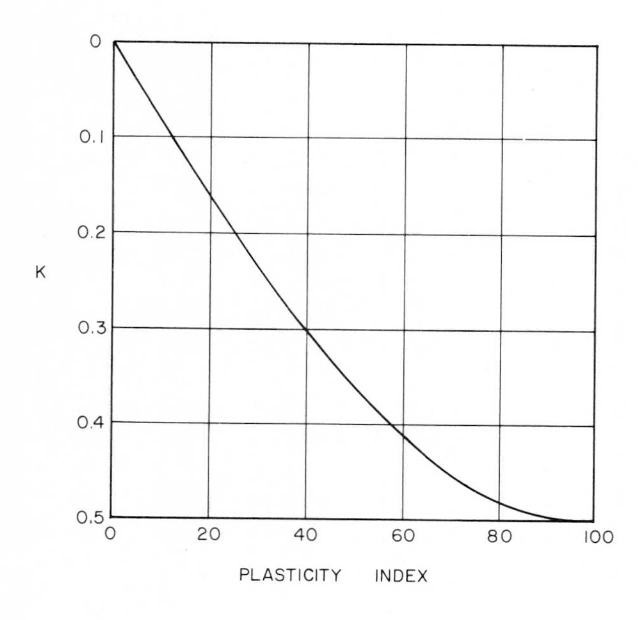

Figure 6-15 Value of coefficient K.

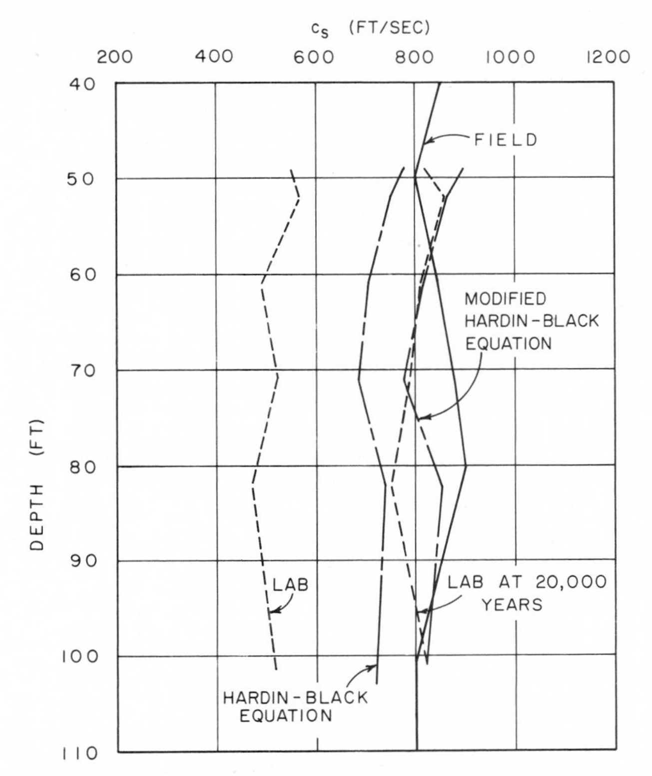

Figure 6-16 Comparison of shear wave velocity of Boston blue clay measured in the field, computed empirically, and measured in the laboratory.[22]

The equation is a very useful tool for estimating the in situ initial values of shear modulus and should be used in any such evaluation, even if only as a check on other methods.

Since the engineer really wishes to have a value of the shear modulus for the conditions in the ground, it is attractive to use some form of in situ test. The most successful of these is the shear wave velocity measurement by cross-hole seismic techniques. In this approach, a source of seismic energy, e.g., an explosion or a blow on a drill rod, is set off in one borehole while geophones or other devices sensitive to seismic waves are located in other boreholes some distance from the first but at the same elevation in the ground. The investigator measures the time of arrival of the shear waves at each of the other boreholes. As he already knows the time at which the seismic energy was made to go off, he is able to calculate the time for the shear wave to travel from the source to the receivers and can compute the shear wave velocity. It is then a simple matter to compute the shear modulus, especially since the unit weight and mass density of soils do not vary significantly.

A great deal of skill is required to make this measurement, and there is considerable controversy over the best procedures to use. One difficulty is that the compressive waves arrive before the shear waves, and it is therefore necessary to identify the shear wave arrival in an already strongly excited record. Nevertheless, the technique has been applied successfully, particularly in the nuclear power-plant construction industry, to determine the in situ properties of soil and rock. Trudeau et al.[22] used the in situ shear wave velocity technique, the Hardin and Black formula, and the resonant column test to determine the shear wave velocity and in situ shear modulus of Boston blue clay. Figure 6-16 shows the results of this study. There is considerable scatter in the initial measurements. However, when the revised Hardin and Black formula is used, and when the resonant column tests are corrected for the effects of secondary compression, the shear wave velocities are found to be in very close agreement by all three techniques. This suggests that the techniques can be used to measure the shear wave velocity, but two or more of them should be used simultaneously as checks on each other. The engineer must be very careful to evaluate the effects of secondary compression and of sample disturbance. The magnitude of the influence of secondary compression can be illustrated by the observation that for a recent postglacial clay the secondary compression accounts for approximately 250 ft/s of the shear wave velocity.[22]

SHANSEP Approach

Ladd[25] has developed a comprehensive procedure for accounting for the effects of sample disturbance when soil properties are to be evaluated. He calls this procedure SHANSEP, an acronym for *s*tress *h*istory *a*nd *n*ormalized *s*oil *e*ngineering *p*roperties. The technique is based on the observation that many properties of soils are proportional to the consolidation stress so long as the overconsolidation ratio is constant. In other words, for all normally consolidated samples of a soil, the same ratio of, say, undrained shear strength to vertical effective consolidation stress applies, and for all samples of the same soil with overconsolidation ratios of 2, another ratio applies.

It is essential in using this procedure to eliminate the effect of sample disturbance. This is done by consolidating the undisturbed samples of the soil to stresses much higher than those experienced in the ground so as to ensure that they are in fact normally consolidated. Then the properties of overconsolidated material are determined by rebounding these samples to the desired overconsolidation ratios.

Figure 6-17 shows results for a number of clays. The engineer wishing to know the shear strength of Boston blue clay at an overconsolidation ratio of 6 would find from this figure that the undrained shear strength is 0.8 times the vertical effective consolidation stress.

The approach can be extended to the evaluation of other properties. Figures 6-18 and 6-19 show families of curves for the properties of two soils used by Simon et al.[26] in the study of the behavior of some embankments. From this figure the engineer can evaluate not only the undrained shear strength but also the anisotropy of the shear strength, the initial shear modulus, and the coefficient of lateral earth pressure at rest.

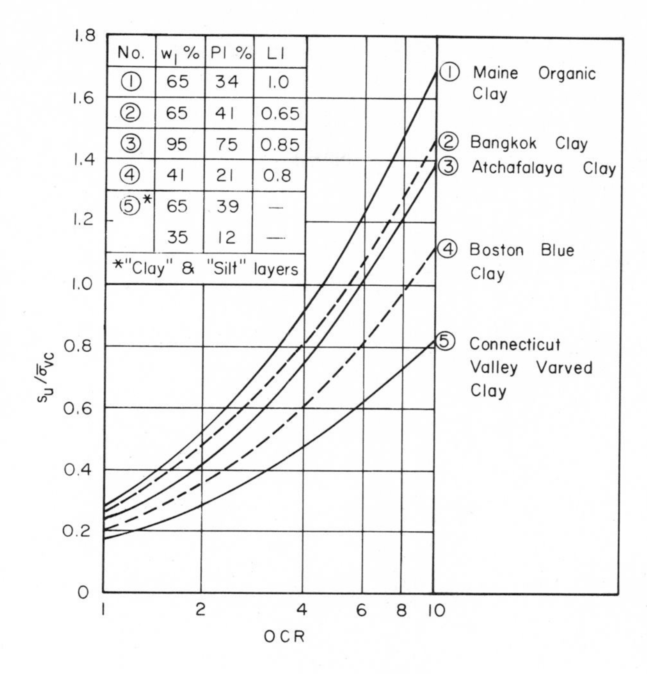

Figure 6-17 Shear-strength ratio vs. overconsolidation ratio for several clays.[25]

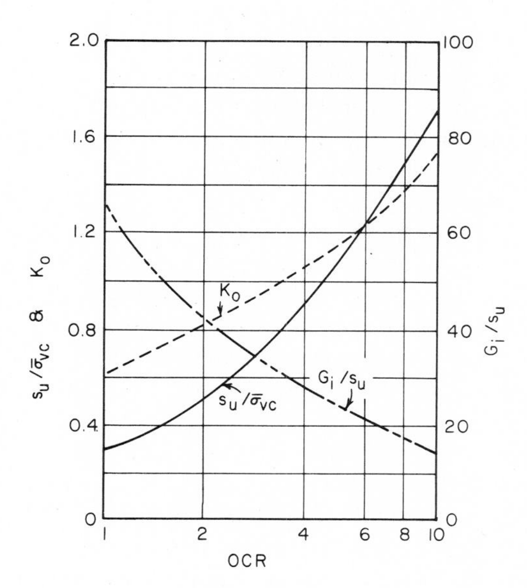

Figure 6-18 Normalized soil properties for Portland organic clay.[26]

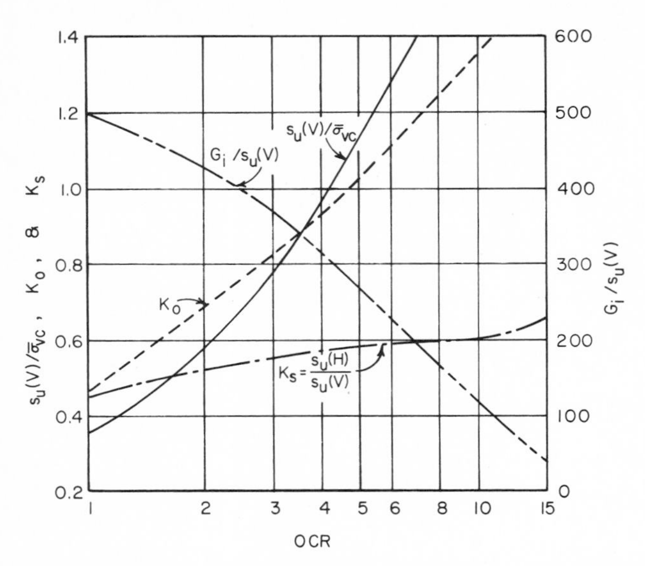

Figure 6-19 Normalized soil properties for Portsmouth sensitive clay.[26]

The shear strength should be measured in a test that conforms as closely as possible to the stress conditions expected in the field. Thus, axial compression, axial extension, and simple shear tests may all be used, and Ladd has done so for several projects. This gives a very comprehensive picture of the isotropic shear strength of the soil as a function of the confining stress and the overconsolidation ratio.

All the parameters in Figs. 6-18 and 6-19 were derived from laboratory tests on the particular soils except for the coefficient of lateral earth pressure at rest K_0, which is based on the published results of Brooker and Ireland.[27] Laboratory tests carried over a number of years indicate that these published results for K_0 are reasonably accurate.

The SHANSEP approach can be useful in finite element analyses. The engineer first evaluates the initial state of stress in the ground as best he can. He then develops a figure or figures similar to Figs. 6-17 to 6-19 for the particular soils with which he has to deal. The next step is to set up a finite element mesh representing the geometry of the problem. For each element in the mesh, the soil properties can be evaluated from the initial state of stress and from the normalized soil properties.

The loads are then applied incrementally, and nonlinear stress-strain laws are used to develop the stiffness of the elements. This is a fairly conventional procedure for finite element analysis. The normalized soil properties, i.e., the ratios between, say, shear modulus and vertical-effect stress, can be used directly in the computer program, so that at any stage in the analysis updated soil properties are used. This procedure is built into the computer program FEECON described by Simon et al.[26]

If at any point in the course of the loading a change occurs in the overconsolidation ratios or in the normalized soil properties, a change must be made in the description of materials. Such changes can occur because of consolidation, drainage, creep, and so on. The engineer can change the material types, i.e., simply use two different material numbers for the same soil, one material for the early stage of the problem and the other for the later. It should be clear that additional variations on the same procedure are possible.

The essential point to remember in carrying out this analysis is that it is very easy to become infatuated with the sophistication of numerical techniques and to lose sight of the complexities of soil behavior. Many finite element programs have been written with very comprehensive descriptions of material properties that not only have little or no relation to observed soil behavior but also require the engineer to use values of soil properties that no one knows how to measure. Actual soils tend to have properties that are quite heterogeneous and are at best statistical averages of rather badly scattered data. Therefore, the description of the material for a finite element or finite difference analysis should include the possibility of accounting for material nonlinearity in a way that is reasonable for soil, including the dependence of the soil properties on the past stress history and on the values of the consolidation stresses. Much further work is needed to improve such techniques and to determine their validity by comparing predictions with observed field behavior. Nevertheless, a promising start has been made in this direction.

Other Techniques

In addition to the techniques described above, two others show particular promise in evaluating the in situ properties of soils. Eisenstein and Morrison[28] describe how the data from a pressuremeter test can be used to provide data for finite element analyses. Increasing use of the pressuremeter, together with analyses and observations, could develop the technology to the point where it will be a reliable tool for settlement analysis.

The second technique is the hydraulic fracturing test, in which the fluid pressure in a boring or probe is increased until there is a fracture in the soil and fluid flows rapidly into the hole. This indicates that one of the principal stresses has been reduced to zero, and this usually is the horizontal stress. Thus, the method provides a direct measurement of the in situ horizontal stress, a quantity that is otherwise very hard to determine. Bjerrum et al.[29] give a detailed description of the theory and practice of hydraulic fracturing.

6-4 DESIGN AND ANALYSIS

As discussed in Secs. 6-2 and 6-3, if adequate soil parameters can be determined, it is possible to predict the full stress-deformation behavior of footings located at ground level and at shallow depths. For problems involving such complexities as nonlinear behavior, layered foundations, influence of initial stresses, and anisotropy, the finite element method can be used with considerable confidence. It permits computation of the progressive development of stresses and deformations before and after yield. From the viewpoint of applications, these results permit the location of critical stresses and a check on tolerable displacements and stresses.

It is also possible to study the effects of flexibility of footings on stress-deformation behavior. One can include the footing in the analysis or assume different loading conditions to approximate the deformation characteristics of the footing. For example, Desai[2,4] simulated flexible and rigid footings by approximating their effect through application of uniform pressure and rigid displacements (Figs. 6-5 and 6-6). It can be seen from these figures that these assumptions do not significantly influence the load-deformation behavior and ultimate loads. However, the stresses from the two assumptions can be significantly different. Figure 6-20 shows computed distributions of contact pressures at different loads for the two assumptions. The distributions for the two cases are different, and the

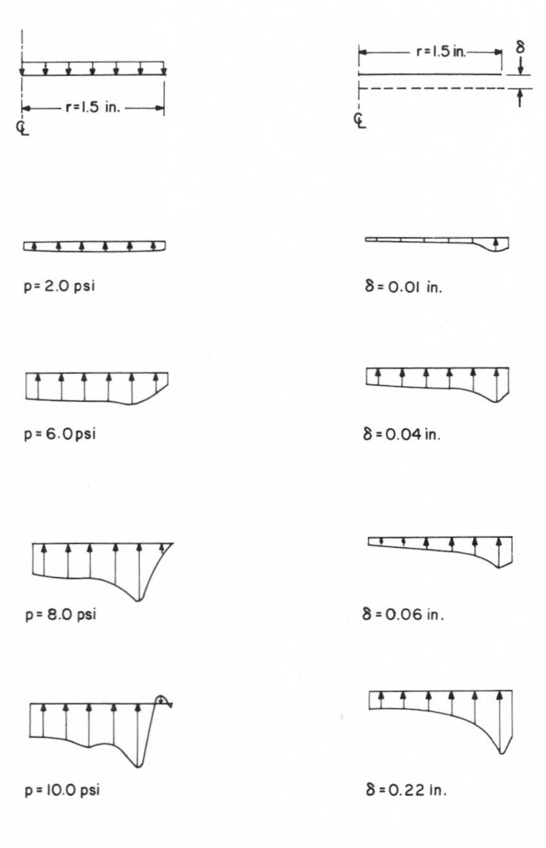

Figure 6-20 Pressure distribution below footing in two-layered system: (*a*) uniform pressure; (*b*) rigid displacements.[2,4]

uniform pressure case indicates separation of the edges of the footing from the foundation, whereas the rigid case does not show such effects.

One of the reasons for use of these techniques is to find the bearing capacity of the foundation. As stated previously, the prediction of load-settlement behavior may not always show a distinct yield point. In general, therefore, it is convenient to use the concept of critical or tolerable displacement to define the bearing capacity (see Chap. 7). Accordingly, the designer can choose a critical displacement and then read from the numerically predicted load-settlement curve (Fig. 6-5) the corresponding value of bearing load. Alternatively, the load corresponding to the intersection of the tangents at the initial and ultimate portions of the curve (Fig. 6-5*a*) can be used for the bearing value. For example, for the model footing with radius equal to 1.5 in, we can adopt a critical displacement of 0.225 in, which is 15 percent of the radius of the footing. At this displacement Fig. 6-5*a* shows a bearing load of about 9.9 lb/in^2 as computed from the plate-loading test. The corresponding values from the curves for uniform-pressure and rigid-displacement approximations are about 10.00 and 10.15 lb/in^2, respectively. The bearing-capacity values found by using the Terzaghi, Skempton, and Meyerhof formulas were 7.95, 9.92, and 9.90 respectively.[2]

REFERENCES

1. Höeg, Kaare, John T. Christian, and Robert V. Whitman: Settlement of Strip Load on Elastic Plastic Soil, *J. Soil Mech. Found. Div. ASCE*, vol. 94, no. SM2, pp. 431–445, March 1968.
2. Desai, C. S.: Solution of Stress-Deformation Problems in Soil and Rock Mechanics Using the Finite Element Methods, Ph.D. dissertation, University of Texas, Austin. 1968.
3. Girijavallabhan, C. V., and Lyman C. Reese: Finite-Element Method Applied to Some Problems in Soil Mechanics, *J. Soil Mech. Found. Div. ASCE*, vol. 94, no. SM2, pp. 473–496, March 1968.
4. Desai, C. S., and L. C. Reese: Analysis of Circular Footings on Layered Soils, *J. Soil. Mech. Found. Div. ASCE*, vol. 96, no. SM4, July 1970.
5. D'Appolonia, D. J., and T. W. Lambe: Method for Predicting Initial Settlement, *J. Soil Mech. Found. Div. ASCE*, vol. 96, no. SM2, pp. 523–545, March 1970.
6. Duncan, J. M., and C. Y. Chang: Non-linear Analysis of Stress and Strain in Soils, *J. Soil Mech. Found. Div. ASCE*, vol. 96, no. SM5, pp. 1629–1653, September 1970.
7. Focht, J. A., F. R. Khan, and J. P. Gemeinhardt: Performance of Deep Mat Foundation of 52-Story One Shell Plaza Building, *ASCE Annu. Nat. Environ. Eng. Meet., St. Louis, October 1971.*
8. Palmer, J. H. LaVerne, and T. Cameron Kenney: Analytical Study of a Braced Excavation in Weak Clay, *Can. Geotech. J.*, vol. 9, no. 2, pp. 145–164, May 1972.
9. Hagmann, Alfred J.: Prediction of Stress and Strain under Drained Loading Conditions, *Tech. Dept. Civ. Eng. Rep.* R71-3, 1971.
10. Desai, C. S.: Finite Element Method for Analysis and Design of Deep Foundations, draft rep., U.S. Army Engineers Waterways Experiment Station, Vicksburg, Miss., 1973.
11. Prediction of the Deformation of a Levee on a Soft Foundation, *MIT Dept. Civ. Eng. Rep.* R69-18, 1969.
12. Marr, W. Allen, Jr., and John T. Christian: Finite Element Analysis of Elasto-plastic Soils, *MIT Dept. Civ. Eng. Rep.* R72-21, 1972.
13. Christian, John T.: Use of Incremental Plasticity Relations in Finite Element Analysis of Deformation of Frictional Soils, *Meet. Soc. Eng. Sci., Duke Univ., October 1974.*
14. Desai, C. S.: Nonlinear Analyses Using Spline Functions, *J. Soil Mech. Found. Div. ASCE*, vol. 97, no. SM10, October 1971.

15. D'Appolonia, D. J.: Prediction of Stress and Deformation for Undrained Loading Conditions, Ph.D. thesis, Department of Civil Engineering, Massachusetts Institute of Technology, Cambridge, Mass., 1968.
16. Christian, John T., and Brian J. Watt: Undrained Visco-elastic Analysis of Soil Deformations, *Proc. Symp. Appl. Finite Element Method Geotech. Eng., Vicksburg, Miss., May 1972*, pp. 533–579.
17. Höeg, Kaare, O. B. Andersland, and E. N. Rolfsen: Undrained Behaviour of Thick Clay under Load Tests at Asrum, *Geotechnique*, vol. 19, no. 1, pp. 101–115, March 1969.
18. D'Appolonia, D. J., H. G. Poulos, and C. C. Ladd: Initial Settlement of Structures on Clay, *J. Soil Mech. Found. Div. ASCE*, vol. 97, no. SM10, pp. 1359–1377, October 1971.
19. Hardin, B. O., and V. P. Drnevich: Shear Modulus and Damping in Soils: Measurement and Parameter Effects, *J. Soil Mech. Found. Div. ASCE*, vol. 98, no. SM6, pp. 603–624, June 1972.
20. Hardin, B. O., and V. P. Drnevich: Shear Modulus and Damping in Soils: Design Equations and Curves, *J. Soil Mech. Found. Div. ASCE*, vol. 98, no. SM1, pp. 667–692, July 1972.
21. Marcuson, W. F., III, and H. E. Wahls: Time Effects on Dynamic Shear Modulus of Clays, *J. Soil Mech. Found. Div. ASCE*, vol. 98, no. SM12, pp. 1359–1373, December 1972.
22. Trudeau, Paul J., Robert V. Whitman, and John T. Christian: Shear Wave Velocity and Modulus of a Marine Clay, *J. Boston Soc. Civ. Eng.*, vol. 61, no. 1, pp. 12–25, January 1974.
23. Hardin, B. O., and W. L. Black: Vibration Modulus of Normally Consolidated Clays, *J. Soil Mech. Found. Div. ASCE*, vol. 94, no. SM2, pp. 353–369, March 1968.
24. Hardin, B. O., and W. L. Black: closure to Vibration Modulus of Normally Consolidated Clays, *J. Soil Mech. Found. Div. ASCE*, vol. 95, no. SM6, pp. 1531–1537, November 1969.
25. Ladd, Charles C., and Roger Foott: New Design Procedure for Stability of Soft Clays, *J. Geotech. Eng. Div. ASCE*, vol. 100, no. GT1, pp. 763–786, July 1974.
26. Simon, Richard M., John T. Christian, and Charles C. Ladd: Analysis of Undrained Behavior of Loads on Clay, *Proc. Conf. Anal. Des. Geotech. Eng. ASCE, Austin, Tex., June 1974*, pp. 51–84.
27. Brooker, Elmer W., and Herbert O. Ireland: Earth Pressures at Rest Related to Stress History, *Can. Geotech. J.*, vol. 2, no. 1, pp. 1–15, February 1965.
28. Eisenstein, Z., and N. A. Morrison: Prediction of Foundation Deformations in Edmonton Using an *in situ* Pressure Probe, *Can. Geotech. J.*, vol. 10, no. 2, pp. 193–210, May 1973.
29. Bjerrum, L., J. K. T. L. Nash, R. M. Kennard, and R. E. Gibson: Hydraulic Fracturing in Field Permeability Testing, *Geotechnique*, vol. 22, no. 2, pp. 319–332, June 1972.

CHAPTER

SEVEN

DEEP FOUNDATIONS

Chandrakant S. Desai

7-1 INTRODUCTION

Conventional methods for computing deformations and bearing capacities of deep (pile) foundations cannot generally account for such factors as in situ stress and its spatial variation, stresses and disturbances caused by driving, variation in strength of soils and interfaces, size and length of embedment, geometrical changes, consolidation and negative skin friction, group and interaction effects, cyclic loading, and physical phenomenon like arching. The finite element and finite difference methods have shown considerable promise in handling many of these factors.

Conventionally, the theory of elasticity is used for evaluation of deformations in piles and soils before the ultimate loading condition. The ultimate bearing capacity is evaluated from formulas based on limit equilibrium.[1,2] These approaches require many simplifying assumptions regarding the properties of soils, structure, and loading. Bearing capacity is often obtained by using a dynamic formula. Such a value of bearing capacity can be in error as much as 100 percent compared with field data.[3] Moreover, this method can provide values of bearing capacity but not the history of deformations.

Analysis of piles can also be made by using the wave equation, covered in Chap. 8. This procedure can be extremely useful for performing parametric studies for evaluation of such factors as efficiency of hammer and types of cushions.

Its usefulness for computing the bearing capacity and in accounting for interaction effects, however, is limited.

Field and laboratory load tests can be expensive and are feasible only for selected sites and for localized soil conditions. It is therefore desirable to evolve general procedures that permit computation of both the history of deformations and bearing capacities by releasing the restrictions concerning the factors stated previously. Numerical methods show considerable promise of developing such general procedures.

Scope

We shall cover essentially the finite element and finite difference approaches. Since most of the work in the past has been done for single piles, a major portion of the chapter will be devoted to analysis of single piles. Numerical methods, particularly the finite element method, however, can be readily applied to such three-dimensional problems as pile groups, pile-pile cap and superstructure combinations, and laterally loaded piles. We shall cover briefly the basic theory of three-dimensional analysis; a number of related factors are also discussed in Chap. 10.

Procedures based on the theory of elasticity using Mindlin's equations for analysis of single and group piles are described in Chap. 10. These procedures provide solutions to many practical problems, but they may become difficult and require modification if such factors as nonlinear soil behavior, layered systems, and nonlinear interaction effects between pile and soil must be included.

A finite difference procedure satisfying compatibility of loads and deformations has been used by various investigators[4–6] following a step-by-step integration scheme. The method requires use of a function to define the transfer of load from the pile to the surrounding soil; the function is usually determined from semiempirical relations and from laboratory tests. We shall cover this method and discuss its limitations.

Since the finite element method is perhaps the most general procedure available at this time, major attention in this chapter will be devoted to it.

Load-Deformation Analysis

A detailed discussion of the theory of elasticity used for computation of stresses and deformations before the ultimate loading can be found in various publications.[1,2] We shall present here some basic aspects of the mechanism of load transfer in pile foundations and of bearing capacity that are used in the subsequent finite element analysis.

The total load Q_T that can be borne safely by a pile without undergoing undue deformation and stresses can be considered to be carried jointly by the point or base resistance Q_p and by the wall or skin friction Q_w (Fig. 7-1). Therefore

$$Q_T = Q_p + Q_w \tag{7-1}$$

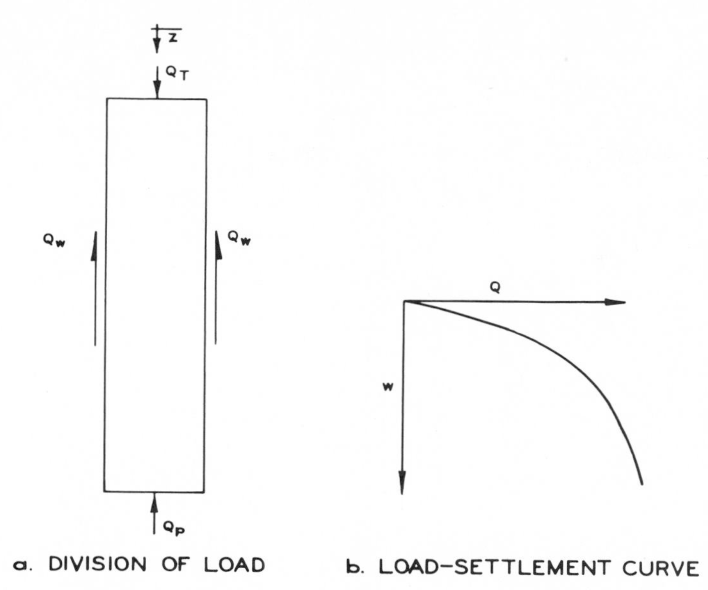

Figure 7-1 Mechanism of load transfer.

The mechanism of transfer of these components will depend upon the properties of soils and interfaces and on other factors introduced by construction and geological characteristics of the site.

Resistance by wall friction can be expressed as

$$Q_w = \sum f_0 A_w \tag{7-2a}$$

where A_w is the circumferential area of the pile and f_0 is the amount of skin resistance, given by

$$f_0 = K\bar{\sigma}_v \tan \delta + c_a \tag{7-2b}$$

where K = coefficient of lateral earth pressure
$\bar{\sigma}_v$ = effective vertical stress at point on pile
δ = angle of frictional resistance between pile and soil
c_a = adhesion between pile and soil

The parameter K that defines the in situ stress condition is an important factor. Its value may vary from less than unity to greater than unity, depending upon the geology at the site and driving conditions. The normal range of the values of K in the vicinity of a driven pile is about 1 to 3. The magnitudes of δ and c_a are usually less than the angle of internal friction ϕ and undrained strength c_u of the surrounding soils, respectively.

The point resistance is expressed as

$$Q_p = q_0 A_p \tag{7-3a}$$

where A_p is the cross-sectional area of the base and q_0 is the ultimate point resistance. The limit-equilibrium approaches[1,2,7–9] lead to the following form for q_0:

$$q_0 = cN_c + qN_q \tag{7-3b}$$

where q is the overburden pressure and N_c and N_q are the bearing-capacity factors.[1]

7-2 GEOTECHNICAL CONSIDERATIONS

One of the important factors in the analysis of pile behavior is the constitutive or stress-strain behavior of soils and interfaces. For clayey soils, stress-strain curves and strength parameters determined from triaxial tests may be adequate for characterization of the soils as nonlinear (piecewise linear) elastic behavior. Behavior at the interface can be derived on the basis of direct shear tests simulating the junction between pile and soil. Generally for undrained conditions, adhesion c_a can be expressed as a fraction $c_a = \alpha_c c_u$ of the undrained strength c_u. The values of α_c can vary[10] from about 0.3 to 0.64; for some soils the value may be equal to or greater than unity.

In an alternative method[11,12] shaft friction f is expressed in terms of the effective overburden pressure p as

$$f = K \tan \delta p \tag{7-4a}$$

in which $K \tan \delta$ is expressed as

$$\beta = \frac{f}{p} = K \tan \delta \tag{7-4b}$$

where β is similar to α_c. The difference between the two is that β is related[12] to the fundamental effective stress parameters δ and K. Some preliminary work toward comparisons between finite element results using α_c and β has been initiated.[13]

If the long-range load-deformation behavior is desired, it will be necessary to incorporate consolidation effects into the analysis. Moreover, if the soil surrounding a pile experiences (consolidation) settlement, the question of negative skin friction needs consideration.

Drained triaxial tests and interface direct shear tests can provide data for simulating the behavior of sands and interfaces between pile and sands, respectively. For cohesionless media, the angle δ is usually less than ϕ and can be expressed as $\delta = \alpha_s \phi$; α_s can be about 0.7 for smooth interface and about 0.85 for rough contact.[10]

Some soils may exhibit such special characteristics as strain softening, stress release, dilatation, and arching. Moreover, the behavior of soils under cyclic and dynamic loading, in which significant variations in the path of loading can occur, is different from static loading, in which essentially monotonic loads are applied.

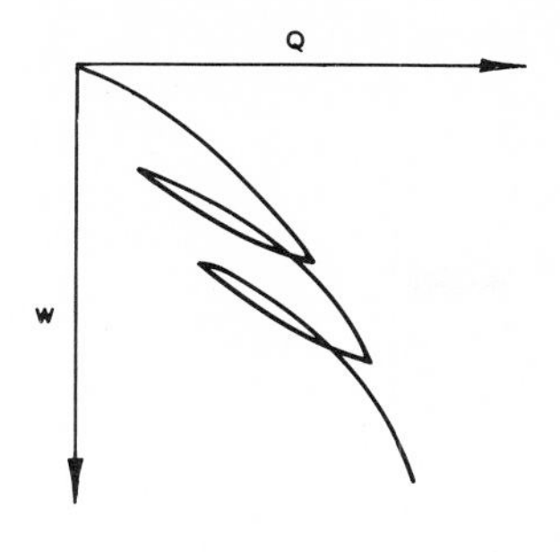

a. TYPICAL LOAD-SETTLEMENT DATA FROM FIELD TESTS

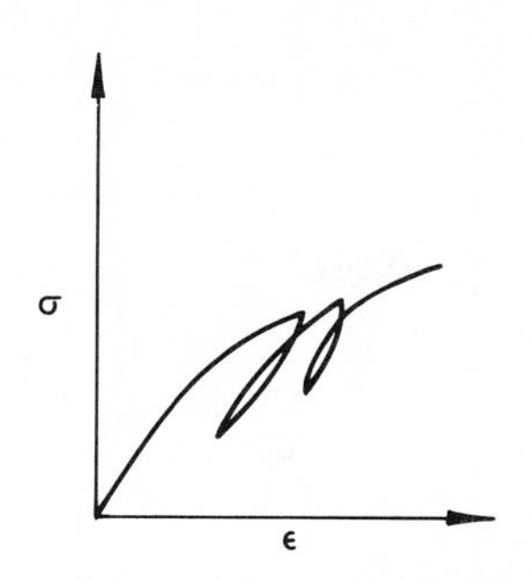

b. TYPICAL STRESS-STRAIN CURVE FOR SOILS

Figure 7-2 Load-settlement and stress-strain curves.

The assumption of nonlinear elastic behavior defined by pairs of parameters such as tangent modulus E_t and Poisson's ratio v_t, and shear modulus G_t and bulk modulus K_t do not account adequately for these factors. Improved material models based on higher-order elasticity and plasticity (critical-state concepts and cap models) become necessary.[14–16] A detailed description of various models is available in Ref. 16 and Chap. 2.

It has been found[13,17,18] that for many problems involving monotonic loading, the nonlinear elastic model may be adequate for practical applications. Although the model may introduce local errors, the gross predictions can be adequate for practical purposes. We shall discuss some of the foregoing special aspects subsequently.

We can also use, in an approximate manner, the nonlinear elastic approach for cyclic loading (Fig. 7-2) involving loading and unloading cycles.

7-3 NUMERICAL METHODS

Finite Difference Method

The basic idea underlying the finite difference approximation for the behavior of (axially loaded) piles is rooted in the satisfaction of compatibility between loads and deformations. Seed and Reese[4] and Reese[5] have presented the following differential equation to account for the compatibility requirement:

$$\frac{d^2w}{dz^2} - \frac{C\beta'}{AE}w = 0 \tag{7-5}$$

where w = axial deformation (Fig. 7-3) of element of pile
z = coordinate direction (Fig. 7-3a)
C = circumference of pile
A = cross-sectional area

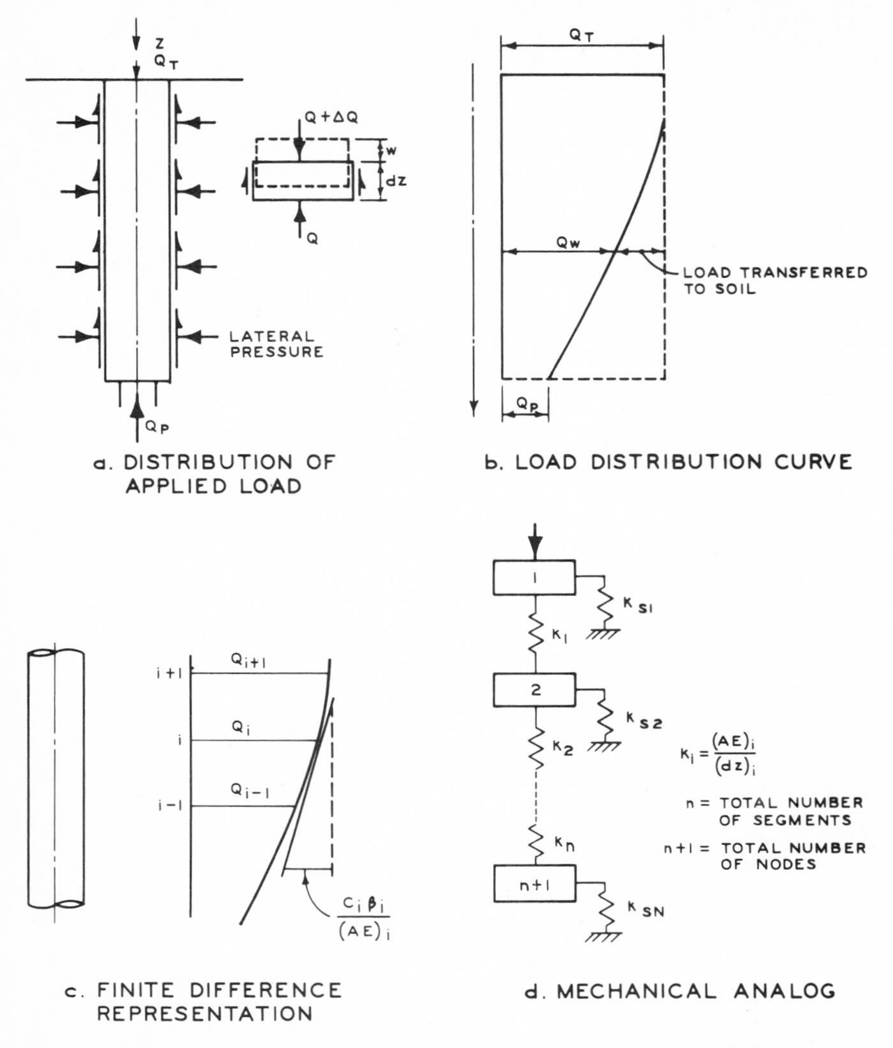

Figure 7-3 Finite difference and mechanical analogs for axially loaded piles.[4-6]

The parameter β' defines a transfer relation between the load (Fig. 7-3*b*) and the shear force as

$$\frac{dQ}{dz} = \tau C = (\beta' w)C \tag{7-6}$$

where τ is the shear force per unit area. Closed-form solutions to Eq. (7-5) can be obtained on the basis of known values of β' and (two) boundary conditions. These solutions, however, are possible only for simple variations of β'. Realistic variations of β' for nonlinear conditions usually call for numerical procedures; the finite difference method has been used as one such numerical procedure.[4–6] For instance, we can write Eq. (7-5) in the difference form (Fig. 7-3*c*) as

$$\frac{(dw/dz)_{i+1} - (dw/dz)_{i-1}}{2\,dz} - \frac{C_i \beta_i'}{(AE)_i} w_i = 0 \tag{7-7a}$$

or

$$Q_{i+1} - Q_{i-1} - 2\,dz \frac{C_i \beta_i'}{(AE)_i} w_i = 0 \tag{7-7b}$$

or

$$Q_{i+1} - Q_{i-1} = \lambda S_i \tag{7-7c}$$

Figure 7-3*d* shows a mechanical analog to Eq. (7-5). The pile segments are replaced by (linear) elastic springs with stiffness equal to k_{pi}, and the soil resistance is simulated by springs (k_{si}) attached to the pile segments.

The finite difference equations can be solved by using a recursive procedure. If we assume a deflection value w_{n+1}^j for the node $n + 1$, we can evaluate the shear force λS_{n+1} from that deflection, which in turn can lead to load Q_n [Eq. (7-6)]. Deflection w_n can now be obtained from the computed value of Q_n as k_{pn} is known. The procedure is continued to evaluate butt load Q_T^j and displacement w_1^j corresponding to the assumed value of w_{n+1}^j, thus yielding points on the load-settlement curve (Fig. 7-1*b*). The procedure can be cast in a recursive form as

$$w_n^j = w_{n+1}^j + \frac{Q_n\,dz}{(AE)_n} \qquad j = 1, 2, \ldots, m \tag{7-8}$$

where m is the number of the values of w_{n+1}^j chosen to compute the corresponding value of Q_T^j to obtain the entire load-settlement plot. This procedure has been used by Reese[5] and Coyle and Reese[6] for solution of practical problems. The finite difference approach can also be employed for pile and soil on the basis of elastic or piecewise elastic soil response.[19]

The foregoing method replaces soil reaction by independent springs. Consequently, it is assumed that shear at one point does not influence deformations at the adjacent points; thus interaction effects are not accounted for precisely. Perhaps a higher-order relation for β' can include interaction effects. Moreover, since the pile and soil are assumed elastic and are replaced by one-dimensional elements, the size effects are not included.

Comments

Solutions based on compatibility of load and deformation and on elasticity approaches are covered in Chap. 10. As shown there, those procedures and the foregoing finite difference scheme can provide solutions for a number of problems. However, the basic assumption of linear elastic behavior for geologic media may not be realistic. Moreover, no provision is presently available for nonlinear behavior at the pile-soil interface. At present neither the finite difference nor the elasticity approach can account for such features as tensile stress conditions, arching, dilatation, and strain softening. The finite element method is perhaps the most general procedure available at this time. We shall devote the remainder of this chapter to the finite element method and its applications.

Finite Element Method

Formulations based on the finite element method have been developed for two- and three-dimensional pile problems. Approximate procedures to reduce a three-dimensional problem to an equivalent two-dimensional problem are obtained for pile groups. Procedures based on Fourier series expansion are possible and can provide an intermediate step between two- and three-dimensional analysis.

Elements used A quadrilateral isoparametric element (Fig. 1-14) either with four primary nodes or with four primary nodes and one internal node is perhaps the most suitable element for this class of problems. Details of stiffness matrices for the quadrilateral element are given in Chap. 1. For three-dimensional problems,

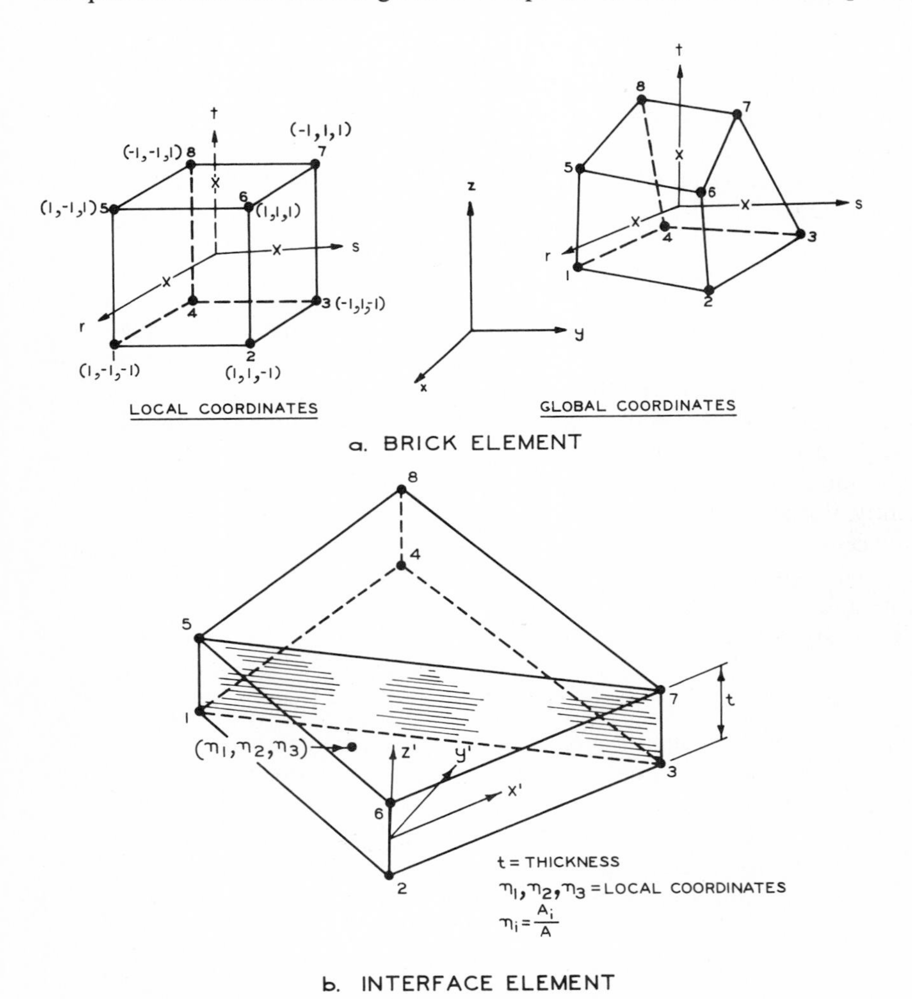

Figure 7-4 Three-dimensional elements.

the corresponding isoparametric brick element (Fig. 7-4*a*) with eight primary nodes or with eight primary and one internal node may be suitable. For the case of the element with eight nodes, the displacement at a point can be expressed as[20–22]

$$\{u\} = \begin{bmatrix} N_1 & 0 & 0 & N_2 & 0 & 0 & \cdots & N_8 & 0 & 0 \\ 0 & N_1 & 0 & 0 & N_2 & 0 & \cdots & 0 & N_8 & 0 \\ 0 & 0 & N_1 & 0 & 0 & N_2 & \cdots & 0 & 0 & N_8 \end{bmatrix} \{q\} \qquad (7\text{-}9)$$

where $\{u\}^T = [u \quad v \quad w]$

$$\{q\}^T = [u_1 \quad v_1 \quad w_1 \quad u_2 \quad v_2 \quad w_2 \quad \cdots \quad u_8 \quad v_8 \quad w_8]$$

and $N_i = \frac{1}{8}(1 + rr_i)(1 + ss_i)(1 + tt_i)$

Details of derivation of the stiffness relation for this element are given elsewhere;[20,21] we shall state here the expression for the stiffness as

$$[k] = \int_{-1}^{1} \int_{-1}^{1} \int_{-1}^{1} [B]^T[C][B]|J| \, dr \, ds \, dt \qquad (7\text{-}10)$$

Interface elements (Chap. 4) are usually employed to simulate behavior at the junction between pile and surrounding medium. For two-dimensional cases, such an element can be a line segment with or without thickness, and for three-dimensional problems, a planar (quadrilateral) element (Fig. 7-4*b*) is suitable. Details of derivation of stiffnesses for the line element for the two-dimensional case are available in Refs. 17 and 18 and in Chap. 4.

Interface element Figure 7-4*b* shows an interface element that can be inserted between a structure and soil idealized as three-dimensional. The element relations can be formulated for the "quadrilateral" with eight nodes and zero or nonzero thickness. For convenience in using the element for different planes, the stiffness relations for the quadrilateral can be obtained as the sum of the stiffnesses of the two component triangles (Fig. 7-4*b*). We shall cover here a brief description of the interface element; further details and additional descriptions are available in Chap. 4 and Refs. 21 to 23.

The three displacement components at any point in the quadrilateral can be expressed as

$$\{u\} = [N_u]\{q\} \qquad (7\text{-}11)$$

where $\{u\}$ is the vector of the displacement components, u, v, and w, $\{q\}$ is the vector of nodal displacements

$$\{q\}^T = [u_1 \quad v_1 \quad w_1 \quad u_2 \quad v_2 \quad w_2 \quad u_3 \quad v_3 \quad w_3 \quad u_4 \quad v_4 \quad w_4]$$

and the interpolation functions in matrix $[N_u]$ are given by

$$N_i = \tfrac{1}{4}(1 + ss_i)(1 + tt_i)$$

For isoparametric formulation, the coordinates of the point can be similarly expressed as

$$\{x\} = [N_c]\{x_n\} \tag{7-12}$$

where $\{x\}^T = [x \quad y \quad z]$ and $\{x_n\}^T = [x_1 \quad x_2 \quad x_3 \quad \cdots \quad z_3 \quad z_4]$

For the triangular element (Fig. 7-4*b*) the displacement components can be expressed as

$$\{u\} = \begin{bmatrix} L_1 & 0 & 0 & L_2 & 0 & 0 & L_3 & 0 & 0 \\ 0 & L_1 & 0 & 0 & L_2 & 0 & 0 & L_3 & 0 \\ 0 & 0 & L_1 & 0 & 0 & L_2 & 0 & 0 & L_3 \end{bmatrix} \{q\} \tag{7-13}$$

where $L_i = A_i/A$; A_1, A_2, and A_3 denote the areas of the three component triangles; and A is the total area of the element.

Now we express relative displacements between the top and bottom of the element. For the quadrilateral element, the relative displacement is[21,23]

$$\{u_r\} = \begin{Bmatrix} u_r \\ v_r \\ w_r \end{Bmatrix} = [[N]\{q\}_{\text{top}} - [N]\{q\}_{\text{bottom}}] = [B]\{q_r\} \tag{7-14}$$

where the matrix $[B]$ is given by

$$[B]' = [[N_1] \quad [N_2] \quad [N_3] \quad [N_4] \quad [N_1^m] \quad [N_2^m] \quad [N_3^m] \quad [N_4^m]]$$

$$\{q\}^T = [u_1 \quad v_1 \quad w_1 \quad \cdots \quad u_4 \quad v_4 \quad w_4 \quad u_5 \quad v_5 \quad w_5 \quad \cdots \quad u_8 \quad v_8 \quad w_8]$$

and $[N_i] = \begin{bmatrix} N_i & 0 & 0 \\ 0 & N_i & 0 \\ 0 & 0 & N_i \end{bmatrix}$ and $[N_i^m] = \begin{bmatrix} -N_i & 0 & 0 \\ 0 & -N_i & 0 \\ 0 & 0 & -N_i \end{bmatrix}$

By minimizing the potential energy of the element, the stiffness matrix in terms of the local coordinates s, t (or L_i) becomes

$$[k_l] = \int_{-1}^{1} \int_{-1}^{1} [B][R][B]\,|J|\; ds\, dt \tag{7-15}$$

which can be integrated numerically, as explained in Chap. 1. The subscript l in Eq. (7-15) indicates that the matrix relates to the local cartesian system x, y, z. The local stiffness matrix can be transformed to the global system x', y', z' as

$$[k_g] = [T]^T[k_l][T] \tag{7-16}$$

where $[T]$ is the transformation matrix composed of the direction cosines between the coordinates x, y, z and x', y', z'. The stiffness matrix for the triangular element can be similarly formulated.

The matrix R in Eq. (7-15) contains shear, k_x, k_y, and normal, k_z, stiffness moduli as

$$[R] = \begin{bmatrix} k_x & 0 & 0 \\ 0 & k_y & 0 \\ 0 & 0 & k_z \end{bmatrix} \tag{7-17}$$

Nonlinear Behavior

The finite element procedure[21] incorporates the three-dimensional solid element (Fig. 7-4*a*) for soil or rock and structure and the interface element (Fig. 7-4*b*) for junction between structure and soil. The formulation has a provision for including nonlinear behavior of both soils and interfaces. As described previously and in Chap. 2, the values of tangent modulus E_t and tangent Poisson's ratio ν_t (or tangent shear modulus G_t and tangent bulk modulus K_t) can be computed by using hyperbolic simulation. The moduli are evaluated as functions of such factors as state of stress, confining pressure, and stress paths. In a similar manner, the tangent shear stiffnesses, k_{xt} and k_{yt}, for interfaces are computed as nonlinear functions of the induced shear and normal stresses. The normal stiffness k_z is set equal to a high value ($= 10^{10}$ lb/ft^3) during loading. If, during loading, the interface fails in shear, the values of shear stiffness are reduced to a small value ($= 10^{-2}$ lb/ft^3); if tensile conditions develop, both shear and normal stiffnesses are reduced to the small value.

The nonlinear behavior can also be handled by using the plasticity approaches described in Chaps. 2 and 3.

Nonlinear techniques used Incremental, incremental-iterative, and residual (initial) stress and strain methods have been used for the problems described in this chapter. The incremental-iterative method was used most commonly; it was combined with the residual method for the softening problem.[24]

Comment

With the current generation of computers, procedures based on two-dimensional, idealized two-dimensional, and series expansion can provide convenient and economical solutions. Use of three-dimensional finite element analyses may involve a significant amount of effort and cost; however, they are necessary for certain types of problems. Moreover, they provide a very useful means of verifying the difference between accuracies provided by solutions from two-dimensional idealizations and from three-dimensional analyses.

7-4 APPLICATIONS AND DESIGN ANALYSIS

In this section, we shall describe typical examples of applications for axially loaded piles in sands and clays and pile groups. A number of special factors that can influence behavior of pile foundations and applications of three-dimensional analysis will also be discussed.

Piles in Sands

Figure 7-5 shows details of soil layers and typical piles driven at the site of Arkansas Lock and Dam number 4 (LD4). Further details are available in Refs. 13, 17, 25, and 26. A number of piles, numbered as 2, 3, and 10, were analyzed.[13,18]

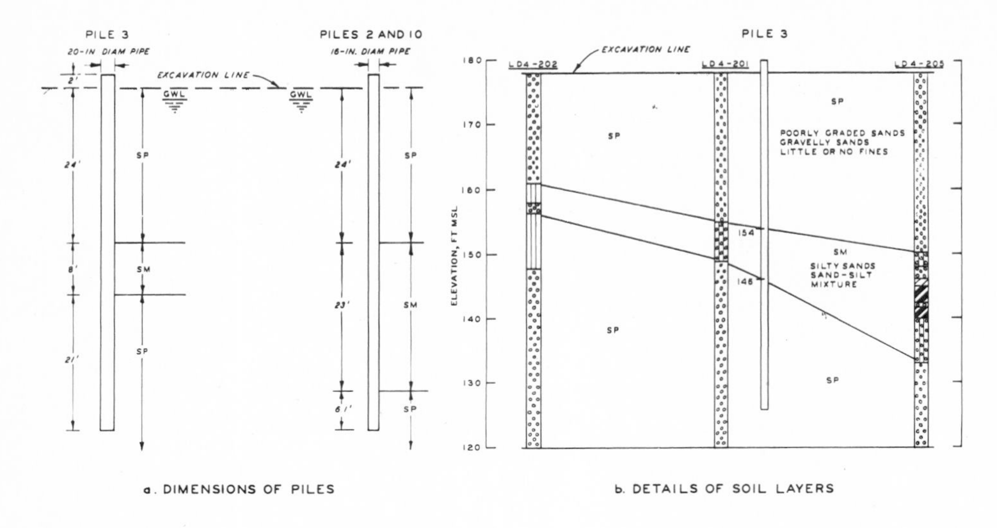

Figure 7-5 Dimensions of piles and soil details.[25]

In the finite element analyses, the pipe piles were replaced by equivalent solid cylinders retaining the axial stiffness of the original piles. The finite element mesh showing layering of soil and interfaces is illustrated in Fig. 7-6. The load on the pile was applied in increments of 17.5 tons/ft^2 for pile 3, and 14.3 tons/ft^2 for piles 2 and 10, which corresponded with the load increments applied during the field load tests. Before applying the load, the in situ stresses corresponding to the value of K_c equal to 1.29, 1.17, and 1.23 for piles 2, 3, and 10, respectively, were introduced into the soil, interfaces, and pile elements; the subscript indicates value of K in compression loading. Initial stresses in the interface and the pile elements were set equal to those in the adjacent soil elements; initial shear stresses were set equal to zero in all elements for horizontal ground surface.

Nonlinear behavior of the soil and interfaces was obtained from drained triaxial tests at different confining pressures and relative densities and drained direct shear tests at different normal loads and relative densities, respectively.[13,17,18] The nonlinear curves were simulated using hyperbolic functions. Values of tangent (elastic) modulus E_t, tangent Poisson's ratio or tangent shear modulus G_t, and tangent bulk modulus K_t were computed for forming the tangent stiffness matrices during incremental loading. Table 7-1 gives material parameters for soils, piles, and interfaces. These values were adopted from tests on the sands and from tests on similar sands at other sites.[18] The applied-load-vs.-butt-displacement curve computed from incremental finite element analysis is compared in Fig. 7-7*a* with that observed in the field for pile 10, which was driven by vibrating hammers. Figure 7-7*b* also shows the comparison between computed and observed point and wall friction loads.

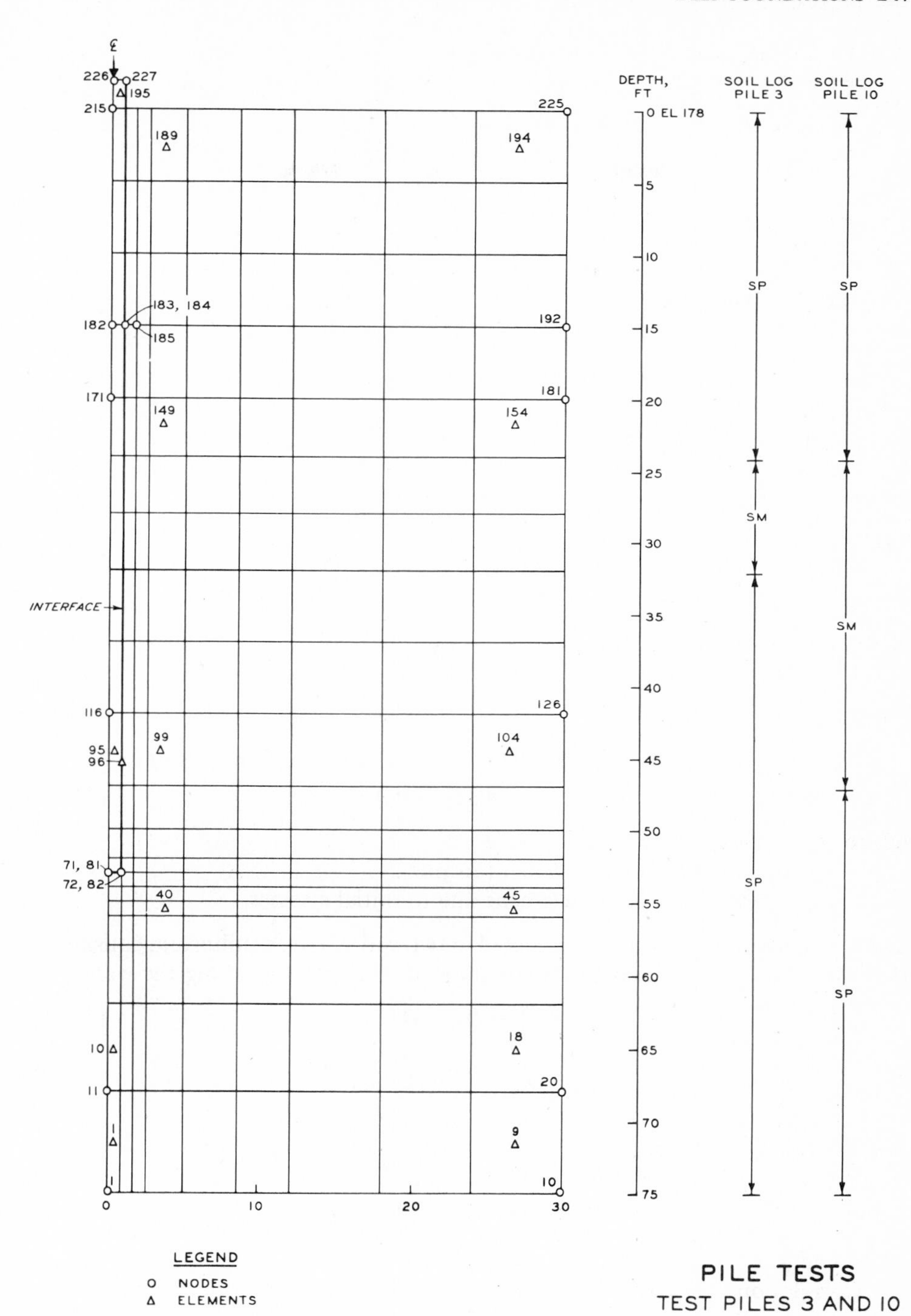

Figure 7-6 Finite element mesh.[18]

Table 7-1 Material parameters for soils at LD4 site†

Sands,
for tangent modulus E_t

Layer	K	n	R_f	ϕ, deg
I and III	1500	0.6	0.9	32
II	1200	0.5	0.8	31

for tangent Poisson's ratio ν_t

Layer	G	F	d
I–III	0.54	0.24	4.00

Interfaces,
for shear stiffness k_{st}

Layer	K_j	n	R_f	δ, deg
I and III	25,000	1.00	0.87	25
II	20,000	1.20	0.88	27

† See Ref. 18 and Chaps. 2 and 16 for details and explanation of symbols.

In the finite element analysis, the point load Q_p was computed as

$$Q_p = A_p(\sigma_p - \sigma_0) \tag{7-18a}$$

where A_p = cross-sectional area of pile tip
σ_p = axial stress in last element in pile
σ_0 = axial stress in last element due to initial stresses

The wall friction load Q_w was evaluated in two ways; from equilibrium considerations $Q_w^1 = Q_T - Q_p$ and as the summation of shear stresses along the wall Q_w^2:

$$Q_w^1 = Q_T - Q_p \tag{7-18b}$$

and

$$Q_w^2 = \sum_{i=1}^{m} \sigma_{ti} C_i \tag{7-18c}$$

where σ_{ti} is the computed shear stress in interface element i and

$$C_i = A_i \pi D_i \tag{7-18d}$$

in which A_i and D_i are the (mean) area and (mean) diameter of the pile element, respectively.

Correlations between computations and observation are satisfactory. In the present finite element analysis, effect of (residual) stresses due to driving is not accounted for. Hence, the correlation (Fig. 7-8) between computed load in pile

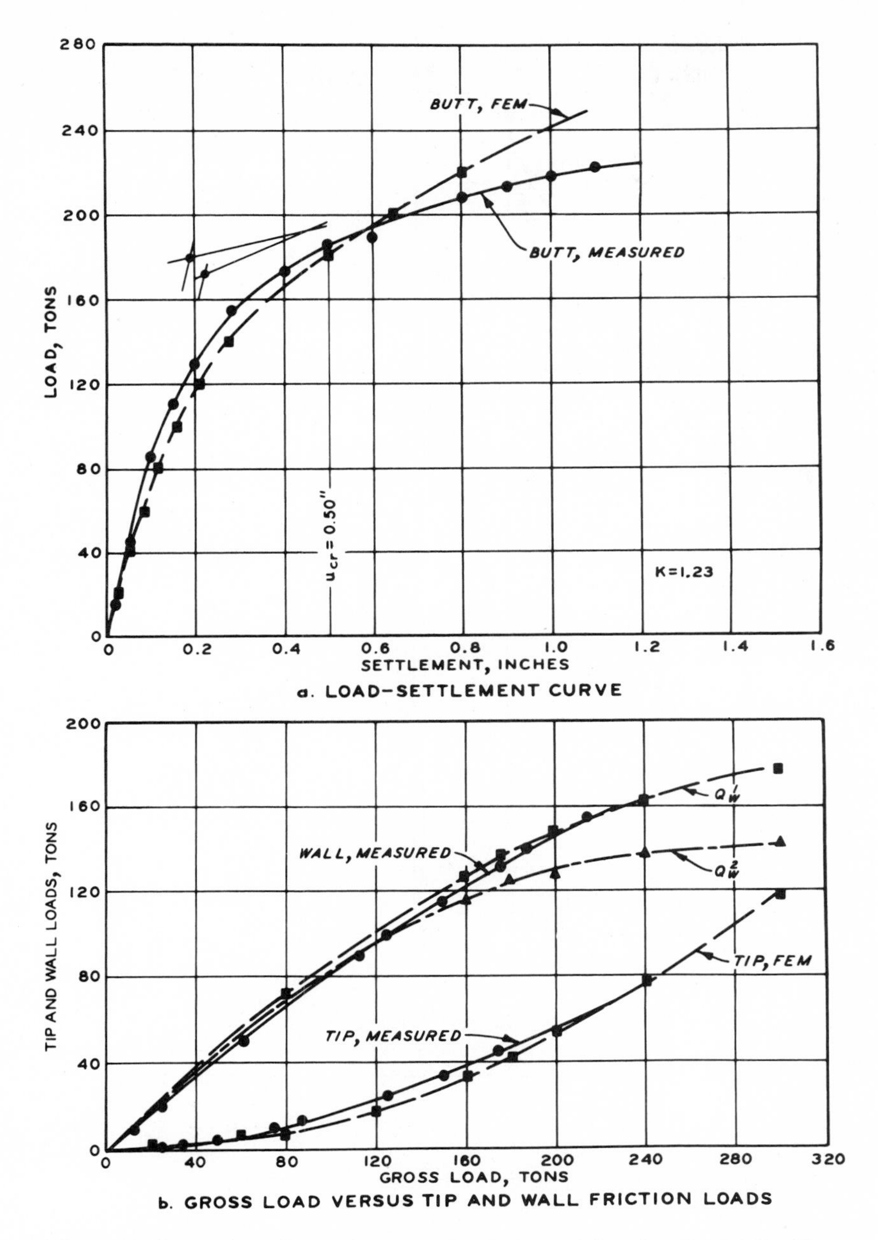

Figure 7-7 Comparisons between computed and observed data for pile 10, LD4.[18]

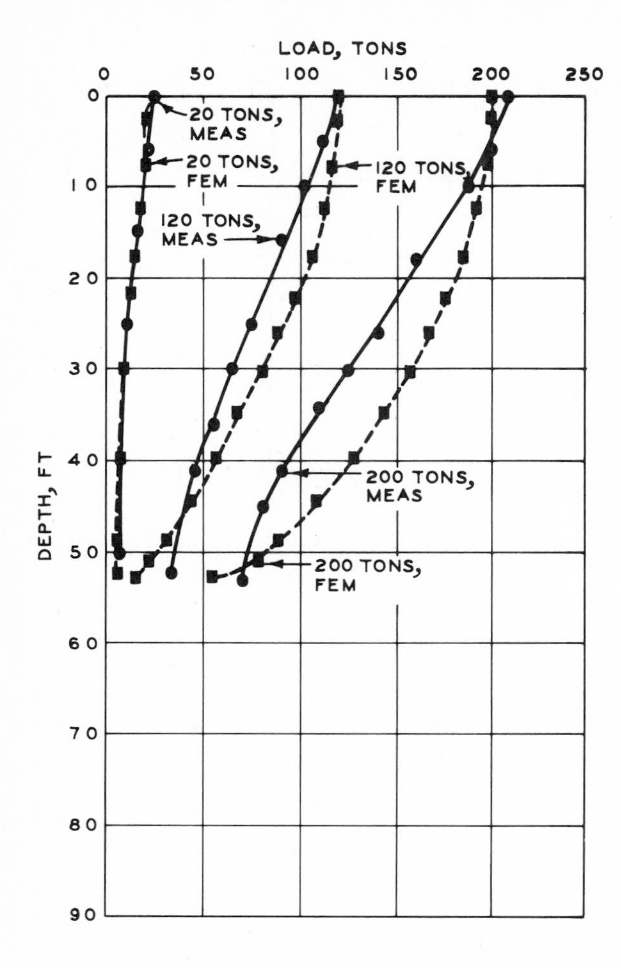

Figure 7-8 Distribution of load in pile 3.

Table 7-2 Comparisons between computed and observed bearing capacities in tons for various piles[18]

	Site LD4						Site JL	
	Pile 2		Pile 3		Pile 10			
	i†	ii	i	ii	i	ii	i	ii
Field	205	210	210	215	180	186	187	160
Finite element method	195	192	192	215	176	180	190	155

† Criterion i = tangent procedure; ii = critical displacement (butt) equal to 0.5 in; this value was adopted since it was used for finding the bearing load from field curves.[25]

2 (which was driven by using a hammer) and observation, particularly at higher loads, did not show good agreement. It is possible to introduce such residual stresses as an initial stress condition, which may possibly improve the correlation.

Values of bearing capacity obtained for three piles at LD4 and a pile at Jonesville lock (JL) site from computed load-settlement curves are compared in Table 7-2 with those values obtained from observed load-settlement curves. In the displacement criterion, a critical value of allowable (butt) displacement is chosen, and the corresponding load is adopted as the bearing capacity. In the tangent procedure, the point of intersection of tangents drawn at initial and ultimate portions of the curve gives the bearing capacity.

Piles in (Stiff) Clay

It is easier to characterize normally consolidated and homogeneous clays than overconsolidated stiff fissured clays. It is difficult to determine material parameters for fissured clays from laboratory tests because the disturbance due to sampling can significantly alter the behavior of the soil. The initial modulus E_i determined from laboratory triaxial tests may highly underestimate the field value, whereas the (undrained) strength may be overestimated.[27–29] Moreover, overconsolidated soils usually exhibit strain-softening behavior. The behavior at the junction between the structure and such soils can also experience softening behavior.

We shall describe an example of the application of finite element method to the analysis of piles in Beamont clay, which is stiff overconsolidated soil similar in behavior to London clay.[27]

Figure 7-9*a* shows some of the bored piles tested by O'Neill[30,31] in Beamont clay. Details of the pile considered herein and the log of the soil borings are also shown in Fig. 7-9*a*. A comprehensive series of triaxial tests were performed[30] to establish the behavior of soils in the various layers. A series of special direct shear tests were performed to simulate behavior at and near the interface. These data were consulted for determination of input parameters in the finite element analysis. However, the parameters such as E_i determined from laboratory tests did not yield satisfactory computations of load-settlement curves when compared with observations in the field.

A parametric finite element study was performed in which the values of λ and α in the following equations were varied:

$$E_i = \lambda c_u \qquad \text{and} \qquad c_a = \alpha_c c_u \tag{7-19}$$

where c_u = undrained strength

c_a = adhesion at interface

α_c = adhesion factor

Values of λ and α_c equal to about 200 and 0.42, respectively, were reported in Refs. 30 and 31. From the finite element analysis it was found that the value of λ should be in the range of 800 to 1600 to obtain satisfactory correlation with initial load-settlement curve (Fig. 7-10*a*) with the value of α_c equal to about 0.4 to 0.44.

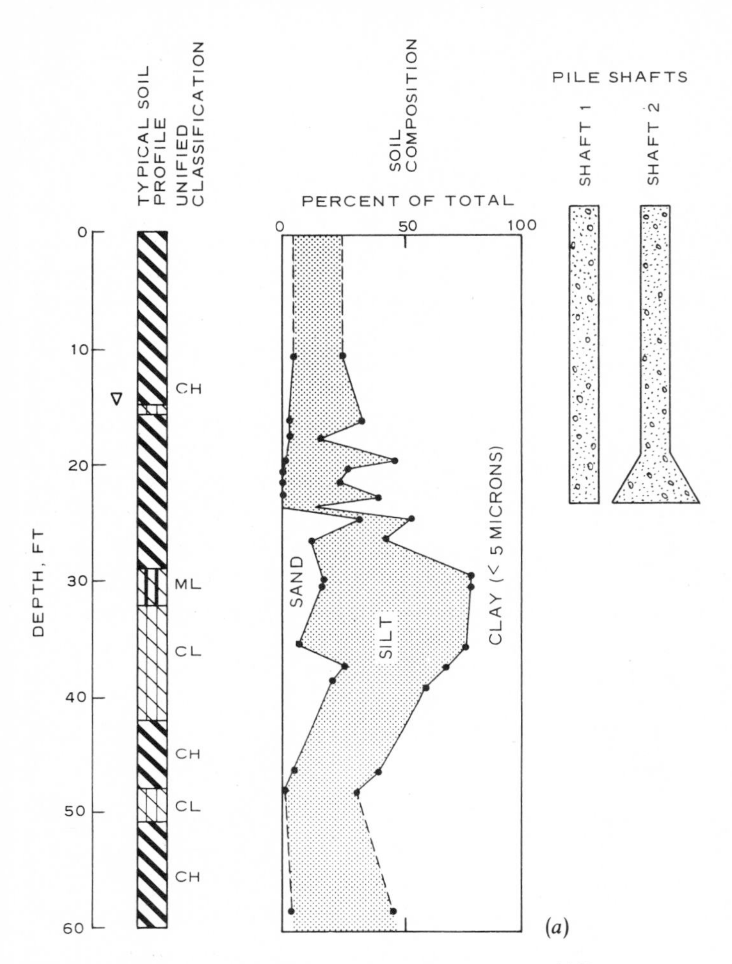

Figure 7-9 Details of piles and soils in stiff fissured clay.[30,31]

Similar values of λ have been reported by D'Appolonia et al.[32] Variation of c_u with depth adopted for the analysis is shown in Fig. 7-9*b* and typical direct-shear-test results for the behavior of soil near the interface are shown in Fig. 7-10*b*.

Strain-softening Effects

Field observations for the wall-friction load Q_w showed work-softening behavior.[30] The laboratory direct-shear tests for the clay also showed strain-softening (Fig. 7-10*b*), indicating that the failure may have occurred in the soil in the vicinity of the pile.

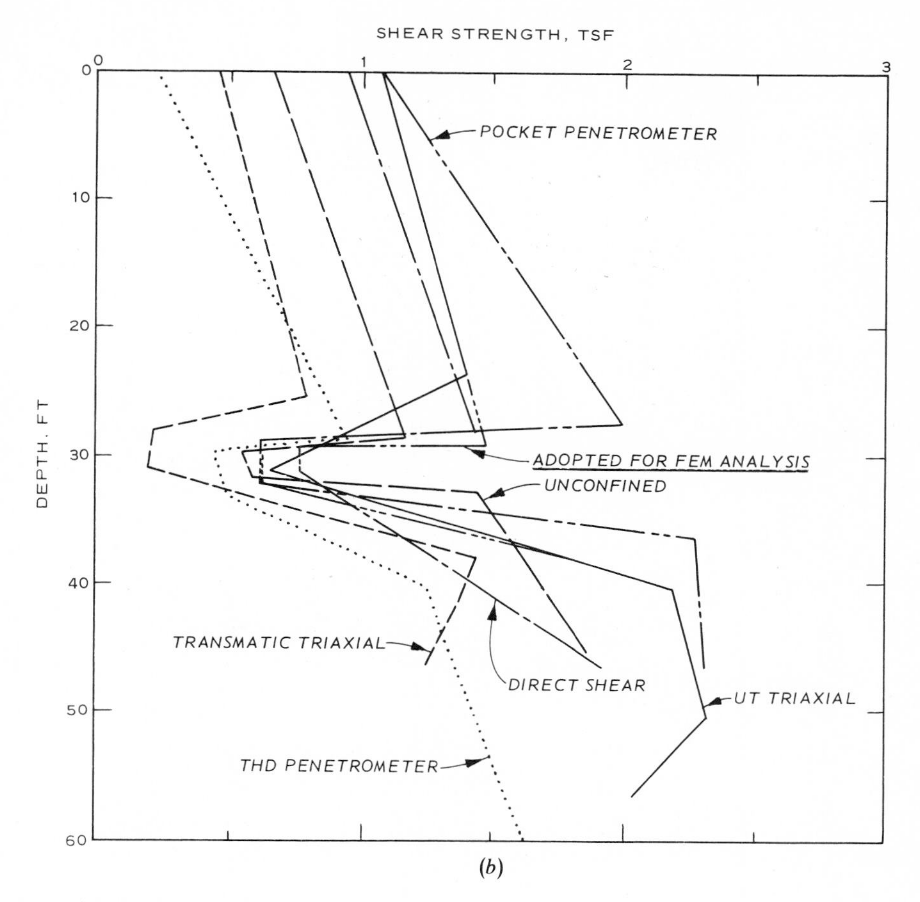

(*b*)

Figure 7-9 (Continued).

A consistent finite element technique[24] based on a combination of residual (initial) stress, relaxation, and iteration was developed to account for the softening behavior. Here during the incremental loading, the excess load according to the shear-stress–relative-displacement curve (Fig. 7-10*b*) was released and transferred to the adjacent elements that had not reached yield conditions. Figure 7-10*c* shows comparisons for Q_w between computed and field observation for straight-sided shaft (Fig. 7-9*a*) after allowance was made for the softening behavior. Use of this technique[24] permits consideration of the strain-softening behavior. These results are considered significant in view of the fact that not many quantitative correlations for strain-softening behavior had been obtained previously.

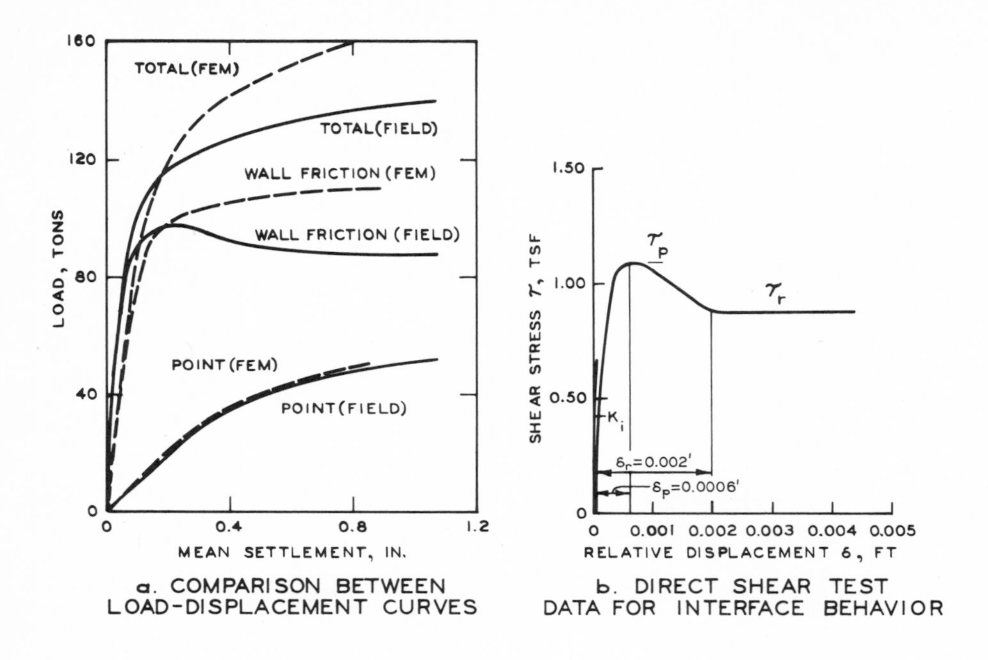

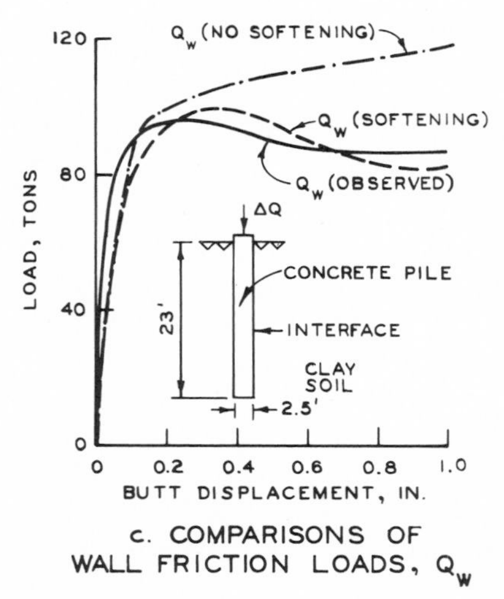

Figure 7-10 Results for piles in stiff clay and effect of softening.[24]

Tensile Stresses

Tensile stresses, particularly at higher loads, occur near the tip of the pile. Sand does not usually possess tensile strength, while clays can possess a small amount of tensile strength. The excess tensile stresses can be relieved and distributed to adjoining elements by using the foregoing technique for work softening. Thus, a realistic condition conforming to the tensile strength of the medium can be restored.

Tension Loading

It is possible to use the finite element procedure for piles loaded in tension during pullout tests. Figure 7-11 shows the observed load-displacement curve for pile 10.[25] In the finite element analysis, the same mesh as in Fig. 7-6 was used, and the incremental load was applied in the upward direction. A value of K_t (in tension) equal to 0.79, as reported in Ref. 25, was used. The computed load-settlement curve can be obtained up to the point ⊗ (Fig. 7-11), beyond which the computed settlements grow very fast. However, the correlation in the zones of interest during initial loading and around the "ultimate" load state seems satisfactory. This is a preliminary analysis; additional investigations will be necessary for tension loading.

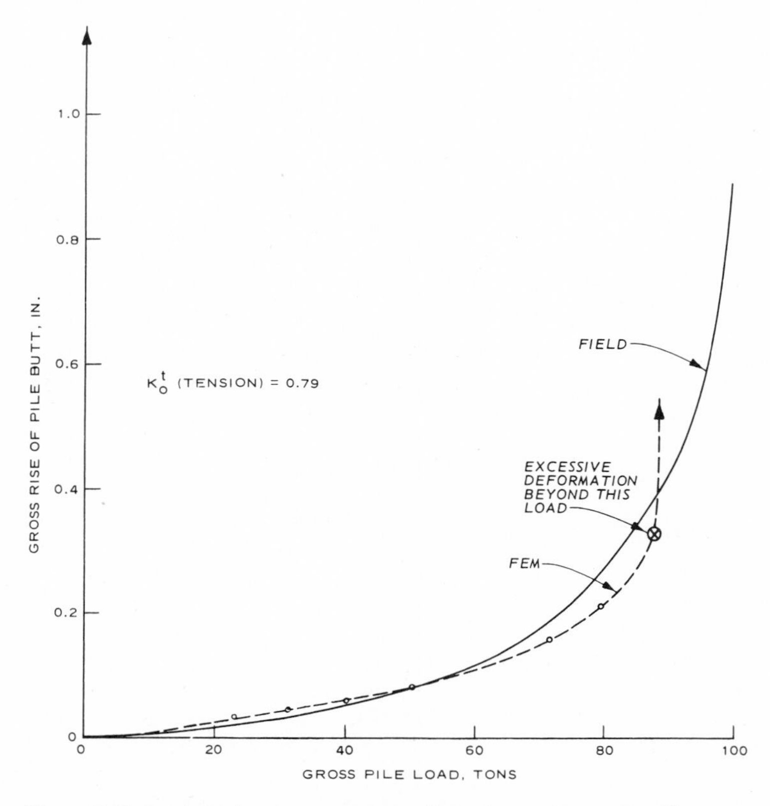

Figure 7-11 Load-displacement curve for pile under tension loading.[13]

Pile Groups

Pile groups or clusters are analyzed conventionally by using criteria derived from the principle of superposition, the theory of elasticity, the equilibrium of forces, and the reciprocal theorem.[33,34] Distribution of load in the individual piles is computed from static equilibrium and elastic theories; the commonly used Hrennikoff's method[35] is an example of this concept. Other procedures based on elasticity solutions are covered in Chap. 10. Two design parameters commonly used are group efficiency and the settlement factor.[33,34] Efficiency is defined as the ratio between the ultimate capacity of the group and sum of the ultimate capacities of individual piles in the group. The efficiency is less than unity for clays but can be greater than unity for sands. A number of formulas and empirical and semitheoretical rules are proposed to compute efficiency.[33,34,36] The settlement ratio is defined as the ratio between settlement at a fraction of the ultimate group capacity and the settlement of a single pile at the same fraction of the ultimate capacity of the single pile. Investigations toward settlement analysis, usually based on model tests, are reported by various workers.[37–40]

Most of the foregoing procedures for design usually ignore such important factors as nonlinear behavior, in situ stresses, interaction, influence of pile cap, and effect of spacing. It is believed that three-dimensional finite element analysis can provide more rational and general solutions. As described earlier in Sec. 7-3, three-dimensional codes based on an arbitrary hexahedron for solid elements and quadrilateral interface elements are possible and available.[21]

Two-dimensional Approximation

It is often possible to approximate a three-dimensional problem by an equivalent two-dimensional system. One of the theoretical procedures based on the use of separation of variables is described in Chap. 9. A scheme for obtaining a physically equivalent system is described below.

Battered-Pile Foundation

Figure 1-11 shows the finite element mesh for the battered-pile foundation of Columbia lock.[41,42] The foundation poses a three-dimensional problem. It will be difficult and costly to solve it with three-dimensional elements for soils, lock, H piles and interfaces between lock and soils and piles and soil. As an approximation, the three-dimensional system was idealized as an equivalent two-dimensional plane strain system. As shown in Fig. 7-12*a*, the total axial stiffness of a monolith was divided among strips of unit length and width, and the equivalent value of E was computed for each pile in the strip.

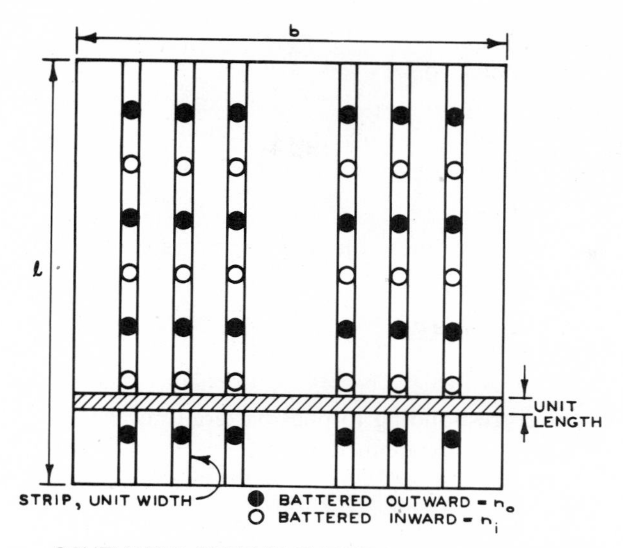

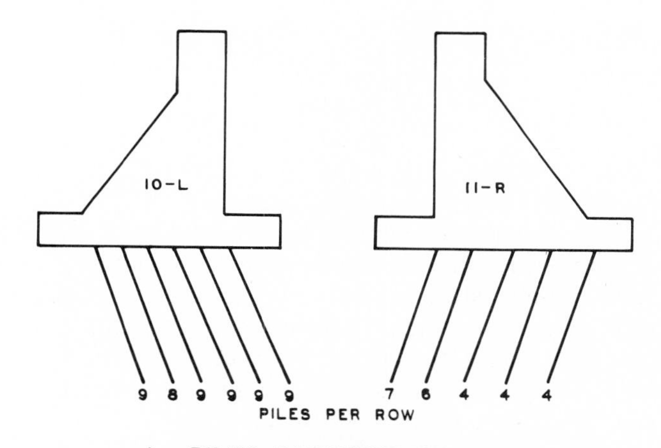

Figure 7-12 Two-dimensional idealization of lock problem, Fig. 1-11.[41, 42]

Two-dimensional Idealization

If we assume that the major response of the foundation is provided by axial stiffness, the total stiffness S_t can be expressed as

$$S_t = \sum_{j}^{n} \frac{A_j E_j}{L_j} = (n_i + n_o) \frac{AE}{L} \tag{7-20}$$

where n_i = number of piles battered inward
n_o = number of piles battered outward
A = projected area of pile
L = length of pile
E = modulus of elasticity
n = total number of piles

If a number of strips (Fig. 7-12*a*) are assumed, the stiffness per strip for piles battered inward is

$$s_i = \frac{n_i}{m_i}\frac{AE}{L} \tag{7-21}$$

where m_i are the number of strips corresponding to piles battered inward.

The equivalent stiffness of a strip is now defined as

$$s_{ei} = \frac{A_{ei}E_{ei}}{L_{ei}} \tag{7-22}$$

Equating expressions in Eqs. (7-20) and (7-21) leads to equivalent modulus for the monolith (Fig. 7-12*b*) as

$$E_{ei} = \frac{n_i AE}{m_i A_{ei}} = \frac{(53 \times 0.149)(4.04 \times 10^9)}{6 \times 40}$$

$$= 1.3 \times 10^8 \text{ lb/ft}^2 (= 6.4 \times 10^6 \text{ kN/m}^2) \tag{7-23}$$

Here the area of a strip is 40 by 1 ft, where 40 is the length in feet of the monolith and 1 denotes the unit width of the strip and $m_i = 6$ (Fig. 7-12*b*). The number of piles n_i in the monolith is 53. Similar computations lead to equivalent moduli for piles battered outward. Further details are provided in Refs. 41 and 42.

The finite element analysis simulated various sequences of construction such as in situ stresses, dewatering, excavation, buildup, installation of piles, backfilling, filling water in locks, and uplift pressures. The soils and interfaces were considered nonlinear. Figure 7-13 shows distributions of loads in the piles in comparison with field observations and those from Hrennikoff's method. In Fig. 7-14 are compared the computed values of the progressive displacements with those observed in the field.

Overall, the finite element predictions showed satisfactory correlations with observations and provided improved evaluation of loads in the piles in comparison with those from Hrennikoff's method. For certain problems, therefore, considerable effort and computer time can be saved by reducing a three-dimensional system to an equivalent two-dimensional system, and at the same time satisfactory solutions can be obtained for practical applications. Moreover, it may be possible to use the finite element procedure to evolve a criterion to modify the conventional design procedures. Such a modification would allow for some of the factors previously ignored in the conventional methods.

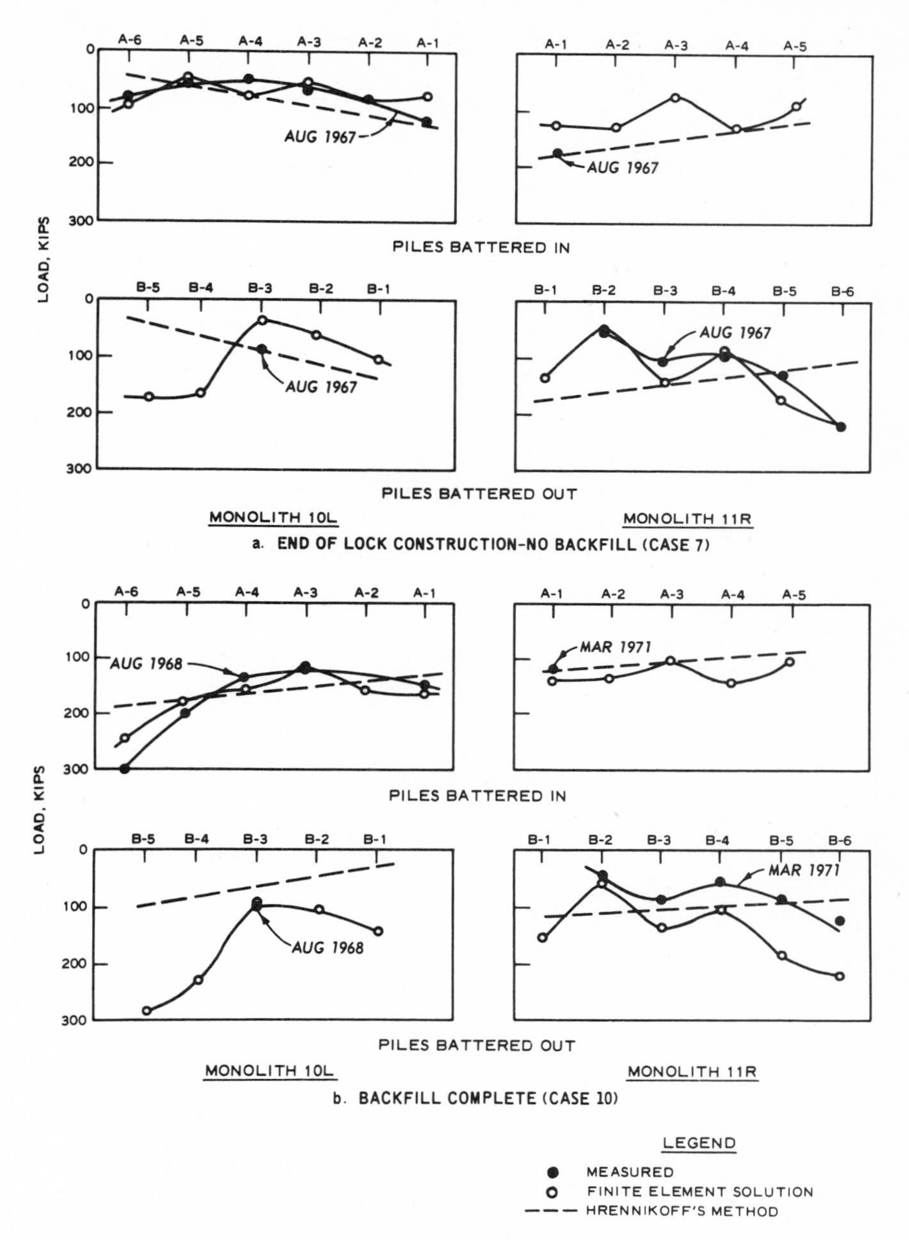

Figure 7-13 Comparisons for distribution of load in piles.[41,42]

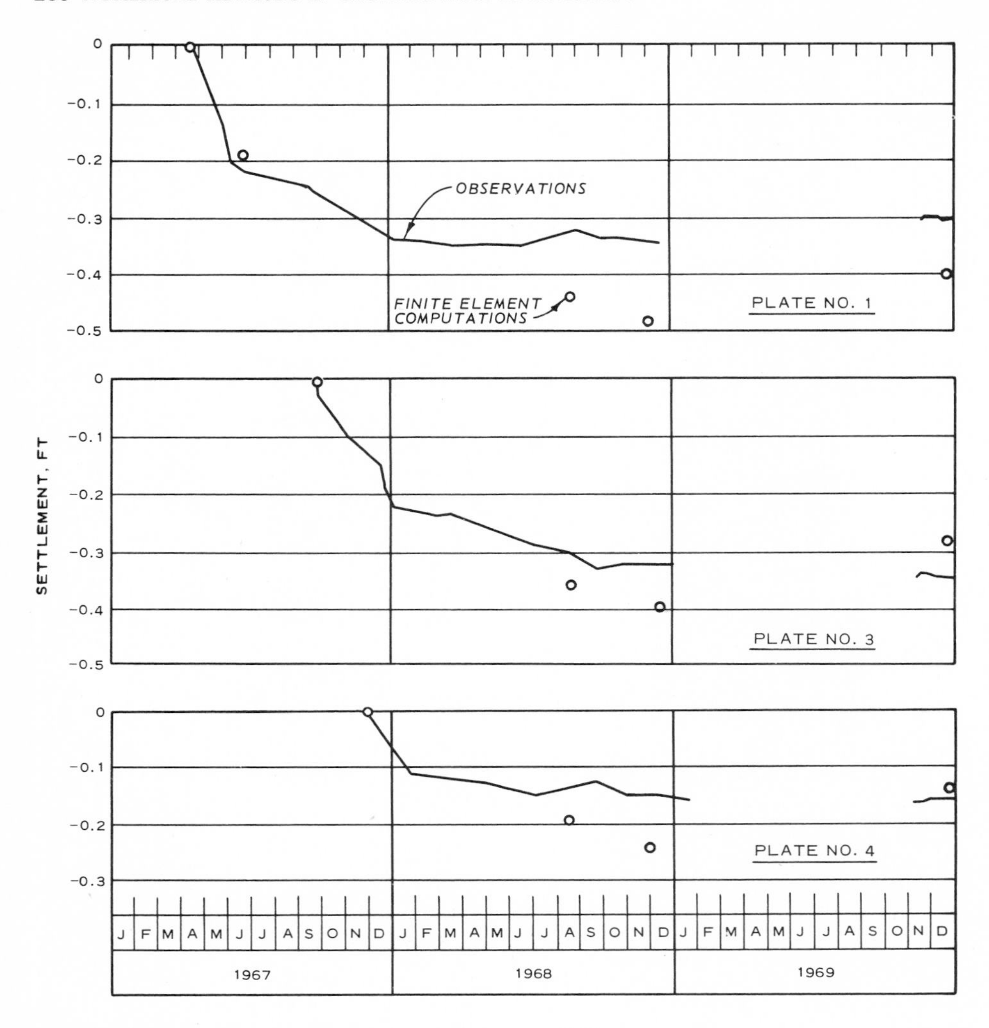

Figure 7-14 Comparisons for progressive displacements.[41,42]

Factors Affecting Pile Behavior

Factors such as magnitudes and spatial variations of K, stress paths, driving and residual stresses, tensile-stress conditions, arching, negative skin friction, and the level of the water table can influence behavior of piles.[43] We shall consider some of these factors.

Coefficient of Lateral Earth Pressure K

This coefficient is used essentially to introduce the in situ stresses as

$$\sigma_x = K\sigma_y \tag{7-24a}$$

If the pile is loaded in compression, we denote K as K_c, and if in tension, by K_t. The magnitude of K will depend upon the geological properties of soil. For normally consolidated linear elastic media, we can compute K from the following formula, which should be used with caution:

$$K = \frac{\nu}{1 - \nu} \tag{7-24b}$$

For overconsolidated soils, with an overconsolidation ratio higher than 4, the value of K is usually greater than unity. Procedures described by Brooker and Ireland[44] can be used to evaluate K. The best way to find K would be to conduct adequate field (or laboratory) tests. Some laboratory procedures have been used, and certain field techniques have recently been proposed[45] for evaluation of K. However, very often K is guessed, varied, or backfigured based on field data and experience. Future research will be needed to obtain realistic values and the spatial distribution of K in geologic media. The magnitude of K can have significant effects on the behavior of a foundation under load. Figure 7-15 shows load-displacement curves for an 18-in prestressed concrete pile at the JL site for various values of K. The actual value backfigured from field load test was found to be about 0.6.[46] In these analyses K was assumed to be uniform over the entire discretized region.

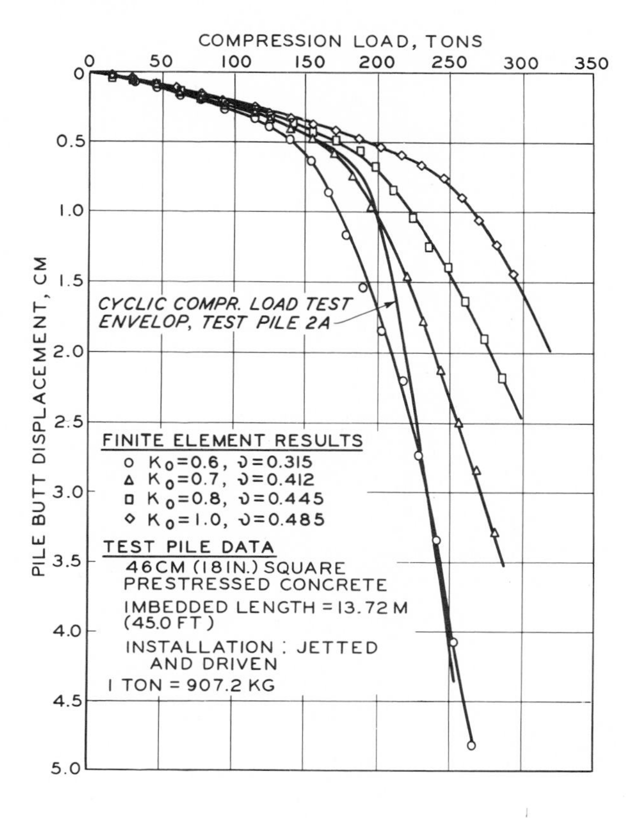

Figure 7-15 Influence of magnitude of K.[18,43]

Spatial Variation of K

The value of K used in the foregoing examples was backfigured from field pile-load tests. The high value, such as 1.29 for pile 3, may be valid only in the vicinity of the driven pile, whereas K can decrease as the distance from the pile increases.

A finite element study was performed by using the variation of K as shown in Fig. 7-16*a*; the zones of high K near the wall and tip of the pile were chosen as suggested by Broms.[47] Beyond such zones, the value of K was assumed to be uniform, as given by Eq. (7-24*b*).

It was found that the behavior from this analysis and from the previous one in which K was assumed to be uniform are not significantly different (Fig. 7-16*b*). However, if K is assumed to be uniform at a low value of 0.48, the behavior will be significantly different (Fig. 7-16*b*). For this problem the finite element computations indicate the extents of zones similar to those suggested by Broms.[47] Variation of K (Fig. 7-16*a*) can be considered to include indirectly the influence of driving on soil properties.

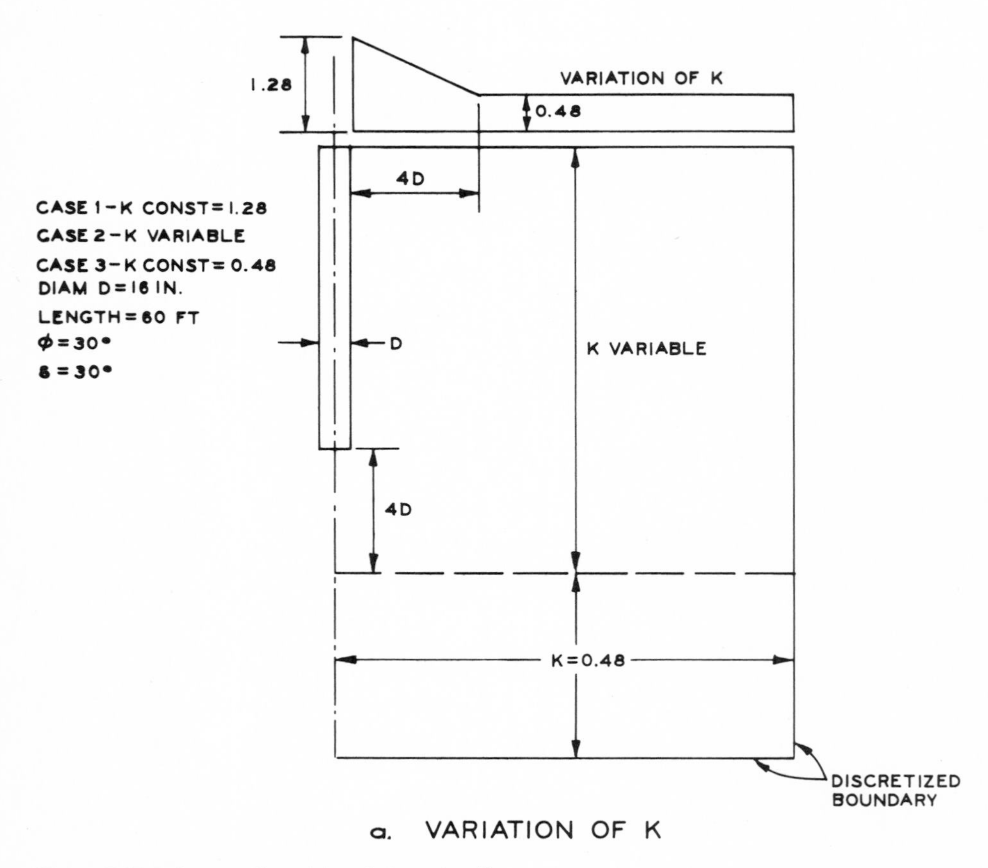

Figure 7-16 Influence of spatial variation of K.[43]

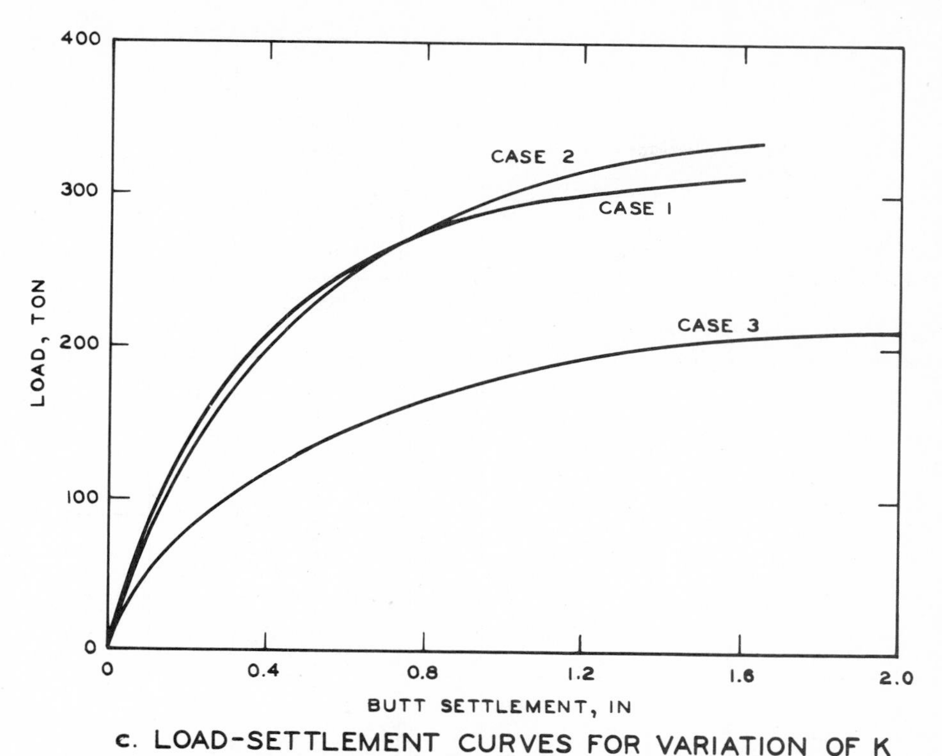

Figure 7-16 (Continued).

Stress Paths

The stress paths followed by various elements during incremental loading may not follow the path of laboratory test samples from which the stress-strain law was derived. Figure 7-17*a* shows comparisons between the stress paths followed by laboratory triaxial samples of the sand[18] and those for typical finite elements in various zones in the foundation for the pile at the JL site. It can be seen that the computed paths differ from the laboratory paths; only at some distance below the pile, elements such as 12, Fig. 7-6, follow triaxial stress paths. Near and just above the tip, the elements load under constant pressure and then unload, indicating stress relief and arching effects. Thus, we can see that at local levels the stress-strain law derived from triaxial tests alone may not provide satisfactory results. However, for the particular case of static and monotonic loading, the gross load-settlement behavior is satisfactory (Fig. 7-7).

If it is important to obtain precise values of stresses and deformations at local levels and for general formulations, it will be desirable to include the effect of stress paths. In a simple constitutive model involving only two parameters, G and K or E and ν, it can be possible to incorporate approximately the effect of

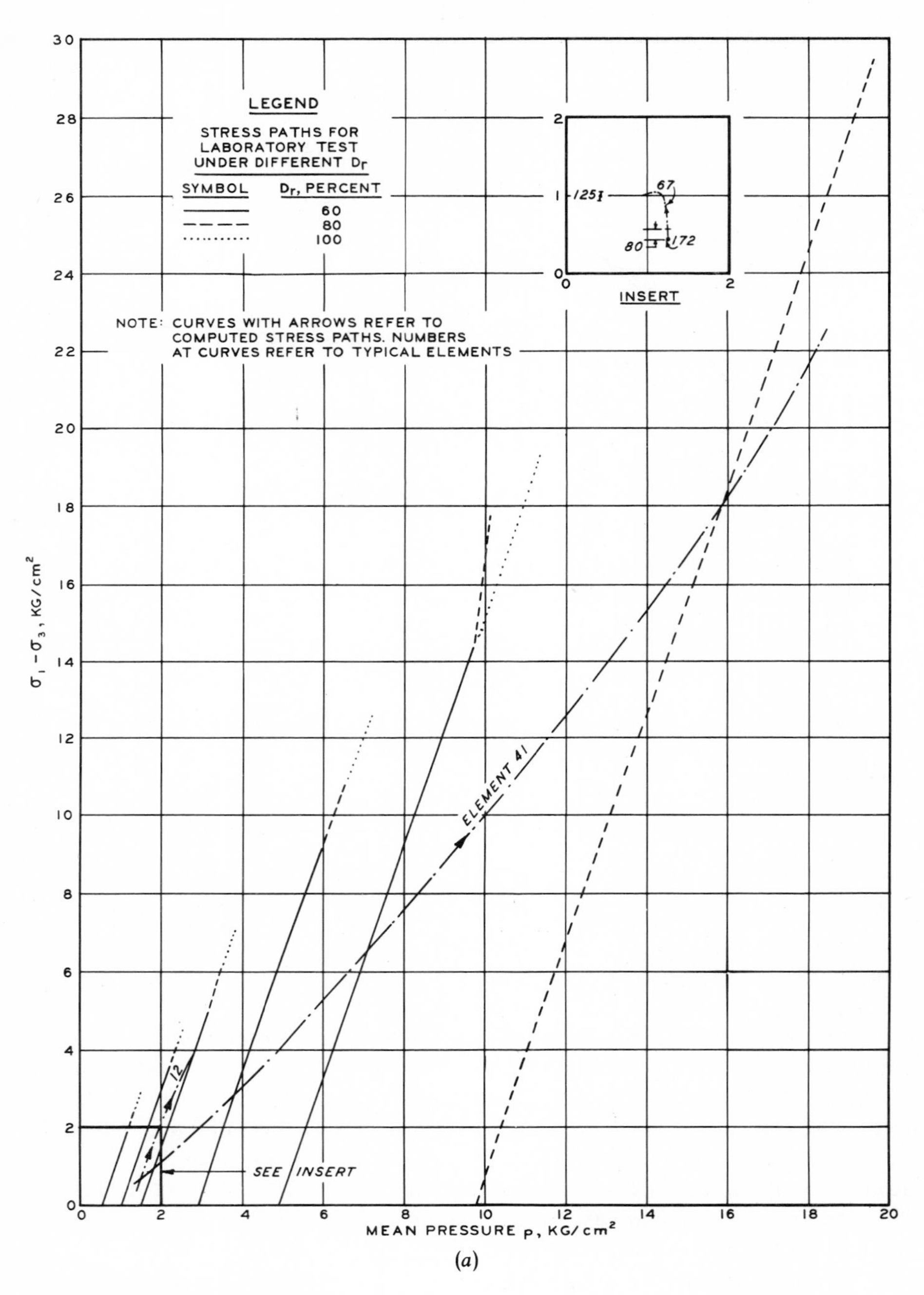

(a)

Figure 7-17 Effect of stress paths.[18,48]

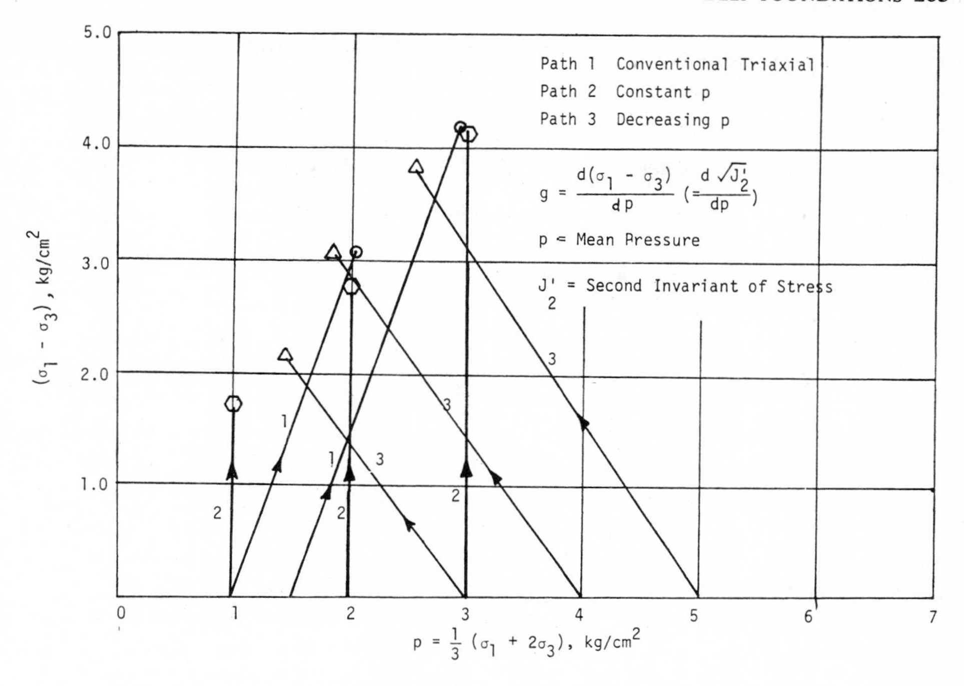

b. LABORATORY LOADING PATHS

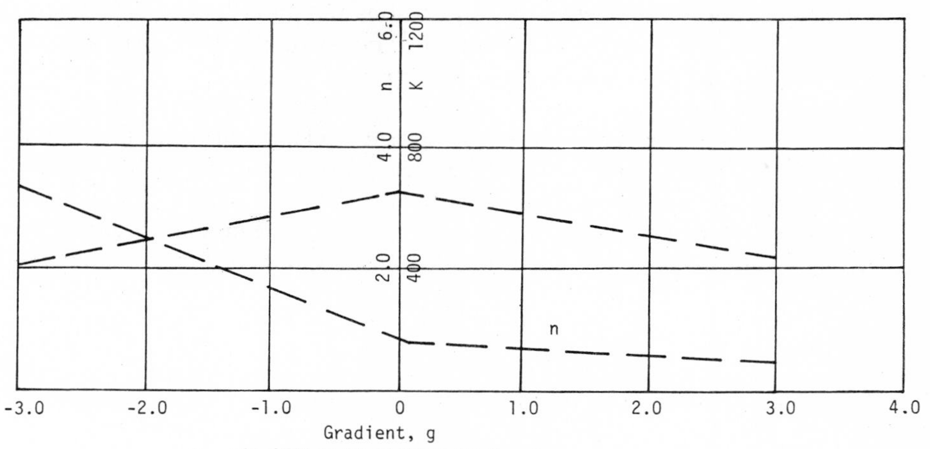

c. VARIATIONS OF PARAMETERS K and n

stress paths. For example E_t can be made a function of the gradient of stress path, $g = d(\sqrt{J'_2})/dp$, in addition to various other parameters such as c, $\sigma_1 - \sigma_3$, σ_3, and D_r

$$E_t = E_t(E_i, c, \phi, \sigma_1 - \sigma_3, \sigma_3, D_r, g) \tag{7-25a}$$

where E_i is the initial modulus. The gradient g can be expressed as a function of such factors as

$$g = g(E_i, \sigma_1 - \sigma_3, \sigma_3) \tag{7-25b}$$

In the finite element analysis, the value of E_t can then be computed as a function of g for each element at the induced state of stress ($\sigma_1 - \sigma_3$, σ_3) during incremental loading. For example, if the hyperbolic formula (Chap. 2) is used, a relation between the parameter K and n (Eq. 2-63b) and gradient for a sand tested (Fig. 7-17b) in true triaxial test equipment[48] is given in Fig. 7-17c. Similar work for clays using standard triaxial apparatus is reported.[49]

A possible procedure using a general function similar to the Ramberg-Osgood model (Chap. 2) for incorporation of stress paths in an approximate manner is given by Desai and Wu (see Chap. 2).

Use of higher-order elastic or elastoplastic cap[16] models that can include effect of stress paths, however, require determination of a large number of parameters. This will usually require more than one type of laboratory tests and will involve additional complexities.

Driving Stresses

Stresses caused by driving may not significantly influence the bearing capacity of a pile foundation, but they affect the magnitude and distribution of load in the pile.[18,25] The results in Fig. 7-8 do not include effect of driving stresses, and it can be seen that the correlation between computations and observation for load in the pile is not satisfactory. Because of the nonlinear behavior, the driving stresses may be one of the main factors causing this discrepancy.

It is possible to add driving stresses to in situ stresses. One of the ways of obtaining driving stresses has been to instrument a pile and measure the residual stresses after the pile is driven and before it is loaded; observed distributions of driving stresses are reported in Refs. 25 and 50.

In the finite element analysis, the in situ stresses $\{\sigma_0\}$ are first introduced. Then the given driving stresses $\{\sigma_d\}$ can be divided into a number of increments m

$$\{\sigma_d\} = \sum_{i=1}^{m} \{\Delta\sigma_d\}_i \tag{7-26a}$$

For each increment $\{\Delta\sigma_d\}_i$ an equivalent (residual) load

$$\{Q_0\}_i = \iiint [B]^T \{\Delta\sigma_d\}_i \, dV \tag{7-26b}$$

is computed and applied at the nodes of the pile element. The initial state of stress is then modified to yield

$$\{\sigma_{0d}\} = \{\sigma_0\} + \sum \{\Delta\sigma\}_i \tag{7-26c}$$

where $i = 1, 2, \ldots, m$. The external load is then applied to the system which already has a state of stress $\{\sigma_{0d}\}$ built into it.

Laterally Loaded Piles

An example of three-dimensional finite element analysis (Sec. 7-3) of a laterally loaded pile is given in Chap. 9.

Design

In addition to case studies and parametric analysis, it is essential from the viewpoint of the user to evaluate the potential of numerical methods for practical designs. One of the ways is to generate design charts. It must be recognized that it is cumbersome and time-consuming to generate such charts since a large number of finite element analyses are required. Nonetheless, it is possible to produce design charts, and we shall consider an example for piles in sands.

Three different piles with diameters of 12, 16, and 20 in were chosen. For each diameter, finite element analyses were performed with five lengths of embedment 20, 40, 60, 80, and 100 ft. The following soil properties were adopted:

$$\begin{gathered}\phi = 31.5^\circ \qquad N_q = 26.84 \qquad c = 0 \qquad K_c = 1.29 \\ \gamma' = 62.8 \text{ lb/ft}^3 \qquad \text{and} \qquad \delta = 31.5^\circ\end{gathered} \tag{7-27}$$

Here γ' is the submerged unit weight of sand and δ is the angle of friction between pile and soil. The value of δ is usually less than ϕ, but for convenience it was assumed to be equal to that of ϕ. The soil mass was assumed to be a homogeneous layer. The nonlinear properties for soil and interface were the same as those in the first layer of pile 10 (Fig. 7-5 and Table 7-1).

Bearing capacities were determined for each pile from each load-settlement curve by adopting both tangent and critical-displacement criteria; values of critical displacements u_{cr} equal to 0.5, 1.0, and 2.0 in were adopted. Figure 7-18 shows the variation of bearing capacity with length of embedment for the three piles. As a comparison, the variation as computed from limit-equilibrium formulas[1] is also shown in Fig. 7-18. It can be seen that the computed bearing capacity does not increase at the same rate as shown by the limit-equilibrium methods. This is consistent with observations.[8,9,51,52] One of the main reasons can be that the lateral pressure does not vary linearly with depth as assumed in the limit-equilibrium concepts but remains uniform or increases at a decreasing rate beyond a certain depth of pile or length-to-width (diameter) ratio of the pile. A factor contributing to this effect can be the occurrence of tensile-stress conditions and unloading (Fig. 7-17) with consequent stress relief and arching effects near the

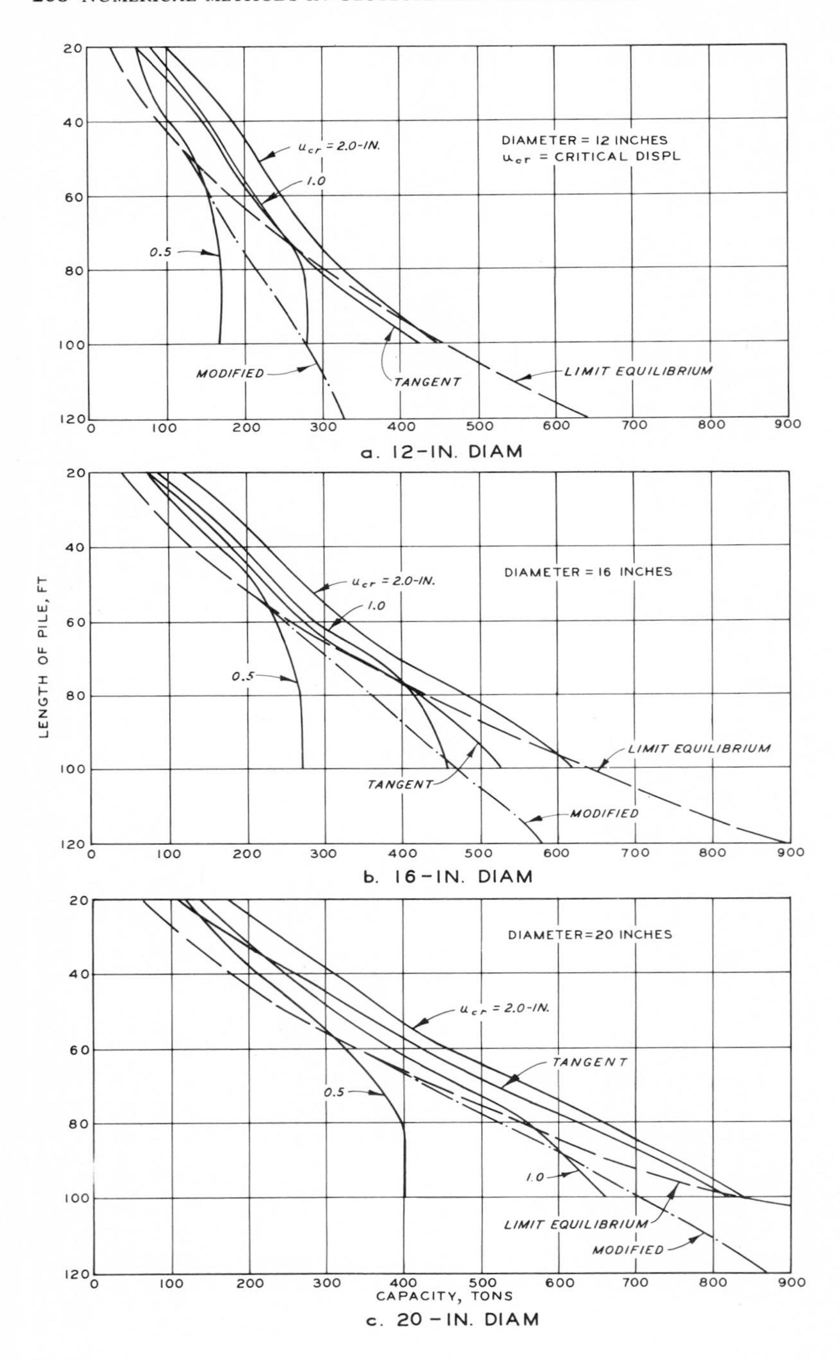

Figure 7-18 Variations of load-carrying capacities and comparisons.[18]

tip of the pile.[16–18,53] This aspect can be illustrated from finite element results by plotting variations of shear stress (lateral pressure) with length of embedment; the numerically computed values do not increase linearly with depth but remain essentially uniform or decrease; see Ref. 18.

The foregoing influence can be introduced by modifying such factors as N_q, K_c, $K_c \tan \delta$, $L\gamma'$ and/or $K_c \gamma' L \tan \delta$ in the conventional formulas. For instance, the conventional bearing-capacity formulas[1,7] can be modified to

$$Q_p = A_p q_0^* N_q \qquad q_0^* = \begin{cases} \gamma' L & \text{for} \quad L < \lambda D \\ \gamma' \lambda D & \text{for} \quad L > \lambda D \end{cases} \tag{7-28a}$$

$$Q_w = A_c K_c \tan \delta q_w^* (0.5L_1 + L_2) \qquad q_w^* = L_1 \gamma'$$

where

$$\begin{aligned} L_1 &= L, \quad L_2 = 0 && \text{for} \quad L < \lambda D \\ L_1 &= \lambda D, \quad L_2 = L - L_1 && \text{for} \quad L > \lambda D \end{aligned} \tag{7-28b}$$

where $\lambda = L/D$ is the ratio beyond which the variations may be assumed to be essentially uniform. Figure 7-18 also shows the variation of bearing capacity with depth for the modified formulas with $\lambda = 40$. Thus we can use the finite element method and generate design charts for a given site, soil, and pile properties and can use them for computing bearing capacity for the required criteria of failure consistent with the requirements for a safe structure.

REFERENCES

1. Terzaghi, K., and R. B. Peck: "Soil Mechanics in Engineering Practice," John Wiley & Sons, Inc., New York, 1967.
2. Scott, R. F.: "Principles of Soil Mechanics," Addison-Wesley Publishing Company, Inc., Reading, Mass., 1963.
3. Mansur, C. I., and J. A. Focht: Pile Loading Tests, Morganza Floodway Control Structure, *Proc. ASCE*, vol. 79, no. 324, pp. 1–31, 1953.
4. Seed, H. B., and L. C. Reese: The Action of Soft Clay along Friction Piles, *Trans. ASCE*, vol. 122, pp. 731–764, 1957.
5. Reese, L. C.: Load versus Settlement for an Axially Loaded Pile, *Proc. Symp. Bearing Capacity Piles*, pt. 2, Central Building Research Institute, Roorkee, India, February 1964.
6. Coyle, H. M., and L. C. Reese: Load Transfer for Axially Loaded Piles in Clay, *J. Soil Mech. Found. Div. ASCE*, vol. 92, no. SM2, pp. 1–26, March 1966.
7. Meyerhof, G. G.: The Ultimate Bearing Capacity of Foundations, *Geotechnique*, vol. 2, pp. 301–332, 1951.
8. Vesic, A. S.: A Study of Bearing Capacity of Deep Foundations, *Georgia Inst. Tech. Rep. Proj.* B-189, Atlanta, March 1967.
9. Vesic, A. S.: Ultimate Loads and Settlements of Deep Foundations in Sand, *Proc. Symp. Bearing Capacity Settlement Found., Duke Univ., Durham, N.C., 1967.*
10. Woodward, R. J., W. S. Gardner, and D. M. Greer: "Drilled Pier Foundations," McGraw-Hill Book Company, New York, 1972.
11. Chandler, R. J.: The Shaft Friction of Piles in Cohesive Soils in Terms of Effective Stress, *Civ. Eng. Publ. Works Rev.*, vol. 63, pp. 48–51, 1968.
12. Burland, J.: Shaft Friction of Piles in Clay, *Ground Eng.*, vol. 6, no. 3, May 1973.
13. Desai, C. S.: Finite Element Method for Analysis and Design of Piles, *U.S. Army Eng. Waterw. Expt. Stn. Misc. Paper* S-76-21, Vicksburg, Miss., 1976.

14. Scofield, A., and P. Wroth: "Critical State Soil Mechanics," McGraw-Hill Book Company, New York, 1968.
15. DiMaggio, F. L., and I. S. Sandler: Material Model for Granular Soils, *J. Eng. Mech. Div. ASCE*, vol. 97, no. EM3, June 1971.
16. Desai, C. S.: Overview, Trends and Projections: Theory and Applications of the Finite Element Method in Geotechnical Engineering, *Proc. Symp. Appl. Finite Element Method Geotech. Eng., Vicksburg, Miss., 1972.*
17. Desai, C. S., and D. M. Holloway: Load-Deformation Analysis of Deep Pile Foundations, *Proc. Symp. Appl. Finite Element Methods Geotech. Eng., Vicksburg, Miss., 1972.*
18. Desai, C. S.: Numerical Design-Analysis for Piles in Sands, *J. Geotech. Eng. Div. ASCE*, vol. 100, no. GT6, pp. 613–635, June 1974.
19. Lee, I. K. (ed.): "Soil Mechanics: Selected Topics," Elsevier, Amsterdam, 1968.
20. Desai, C. S., and J. F. Abel: "Introduction to the Finite Element Method," Van Nostrand Reinhold Company, New York, 1972.
21. Desai, C. S.: "A Three-dimensional Finite Element Procedure and Computer Program for Nonlinear Soil-Structure Interaction Problems," *VPI State Univ. Dept. Civ. Eng. Rep.* VPI-E-75.27, Blacksburg, Va., June 1975.
22. Desai, C. S., and G. C. Appel: 3-D Analysis of Laterally Loaded Structures, *Proc. 2d Int. Conf. Numer. Methods Geomech., Blacksburg, Va., June 1976.*
23. Mahtab, M. A., and R. E. Goodman: Three-dimensional Finite Element Analysis of Jointed Rock Slopes, *Proc. 2d Cong., Int. Soc. Rock Mech., Belgrade, September 1970*, vol. 3.
24. Desai, C. S.: A Consistent Finite Element Technique for Work-softening Behavior, *Proc. Int. Conf. Comput. Methods Nonlinear Mech., Univ. Texas, Austin, September 1974.*
25. Fruco and Associates: Pile Driving and Loading Tests: Lock and Dam No. 4, Arkansas River and Tributaries, Arkansas and Oklahoma, U.S. Army Engineers District, Little Rock, Ark., September 1964.
26. Mansur, C. I., and A. H. Hunter: Pile Tests: Arkansas River Project, *J. Soil Mech. Found. Div. ASCE*, vol. 96, no. SM5, pp. 1545–1582, September 1970.
27. Skempton, A. W.: Cast in-situ Piles in London Clay, *Geotechnique*, vol. 9, pp. 153–73, 1959.
28. Burland, J. B., F. G. Butler, and P. Dunican: The Behavior and Design of Large Diameter Bored Piles in Stiff Clay, *Proc. Symp. Large Bored Piles, Inst. Civ. Eng., London, 1966.*
29. Parry, R. H. G.: personal communication.
30. O'Neill, M. W., and L. C. Reese: Behavior of Axially Loaded Drilled Shafts in Beamont Clay, *Cent. Highw. Res., Res. Rep.* 89-7 to 89-10, Univ. Texas, Austin, December 1970.
31. O'Neill, M. W., and L. C. Reese: Behavior of Bored Piles in Beamont Clay, *J. Soil Mech. Found. Div. ASCE*, vol. 98, no. SM2, February 1972.
32. D'Appolonia, D. J., H. G. Poulos, and C. C. Ladd: Initial Settlement of Structures on Clay, *J. Soil Mech. Found. Div. ASCE*, vol. 97, no. SM10, October 1971.
33. Chellis, R. D.: "Pile Foundations," McGraw-Hill Book Company, New York, 1962.
34. Leonards, G. A. (ed.): "Foundation Engineering," McGraw-Hill Book Company, New York, 1962.
35. Hrennikoff, A.: Analyses of Pile Foundations with Batter Piles, *Trans. ASCE*, vol. 115, pp. 351–381, 1950.
36. Whitaker, T.: Experiments with Model Piles in Groups, *Geotechnique*, vol. 7, pp. 147–167, 1957.
37. Sowers, G. F., et al.: The Bearing Capacity of Friction Pile Groups in Homogeneous Clay from Model Studies, *Proc. 5th Int. Conf. Soil Mech. Found. Eng., Paris, 1961*, vol. 2, pp. 155–159.
38. Kezdi, A.: Deep Foundations, *Proc. 6th Int. Conf. Soil Mech. Found. Eng., Montreal, 1965*, vol. 2, pp. 265–284.
39. Berezantzev, V. G., V. S. Khristoforov, and V. S. Golubkov: Load Bearing Capacity and Deformation of Piled Foundations, *Proc. 5th Int. Conf. Soil Mech. Found. Eng., Paris, 1961*, vol. 3, pp. 11–15.
40. Hanna, T. H.: Model Studies of Foundation Groups in Sand, *Geotechnique*, vol. 13, no. 4, pp. 334–351, 1963.
41. Desai, C. S., L. D. Johnson, and C. M. Hargett: Analysis of Pile-Supported Gravity Lock, *J. Geotech. Eng. Div. ASCE*, vol. 100, no. GT9, 1974.

42. Desai, C. S., et al.: Finite Element Analysis of Columbia Lock-Pile-Foundation System, *U.S. Army Eng. Waterw. Expt. Stn. Tech. Rep.* S-74-61, Vicksburg, Miss., 1974.
43. Desai, C. S.: Soil-Structure Interaction and Some Relevant Topics, *Proc. Symp. Numer. Methods Soil Rock Mech., Univ. Karlsruhe, September 1975.*
44. Brooker, E. W., and H. O. Ireland: Earth Pressures at Rest Related to Stress History, *Can. Geotech. J.*, vol. 2, no. 1, February 1965.
45. Wroth, P., and J. M. O. Hughes: An Instrument for the in situ Measurement of the Properties of Soft Clay, *Proc. 8th Int. Conf. Soil Mech. Found. Eng., Moscow, 1973.*
46. Furlow, C. R.: Pile Tests, Jonesville Lock and Dam, Ouachita and Black Rivers, Arkansas and Louisiana, *U.S. Army Eng. Waterw. Expt. Stn. Tech. Rep.* S-68-10, Vicksburg, Miss., December 1968.
47. Broms, B. B.: Methods of Calculating the Ultimate Bearing Capacity of Piles, *Swed. Geotech. Inst., Stockholm, Sols Soils*, vol. 18–19, pp. 21–31, 1966.
48. Ko, H. Y., and C. S. Desai: Effect of Stress Paths on Behavior of Sands at Jonesville Lock Site, unpublished.
49. Varadrajan, A.: Effect of Overconsolidation and Stress-Path on the Behavior of Saturated Clays during Shear," Ph.D. thesis, Indian Institute of Technology, Kanpur, 1974.
50. Sherman, W. C., et al.: Analysis of Pile Tests, *U.S. Army Eng. Waterw. Expt. Stn. Tech. Rep.* S-74-3, Vicksburg, Miss., April 1974.
51. Kerisel, J.: Deep Foundations: Basic Experimental Facts, *Proc. N. Am. Conf. Deep Found., Mexico City, 1964.*
52. Robinshky, G. I., and C. F. Morrison: Sand Displacement and Compaction around Model Friction Piles, *Can. Geotech. J.*, vol. 1, no. 2, March 1964.
53. Ellison, R. D., et al.: Load-Deformation Mechanism of Bored Piles, *J. Soil Mech. Found. Div. ASCE*, vol. 97, no. SM4, pp. 661–678, April 1971.

CHAPTER

EIGHT

WAVE-EQUATION ANALYSIS OF PILING BEHAVIOR

Harry M. Coyle, Lee L. Lowery, Jr., and Teddy J. Hirsch

8-1 INTRODUCTION

Description of the Problem

The rapidly increasing use of pile foundations and the appearance of new pile-driving techniques have led engineers to look for more reliable methods of pile analysis and design. In the past, the use of piles in foundations has always been one of the less certain and variable aspects of foundation engineering. The diversity and variability of piles, types of driving hammers, and soils encountered have always been major problems in building construction and offshore structures. Until recently, it was not possible to formulate a general method of analysis capable of solving such a complex problem, and empirical solutions with severe limitations were the only tools generally available. The use of simple pile-driving formulas to predict bearing capacity or impact stresses for so complex a problem was indeed risky.

Historical Review and Current Trends

In 1950, E. A. L. Smith[1] proposed a numerical solution which could be used to solve extremely complex pile-driving problems if it was possible to quantify the numerous input variables involved. In 1960, Smith[2] published a second paper dealing exclusively with the application of wave theory to the behavior of driven piles and recommended values for certain of the required input variables based on his past observations. From that time to the present, the authors have been engaged in continuing research to determine values for the required input quantities by conducting both laboratory and field tests. At present many large construction firms and oil companies are using the wave equation with great success for the design of pile foundations. Some states, including the Texas Highway Department, have written it into their foundation manuals. The purpose of this chapter is to familiarize the reader with how the wave equation is formed and utilized and to present its practical applications.

8-2 NUMERICAL METHOD OF ANALYSIS

Smith's proposed solution to the wave equation was based on dividing the distributed mass of the pile, as shown in Fig. 8-1, into a number of concentrated weights $W(1)$ through $W(p)$, which are connected by weightless springs $K(1)$ through $K(p-1)$, with the addition of soil resistance acting on the masses. Time is also divided into small increments.

For the idealized system, Smith set up a series of simple equations of motion in the form of finite difference equations, which were easily solved using high-speed digital computers. He extended his original method of analysis to include various nonlinear parameters such as elastoplastic soil resistance.

A derivation of the one-dimensional wave equation and calculation of the critical time interval for use therein is presented in Appendixes 8A and 8B, respectively.

Hammer-Pile Simulation

Figure 8-1 illustrates the idealization of the pile system. In general, the system is considered to be composed of:

1. A ram, to which an initial velocity is imparted by the pile driver
2. A cap block (cushioning material)
3. A pile cap
4. A cushion block (cushioning material)
5. A pile
6. The supporting medium, or soil

In Fig. 8-1*b* are shown the idealizations for various components of the actual pile. The ram, cap block, pile cap, cushion block, and pile are pictured as appropriate discrete weights and springs. The frictional soil resistance on the side of the pile is represented by a series of side springs and dashpots; the point resistance is accounted for by a single spring and dashpot under the point of the pile. The characteristics of these various components will be discussed in greater detail

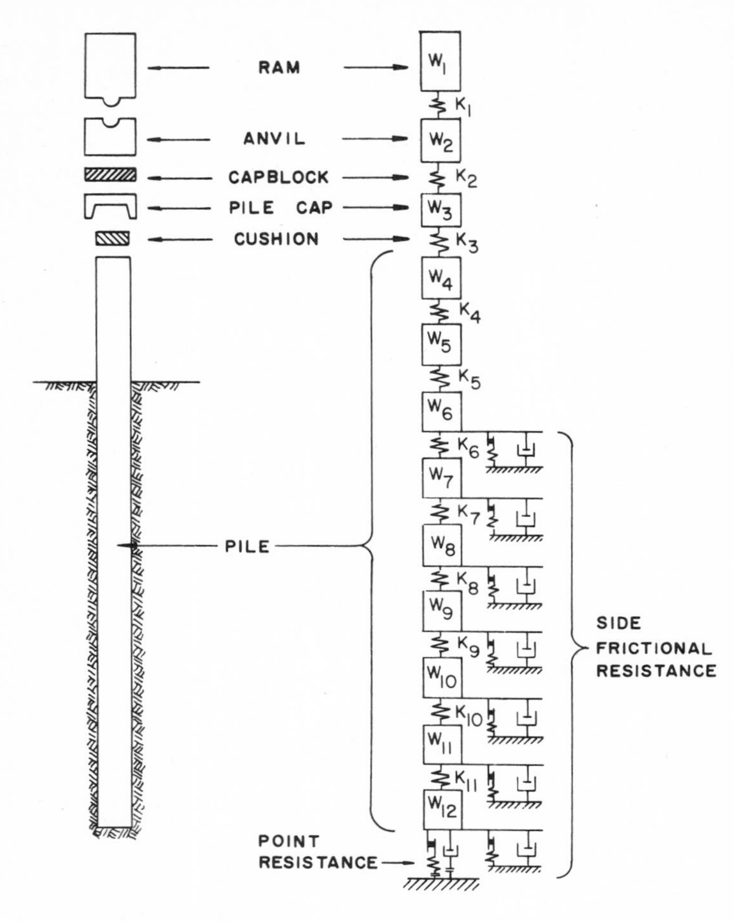

Figure 8-1 Simulation of a typical Hammer-pile-soil system.[2]

later. Actual situations may deviate from that illustrated in Fig. 8-1. For example, a cushion block may not be used, or an anvil may be placed between the ram and cap block. However, such cases can readily be accommodated.

The ram, cap block, pile cap, and cushion block may in general be considered to consist of "internal" springs, although in the representation of Fig. 8-1*B* the ram and the pile cap are assumed rigid, which is a reasonable assumption for many practical cases. However, in the case of a long slender ram, it too can be broken into a series of weights and springs.

Figure 8-2*A* and *B* suggests different possibilities for representing the load-deformation characteristics of the internal springs. In Fig. 8-2*A*, the material is considered to experience no internal damping. In Fig. 8-2*B*, the material is assumed to have internal damping according to the linear relationship shown.

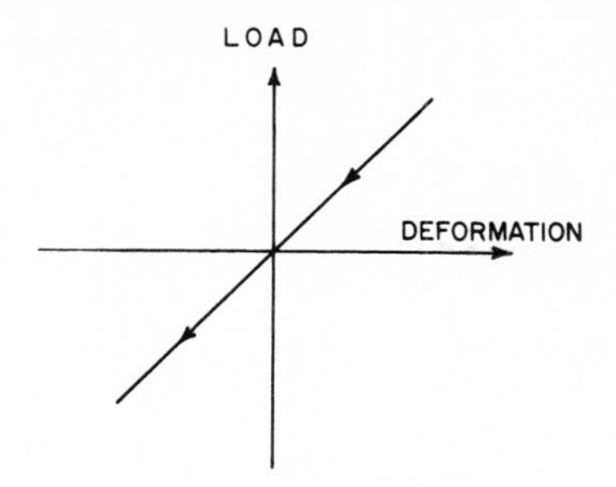

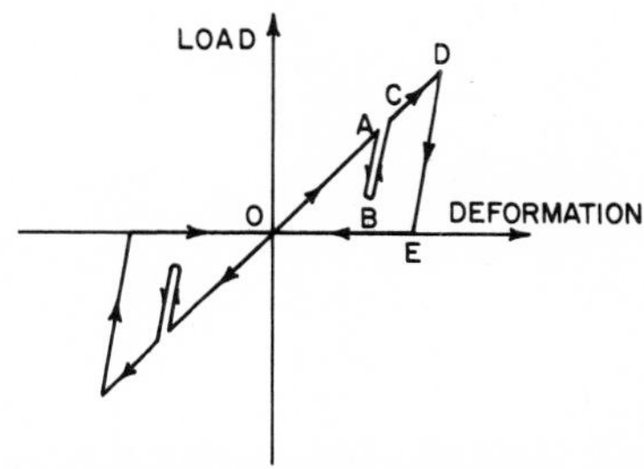

Figure 8-2 Load-deformation relationships for internal springs.

Theory

The following equations, developed by Smith,[2] are the controlling equations on which the method is based:

$$D(m, t) = D(m, t-1) + 12\,\Delta t\,V(m, t-1) \tag{8-1}$$

$$C(m, t) = D(m, t) - D(m+1, t) \tag{8-2}$$

$$F(m, t) = C(m, t)K(m) \tag{8-3}$$

$$R(m, t) = [D(m, t) - D'(m, t)]K'(m)[1 + J(m)V(m, t-1)] \tag{8-4}$$

$$V(m, t) = V(m, t-1) + [F(m-1, t) - F(m, t) - R(m, t)]\frac{g\,\Delta t}{W(m)} \tag{8-5}$$

where m = element number
t = time interval number
Δt = size of time interval, s
$C(m, t)$ = compression of internal spring m in time interval t, in
$D(m, t)$ = displacement of element m in time interval t, in
$D'(m, t)$ = plastic displacement of external soil spring m in time interval t, in
$F(m, t)$ = force in internal spring m in time interval t, lb
g = acceleration due to gravity, ft/s^2
$J(m)$ = damping constant of soil at element m, s/ft
$K(m)$ = spring constant associated with internal spring m, lb/in
$K'(m)$ = spring constant associated with external soil spring m, lb/in

$R(m, t)$ = force exerted by external spring m on element m in time interval t, lb
$V(m, t)$ = velocity of element m in time interval t, ft/s
$W(m)$ = weight of element m, lb

As usual, the parentheses indicate functions.

The use of a spring constant $K(m)$ implies a load-deformation behavior of the type shown in Fig. 8-2*a*, in which $K(m)$ is the slope of the straight line. Smith also developed special relationships to account for internal damping in the cap block and the cushion block. Instead of Eq. (8-3) he obtained

$$F(m, t) = \frac{K(m)}{[e(m)]^2} C(m, t) - \left\{ \frac{1}{[e(m)]^2} - 1 \right\} K(m) C(m, t)_{\max} \qquad (8\text{-}6)$$

where $e(m)$ is the coefficient of restitution of internal spring m and $C(m, t)_{\max}$ is the temporary maximum value of $C(m, t)$. With reference to Fig. 8-1, Eq. (8-6) would be applicable in the calculation of the forces in internal springs $m = 2$ and $m = 3$. The load-deformation relationship characterized by Eq. (8-6) is illustrated by the path *OABCDEO* in Fig. 8-2*b*. Tensile forces cannot exist in a pile cap or a cushion block; consequently, only the compressive part of the diagram applies. Intermittent unloading and loading is typified by the path *ABC*, established by control of the quantity $C(m, t)_{\max}$ in Eq. (8-6). The slopes of lines *AB*, *BC*, and *DE* depend upon the coefficient of restitution $e(m)$.

Computations

The computations proceed as follows:

1. The initial velocity of the ram is determined from the properties of the pile driver. Other time-dependent quantities are initialized either at zero or to satisfy static equilibrium conditions.
2. Displacements $D(m, 1)$ for each mass are first calculated by using Eq. (8-1). Note that $V(1, 0)$ is the initial velocity of the ram.
3. Compressions $C(m, 1)$ for each mass are next calculated by using Eq. (8-2).
4. Internal spring forces $F(m, 1)$ for each spring are calculated by using Eq. (8-3) or (8-6).
5. External spring forces $R(m, 1)$ are calculated by using Eq. (8-4).
6. Velocities $V(m, 1)$ are calculated by using Eq. (8-5).
7. The cycle is repeated for successive time intervals.

In Eq. (8-4), the plastic soil deformation $D'(m, t)$ for a given external spring follows Fig. 8-3*a* and can be determined by special routines. For example, when $D(m, t)$ is less than $Q(m)$, $D'(m, t)$ is held at zero; when $D(m, t)$ is greater than $Q(m)$ along line *AB* (Fig. 8-3*a*), $D'(m, t)$ is set equal to $D(m, t) - Q(m)$. $Q(m)$ is called the soil quake.

Smith noted that Eq. (8-4) produces no damping when $D(m, t) - D'(m, t)$ becomes zero at point C in Fig. 8-3*a*. He suggested an alternative equation to be

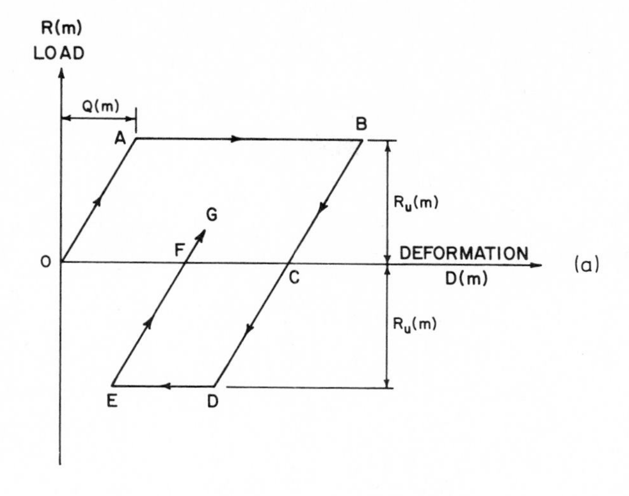

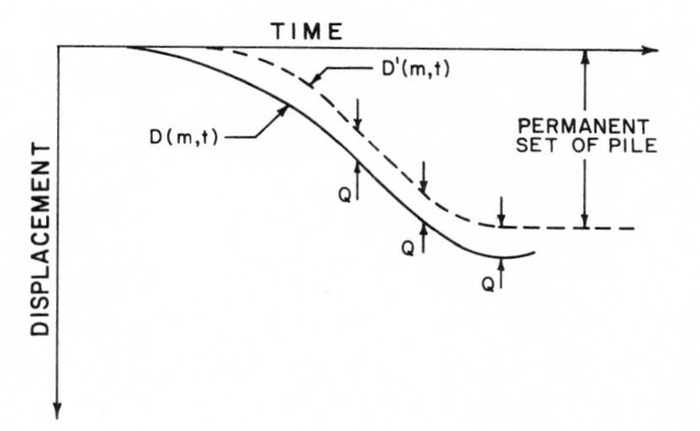

Figure 8-3 (*a*) Load-deformation characteristics assumed for soil spring *m*. (*b*) Time-displacement curve for pile point.

used after $D(m, t)$ first becomes equal to $Q(m)$, as follows:

$$R(m, t) = [D(m, t) - D'(m, t)]K'(m) + J(m)K'(m)Q(m)V(m, t - 1) \qquad (8\text{-}7)$$

Most computer programs are programmed to stop automatically when the following conditions are satisfied:

1. The pile point has penetrated the soil to some maximum value and is now rebounding upward.
2. The velocities $V(1)$ through $V(MP)$ are all simultaneously negative or equal to zero.

This ordinarily stops the calculations soon after the driving is complete, as shown in Fig. 8-3*b*. Also, as noted in this figure, D', the plastic or permanent set of the soil, lags behind the total displacement by a distance Q. Thus, the permanent set of the pile is assumed to be equal to the maximum point displacement recorded less the soil quake at the point.

When soil quake loading and unloading values vary along the pile length, it may prove more accurate to let the computer program run until the pile settles down at some depth and use this computed value as the permanent set.

Care must be taken to satisfy conditions at the head and point of the pile when applying Eqs. (8-1) to (8-5). Consider Eq. (8-3). When $m = p$, where p is the number of the last element of the pile, $K(p)$ must be set equal to zero since there is no force $F(p, t)$. Beneath the point of the pile, the soil spring must be prevented from exerting tension on the pile point. In applying Eq. (8-5) to the ram ($m = 1$), one should set $F(0, t)$ equal to zero.

For the idealization of Fig. 8-1, it is apparent that the spring associated with $K(3)$ represents both the cushion block and the top element of the pile. Its spring rate can be obtained from

$$\frac{1}{K(3)} = \frac{1}{K(3)_{\text{cushion}}} + \frac{1}{K(3)_{\text{pile}}} \tag{8-8}$$

where the spring rate of each element is computed by

$$K(I) = A(I)\frac{E(I)}{L(I)} \tag{8-9}$$

Table 8-1 Typical hammer properties

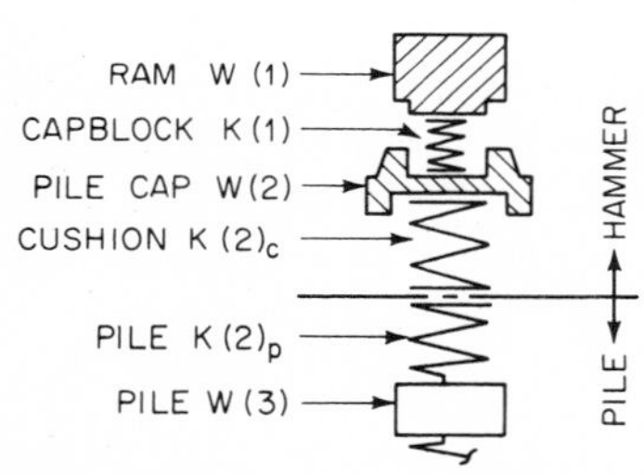

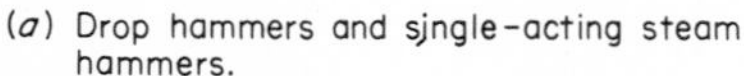
(a) Drop hammers and single-acting steam hammers.

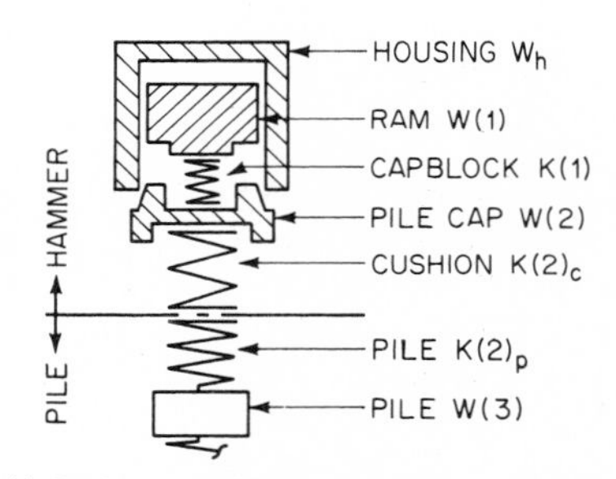

(b) Double and differential-acting steam hammers.

$K(1)$ and $K(2)_c$ depend on material properties and dimensions; $K(2)_p = [A(2)E(2)]/L(2)$

Hammer	Type	$W(1)$, lb	$W(2)$,† lb	$W(h)$, lb	Stroke h, ft	P rated, lb/in^2	Eff e_f
Mkt S3	*a*	3,000	···	···	3.00	···	0.80
Mkt S5	*a*	5,000	···	···	3.25	···	0.80
Vulcan 1	*a*	5,000	1000	···	3.00	···	0.80
Vulcan 2	*a*	3,000	1000	···	2.42	···	0.80
Vulcan 30C	*b*	3,000	1000	4,036	1.04	120	0.85
Vulcan 50C	*b*	5,000	1000	6,800	1.29	120	0.85
Vulcan 80C	*b*	8,000	2000	9,885	1.38	120	0.85
Vulcan 140C	*b*	14,000	···	13,984	1.29	140	0.85

† Representative values for pile normally used in highway construction.

where $A(I)$ = cross-sectional area of pile at element I
$E(I)$ = modulus of elasticity of element I
$L(I)$ = length of element I

Only a few input quantities for the hammer-pile simulation are included in Table 8-1 because of space limitation. The reader should refer to Lowery et al.[3] for additional tabulated numerical quantities recommended for use with wave equation analysis.

Program Limitations

The only basic limitations which are inherent to the program are the exclusion of residual stresses (either during manufacture of the pile or from previous hammer blows) and flexural vibrations. Other parameters such as damping in the pile material are normally handled by special subroutines.[3] Piles of variable cross section are easily handled by simply computing the actual spring rate and weight for each element selected.

8-3 GEOTECHNICAL ASPECTS OF THE PROBLEM

Pile-Soil Simulation

The pile-soil system simulation used in the wave equation analysis is given in Fig. 8-4. The input parameters needed for the pile-soil simulation include the static soil resistance, the soil quake, and the soil damping. The static soil resistance is input as point resistance *RUP* and the side resistance *RU*, which can be distributed in accordance with any variation in the soil profile. The total soil resistance *RUT* is the sum of the side resistance and the point resistance. The soil quake Q, which is the amount of static deformation the soil will experience before failure, is shown in Fig. 8-5*a*. It is input both along the side and at the point of the pile. The magnitude of the quake can vary for different types of soils, and the loading and unloading values can also differ. The soil damping, which accounts for the dynamic behavior of the soil, is input as side damping J' and point damping J. The magnitude of the soil damping can vary for different types of soils and the values of side damping and point damping can also differ.

The soil load-deformation relationship used in wave equation analysis is shown in Fig. 8-5*a* and the pile-soil model is shown in Fig. 8-5*b*. The soil resistance mobilized during dynamic loading is determined by using the equation given in Fig. 8-5*b*. The terms appearing in this equation are defined as follows:

R_u = dynamic or static soil resistance, lb
J = damping constant for the soil at point of pile, s/ft
J' = damping constant for soil along side of pile, s/ft
V = instaneous velocity of segment of pile at given time, ft/s

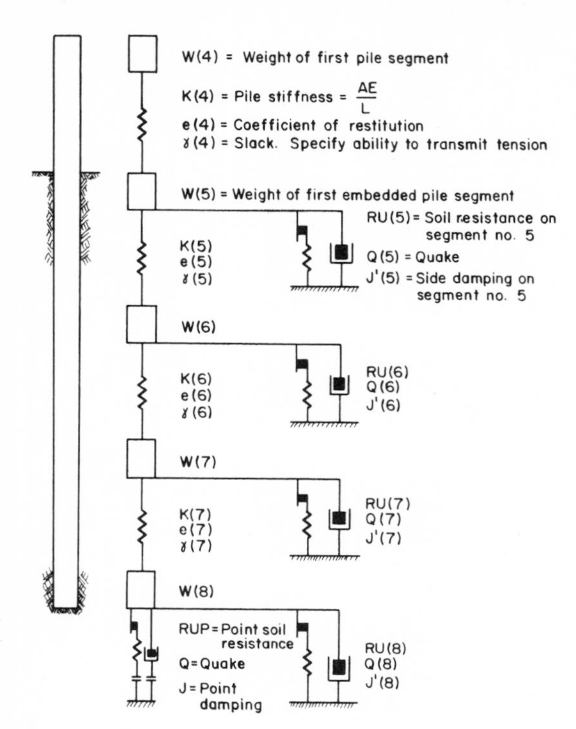

Figure 8-4 Pile-soil system.

Soil Parameters

It is important to note that the reliability of predicted bearing capacity and predicted driving stresses is primarily a function of the accuracy of the soil parameters. Considerable research effort was expended at Texas A & M University from 1967 to 1973 toward determining appropriate soil parameters. Data from static tests on five instrumented piles were used to determine appropriate values of soil quake for sands and clays. By using measured dynamic force-time data as input at the pile head from the five test piles and the appropriate static soil quake values a satisfactory combination of side and point damping was determined. The correctness of the soil parameters was verified by comparing computed bearing capacities and pile stresses with measured data. Details on the analysis of the pile test data are given in Ref. 4.

The testing sequence involved initial driving of the pile, static load testing as soon as possible after initial driving, static load testing after an elapsed time of 8 to 11 days, and redriving after the final static load test. Dynamic data were recorded from strain gage bridge outputs during the last 3 to 5 ft (0.9 to 1.5 m) of initial driving and the first 3 to 5 ft (0.9 to 1.5 m) of redriving. The outputs of the strain gage bridges during the dynamic tests were mechanically recorded during driving and redriving. This provided a permanent record of the load in the pile as a function of time at each bridge location for every blow of the hammer throughout the time the data were being recorded. The force-time data at the top of the

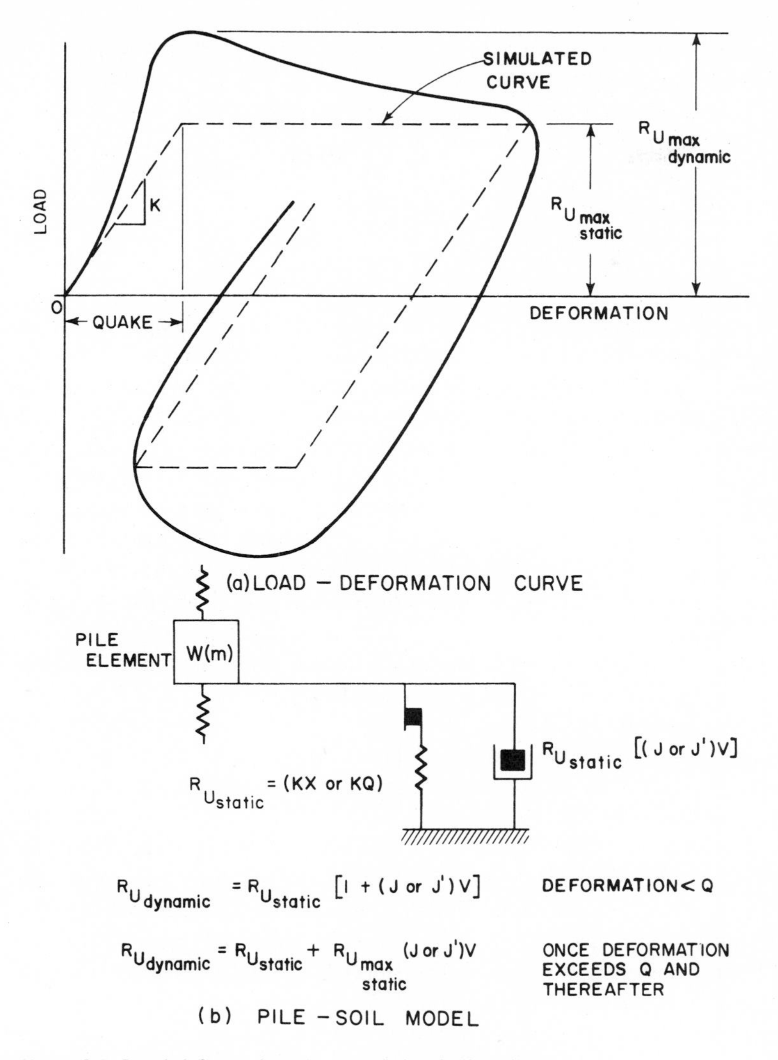

Figure 8-5 Load-deformation characteristic of pile-soil model.

pile are particularly valuable from a wave equation analysis standpoint because these data can be used as input to the computer program. In so doing, all the uncertainties connected with simulating the hammer-pile system can be eliminated.

The static load tests included the use of strain gage bridges located along the length of the test pile. The bridge at the head of the pile, along with a calibrated load cell, measured the total static force being applied to the pile during any load increment. The bridge at the point of the pile measured the load being supported by the pile in end bearing. The ratio of point load to total load must be known for wave equation analysis purposes. Interior strain gage bridges were placed so that they would be as close as possible to the interface of major soil stratum changes at the end of the initial driving. In this manner, it was possible to determine the load transfer for each stratum. This information is useful for wave equation analysis purposes since it is possible to input soil distribution in accordance with the actual (measured) distribution for each layer of soil.

Soil Quake

The soil quake Q is defined as the amount of soil deformation or movement which must occur before the soil reaches a state of plastic failure. This concept is shown graphically in Fig. 8-5*a*. The deliberate placement of strain gage bridges and the incremental nature of the static load tests made it possible to develop load transfer vs. movement curves for each soil stratum between bridges. Similarly, the data were used to construct a tip load vs. tip movement curve for the soil beneath the tip of the pile. The details of the procedure used to develop these curves can be found in the paper by Coyle and Reese.[5] By developing these curves from the static load test data it was possible to evaluate the soil quake Q. Based on this research,[4] the loading quake values given in Table 8-2 are recommended for use in sands and clays.

The recommended quake values given in Table 8-2 are designated loading quakes because the values were derived from loading tests on piles. From the standpoint of a wave equation analysis the unloading quake is also significant because under dynamic loading the motion of the soil with respect to the pile may change directions several times before the energy causing the displacement of the pile-soil system has dissipated.

It was also possible to obtain[4] an indication of the unloading quake from the static field load test data. Gross settlement at the head of each test pile was recorded with maximum load applied. The pile was unloaded and allowed to rebound, whereupon gross settlement at the head of the test pile was again recorded. The difference between these gross settlements from maximum load to no load gives the total rebound of the pile head. The unloading quake of the soil can then be estimated by subtracting the elastic compression of the pile under maximum load from the total rebound of the pile head. Using this procedure, an

Table 8-2 Recommended loading soil quake values

Soil type	Side quake Q, in†	Point quake Q, in†
Sand	0.2	0.4
Clay	0.1	0.1

† 1 in = 25.4 mm.

Table 8-3 Recommended soil-damping values

Soil type	Friction damping J', s/ft†	Point damping J, s/ft†
Sand	0.5	0.15
Clay	0.2	0.01

† 1 ft = 0.305 m.

unloading quake of 0.1 in (2.54 mm) was determined for both sands and clays. It should be noted that most computer programs currently in use would have to be modified to enable them to use an unloading quake which differs from the loading quake.

Damping Parameters

Once the proper soil quake values for both sands and clays had been established,[4,5] the next step in the study was to determine the proper soil damping values. This step was accomplished by conducting a parametric study using known force-time data as input along with the established soil quake values. All reasonable combinations of soil damping values were used as input to the computer program, and the calculated pile stresses and blow counts were correlated with the measured test pile stresses and blow counts. The soil damping values given in Table 8-3 are recommended for use in sands and clays. The damping values given in Table 8-3 for sand are considered to be applicable for sands which are completely saturated. A smaller friction damping value may be more appropriate for partially saturated sands.

8-4 APPLICATIONS AND DESIGN ANALYSIS

Early in 1960 a number of field tests were performed to compare the results of the wave equation with full-scale piles. Several steel and concrete piles 85 ft (25.9 m) long were instrumented and suspended above the ground. This eliminated the soil resistance and made it possible to check the method's ability to predict accurately driving stresses as influenced by the hammer-pile configuration. Typical results for a steel H pile are shown in Fig. 8-6. The method gives relatively accurate results, especially during the first few milliseconds.

Selection of Driving Equipment

Pile driving hammers One of the most important questions to be answered in the installation of piles is the selection of a hammer to drive the pile to the required penetration successfully. Unfortunately, the driving capability of any hammer is greatly influenced by numerous factors. Many cases have been reported in which the hammer selected was unable to drive the piles to the design penetration.

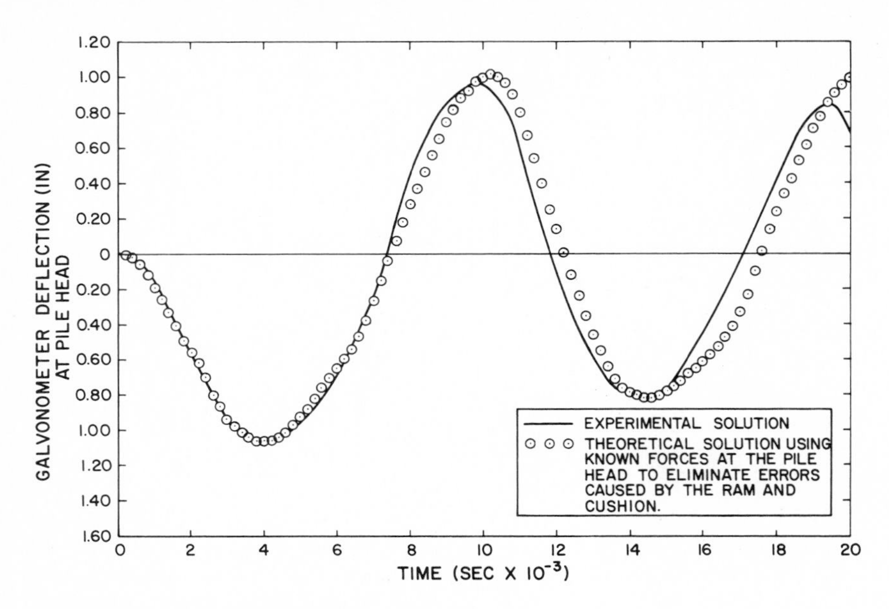

Figure 8-6 Theoretical vs. experimental solution.

The wave equation was used to analyze three offshore pile driving hammers, the Vulcan 020, 040, and 060 hammers. The results of this analysis are shown in Fig. 8-7. In each case, all variables such as the hammer efficiency, cushion stiffness, and other input data required for the wave equation solution were held constant.

As shown in Fig. 8-7, although the energy output of the Vulcan 040 and 060 hammers are respectively two and three times that of the Vulcan 020 hammer, in

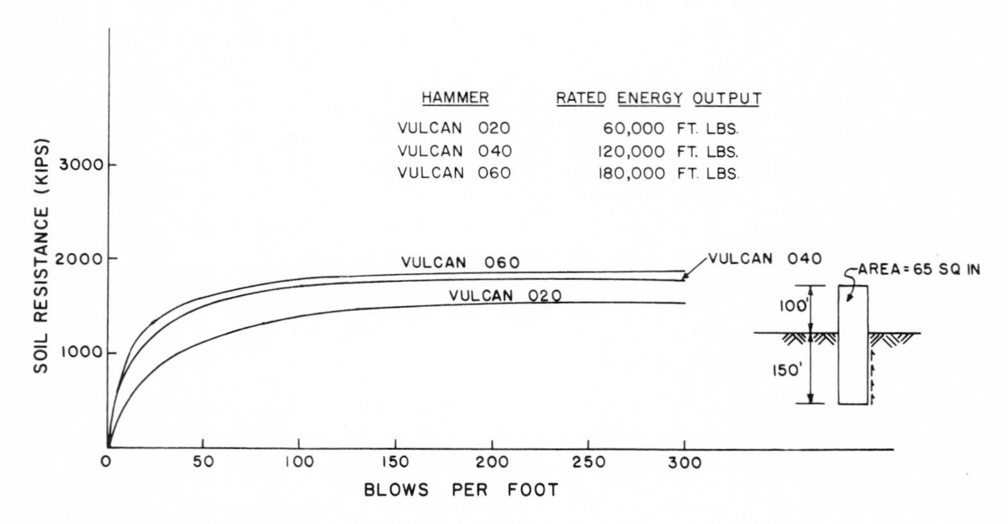

Figure 8-7 A typical comparison of pile driving hammers.

Table 8-4 Recommended cushion properties

Material	E, lb/in^2	e
Pine plywood	21,200	0.27
Gum	25,000	0.20
Fir plywood	32,100	0.43
Oak	43,000	0.50
Micarta-aluminum	400,000	0.90
Garlock asbestos	40,000	0.50

no case was the driving capacity of the 020 doubled or tripled by the use of the larger hammers. Actually, the 200 percent increase in energy from a Vulcan 020 to an 060 hammer increased the driving resistance by only 20 percent.

The demands of present offshore construction often require that these piles be driven to resistances of 3000 tons (2.72×10^6 kg) or more. These requirements can barely be achieved with existing hammers, and it is obvious that larger hammers will eventually have to be built.

Cushions Cushions are normally used to reduce the stresses in both the hammer and pile during driving. Common cushioning materials include hard- and softwoods, plywood, asbestos, rope, and micarta and aluminum disks. The most significant properties of a cushion include its stiffness and coefficient of restitution. Typical cushion properties are given in Table 8-4.

Figure 8-8 shows the effect of varying the cushion stiffness. For the pile shown, an increase in the cushion stiffness increased the ability to drive the pile, especially at high soil resistances.

The effect of variations in coefficient of restitution of the cushion is illustrated in Fig. 8-9. As might be expected, the more efficient cushion increases the ability to drive the pile at all levels of resistance since less energy is absorbed in the cushion. However, as shown in Fig. 8-9, the effect of coefficient of restitution is far less than might be expected.

Driving accessories The weight of the pile cap can have a significant influence on the ultimate soil resistance to which a hammer can drive a pile. Figure 8-10 gives the wave equation results for two hammers driving the same pile. The pile cap weight was varied as shown.

For the 020 hammer, increasing the weight of the pile cap from 5 to 25 kips (2265 to 11,325 kg) resulted in a 12 percent decrease in the ultimate resistance to which the pile could be driven, whereas an increase to 45 kips (20,385 kg) resulted in a 26 percent loss in ultimate resistance. This phenomenon can be explained by the fact that the energy available to drive the pile was reduced by the work done in accelerating the pile cap. The results obtained for the Vulcan 060 are also indicated in Fig. 8-10.

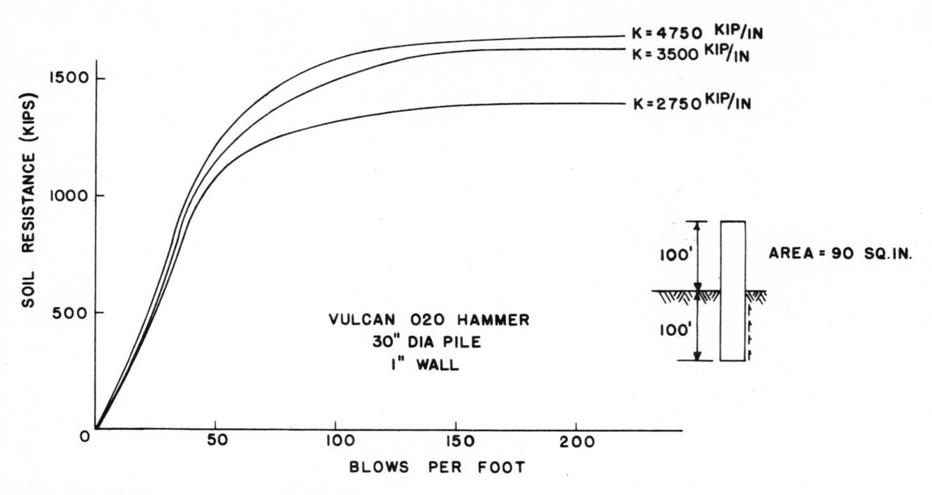

Figure 8-8 Effect of cushion stiffness.

Pile configuration One of the more effective methods of increasing the ability of a pile to be driven is to increase its stiffness by increasing its cross-sectional area. Figures 8-11 and 8-12 illustrate typical results found by increasing the wall thicknesses of given piles.

As noted in Fig. 8-7, tripling the energy of a Vulcan 020 hammer increased the ability to drive the pile only 20 percent. In contrast, as illustrated in Figs. 8-11 and 8-12, tripling the cross-sectional area of the pile increases the ability of the 020 hammer to drive the pile by approximately 60 percent and increases the ability of the 060 hammer to drive the pile by 100 percent.

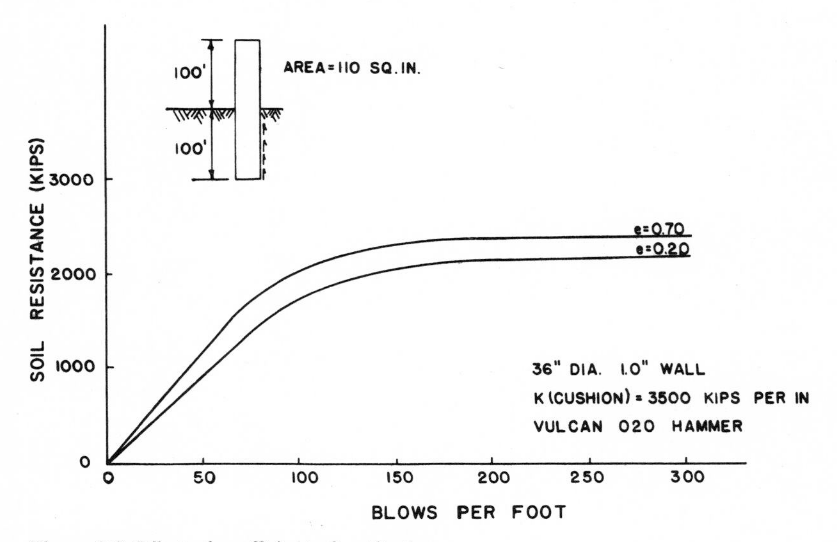

Figure 8-9 Effect of coefficient of restitution *e*.

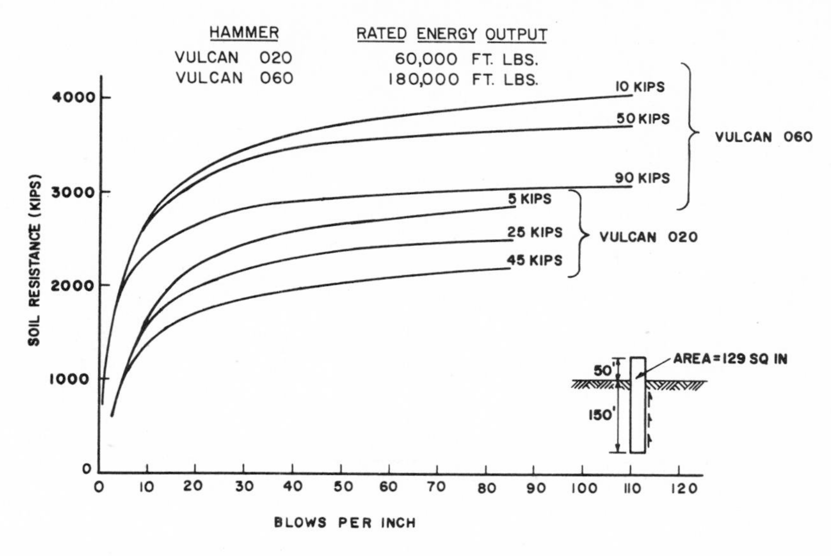

Figure 8-10 Effect of pile cap weight.

It is interesting to note in Fig. 8-11 that at low soil resistances (less than 1200 kips) the lighter pile drives at a lower blow count than the heavier pile. This is because at low resistances the pile moves a larger distance per blow, and the increased mass of the heavier pile traveling through this displacement absorbs much of the energy output of the hammer. However, at a higher resistance, the number of blows per foot becomes sufficiently large, i.e., the penetration per blow becomes so small that the additional capacity of the heavy pile to transmit the stress wave through the soil far exceeds the inertial effects, thus making it easier to drive.

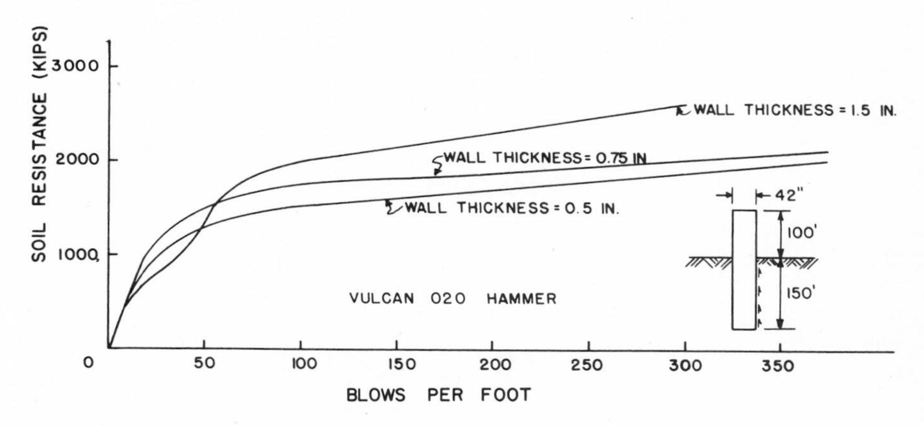

Figure 8-11 Effect of wall thickness, Vulcan 020 hammer.

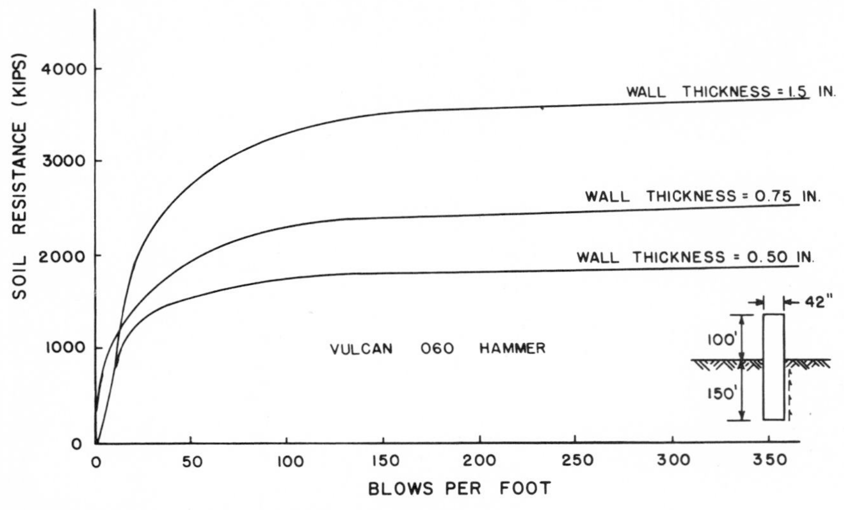

Figure 8-12 Effect of wall thickness, Vulcan 060 hammer.

Prediction of Bearing Capacity

Another application of wave equation analysis is the prediction of bearing capacity either by use of a bearing graph, which relates soil resistance to blow count, or by development of a bearing capacity-vs.-depth relationship. The wave equation computer program can be used to compute a soil resistance vs. blow count curve (bearing graph) for selected intervals of depth by using the ratio of point to total load encountered throughout the particular stratum under investigation. The ratio can be predicted on the basis of soil shear strength data, or an approximate choice can be made from a knowledge of the general character of the soil profile down to that point. For each stratum investigated, a family of curves can be generated with a different hammer efficiency for each curve if it is expected that the hammer efficiency will vary over a relatively wide range.

This procedure is illustrated in Figure 8-13. Figure 8-13*a* shows a typical driving record for a 100-ft (30.5-m) pile. Figure 8-13*b* shows the bearing graphs obtained from a wave equation analysis, where it has been assumed that the hammer efficiency is expected to vary between 75 and 95 percent. For this example, it is further assumed that the variation in the ratio of point to total load during driving will not significantly affect the bearing graph. The plot of predicted bearing capacity vs. depth (Fig. 8-13*c*) from 0 to 60 ft (18.3 m) penetration was obtained from the 75 percent hammer-efficiency bearing graph; the 95 percent efficiency bearing graph was used to develop the curve from 60 to 100 ft (18.3 to 30.5 m) penetration. For example, while driving from 56 to 60 ft (17.1 to 18.3 m) of penetration, the average blow count was 40 blows per foot (131 blows per meter), which yields a predicted bearing capacity of 160 tons (1.45×10^5 kg) with the hammer operating at 75 percent efficiency. Other points are obtained in a similar manner.

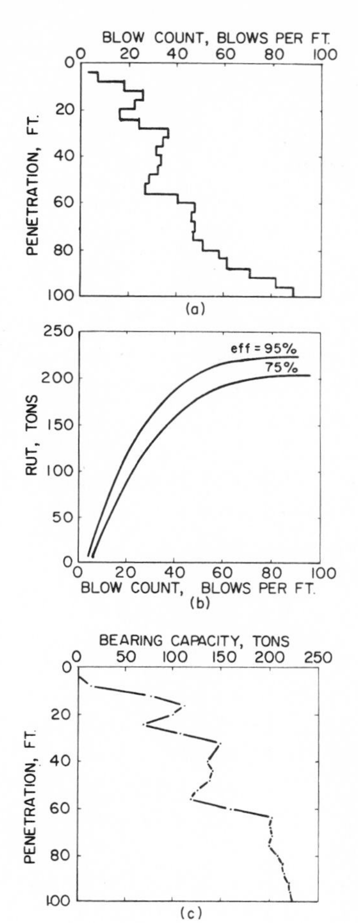

Figure 8-13 Determination of bearing capacity vs. depth from driving record.

Although the data in Fig. 8-13 are hypothetical and are not intended to be typical of all driving conditions, they are helpful in describing an important detail regarding the interpretation of a driving record and bearing graph. A rather large increase in blow count is not necessarily indicative of a correspondingly large increase in static bearing capacity. As shown in Fig. 8-13*a*, the blow count increased from 48 to 88 blows per foot (157 to 289 blows per meter), or nearly doubled, between 64 and 100 ft (19.5 and 30.5 m) penetration. Figure 8-13*c* shows that the predicted bearing capacity increased by only 20 tons (1.81×10^4 kg), from 200 to 220 tons (1.81×10^5 to 1.99×10^5 kg), as the driving capacity of the hammer was approached.

In cases where the ratio of point to total load does not change significantly after the pile is driven, the bearing graph which is valid at the time of driving can be used to predict the bearing capacity after soil setup has occurred. Soil setup is defined as increase in soil shear strength around a pile with time. Setup is not likely to occur for piles in sand, but an appreciable amount of setup may develop in some clays. Stiff or overconsolidated and normally consolidated clays may exhibit considerable setup, while underconsolidated clays may exhibit very little, if any, setup. To predict an after setup bearing capacity, a pile should be redriven an additional 1 to 3 ft (0.3 to 0.9 m) or until a relatively constant redriving blow count is obtained and the pile is moving relative to the soil. The blow count then obtained can be used to determine the bearing capacity from the bearing graph in the usual manner. For piles which are driven in clay, the point load generally will not vary greatly, as these piles are predominantly friction piles. For a pile driven through soft clay to bearing on a dense sand or a pile driven through dense sand into a soft clay, the point load can vary considerably. For these conditions, the initial driving bearing graph may not be valid for redriving data. The wave equation should be used to determine setup where possible, especially on large jobs, as a substantial saving in cost can be realized from a reduction in pile length which can be obtained by utilizing the increased capacity due to setup. For the majority of the test piles analyzed during the Texas A & M study,[4] the predicted bearing capacity was within ± 10 percent of the measured capacity.

Field Control

A good example of how wave equation analysis can be used for field control is the present procedure used by the Texas Highway Department to prevent tension cracks from occurring in prestressed concrete piles. Generally speaking, there are five basic causes of tension cracks which can be summarized briefly as follows:

1. Stress waves of high magnitude and short duration caused by an insufficient amount of cushion material
2. High magnitude stress waves caused by high ram impact velocities or a very high ram drop
3. Tensile strength of concrete too low
4. Little or no soil resistance at the point of long piles, causing critical tensile stresses near the bottom or middle of the pile
5. Hard driving at the point of long piles, causing critical tensile stresses in the upper half of the pile due to reflected tensile stresses from the pile head

Generally speaking, the probability of critical tensile stresses existing in short concrete piles is small in comparison with long piles. Provided that adequate cushioning material is used and reasonable precautions are taken to reduce driving stresses during easy driving (by reducing ram velocity or using a smaller stroke) tension cracks will generally not be much of a problem except when little or no resistance is present at the point of a long pile.

To illustrate how wave equation analysis can be applied to the problem of field control of driving stresses, assume that a concrete pile 100 ft (30.5 m) long is to be driven through clay with a double acting diesel hammer. This problem was chosen for illustrative purposes for two reasons: (1) compared with a single acting steam hammer of comparable energy rating, the double acting diesel hammer produces a high magnitude, short duration stress wave due to the relatively high impact velocity of the comparatively lightweight ram and (2) very little point soil resistance is encountered throughout the entire driving operation when driving through soft clay. These two conditions are most likely to cause a potential tensile crack problem.

The problem was analyzed by the wave equation to determine the relationship between pile penetration, blow count, and maximum tensile stress for pile penetrations of 10, 50, and 90 percent. At a pile penetration of 10 percent, the ratio of point to total soil resistance (RUP/RUT) was assumed to be 90 percent; at a penetration of 50 percent, the ratio was taken as 50 percent; and a ratio of 10 percent was assumed at 90 percent penetration. Maximum soil resistances of 50, 100, and 200 tons were assumed for penetrations of 10, 50, and 90 percent, respectively. The data obtained are presented in Fig. 8-14. The maximum tensile stress was plotted vs. soil resistance. The maximum tensile stress allowable in the concrete was assumed to be 1500 lb/in^2. To determine the blow count at which critical tensile stresses may occur, the tensile stress vs. blow count curve was entered with the allowable stress, and the corresponding soil resistance was determined. The soil resistance value thus determined was used to enter the soil resistance vs. blow count curve for the same penetration, and the corresponding blow count was determined. Proceeding in a similar manner for the two remaining values of penetration, three points were obtained which were then used to plot the curve shown in Fig. 8-15. This curve can be used to determine the blow count, for any penetration, below which critical tensile stresses are most likely to occur. For example, if the blow count becomes equal to or less than 23 blows per foot when

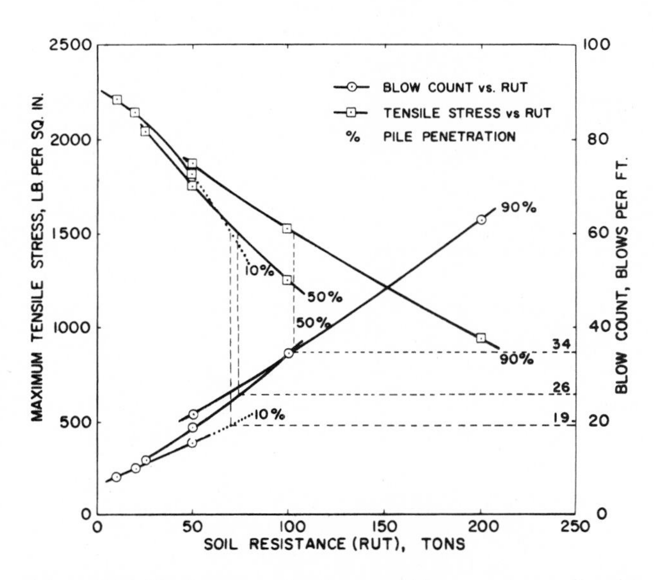

Figure 8-14 Maximum tensile stress and blow count vs. soil resistance.

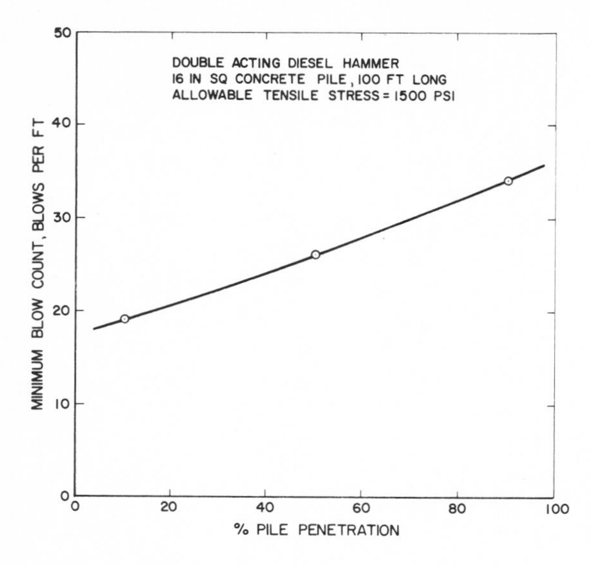

Figure 8-15 Minimum allowable blow count for prevention of critical tensile stress.

the pile is one-third of the way into the ground, the driving operation ought to be altered in some manner, e.g., reduce ram velocity, to prevent pile damage.

An actual case of the use of wave equation analysis to stop excessive driving stresses from breaking long concrete piles occurred at a construction site on the Texas Gulf coast. Approximately every tenth pile cracked during driving, requiring pulling and redriving of new piles—a costly operation. The piles were prestressed concrete 90 ft (27.5 m) long, 20 in (508 mm) square, with an 11-in (279-mm) central circular void running through the length of the pile. Soil conditions were such that jetting through a stiff clay lens at 20 ft (6.1 m) embedment was necessary. Final penetration was around 60 ft (18.3 m).

It was decided to run a parametric study by the wave equation to determine exactly which parameters could be expected to reduce the driving stresses. The variables studied included the cushion stiffness, hammer efficiency, and type of driving hammer. The results of this analysis are shown in Fig. 8-16, on the basis of which it was recommended that new 6- to 12-in cushions be used and the hammer efficiency be reduced after jetting until firm point resistance was again encountered. No further breakage was experienced. Total cost of piles lost before the analysis was $100,000 or more. Total cost of the computer analysis was around $1000.

Because of space limitations this chapter has presented only the briefest summary of the use and applications of the wave equation for pile driving analysis. During the past 14 years the method has been applied to hundreds of pile driving problems, and considerable sums of money have been saved by its application. Most of the information presented in this chapter has been summarized from two state of the art reports,[3,4] the result of the wave equation research conducted at Texas A & M University. A selected additional list of references at the end of the chapter includes published work by other researchers in this area.

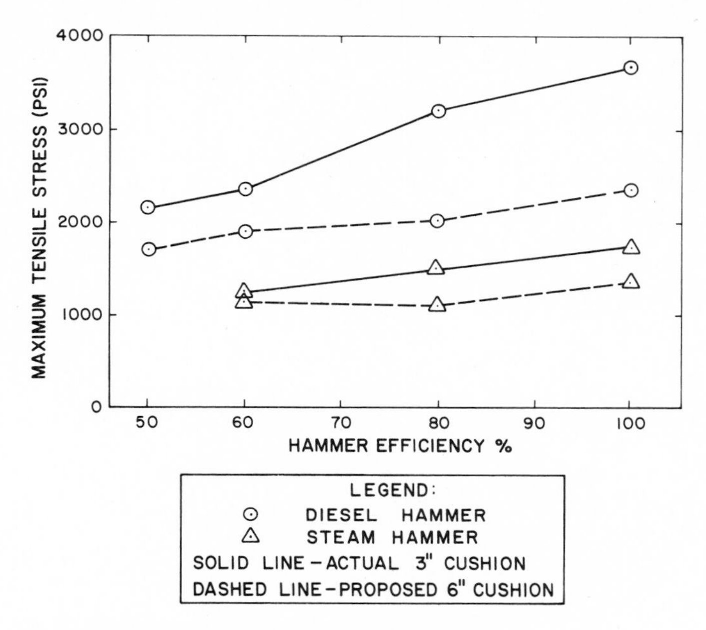

Figure 8-16 Stress after jetting vs. hammer efficiency.

8-5 COMPUTER CODES

Availability of various computer codes is widespread, ranging from no cost[6] for extremely limited programs, to several thousand dollars for full capability proprietary codes. Such computer codes are regularly used to solve a myriad of impact problems. The user can expect to invest approximately 3 man-months to write his own codes, which is probably the most efficient approach, as later modifications are easily incorporated.

APPENDIX 8A: DERIVATION OF ONE-DIMENSIONAL WAVE EQUATION

When a pile driving hammer impacts the head of a pile, it sends a compressive stress wave down the pile at a velocity equal to that of the speed of sound in the material, given by

$$C = \sqrt{\frac{E}{\rho}}$$

where C = velocity of stress wave
E = modulus of elasticity of pile material
ρ = mass density of pile material

A typical stress wave produced by a Menck 7000 single-acting steam hammer driving a 54-in-OD pipe pile is shown in Fig. 8-17.

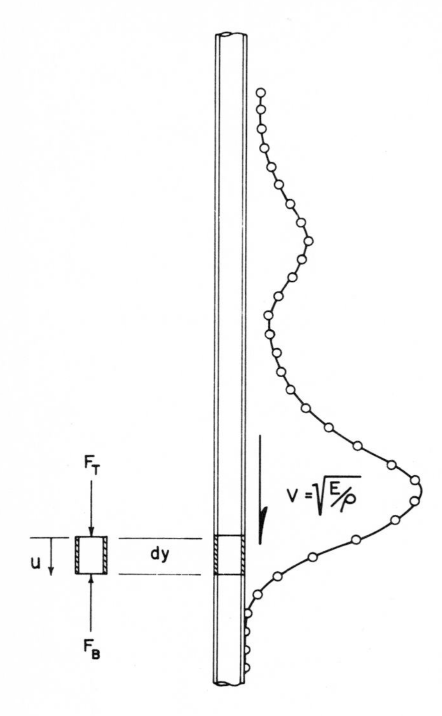

Figure 8-17 Typical shear wave produced by Menck 7000 steam hammer driving a 54-in-OD pipe pile.

Isolating a short segment of dy length from the pile reveals forces on the top and bottom of the segment (F_T and F_B, respectively) due to the corresponding stresses at these points. When the motion of any point is denoted by u, these forces can be expressed in terms of strains in the pile by

$$F_T = AE\epsilon_T = AE\frac{\partial u}{\partial y} \qquad \text{and} \qquad F_B = AE\left(\frac{\partial u}{\partial y} + \frac{\partial^2 u}{\partial y^2}\,dy\right)$$

where A is the cross-sectional area of the element and $\epsilon = \partial u/\partial y =$ the strain at a point on the element. Dynamic equilibrium requires that the unbalanced force on the element be balanced by the equation

$$\Delta F = Ma \qquad F_B - F_T = (A\rho\,dy)\frac{\partial^2 u}{\partial t^2}$$

$$AE\left(\frac{\partial u}{\partial y} + \frac{\partial^2 u}{\partial y^2}\,dy\right) - AE\frac{\partial u}{\partial y} = A\rho\,dy\frac{\partial^2 u}{\partial t^2}$$

or

$$\frac{E}{\rho}\frac{\partial^2 u}{\partial y^2} = \frac{\partial^2 u}{\partial t^2}$$

and since $c = \sqrt{E/\rho}$,

$$\frac{\partial^2 u}{\partial t^2} = c^2\frac{\partial^2 u}{\partial y^2} \tag{8-10}$$

Equation (8-10) is the controlling differential equation for the longitudinal transmission of stress in a rod.

APPENDIX 8B: CRITICAL TIME INTERVAL

The accuracy of the discrete element solution is related to the size of the time increment Δt. For free longitudinal vibrations in a continuous elastic bar, the discrete element solution is an exact solution of the partial differential equation when

$$\Delta t = \frac{\Delta L}{\sqrt{E/\rho}}$$

where ΔL is the segment length. Smith draws a similar conclusion and has expressed the critical time interval as

$$\Delta t = \frac{1}{19.648}\sqrt{\frac{W_{m+1}}{K_m}} \qquad \text{or} \qquad \Delta t = \frac{1}{19.648}\sqrt{\frac{W_m}{K_m}}$$

If a time increment larger than that given is used, the discrete element solution will diverge and no valid results can be obtained. As pointed out by Smith, in this case the numerical calculation of the discrete element stress wave does not progress as rapidly as the actual stress wave. Consequently the value of Δt given is called the *critical value*.

Also, when

$$\Delta t < \frac{\Delta L}{\sqrt{E/\rho}}$$

is used in a discrete element solution, a less accurate solution is obtained for the continuous bar. As Δt becomes progressively smaller, the solution approaches the actual behavior of the discrete element system (segment lengths equal to ΔL) used to simulate the pile.

This in general leads to a less accurate solution for the longitudinal vibrations of a slender continuous bar. If, however, the discrete element system were divided into a large number of segments, the behavior of this simulated pile would be essentially the same as that of the slender continuous bar irrespective of how small Δt becomes provided that

$$\frac{\Delta L}{\sqrt{E/\rho}} \geqslant \Delta t > 0$$

This means that if the pile is divided into only a few segments, the accuracy of the solution will be more sensitive to the choice of Δt than if it is divided into many segments. For practical problems, a choice of Δt equal to about one-half the critical value appears suitable since inelastic springs and materials of different densities and elastic moduli are usually involved.

REFERENCES

1. Smith, E. A. L.: Pile Driving Impact, *Proc. Ind. Comput. Semin.*, International Business Machines Corp., New York, September 1950, p. 44.
2. Smith, E. A. L.: Pile Driving Analysis by the Wave Equation, *J. Soil Mech. Found. Div. ASCE*, vol. 86, no. SM4, pap. 2574, pp. 35–61, August 1960.
3. Lowery, L. L., T. J. Hirsch, T. C. Edwards, H. M. Coyle, and C. H. Samson: Pile Driving Analysis: State of the Art, *Texas A & M Univ. Tex. Transp. Inst. Res. Rep.* 33-13(F), January 1969.
4. Coyle, H. M., R. E. Bartoskewitz, and W. J. Berger: Bearing Capacity Prediction by Wave Equation Analysis: State of the Art, *Texas A & M Univ. Tex. Transp. Inst. Res. Rep.* 125-8(F), August 1973.
5. Coyle, H. M., and L. C. Reese: Load Transfer for Axially Loaded Piles in Clay, *J. Soil Mech. Found. Div. ASCE*, vol. 92, no. SM2, pap. 4702, pp. 1–26, March 1966.
6. Bowles, Joseph E.: "Analytic and Computer Methods in Foundation Engineering," McGraw-Hill Book Company, New York, 1974.

SELECTED ADDITIONAL REFERENCES

Bender, C. H., Jr., C. G. Lyons, and L. L. Lowery, Jr.: Applications of Wave Equation Analysis to Offshore Pile Foundations, *Proc. 1st Annu. Offshore Technol. Conf., May 1969*, pap. 1055, pp. 575-586.

Davisson, M. T., and V. J. McDonald: Energy Measurements for a Diesel Hammer, pp. 295–337 in *Performance of Deep Foundations, ASTM Spec. Tech. Publ.* 444, March 1969.

Davisson, M. T.: Design Pile Capacity, *Proc. Conf. Des. Install. Pile Found. Cell. Struct., Lehigh Univ., April 1970.*

Forehand, P. W., and J. L. Reese: Predictions of Pile Capacity by the Wave Equation, *J. Soil Mech. Found. Div. ASCE*, vol. 90, no. SM2, pap. 3820, pp. 1–25, March 1964.

Goble, G. G., and F. Rausche: Pile Load Test by Impact Driving, Pile Foundations, *Highw. Res. Rec.* 333, Highway Research Board, Washington, D.C., 1970.

Goble, G. G., R. H. Scanlan, and J. J. Tomko: Dynamic Studies on the Bearing Capacity of Piles, Bridges and Structures, *Highw. Res. Rec.* 167, Highway Research Board, Washington, D.C., 1967.

McClelland, B., J. A. Focht, Jr., and W. J. Emrich: Problems in Design and Installation of Offshore Piles, *J. Soil Mech. Found. Div. ASCE*, vol. 95, no. SM6, pap. 6913, pp. 1491–1514, November 1969.

Mosley, E. T.: Test Piles in Sand at Helena Arkansas, Wave Equation Analysis, *Found. Facts*, vol. 3, no. 2, Raymond International, New York, 1967.

Rausche, F., F. Moses, and G. G. Goble: Soil Resistance Predictions from Pile Dynamics, *J. Soil Mech. Found. Div. ASCE*, vol. 98, no. SM9, pap. 9220, pp. 917–937, September 1972.

Sulaiman, I. H., H. M. Coyle, and T. J. Hirsch: Static versus Dynamic Resistance of Piles in Clay, *Proc. 4th Annu. Offshore Technol. Conf., May 1972*, pap. 1601, pp. 823–838.

CHAPTER

NINE

LATERALLY LOADED PILES

Lymon C. Reese and Chandrakant S. Desai

9-1 INTRODUCTION

Definition of Problem

The problem of a pile subjected to lateral loading is one of a class of problems concerned with the interaction of soils and structures. The solution of such problems usually involves the use of iterative techniques because the soil response is a nonlinear function of the deflection of the structure.

Figure 9-1 presents examples of the kinds of loading that are frequently encountered in practice. Figure 9-1*a* shows an anchor pile subjected to a horizontal load from a cable attached at the ground surface. Figure 9-1*b* shows a pile that supports an overhead structure such as an advertising sign. Figure 9-1*c* shows a pile that is fixed against rotation at the groundline such as a pile supporting a reinforced concrete retaining wall. Figure 9-1*d* shows a pile that extends upward and becomes a part of a flexible structure such as a pile supporting an offshore drilling platform. Other loadings can also be analyzed.

In all the cases illustrated the pile could also be subjected to an axial load; however, the presently available methods of analysis do not allow for a torque on the pile. Furthermore, the applied shear and moments and the resulting deflection

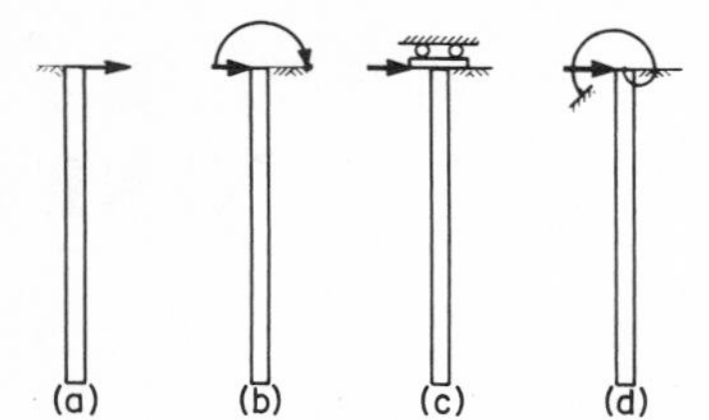

Figure 9-1 Examples of loadings.

must be in the same plane. It is assumed that the pile was straight before and after installation.

The methods of analysis presented here are applicable to short-term static loading and to repeated loading. Adjustments in the method would allow the analysis of the case of sustained loading and to some cases of earthquake loading; however, inertia effects are not taken into account. The method of analysis is applicable to piles with mild batter, and modifications would allow the consideration of piles with large batter.

Conditions to be Satisfied

A valid solution of the problem of the laterally loaded pile requires that, as for other boundary value problems, the conditions of equilibrium and compatibility be satisfied. These conditions are satisfied in the method that is presented; the limitations are concerned with the ability to predict soil properties and soil response. As more knowledge is gained about the prediction of soil behavior, it can readily be incorporated into the analytical procedures.

In the past, a number of proposals have been made for analyzing piles under lateral loading that violate basic principles. One such proposal is the point of fixity method. The soil is assumed to be removed and the pile is fixed against rotation at some distance below the groundline. The distance from the ground surface to the point of fixity is a measure of the strength of the soil; the greater the strength the less the distance to the point of fixity. The point of fixity method not only violates the principles of equilibrium and compatibility, but it does so in such a gross way that even volumes of field experience would be virtually uninstructive in suggesting improvements.

Nature of Solution

Of most practical interest to the engineer is a knowledge of the deflection of the pile and the bending moment in the pile, both as functions of depth. The bending moment is required in sizing the pile, and the deflection would be important with regard to the serviceability of the supported structure. Furthermore, the magnitude of the deflection would be instructive in regard to the ability of the soil to

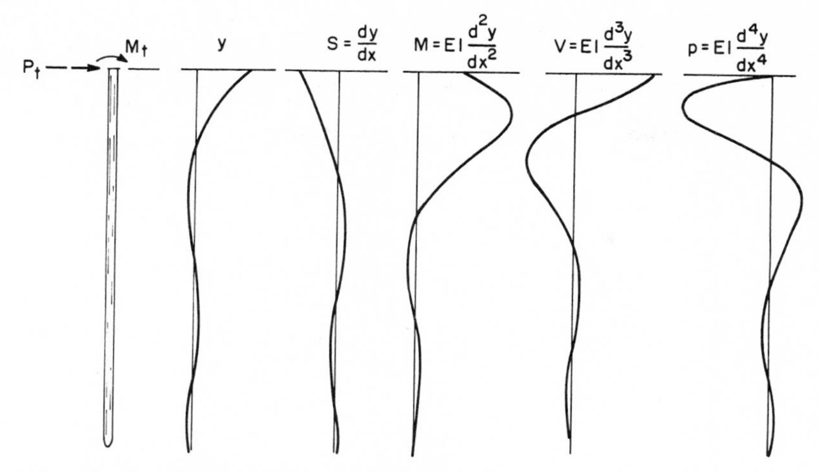

Figure 9-2 Form of results from a complete solution.

respond to somewhat higher loads. The family of curves shown in Fig. 9-2 indicates the form of a complete solution. The curves show deflection y, slope S, moment M, shear V, and soil reaction p, all as a function of depth x.

While a single set of curves is shown in Fig. 9-2, in practice solutions should be obtained for a number of loadings, ranging below and above the working loads. Because the problem is nonlinear, the solution for a single set of loads would be insufficient to allow the appropriate judgment concerning the stability and serviceability of the pile.

Differential Equation

The following differential equation for the problem of the laterally loaded pile is well known, and its solution has been discussed by a number of authors:[1–3]

$$EI\frac{d^4y}{dx^4} + P_x\frac{d^2y}{dx^2} + E_s y = 0 \tag{9-1}$$

where P_x = axial load
y = deflection
x = length along pile
EI = flexural stiffness of pile
E_s = soil modulus

A number of methods proposed for the solution of the differential equation will be discussed later.

The solution of the differential equation, employing appropriate boundary conditions and with the appropriate representation of the soil response through the selection of appropriate values of soil modulus, ensures the satisfaction of the conditions of equilibrium and compatibility and results in a set of curves of the form as shown in Fig. 9-2. Such a solution of the differential equations is a highly desirable procedure for the engineer in that judgments are possible concerning the effects of important parameters.

While the discussions have principally been concerned with the behavior of a single pile, the method can be applied to the study of the behavior of the pile group. For a closely spaced pile group the soil would be modified to account for the interaction between the closely spaced piles. For a widely spaced pile group, procedures are employed to ensure that the structure supported by the piles is in equilibrium and that compatibility is achieved.[4,5]

9-2 GEOTECHNICAL ASPECTS

Soil Response

The soil response is given by a family of curves that shows the soil resistance p as a function of deflection y and depth below the ground surface x. The idea of p-y curves is presented in Fig. 9-3. Figure 9-3a shows a section through a pile at depth below the ground surface. The behavior of the thin stratum of soil at a depth x_1

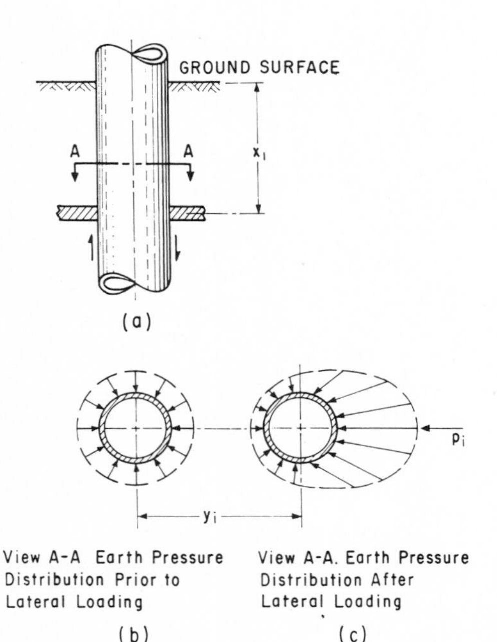

Figure 9-3 Distribution of p and y.[6]

below the ground surface will be discussed. Figure 9-3*b* shows a possible earth-pressure distribution around the pile after it has been installed and before it has been loaded laterally. The earth-pressure distribution in Fig. 9-3*c* assumes that the pile was perfectly straight before driving and that there was no bending of the pile during driving. While neither of the conditions is precisely met in practice, it is believed that in most instances the assumptions can be made without serious error. The deflection of the pile through a distance y_i, as shown in Fig. 9-3*c*, generates unbalanced soil pressures against the pile, perhaps as indicated in the figure. Integration of the soil pressures around the pile yields an unbalanced force p_i per unit of length of the pile.

The deflection of the pile could generate a soil resistance parallel to the axis of the pile; however, it is assumed that such soil resistance is quite small and can be ignored in the analysis.

As shown in Fig. 9-3, the deflection of y_i is the distance the pile deflects laterally on being subjected to a lateral load. The soil resistance p_i is the force per unit length from the soil against the pile which develops as a result of the pile deflection.

For the solution of the problem of laterally loaded pile, it is desirable to be able to predict a set of *p-y* curves like those shown in Fig. 9-4. If such a set of curves can be predicted, Eq. (9-1) can readily be solved to yield pile deflection, pile rotation, bending moment, shear, and soil reaction for any load capable of being sustained by the pile.

The set of curves shown in Fig. 9-4 seems to imply that the behavior of the soil at a particular depth is independent of the soil behavior at all other depths. That assumption, of course, is not strictly true. However, it has been found by experiments that for the patterns of pile deflections which can occur in practice the soil reaction at a point is dependent essentially on the pile deflection at that point and not on pile deflections above and below. Thus, for purposes of analysis the soil can be removed and replaced by a set of discrete mechanisms with load-deflection characteristics of the character shown in Fig. 9-4.

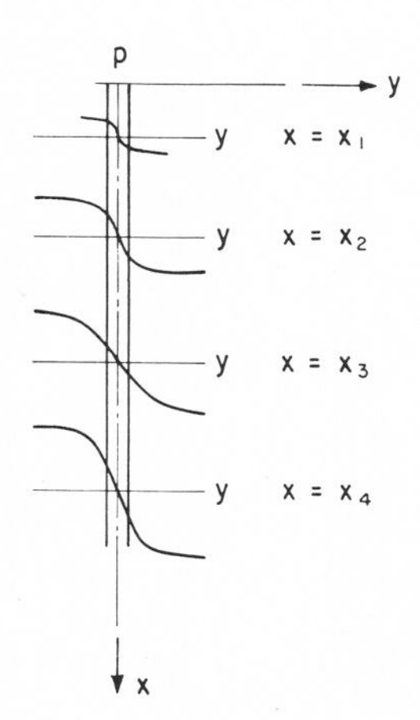

Figure 9-4 Set of *p-y* curves.

Determination of Soil Behavior

A number of field experiments have been performed and analyzed to obtain experimental *p-y* curves.[6–9] The experiments entail the application of known lateral loads to piles that were instrumented for the measurement of bending moment along the length of the piles. Measurements were also made of pile head deflection and rotation. In most of the experiments two types of loading were employed, static and cyclic.

From the sets of experimental bending-moment curves, values of p and y at points along the pile can be obtained by solving the equations

$$y = \iint \frac{M(x)}{EI} \tag{9-2}$$

and

$$p = \frac{d^2M(x)}{dx^2} \tag{9-3}$$

Appropriate boundary conditions must be used, and the equations must be solved numerically.

The solution of Eq. (9-2) for values of y can normally be accomplished with appropriate accuracy. However, analytical difficulty is encountered in the solution of Eq. (9-3). If extremely accurate moment values are available, the double differentiation can be performed numerically.[10,11] Another procedure for obtaining the soil resistance curves involves the prior assumption that the soil modulus can be described as a function of depth by a two-parameter nonlinear curve. The two parameters can be computed from experimental data, allowing the soil reaction curve to be computed analytically.[6]

Boundary Conditions

In the analyses, the boundary conditions employed at the top of the pile depend on the loading of the pile in the field. At the bottom of the pile, there will be deflection and rotation, but the moment can be assumed to be zero. The shear can be taken equal to zero even though the pile is driven to rock. The possible horizontal force on the base of the pile can be treated without error as a horizontal force just above the base of the pile.

Recommendations for the prediction of *p-y* curves are developed by using the theories of elasticity and plasticity to derive analytical soil behavior to be compared with experimentally determined behavior. The details of the development of such prediction methods are not presented here, but brief summaries of some of the methods are given in the next section.

Summary of Recommendations for Predicting *p-y* Curves

In this section, details of procedures will be presented for predicting *p-y* curves for soft clay, stiff clay, and sand. These methods are believed to be the best now available, representing the current state of the art.

Soft clay ***Short-term static loads*** Matlock's[7] criteria are for soft clays. For short-term static loading, the step-by-step procedure for computing values for p-y curves is given below.

1. Obtain the best possible estimate of the variation of shear strength and effective unit weight with depth. Also obtain the value of ϵ_{50}, the strain corresponding to one-half the maximum principal stress difference. If no values of ϵ_{50} are available, typical values suggested by Skempton[12] are given in Table 9-1.
2. Compute the ultimate soil resistance per unit length of shaft p_u, using the smaller of the values given by

$$p_u = \left(3 + \frac{\gamma}{c}x + \frac{0.5}{b}x\right)cb \tag{9-4}$$

and

$$p_u = 9cb \tag{9-5}$$

where γ = average effective unit weight from ground surface to p-y curve
x = depth from ground surface to p-y curve
c = shear strength at depth x
b = width of pile

3. Compute the deflection y_{50} at one-half the ultimate soil resistance from

$$y_{50} = 2.5\epsilon_{50}b \tag{9-6}$$

4. Points describing the p-y curve are now computed from

$$\frac{p}{p_u} = 0.5\left(\frac{y}{y_{50}}\right)^{1/3} \tag{9-7}$$

Cyclic loads The effect of cyclic loading as presented by Matlock[7] is as follows:

1. Construct the p-y curve in the same manner as for short-term static loading for values of p less than $0.72p_u$.
2. Solve Eqs. (9-4) and (9-5) simultaneously to find the depth x_r where the transition occurs. If the unit weight and shear strength are constant in the upper zone, then

$$x_r = \frac{6cb}{\gamma b + 0.5c} \tag{9-8}$$

Table 9-1

Consistency of clay	ϵ_{50}	E/c
Soft	0.020	50
Medium	0.010	100
Stiff	0.005	200

3. If the depth to the p-y curve is greater than or equal to x_r, p is equal to $0.72p_u$ for all values of y greater than $3y_{50}$.
4. If the depth to the p-y curve is less than x_r, the value of p decreases from $0.72p_u$ at $y = 3y_{50}$ to the value given by the expression below at $y = 15y_{50}$:

$$p = 0.72p_u \frac{x}{x_r} \tag{9-9}$$

The value of p remains constant beyond $y = 15y_{50}$.

Stiff clay ***Short-term static loads*** The correlations developed by Reese and Welch[13] provide the basis for the method for predicting behavior in stiff clay.

1. Obtain the best possible estimate of the variation of shear strength and effective unit weight with depth. Also obtain the value of ϵ_{50}, the strain corresponding to one-half the maximum principal stress difference. If no value of ϵ_{50} is available, use a value of 0.005 or 0.010, the larger value being more conservative.
2. Compute the ultimate soil resistance per unit length of shaft p_u, using the smaller of the values given by Eqs. (9-4) and (9-5). [In the use of Eq. (9-4), the shear strength is taken as the average from the ground surface to the depth being considered.]
3. Compute the deflection y_{50} at one-half the ultimate soil resistance from Eq. (9-6).
4. Points describing the p-y curve can be computed from

$$\frac{p}{p_u} = 0.5\left(\frac{y}{y_{50}}\right)^{1/4} \tag{9-10}$$

5. Beyond $y = 16y_{50}$, p is equal to p_u for all values of y.

Cyclic loads The effect of repeated loading is predicted by the following procedure:

1. Determine the p-y curve for short-term static loading by the procedure previously given.
2. Determine the number of times the design lateral load will be applied to the shaft.
3. For several values of p/p_u obtain the value of C, the parameter describing the effect of repeated loading on deformation, from a relationship developed by laboratory test or, in the absence of tests, from

$$C = 9.6\left(\frac{p}{p_u}\right)^4 \tag{9-11}$$

4. At the values of p corresponding to the values of p/p_u selected in step 3, compute new values of y for cyclic loading from

$$y_c = y_s + y_{50} C \log N \tag{9-12}$$

where y_c = deflection under N cycles of load
y_s = deflection under short-term static load
y_{50} = deflection under short-term static load at one-half ultimate resistance
N = number of cycles of load application

5. The p-y_c curve defines the soil response after N cycles of load.

Sand ***Short-term static and cyclic loads*** The criteria proposed by Reese, Cox, and Koop[14] give detailed procedures for computing p-y curves for sand. The step-by-step procedure is as follows:

1. Obtain values for significant soil properties and pile dimensions, ϕ, γ, and b.
2. Use the following parameters for computing soil resistance.

$$\alpha = \frac{\phi}{2} \qquad \beta = 45 + \frac{\phi}{2} \qquad K_0 = 0.4 \qquad \text{and} \qquad K_a = \tan^2\left(45 - \frac{\phi}{2}\right)$$

3. Use the following equations for computing soil resistance:

   Ultimate resistance near ground surface:

$$p_{ct} = \gamma H\left[\frac{K_0 H \tan\phi \sin\beta}{\tan(\beta - \phi)\cos\alpha} + \frac{\tan\beta}{\tan(\beta - \phi)}(b + H\tan\beta\tan\alpha) + K_0 H \tan\beta(\tan\phi\sin\beta - \tan\alpha) - K_a b\right] \tag{9-13}$$

   Ultimate resistance well below the ground surface:

$$p_{cd} = K_a b\gamma H(\tan^8\beta - 1) + K_0 b\gamma H \tan\phi\tan^4\beta \tag{9-14}$$

4. Find the intersection X_t of the equations for the ultimate soil resistance near the ground surface and the ultimate soil resistance well below the ground surface. Above this depth use Eq. (9-13). Below this depth use Eq. (9-14).
5. Select a depth x at which a p-y curve is desired (Fig. 9-5).
6. Establish y_u as $3b/80$. Compute p_u by

$$p_u = Ap_c \tag{9-15}$$

   Use the appropriate value of A from Fig. 9-6, for the particular nondimensional depth and for either the static or cyclic case. Use the appropriate equation for p_c [Eq. (9-13) or (9-14)] by referring to the computation in step 4.
7. Establish y_m as $b/60$. Compute p_m by

$$p_m = Bp_c \tag{9-16}$$

   Use the appropriate value of B from Fig. 9-7 for the particular nondimensional depth and for either the static or cyclic case. Use the appropriate equation for p_c.
8. Establish the slope of the initial portion of the p-y curve by selecting the appropriate value of k from the Table 9-2.

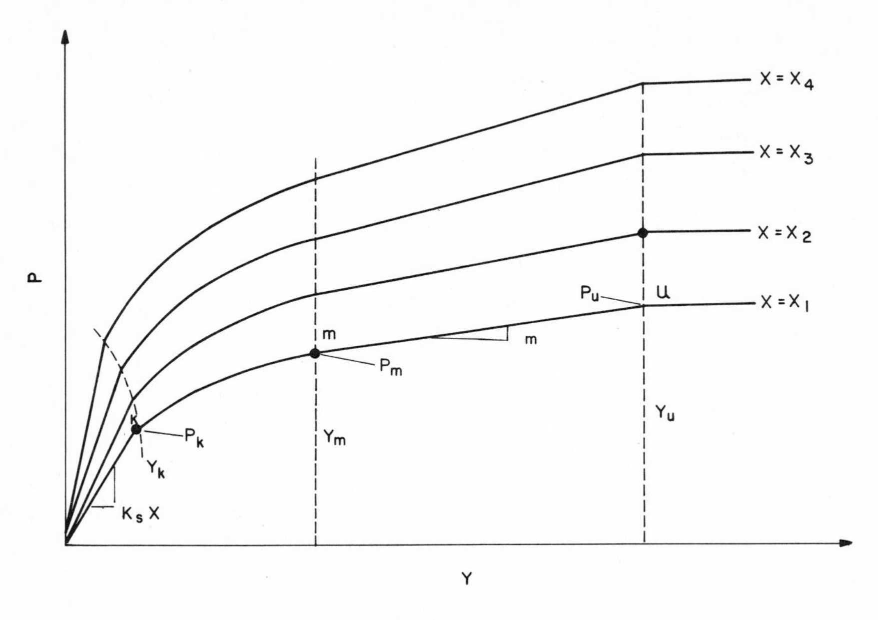

Figure 9-5 *p-y* curves for sands.

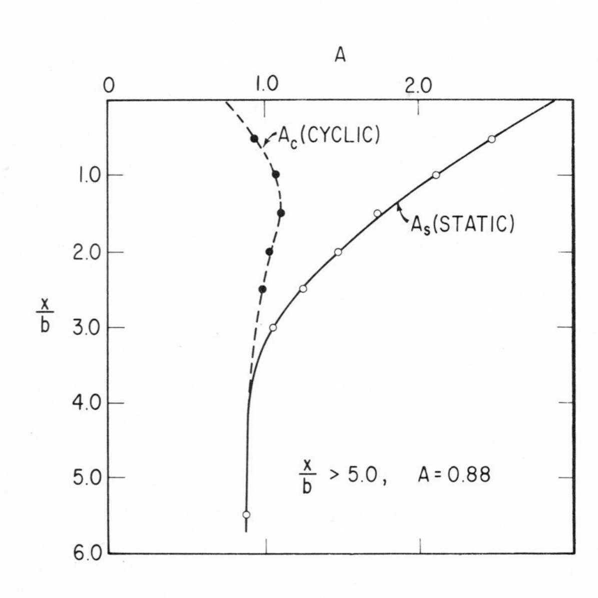

Figure 9-6 Nondimensional coefficient *A* for ultimate soil resistance vs. depth.

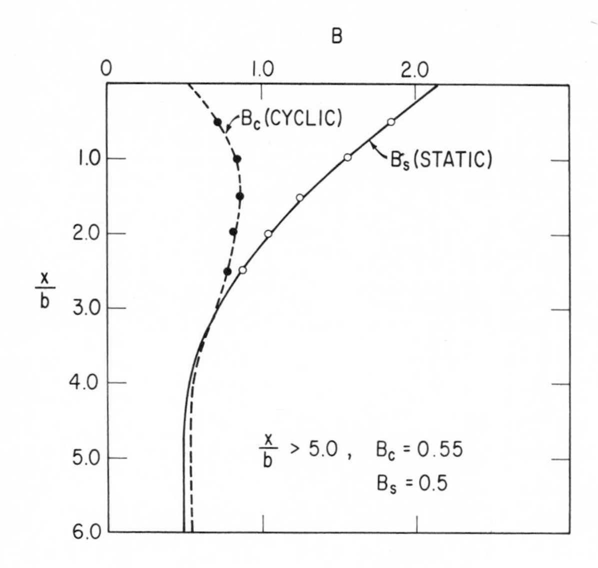

Figure 9-7 Nondimensional coefficient *B* for soil resistance vs. depth.

9. Select the following parabola to be fitted between points k and m:

$$p = Cy^{1/n} \tag{9-17}$$

10. Fit the parabola between points k and m as follows (Fig. 9-5): Compute the slope of the line between points m and u (Fig. 9-5) by

$$m = \frac{p_u - p_m}{y_u - y_m} \tag{9-18}$$

Obtain the power of the parabolic section by

$$n = \frac{p_m}{my_m} \tag{9-19}$$

Obtain the coefficient C as follows:

$$C = \frac{p_m}{y_m^{1/n}} \tag{9-20}$$

Determine point k as

$$y_k = \left(\frac{C}{kx}\right)^{n/(n-1)} \tag{9-21}$$

Table 9-2 Recommended values of k for submerged sand

Relative density	Loose	Medium	Dense
Recommended k, lb/in^3	20	60	125

Compute the appropriate number of points on the parabola by using Eq. (9-17).

This completes the development of the p-y curve for the desired depth. Any number of curves can be developed by repeating the steps above for each depth desired.

It is important to note that Matlock's[7] tests with soft clay were performed with the water table along the ground surface and the tests described by Reese and Welch[13] were performed with the water table below the ground surface. Therefore, the recommendations for Reese and Welch are not applicable to the design of offshore foundations subjected to cyclic loading.

9-3 NUMERICAL METHODS

Behavior of laterally loaded piles can be considered analogous to that of long beams on elastic foundations.[1,15] A form of the equation governing this problem is given in Eq. (9-1). Often, the value of axial load P_x can be much smaller than the axial load that can cause buckling, and hence such a load can be ignored. In absence of P_x, Eq. (9-1) reduces to

$$EI \frac{d^4 y}{dx^4} + E_s y = 0 \tag{9-1a}$$

This equation has the same form as Eq. (1-30); here we have used different symbols and coordinate labels in order to be consistent with those used by various investigators of the subject of laterally loaded piles.

Closed-Form Solutions and Computational Schemes

By making simplifying assumptions concerning properties of pile and soils, it is possible to obtain closed-form solutions to Eq. (9-1a); such assumptions are that E_s varies linearly with depth and that the pile geometry can be defined by simple functions. The closed-form solutions can be programmed for hand calculations or for digital computer.

If the stiffness of a pile is much higher than that of the surrounding soils, the pile may be considered rigid. Various procedures have been proposed (Davisson and Prakash,[16] Broms,[17,18] and others) for analysis of such piles and for defining mechanisms of failure. In this chapter, we shall consider essentially deformable piles. Separation between deformable and rigid piles is often based on relative stiffness factors expressed in terms of stiffness EI of pile and stiffness E_s of soil. For the soil modulus constant with depth, the stiffness factor R is defined as[19,20]

$$R = \sqrt[4]{\frac{EI}{E_s}} \tag{9-22}$$

where E_s is in units of force/length^2 and R is in the units of length. If the soil modulus varies linearly with depth, the stiffness factor $T(L)$ is defined as

$$T = \sqrt[5]{\frac{EI}{k}} \tag{9-23}$$

where k is the coefficient of horizontal subgrade reaction and is given by $E_s = kx$.

The closed-form solutions can be expressed either as a combination of exponential and trigonometric functions or in a series form; typical closed-form and series solution are described in Appendix 9A. A final equation for displacement y resulting from the first procedure is

$$y = \frac{P_t T^3}{EI} A_y + \frac{M_t T^2}{EI} B_y \tag{9-24a}$$

in which M_t and P_t are applied moment and lateral load, respectively, and A_y and B_y are dimensionless constants.[19,21] Other design quantities such as slope S, moment M, shear V, and soil reaction p can be obtained by successive differentiations as

$$S = \frac{P_t T^2}{EI} A_s + \frac{M_t T}{EI} B_s \tag{9-24b}$$

$$M = (P_t T)A_m + (M_t)B_m \tag{9-24c}$$

$$V = P_t A_v + \frac{M_t}{T} B_v \tag{9-24d}$$

$$p = \frac{P_t}{T} A_p + \frac{M_t}{T^2} B_p \tag{9-24e}$$

Figure 9-8 shows typical variations of A's and B's; these are dimensionless coefficients and are dependent on the parameters $Z = x/T$, $Z_{max} = L/T$, and T.

If the pile is long, with $L \geqslant 4T$, the solution procedure can be simplified as[19,21,22]

$$y = C_y \frac{P_t T^3}{EI} \qquad \text{where} \quad C_y = A_y + \frac{M_t}{P_t T} By \tag{9-24f}$$

The factor C_y has been used for solution of long piles. Here the value of M_t is obtained by equating the expression for slope [Eq. (9-24b)] to the given fixity at the pile head; the results are expressed in terms of $M_t/P_t T$ and Z.

For nonlinear analysis, the foregoing procedure follows iterative cycles in which successive values for T are assumed for various values of E_s until convergence is achieved. The procedure can be briefly described in the following steps:

1. Assume a value of $E_s = kx$. Evaluate T and $Z_{max} = L/T$.
2. Obtain dimensionless coefficients from the plots in Fig. 9-8.

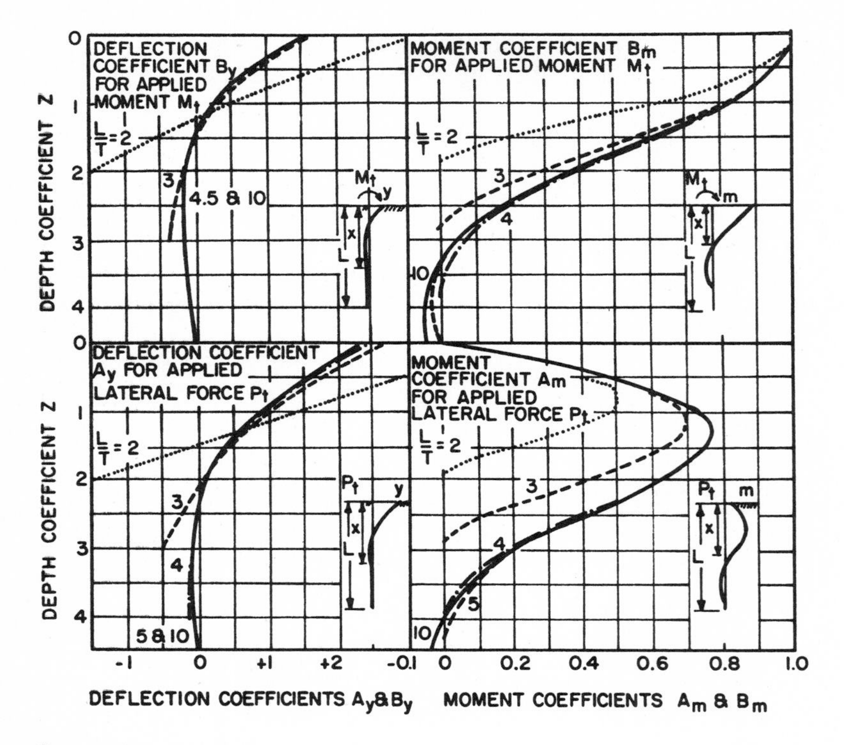

Figure 9-8 Variations of A_s and B_s.[21]

3. Evaluate deflection y [Eq. (9-24*a*)].
4. Revise the value of E_s if necessary and continue until

$$T_i \approx T_{i-1}$$

where i denotes the stage of iteration.

Finite Difference Method

The closed-form solution can become cumbersome and difficult if such factors as arbitrary variation of E_s and irregular geometry must be considered. The finite difference method has been one of the main numerical schemes used in the past.[19–22] We shall cover details of the finite difference method.

A finite difference analog to various derivatives in the general equation (9-1) for a generic point in Fig. 9-9 is

$$\left[\frac{d^2}{dx^2}\left(R\frac{d^2y}{dx^2}\right)\right] = \left(\frac{d^2M}{dx^2}\right)_m = \frac{(dM/dx)_{m-1/2} - (dM/dx)_{m+1/2}}{h} \tag{9-25}$$

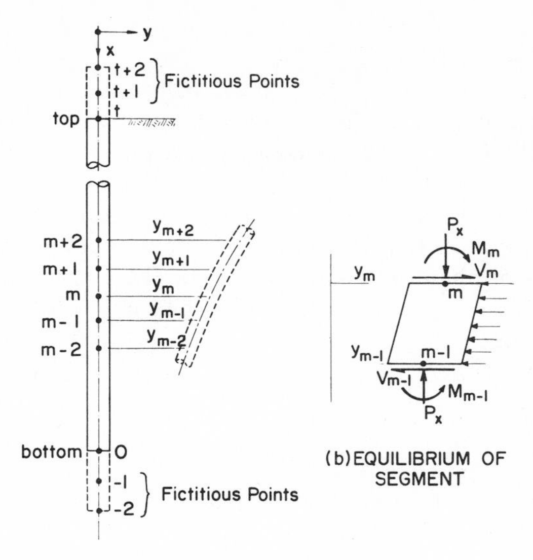

Figure 9-9 Finite difference approximation.

where h is the length of a segment and $R = EI$. Equation (9-25) is based on equilibrium of forces on a pile segment (Fig. 9-9b). Substitution of the finite difference forms for dM/dx and d^2y/dx^2 in Eq. (9-25) leads to

$$\left(\frac{d^2M}{dx^2}\right)_m \approx \frac{1}{h^4}[R_{m-1}y_{m-2} + (-2R_{m-1} - 2R_m)y_{m-1} + (R_{m-1} + 4R_m + R_{m+1})y_m + (-2R_m - 2R_{m+1})y_{m+1} + R_{m+1}y_{m+2}] \quad (9\text{-}26a)$$

Similarly

$$P_x\frac{d^2y}{dx^2} \simeq \frac{P_x(y_{m-1} - 2y_m + y_{m+1})}{h^2} \quad (9\text{-}26b)$$

Here y_m is the deflection at point m, and P_x is assumed to be constant over the length of the pile; P_x is positive if it is compressive. The difference equation for the entire differential equation (9-1) is

$$R_{m-1}y_{m-2} + (-2R_{m-1} - 2R_m + P_xh^2)y_{m-1} + (R_{m-1} + 4R_m + R_{m+1} - 2P_xh^2 + E_sh^4)y_m + (-2R_m - 2R_{m+1} + P_xh^2)y_{m+1} + R_{m+1}y_{m+2} = 0 \quad (9\text{-}27)$$

This equation can be written recursively for each point 0, 1, 2, ..., t (Fig. 9-9), resulting in a set of simultaneous equations in y.

The foregoing derivation is based on the assumption that the pile is initially straight and carries no bending moment due to driving.

Boundary Conditions

The boundary conditions must be introduced in Eqs. (9-27) before they can be solved by using any one of the procedures in Chap. 1 for solving simultaneous equations. Various physical conditions at the top and base of a pile can be introduced to modify the equations (Chap. 1).

In the finite difference method, boundary conditions in terms of gradients of y such as d^3y/dx^3 (shear), d^2y/dx^2 (moment), and dy/dx (slope) are expressed in difference forms by using fictitious or phantom points (Fig. 9-9). These forms are then introduced into the general set of equations (9-27). We shall now consider some examples of boundary conditions.

Boundary Condition at the Base[24]

If the pile is long, moment and shear at the base (Fig. 9-9*a*) are small and can be assumed to be zero. Thus for zero moment, we obtain

$$y_{-1} - 2y_0 + y_1 \approx 0 \tag{9-28a}$$

and for zero shear

$$R_0 \frac{d^3y}{dx^3} + P_x \frac{dy}{dx} \approx 0 \tag{9-28b}$$

which leads to

$$y_{-2} - 2y_{-1} + 2y_1 - y_2 + \frac{P_x h^2}{R_0}(y_{-1} - y_1) = 0 \tag{9-28c}$$

Here, as an approximation, we assumed that $R_{-1} = R_0 = R_1$. Now by expressing Eq. (9-27) at point 0 and by using Eqs. (9-28*a*) and (9-28*b*) we obtain a recursive relation[24]

$$y_m = a_m y_{m+1} - b_m y_{m+2} \tag{9-29}$$

where $\qquad a_m = \dfrac{d_m}{c_m} \qquad b_m = \dfrac{R_{m+1}}{c_m}$

$$d_m = -2b_{m-1}R_{m-1} + a_{m-2}b_{m-1}R_{m-1} + 2R_m - 2b_{m-1}R_m + 2R_{m+1} + P_x h^2(1 - b_{m-1}) \tag{9-30a}$$

and

$$c_m = R_{m-1} - 2a_{m-1}R_{m-1} - b_{m-2}R_{m-1} + a_{m-2}a_{m-1}R_{m-1} + 4R_m - 2a_{m-1}R_m + R_{m+1} + E_s h^4 + P_x h^2(2 - a_{m-1}) \tag{9-30b}$$

The coefficients a and b can be readily computed by using Eqs. (9-30), with specialized form for points 0 and 1.

Boundary Conditions at the Top[24]

Three different conditions are possible at the top, depending upon the nature of superstructure:

1. Applied lateral load (shear) P_t and moment M_t
2. Applied lateral load P_t and slope S_t
3. Applied lateral load P_t and rotational restraint M_t/S_t

In the following, we state how the expressions for deflections are obtained at the top node t and the two fictitious nodes $t = 1$ and $t = 2$.

Case 1: P_t and M_t The difference equations corresponding to the differential equations for moment and shear are

$$y_{t-1} - 2y_t + y_{t+1} = \frac{M_t h^2}{R_t} = J_2 \tag{9-31a}$$

and

$$(y_{t-2} - 2y_{t-1} + 2y_{t+1} - y_{t+2}) + \frac{P_x h^2}{R_t}(y_{t-1} - y_{t+1}) = \frac{2P_t h^3}{R_t} = J_3 \tag{9-31b}$$

Case 2: P_t and S_t The finite difference expression for S_t is

$$y_{t-1} - y_{t+1} \approx 2hS_t \tag{9-32a}$$

Case 3: P_t and M_t/S_t The finite difference form for M_t/S_t is

$$\frac{y_{t-1} - 2y_t + y_{t+1}}{y_{t-1} - y_{t+1}} = \frac{M_t h}{2R_t S_t} \tag{9-33}$$

Equations (9-31) to (9-33) permit computation of values of y_t, y_{t+1}, and y_{t+2} that can be substituted in the set of equations (9-27).

Solution of Equations for Deflections

The set of equations (9-27) is modified by introducing the special forms for deflections at top and bottom derived from given boundary conditions. The resulting equations are solved simultaneously for y by using one of the available solution procedures; gaussian elimination has been the common procedure.

Computation of derived quantities The deflections y_m are the primary quantities. Other design quantities such as slope, moment, shear, and soil reaction can be derived from known values of y_m. For example, moment at point m is

$$M_m \approx \frac{R_m(y_{m-1} - 2y_m + y_{m+1})}{h^2} \tag{9-34}$$

Finite Element Method

The foregoing approach that uses beam-on-elastic-foundation theory and the finite difference method is based essentially on the assumption that the pile can be replaced by a (line) segment and that the stiffness can be lumped in the segment. The soil resistance is represented by individual springs, as in the case of Winkler hypothesis. Since the springs are independent, influence of (shear) stress in adjacent elements is normally not accounted for. The soil and pile are assumed to move together; hence large relative displacements between the pile and soil cannot be simulated.

The finite element method, on the other hand, can permit inclusion of many of the above mentioned factors and can provide a general scheme. Although the three-dimensional finite element formulations and codes are possible and available (see Chap. 7), their use (at the present time) for practical applications may involve considerable human and computer effort. However, for certain problems it is essential to use three-dimensional idealization, and in many cases only a three-dimensional formulation can allow study of such design factors as group effects and efficiency, interaction behavior, and distribution of load in pile groups.

A description of a three-dimensional solid element for soil and pile and an interface element and of a code for three-dimensional nonlinear analysis is given in Refs. 25 and 26 and in Chap. 7. That formulation and code can be used for laterally loaded piles. Preliminary applications of three-dimensional finite element analysis based on linear elastic approach have been obtained;[27,28] an application based on a two-dimensional approximation is given in Ref. 29. Applications including effects of interface elements are described by Desai and Appel.[26]

A procedure intermediate between three- and two-dimensional idealizations that is possible for laterally loaded pile problems can be formulated by using the concept of separation of variables. As shown in Fig. 9-10*a*, the axially symmetric pile is subjected to an asymmetric lateral loading. A periodic function can be expanded into a Fourier series if it satisfies certain mathematical conditions;[30,31]

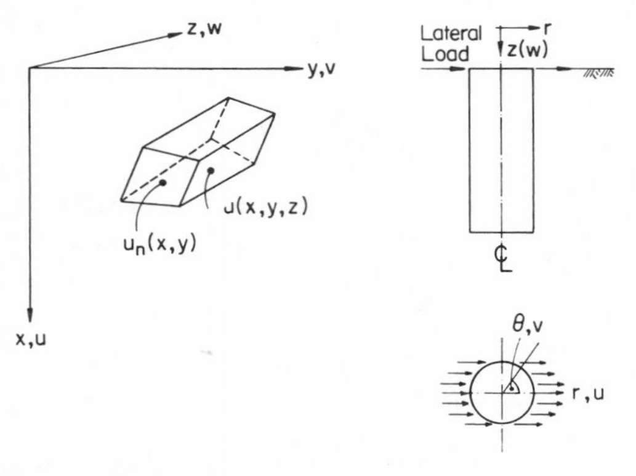

Figure 9-10 Separation of variables for laterally loaded structures.

then the three displacement components can be expressed as

$$u(x, y, z) = \sum_{n=0}^{N} u_n(x, y) \cos nz$$
$$v(x, y, z) = \sum_{n=0}^{N} v_n(x, y) \cos nz \tag{9-35}$$
$$w(x, y, z) = \sum_{n=1}^{N} w_n(x, y) \sin nz$$

where N is the number of finite element solutions. The applied forces can be expanded similarly. The stiffness equations can be obtained by using the minimum-energy principle. The displacements $u(x, y, z)$ are then computed as the sum of the series once $u_n(x, y)$, etc., are known in two-dimensional space (x, y). The procedure can be specialized for axisymmetric piles loaded asymmetrically (Fig. 9-10*b*). Complete details of this procedure are given by Desai and Abel,[31] and additional details and applications to laterally loaded piles are described by Desai and Patil[32] and Patil and Desai.[33]

The concept of separation of variables is valid for linear problems. As an approximation, however, nonlinear analysis can be done by using the incremental load method. For example, the element equations can be written in incremental form as

$$[k]_n\{\Delta q\}_n = \{\Delta Q\}_n + \{\Delta Q_0\}_n \tag{9-36}$$

where $n = 1, 2, \ldots, N$
$n =$ nth harmonic
$\{Q_0\} =$ incremental residual (initial) load vector

The incremental stiffness matrix can be formed on the basis of nonlinear elastic or elastoplastic constitutive laws (Chaps. 2 and 3).

9-4 APPLICATIONS

A number of studies have considered applications of the finite difference method[6,10,22,24] and the finite element method[25,26,32,33] for the problem of laterally loaded piles. We consider here two typical examples.

Example 9-1 Laterally loaded pile in sand using the finite difference method The finite difference method described in the preceding pages for the analysis is applied to the computation of deflections and bending moments for a pile installed in sand. The site is on Mustang Island, Texas.[34] The sand varies from clean fine sand to silty fine sand, both having high relative densities. The particles are subangular, with a large percentage of flaky grains. The angle of internal friction ϕ was 39° and the submerged unit weight γ' was 66 lb/ft^3.

Two piles of identical geometry were tested. They were 24 in OD and were driven open-ended to a penetration of 70 ft below the ground surface. Horizontal loads were applied near the ground surface. One of the piles was subjected to short-term static loading and the other to cyclic loading. The piles were instrumented so that bending moment could be measured along the length of the piles. Deflection and rotation at the ground surface were measured, as was the applied load.

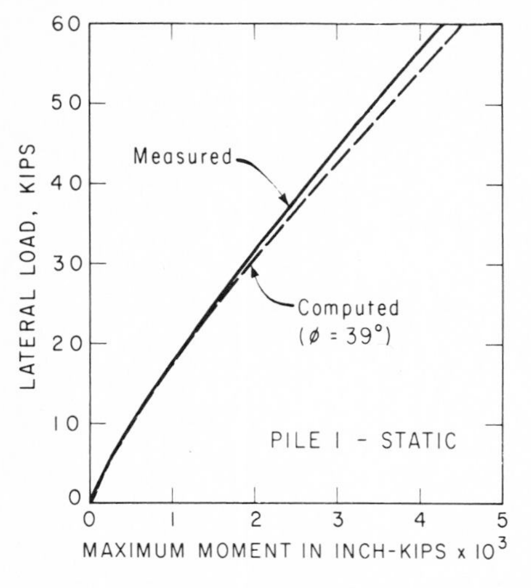

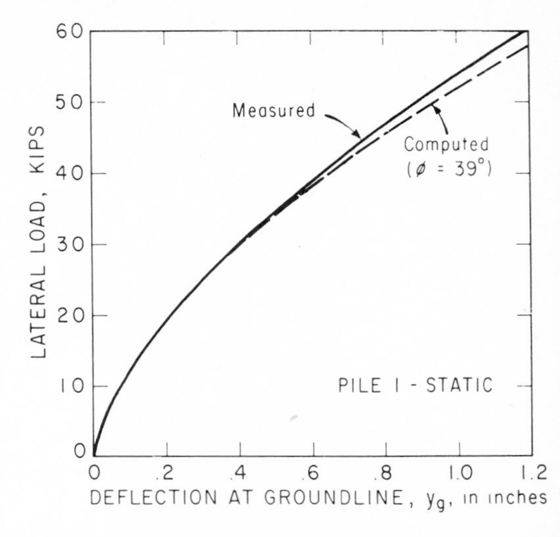

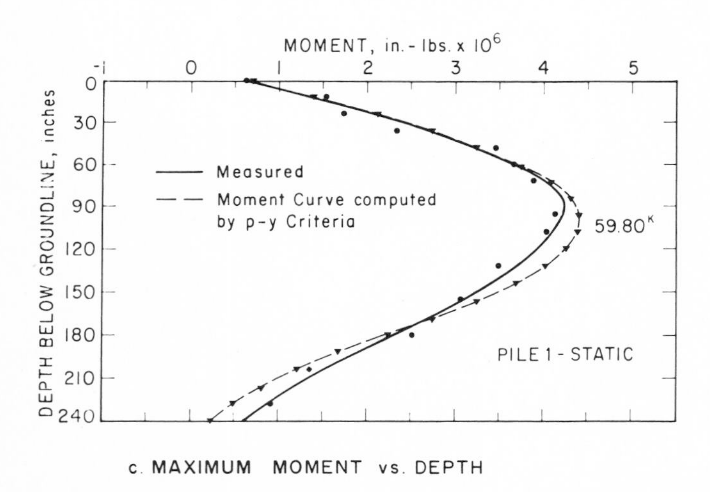

Figure 9-11 Comparisons between measured results of Mustang Island tests and computed results from proposed criteria, pile 1: short-term loading.[14]

The criteria for sand, described above, were used to predict *p-y* curves, and the differential equation was solved by computer to compute the behavior of the two piles. The results from the analyses compare very favorably with the experimental results. Figure 9-11*a* to *c* shows comparisons for the pile that was tested using short-term static loading. Figure 9-12*a* and *b* shows comparisons for the pile that was tested using cyclic loading.

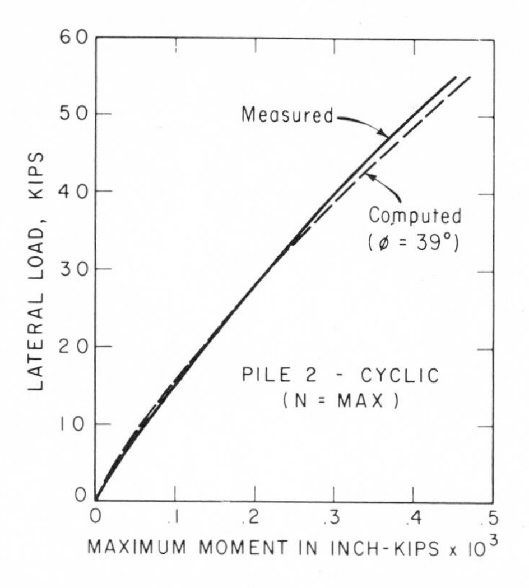

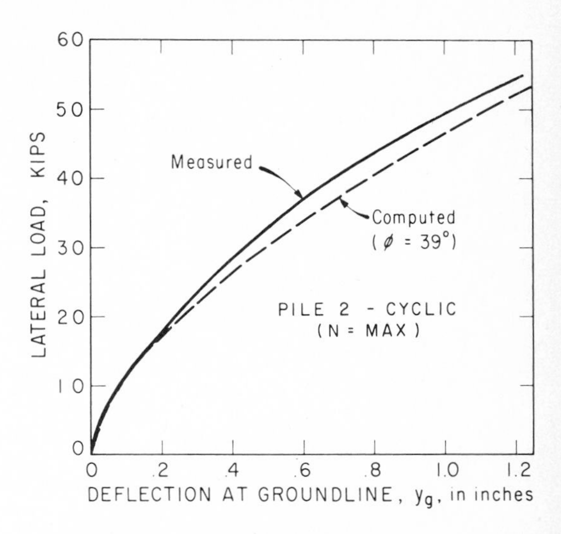

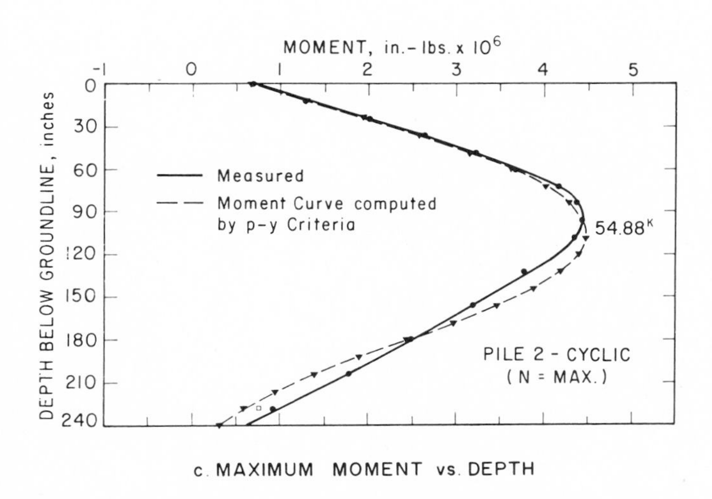

Figure 9-12 Comparisons between measured results of Mustang Island tests and computed results with proposed criteria, pile 2 (N = max): cyclic loading.[14]

Example 9-2 Laterally loaded piles using the finite element method Figure 9-13 shows a finite element mesh including interface elements around the pile[25,26] (see also Chap. 7) for a model pile of square cross section, 0.5 by 0.5 in (1.3 by 1.3 cm), driven in sandy foundation. The foundation was laid in a test tank (Fig. 9-14); details are given in Ref. 35. A lateral load applied at the butt of the pile was simulated in the three-dimensional finite element analysis, and displacements and stresses were obtained in the discretized mass.

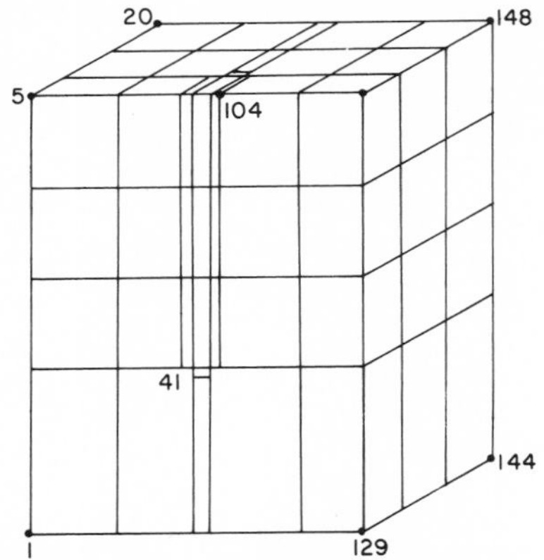

(a) FE MESH OF PILE AND SOIL CONTINUUM WITH JOINT ELEMENTS

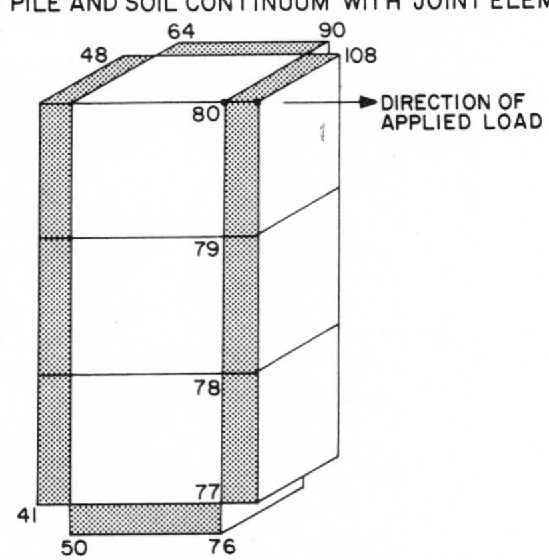

(b) DETAILS OF JOINT ELEMENTS (JOINT ELEMENTS SHOWN WITH ONE EDGE SHADED.)

Figure 9-13 Details of finite element mesh.[26]

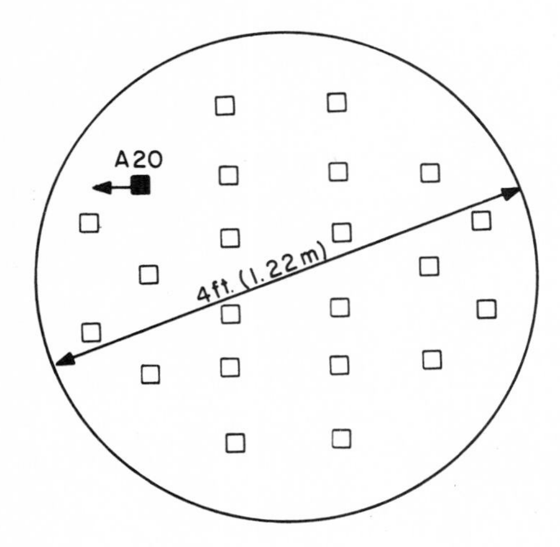

Figure 9-14 Plan view of test tank and model piles. Arrow on pile A20 shows direction of applied load.[35]

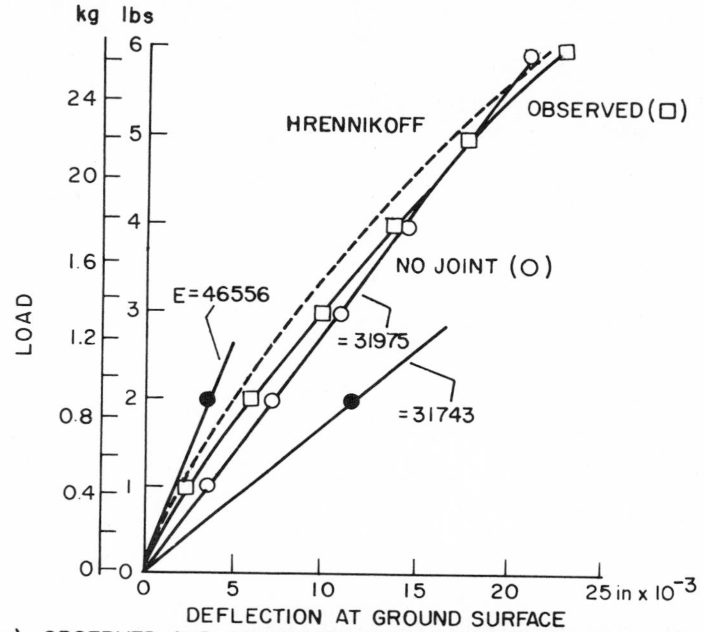

(a) OBSERVED AND COMPUTED LOAD – DISPLACEMENT RELATIONSHIPS

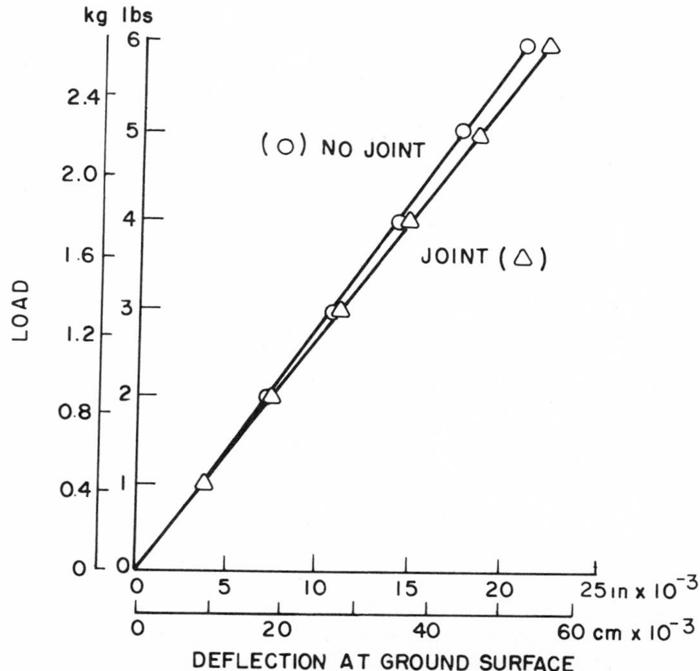

(b) BUTT DISPLACEMENT OF PILE A20 UNDER LATERAL LOAD.

Figure 9-15 Results and comparisons.[26]

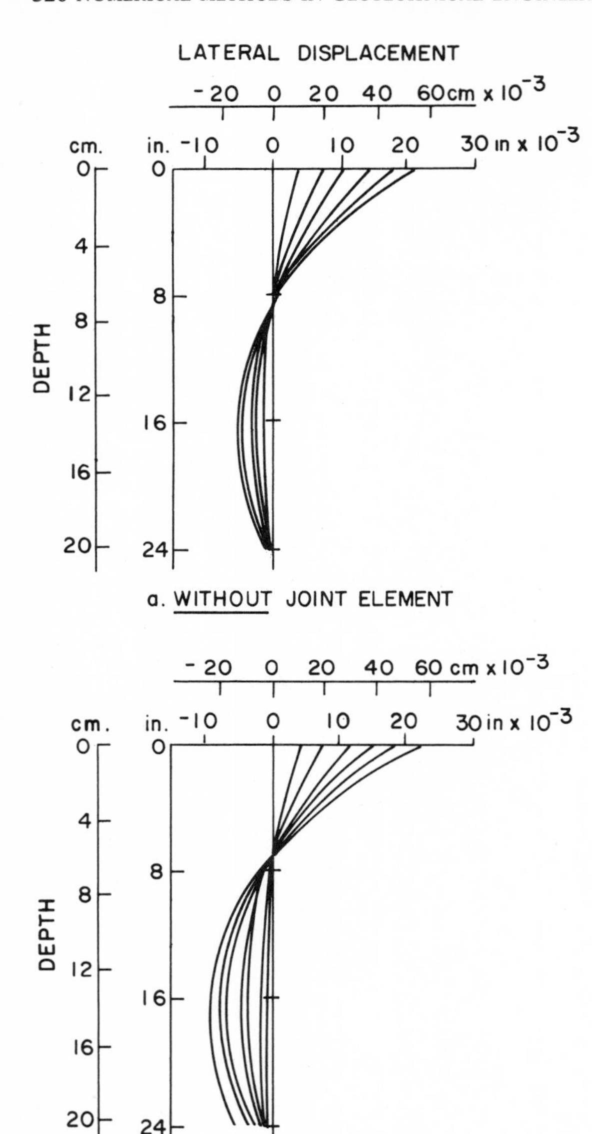

Figure 9-16 Displacement profiles of pile under successive load increments.

The values of elastic modulus and Poisson's ratio for the steel pile were adopted as 4.3×10^9 lb/ft^2 (2×10^{11} Pa) and 0.2, respectively. Since sufficient details of soil properties were not available, a parametric study was performed in which various values of elastic modulus E were used: 31,743 lb/ft^2 (1.5×10^6 Pa), 35,975 lb/ft^2 (1.7×10^6 Pa), and 46,556 lb/ft^2 (2.2×10^6 Pa). A value of the Poisson's ratio equal to 0.3 was used.

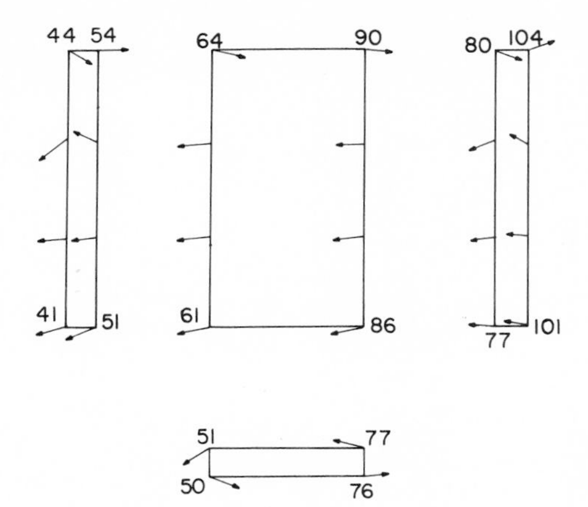

Figure 9-17 Displacements of nodes in joint elements (Fig. 9-13).

The shear stiffness k_{sx} and k_{sy} for the interfaces were adopted as 20,000 lb/ft^3 (3.2 × 10^5 kg/m^3). The normal stiffness k_{nz} was taken as equal to 10^9 lb/ft^3 (1.6 × 10^{10} kg/m^3). The material properties for sands and interfaces were based on laboratory test results on similar sands (Chap. 7).

Figure 9-15*a* shows comparisons between results for load-displacement curves from Hrennikoff's method, numerical predictions, and laboratory observations. It can be seen that the predictions agree most closely with the observations for the value of $E = 35{,}975$ lb/ft^2. This value compares approximately to the average value of the initial modulus at low confining pressures and relative density equal to about 70 percent; these conditions can be considered to simulate the situation in the test tank. Figure 9-15*b* shows the influence on the load-deformation behavior of providing the interface elements around the pile.

The effect of interface elements is further illustrated in Fig. 9-16*a* and *b*, where it can be seen that at higher loadings the existence of interface elements permits large relative displacements between the pile and the soil. Figure 9-17 shows vectors of displacements at typical nodes and illustrates the capability of the analysis in allowing computations of stresses and deformations around the pile.

Comments

There are a number of advantages of using the finite element method. In contrast to the finite difference method in conjunction with the concept of *p-y* curves, the finite element method can permit realistic three-dimensional effects and computation of stresses and deformations in and around the piles. It is also possible to study progressive development of stresses and deformations leading to demarcation of failure zones and wedges. Moreover, the generalized three-dimensional description of constitutive behavior can be included, and factors such as stress paths, strain softening, stress release, and interaction effects can be considered.

At this time, use of three-dimensional finite element analysis can be relatively expensive. However, with the introduction of the new generation of computers and development of efficient solving and data-storage routines, its use will become competitive.

APPENDIX 9A

9A-1 DERIVATION OF CLOSED-FORM AND SERIES SOLUTIONS

Let us consider Eq. (9-1*b*). The simplified equation can be expressed as

$$\frac{d^4y}{dx^4} + 4\beta^4 y = 0 \tag{9-37a}$$

where $\beta^4 = E_s/4EI$. In operator form the equation becomes

$$(D^4 + 4\beta^4)y = 0 \tag{9-37b}$$

in which $D \equiv d/dx$. From our knowledge of the solutions of ordinary linear differential equations[1,36] we can write the solution to Eq. (9-37*b*) as

$$y = e^{\beta x}(C_1 \cos \beta x + C_2 \sin \beta x) + e^{-\beta x}(C_3 \cos \beta x + C_4 \sin \beta x) \tag{9-38}$$

Here C_i are the constants to be determined from the given boundary conditions.

If the pile is long and can be assumed to be of "infinite" length, the deflection at the base can be considered to be essentially zero; hence, at large values of x, $y \approx 0$ and $C_1 = C_2 = 0$. If M_t and V_t are the applied moment and shear at the top, at $x = 0$

$$\frac{d^2y}{dx^2} = \frac{M_t}{EI}$$

hence

$$C_4 = -\frac{M_t}{2EI\beta^2} \tag{9-39a}$$

and

$$\frac{d^3y}{dx^3} \approx \frac{V_t}{EI}$$

hence

$$C_3 + C_4 = \frac{V_t}{2EI\beta^3} \tag{9-39b}$$

Substitution of C_i in Eq. (9-38) yields the required solution. The above formulation can be satisfactory for lengths given by[15] $BL \geqslant 4$. The sign convention relevant to this derivation is shown in Fig. (9-18).

If the pile is of finite length, at $x = 0$

$$\frac{d^2y}{dx^2} = \frac{M_t}{EI}$$

hence

$$2\beta^2(C_2 - C_4) = \frac{M_t}{EI} \tag{9-40a}$$

and

$$\frac{d^3y}{dx^3} = \frac{V_t}{EI}$$

hence

$$2\beta^3(-C_1 + C_2 + C_3 + C_4) = \frac{V_t}{EI} \tag{9-40b}$$

At $x = L$

$$EI\frac{d^2y}{dx^2} = 0$$

hence

$$2\beta^2 e^{\beta L}(C_2 \cos \beta L - C_1 \sin \beta L) + 2\beta^2 e^{-\beta L}(C_3 \sin \beta L - C_4 \cos \beta L) = 0 \tag{9-40c}$$

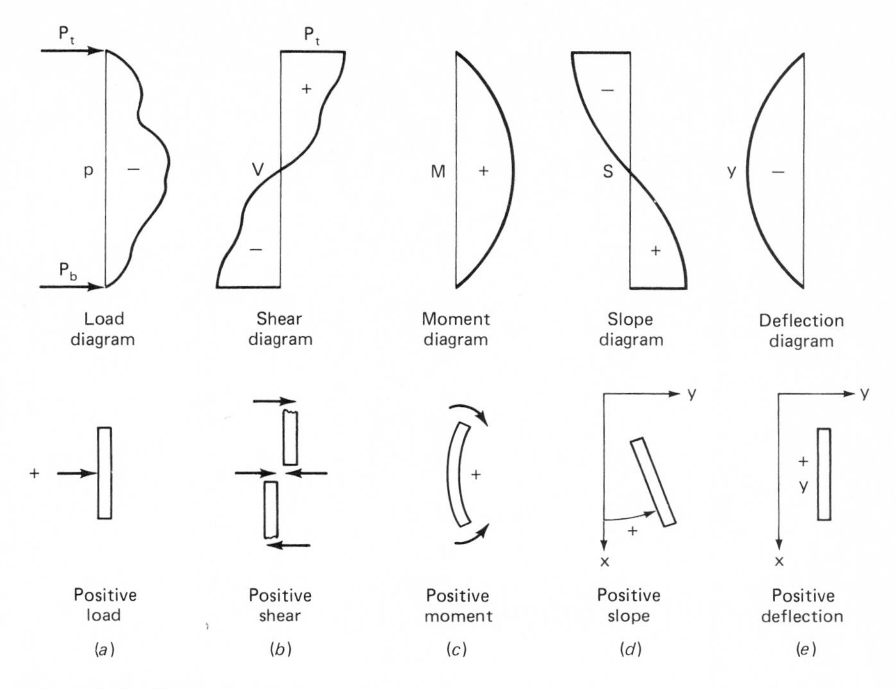

Figure 9-18 Sign convention.

and $EI\ d^3y/dx^3 = 0$; hence

$$2\beta^3 e^{\beta L}(C_2 \cos \beta L - C_1 \sin \beta L - C_2 \sin \beta L - C_1 \cos \beta L)$$

$$+ 2\beta^3 e^{-\beta L}(-C_3 \sin \beta L + C_4 \cos \beta L + C_3 \cos \beta L + C_4 \sin \beta L) = 0 \qquad (9\text{-}40d)$$

Substitution of values of C_i from Eqs. (9-40) leads to the solution for a pile with finite length.

9A-2 SERIES SOLUTIONS

The solutions to Eq. (9-1) can be expressed in an infinite series. Approximate solutions can be obtained by truncating the series. A form of series solution can be

$$y = \sum_{i=0}^{\infty} A_i x^i \qquad (9\text{-}41)$$

where A_i are the constants to be determined. The computer can be used to evaluate the series solutions. Solutions in series form for piles in elastic and elastoplastic soils have been presented by various investigators.[37,38] Like the previous closed-form solutions [Eq. (9-38)] the series solution can generally handle simple geometries and boundary conditions; numerical techniques become convenient when these factors become complex.

REFERENCES

1. Hetenyi, M.: "Beams on Elastic Foundation," University of Michigan Press, Ann Arbor, Michigan, 1946.
2. Feagin, L. B.: Lateral Pile Loading Tests, *Trans. ASCE*, vol. 63, no. 8, pt. 2, pp. 236–254, 1937.
3. Focht, John A., and Bramlette McClelland: Analysis of Laterally Loaded Piles by Difference Equation Solution, *Tex. Eng.* Texas Section *ASCE*, September–November, 1955.
4. Matlock, Hudson, and Lymon C. Reese: Foundations Analysis of Offshore Pile-supported Structures, *Proc. 5th Int. Conf. Int. Soc. Soil Mech. Found. Eng., Paris, 1961*, vol. 2, p. 91.
5. Reese, Lymon C., Michael W. O'Neill, and Robert E. Smith: Generalized Analysis of Pile Foundations, *Proc. ASCE*, vol. 96, no. SM1, pp. 235–250, January 1970.
6. Reese, Lymon C., and William R. Cox: Soil Behavior from Analysis of Tests of Uninstrumented Piles under Lateral Loading, *ASTM Spec. Tech. Pub.* 444, 1969.
7. Matlock, Hudson: Correlations for Design of Laterally Loaded Piles in Soft Clay, *2d Offshore Technol. Conf., Houston, Tex., 1970.*
8. McClelland, B., and J. A. Focht: Soil Modulus for Laterally Loaded Piles, *Trans. ASCE*, vol. 123, pp. 1071–1074, 1958.
9. Parker, Frazier, Jr., and Lymon C. Reese: Experimental and Analytical Studies of Behavior of Single Piles in Sand under Lateral and Axial Loading, *Univ. Texas Cent. Highw. Res., Res. Rep.* 117-2, Austin, 1971.
10. Matlock, Hudson, and E. A. Ripperger: Procedures and Instrumentation for Tests on a Laterally Loaded Pile, *Proc. 8th Tex. Conf. Soil Mech. Found. Eng., Austin, 1956.*
11. Matlock, Hudson, and E. A. Ripperger: Measurements of Soil Pressure on a Laterally Loaded Pile, *Proc. ASTM*, 1958.
12. Skempton, A. W.: The Bearing Capacity of Clays, *Proc. Build. Res. Congr. Div. I, London, 1951.*
13. Reese, Lymon C., and Robert C. Welch: Lateral Loading of Deep Foundations in Stiff Clay, *J. Geotech. Eng. Div. ASCE*, vol. 101, no. GT7, July 1975.
14. Reese, Lymon C., William R. Cox, and Francis D. Koop: Analysis of Laterally Loaded Piles in Sand, *6th Offshore Technol. Conf., Houston, Tex., 1974.*
15. Timoshenko, S.: "Strength of Materials," pt. 2, D. Van Nostrand Company, Inc., New York, 1941.
16. Davisson, M. T., and S. Prakash: A Review of Soil-Pile Behavior, *Highw. Res. Rec.* 39, 1963.
17. Broms, B. B.: Lateral Resistance of Piles in Cohesionless Soils, *J. Soil Mech. Found. Div. ASCE*, vol. 90, no. SM3, 1964.
18. Broms, B. B.: Design of Laterally Loaded Piles, *J. Soil Mech. Found. Div. ASCE*, vol. 91, no. SM3, 1965.
19. Reese, L. C., and H. Matlock: Mechanics of Laterally Loaded Piles, lecture notes, University of Texas, Austin (unpublished).
20. Woodward, R. J., W. S. Gardner, and D. M. Greer: "Drilled Pier Foundations," McGraw-Hill Book Company, New York, 1972.
21. Matlock, H., and L. C. Reese: Generalized Solutions for Laterally Loaded Piles, *Trans. ASCE*, vol. 127, pt. 1, 1962.
22. Reese, L. C., and H. Matlock: Non-dimensional Solutions for Laterally Loaded Piles with Soil Modulus Assumed Proportional to Depth, *Proc. 8th Tex. Conf., Soc. Mech. Found. Eng. Univ. Texas, 1956.*
23. Glesser, S. M.: Lateral Load Tests on Vertical Fixed-Head and Free-Head Piles, *Symp. Lateral Load Tests Piles ASM*, Spec. pap. 154, pp. 75–101, 1953.
24. Reese, L. C.: Program Documentation: Analysis of Laterally Loaded Piles by Computer, *Univ. Texas Dept. Civ. Eng.*, Austin, July 1972.
25. Desai, C. S.: A Three-dimensional Finite Element Procedure and Computer Program for Nonlinear Soil-Structure Interaction, *VPI & State Univ. Dept. Civ. Eng. Rep.* VPI-E75.27, Blacksburg, Va., June 1975.
26. Desai, C. S., and G. C. Appel: 3-D Analysis of Laterally Loaded Structures, *Proc. 2d Int. Conf. Numer. Methods Geomech., Blacksburg, Va., June 1976.*

27. Ruser, J. R., and W. P. Dawkins: Three Dimensional Finite Element Analysis of Soil-Structure Interaction, *Proc. Symp. Appl. Finite Element Method Geotech. Eng., Vicksburg, Miss., September 1972.*
28. Wittke, W., et al.: Bemessung von horizontal belasteten Grossbohrfahlen nach Methode finiter Elemente, *Bauingenieur*, vol. 49, pp. 219–226, 1974.
29. Yegian, M., and Wright, S. G.: Lateral Soil Resistance–Displacement Relationships for Pile Foundations in Soft Clays, *5th Annu. Offshore Technol. Conf., Houston, 1973.*
30. Churchill, R. V.: "Fourier Series and Boundary Value Problems," McGraw-Hill Book Company, New York, 1941.
31. Desai, C. S., and J. F. Abel: "Introduction to the Finite Element Method," Van Nostrand Reinhold Company, New York, 1972.
32. Desai, C. S., and U. K. Patil: Finite Element Analysis of Some Soil-Structure Interaction Problems, *Proc. Intl. Symp. on Soil Structure Interaction, Univ. of Roorkee, India*, January 1977.
33. Patil, U. K., and C. S. Desai: Static and Dynamic Finite Element Analysis of Laterally Loaded Structures By Separation of Variables, *Report no. VPI-E-76-20*, Dept. of Civil Eng., V.P.I. & S.U., Blacksburg, July 1976.
34. Cox, William R., Lymon C. Reese, and B. R. Grubbs: Field Testing of Laterally Loaded Piles in Sand, *6th Offshore Technol. Conf., Houston, 1974.*
35. Fruco and Associates: Pile Driving and Loading Tests: Lock and Dam No. 4, Arkansas River and Tributaries, Arkansas and Oklahoma, U.S. Army Corps of Engineers District, Little Rock, September 1964.
36. Sokolnikoff, I. S., and R. M. Redheffer: "Mathematics of Physics and Modern Engineering," McGraw-Hill Book Company, New York, 1958.
37. Venkata Ratnam, M., G. Subrahmanyam, and Z. H. Mazindrani: Behavior of a Vertical Pile under Oblique Load, *Ind. Geotech. J.*, vol. 1, no. 1, pp. 11–26, January 1971.
38. Valsangkar, A. J., N. S. V. Kameswara Rao, and P. K. Basudhar: Generalized Solutions of Axially and Laterally Loaded Piles in Elasto-plastic Soil, *Soils Found.*, vol. 13, no. 4, December 1973.

CHAPTER

TEN

SETTLEMENT OF PILE FOUNDATIONS

H. G. Poulos

10-1 INTRODUCTION

Until relatively recently, predictions of the settlement of pile foundations, if made at all, have generally been based on empirical data or simplified one-dimensional consolidation approaches. With the development of numerical techniques and the increased use of computers, increased efforts have been directed toward making more rational analyses of pile settlement behavior. The currently available theoretical approaches may be classified broadly into three categories:

1. Methods based on the theory of elasticity, which employ the equations of Mindlin[1] for subsurface loading within a semi-infinite mass
2. Step-integration methods, which use measured relationships between pile resistance and pile movement at various points along the pile
3. Numerical methods and, in particular, the finite element method

Methods in the first category have been described by several investigators, e.g., D'Appolonia and Romualdi,[2] Thurman and D'Appolonia,[3] Salas and Belzunce,[4] Poulos and Davis,[5] and Mattes and Poulos.[6] All rely initially on the assumption of the soil as a linear elastic material, although more realistic soil behavior can be incorporated readily into the analyses in an approximate manner.

Such methods also provide a relatively rapid means of carrying out parametric analyses of the effects of pile and soil characteristics and of preparing series of solutions which can be used for design purposes. Moreover, the settlement of pile groups can be analyzed by a relatively simple extension of the single pile analysis.

The step-integration approach was first described by Seed and Reese[7] and then developed by Coyle and Reese[8] and Coyle and Sulaiman.[9] This method uses measured pile load-transfer data and hence requires no assumptions regarding linearity of soil behavior. However, it inherently assumes that the movement of a point on the pile is related only to the shear stress at the point and is independent of the stresses elsewhere along the pile. This assumption is analogous to that used in analyses employing the theory of subgrade reaction or Winkler's hypothesis, and, since no account is taken of the continuity of the soil, the method cannot be used for analyzing pile groups.

The finite element method is potentially the most powerful method, as it can take into account such factors as soil layering and nonlinear stress-strain behavior of the soil. However, although a single pile is readily analyzed as a radially symmetrical problem, a pile group can present difficulties, as the loss of axial symmetry will require a time-consuming three-dimensional analysis. In addition, suitably accurate values of the soil properties required for input may be difficult to determine.

In this chapter, attention will be concentrated on the elastic approach using Mindlin's equations. A basic analysis for a single pile will be described in detail, and an examination of the accuracy of the solutions will be made, including comparisons with finite element solutions. Modifications of the basic analysis to account for more realistic cases, including pile groups, will be outlined, and a series of parametric solutions will be presented. A number of cases will then be described in which measured and theoretical settlements are compared.

10-2 GEOTECHNICAL ASPECTS OF THE PROBLEM

10-2.1 Idealization of the Problem

In the most general case, the geotechnical engineer is required to predict the behavior of a group of piles subjected to various types of loading such as vertical, horizontal, and moment, in a soil profile consisting of several different layers. Attention will be confined to the response of piles subjected to vertical loads, as the behavior of piles subjected to horizontal load and moment was discussed in Chap. 9. In order to make a systematic analysis of pile behavior, it is advantageous to idealize the real problem and reduce it to a mathematically tractable model. A relatively simple idealization can be attempted first, and when experience has been gained regarding the mathematical behavior of this simple model, the idealization can be progressively refined to make it approach the real problem more closely.

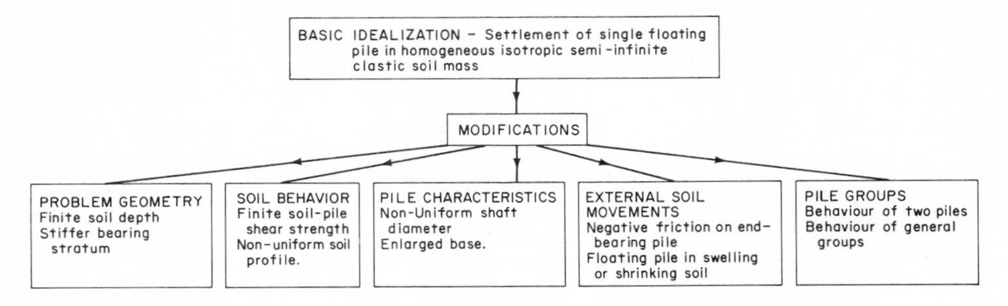

Figure 10-1 Progressive idealization of the pile settlement problem.

The simplest problem to consider is a single floating pile in a homogeneous, isotropic, semifinite elastic mass. From this basic starting point, refinements can be made to the problem idealization, as shown in Fig. 10-1. Successive idealizations involve building on to the analyses of the simpler idealizations. These analyses are described in Sec. 10-3.

10-2.2 Idealization of Soil Behavior

For mathematical simplicity, the soil is assumed to be an isotropic elastic material in the basic analysis. It is possible to adopt the elastic approach to analyze approximately a soil whose stress-strain behavior is represented by a piecewise linear relationship. However, in order to obtain a general appreciation of the factors influencing pile settlement behavior, it is convenient to consider the soil as having the simplified stress-displacement relationship illustrated in Fig. 10-2*a*, consisting of a linear relationship between shear stress and displacement up to a limiting stress τ_a. This soil model has been found to reproduce the primary behavioral characteristics of piles under axial loading, and, although more refined representations of soil stress-strain behavior may modify some of the predicted details, they are not expected to alter the broad conclusions given by the simpler model. As shown later in this chapter, a purely elastic analysis can often give an adequate solution to practical problems.

A similar modified elastic model has been used successfully to examine the settlement of shallow foundations[10] and laterally loaded piles[11] provided that a suitable elastic "modulus" can be chosen for the soil. Some of the problems involved in choosing such a modulus are described in Sec. 10-4. Further advantages of the use of the simple soil model are:

1. The immediate and consolidation settlements can readily be calculated, as described in Sec. 10-3.6.
2. Parametric studies can readily be carried out to determine the significant factors affecting pile settlement.
3. The analysis of pile groups is a simple extension of the single pile analysis.

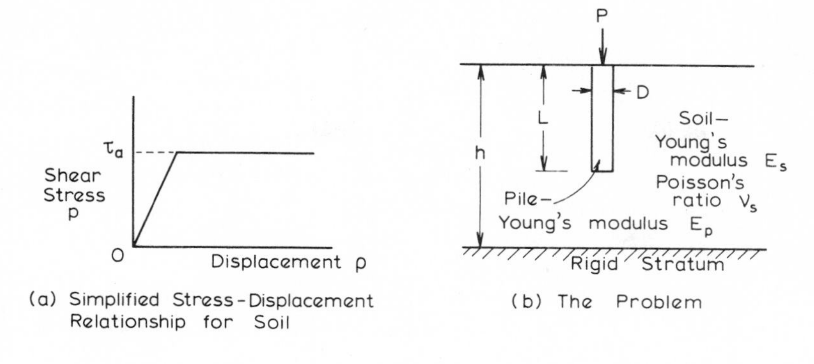

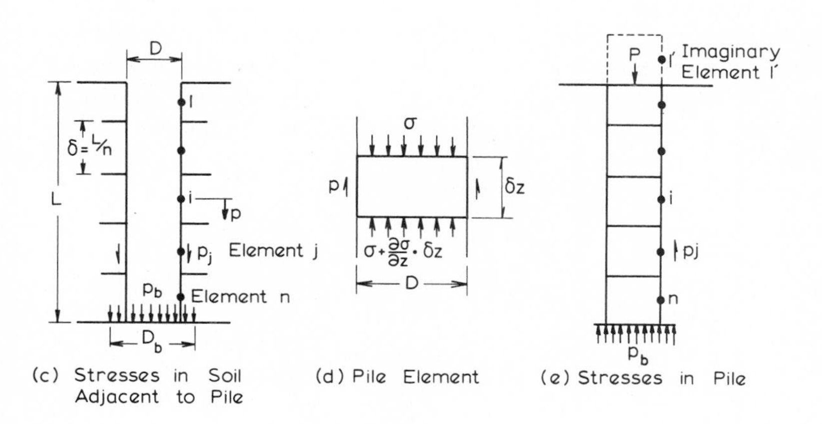

Figure 10-2 Floating pile problem.

4. The theoretical *relative* behavior of different types of piles and of single piles and pile groups is generally in very good agreement with that observed in field and model tests (Sec. 10-4).

10-3 NUMERICAL ANALYSIS

A basic analysis for a single pile in a uniform semi-infinite elastic mass will be considered first. Modifications to the basic analysis (Fig. 10-1) will then be described.

10-3.1 Analysis for Single Floating Pile

The pile is considered to be cylindrical with length L, shaft diameter D, and base diameter D_b and loaded with an axial force P at the ground surface. For the purposes of the analysis, the pile is acted upon by a system of uniform shear stresses p around the periphery, and the base is acted upon by a uniform vertical stress p_b (Fig. 10-2). The sides of the pile are assumed to be perfectly rough, but the base is assumed perfectly smooth. The soil is initially considered to be an ideal homogeneous isotropic elastic half-space, having Young's modulus E_s and Poisson's ratio ν_s, which are not influenced by the presence of the pile. If conditions at the pile-soil interface remain elastic and no slip occurs, the movements of the pile and the adjacent soil must be equal. The correct values of the stress system p† and base stress p_b will be those which satisfy this condition of displacement compatability.

Ideally, consideration should be given to compatability of both vertical and radial displacements, and a normal stress system should also then be imposed on the pile elements. However, this more complete analysis with additional compatability conditions gives solutions which are generally almost identical with those from a simpler analysis considering only vertical displacement compatability (Sec. 10-3.3). Details of the more complete analysis are given in Refs. 12 and 13.

A solution for the values of p and p_b and the displacement of the pile can be obtained by deriving expressions for the vertical displacement of the pile and the soil at each element in terms of the unknown stresses on the pile, imposing the compatability condition, and solving the resulting equations.

Soil displacement equations Considering a typical element i in Fig. 10-2, the vertical displacement of the soil adjacent to the pile at i due to the stress p_j on an element j can be expressed as

$$ {}_s\rho_{ij} = \frac{D}{E_s} I_{ij} p_j \tag{10-1} $$

where I_{ij} is the vertical-displacement factor for i due to shear stress at element j.

The soil displacement at i to all n elements and due to the base is

$$ {}_s\rho_i = \frac{D}{E_s} \sum_{j=1}^{n} I_{ij} p_j + \frac{D}{E_s} I_{ib} p_b \tag{10-2} $$

where I_{ib} is the vertical displacement factor for i due to uniform stress on the base.

A similar expression can be written for the base, and for all elements on the pile the soil displacements can be written

$$ \{{}_s\rho\} = \frac{D}{E_s} [I_s]\{p\} \tag{10-3} $$

† The shear stresses p are fictitious in that they represent tractions applied to the boundaries of the imaginary surface in the half-space representing the pile surface and are not necessarily the actual stresses acting on the real pile surfaces. Once the values of p are determined, the actual stresses and displacements they produce anywhere in the half-space, including the real pile boundaries, can be calculated. Mattes[12] shows that an analysis in which the difference between fictitious and real stresses is taken into account gives similar solutions to those from the simple analysis.

where $\{_s\rho\}$ = soil displacement vector
$\{p\}$ = pile stress vector
$[I_s]$ = $n+1$ square matrix of soil displacement factors, given by

$$[I_s] = \begin{bmatrix} I_{11} & I_{12} & \cdots & I_{1n} & I_{1b} \\ I_{21} & I_{22} & \cdots & I_{2n} & I_{2b} \\ \cdots & \cdots & \cdots & \cdots & \cdots \\ I_{n1} & I_{n2} & \cdots & I_{nn} & I_{nb} \\ I_{b1} & I_{b2} & \cdots & I_{bn} & I_{bb} \end{bmatrix}$$

Evaluation of the elements of $[I_s]$ is most conveniently carried out by integration of the Mindlin equations[1] for the displacement due to a point load within a semi-infinite mass. The geometry of a typical cylindrical pile element is shown in Fig. 10-3. For point i at midheight of the ith element on the periphery of the pile, the value of I_{ij} is

$$I_{ij} = 2\int_{(j-1)\delta}^{j\delta}\int_0^{\pi/2} {}_pI \, d\theta \, dc \tag{10-4}$$

where ${}_pI$ = influence factor for vertical displacement due to vertical point load
δ = length of element = L/n
L = total length of pile
n = number of elements

From Mindlin's equation, ${}_pI$ is given by

$${}_pI = \frac{1+\nu}{8\pi(1-\nu)}\left[\frac{z_1^2}{R_1^3} + \frac{3-4\nu}{R_1} + \frac{5-12\nu+8\nu^2}{R_2} + \frac{(3-4\nu)z^2 - 2cz + 2c^2}{R_2^3} + \frac{6cz^2(z-c)}{R_2^5}\right] \tag{10-5}$$

where $\qquad z = (i-0.5)\delta + c \qquad z_1 = (i-0.5)\delta - c$

$$R_2^2 = D^2\cos^2\theta + z^2 \qquad R_1^2 = D^2\cos^2\theta + z_1^2$$

The integral with respect to c in Eq. (10-4) can be evaluated analytically, but that with respect to θ is more conveniently evaluated numerically.

The geometry of the pile base is also shown in Fig. 10-3. To allow for an enlarged base, a base radius R_b $(= D_b/2)$ different from the pile-shaft radius R $(= D/2)$ is considered. For point i on the shaft,

$$I_{ib} = \frac{1}{D}\int_0^{2\pi}\int_0^{R_b} {}_pIr \, dr \, d\theta \tag{10-6}$$

where ${}_pI$ is given in Eq. (10-5) and, for this case,

$$c = n\delta \qquad R_2^2 = z^2 + R^2 + r^2 - 2rR\cos\theta$$

$$R_1^2 = z_1^2 + R^2 + r^2 - 2rR\cos\theta$$

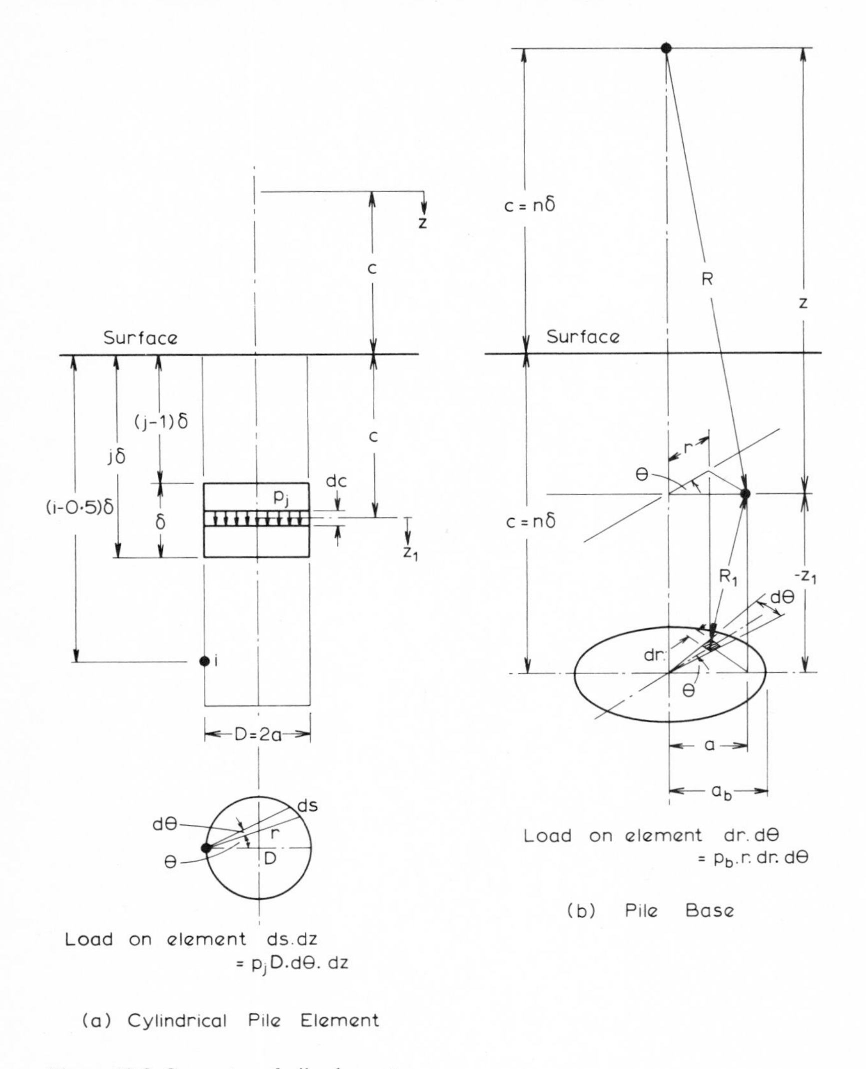

Figure 10-3 Geometry of pile elements.

The integration with respect to r can be performed analytically, but again the integration with respect to θ is most readily evaluated numerically.

In evaluating the integrals with respect to θ in Eqs. (10-4) and (10-6), intervals of $\pi/50$ are usually adequate. To avoid the singularity which occurs when $i = j$, it is most expedient to calculate ordinates at the midpoint of each interval and then apply the simple rectangular rule for integration to evaluate the complete double integral. The value of the integral converges to a constant value as the number of intervals in θ is increased.

For the center of the base due to shear stress on element j,

$$I_{bj} = \pi \int_{(j-1)\delta}^{j\delta} {}_pI \, dc \tag{10-7}$$

where ${}_pI$ is given in Eq. (10-5) with

$$z = L + c \qquad z_1 = L - c \qquad R_2^2 = z^2 + R^2 \qquad R_1^2 = z_1^2 + R^2$$

The integration in Eq. (10-7) can be performed analytically.

For the vertical displacement of the base due to the base loading itself, it is desirable to make an approximate allowance for the effect of the rigidity of the base by multiplying the displacement of the center of the uniformly loaded circular base by a factor of $\pi/4$. (This is the ratio of the surface displacement of a rigid circle on the surface of a half-space to the the center displacement of a corresponding uniformly loaded circle and may be assumed to apply approximately to embedded areas.) Thus

$$I_{bb} = \frac{\pi}{4}\frac{\pi}{D}\int_0^{R_b} {}_pIr \, dr \tag{10-8}$$

where ${}_pI$ is given in Eq. (10-5) and now

$$z = 2L \qquad z_1 = 0 \qquad c = n\delta \qquad R_2^2 = 4c^2 + r^2 \qquad R_1 = r$$

The integral in Eq. (10-8) can readily be evaluated analytically.

For larger diameter piles or piles with pronounced enlarged bases, greater accuracy can be achieved by dividing the base into a number of uniformly loaded concentric rings and considering the displacements due to each ring. However, as will be shown in Sec. 10-3.3, the approximate approach adopted in Eq. (10-8) generally gives very satisfactory solutions for pile displacement and portion of load taken by the base.

Pile displacement equations The pile material is assumed to have a constant Young's modulus E_p and area of pile section A_p. It is convenient to define the area ratio R_A as

$$R_A = \frac{A_p}{\pi d^2/4}$$

= ratio of area of pile section to area bounded by outer circumference of pile (10-9)

For a solid pile, $R_A = 1$.

In calculating the displacement of the pile elements, only the axial compression of the pile is considered. Referring to Fig. 10-3b, consideration of vertical equilibrium of a small cylindrical pile element yields

$$\frac{\partial \sigma}{\partial z} = \frac{-4p}{R_A D} \tag{10-10}$$

where σ is the normal stress in the pile and p is the shear stress on the pile surface.

The axial strain of this element is

$$\frac{\partial_p \rho}{\partial z} = \frac{-\sigma}{E_p} \tag{10-11}$$

where ${}_p\rho$ is the displacement of the pile. From Eqs. (10-10) and (10-11)

$$\frac{\partial^2_p \rho}{\partial z^2} = \frac{4p}{D} \frac{1}{E_p R_A} \tag{10-12}$$

This equation can be written in finite difference form and applied to the points $i = 1$ to n. Equation (10-10) can similarly be applied to the top of the pile, where $\sigma = P/A_p$, and Eq. (10-11) to the base of the pile, where $\sigma = p_b$. The following relationship is then obtained for the pile displacements:[6]

$$\{p\} = \frac{D}{4\delta^2} E_p R_A [{}_pI]\{{}_p\rho\} + \{Y\} \tag{10-13}$$

where $\{p\} = n + 1$ shear-stress vector
$\{{}_p\rho\} = n + 1$ pile displacement vector
$[{}_pI] =$ pile-action matrix, $(n + 1) \times (n + 1)$, given by

$$[{}_pI] = \begin{bmatrix} -1 & 1 & 0 & 0 & \cdots & 0 & 0 & 0 & 0 \\ 1 & -2 & 1 & 0 & \cdots & 0 & 0 & 0 & 0 \\ 0 & 1 & -2 & 1 & \cdots & 0 & 0 & 0 & 0 \\ \cdots & \cdots & \cdots & \cdots & \cdots & \cdots & \cdots & \cdots & \cdots \\ 0 & 0 & 0 & 0 & \cdots & 0 & 1 & -2 & 1 \\ 0 & 0 & 0 & 0 & \cdots & 2 & 2 & -5 & 3.2 \\ 0 & 0 & 0 & 0 & \cdots & 0 & \frac{-4f}{3} & 12f & \frac{-32f}{3} \end{bmatrix}$$

$$f = \frac{L/D}{nR_A} \qquad \text{and} \qquad \{Y\} = \begin{Bmatrix} \frac{P}{\pi D^2} \frac{n}{L/D} \\ 0 \\ 0 \\ \vdots \\ 0 \\ 0 \\ 0 \end{Bmatrix}$$

Displacement compatability When purely elastic conditions prevail at the pile-soil interface, i.e., no slip, the displacements of adjacent points along the interface are equal, i.e.,

$$\{{}_p\rho\} = \{{}_s\rho\} \tag{10-14}$$

Use of Eqs. (10-3) and (10-14) in Eq. (10-13) yields

$$\{p\} = \left[[I] - \frac{n^2 K}{4(L/D)^2} [{}_pI][I_s]\right]^{-1} \{Y\} \tag{10-15}$$

where $[I]$ is the unit matrix of order $n + 1$ and K the pile stiffness factor, given by

$$K = \frac{E_p}{E_s} R_A \tag{10-16}$$

The factor K is a measure of the relative compressibility of the pile and the soil. The more (relatively) compressible the pile, the smaller the value of K.

Equation (10-15) can be solved by computer to obtain the unknown stress distribution along the pile. The displacement distribution can then be calculated from Eq. (10-13).

Approximate treatment for nonuniform soil The foregoing analysis assumes constant soil deformation parameters at all points within the soil. An approximate allowance can be made for the effects of varying soil deformation moduli along the length of the pile by assuming that the soil displacement at a point adjacent to the pile is given by the Mindlin equation but that E_s and ν_s in the equation are the values at that point. This assumption implies that the stress distribution in the nonuniform soil is the same as that in a uniform soil. Equation (10-3) then becomes

$$\{_s\rho\} = D\left[\frac{1}{E_s}\right][I_s]\{p\} \tag{10-17}$$

where $[1/E_s]$ is the $n + 1$ diagonal matrix of reciprocal values of Young's modulus for soil along the pile.

The pile displacement equation (10-13) remains unchanged, so that combination of Eqs. (10-17) and (10-13) yields the equations analogous to (10-15), which can be solved for stresses and displacements along the pile. Although the above approach is only approximate, it can give sufficiently accurate solutions for practical purposes unless sudden large variations in soil moduli occur along the pile length.

Approximate treatment for layer of finite depth The elements of $[I_s]$ calculated as previously described apply only for a soil mass of infinite depth. For soil layers of finite depth, the elements of $[I_s]$ can be obtained approximately by employing the Steinbrenner approximation.[14] For a point i in a layer of depth h, the displacement-influence factor I_{ij} is

$$I_{ij(h)} = I_{ij(\infty)} - I_{hj(\infty)} \tag{10-18}$$

where $I_{ij(\infty)}$ is the displacement-influence factor for i due to stress on element j in a semi-infinite mass and $I_{hj(\infty)}$ is the displacement-influence factor for a point within the semi-infinite mass directly beneath i at a depth h below the surface due to stress on element j. With these adjusted elements of $[I_s]$, Eq. (10-15) can be solved for the stress and displacement distributions along the pile.

For the case of $h = L$, that is, an end bearing pile resting on a rigid or stiffer stratum, an alternative (and probably more reliable) approach is described in Sec. 10-3.2.

10-3.2 Pile Resting on a Stiffer Stratum

A great number of piles are installed so that the tip bears onto a stratum which is stiffer than the soil along the shaft of the pile. Such piles are often designated as *end bearing* or *point bearing piles*, but the results of several analyses and field and laboratory measurements have shown that a significant proportion of the load may be transferred from the pile shaft to the surrounding soil.

To analyze the behavior of such piles, the analysis described in the preceding section for a floating pile must be modified to allow for the effect of the stiffer bearing stratum. The same assumptions are made again for the pile and soil behavior, but, in addition, the bearing stratum is assumed to be an ideal elastic half-space with constant parameters E_b and ν_b. The problem is defined in

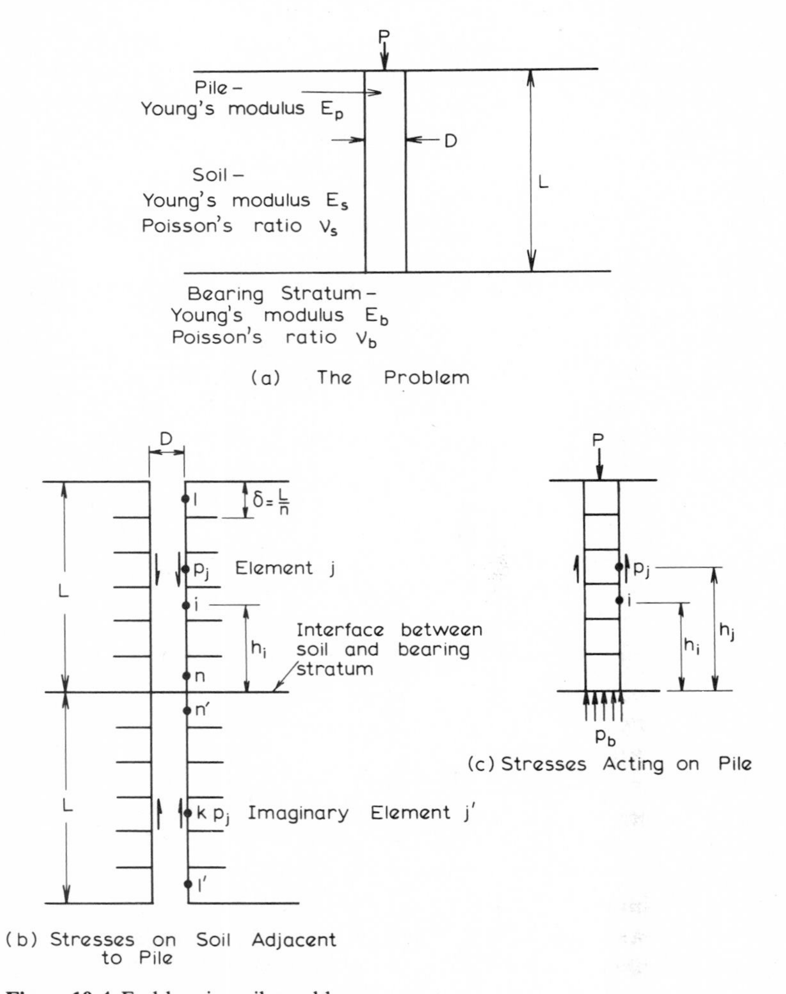

Figure 10-4 End bearing pile problem.

Fig. 10-4*a*. To obtain the solution for the unknown stresses on the pile shaft and tip and the corresponding pile movements, compatability of the vertical displacements of the pile and adjacent soil is again assumed.

Soil displacement equations To determine properly the displacement in the soil surrounding the pile, it would be necessary to use equations for loading within a two-layer elastic system. Since suitable analytical solutions to this problem are not available, Mindlin's equations for displacements due to loading within a half-space can be utilized in an approximate manner. To allow for the reduction in soil displacements due to the presence of the bearing stratum, a correction, based on the mirror image approximation suggested by D'Appolonia and Romualdi,[2] is used. Referring to Fig. 10-4*b*, pile element j is mirrored in the soil bearing stratum interface by an imaginary pile element j, which is acted on by stress kp_j in the opposite direction to the stress p_j on the real element j. The limiting values of k are $k = 0$ for a floating pile ($E_b = E_s$), where the stratum has no effect on soil displacement, and $k = 1$ for a pile resting on a rigid stratum ($E_b = \infty$), in which case the condition of zero tip displacement is satisfied.

If we take downward displacements of the soil as positive, the displacement ${}_s\rho_{ij}$ of the soil at i due to the stress on the real element j and the imaginary element j' is

$$ {}_s\rho_{ij} = \frac{D}{E_s} p_j(I_{ij} - kI'_{ij}) \tag{10-19} $$

where I_{ij} is the vertical displacement factor for i due to shear stress on element j, as before, and I'_{ij} is the vertical displacement factor for i due to shear stress on imaginary element j' (calculated for a distance $L + h_i$ from the imaginary soil surface). If the simplifying assumption is made that the influence of the stress on the pile tip has negligible effect on the soil displacement ${}_s\rho_i$ at i, then

$$ {}_s\rho_i = \frac{D}{E_s} \sum_{j=1}^{n} (I_{ij} - kI'_{ij}) \tag{10-20} $$

and for all n elements along the pile shaft

$$ \{_s\rho\} = \frac{D}{E_s} [I_s - kI'_s]\{p\} \tag{10-21} $$

where $\{_s p\}$ and $\{p\}$ are vectors of soil displacement and shear stress, respectively (of order n), and $[I_s - kI's]$ is the $n \times n$ matrix of values of $I_{ij} - kI'_{ij}$. Equation (10-21) is analogous to Eq. (10-3) for a floating pile.

The values of I_{ij} and I'_{ij} are evaluated from the Mindlin equation, as described in the previous section.

Pile displacement equation The displacement of each element of the pile itself can be divided into three components: displacement due to shear stresses along the pile, displacement due to applied axial load P, and displacement due to the finite compressibility of the bearing stratum. Assuming again axial compression of the pile, it can be shown[15] that the displacement vector $\{_p\rho\}$ for the n elements along the shaft is

$$\{_p\rho\} = -\left[\frac{1}{E_p R_A}[D_p] + \frac{\pi(1-v_b^2)}{E_b}\frac{L}{n}\frac{D}{D_b}[X]\right]\{p\} + \frac{P}{E_p R_A}\{h\} + \frac{P}{D}\frac{(1-v_b^2)}{E_b}\frac{D}{D_b}\{W\} \tag{10-22}$$

where $[D_p] = n \times n$ matrix of pile displacement factors, with

$$D_{pij} = \begin{cases} \dfrac{4\delta h_j}{D} & \text{for } i \leqslant j \\ \dfrac{4\delta h_i}{D} & \text{for } i \geqslant j \end{cases}$$

$\delta = L/n$

h_i, h_j = distances from bearing stratum to points i and j (see Fig. 10-4c)

$\{h\}$ = n column vector of values of h_i

$[X]$ = $n \times n$ matrix, every term of which is unity

$\{p\}$ = n column vector of p_j values

$\{W\}$ = n column vector of values of unity

R_A = area ratio of pile [Eq. (10-9)]

Displacement compatability Assuming no slip at the pile-soil interface gives $\{_p\rho\} = \{_s\rho\}$, so that from Eqs. (10-21) and (10-22)

$$\left[\frac{1}{KD}[D_p] + \frac{\pi(1-v_b^2)}{n}\frac{L}{D_b}\frac{E_s}{E_b}[X] + [I - kI']\right]\{p\} = \frac{P}{D^2}\left[\frac{4}{\pi KD}\{h\} + (1-v_b^2)\frac{E_s}{E_b}\frac{D}{D_b}\{W\}\right] \tag{10-23}$$

where K is the pile stiffness factor, as defined in Eq. (10-16).

For a chosen initial value of k, Eq. (10-23) can be solved to give the n unknown stresses p_j. The stress acting on the pile base can then be evaluated from the equilibrium equation

$$P = \sum_{j=1}^{n} p_j \pi D \frac{L}{n} + p_b \frac{\pi D_b^2}{4} \tag{10-24}$$

Once the solutions for the chosen value of k have been obtained, a closer estimate of k can be found by examining compatability between the displacement of the soil and the bearing stratum at the pile base. Poulos and Mattes[15] show that the next approximation for the value of k is then given by

$$k = 1 - \frac{\pi}{4}\frac{E_s}{E_b}\frac{D_b}{D}\frac{(1-v_b^2)p_b}{\sum_{j=1}^{n} p_j I_{bj}} \tag{10-25}$$

Equations (10-23) and (10-24) can now be solved again, using this new value of k, and the process repeated until convergence in the value of k is obtained. In many cases, only two or three iteration cycles give adequate convergence of k.

The above analysis involves a number of approximations but has nevertheless been found to give solutions for the settlement of a floating pile ($E_b/E_s = 1$) which are within 10 percent of the values from the floating pile analysis in the previous sections. The errors involved in applying the analysis to piles bearing on stiffer strata ($E_b/E_s > 1$) should be considerably less.

10-3.3 Accuracy of Elastic Single Pile Solutions

The accuracy of solutions from the preceding analyses for single piles will now be considered in relation to:

1. The effect of the number of elements
2. The effect of ignoring the requirements of compatability of radial displacements
3. Comparisons with solutions from the finite element method

Effect of number of elements For a floating pile in a semi-infinite mass, the effect of the number of elements on the head displacement and proportion of load at the pile tip is shown in Table 10-1. For the compressible pile ($K = 100$), the solutions are reasonably sensitive to the number of elements, but for $K \geqslant 1000$, even the use of only five elements along the shaft gives a remarkably good solution. In general, unless the pile is highly compressible, 10 elements should be ample to ensure solutions of acceptable accuracy.

Table 10-1 Effect of number of elements

$\frac{L}{D} = 50 \qquad \nu_s = 0.5$

Pile-stiffness factor K	No. of elements n	Displacement influence factor I_s	Proportion of load taken by pile base
100	5	0.1167	0.0069
	10	0.0953	0.0059
	20	0.0858	0.0053
1,000	5	0.0357	0.028
	10	0.0339	0.025
	20	0.0330	0.023
20,000	5	0.0226	0.038
	10	0.0225	0.034
	20	0.0224	0.031

Table 10-2 Effect of number of base elements

$\frac{L}{D} = 5 \quad \nu_s = 0.5 \quad$ incompressible pile ($K = \infty$)

displacement $\rho = \frac{P}{DE_s} I_\rho$

No. of base annuli	Displacement factor I_ρ	Proportion of base load	Remarks
1	0.1107	0.126	Each annulus assumed uniformly loaded
2	0.1100	0.164	
3	0.1098	0.173	
5	0.1096	0.181	
1	0.1092	0.185	Factors of $\pi/4$ applied to influence factor I_{bb} [Eq. (10-8)]

The effect of using a number of annular elements instead of a simple circular element to represent the pile base is illustrated in Table 10-2 for $L/D = 5$. If the annuli are treated as being uniformly loaded, the number of annuli used has a relatively large effect on the proportion of base load but only a small effect on displacement. If, however, only a single base annulus is considered but allowance is made for rigidity by introducing a factor of $\pi/4$ [Eq. (10-8)], the resulting solutions are almost identical with those using five annuli. The recommended procedure of using a single base element and applying a rigidity correction factor therefore appears to be quite adequate.

Effect of ignoring radial displacement compatability Complete solutions for the settlement of a pile, in which both vertical and radial displacement compatability are considered, have been presented by Butterfield and Banerjee[13] and Mattes.[12,16] Comparisons between the complete solutions and solutions in which only vertical displacement compatability is considered are shown in Table 10-3 for the top displacement of the pile.[12] Only for relatively short piles ($L/D < 25$) does the inclusion of radial displacement compatability have any effect on the solutions, and even in such cases the effect is unimportant from a practical point of view. It therefore appears quite adequate to employ analyses in based on compatability of vertical displacement only.

Comparison with finite element solutions A parametric study of the settlement of a pile in a two-layer system based on a finite element analysis has been presented by Lee.[17] Table 10-4 gives a comparison between Lee's solutions for a floating pile in a uniform mass and the corresponding solutions from the analysis presented herein. There is close agreement between the two series of solutions, and the

Table 10-3 Effect on pile displacement of considering radial displacement compatability[18]

Displacement $\rho = \frac{P}{LE_s} I_\rho$

$\frac{L}{D}$	K	Top displacement-influence factor I_ρ	
		Vertical displacement compatability only	Vertical and radial displacement compatability
10	100	1.793	1.782
	1,000	1.378	1.448
25	100	3.559	3.542
	1,000	3.181	3.160
100	100	10.670	10.488
	1,000	5.220	5.140
	20,000	2.758	2.712

difference that does exist may well arise from numerical inaccuracies in one or both of the solutions.

A further comparison with Lee's solutions is shown in Table 10-4 for a pile bearing on a stiffer stratum. The agreement is again reasonable, suggesting that the analysis based on the Mindlin equation should give results of adequate accuracy in practical situations provided that severe variations in subsoil conditions do not occur along the pile.

Table 10-4 Comparisons between elastic and finite element solutions for pile settlement

$K = 1000 \qquad \nu_s = 0.4 \qquad p = \frac{P}{DE_s} I_\rho$

Floating pile in semi-infinite mass			Endbearing pile			
	I_ρ				I_ρ	
$\frac{L}{D}$	Lee[17]	Author	$\frac{L}{D}$	$\frac{E_b}{E_s}$	Lee[17]	Author
3.5	0.267	0.258	5	10	0.078	0.075
5.0	0.211	0.205		100	0.014	0.016
10.5	0.115	0.112	15	100	0.020	0.020
15.0	0.103	0.100				
19.5	0.094	0.092				

10-3.4 Modification to Basic Analyses for Single Pile

Pile-soil slip The analyses presented so far require that no slip occurs at the pile-soil interface. However, since real soils have a finite shear strength and the pile-soil interface has a finite adhesive strength, slip or local yield will occur when the shear stress reaches the adhesive (or yield) strength. By using a method similar to that described by D'Appolonia and Romualdi,[2] Salas,[18] and Poulos and Davis,[5] the elastic analyses can be modified to take account of possible slip provided that the following assumptions are made:

1. Slip occurs at the pile-soil interface when the average shear stress on any element reaches the limiting value τ_a.
2. Displacements anywhere in the soil due to the limiting stress τ_a are still given by elastic theory.
3. Failure of the tip or base of the pile occurs when the base pressure reaches the ultimate bearing capacity of the base.

In carrying out the modified analysis, it is convenient to restate the elastic equation governing pile behavior in terms of displacement rather than shear stress. Thus using the floating pile analysis as an example, Eq. (10-15) in terms of the shear-stress vector $\{p\}$ alters to the following form in terms of the displacement vector $\{\rho\}$:

$$\frac{E_s}{D}\left[[I_s]^{-1} - \frac{KD^2n^2}{4L^2}[I_p]\right]\{\rho\} = \{Y\} \tag{10-26a}$$

or

$$[Z]\{\rho\} = \frac{D}{E_s}\{Y\} \tag{10-26b}$$

where $[I_s]^{-1}$ is the inverse of matrix $[I_s]$ and

$$[Z] = [I_s]^{-1} - \frac{KD^2n^2}{4L^2}[I_p]$$

The analysis is carried out incrementally, increasing the applied load P successively. For a given value of P, Eq. (10-26b) is solved on the assumption that all elements are elastic. From the resulting solution for $\{\rho\}$, the shear stresses p are calculated from Eq. (10-13) or (10-3) and are then compared with the specified limiting stresses τ_a. At an element where the computed stress exceeds τ_a, the displacement-compatability equation for that element, i.e., the appropriate row of the matrix $[Z]$ in Eq. (10-26b), is replaced by the pile displacement equation for that element, i.e., the appropriate row of the matrix in Eq. (10-13), setting the shear stress at that element equal to τ_a. For example, if an element i has slipped, the elements Z_{ij} in row i of the matrix Z in Eq. (10-26b) are replaced by the elements ${}_pI_{ij}$ of matrix $[{}_pI]$, while the element $(D/E_s)Y_i$ on the right-hand side is replaced by $(4L^2/E_p R_A Dn^2)(\tau_{ai} - Y_i)$, where τ_{ai} is the value of τ_a at element i. The modified system of equations is now re-solved. and the procedure is repeated until the computed values of shear stress do not exceed the limiting values τ_a. By successively increasing the applied load P until all elements have failed, a load settlement curve to failure can be obtained.

Analysis of negative friction effects If a pile is situated in a soil mass subjected to consolidation or settlement from some source, additional forces are induced in the pile. This downdrag effect is commonly termed *negative friction* and can have very serious effects on the performance of a pile foundation if not allowed for in design.

The effects of negative friction can readily be analyzed by the approach employed in this chapter.[19] The problem is illustrated in Fig. 10-5 for an end bearing pile. For the initial elastic analysis, before pile-soil slip effects are considered, compatability of soil and pile displacement at each element of the pile is imposed. The pile displacement equation remains the same as that for a pile subjected to axial load [Eq. 10-22]. The soil displacements, however, must now include the external soil movements due to consolidation as well as the movements due to the shear stresses acting on the interface.

With downward displacements taken as positive, the soil displacement equation becomes [in lieu of Eq. (10-21)]

$$\{_s\rho\} = \{S_0\} - \frac{D}{E_s}[I_s - kI_s']\{p\} \tag{10-27}$$

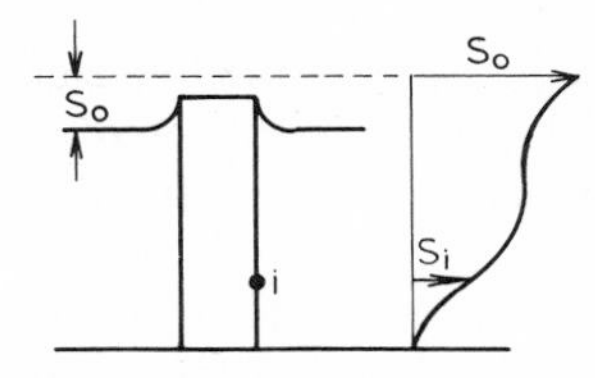

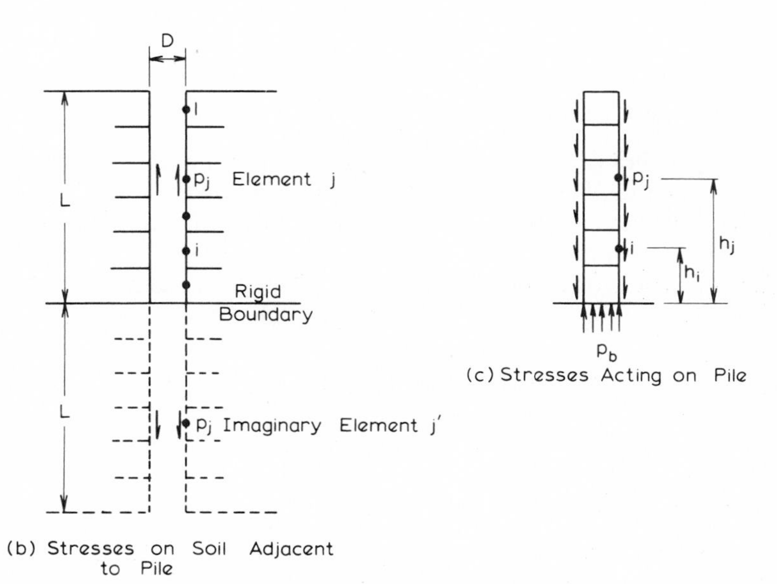

Figure 10-5 Problem of negative friction on end bearing pile.

where $\{S_0\}$ is the vector of soil movements due to external source, e.g., consolidation.

Because most negative friction problems occur with piles in soft soils, the effects of pile-soil slip should always be included in the analysis. The resulting solution will give the displacements and shear stresses (and hence loads) along the pile due to negative friction for a given set of input soil movements. Simultaneous axial loading may be considered, and by combining the pile analysis with a consolidation analysis to obtain the variation of $\{S_0\}$ with time the development of downdrag forces and movements in the pile with time can be computed. Also by considering a floating pile in a soil mass subjected to upward movements, the case of a pile or pier in a swelling soil can be investigated.[20]

Other modifications The basic analyses have been formulated in terms of a uniform pile with provision for a thin enlarged base, but extensions can readily be made to allow for cases where the shaft is not of uniform diameter or the pile is attached to a pile cap resting on the soil surface.

For piles having a nonuniform shaft diameter, the relative diameters of the various shaft elements are considered when calculating the pile and soil displacements. In cases where the shaft diameter of an element is less than that of the element above it, the stress on the annular area at the junction of the two elements must be included as an additional variable. Examples of the analysis of underreamed and step-taper piles using the above approach have been given by Poulos.[21]

For a pile with a rigid cap resting on the soil surface, uniformly loaded annular elements are included in the analysis to represent the pile cap. Compatibility of pile and soil displacements is considered at these cap elements as well as along the pile. Details of such an analysis are given by Poulos[22] and Butterfield and Banerjee.[23]

10-3.5 Analysis of Pile Groups

The analyses for single piles can readily be extended to pile groups if the simplifying assumption is made that the reinforcing effect of the piles on the soil displacements is negligible. The analysis will be described for uniform floating piles but can be adapted to end bearing piles and piles connected by a raft.

For a group of m identical piles, each divided into n elements, the soil displacement equation, analogous to Eq. (10-3) for a single floating pile, is

$$\{_s\rho\} = \frac{D}{E_s}[IG]\{p\} \tag{10-28}$$

where $\{_s\rho\} = m(n+1)$ soil displacement vector for all elements of all piles
$\{p\} = m(n+1)$ shear-stress vector
$[IG] = m(n+1)$ square matrix of soil displacement factors

Each term of matrix $[IG]$ is evaluated by double integration of the Mindlin point load equation [Eq. (10-5)], in a manner similar to that described previously for a single pile.

The pile displacement equation for each pile in the group is identical with the single pile equation (10-15) for a floating pile. By equating soil and pile displacements at each element for all piles a system of $m(n + 1)$ equations is obtained. In addition, further conditions regarding the pile group behavior must now be specified since the load on each pile is generally not known. The simplest conditions of practical interest are either that all piles settle equally, i.e., a rigid pile cap connects the pile heads above the surface, or that all piles have known loads, i.e., a flexible pile cap connects the pile heads. Once such conditions are specified, an overall system of $m(n + 2)$ equations is obtained and can be solved for the $m \times n$ unknown shear stresses, m unknown base pressures, and m unknown pile loads. Detailed treatment of the above type of approach has been presented by Pichumani and D'Appolonia[24] and Butterfield and Banerjee.[23]

Simplified approach Use can be made of symmetry (if applicable) to reduce the number of equations to be solved for a group, but even then a large number of equations may remain to be solved. In a simpler approach adopted by Poulos[25] and Poulos and Mattes[26] the above analysis need only be carried out for a group of two identical equally loaded piles. From this analysis, an *interaction factor* α can be derived

$$\alpha = \frac{\text{increase in settlement of each pile}}{\text{settlement of single isolated pile}} \tag{10-29}$$

By varying the spacing between the two piles, a relationship between α and the pile spacing, expressed in dimensionless form, can be obtained. Series of such relationships have been presented[25,26] for various values of L/D and K.

By analyzing larger groups with circular symmetry, it was demonstrated that the interaction factors could be superimposed to give the increase in settlement (and hence the settlement) of a pile due to the other piles in the group. For example, for a four pile square group with a center-to-center spacing s, the settlement of each pile is given by

$$\rho = P\rho_1(1 + 2\alpha_1 + \alpha_2) \tag{10-30}$$

where ρ_1 = settlement of a single isolated pile under *unit* load
P = load on each pile
α_1 = interaction factor for spacing s between two piles
α_2 = interaction factor for spacing $s\sqrt{2}$ between two piles

Assuming that superposition can be applied to any group,† the settlement ρ_k of any pile k in a group of m piles can be written in general form as

$$\rho_k = \rho_1\left(\sum_{\substack{j=1 \\ j \neq k}}^{m} P_j\alpha_{kj} + P_i\right) \tag{10-31}$$

† Superposition does not apply a priori to all groups, as the presence of additional piles in a mass changes the overall elastic system.

where P_j is the load on pile j and α_{kj} is the interaction factor for two piles corresponding to spacing between piles k and j. Equation (10-31) can be applied to all m piles on the group, giving

$$\{\rho\} = \rho_1[\alpha]\{P_j\} \tag{10-32}$$

where $\{\rho\}$ = m vector of pile settlements
$\{P_j\}$ = m vector of pile loads
$[\alpha]$ = m square matrix of interaction factors (note that $\alpha_{kk} = 1$)

In addition, the vertical load-equilibrium equation is

$$P_G = \sum_{j=1}^{m} P_j \tag{10-33}$$

where P_G is the total group load. Equations (10-32) and (10-33) can be solved for the simple conditions of either equal settlement of all piles or known load in all piles. The settlement and load distribution can thus be obtained.

A convenient way of expressing the results of a pile-group analysis is in terms of the settlement ratio R_s, where

$$R_s = \frac{\text{group settlement}}{\text{settlement of single pile at same } \textit{average} \text{ load as group}} \tag{10-34}$$

An alternative expression is in terms of the group reduction factor R_G, where

$$R_G = \frac{\text{group settlement}}{\text{settlement of single pile at same } \textit{total} \text{ load as group}} \tag{10-35}$$

For example, if the group settlement from the above analysis is found to be ρ_G, R_s and R_G are calculated as follows:

$$R_s = \frac{m\rho_G}{\rho_1 P_G} \tag{10-36}$$

and

$$R_G = \frac{\rho_G}{\rho_1 P_G} \tag{10-37}$$

R_s expresses the increase in settlement of a single pile due to effects of interaction, whereas R_G expresses the decrease in settlement due to the use of a group rather than a single pile. R_G is strictly meaningful only for an elastic soil where there is a linear relationship between load and settlement and failure of the single pile under the group load does not develop. R_G is, however, a useful quantity as it always lies in the range $1/m \leqslant R_G \leqslant 1$; also, as is evident from Eqs. (10-36) and (10-37), R_G and R_s are simply related:

$$R_s = mR_G \tag{10-38}$$

While R_s is likely to be the more useful and familiar quantity for practical problems, there is some advantage in using R_G for examining the comparative behavior of pile groups, since R_G represents the settlement of a group for a single pile settlement of unity and thus gives a direct measure of the relative settlement of groups containing different numbers of piles and subjected to the same total load.

The group analysis described above can readily be applied to groups of end bearing piles and to piles connected by a cap or raft which rests on the surface.[23,27]

10-3.6 Typical Numerical Results

In this section, some parametric solutions will be presented to illustrate the use of the analyses in defining the important factors influencing the settlement behavior of piles and pile groups. A numerical example will be given in Sec. 10-4 to illustrate the use of these parametric solutions for rapid preliminary estimates of settlement.

Parametric solutions for a single pile Parametric solutions for the settlement at the top of a compressible floating pile in a half-space were presented by Mattes and Poulos[6] and are shown in Fig. 10-6. This figure shows the increase in settlement

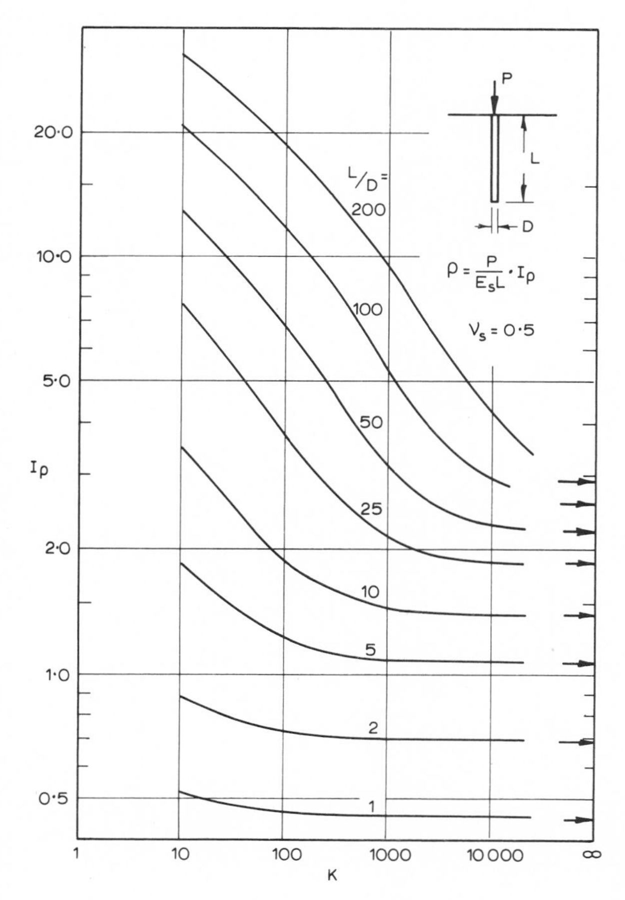

Figure 10-6 Top displacement factors for single floating pile.

with increasing pile compressibility (decreasing K) and the increase in settlement if, for a constant length L, the slenderness ratio L/D is increased.

For an end bearing pile resting on a stiffer stratum, solutions for the settlement of the tip and top of the pile have been obtained by Poulos and Mattes[15] and Mattes,[12] and the solutions for the top are shown in Fig. 10-7, in terms of an influence factor I_ρ, which multiplies the elastic shortening of the pile as if it were a simple column subjected to the applied load. Thus, I_ρ represents the *movement ratio* referred to by Focht,[28] who reported measured values between 0.4 and 2.2 from a number of pile load tests. Figure 10-7 reveals that the movement ratio I decreases with decreasing pile stiffness factor K and with increasing L/D and modular ratio E_b/E_s.

An example of the influence of pile-soil slip is shown in Fig. 10-8 for a floating pile having $L/D = 25$. This figure expresses the ratio M_s of the "elastic" settlement, assuming that slip does not occur, to the actual settlement allowing for the

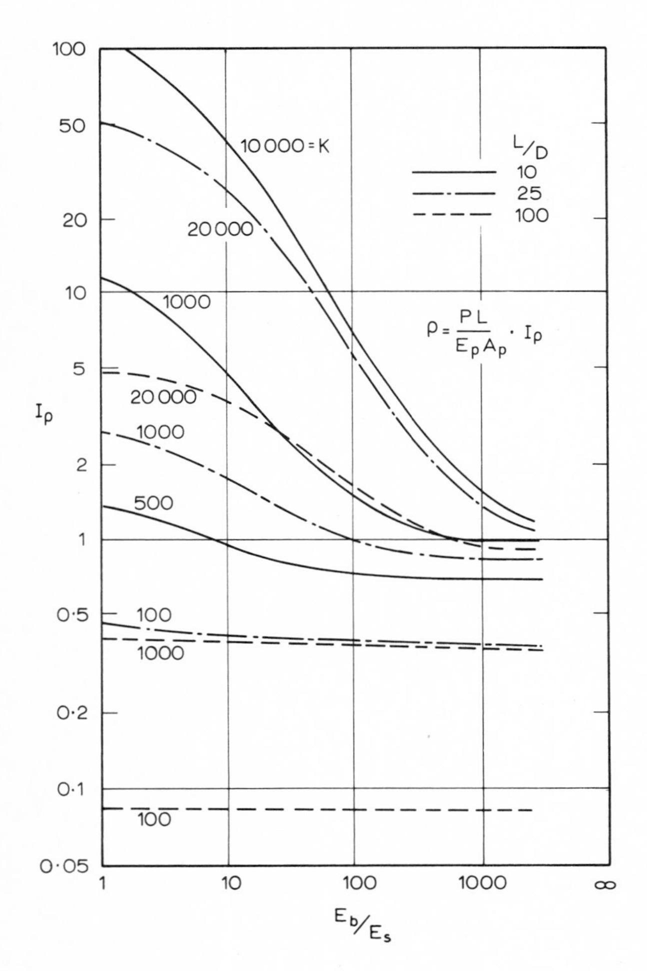

Figure 10-7 Top displacement factors for single end bearing pile.

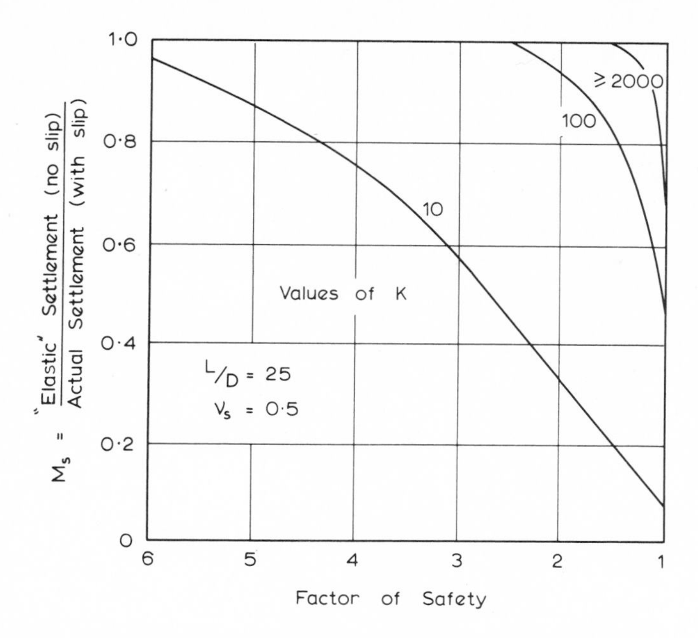

Figure 10-8 Settlement modification factor M_s for pile-soil slip.

effects of slip as a function of the factor of safety against undrained failure. Unless the pile is highly compressible ($K < 100$), M_s is not much less than 1.0 at factors of safety even as low as 1.5; that is, pile-soil slip does not produce a significant increase in settlement at normal working loads. Figure 10-8 therefore has the important implication that, for most normal piles, a purely elastic analysis should give an adequate prediction of settlement.

For larger diameter piers, in which a large proportion of the load is transmitted to the base, the effects of pile-soil slip may be more significant. For such cases, a simplified procedure for constructing the load settlement curve to failure has been proposed by Poulos[29] utilizing a series of parametric elastic solutions for settlement and base load of the pile.

A significant finding from the elastic solutions is that most of the settlement of a pile is immediate settlement, occurring as soon as the load is applied, and only a relatively small proportion of the final settlement, generally between 5 and 30 percent, can be attributed to consolidation. This result can be derived by assuming the soil to be an ideal two-phase elastic material. The immediate settlement ρ_i is calculated using the undrained Young's modulus of the soil E_u and the settlement influence factor for the undrained Poisson's ratio ν_u (= 0.5 for a saturated soil). The total final settlement ρ_{TF} is calculated for the drained Young's modulus $\bar{E}_s$ of the soil and the settlement influence factor for the drained Poisson's ratio $\bar{\nu}_s$. For an ideal two-phase soil, $E_u = 3\bar{E}_s/2(1 + \bar{\nu}_s)$, so that the ratio of immediate to total final settlement can be determined. These theoretical conclusions regarding the importance of immediate settlement are supported by a considerable number of field data from maintained loading tests.

Parametric solutions for pile groups An extensive series of solutions for the group reduction factor R_G [Eq. (10.37)] as a function of relative pile spacing s/D has been obtained[26] both for floating piles in a deep soil layer and for end bearing piles resting on a rigid bearing stratum. Some of these solutions are plotted in Fig. 10-9,

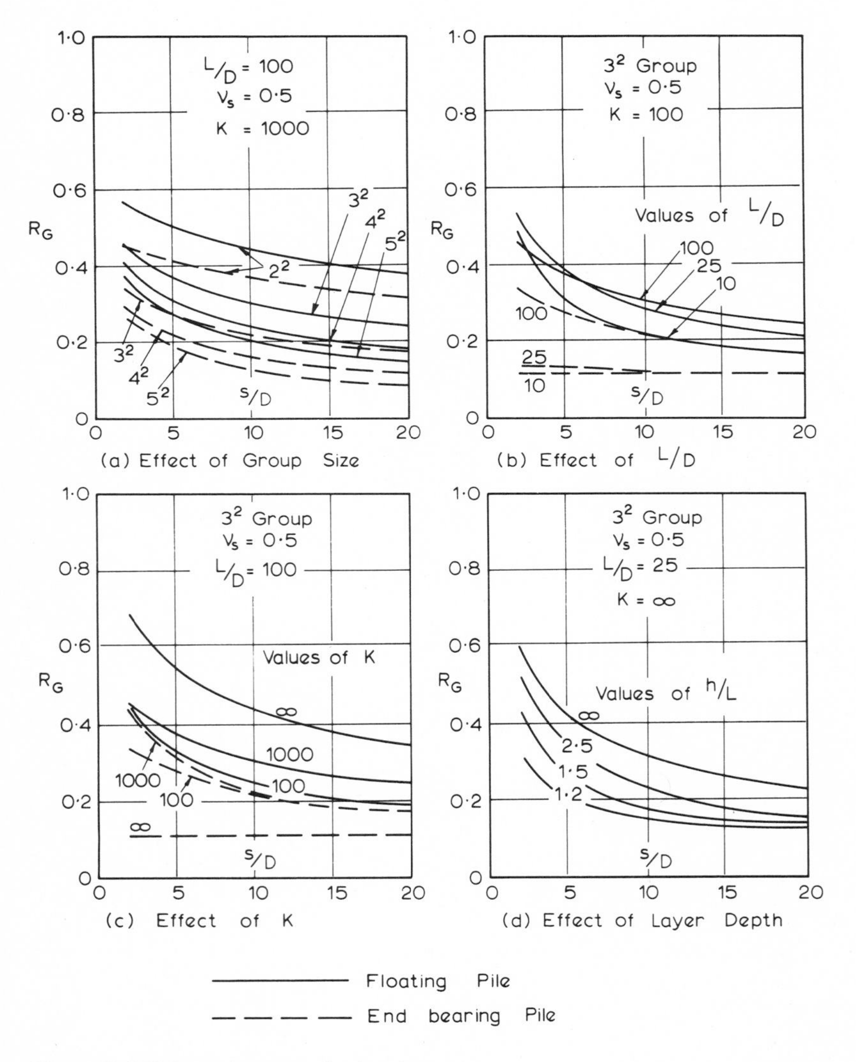

Figure 10-9 Values of group reduction factor R_G.

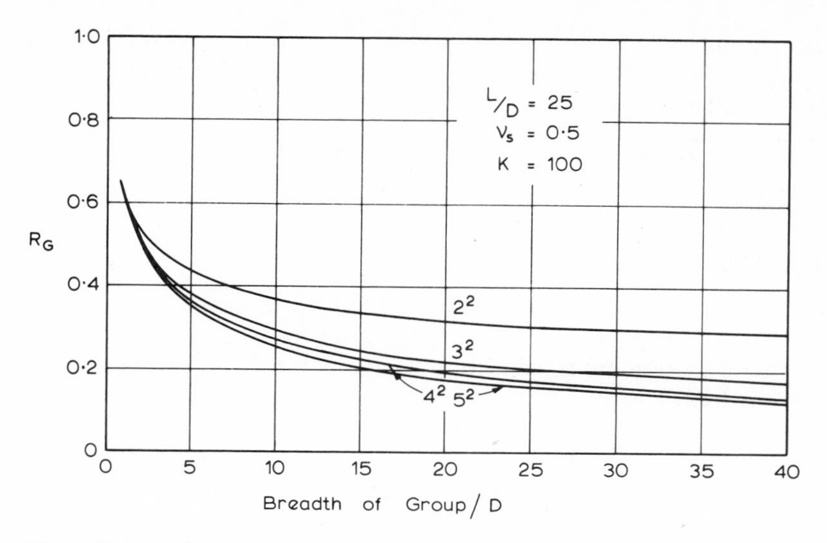

Figure 10-10 R_G vs. group breadth; floating pile groups.

showing the effect of the significant variables on R_G. The following observations can be made:

1. R_G is generally less for end bearing piles than for friction piles.
2. R_G decreases as the number of piles in the group increases.
3. R_G generally increases as L/D increases.
4. R_G for floating piles increases as K increases, i.e., as the piles become relatively less compressible, whereas R_G for end bearing piles decreases as K increases.
5. The effect of a finite layer is to decrease interaction between a group of floating piles.

If R_G is plotted against the relative breadth of a group B/D instead of the relative pile spacing s/D, the effect of the number of piles is greatly reduced. Such a plot is given in Fig. 10-10, and it appears that for groups containing more than 25 piles a common limiting curve of R_G vs. B/D, coincident with the curve for the 5^2 group, can be used over a practical range of group breadths. Figure 10-10 also implies that a reduction in the settlement of a pile group may be more economically attained by increasing the pile spacing and hence the group breadth rather than increasing the number of piles in the group and maintaining the same group breadth.

10-4 APPLICATIONS AND DESIGN ANALYSIS

Attention will now be turned to the application of the theory to practical problems involving real soils. The first application to be described is the use of the theory for determining the average soil modulus from the results of a pile load test.

This procedure, and the subsequent use of the modulus so deduced to calculate the settlement of a pile group will be illustrated by an example. To demonstrate the applicability of the theory to field situations, some comparisons will then be described between pile behavior observed from field tests and that predicted from the theory.

10-4.1 Determination of Soil Modulus from Pile Load Tests

In using the theoretical solutions to predict pile settlements the choice of an appropriate value of Young's modulus of the soil E_s is of prime importance. It has been found[12] that conventional triaxial tests give values of E_s which are far too low, typically one-fourth to one-tenth of the value required to give agreement between measured and predicted settlements. It may be possible to obtain more appropriate values of E_s in the laboratory by using an apparatus similar to that described by Coyle and Reese.[8] However, because the complex initial and final stress states within the soil surrounding the pile cannot easily be reproduced in the laboratory, the most reliable means of obtaining E_s is to carry out a pile load test and backfigure E_s from the measured settlement, using the theoretical solutions. The settlement of piles of different proportions and of pile groups can then be estimated.

The pile to be used in the test should preferably be of length similar to the foundation piles for which predictions are required, but it may be of smaller diameter, especially if the foundation is to consist of large-diameter bored piers. Ideally, two piles of different proportions should be tested, so that both E_s and Poisson's ratio ν_s can be backfigured; however, the influence of ν_s on the theoretical solutions is relatively small, and ν_s can generally be assumed. Typical values of ν_s are 0.3 for sands, 0.5 for clays under undrained conditions, and 0.2 to 0.4 for clays under drained conditions, the higher values for softer soils. If a maintained-load test is carried out, the immediate settlement together with the theoretical solutions for undrained Poisson's ratio ν_u are used to deduce the undrained modulus E_u. The final settlement together with the theoretical solutions for drained Poisson's ratio $\bar{\nu}_s$ are used to deduce the drained modulus $\bar{E}_s$. The method of using the theory to deduce these values will now be described.

Considering the case of a floating pile, the theoretical solutions, e.g., Fig. 10-6, for the settlement at the head of a pile of length L under a load P can be expressed in dimensionless form as

$$\rho = \frac{P}{LE_s} I_\rho \tag{10-39}$$

where I_ρ is the settlement influence factor. I_ρ is a function of pile geometry, ν_s, and the pile stiffness factor K (and hence E_s). Equation (10-39) can be rewritten

$$E_s = I_\rho \frac{P}{L\rho} = f(K) \frac{P}{L\rho} \tag{10-40}$$

Also, from the definition of K [Eq. (10-16)],

$$E_s = E_p R_A \frac{1}{K} \tag{10-41}$$

Equations (10-40) and (10-41) can be solved (most conveniently by graphical means) for the two unknowns K and E_s. For end bearing piles, a similar approach is followed except that the settlement influence factor or movement ratio is now also a function of E_b/E_s.

Poulos[29] has analyzed a number of pile load tests and has produced rough correlations between E_s and undrained cohesion c_u for driven and bored piles in clay and ranges of values of E_s for piles driven into sand.

To illustrate the application of the procedure described above, a simple example is presented below. To further illustrate the use of the pile settlement theory, the prediction of the settlement of a pile group using the deduced values of E_s is then described.

Example 10-1 A concrete test pile 0.3 m in diameter and 15 m long is driven into a deep clay layer. Under the average working load of 50 t an immediate settlement of 10 mm and a total final settlement of 14 mm was recorded. It is desired to backfigure the average value of E_s for the clay, for both undrained and drained conditions, assuming that Young's modulus of the pile is 2×10^5 kg_f/cm^2.

As the second part of the example, the backfigured value of E_s from the pile load test will be used to predict the immediate and final settlements of a 3^2 group of steel tube piles 20 m long and 0.5 m OD with 15 mm wall thickness subjected to a total load of 450 t. The piles are spaced at 1.5 m center to center and are set into a rigid concrete cap.

Interpretation of pile load test Considering first the undrained case, from Eq. (10-40)

$$E_u = \frac{(50)(1000)}{(1500)(1.0)} I_\rho \qquad kg_f/cm^2$$

$$= 33.33 I_\rho \qquad kg_f/cm^2 \tag{10-42}$$

From Eq. (10-41)

$$E_u = \frac{(2 \times 10^5)(1)}{K} \qquad kg_f/cm^2 \tag{10-43}$$

From Fig. 10-6, values of I_ρ for $\nu_s = \nu_u = 0.5$ and $L/D = 15/0.3 = 50$ as a function of K (and hence E_u) can be obtained; they are tabulated in Table 10-5 and plotted in Fig. 10-11. Equation (10-43) can also be plotted, and the intersection of the two plots gives the required value of E_u, 87 kg_f/cm^2. For the drained case, Eq. (10-40) becomes

$$\bar{E}_s = 23.8 I_\rho \qquad kg_f/cm^2 \tag{10-44}$$

where I_ρ is now for $\nu_s = \bar{\nu}_s$, assumed here to be 0.3. The plot of $\bar{E}_s$ vs. K derived from Eq. (10-44) is shown in Fig. 10-11. Equation (10-43) remains the same as for the undrained case, and $\bar{E}_s$ is thus found to be 54 kg_f/cm^2.

Table 10-5 Modulus calculation for Example 10-1

		Undrained case		Drained case	
K	E_s kg_f/cm^2 [Eq. (10-43)]	I_ρ ($\nu_s = 0.5$)	E_u kg_f/cm^2 [Eq. (10-42)]	I_ρ ($\nu_s = 0.3$)	$\bar{E}_s$ kg_f/cm^2 [Eq. (10-44)]
10,000	20	2.26	75.3	2.10	50.0
5,000	40	2.39	79.7	2.22	52.8
2,000	100	2.68	89.3	2.52	60.0
1,000	200	3.06	102.0	2.91	69.3
500	400	3.76	125.3	3.60	85.7

Prediction of pile-group settlement For the individual piles,

$$\frac{L}{D} = \frac{20}{0.5} = 40 \qquad R_A = \frac{(\pi)(0.5)(0.015)}{(\pi)(0.25)^2} = 0.12$$

Considering, first, immediate settlement and assuming that $E_p = 2 \times 10^6$ kg_f/cm^2 gives

$$\begin{aligned} K &= \frac{E_p}{E_u} R_A \\ &= \frac{(2 \times 10^6)(0.12)}{87} \\ &= 2760 \end{aligned}$$

For the single pile, from Fig. 10-6

$$\rho_i = \frac{P}{LE_u} I_\rho$$

and for $L/D = 40$ and $K = 2760$, $I_\rho = 2.32$. Substitution into the above equation, using the average pile load of $P = 50$ t, gives $\rho_i = 6.7$ mm. Group action must now be taken into account. From Fig. 10-9, by interpolation, the group reduction factor R_G is estimated to be 0.48. Thus, the settlement ratio R_s is $R_s = (9)(0.48) = 4.32$, and the immediate settlement of the group ρ_{iG} is thus

$$\begin{aligned} \rho_{iG} &= 4.32\rho_i = (4.32)(6.7) \\ &= 29.0 \text{ mm} \end{aligned}$$

For the total final settlement,

$$\begin{aligned} K &= \frac{E_p}{E_s} R_A \\ &= \frac{(2 \times 10^6)(0.12)}{54} \\ &= 4450 \end{aligned}$$

Following the same procedure as for ρ_i, I_ρ is found to be 2.21, and using $E_s = \bar{E}_s$, the total final settlement of the single pile ρ_{TF} is computed to be 10.2 mm.

The value of R_G is found to be 0.50, so that $R_s = (0.50)(9) = 4.50$, and the total final settlement of the group ρ_{TFG} is

$$\rho_{TFG} = (4.50)(10.2) = 46.0 \text{ mm}$$

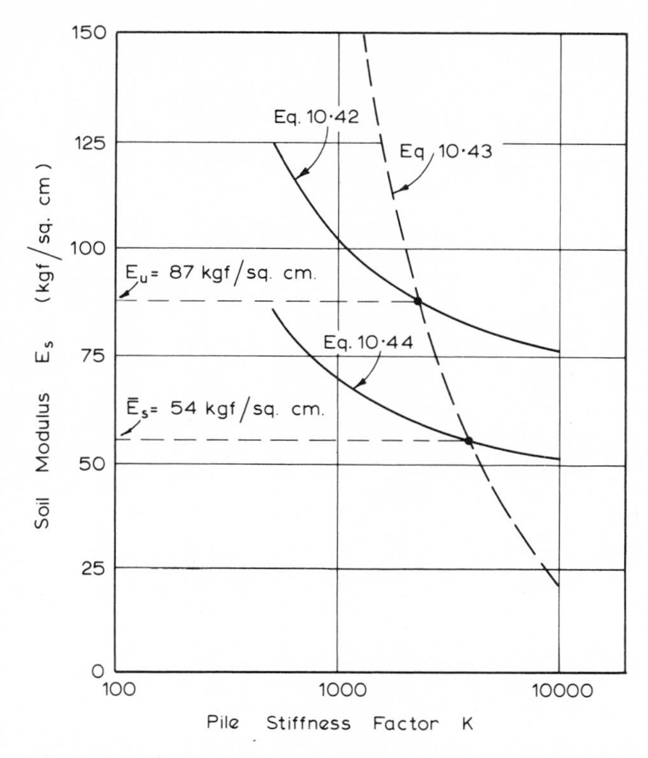

Figure 10-11 Example of determination of soil modulus from pile load test.

10-4.2 Observed and Predicted Behavior of Single Piles

Tests by Darragh and Bell[29a] Mattes[12] has analyzed an interesting series of pile tests carried out by Darragh and Bell[29a] at the site of Gulf Oil Corporation's Faustina Works, on the banks of the Mississippi River. Brief details of the piles driven and of the site subsurface conditions are given in Fig. 10-12. Two pairs of step-taper piles and one pair of steel-tube piles were driven, and the site involved about 120 ft of natural levee and backswamp deposits consisting of layers and laminations of clays, silts, and fine sands, which overlay a 70-ft-deep layer of fine silt grading to sandy gravel with depth. Piles 1*B* and 3 gave very similar load-test results and were analyzed as floating piles in a finite layer to derive a backfigured soil modulus for the backswamp deposits. The relevant details of these piles are

Pile length $L = 108$ ft Average diameter $D = 13$ in

$$\frac{L}{D} \approx 100 \qquad \text{Depth of founding layer } h = 120 \text{ ft} \qquad \frac{h}{L} \approx 1.1$$

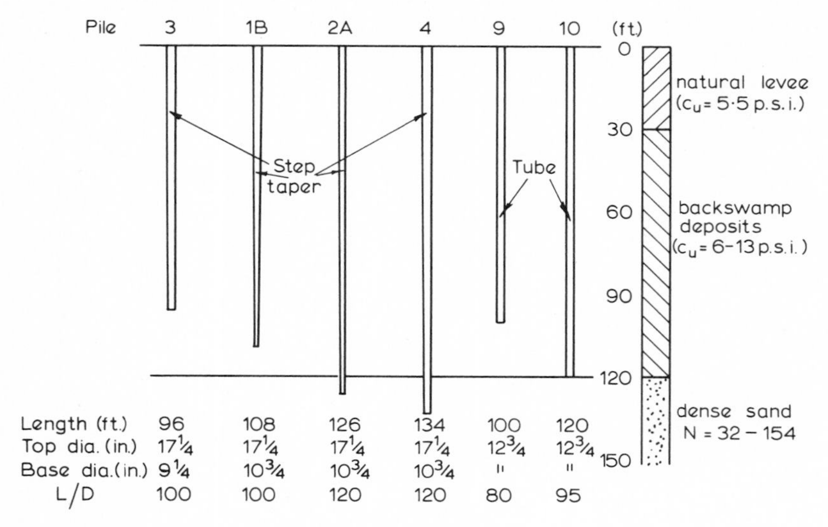

Figure 10-12 Details of tests by Darragh and Bell.[29a]

$$\text{Ultimate load} = 160 \text{ tons} \qquad \text{Settlement at 80 tons} = 0.12 \text{ in}$$

$$\text{Soil modulus } E_s \text{ (backfigured)} = 6500 \text{ lb·ft/in}^2$$

$$\text{Pile stiffness factor } K = 460 \text{ to } 500$$

For the purpose of predicting the performance of the end bearing piles 2*A*, 4, and 10, it was assumed that the modular ratio E_b/E_s was 2.

In Table 10-6, the observed settlements of piles 2, 4*A*, 9, and 10 are compared with those predicted by the elastic theory using the soil modulus derived from the floating pile tests 1*B* and 3. It can be seen that the settlement performance of both the floating and end bearing piles has been closely predicted.

Tests by Mansur and Kaufman[30] Six instrumented piles were driven into a fairly deep layered system of silts, sandy silts, and silty sands with interspersed clay strata underlain by a deep layer of dense fine sand. All except pile 5 were driven to end bearing in the dense fine sand; pile 5 was a floating pile. One of the end bearing piles (pile 3) was an H pile with a rectangular baseplate attached, and because of the disturbing effect of the plate during driving, this test was not considered. The test results were analyzed as follows:

1. Using the single pile theory, pile 5 was analyzed as a floating pile in a finite layer, and a soil modulus E_s of 10,000 lb_f/in^2 was backfigured from the pile test results.

Table 10-6 Predicted pile performance against test by Darragh and Bell[29a]

Pile†	Length, ft	$\frac{L}{D}$, approx.	Pile type	E_s, lb/in²	$\frac{E_b}{E_s}$	K	Applied load, tons	Top settlement Pred.	Top settlement Meas.	Base settlement Pred.	Base settlement Meas.
2A, EB	126	120	Steel, step-taper, closed	6500	2	500	120	0.25	0.20	0.05	0.04
4, EB	132	120	Steel, step-taper, closed	6500	2	500	120	0.25	0.24	0.05	0.06
9, F	100	80	Closed steel tube, 0.188-in wall	6500	1	270	40	0.09	0.10	0.01	0.015
10, EB	120	95	Closed steel tube, 0.188-in wall	6500	2	270	80	0.16	0.17	0.016	0.02

† EB = end bearing; F = floating.

2. From the standard penetration test blow counts for the silty soils and the dense fine sand, it was deduced that a ratio of soil to bearing stratum moduli E_b/E_s of about 3 would be applicable for the end bearing piles.
3. The analysis for a single end bearing pile was used to evaluate the load distribution along piles 1, 2, 4, and 6 and the settlement of these piles. In Table 10-7 the details of pile properties, settlements, and settlement predictions are given, while in Fig. 10-13 the predicted and measured load distributions along the piles are compared.

Figure 10-13 and Table 10-7 show that quite low values of K are possible when steel tubes or H sections are used as piles. In such cases, it is likely that very little load does in fact reach the pile base, even in nominally end bearing piles. In the case described here, the results of a floating pile test combined with the results of a routine borehole test allowed the accurate prediction of the load distribution along end bearing piles on the same site and the settlement of these piles.

Tests by D'Appolonia and Romualdi[2] Tests on two instrumented H piles were reported; the piles were about 45 ft long, and passed through layers of fill, sandy silt, sand and gravel, fine to medium sand, sand and gravel, and sandy silt to end bearing in shale. Since no satisfactory soil data were available, a K value for solid-steel piles of 3000 was adopted, based on an E_s value from Poulos.[29] The bearing stratum was assumed to be rigid. In Table 10-8 the pile properties and

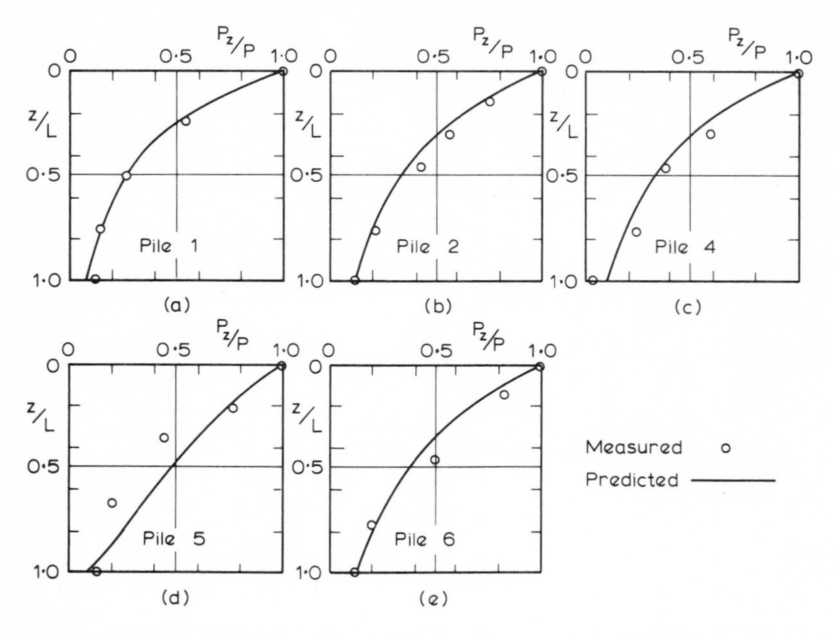

Figure 10-13 Comparisons between predicted and observed load distributions.[30]

settlement details are listed, and comparisons based on the assumed soil properties are made. In Fig. 10-14, the load distributions within the piles are compared with the calculated distributions. In each case, reasonable agreement between prediction and observation is obtained.

Table 10-7 Tests by Mansur and Kaufman[30]

Soils alternating strata of silts, silty sands, sandy silts with interspersed clay strata; bearing stratum: dense fine sand

Pile no. and type†	Size, in.	Type	Length, ft	$\frac{L}{D}$	K	$\frac{E_b}{E_s}$	Load, tons	Settlement Obs., in	Pred., in	Pred./Obs.
1, EB	14	H beam	81	70	470	3	125	0.13	0.144	1.10
2, EB	21	Pipe	65	37	250	3	125	0.13	0.130	1.00
4, EB	17	Pipe	66	47	350	3	125	0.16	0.142	0.89
5, F	17	Pipe	45	32	350	1‡	75	0.10	0.100‡	1.00‡
6, EB	19	Pipe	65	41	350	3	125	0.13	0.150	1.15

† EB = end bearing; F = floating.
‡ Pile 5 used as a control pile for predictions.

Table 10-8 Tests by D'Appolonia and Romualdi[2]

Soil: layers of fill, sandy silt, sand and gravel, fine to medium sand, sand and gravel, and sandy silt; bearing stratum: shale

Pile no.	Type	Length, ft	$\frac{L}{D}$, assumed	Area ratio R_A, assumed	K	$\frac{E_b}{E_s}$	Load, tons	Settlement Obs., in	Pred., in	Pred./Obs.
1	14BP89 H pile	44	33	0.143	430	∞	75	0.07	0.06	0.86
2	14BP119 H pile	45	34	0.186	560	∞	100	0.11	0.09	0.82

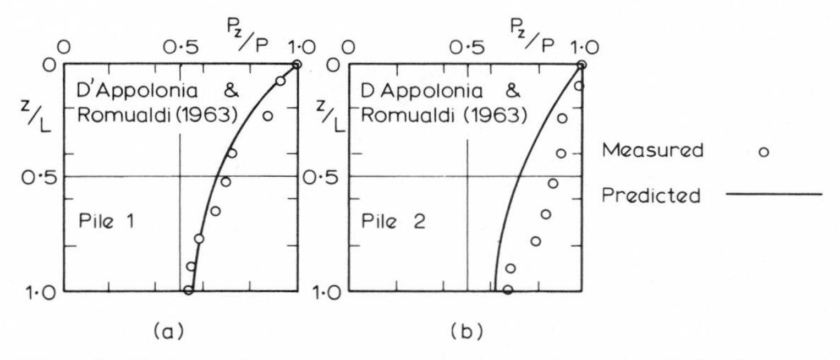

Figure 10.14 Comparisons between predicted and observed load distributions.[2]

10-4.3 Observed and Theoretical Group Behavior

Settlements A number of comparisons between measured and theoretical values of settlement ratio for floating pile groups were made by Poulos and Mattes.[26] A summary of the cases considered is given in Table 10-9, and the comparisons are shown in Fig. 10-15. In all cases, the load level corresponds to a factor of safety of at least 2 against ultimate failure of the group. With the exception of the model tests by Hanna[31] in loose sand, the agreement is generally satisfactory for both large and small values of K. The poor agreement for the tests in loose sand may be attributed to the effects of the greater densification of the loose sand by the pile group compared with the single pile. These comparisons therefore indicate that the theoretical approach should be satisfactory in practical cases except for pile groups in loose sand.

Group behavior predicted from single pile test results A series of full-scale tests by Koizumi and Ito[36] was studied by Mattes[12] in an attempt to predict the performance of a pile group from the results of a single pile test. A single floating pile

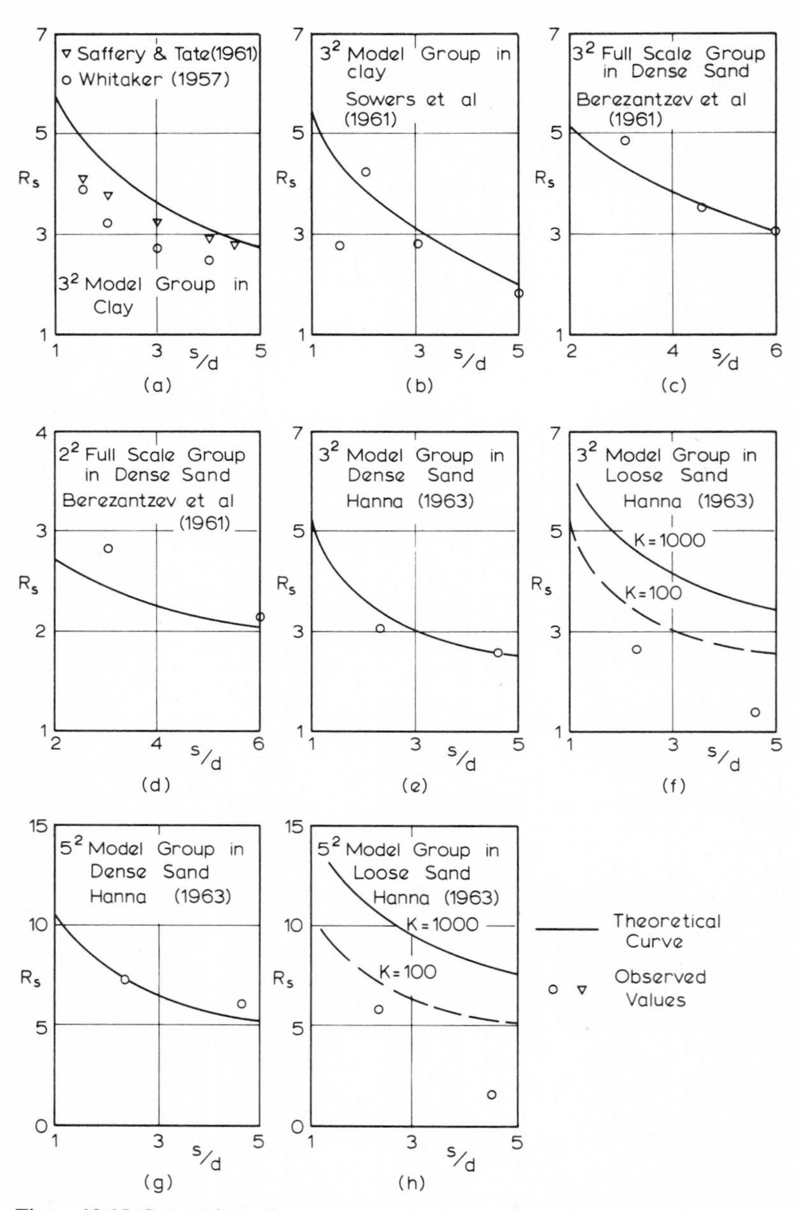

Figure 10-15 Comparisons between theoretical and observed group settlement ratios.

and a nine-pile rigid-capped group of similar piles were founded in a thick uniform layer of silty clay overlain by a thin layer of sandy silt. The piles were closed-end steel tubes, length 5.5 m, diameter 0.3 m, wall thickness 3.2 mm, and were instrumented to allow pile loads, earth pressures, and pore pressures to be

Table 10-9 Summary of data on floating pile group tests

Test	Pile material	Soil type	Assumed parameters for comparisons L/D	K	Layer depth/L	Remarks
Whitaker[32]	Brass	Remolded London clay	24	∞	2	Model tests
Saffery and Tate[33]	Stainless steel	Remolded clay	20	∞	2	Model tests
Sowers et al.[34]	Aluminum tube	Remolded bentonite	24	2000	2	Model tests
Berezantzev et al.[35]	Concrete	Dense sand	20	1000	∞	Field tests; K estimated for quoted values of E_s
Hanna[31]	Wood	Dense sand	33	100	2	Model tests; K estimated from Poulos and Mattes[15]
Hanna[31]	Wood	Loose sand	33	1000	2	

measured. Provision was also made for measuring displacements and pressures in the soil remote from the piles.

By using the single pile load-test results, a soil modulus of 2500 lb_f/in^2 was backfigured, corresponding to a pile stiffness factor of 500 at a load factor against failure of approximately 2.5. From theory, the settlement of the rigid-capped group (at the same load factor) was calculated and compared with the measured settlement, as follows:

$$\text{Group} = 3^2\text{, floating, rigid cap}$$

$$L/D = 18.5 \text{ (individual piles)}$$

$$\text{Spacing} = 3 \text{ pile diameters center-to-center}$$

$$\text{Group load } P_G = 90 \text{ t}$$

$$\text{Group reduction factor } R_G = 0.40$$

$$\text{Single pile settlement for unit load } \rho_1 = 0.20 \text{ mm/t}$$

$$\text{Predicted settlement of group } P_G \rho_1 R_G = 7.2 \text{ mm}$$

$$\text{Measured settlement of group} = 7.1 \text{ mm}$$

Although the cap of this group was in contact with the soil, the effect of the cap on group settlement is negligible.[27] It can be seen that there is excellent agreement between predicted and measured group settlement. The measured and theoretically predicted load distributions within the group also agree closely, as shown in Table 10-10.

Table 10-10 Theoretical and measured load-distribution tests of Koizumi and Ito[13]

	Pile load/average pile load	
Pile location	Theoret.	Meas.
Center	0.35	0.46
Corner	0.82	0.86
Midside	1.35	1.20

† Group load 120 t.

REFERENCES

1. Mindlin, R. D.: Force at a Point in the Interior of a Semi-Infinite Solid, *Physics*, vol. 7, p. 195, 1936.
2. D'Appolonia, E., and J. P. Romualdi: Load Transfer in End-bearing Steel H-Piles, *J. Soil Mech. Found. Div., ASCE*, vol. 89, no. SM2, pp. 1–25, 1963.
3. Thurman, A. G., and E. D'Appolonia: Computed Movement of Friction and End-bearing Piles Embedded in Uniform and Stratified Soils, *Proc. 6th Int. Conf. Soil Mech. Found. Eng., 1965*, vol. 2, pp. 323–327.
4. Salas, J. A. J., and J. A. Belzunce: Resolution théorique de la distribution des forces dans les pieux, *Proc. 6th Int. Conf. Soil Mech. Found. Eng., 1965*, vol. 2, pp. 309–313.
5. Poulos, H. G., and E. H. Davis: The Settlement Behavior of Axially-Loaded Incompressible Piles and Piers, *Geotechnique*, vol. 18, pp. 351–371, 1968.
6. Mattes, N. S., and H. G. Poulos: Settlement of Single Compressible Pile, *J. Soil Mech. Found. Div., ASCE*, vol. 95, no. SM1, pp. 189–207, 1969.
7. Seed, H. B., and L. C. Reese: The Action of Soft Clay along Friction Piles, *Trans. ASCE.*, vol. 122, 1957.
8. Coyle, H. M., and L. C. Reese: Load Transfer for Axially-Loaded Piles in Clay, *J. Soil Mech. Found. Div. ASCE*, vol. 92, no. SM2, pp. 1–26, 1966.
9. Coyle, H. M., and T. Sulaiman: Skin Friction for Steel Piles in Sand, *J. Soil Mech. Found. Div. ASCE*, vol. 93, no. SM6, 1967.
10. D'Appolonia, D. J., H. G. Poulos, and C. C. Ladd: Initial Settlement of Structures on Clay, *J. Soil Mech. Found. Div., ASCE*, vol. 97, no. SM10, 1971.
11. Poulos, H. G.: Load-Deflection Prediction for Laterally Loaded Piles, *Univ. Sydney Civ. Eng. Res. Rep.* R208, 1973.
12. Mattes, N. S.: The Analysis of Settlement of Piles and Pile Groups in Clay Soils, Ph.D. thesis, University of Sydney, 1972.
13. Butterfield, R., and P. K. Banerjee: The Elastic Analysis of Compressible Piles and Pile Groups, *Geotechnique*, vol. 21, pp. 43–60, 1971.
14. Steinbrenner, W.: Tafeln zur Setzungberechnung, *Strasse*, vol. 1, p. 221, 1934.
15. Poulos, H. G., and N. S. Mattes: The Behavior of Axially-Loaded End-bearing Piles, *Geotechnique*, vol. 19, pp. 285–300, 1969.
16. Mattes, N. S.: The Influence of Radial Displacement Compatability on Pile Settlements, *Geotechnique*, vol. 19, pp. 157–159, 1969.
17. Lee, I. K.: Application of Finite Element Method in Geotechnical Engineering, I: Linear Analysis, Chap. 17 in "Finite Element Techniques: A Short Course of Fundamentals and Applications," University of New South Wales, 1973.
18. Salas, J. A. J.: Discussion on Div. 4, *Proc. 6th Int. Conf. Soil Mech. Found. Eng., 1965*, vol. 3, pp. 489–492.

19. Poulos, H. G., and N. S. Mattes: The Analysis of Downdrag in End-bearing Piles, *Proc. 7th Int. Conf. Soil Mech. Found. Eng., 1969*, vol. 2, pp. 203–209.
20. Poulos, H. G., and E. H. Davis: Theory of Piles in Swelling and Shrinking Soils, *Proc. 8th Int. Conf. Soil Mech. Found. Eng., Moscow, 1973.*
21. Poulos, H. G.: The Settlement of Under-reamed and Step-Taper Piles, *Civ. Eng. Trans. Inst. Eng. Aust.*, vol. CE11, pp. 93–87, 1969.
22. Poulos, H. G.: The Influence of a Rigid Pile Cap on the Settlement Behavior of an Axially-Loaded Pile, *Civ. Eng. Trans. Inst. Eng. Aust.*, vol. CE10, no. 2, pp. 206–208, 1968.
23. Butterfield, R., and P. K. Banerjee: The Problem of Pile Group–Pile Cap Interaction, *Geotechnique*, vol. 21, pp. 135–142, 1971.
24. Pichumani, R., and E. D'Appolonia: Theoretical Distribution of Loads among the Piles in a Group, *Proc. 3d Pan-Am. Conf. Soil Mech. Found. Eng., Caracas, 1967.*
25. Poulos, H. G.: Analysis of the Settlement of Pile Groups, *Geotechnique*, vol. 18, pp. 449–471, 1968.
26. Poulos, H. G., and N. S. Mattes: Settlement and Load Distribution Analysis of Pile Groups, *Aust. Geomech. J.*, vol. G1, no. 1, pp. 18–28, 1971.
27. Davis, E. H., and H. G. Poulos: The Analysis and Design of Pile-Raft Systems, *Aust. Geomech. J.*, vol. G2, no. 1, pp. 21–27, 1972.
28. Focht, J. A.: Discussion of Ref. 8, *J. Soil Mech. Found. Div., ASCE*, vol. 93, no. SM1, pp. 133–138, 1967.
29. Poulos, H. G.: Load-Settlement Prediction for Piles and Piers, *J. Soil Mech. Found. Div., ASCE*, vol. 98, no. SM9, pp. 879–897, 1972.
29a. Darragh, R. D., and R. A. Bell: Load Tests on Long Bearing Piles, *ASTM Spec. Tech. Pap.* 444, pp. 41–67, 1969.
30. Mansur, C. I., and R. I. Kaufman: Pile Tests, Low-Sill Structure, Old River La., *J. Soil Mech. Found. Div., ASCE*, vol. 82, no. SM4, Proc. Pap. 1079, 1956.
31. Hanna, T. H.: Model Studies of Foundation Groups in Sand, *Geotechnique*, vol. 13, pp. 334–351, 1963.
32. Whitaker, T.: Experiments with Model Piles in Groups, *Geotechnique*, vol. 7, pp. 147–167, 1957.
33. Safferey, M. R., and A. P. K. Tate: Model Tests on Pile Groups in a Clay Soil with Particular Reference to the Behaviour of the Group When It is Loaded Eccentrically, *Proc. 5th Int. Conf. Soil Mech. Found. Eng., 1973*, vol. 2, pp. 129–134.
34. Sowers, G. F., C. B. Martin, L. L. Wilson, and M. Fausold: The Bearing Capacity of Friction Pile Groups in Homogeneous Clay from Model Studies, *Proc. 5th Int. Conf. Soil Mech. Found. Eng., 1961*, vol. 2, pp. 155–159.
35. Berezantzev, V. G., V. Khristoforov, and V. Golubkov: Load Bearing Capacity and Deformation of Piled Foundations, *Proc. 5th Int. Conf. Soil Mech. Found. Eng. 1961*, vol. 2, pp. 11–15.
36. Koizumi, Y., and K. Ito: Field Tests with Regard to Pile Driving and Bearing Capacity of Piled Foundations, *Soil Found.*, vol. 7, no. 3, pp. 30–53, 1967.

CHAPTER

ELEVEN

ONE-DIMENSIONAL CONSOLIDATION

Robert L. Schiffman and Santosh K. Arya

11-1 INTRODUCTION†

The theory of consolidation[27,32] is traditionally used to predict the progress of settlement of loaded clay layers. This theory is based upon the hypothesis that the progress of compression is governed by the dissipation of the water pressure generated by an external loading.

Historical Background

That water plays an important role in the time behavior of soil systems has long been recognized. Skempton[21] has summarized this history as follows: "The concept of effective stress was first explicitly stated by Terzaghi in relation to the consolidation of clays. Geologists and civil engineers had long recognized that clay under load gradually consolidates as water escapes from the voids." The earliest record of consolidation is by Telford,[25] who in 1809 preloaded a thick

† Acknowledgment is made to the National Science Foundation, the Office of Naval Research, the U.S. Bureau of Reclamation, and the Control Data Corporation for contributing to the support of the research which led to this chapter. Particular acknowledgment is made to Messrs. Richard A. Jones and Jack R. Stein, who assisted in the development of some of the computer programs.

stratum of clay and permitted settlement to occur for about 9 months "... for the purpose of squeezing out the water and consolidating the mud." At the end of the nineteenth century, experiences in Chicago[22,20] indicated a realization that settlements were delayed in time due to the expulsion of water. Sorby[23] attempted to explain the consolidation phenomenon by considering porosity changes. This attempt, however, failed. Sorby neglected consideration of the buoyancy of saturated submerged soils.

The first experimental work was recorded by Frontard.[10] In this study a one-dimensional test was performed, and the results were recorded in terms of curves of water content vs. pressure. Forchheimer[9] attempted a theoretical treatment of consolidation, but this failed due to its oversimplification. It neglected pore-pressure dissipation and compressibility. In 1916 to 1922, the Swedish State Railway Commission[24] studied the compressive behavior of soft soil contributing to landslides. In their report the mechanical coupling of soil and water was represented by a model similar to the famous Terzaghi spring-piston model.

Terzaghi[26–31] used both experimental and analytical means to develop a single consistent theory for predicting the rate of settlement of structures founded on clay. In the development of this theory, Terzaghi described the consolidation process as follows: "A change in pressure at any point in a clay causes an alteration in water content. But for a change in water content to occur, some water must flow out, and this flow must be the consequence of a gradient in the pore water pressure. Due to the low permeability of clays, the rate of flow will be correspondingly small but finally the pore pressures will disappear."

Theory of Consolidation

The theory of consolidation is a rational theory which predicts the progress of development of effective stress in a soil. It is based upon the effective stress equation

$$\sigma = \sigma' + u_w \tag{11-1}$$

where σ = total stress
σ' = effective stress
u_w = pore water pressure (pore pressure or neutral pressure)

Equation (11-1) states that the stress at a point in a porous medium due to boundary tractions or body forces is the sum of the stress in the water (or other fluid) at that point and the stress in the solid portion (soil skeleton or soil structure) of the porous medium at that point.

The process of consolidation is the process of generation and dissipation of a pore water pressure established by a foundation loading or change in water conditions. Consider the establishment of a total stress increment $\Delta\sigma$ by means of a foundation loading or other mechanism. At the instant of loading, the total stress increment $\Delta\sigma$ is transmitted to the soil as an excess pore pressure u. This is the pore water pressure which is in excess of the static water pressure. The effective stress increment $\Delta\sigma'$ is governed by the effective stress equation (11-1) as

$$\Delta\sigma' = \Delta\sigma - u \tag{11-2}$$

Initially, the excess pore pressure is equal to the total stress increment. Thus, the effective stress increment is initially zero. With time, however, the pore water escapes from the soil, dissipating the excess pore pressure. In the one-dimensional geometry, the total stress increment is fully defined by the external foundation loading. Thus the progress of generation of the effective stress increment is equivalent to the dissipation of excess pore pressure. The pore water is incompressible. Thus the deformations are controlled by the effective stress increment. As a result, the progress of settlement can be predicted from the dissipation of the excess pore pressure.

The mathematical model of one-dimensional consolidation is a parabolic differential equation of the heat conduction type.

Numerical Analysis

Numerical analysis can be defined as the solution of a physical problem by arithmetic means. A physical problem is usually modeled by a symbolic mathematical relationship. The numerical analysis is a means by which arithmetic operations are used to provide numerical results for engineering use. For one-dimensional consolidation the symbolic form is a partial differential equation governing the behavior of the excess pore pressure u as a function of space and time. For a given set of physical parameters and boundary and initial conditions, the numerical analysis provides numerical values for the excess pore pressure.

Three types of numerical analysis can be applied to one-dimensional consolidation problems. The first concerns the evaluation of analytical solutions by numerical means. This involves the evaluation of the roots of a characteristic equation and the summation of a series over the characteristic values. The second numerical technique is a finite difference procedure in which the governing differential equation is represented by a set of difference equations, which are then solved algebraically. The third type of analysis is by finite elements, in which the consolidating mass is discretized by using compatible elements. A solution is accomplished by application of a variational principle. Numerically, the finite element procedure involves the solution of a set of simultaneous linear algebraic equations. All these numerical methods require computation by machine procedures.

11-2 THEORY

Settlements are controlled by the effective stress increments generated in a soil mass. If the total stresses in a soil mass are known at all times, the progress of settlement is directly related to the dissipation of excess pore pressure.[27] Excess pore pressures are also generated by changes in drainage conditions and changes in groundwater conditions. The transient behavior of the excess pore pressure is governed by the theory of consolidation.

Formulation of Theory

The one-dimensional theory of primary consolidation[27,32] predicts the progress of changes of effective stress increments within loaded clay layers. This theory is based upon the following constitutive assumptions:[18,11]

1. The soil is completely saturated with water.
2. The soil particles and the pore water are incompressible.
3. The fluid flow equations follow Darcy's law.
4. The strains in the soil skeleton are controlled exclusively by the effective stresses via a linear relationship.
5. The soil skeleton is nonhomogeneous; i.e., a spatially dependent stress-strain and velocity–pressure gradient relationship governs the soil mass.
6. The strains, velocities, and stress increments are small, and the theory is quasistatic.

The continuity relationship of a compressible porous mass in which the compression and the fluid flow are in the z direction is

$$\frac{\partial v}{\partial z} = \frac{\partial e}{\partial t} \tag{11-3}$$

where v is the velocity of the pore fluid relative to the soil particles and e is the dilatation of the soil skeleton. From the effective stress principle and the assumptions given above, the dilatation is

$$e(z, t) = \int_0^t m_v(z, t - \tau) \frac{\partial \sigma'(z, \tau)}{\partial \tau} \, d\tau \tag{11-4}$$

where m_v is the constrained compressibility of the soil skeleton.

In general, the pore water pressure u_w is related to the excess pore water pressure u by

$$u_w = u + \gamma_w(z - z_0) \tag{11-5}$$

where γ_w is the unit weight of water and z_0 is an arbitrary constant.

Darcy's law[5] can be expressed as

$$v = -\frac{k(z)}{\gamma_w} \frac{\partial u}{\partial z} \tag{11-6}$$

where k is the coefficient of permeability.

From Eqs. (11-3) to (11-6), assuming that the total stress σ in the soil mass is time dependent but independent of depth, it follows that

$$\frac{\partial}{\partial z}\left[\frac{k(z)}{\gamma_w}\frac{\partial \sigma'}{\partial z}\right] = m_v(z, 0)\frac{\partial \sigma'}{\partial t} + \int_0^t \frac{\partial \sigma'(z, \tau)}{\partial \tau}\frac{\partial m_v(z, t-\tau)}{\partial t}\,d\tau \tag{11-7}$$

where the change in total stress as a function of time is reflected in the boundary conditions.

Equation (11-7) with the appropriate boundary and starting conditions† constitutes a complete mathematical statement of the theory of one-dimensional consolidation as it relates to the generation of effective stress.

Several special cases of Eq. (11-7) provide solutions to practical engineering problems. In the first, primary consolidation of a nonhomogeneous clay stratum,[18] the constrained compressibility m_v is time independent. Then

$$\frac{\partial}{\partial z}\left[\frac{k(z)}{\gamma_w}\frac{\partial u}{\partial z}\right] = m_v(z)\left\{\frac{\partial u}{\partial t} - \frac{\partial}{\partial t}[\Delta\sigma(z, t)]\right\} \tag{11-8}$$

The last term on the right-hand side of Eq. (11-8) defines a total stress history which varies with depth.

If the clay stratum is homogeneous, both m_v and k are constants. Then

$$c_v \frac{\partial^2 u}{\partial z^2} = \frac{\partial u}{\partial t} - \frac{\partial}{\partial t}(\Delta\sigma) \tag{11-9a}$$

where the coefficient of consolidation c_v is

$$c_v = \frac{k}{\gamma_w m_v} \tag{11-9b}$$

The Primary Consolidation of a Homogeneous Multilayered Stratum

It is assumed that the soil profile consists of n contiguous layers, as shown in Fig. 11-1. The arbitrary layer is indexed l with thickness h^l. The soil properties of the lth layer are the coefficient of consolidation c_v^l, the constrained compressibility m_v^l, and the coefficient of permeability k^l. The compressible stratum is the system of n compressible layers and has a total thickness H.

In the analysis which follows, a consistent notation will be used. Indices which refer to a layer are written as superscripts. The superscript l refers to an arbitrary layer. The specific value is superscripted accordingly. Thus all soil properties are carried with superscripts.

The space coordinate z is a global coordinate and has its origin at the surface $z = 0$. All indices which depend on the global space coordinate z are written as

† For clarity of presentation, the term *starting condition* will be used instead of the more general term *initial condition*, which is commonly used in the mathematics literature.

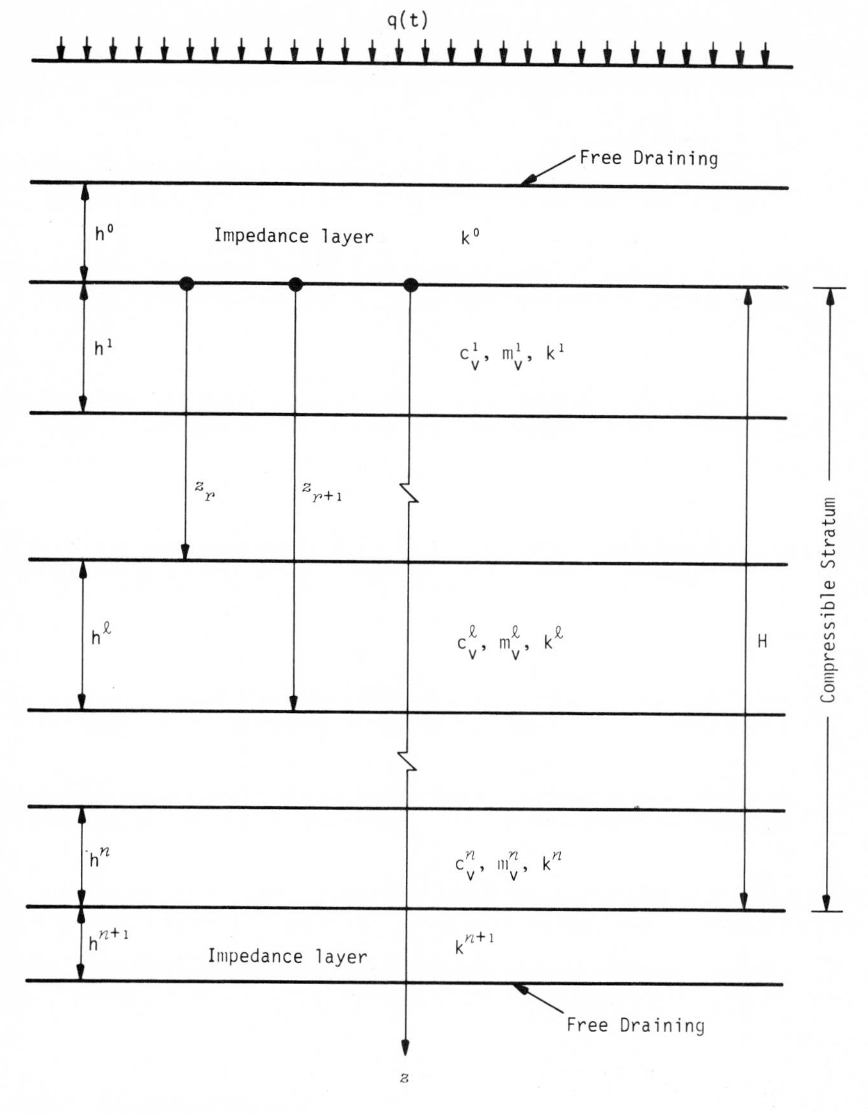

Figure 11-1 Multilayered system.

subscripts. Similarly, all indices which are dependent on time are written as subscripts.

As a general rule the superscript l refers to a layer number; the subscript i refers to a z coordinate point, and the subscript j refers to a value at a particular time. The value of the superscript l runs from (1) for the uppermost layer to n for the lowest layer. The value of i runs from 0 at the surface ($z = 0$) to a value p which is the lower boundary point ($z = H$). The subscripted space index at the layer interface is r. The subscripted time index j runs from 0 at $t = 0$ in an arithmetic progression (0, 1, 2, ...).

The process of consolidation is the process of generation and dissipation of the excess pore pressure u^l of each layer. The dissipation is governed by

$$c_v^l \frac{\partial^2 u^l}{\partial z^2} = \frac{\partial u^l}{\partial t} - \frac{\partial}{\partial t}(\Delta\sigma) \qquad l = 1, 2, \ldots, n \tag{11-10a}$$

There are n equations, which must be solved in order to determine the excess pore pressure at any point in space and time.

If it is assumed that the total stress increment $\Delta\sigma$ is spatially independent and equal to the applied load q, Eq. (11-10a) becomes

$$c_v^l \frac{\partial^2 u^l}{\partial z^2} = \frac{\partial u^l}{\partial t} - \frac{dq}{dt} \qquad l = 1, 2, \ldots, n \tag{11-10b}$$

Equations (11-10) define the generation and dissipation of the excess pore pressure u in the lth layer at a particular depth z and time t.

The three types of time-independent boundary conditions that can apply to the stratum boundaries $z = 0$ and $z = H$ can be expressed in general form as

$$a^1 \frac{\partial u^1}{\partial z}(0, t) - b^1 u^1(0, t) = -c^1 \tag{11-11}$$

and

$$a^n \frac{\partial u^n}{\partial z}(H, t) + b^n u^n(H, t) = c^n \tag{11-12}$$

where the coefficients a^1, b^1, c^1, a^n, b^n, and c^n take on specific values for specific conditions. Table 11-1 presents the particular values of these coefficients for

Table 11-1 Drainage conditions†

Boundary condition	Upper boundary			Lower boundary		
	a^1	b^1	c^1	a^n	b^n	c^n
Excess pore pressure	0	1	$-\phi_0$	0	1	ϕ_p
Free draining	0	-1	0	0	1	0
Velocity	1	0	$\frac{\gamma_w}{k^1 v_0^1}$	1	0	$\frac{\gamma_w}{k^n} v_p^n$
Impervious	1	0	0	1	0	0
Impeded	h^1	$\lambda^1 = \frac{k^0 h^1}{k^1 h^0}$	0	h^n	$\lambda^n = \frac{k^{n+1} h^n}{k^n h^{n+1}}$	0

† ϕ_0, ϕ_p = specified excess pore pressure at upper and lower boundaries, respectively; v_0^1, v_p^n = specified flow velocities at upper and lower boundaries, respectively; k^0, h^0 = coefficient of permeability and thickness, respectively, of impedance layer above the upper boundary; k^{n+1}, h^{n+1} = coefficient of permeability and thickness, respectively, of impedance layer below the lower boundary.

excess pore pressure, free draining, velocity, impervious, and impeded boundaries.

In addition to the boundary conditions for the compressible stratum, it is assumed that there is full continuity between the clay layers. This assumption requires that the excess pore pressures and flow velocities in adjacent layers are equal at the common layer interfaces. These conditions are formulated as

$$u^l(z_r, t) = u^{l+1}(z_r, t) \qquad \text{and} \qquad k^l \frac{\partial u^l}{\partial z}(z_r, t) = k^{l+1} \frac{\partial u^{l+1}}{\partial z}(z_r, t) \quad (11\text{-}13)$$

where, as shown in Fig. 11-1, the distance z_r is the distance from the surface to the layer interface separating the lth and $(l + 1)$st layer.

If the clay stratum is subjected to a sequence of loadings and unloadings, conditions of swell and consolidation will be occurring within a layer. The boundary between regions of swell and consolidation can be modeled as a layer interface governed by Eqs. (11-13). The interface z_r in this case is time dependent.

Excess pore pressures can be generated by the stress history and the starting excess pore pressure. The stress history generates an excess pore pressure during the process of consolidation. The starting excess pore pressures are generated at the starting time t_0. This is the time at the beginning of consolidation.

The stress history $q(z, t)$ is constructed as a sequence of linear load-time curves, as shown in Fig. 11-2. It is assumed that the stress history $q(z, t)$ is equivalent to the history of the total stress increment $\Delta\sigma(z, t)$. It is further assumed that the stress history generates an excess pore pressure which may vary throughout the clay stratum.

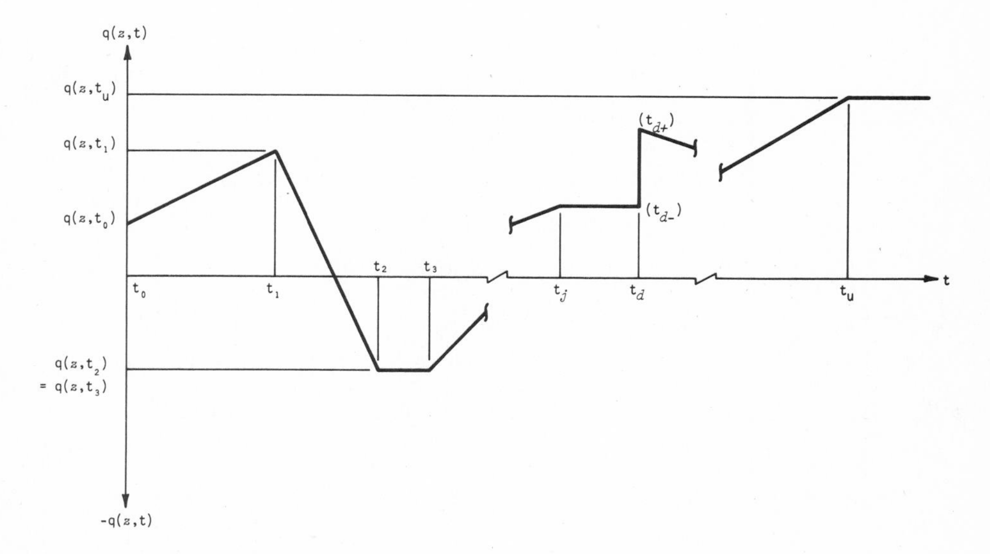

Figure 11-2 Stress history.

11-3 ANALYTICAL PROCEDURES

This section presents an analytical solution for the problem of the consolidation of a multilayered soil stratum.[19] The solution presented is applicable to compressible strata with free draining, impervious, or impeded boundaries.

A solution is sought to the governing equation (11-10*b*) with boundary and interface conditions as given by Eqs. (11-11) to (11-13). The starting condition is

$$u^l(z, 0) = f^l(z) \tag{11-14}$$

where $f^l(z)$ is the net starting condition.

Formal Solution

A formal solution to the boundary-value problem posed above is established by separating variables and expanding starting conditions in orthogonal functions.[2,33,15] The resulting expression for the excess pore pressure is

$$u^l(z, t) = \sum_{m=1}^{\infty} A_m Z_m^l(z) \exp\left[-c_v^l(\beta_m^l)^2 t\right]$$

$$+ \sum_{m=1}^{\infty} D_m Z_m^l(z) \int_0^t \frac{dq}{dt} \exp\left[-c_v^l(\beta_m^l)^2(t-\tau)\right] d\tau \tag{11-15a}$$

where

$$A_m = \frac{\sum_{l=1}^{n} m_v^l \int_{h^l} f^l(\xi) Z_m^l(\xi)\, d\xi}{\sum_{l=1}^{n} m_v^l \int_{h^l} [Z_m^l(\xi)]^2\, d\xi} \tag{11-15b}$$

$$D_m = \frac{\sum_{l=1}^{n} m_v^l \int_{h^l} Z_m(\xi)\, d\xi}{\sum_{l=1}^{n} m_v^l \int_{h^l} [Z_m^l(\xi)]^2\, d\xi} \tag{11-15c}$$

$$Z_m^l(z) = B_m^l \cos \beta_m^l z + C_m^l \sin \beta_m^l z \qquad l = 1, 2, \ldots, n \tag{11-15d}$$

where the coefficients B_m^l and C_m^l are obtained from

$$-b^1 B_m^1 + a^1 \beta_m^1 C_m^1 = 0 \tag{11-15e}$$

$$B_m^l \cos \beta_m^l z_r + C_m^l \sin \beta_m^l z_r - B_m^{l+1} \cos \beta_m^{l+1} z_r - C_m^{l+1} \sin \beta_m^{l+1} z_r = 0$$

$$r = 1, 2, \ldots, n-1 \tag{11-15f}$$

$$\alpha^l \beta_m^l - B_m^l \sin \beta_m^l z_r + C_m^l \cos \beta_m^l z_r + \beta_m^{l+1} B_m^{l+1} \sin \beta_m^{l+1} z_r - C_m^{l+1} \cos \beta_m^{l+1} z_r = 0$$

$$r = 1, 2, \ldots, n-1 \tag{11-15g}$$

$$B_m^n b^n \cos \beta_m^n H - a^n \beta_m^n \sin \beta_m^n H + C_m^n b^n \sin \beta_m^n H + a^n \beta_m^n \cos \beta_m^n H = 0 \tag{11-15h}$$

where

$$\alpha^l = \frac{k^l}{k^{l+1}} \qquad l = 1, 2, \ldots, n-1 \tag{11-15i}$$

From the conditions of continuity of excess pore pressure it is noted that the excess pore pressure history must be the same at a layer interface when approached from either side. Thus

$$c_v^l(\beta_m^l) = c_v^{l+1}(\beta_m^{l+1}) \qquad l = 1, 2, \ldots, n-1 \tag{11-16}$$

which relates the n sets of eigenvalues β_m^l in terms of a single set of eigenvalues such as β_m^1.

The eigenvalues β_m^1 are evaluated by setting the determinant of the coefficients of B_m^l and C_m^l in Eqs. (11-15*e*) to (11-15*i*) equal to zero. This forms the characteristic equation, from which the values of β_m^1 are obtained. The values of β_m^l, where $l \neq 1$, are determined successively from Eq. (11-16).

Equations (11-15*h*) and (11-15*i*) are a set of $2n$ homogeneous equations. The solution of this set of equations will relate the $2n$ coefficients B_m^l and C_m^l to a single base coefficient such as B_m^1, C_m^1, B_m^n, or C_m^n. The choice of a base coefficient depends on the particular boundary condition. If, as an example, the upper boundary is free draining, a^1 is zero. This is equivalent to setting $B_m^1 = 0$. An appropriate coefficient is C_m^1.

Numerical Procedure

The numerical procedure required to evaluate u^l at an arbitrary point z and an arbitrary time t is as follows:

1. Calculate the eigenvalues β_m^l for each layer.
2. Calculate the relationships between the coefficients B_m^l and C_m^l in terms of a base coefficient for each layer.
3. Calculate the coefficients A_m and D_m from Eqs. (11-15*b*) and (11-15*c*), respectively.
4. Calculate the excess pore pressure $u^l(z, t)$ by summing the series given by Eq. (11-15*a*).

The critical computations from the viewpoint of machine efficiency are those involved with the computation of the eigenvalues β_m^l and the relationship between the coefficients B_m^l and C_m^l in terms of a base coefficient. The eigenvalue computation involves the calculation of the eigenvalues of a $2n \times 2n$ matrix. The coefficient relationships involve the solution of a $2n \times 2n$ set of simultaneous equations. The machine procedures for these calculations can be quite tedious.

11-4 FINITE DIFFERENCE SOLUTION PROCEDURE FOR MULTILAYERED SYSTEMS

The Crank-Nicholson procedure[4] is used to develop a solution for the excess pore pressure in a consolidating multilayered stratum. This method provides a small, uniform truncation error throughout the clay stratum. Furthermore, it is unconditionally stable. The time-space mesh is shown in Fig. 11-3. The spatial mesh

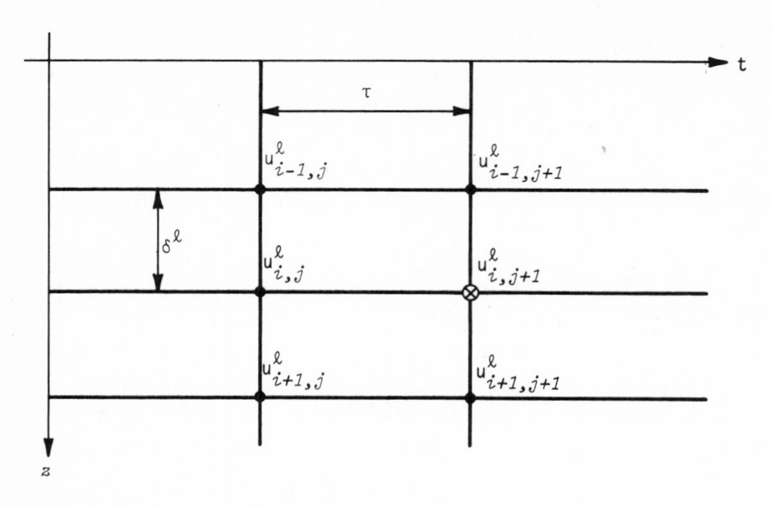

Figure 11-3 Crank-Nicholson scheme.

points are designated by the subscript i, and the time steps are designated by the subscript j. The time increment τ is constant throughout the stratum, while the spatial increment δ^l is constant only within a layer. It may vary from layer to layer.

Difference Equations

Difference equations are required for points interior to a layer, layer interfaces, and stratum boundaries. The difference equation for interior points is

$$u_{i,j+1} - u_{i,j} - q_{j+1} + q_j = \frac{c_v^l \tau}{2}[(\delta^2 u^l)_{i,j+1} + (\delta^2 u^l)_{i,j}] \tag{11-17a}$$

where

$$(\delta^2 u^l)_{i,j} = \frac{1}{(\delta^l)^2}[u^l_{i-1,j} - 2u^l_{i,j} + u^l_{i+1,j}] \tag{11-17b}$$

with a similar expression for $(\delta^2 u^l)_{i,j+1}$.

It is noted that the load history q carries a time subscript j only. This reflects the constancy of q with respect to z.

In expanded form the governing difference equation is

$$F^l u^l_{i-1,j+1} - Q^l u^l_{i,j+1} + F^l u^l_{i+1,j+1} = -E^l_{i,j} \tag{11-18a}$$

where

$$E^l_{i,j} = F^l u^l_{i-1,j} + P^l u^l_{i,j} + F^l u^l_{i+1,j} + \Delta q \tag{11-18b}$$

in which

$$F^l = \frac{R^l}{2} \tag{11-18c}$$

$$Q^l = 1 + R^l \tag{11-18d}$$

$$P^l = 1 - R^l \tag{11-18e}$$

$$\Delta q = q_{j+1} - q_j \tag{11-18f}$$

and
$$R^l = \frac{c_v^l \tau}{(\delta^l)^2} \tag{11-18g}$$

Expanding Eqs. (11-13) in finite differences results in the following difference equation for the layer interface:

$$A^l u_{r-1,j+1}^l - B^l u_{r,j+1}^l + R^{l+1} u_{r+1,j+1}^{l+1} = -G_{r,j}^l \tag{11-19a}$$

where
$$G_{r,j}^l = A^l u_{r-1,j}^l + C^l u_{r,j}^l + R^{l+1} u_{r+1,j}^{l+1} + (1 + \gamma^l \beta^l)\,\Delta q \tag{11-19b}$$

in which

$$A^l = \frac{\alpha^l R^{l+1}}{\beta^l} \qquad B^l = Q^{l+1} + \gamma^l \beta^l Q^l \qquad C^l = P^{l+1} + \gamma^l \beta^l P^l \tag{11-19c}$$

and
$$\alpha^l = \frac{k^l}{k^{l+1}} \qquad \beta^l = \frac{\delta^l}{\delta^{l+1}} \qquad \gamma^l = \frac{m_v^l}{m_v^{l+1}} \tag{11-19d}$$

The Q^l, P^l, and R^l terms are defined by Eqs. (11-18*d*), (11-18*e*), and (11-18*g*), respectively.

It is noted that the coefficients of the difference equations (11-19*a*) and (11-19*b*) are dependent on the ratios α^l and γ^l, as well as the value of c_v^l, which is incorporated in R^l. The difference equations assume a consistency in the interrelationship between c_v^l, k^l, and m_v^l.

The boundary conditions are used to modify the difference equations at the boundaries. In general, they combine the difference equation with the particular boundary condition. At the upper boundary ($z = 0$, $i = 0$), the difference equation is

$$-\left(\frac{\delta^1 b^1 R^1}{a^1} + Q^1\right) u_{0,j+1}^1 + R^1 u_{1,j+1}^1 + \frac{\delta^1 R^1}{a^1} c^1 = -E_{0,j}^1 \tag{11-20a}$$

where

$$E_{0,j}^1 = \left(P^1 - \frac{\delta^1 b^1 R^1}{a^1}\right) u_{0,j}^1 + R^1 u_{1,j}^1 + \frac{\delta^1 R^1}{a^1} c^1 + \Delta q \tag{11-20b}$$

The difference equation at the lower boundary ($z = H$, $i = p$) is

$$R^n u_{p-1,j+1}^n - \left(\frac{\delta^n b^n R^n}{a^n} + Q^n\right) u_{p,j+1}^n + \frac{\delta^n R^n}{a^n} c^n = -E_{p,j}^n \tag{11-21a}$$

where

$$E_{p,j}^n = R^n u_{p-1,j}^n + \left(P^n - \frac{\delta^n b^n R^n}{a^n}\right) u_{p,j}^n + \frac{\delta^n R^n}{a^n} c^n + \Delta q \tag{11-21b}$$

The truncation error $e[u^l]$ of the Crank-Nicholson scheme is

$$e[u^l] = O(\tau^2) + O[(\delta^l)^2] \tag{11-22}$$

Solution to the Difference Equations

The difference equations previously given define a system of simultaneous equations. In matrix form these are

$$[A]\{u(t + \tau)\} = \{E(t)\} \qquad (11\text{-}23)$$

where $[A]$ = coefficient matrix of left-hand side of Crank-Nicholson difference equations
$\{u(t + \tau)\}$ = column vector of $u_{i,\, j+1}$
$\{E(t)\}$ = column vector of right-hand side of difference equations

The solution to the above system of linear simultaneous equations can be accomplished by an equivalent line inversion method.[34,3]

11-5 A COMPUTER PROGRAM TO CALCULATE THE PROGRESS OF CONSOLIDATION USING A FINITE DIFFERENCE PROCEDURE

PROGRS (*pro*gress *o*f *gr*ound *s*ettlement) is a computer program to calculate the progress of consolidation. The program uses the Crank-Nicholson finite difference procedure to perform the calculations.

The logic of PROGRS is presented in flow chart form in Fig. 11-4. This is a conceptual flow chart designed to show the general logic of the program, which is as follows:

1. The input data for a given problem are read. These data consist of the soil profile, the soil properties, the boundary conditions, the starting excess pore pressures (if any), the stress history, the starting time, the problem control data, the printout time, and printout spatial coordinates.
2. The spatial mesh δ^l and the time increment τ are calculated.
3. The starting time t_0 is set.
4. The starting values are calculated. These values are the starting excess pore pressures at the spatial mesh points calculated from the starting excess pore pressures and the stress history.
5. The starting values are printed.
6. The coefficients of the difference equations and the auxiliary matrix coefficients are calculated and stored.
7. The time is incremented.
8. The excess pore pressures at each spatial mesh point are calculated.
9. The consolidation settlement, degree of consolidation, and average excess pore pressures are calculated.
10. The calculated excess pore pressures, degree of consolidation, consolidation settlement, and average excess pore pressures are printed. The excess pore

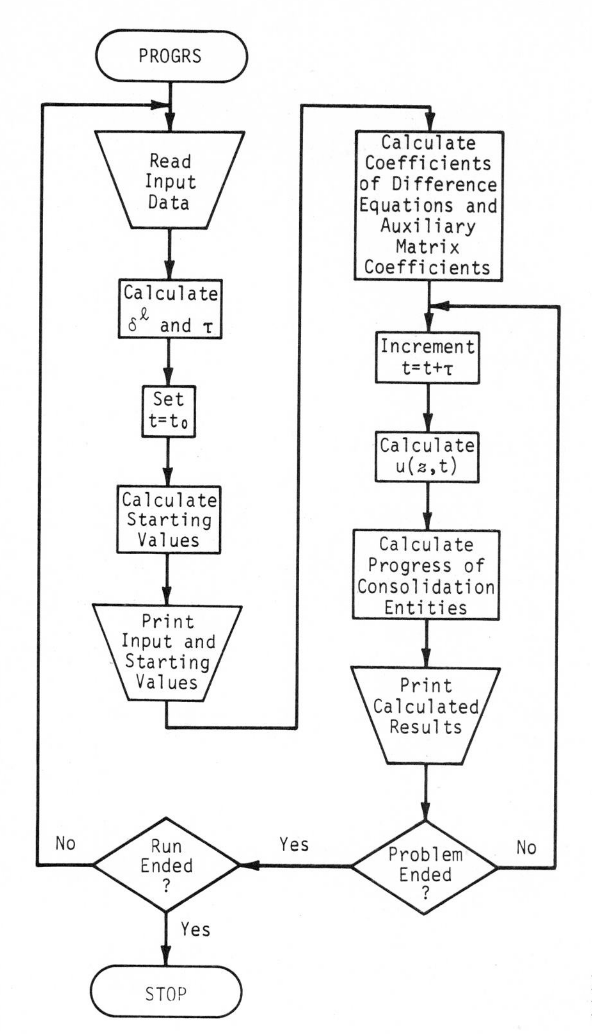

Figure 11-4 Conceptual flow chart for PROGRS.

pressures are printed at the calculation mesh points or at points specified by the input data.

11. The input data specify a termination. This termination is tested. If the problem termination has been reached, the run termination is tested. If not, the time is incremented by τ and calculations for the next time are processed.
12. Once a problem has been terminated, the problem control is tested to determine whether additional problem data are present. If not, the run is terminated. If additional data are to be input, the program starts again with the new input data.

11-6 FINITE ELEMENT FORMULATION

The finite element procedure for one-dimensional consolidation is based upon a discretization of the compressible stratum into a sequence of connecting elements. As an example, consider the compressible stratum shown in Fig. 11-5. In this example the stratum is divided into p elements designated by a square brackets. The element is a line segment of arbitrary length. The arbitrary element in the lth layer is indicated by index k.

Each element is defined by a sequence of geometric nodes indicated in Figure 11-5 by circles. There must be at least two nodes per element. As shown in Figure 11-5, intermediate nodes within an element are permissible. The arbitrary node in the lth layer is indicated by the index i. It is noted that this analysis assumes that every layer boundary is also an element boundary.

In the analysis which follows the element designation will be indicated as a leading superscript. The layer designation will be indicated by a trailing superscript. Thus $\{{}^kW^l\}$ is the column vector $\{W\}$ for the kth element in the lth layer. Similarly the length of an element is designated ${}^k\delta^l$.† The local spatial coordinate for each element is designated $\bar{z}$.

Variational Integral

The variational formulation for one-dimensional consolidation can be determined by applying the calculus of variations.[1,12,13] The partial differential equation to be solved is

$$a\frac{\partial^2 u}{\partial z^2} = m_v\left(\frac{\partial u}{\partial t} - \frac{\partial q}{\partial t}\right) \qquad \text{where } a \equiv \frac{k}{\gamma_w} \tag{11-24}$$

and where it is noted that q can be a function of z as well as t.

It is noted that Eqs. (11-24) do not use the coefficient of consolidation c_v. The above form is preferred because it avoids errors in the assembly of the finite element matrices in a multilayered system when the compressibilities of adjacent layers are unequal.

The possible boundary conditions are

$$u(0, t) = u_0 \qquad u(H, t) = u_1 \tag{11-25}$$

or

$$\frac{\partial u}{\partial z}(0, t) = 0 \qquad \frac{\partial u}{\partial z}(H, t) = 0 \tag{11-26}$$

or

$$\frac{\partial u}{\partial z}(0, t) - \beta_1 u(0, t) = -\mu_1 \qquad \frac{\partial u}{\partial z}(H, t) - \beta_2 u(H, t) = \mu_2 \tag{11-27a}$$

† For simplicity of notation, the superscripts k and l will be dropped when not necessary. Thus the element length in reduced notation is δ.

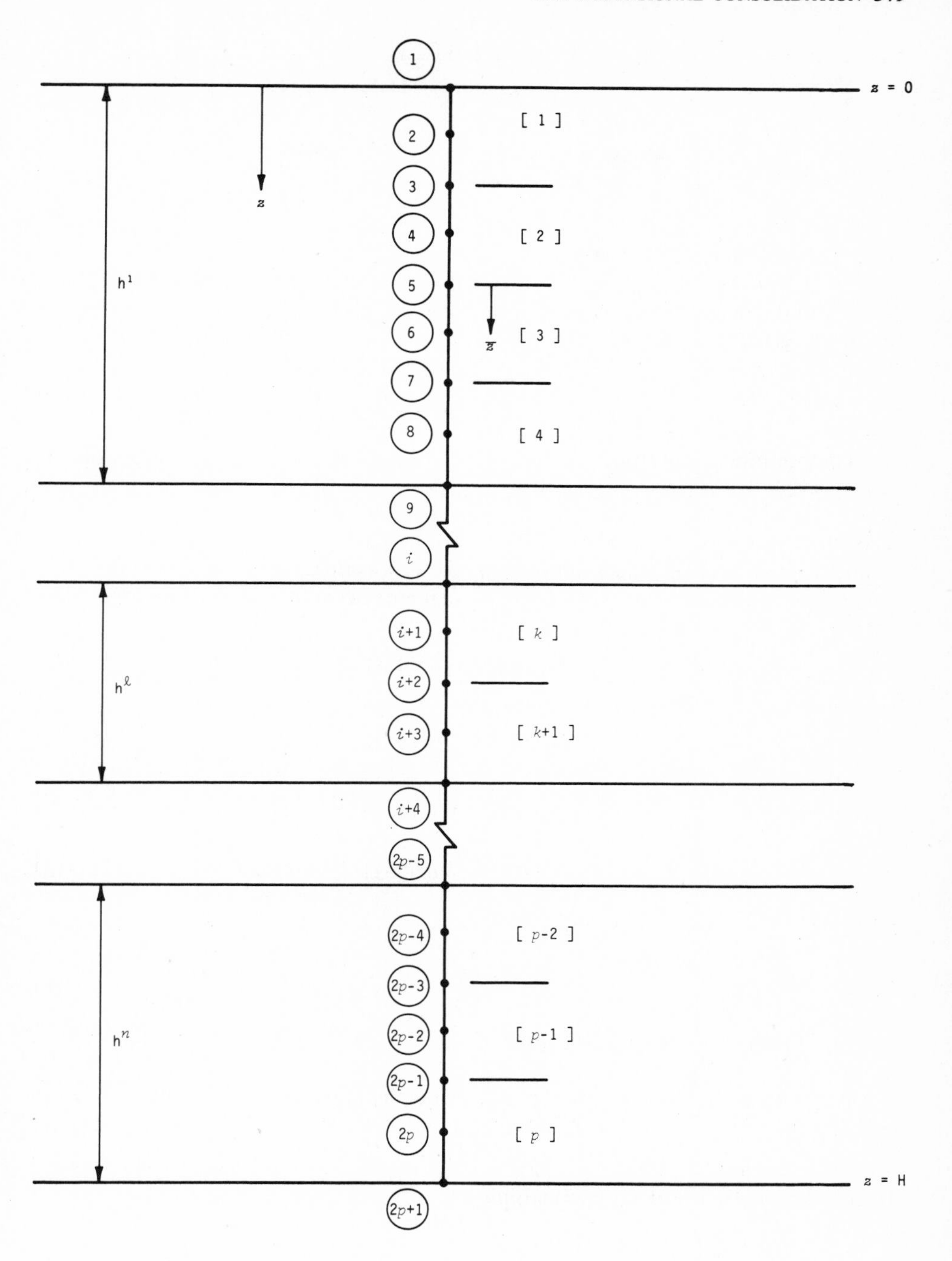

Figure 11-5 Finite element discretization.

where, using the notation defined in Sec. 11-2.2,

$$\beta_1 = \frac{k^0}{k^1 h^0} \qquad \beta_2 = \frac{k^{n+1}}{k^n h^{n+1}} \qquad \mu_1 = \frac{c^1}{h^1} \qquad \mu_2 = \frac{c^2}{h^n} \tag{11-27b}$$

The starting condition is

$$u(z, 0) = g(z) \tag{11-28}$$

where $g(z)$ is known.

The consolidation of each element is governed by Eqs. (11-24) in the domain $(0, \delta)$, where the local coordinate $\bar{z}$ replaces the global coordinate z. The variational integral ${}^kI^l$ for the interior points is

$${}^kI^l = \frac{a}{2}\int_0^\delta \left(\frac{\partial u}{\partial \bar{z}}\right)^2 d\bar{z} + \frac{m_v}{2}\int_0^\delta \frac{\partial(u^2)}{\partial t}\, d\bar{z} - m_v \int_0^\delta u \frac{\partial q}{\partial t}\, d\bar{z} \tag{11-29}$$

The boundary conditions are considered by adding variational expressions ${}^k\bar{I}^l$ as follows:

1. No additional variational expressions are necessary if the element boundary is impervious or if the excess pore pressure is specified.
2. If the excess pore water velocity at an element boundary is specified, the additional required expression is

$${}^k\bar{I}^l = a\mu_1 u^2(0, t) \tag{11-30a}$$

for the upper boundary and

$${}^k\bar{I}^l = a\mu_2 u^2(\delta, t) \tag{11-30b}$$

at the lower boundary.
3. If the boundary of an element is impeded, the additional required expression is

$${}^k\bar{I}^l = \frac{a\beta_1}{2} u^2(0, t) \tag{11-31a}$$

if the upper boundary is impeded and

$${}^k\bar{I}^l = \frac{a\beta_2}{2} u^2(\delta, t) \tag{11-31b}$$

if the lower boundary is impeded.

It is noted that similar variational principles have been established for heat conduction,[35,14] consolidation,[8] and flow problems.[7,36,37]

11-6.2 Basic Element Relationships

Consider the one-dimensional (line) elements shown in Fig. 11-6. If an odd order element is used, as shown in Fig. 11-6*a*, the kth element connects nodal points i and $i + 1$. If an even order element is used, as shown in Fig. 11-6*b*, the kth element connects nodal points i and $i + 2$ with an intermediate nodal point $i + 1$ in the center of the element.

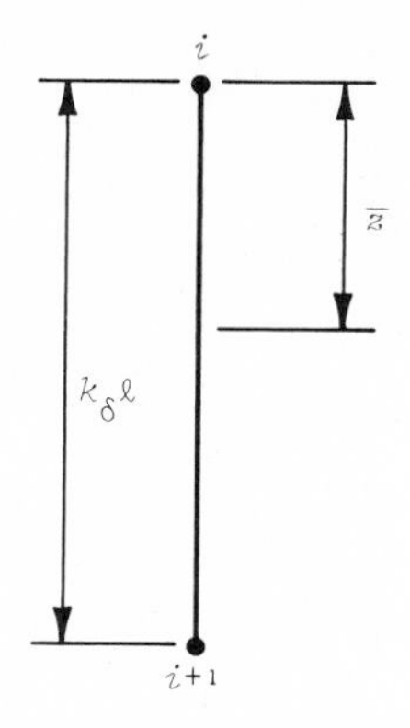

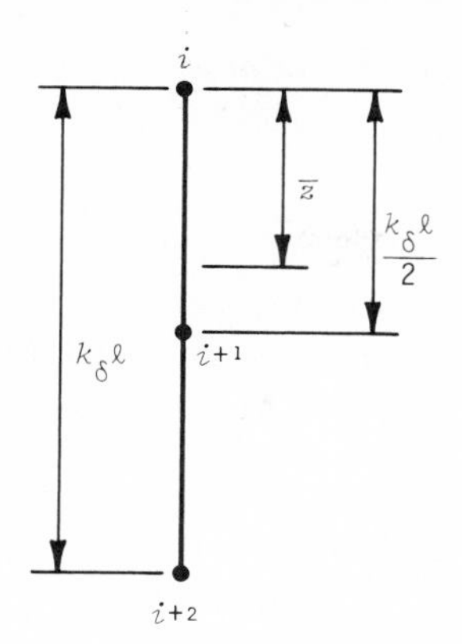

Figure 11-6 kth line element.

It is assumed that the excess pore pressure within the elements shown in Fig. 11-5 varies as a power of $\bar{z}$. If an nth-order element is chosen,

$$u(\bar{z}, t) = \lambda_1(t) + \lambda_2(t)\bar{z} + \lambda_3(t)\bar{z}^2 + \cdots + \lambda_{n+1}(t)\bar{z}^n \tag{11-32}$$

where n is the order of the element. It is noted that the time dependency of u is accounted for by the λ terms.

Each element has $n + 1$ constraints, which can be represented as a column vector $\{^kW^l\}$ (in reduced notation as $\{W\}$).† This vector contains the values of the excess pore pressure and appropriate derivatives taken at each nodal point. For odd order elements the vector $\{W\}$ has the general form

$$\{^kW^l\} \equiv \begin{Bmatrix} u(\bar{z}_i, t) \\ \dfrac{\partial u}{\partial \bar{z}}(\bar{z}_i, t) \\ \vdots \\ \dfrac{\partial^m u}{\partial \bar{z}^m}(\bar{z}_i, t) \\ u(\bar{z}_{i+1}, t) \\ \dfrac{\partial u}{\partial \bar{z}}(z_{i+1}, t) \\ \vdots \\ \dfrac{\partial^m u}{\partial \bar{z}^m}(\bar{z}_{i+1}, t) \end{Bmatrix} \tag{11-33a}$$

† The notation in this chapter may differ from other chapters. For example, the notation for the vector of nodal unknowns is $\{W\}$ in this chapter instead of $\{q\}$. The symbol q is used here to define the load history.

where the highest derivative m is

$$m = \frac{n-1}{2} \tag{11-33b}$$

in which n is the order of the element.

If even order elements are specified, the vector $\{W\}$ has the general form

$$\{^kW^l\} \equiv \left\{ \begin{array}{c} u(\bar{z}_i, t) \\ \dfrac{\partial u}{\partial \bar{z}}(\bar{z}_i, t) \\ \vdots \\ \dfrac{\partial^m u}{\partial \bar{z}^m}(\bar{z}_i, t) \\ u(\bar{z}_{i+1}, t) \\ u(\bar{z}_{i+2}, t) \\ \dfrac{\partial u}{\partial \bar{z}}(\bar{z}_{i+2}, t) \\ \vdots \\ \dfrac{\partial^m u}{\partial \bar{z}^m}(\bar{z}_{i+2}, t) \end{array} \right\} \tag{11-34a}$$

where the highest derivative m is

$$m = \frac{n-2}{2} \tag{11-34b}$$

in which n is the order of the element.

In the establishment of a computer program to develop a finite element solution, a distinction must be made between the geometric node numbering, as shown in Fig. 11-6, and the constraint numbering. Each geometric node must be identified by $m + 1$ nodal quantities, where m is the highest derivative evaluated at the geometric node. For example, consider the first element of a system. The nodal quantity numbering system for the five elements being considered is shown in Fig. 11-7.

A series of five computer programs was written to solve the multilayered one-dimensional consolidation problem. These five programs used first-, second-, third-, fourth-, and fifth-order elements, respectively. Since empirical data, based on bench mark runs, indicate that the most efficient use of machine resources, from the viewpoint of time and accuracy, is achieved with first-order elements, all further discussion will be centered around first-order line elements.

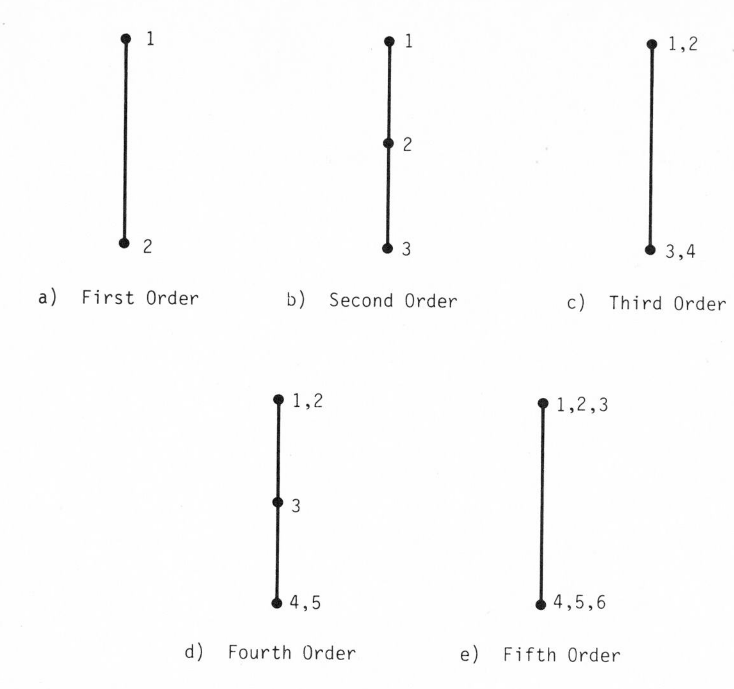

Figure 11-7 Nodal quantity numbering.

From Eq. (11-32) the matrix equation for $\{W\}$ is

$$\{W\} = [A]\{\alpha\} \tag{11-35a}$$

where

$$[A] = \begin{bmatrix} 1 & 0 \\ 1 & \delta \end{bmatrix} \quad \text{and} \quad \{\alpha\} = \begin{Bmatrix} \lambda_1 \\ \lambda_2 \end{Bmatrix} \tag{11-35b}$$

The vector $\{\alpha\}$ is obtained from Eq. (11-35*a*) as

$$\{\alpha\} = [A^{-1}]\{W\} \qquad \text{where} \qquad [A^{-1}] = \begin{bmatrix} 1 & 0 \\ -\dfrac{1}{\delta} & \dfrac{1}{\delta} \end{bmatrix} \tag{11-36}$$

A column vector $\{^kV^l\}$ (or $\{V\}$ in reduced notation) is defined as containing the excess pore pressure and excess pore pressure gradient at an arbitrary point in the element. Thus

$$\{^kV^l\} \equiv \begin{Bmatrix} u(\bar{z}, t) \\ \dfrac{\partial u}{\partial \bar{z}}(\bar{z}, t) \end{Bmatrix} \tag{11-37}$$

From Eq. (11-32),

$$\{V\} = [\phi]\{\alpha\} \tag{11-38a}$$

where
$$[\phi] = \begin{bmatrix} 1 & \bar{z} \\ 0 & 1 \end{bmatrix} \tag{11-38b}$$

The matrix $[\phi]$ can be partitioned to

$$[\phi] = \begin{Bmatrix} \{\phi_1\}^T \\ \{\phi_2\}^T \end{Bmatrix} \tag{11-39}$$

Applying Eq. (11-3) to Eq. (11-38*a*) gives

$$\{V\} = [\phi][A^{-1}]\{W\} \tag{11-40}$$

which relates the excess pore pressure and the velocity at any point inside the element to the excess pore pressure and velocity at the nodal points. When the partitioning of $[\phi]$ is used, the excess pore pressure is

$$u(\bar{z}, t) = \{\phi_1\}^T[A^{-1}]\{W\} \tag{11-41}$$

and the excess pore pressure gradient is

$$\frac{\partial u}{\partial \bar{z}}(\bar{z}, t) = \{\phi_2\}^T[A^{-1}]\{W\} \tag{11-42}$$

Element Pressure Gradient Equations

The element pressure gradient equations, which are analogous to the element stiffness equations in solid mechanics, are obtained by taking the first variation of the functional ${}^kI^l$. Thus,

$$\frac{{}^kI^l}{\partial\{{}^kW^l\}} = 0 \tag{11-43}$$

This relationship in matrix form is

$$[{}^kk^l]\{{}^kW^l\} + [{}^kc^l]\{{}^k\dot{W}^l\} = \{{}^kE^l\} \tag{11-44a}$$

where
$$[k] = a\begin{bmatrix} \frac{1}{\delta} & -\frac{1}{\delta} \\ -\frac{1}{\delta} & \frac{1}{\delta} \end{bmatrix} \qquad [c] = m_v\begin{bmatrix} \frac{\delta}{3} & \frac{\delta}{6} \\ \frac{\delta}{6} & \frac{\delta}{3} \end{bmatrix} \qquad \{E\} = m_v\frac{\partial q}{\partial t}\begin{Bmatrix} \frac{\delta}{2} \\ \frac{\delta}{2} \end{Bmatrix} \tag{11-44b}$$

Specific boundary conditions are incorporated in the above as follows:

1. If the boundary element has the excess pore pressure specified or is impervious, Eqs. (11-44) apply as written above.
2. If the excess pore water velocity is specified at the upper boundary ($z = 0$), the matrix $[{}^1k^1]$ is augmented by adding $a\beta_1$ to the ${}^1k_{11}$th element of that matrix.
3. If the upper boundary ($z = 0$) is impeded, the vector $\{{}^1E^1\}$ is augmented by subtracting $a\mu_1$ from the ${}^1e_1^1$th element of that vector.

4. If the excess pore water velocity is specified at the lower boundary ($z = H$), the matrix $[{}^{p}k^{n}]$ is augmented by adding $a\beta_2$ to the ${}^{p}k^{n}_{22}$ th element of that matrix.
5. If the lower boundary ($z = H$) is impeded, the vector $\{{}^{p}E^{n}\}$ is augmented by subtracting $a\mu_2$ from the ${}^{p}e^{n}_{2}$th element of that vector.

Details of the analysis can be found elsewhere.[17]

It is noted that $[k]$ is a symmetric matrix. This matrix has a counterpart in structural mechanics as the element stiffness matrix. In heat conduction it is known as the element conductance matrix.[14] For consolidation problems it is referred to as the element velocity gradient matrix. The matrix $[c]$ is also symmetric. The heat conduction analog is the element capacitance matrix.[14] Its name for consolidation is the element dilatation rate matrix. The vector $\{E\}$ is the element stress history vector.

11-7 GLOBAL FINITE ELEMENT RELATIONSHIPS

In order to develop the governing finite element equations for the entire stratum, the element relationships are summed over the stratum

$$\sum_{k=1}^{p} \frac{\partial({}^{k}I^{l})}{\partial\{{}^{k}W^{l}\}} = 0 \tag{11-45}$$

From the element pressure gradient equations it can be deduced that the global pressure gradient equations will have the form

$$[K]\{r\} + [P]\{\dot{r}\} = \{R\} \tag{11-46}$$

where $[K]$ = global velocity gradient matrix
$[P]$ = global dilatation rate matrix
$\{R\}$ = global stress history vector
$\{r\}$ = global pressure gradient vector

This is a system of n simultaneous equations, where n is one more than the number of elements.

The assembly process combines the element matrices $[k]$ and $[c]$ to form the global matrices $[K]$ and $[P]$, respectively. It also combines the element vectors $\{E\}$ to form the global vector $\{R\}$. The global vector $\{r\}$ is formed from the element vector $\{W\}$. Details of the assembly process are given elsewhere.[17]

11-8 SOLUTION OF THE GLOBAL FINITE ELEMENT EQUATIONS

The Crank-Nicholson procedure approximates the time derivatives by setting

$$\{r(t + \tau)\} = \{r(t)\} + \frac{\tau}{2}[\{\dot{r}(t + \tau)\} + \{\dot{r}(t)\}] \tag{11-47}$$

where τ is the time increment. The resulting set of simultaneous equations expressed in matrix form is

$$[M]\{r(t+\tau)\} = \{Z(t, \tau)\} \qquad \text{where} \qquad \{Z(t, \tau)\} = [N]\{V(t)\} + \{D\} \quad (11\text{-}48a)$$

and
$$[M] = [P] + \frac{\tau}{2}[K] \qquad [N] = [P] - \frac{\tau}{2}[K] \qquad \{D\} = \tau\{R\} \quad (11\text{-}48b)$$

Calculation Procedures

The procedural steps to accomplish a solution for $\{V(t+\tau)\}$ are basically those required to solve a system of simultaneous linear equations. Two cases are of importance: (1) the procedure when τ is a constant for all time steps and (2) the procedure when τ differs from time step to time step. For clarity the notation is changed to

$$[M_j]\{r_{j+1}\} = \{Z_j\} \qquad \text{where} \qquad \{Z_j\} = [N_j]\{r_j\} + \{D_j\} \quad (11\text{-}49)$$

in which the subscript j indicates that the calculations are occurring at time t_j. The time increment τ is defined as

$$\tau = t_{j+1} - t_j \quad (11\text{-}50)$$

If a constant time increment is used, the matrices $[M]$ and $[N]$ and the vector $\{D\}$ are assembled only once. If variable time steps are used, these matrices and vectors must be assembled whenever the time increment is changed.

11-9 A COMPUTER PROGRAM TO CALCULATE THE PROGRESS OF CONSOLIDATION USING A FINITE ELEMENT PROCEDURE

FECON1 (*f*inite *e*lement *con*solidation, *1* dimension) is a computer program to calculate the progress of consolidation. The program uses a finite element formulation. A Crank-Nicholson technique is used to account for the transient behavior of the excess pore pressure.

The logic of FECON1 is presented in flow chart form in Fig. 11-8. This is a conceptual flow chart designed to show the overall logic of the program. The logic shown in Fig. 11-8 and presented below is independent of the order of the elements.

1. The input data for a given problem are read. These data consist of the soil profile, the soil properties, the nodal point coordinates, the boundary conditions, the starting excess pore pressures (if any), the stress history (if any), the starting time, the time increment, termination controls, the printout times, and the printout spatial coordinates.
2. The input data are checked for compatibility. If incompatible, an error message is produced and the program aborts.

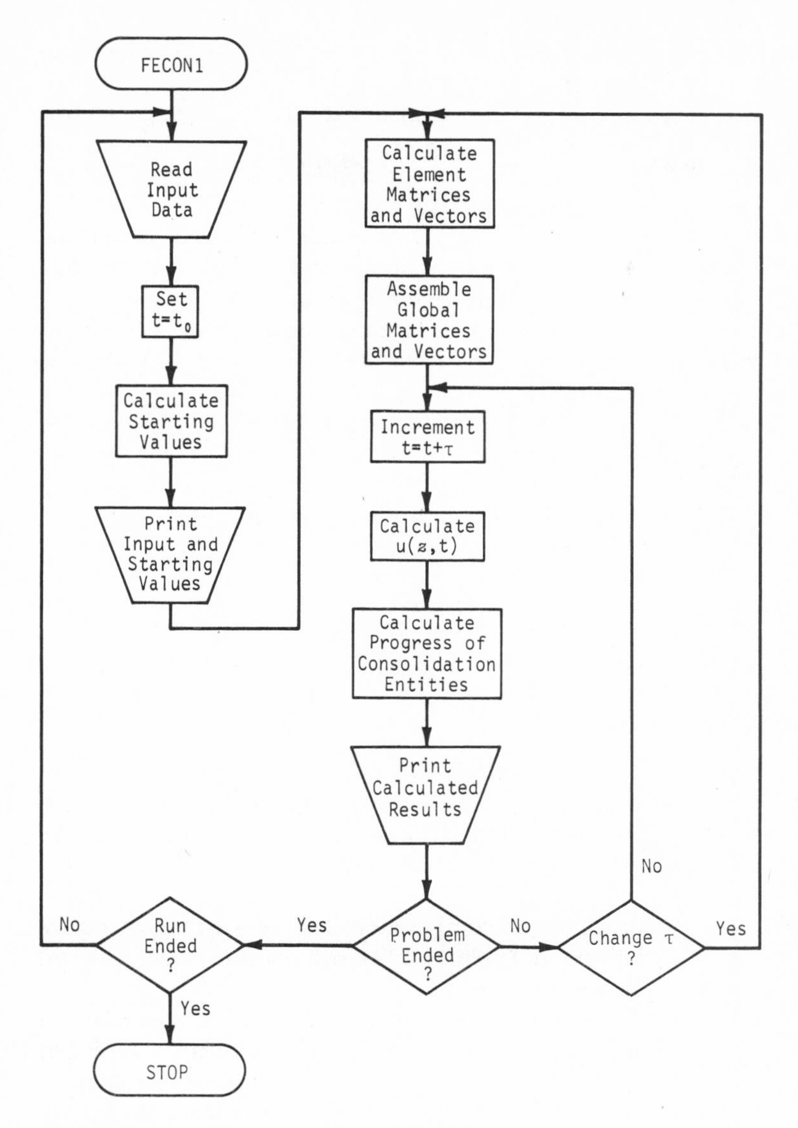

Figure 11-8 Conceptual flow chart for FECON1.

3. The starting time t_0 is set.
4. The starting values are calculated. These values are the starting excess pore pressures at the geometric nodal points and the settlements. They are calculated from the initial excess pore pressure and the stress history.
5. The input and starting values are printed.
6. The components of the element matrices $[k]$ and $[c]$ and the element vector $\{E\}$ are calculated.

7. The global matrices $[K]$ and $[P]$ and the global vector $\{R\}$ are assembled.
8. The time is incremented by adding τ.
9. The excess pore pressure and excess pore pressure gradient at each geometric nodal point are calculated.
10. The consolidation settlement, total settlement, degree of consolidation, and average excess pore pressures are calculated.
11. The calculated excess pore pressures, degree of consolidation, consolidation and total settlement, and average excess pore pressure are printed. The excess pore pressures are printed at the points specified by the input data.
12. The input data specify a termination. This termination is tested. If the problem termination has been reached, the run termination is tested. If not, the time increment τ is tested.
13. If the time increment τ for the next time step changes, the element matrices and vectors must be recalculated starting with step 6. If not, the time is incremented (step 8) and the calculations for the next time are processed.
14. Once a problem has been terminated, the problem control is tested to determine whether additional problem data are present. If not, the run is terminated. If additional data are to be input, the program starts again with the new input data.

11-10 A PRACTICAL EXAMPLE

The following is an example of the use of the preceding analysis and computer programs in solving practical problems concerning the progress of one-dimensional consolidation of multilayered soil systems. This problem concerns the development of excess pore pressures in the three-layer compressible soil system shown in Fig. 11-9.

The consolidation properties of the compressible system which exist before any construction activity are listed in Table 11-2. As will be noted later, the soil properties will change during construction.

The compressible soil system is overlaid by 3 ft of silt with a coefficient of permeability of 0.00175 ft^2/month. The total system of silt and clay is bounded at top and bottom by thick, free draining sand deposits. The groundwater table remains above the top of the silt at all times. The silt and clay layers are saturated. They remain in the saturated state during and after construction.

Table 11-2 In situ soil properties: three-layer system

Depth, ft	Layer no.	Thickness, ft	c_v, ft^2/month	m_v, ft^2/kip
0–5	1	5	6.23	0.014
5–10	2	5	3.40	0.005
10–18	3	8	10.25	0.036

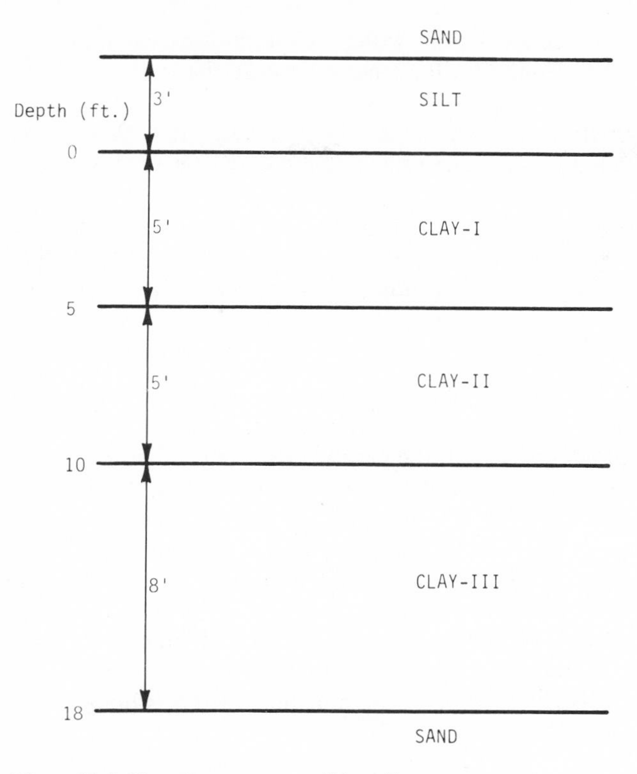

Figure 11-9 Three-layer compressible soil system.

The sequence of events to be analyzed is as follows:

1. Dewatering the site in preparation for construction
2. Erection of the structure
3. Removal of the pumps

The site is dewatered in the first $\frac{1}{2}$ month. It is assumed that the dewatering occurs at a uniform rate in this time. Due to artesian conditions in the lower sand, the total stress increment $\Delta\sigma$ at the end of dewatering is a linear function of depth with a value of 2.503 kips/ft^2 at the top of the clay stratum and -1.079 kips/ft^2 at the bottom of the stratum.

The erection phases commence at the end of dewatering ($t = 0.5$) and consist of the following:

1. Excavation, which occurs in a 2.5-month time span. The net surface load q at the end of excavation ($t = 3$) is -2.57 kips/ft^2.
2. Foundation construction, which occurs in a 1-month period after excavation. The load increment at the surface of the clay due to the foundation construction is 1.37 kips/ft^2.

3. Phase 1 of the superstructure construction, which occurs in a 6-month period after the foundation construction. The load increment at the surface of the clay due to this construction phase is 1.24 kips/ft^2.
4. Phase 2 of the superstructure construction, which occurs in a 9.5-month period after phase 1. The loading at the surface of the clay due to this phase is 2.75 kips/ft^2.
5. Phase 3 of the superstructure construction, which occurs in a 7.5-month period after phase 2. The load increment at the surface of clay due to this phase is 3.07 kips/ft^2. The structure is assumed to be completed at the end of phase 3.

It is assumed that all loadings and unloadings during the erection phases occur at a uniform rate. It is also assumed that the excavation and construction loadings are transmitted uniformly through the clay stratum as total stress increments $\Delta\sigma$. A tabulation of the surface load intensity q is presented in Table 11-3.

The removal of the pumps at 36 months after the commencement of dewatering decreases the total stress increment by 3.0 kips/ft^2 at the top of the stratum and increases the total stress increment by 0.61 kips/ft^2 at the bottom of the stratum.

The total stress increments $\Delta\sigma(z, t)$ at the various critical times are summarized in Table 11-4. The values presented are for the top ($z = 0$) and bottom ($z = 18$) of the stratum. The distribution of $\Delta\sigma$ within the stratum is linear.

Table 11-3 Surface load-intensity schedule

Time, months	Net load intensity q, kips/ft^2	Activity
0	0	
		Dewatering
0.5	0	
		Excavation
3.0	−2.57	
		Foundation construction
4.0	−1.20	
		Phase 1 Superstructure construction
10.0	0.04	
		Phase 2 Superstructure construction
19.5	2.79	
		Phase 3 Superstructure construction
27.0	5.86	
36.0	...	Removal of pumps

Table 11-4 Summary of total stress increments

Time t, months	$\Delta\sigma(0, t)$, kips/ft^2	$\Delta\sigma(18, t)$, kips/ft^2
0+	0	0
0.5	2.503	−1.079
3.0	−0.067	−3.649
4.0	1.303	−2.279
10.0	2.543	−1.039
19.5	5.293	1.711
27.0	8.363	4.781
36.0+	5.363	5.391

The laboratory data have shown that the soil properties c_v and m_v change with effective stress. A series of trial computer runs linearizes the problem by approximating the relationship between soil properties and effective stress with a relationship between soil properties and time. This relationship is shown in Table 11-5.

Using the data described above, a series of eight problems were prepared. Both PROGRS and FECON1 were used. These eight problems were set up for a single computer run, in which data were passed between successive problems. Thus the excess pore pressure calculations proceeded continuously with due regard to the change in soil properties with time.

Figure 11-10 presents the development of the total stress increment and excess pore pressure at the center of the middle layer ($z = 7.5$ ft). This type of result is useful in predicting and monitoring field behavior. Figures 11-11 to 11-13 present the profile of total stress increments and excess pore pressures at various stages of the construction process. Figure 11-11 presents the data at the

Table 11-5 Change of soil properties with time

	Clay I		Clay II		Clay III	
Time, months	c_v, ft^2/month	m_v, ft^2/kip	c_v, ft^2/month	m_v, ft^2/kip	c_v, ft^2/month	m_v, ft^2/kip
0+	6.23	0.014	3.40	0.005	10.25	0.036
0.5	5.27	0.010	2.45	0.004	8.36	0.031
3.0	7.43	0.016	4.10	0.005	12.28	0.042
4.0	6.72	0.023	3.82	0.006	10.97	0.053
10.0	6.42	0.021	3.72	0.004	10.63	0.045
19.5	5.87	0.017	3.00	0.003	9.24	0.038
27.0	5.55	0.012	2.87	0.003	8.76	0.036
36.0+	6.38	0.015	3.64	0.003	11.12	0.042

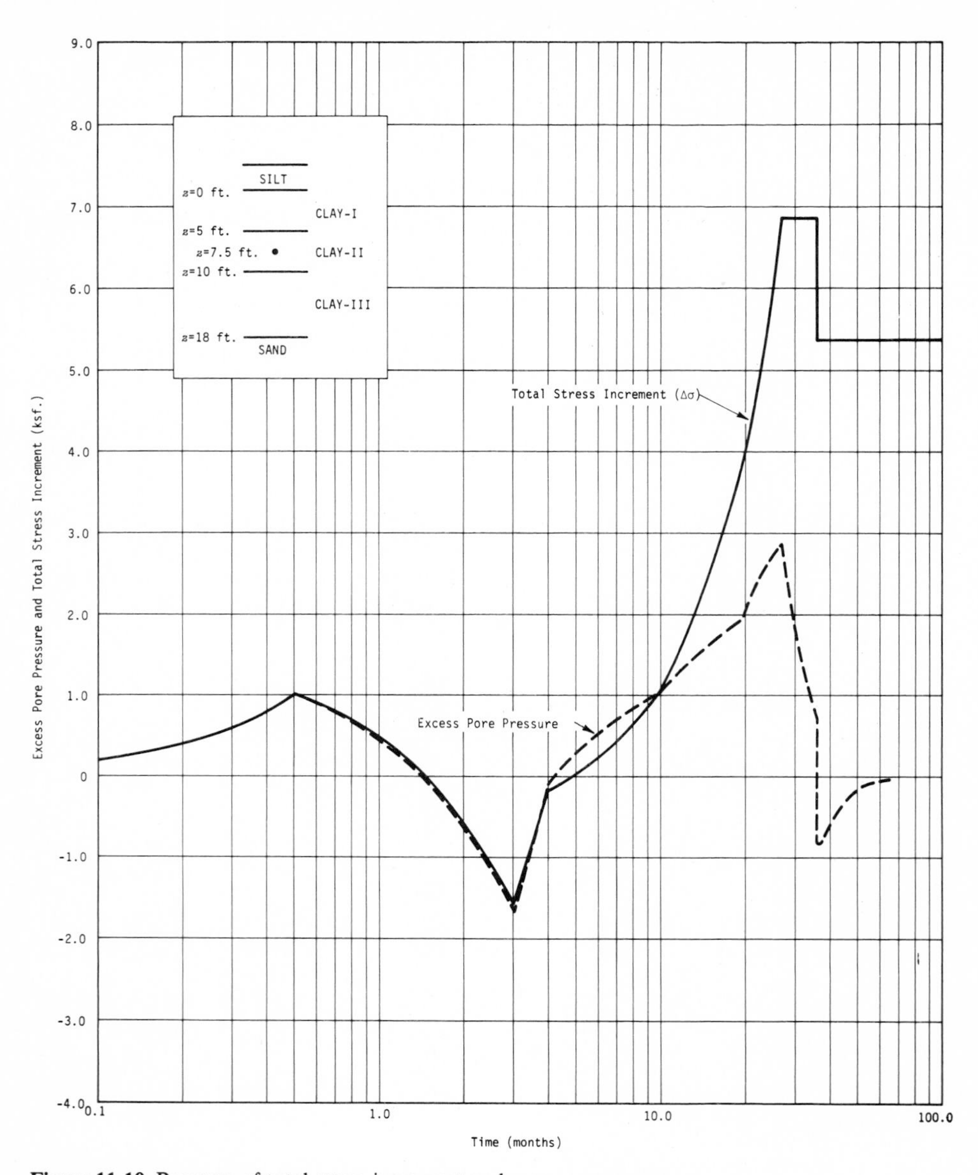

Figure 11-10 Progress of total stress increment and excess pore pressure.

end of 3 months, when the excavation is completed. Figure 11-12 presents the total stress increment which exists between 27 and 36 months, before the groundwater was raised. The excess pore pressure isochrones presented in Figure 11-12 are for two times, 27 months and 36 months. Figure 11-13 presents the total stress increment which exists immediately after the groundwater is restored to its

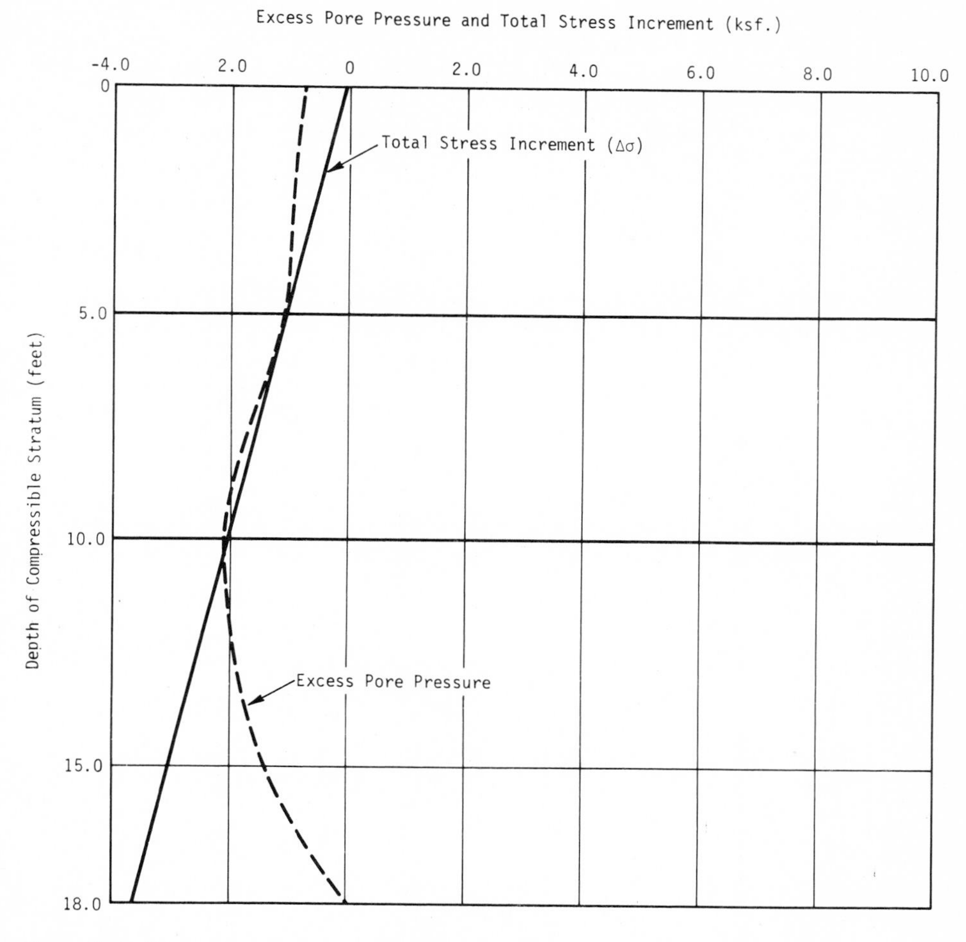

Figure 11-11 Excess pore pressure and total stress increments at the end of three months.

normal condition (36+ months) and which exists from that time forward. It also presents the excess pore pressure isochrones at 36+ and 50 months.

The computations were performed at the University of Colorado Computing Center. The hardware used was a CDC 6400. The software was the RUN 2.3 compiler and the KRONOS 2.0 operating system. The full set of eight interconnected problems were computed in one run on July 7, 1974. Compilation, loading, and execution from a FORTRAN source program was accomplished in 68.486 s of central processor time. Loading and execution from a binary object program was accomplished in 55.034 s of central processor time.

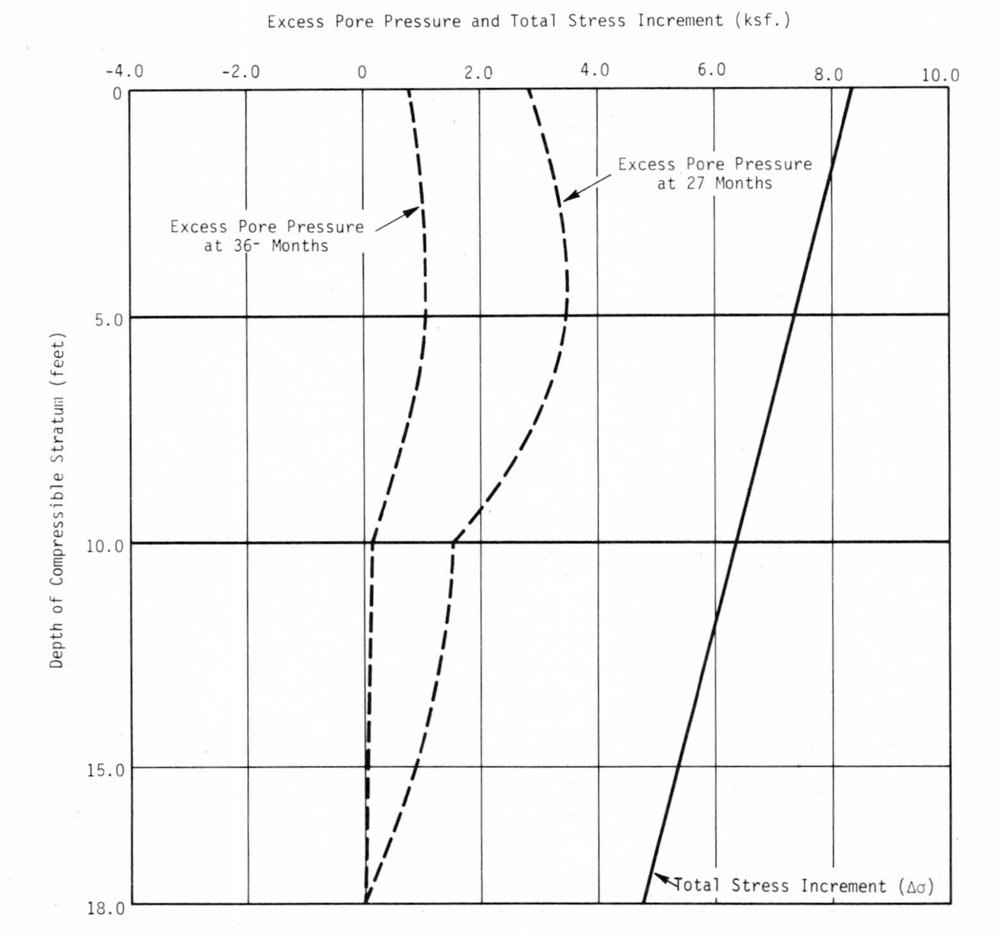

Figure 11-12 Excess pore pressure and total stress increments at the end of 27 and 36 months: before water table restoration.

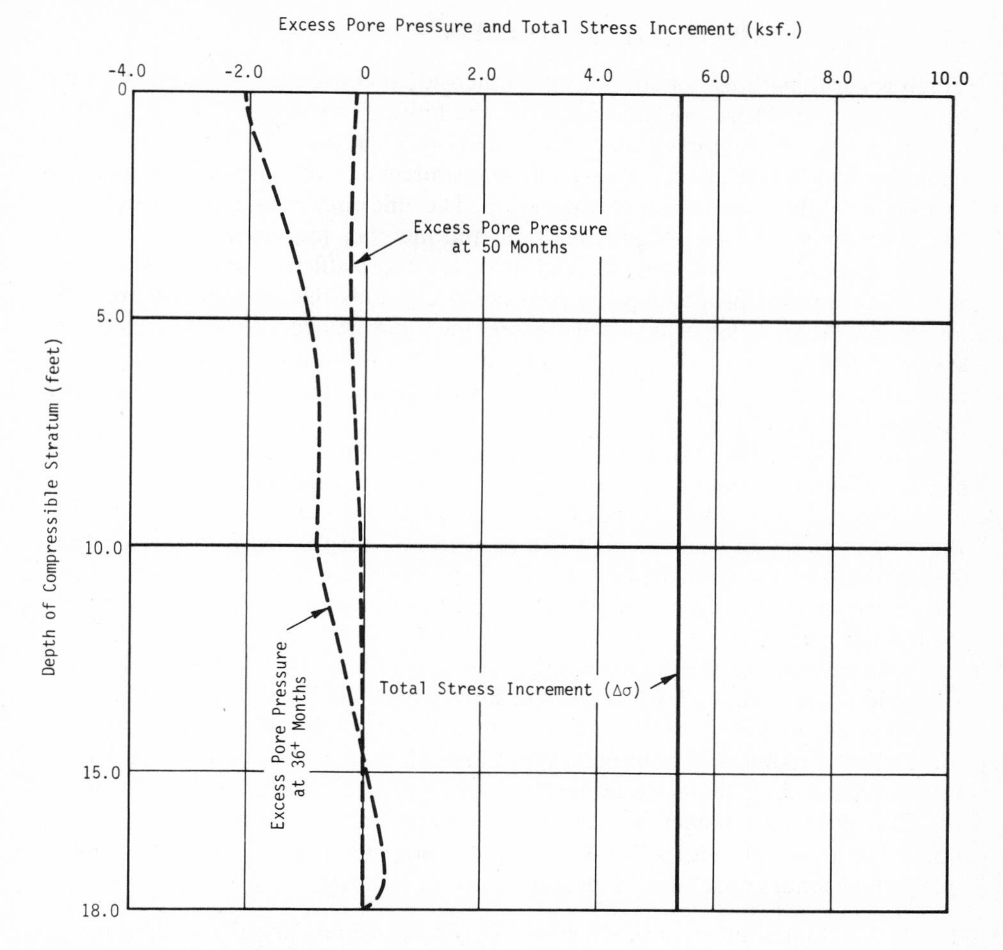

Figure 11-13 Excess pore pressure and total stress increments at the end of 36 and 50 months: after water table restoration.

11-11 COMPARISON OF METHODS

Three general methods for solving one-dimensional consolidation problems have been presented. These are the analytical, the finite element, and the finite difference procedures. A comparison of these methods for this class of problems can be made in terms of the efficiency of machine resources and the ease of implementing certain capabilities in a computer program. The efficiency of machine resources is best judged on the basis of machine time and memory requirements.

In general, the analytical procedure is the most efficient one on all counts when the physical problem being considered contains one or two compressible layers. As the number of layers increases, the efficiency rapidly degrades. If the system contains three or more layers, one of the two numerical approaches is considerably more efficient, both from the viewpoint of computational speed and memory requirements.

The capabilities of both the finite difference and finite element procedures are similar, with two exceptions. First, the finite element procedure is somewhat more amenable to the treatment of stress histories which are spatially dependent, since the stress history of each element can be readily incorporated into the solution. Space dependent stress histories can be solved by finite difference methods. Their use, however, from a user image viewpoint, is not as easy as by finite element techniques.

Finite difference techniques are generally established on the basis of a constant mesh size within a layer. Thus if in a given problem it is desirable to decrease the spatial mesh size in certain critical areas, the mesh size throughout the layer must also be reduced. The finite element procedure is substantially more flexible in this regard since there are no restrictions on the variation of element lengths.

The efficiency of machine resources for a given numerical technique depends upon a number of factors.[16] Some of the more important factors applying to one-dimensional consolidation problems are as follows:

1. The machine architecture, which in many cases will dictate the computational algorithms; e.g., the algorithms used to solve a given problem on pipeline computers may be vastly different from those used on parallel processing machines.
2. The operating system software can substantially affect the computational efficiency. Some machines and their operating systems are more attuned to solving a given class of problems than others. For the class of problems discussed here, the most efficient operating systems are those oriented to scientific calculation.
3. The programming techniques used can be a dominant feature in the efficiency of a computer program; e.g., the computer time to solve a given program can vary by a factor of 15, depending on the skill and ingenuity of the programmer.[6]

As a general rule, capabilities being equal, and considering the above factors, the finite difference procedures are the most efficient methods for solving one-dimensional consolidation problems by numerical means. For example, a series of bench mark problems were run on a CDC 6400 at the University of Colorado Computing Center. The operating system was KRONOS 2.0. The compiler was RUN 2.3. The following results were obtained:

1. The bench mark set was solved by PROGRS in 70 percent of the central processor time required to solve the set by means of FECON1 with first-order elements.
2. The memory requirements for PROGRS in the bench mark set cited in item 1 were approximately one-half those required for FECON1 with first-order elements.
3. The ratio of central processor times for the bench mark set cited in item 1 for FECON1 using first-, second-, third-, fourth-, and fifth-order elements was 1: 2.88: 3.97: 7.26: 8.88.
4. The ratio of the memory requirements in the progression cited in item 3 was 1: 1.27: 1.38: 1.81: 2.08.

From an analysis of the bench mark data two conclusions can be drawn: (1) finite difference procedures are generally more efficient users of machine resources than finite element procedures; (2) when finite element procedures are required, first-order linear elements provide the most efficient solution.

REFERENCES

1. Becker, M.: The Principles and Applications of Variational Methods, *Res. Monogr.* 27, The MIT Press, Cambridge, Mass., 1964.
2. Bulavin, P. E., and V. M. Kascheev: Solution of the Non-homogeneous Heat Conduction Equation for Multilayered Bodies, *Int. Chem. Eng.*, vol. 5, pp. 112–115, 1965.
3. Conte, S. D.: "Elementary Numerical Analysis," McGraw-Hill Book Company, New York, 1965.
4. Crank, J., and P. Nicholson: A Practical Method for Numerical Evaluation and Solution of Partial Differential Equations of the Heat-Conduction Type, *Proc. Camb. Phil. Soc.*, vol. 43, pp. 50–67, 1947.
5. Darcy, H.: "Les Fontaines publiques de la ville de Dijon," Dalmont, Paris, 1856.
6. David, E. E.: The Production of Software for Large Systems, Giant Computers, *INFOTECH State Arts Rep.* 2, INFOTECH Ltd., Maidenhead, England, pp. 403–414, 1971.
7. Desai, C. S., and J. F. Abel: "Introduction to the Finite Element Method," Van Nostrand Reinhold Company, New York, 1972.
8. Desai, C. S., and L. D. Johnson: Evaluation of Two Finite Element Formulations for One-dimensional Consolidation, *Comput. Struct.*, vol. 2, pp. 469–486, 1972.
9. Forchheimer, Ph.: "Hydraulik," pp. 26, 494, 495, Leipzig, 1914.
10. Frontard, J.: Notice sur l'accident de la dique de charmes, *Ann. Ponts Chaussees*, 9th ser., vol. 23, pp. 173–280, 1910.
11. Habetler, G. J., and R. L. Schiffman: A Finite Difference Method for Analyzing the Compression of Poro-viscoelastic Media, *Computing*, vol. 6, pp. 342–348, 1970.
12. Lanczos, C.: "The Variational Principles of Mechanics," 4th ed., University of Toronto Press, Toronto, 1970.
13. Mikhlin, S. G.: "Variational Methods in Mathematical Physics," The Macmillan Company, New York, 1964.

14. Myers, G. E.: "Analytical Methods in Conduction Heat Transfer," McGraw-Hill Book Company, New York, 1971.
15. Ozisik, M. N.: "Boundary Value Problems of Heat Conduction," International Textbook Company, Scranton, Pa., 1968.
16. Schiffman, R. L.: The Efficient Use of Computer Resources, pp. 91–129, in C. S. Desai (ed.), "Applications of the Finite Element Method in Geotechnical Engineering," Soil Mechanics Information Analysis Center, U.S. Army Engineer Waterways Experiment Station, Vicksburg, Miss., 1972.
17. Schiffman, R. L.: The Solution of One-dimensional Consolidation Problems, *Univ. Colo. Comput. Cent. Rep.* 73-22, 1973.
18. Schiffman, R. L., and R. E. Gibson: Consolidation of Non-homogeneous Clay Layers, *J. Soil Mech. Found. Div. ASCE*, vol. 90, no. SM5, proc. pap. 4043, pp. 1–30, 1964.
19. Schiffman, R. L., and J. R. Stein: One-dimensional Consolidation of Layered Systems, *J. Soil Mech. Found. Div. ASCE*, vol. 96, no. SM4, proc. pap. 7387, pp. 1499–1504, 1970.
20. Shankland, E. C.: Steel Skeleton Construction in Chicago, *Min. Proc. Inst. Civ. Eng.*, vol. 128, pp. 1–27, 1896.
21. Skempton, A. W.: Terzaghi's Theory of Effective Stress, pp. 42–53 in "From Theory to Practice in Soil Mechanics," John Wiley & Sons, Inc., New York, 1960.
22. Smith, W. S.: The Building Problem in Chicago from an Engineering Standpoint, *Univ. Ill. Technogr.*, no. 6, pp. 9–19, 1892.
23. Sorby, H. C.: On the Application of Quantitative Methods to the Study of the Structure and History of Rocks, *Q. J. Geol. Soc.*, vol. 64, pp. 171–231, 1908.
24. Statens Jarnvagars Geotekniska Kommission 1914–1922: Slutbetankande Avgivet Till Kungl. Jarnvagsstyrelsen, Den 31 Maj 1922, *Geotek. Medd.*, vol. 2, Stockholm, 1922.
25. Telford, T.: Inland Navigation, pp. 209–315 in "Edinburgh Encyclopedia," vol. 15, 1830.
26. Terzaghi, K.: Die physikalischen Grundlagen der technisch-geologischen Gutachtens, *Osterr. Ing. Archit. Verein Z.*, vol. 73, no. 36/37, pp. 237–241, 1921.
27. Terzaghi, K.: Die Berechnung der Durchlässigkeitsziffer des Tones aus dem Verlauf der hydrodynamischen Spannungserscheinungen, *Akad. Wiss. Wein Sitzungsber. Math-naturwiss Kl.*, pt. IIa, vol. 132, no. 3/4, pp. 125–138, 1923.
28. Terzaghi, K.: Die Beziehungen zwischen Elastizität und Innendruck, *Akad. Wiss. Wein Sitzungsber. Math-naturwiss Kl.*, pt. IIa, vol. 132, pp. 105–124, 1923.
29. Terzaghi, K.: Die Theorie der hydrodynamischen Spannungserscheinungen und ihr erdbautechnisches Anwendungsgebiet, *Proc. 1st Int. Congr. Appl. Mech., Delft, 1924*, pp. 288–294.
30. Terzaghi, K.: "Erdbaumechanik auf bodenphysikalischer Grundlage," F. Deuticke, Vienna, 1925.
31. Terzaghi, K.: Principles of Soil Mechanics, *Eng. News-Rec.*, vol. 95, pp. 742–746, 796–800, 832–836, 874–878, 912–915, 987–990, 1026–1029, 1064–1068, published as "Principles of Soil Mechanics," McGraw-Hill Book Company, New York, 1926.
32. Terzaghi, K., and O. K. Frohlich, "Theorie der Setzung von Tonschichten," F. Deuticke, Leipzig, 1936.
33. Tittle, C. W.: "Boundary-Value Problems in Composite Media: Quasi-orthogonal Functions," *J. Appl. Phys.*, vol. 36, pp. 1486–1488, 1965.
34. Wachspress, E. L.: The Numerical Solution of Boundary Value Problems, pp. 121–127 in A. Ralston and H. S. Wilf (eds.), "Mathematical Methods for Digital Computers," John Wiley & Sons, Inc., New York, 1960.
35. Wilson, E. L., and R. E. Nickell: Application of the Finite Element Method to Heat Conduction Analysis, *Nucl. Eng. Des.*, vol. 4, pp. 276–286, 1966.
36. Zienkiewicz, O. C.: "The Finite Element Method in Engineering Science," 2d ed., McGraw-Hill Publishing Company, Ltd., London.
37. Zienkiewicz, O. C., and C. J. Parekh: Transient Field Problems: Two-dimensional and Three-dimensional Analysis by Isoparametric Finite Elements, *Int. J. Numer. Methods Eng.*, vol. 2, pp. 61–71, 1970.

CHAPTER

TWELVE

TWO- AND THREE-DIMENSIONAL CONSOLIDATION

John T. Christian

The previous chapter discussed the use of numerical methods for problems involving consolidation of soils under conditions of strain and flow in one direction. The extension of the theory of consolidation to circumstances entailing deformation or flow in two or three directions raises the analytical and experimental complexity significantly. This chapter describes the application of numerical techniques to make the analytical problems tractable.

12-1 GEOTECHNICAL BACKGROUND AND MATHEMATICAL DERIVATIONS

Much of the necessary background for the theory of consolidation has been presented in Chap. 11, and we shall not repeat it here. Let us begin by considering the simplest case with the most restrictive assumptions:

1. Small strains and small displacements occur.
2. The skeleton of solid particles is linearly elastic in terms of effective stress.
3. The soil is saturated with incompressible fluid.
4. The flow of the pore fluid obeys Darcy's law.

From these assumptions one version of the Biot[1] theory can be derived.

The assumptions of linear elasticity and of small strain lead to the elastic stress-strain relations described in Chap. 2. For an isotropic material these become

$$\bar{\sigma}_{ij} = \bar{\lambda}\epsilon_{\text{vol}}\delta_{ij} + 2\bar{G}\epsilon_{ij} \tag{12-1}$$

where $\bar{\lambda}$ = Lamé constant = $\bar{E}\bar{v}/(1+\bar{v})(1-2\bar{v})$
ϵ_{vol} = volumetric strain
δ_{ij} = Kronecker delta = 1 for $i = j$ and = 0 for $i \neq j$
$\bar{G}$ = shear modulus = $\bar{E}/2(1+\bar{v})$

This equation is similar to Eqs. (2-53) and (2-59) except that here a bar over a quantity indicates that it is an effective stress or that it relates strain to effective stress. For plane strain conditions Eq. (2-45) specializes to

$$\begin{Bmatrix} \bar{\sigma}_{xx} \\ \bar{\sigma}_{yy} \\ \bar{\sigma}_{xy} \end{Bmatrix} = \frac{\bar{E}}{(1+\bar{v})(1-2\bar{v})} \begin{bmatrix} 1-\bar{v} & \bar{v} & 0 \\ \bar{v} & 1-\bar{v} & 0 \\ 0 & 0 & 1-2\bar{v} \end{bmatrix} \begin{Bmatrix} \epsilon_{xx} \\ \epsilon_{yy} \\ \epsilon_{xy} \end{Bmatrix} \tag{12-2}$$

In matrix notation, Eq. (12-2) becomes

$$\{\bar{\sigma}\} = [\bar{C}]\{\epsilon\} \tag{12-3}$$

Under conditions of plane strain $\bar{\sigma}_{zz} = \bar{v}(\bar{\sigma}_{xx} + \bar{\sigma}_{yy})$.

Now the volumetric stress-strain behavior can be obtained from

$$\bar{\sigma}_{\text{oct}} = \frac{\bar{\sigma}_{xx} + \bar{\sigma}_{yy} + \bar{\sigma}_{zz}}{3} = \frac{\bar{E}}{3(1-2\bar{v})}(\epsilon_{xx} + \epsilon_{yy} + \epsilon_{zz}) = \bar{K}\epsilon_{\text{vol}} \tag{12-4}$$

For plane strain,

$$\frac{\bar{\sigma}_{xx} + \bar{\sigma}_{yy}}{2} = \frac{\bar{E}}{2(1+\bar{v})(1-2\bar{v})}(\epsilon_{xx} + \epsilon_{yy}) = \bar{M}(\epsilon_{xx} + \epsilon_{yy}) = \bar{M}\epsilon_{\text{vol}} \tag{12-5}$$

For the one-dimensional oedometer condition,

$$\bar{\sigma}_{zz} = \frac{\bar{E}(1-\bar{v})}{(1+\bar{v})(1-2\bar{v})}\epsilon_{xx} = \bar{D}\epsilon_{zz} = \bar{D}\epsilon_{\text{vol}} \tag{12-6}$$

where $\bar{D}$, called the *constrained modulus*, is the inverse of m_v, the coefficient of volume compressibility in one-dimensional compression. Note that $\bar{M}$ relates the volumetric strain to an average in plane effective stress and $\bar{D}$ relates the volumetric strain to the vertical effective stress. In either of the last two cases Eq. (12-4) remains valid. Note that $\bar{M}$ is not a bulk modulus; neither is $\bar{D}$.

We now have three expressions relating effective stress to change in volume, one each for the general case, for plane strain, and for one-dimensional strain.

The flow law equation governed by Darcy's law is

$$v = -ki \tag{12-7}$$

where v = approach velocity
i = hydraulic gradient
k = coefficient of permeability (as defined in soil mechanics)

A more general way of stating Eq. (12-7) is

$$v_i = -\sum_{j=1}^{3} k_{ij} \frac{\partial h}{\partial x_j} \qquad \begin{matrix} i = 1, 2, 3 \\ j = 1, 2, 3 \end{matrix} \tag{12-8}$$

where h = total head
k_{ij} = permeability tensor
i, j = coordinate directions

Equation (12-8) can also be expressed in matrix form, which gives a permeability matrix $[R]$. The reader should distinguish between the bulk modulus K or $\bar{K}$, the stiffness matrix $[K]$, and the permeability k, k_{ij}, or $[R]$. If the soil is isotropic, only the diagonal $(i = j)$ terms of $[R]$ exist and

$$v_i = -k \frac{\partial h}{\partial x_i} \tag{12-9}$$

The net rate of flow of fluid out of an infinitesimal element of soil must be $\partial v_x/\partial x + \partial v_y/\partial y + \partial v_z/\partial z$, and for a soil saturated with incompressible fluid this must be the rate of decrease in volume, which is equal in magnitude to the volumetric strain rate but opposite in sign. Equation (12-9) gives

$$\frac{\partial \epsilon_{\text{vol}}}{\partial t} = k\left(\frac{\partial^2 h}{\partial x^2} + \frac{\partial^2 h}{\partial y^2} + \frac{\partial^2 h}{\partial z^2}\right) = k\,\nabla^2 h \tag{12-10}$$

The head is composed of an elevation component h_e plus a pressure head h_p. Obviously

$$\nabla^2 h_e = 0 \tag{12-11}$$

and so we can work with h_p alone. Further, h_p is related to the pore fluid stress p and the unit weight of water γ_w by

$$h_p = \frac{-p}{\gamma_w} \tag{12-12}$$

We use p instead of the usual u to avoid confusion with displacement. We also make p positive in tension to agree with the sign convention for stress. Then

$$\frac{\partial \epsilon_{\text{vol}}}{\partial t} = -\frac{k}{\gamma_w} \nabla^2 p \tag{12-13}$$

From Eq. (12-4)

$$\frac{\partial \epsilon_{\text{vol}}}{\partial t} = \frac{1}{\bar{K}} \frac{\partial \bar{\sigma}_{\text{oct}}}{\partial t} = \frac{1}{\bar{K}}\left(\frac{\partial \sigma_{\text{oct}}}{\partial t} - \frac{\partial p}{\partial t}\right) \tag{12-14}$$

Combining Eqs. (12-13) and (12-14) gives

$$\frac{k\bar{K}}{\gamma_w} \nabla^2 p = \frac{\partial p}{\partial t} - \frac{\partial \sigma_{oct}}{\partial t} \tag{12-15}$$

and the equations of equilibrium are

$$\begin{aligned} \frac{\partial \sigma_{xx}}{\partial x} + \frac{\partial \sigma_{xy}}{\partial y} + \frac{\partial \sigma_{xz}}{\partial z} &= 0 \\ \frac{\partial \sigma_{xy}}{\partial x} + \frac{\partial \sigma_{yy}}{\partial y} + \frac{\partial \sigma_{yz}}{\partial z} &= 0 \\ \frac{\partial \sigma_{xz}}{\partial x} + \frac{\partial \sigma_{yz}}{\partial y} + \frac{\partial \sigma_{zz}}{\partial z} &= \gamma_t \end{aligned} \tag{12-16}$$

Equations (12-15) and (12-16) form a set that must be solved with appropriate boundary and initial conditions. In particular, Eq. (12-15) has both p and σ_{oct} as dependent variables. If only p (or some other parameter) were the dependent variable, the equation would be a simple diffusion equation, which can be solved analytically or numerically for a variety of conditions. As it is, solution of Eq. (12-15) requires the addition of the equations of equilibrium and involves a complicated coupled system. It is not surprising that very few practical solutions have been obtained; Ref. 2 lists most of the readily available ones.

Uncoupled Theory; Rendulic Assumption

It is useful to ask under what circumstances Eq. (12-15) can be reduced to one dependent variable. The best known case is that of one-dimensional consolidation, in which Eq. (12-14) becomes

$$\frac{\partial \epsilon_{vol}}{\partial t} = \frac{1}{\bar{D}}\left(\frac{\partial \sigma_{zz}}{\partial t} - \frac{\partial p}{\partial t}\right) \tag{12-17}$$

However, equilibrium requires that σ_{zz} be constant for a constant applied load; hence

$$\frac{\partial \epsilon_{vol}}{\partial t} = -\frac{1}{\bar{D}} \frac{\partial p}{\partial t} \tag{12-18}$$

and

$$\frac{k\bar{D}}{\gamma_w} \nabla^2 p = \frac{k}{\gamma_w m_v} \nabla^2 p = \frac{\partial p}{\partial t} \tag{12-19}$$

which is Terzaghi's well known equation of one-dimensional consolidation.

The second circumstance under which the equation can be reduced to one dependent variable is that in which σ_{oct} is a constant over time everywhere in the soil. This attractive assumption is the basis of the Rendulic,[3] or pseudo-three-dimensional, theory. It leads to an equation that is identical to that of the diffusion process, an equation that has been extensively treated in the literature of engineering mechanics. This simplified approach has been applied to several practical problems in soil mechanics[4] and was recommended by Terzaghi.[5]

The resulting diffusion equation is usually written

$$c_v \nabla^2 p = \frac{\partial p}{\partial t} \tag{12-20}$$

What should be used for c_v? Expressed in terms of elastic parameters, the one-dimensional c_v is

$$c_{v1} = \frac{k}{\gamma_w m_v} = \frac{\bar{D}}{\gamma_w} = \frac{\bar{E}}{\gamma_w} \frac{1 - \bar{\nu}}{(1 + \bar{\nu})(1 - 2\bar{\nu})} \tag{12-21a}$$

For plane strain conditions it is

$$c_{v2} = \frac{\bar{M}}{\gamma_w} = \frac{\bar{E}}{\gamma_w} \frac{1}{2(1 + \bar{\nu})(1 - 2\bar{\nu})} \tag{12-21b}$$

and for the general three-dimensional case it is

$$c_{v3} = \frac{k\bar{K}}{\gamma_w} = \frac{\bar{E}}{\gamma_w} \frac{1}{3(1 - 2\bar{\nu})} \tag{12-21c}$$

It is clear that even when the simplified theory is used, c_v must be chosen on the basis of the geometry of the problem and is not a simple material constant. Figure 12-1 illustrates the dependence of c_v on Poisson's ratio.

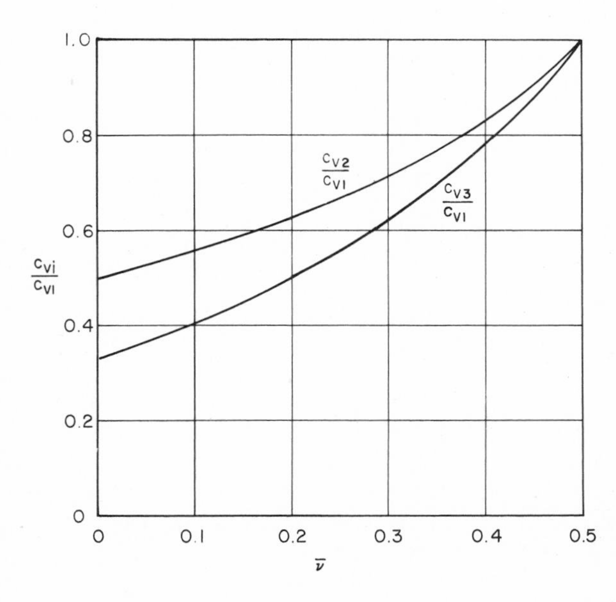

Figure 12-1 Ratio of coefficients of consolidation as functions of Poisson's ratio.

When the full three-dimensional theory is applied to simple problems, it is observed that in certain regions the pore pressure rises before it decays. The phenomenon is called the *Mandel-Cryer effect*, after Mandel,[6] who demonstrated it for the consolidation of a brick shaped body loaded uniaxially under conditions of plane strain, and Cryer,[7] who demonstrated it for the consolidation of a uniformly squeezed sphere drained at its surface. Figure 12-2 shows the time history of pore pressure at the center of the sphere for different Poisson's ratios. Experimental results have confirmed that the Mandel-Cryer effect does occur for overconsolidated soils though not as strongly as theory would predict.

The Rendulic simplified theory does not predict the Mandel-Cryer effect because it involves a pure diffusion and hence a monotonic decrease in pore fluid pressure. This limitation of the simplified theory could be a serious drawback for conditions in which an increase in pore pressure is important. When it is possible to do so, the full theory should be used.

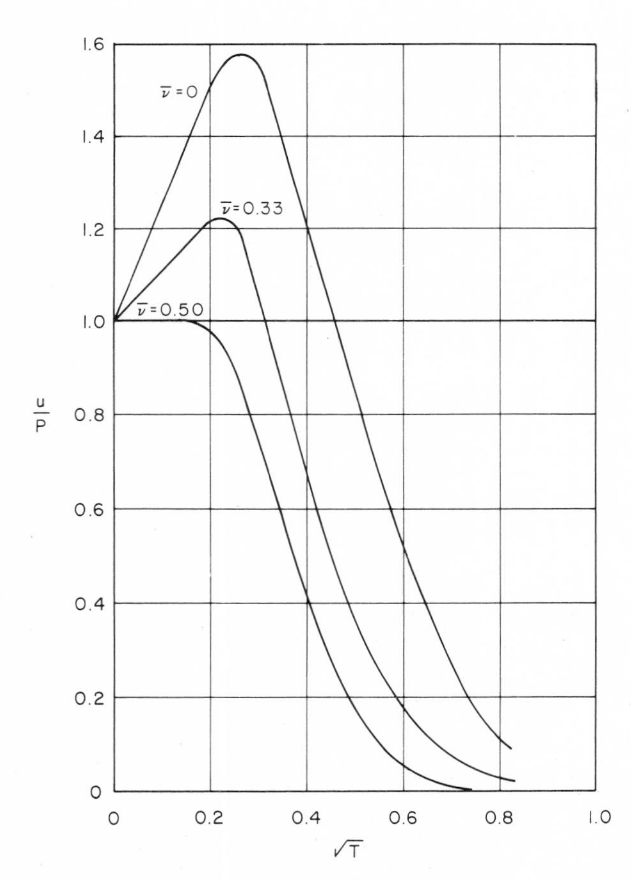

Figure 12-2 Pore pressure vs. time in a spherical sample.

12-2 FINITE ELEMENT FORMULATION: FUNCTIONAL MINIMIZATION

To convert Eqs. (12-15) and (12-16) into a form suitable for solution by finite elements, two approaches are possible. A generalized variational principle can be applied to the static equations of equilibrium (12-15), giving a formulation in terms of unknown displacements and pore pressures. Such a technique has been proposed by Herrmann[8] and used by Christian[9] in the solution of static stress distribution for undrained, incompressible soils. The time-dependent change in the stresses and pore pressures is then introduced by applying Eq. (12-13) in a finite difference form. Either explicit or implicit marching procedures are possible. This procedure has been used by Christian et al.[2,10]

A second and more general approach was proposed by Sandhu and Wilson[11] and has been used by many subsequent authors.[12–16] This method can elegantly incorporate a variety of elements and of marching schemes. Therefore, we use it here as the basis for the finite element approach. We shall consider the physical meaning of the terms later in the chapter.

The analysis for a linearly elastic soil skeleton and incompressible pore fluid begins from the functional

$$\Omega(u, p) = \int_v [\tfrac{1}{2}\bar{\sigma}_{ij} * \epsilon_{ij} - \rho F_i * u_i + p * u_{i,i} - \tfrac{1}{2}g * q_i * (p_{,i} + \rho_w F_i)]\, dV$$

$$- \int_{S_1} (\bar{T}_i * u_i)\, dS + \int_{S_2} (g * \bar{Q} * p)\, dS \quad (12\text{-}22)$$

In this equation repeated subscripts are summed from 1 to 3 and a comma indicates differentiation. Thus

$$u_{i,i} = \frac{\partial u_1}{\partial x_1} + \frac{\partial u_2}{\partial x_2} + \frac{\partial u_3}{\partial x_3} \quad (12\text{-}23)$$

The terms in the functional are defined as follows:

$\bar{\sigma}_{ij}$ = effective stress tensor
ϵ_{ij} = strain tensor = $\frac{1}{2}(u_{i,j} + u_{j,i})$
u_i = displacement vector
ρ = total mass density of the soil
ρ_w = mass density of fluid
F_i = body-force components, usually gravity
p = fluid pressure (tension positive)
$g = 1$
q_i = relative approach velocity of the fluid = $k_{ij}(\rho_{,j} + \rho_w F_i)$, where k_{ij} is the permeability tensor
$\bar{T}_i$ = prescribed boundary tractions on S_1
$\bar{Q}$ = prescribed fluid flow normal to surface S_2

The $*$ indicates convolution and is defined as

$$A * B = \int_0^t A(\ldots, \tau)B(\ldots, t - \tau)\, d\tau \tag{12-24}$$

The functional of Eq. (12-22) can be found from Gurtin's[17] approach, in which the Laplace transform is used on differential equations with time as an independent variable and the resulting equation is divided through by the transform variable s. The inverse of the result gives the functional desired. Alternatively, the Galerkin method of weighted residuals (Chap. 1) could be used to obtain the same result, or the general variational formulations[11,18–21] could be invoked.

Let us write Eq. (12-22) for finite elements in matrix notation:

$$\begin{aligned}\Omega(u, p) = \Bigg(&\int_V \tfrac{1}{2}\{\bar{\sigma}\}^T * \{\epsilon\} - \int_V \{\rho F\}^T * \{u\}\, dV + \int_V p * \epsilon_{\text{vol}}\, dV \\ &- \frac{1}{2}\int_V g * \{q\}^T * (\{p_{,i}\} + \{\rho_w F\}\Bigg)\, dV - \int_{S_1} \{\bar{T}\}^T * \{u\}\, ds \\ &+ \int_{S_2} g * \bar{Q} * ds\end{aligned} \tag{12-25}$$

We can express each of these in terms of the nodal values of displacements and pore pressures. For generality we assume different nodes are used for displacements and pore pressures. Letting $\{q_m\}$ be the nodal displacements, $\{p_m\}$ the nodal pressures, and $\{\bar{T}_m\}$ and $\{\bar{Q}_m\}$ the nodal tractions and fluxes prescribed as boundary conditions, we find

$$\begin{aligned}&\{u\} = [N_u]\{q_m\} & &p = [N_p]\{p_m\} & &\{\epsilon\} = [B_e]\{q_m\} \\ &\{\bar{\sigma}\} = [\bar{C}][B_e]\{q_m\} + \{\bar{\sigma}_0\} & &\{\epsilon_{\text{vol}}\} = [B_\Delta]\{q_m\} & &\{p_{,i}\} = [B_q]\{p_m\} \\ &\{q\} = [k]([B_q]\{p_m\} + \{p_w F\}) & &\{\bar{T}\} = [N_u]\{\bar{T}_m\} & &\bar{Q} = [N_p]\{\bar{Q}_m\}\end{aligned} \tag{12-26}$$

and

$$\begin{aligned}\Omega(u, p) = \sum_{m=1}^{M} \Bigg(&\int_v (\tfrac{1}{2}\{q_m\}^T[B_e]^T[\bar{C}] * [B_e]\{q_m\} + \{p_m\}^T[N_p]^T * [B_\Delta]\{q_m\} \\ &+ \tfrac{1}{2}\{\bar{\sigma}_0\}^T * [B_e]\{q_m\} - \{q_m\}^T[N_u]^T * \{\rho F\} - \tfrac{1}{2}g * \{p_m\}^T[B_q]^T[k] * [B_q]\{p_m\} \\ &- g * \{p_m\}^T[B_q]^T[k] * \{\rho_w F\} - \tfrac{1}{2}g * \{\rho_w F\}^T[k] * \{\rho_w F\})\, dv \\ &- \int_{S_1} \{q_m\}^T[N_u]^T * [N_u]\{\bar{T}_m\}\, ds + \int_{S_2} g * \{p_m\}^T[N_p]^T * [N_p]\{\bar{Q}_m\}\, ds\Bigg)\end{aligned} \tag{12-27}$$

The vector of all nodal displacements for all elements we now write $\{r_q\}$ and the vector of all nodal pore pressures $\{r_p\}$. If we now understand that the summations mean collecting terms consistently with the proper assemblage of the element stiffnesses into a global stiffness, Eq. (12-27) becomes

$$\Omega(u, p) = \tfrac{1}{2}\{q\}^T[K_1] * \{r_q\} + \{q\}^T\{M_1\} * g + \{q\}^T[K_3]\{r_p\} - \{q\}^T\{M_2\} * g$$
$$- \tfrac{1}{2}g * \{r_p\}^T[K_2] * \{r_p\} - g * \{r_p\}^T\{M_3\} * g - \tfrac{1}{2}g * M_4 * g$$
$$- \{r_q\}^T * \{P_1\} + g * \{r_p\}^T * \{P_2\} \tag{12-28}$$

where

$$[K_1] = \sum_{m=1}^{M} \int_V [B_e]^T[\bar{C}][B_e]\, dV \qquad [K_2] = \sum_{m=1}^{M} \int_V [B_q]^T[k][B_q]\, dV$$

$$[K_3] = \sum_{m=1}^{M} \int_V [B_\Delta]^T[N_p]\, dV \qquad \{M_1\} = \sum_{m=1}^{M} \int_V [B_e]^T\{\bar{\sigma}_0\}\, dV$$

$$\{M_2\} = \sum_{m=1}^{M} \int_V [N_u]^T\{\rho F\}\, dV \qquad \{M_3\} = \sum_{m=1}^{M} \int_V [B_q]^T[k]\{\rho_w F\}\, dV$$

$$\{M_4\} = \sum_{m=1}^{M} \int_V \{\rho_w F\}^T[k]\{\rho_w F\}\, dV \qquad \{P_1\} = \sum_{m=1}^{M} \int_{S_1} [N_u]^T[N_u]\{\bar{T}_m\}\, dS$$

$$\{P_2\} = \sum_{m=1}^{M} \int_{S_2} [N_p]^T[N_p]\{\bar{Q}_m\}\, dS$$

Note that some of these terms require moving the $*$ operator among constant terms, which is permissible for a linear operation such as convolution. Taking the first variation of (12-28) with respect to $\{r_q\}$ and setting the result to zero gives

$$[K_1]\{r_q\} + [K_3]\{r_p\} = -\{M_1\} + \{M_2\} + \{P_1\} \tag{12-29a}$$

The same operation with respect to $\{r_p\}$ gives

$$[K_3]^T\{r_q\} - g * [K_2]\{r_p\} = g * \{M_3\} - g * \{P_2\} \tag{12-29b}$$

Time Integration

In Eqs. (12-28) and (12-29) $\{r_q\}$ and $\{r_p\}$ are functions of time, and Eq. (12-29*b*) contains three convolutions. This requires that an approximation be made for the variation of $\{r_p\}$ with time. Each term in the convoluted vectors can be written

$$g * f(t) = \int_0^t f(\tau)g(t - \tau)\, d\tau = \int_0^t f(\tau)\, d\tau \tag{12-30}$$

since $g(t - \tau) = 1$ by definition. In a discrete interval from t_{n-1} to t_n $(= t_{n-1} + \Delta t)$ the integration can be approximated by

$$\int_{t_{n-1}}^{t_n} f(\tau)\, d\tau = \alpha\, \Delta t f(t_n) + (1 - \alpha)\, \Delta t f(t_{n-1}) \tag{12-31}$$

The value of α depends on the way $f(t)$ is assumed to change during the interval. A purely explicit scheme such as that used by Christian et al.[2,10] has $\alpha = 0$. A purely implicit scheme has $\alpha = 1$. Booker and Small[22] have shown that any scheme with $\alpha \geqslant \frac{1}{2}$ is unconditionally stable.

Sandhu[18] suggested that one reasonable assumption is that $f(t)$ varies linearly between t_{n-1} and t_n. This leads to $\alpha = \frac{1}{2}$. He also proposed assuming that $f(t)$ varies logarithmically between t_{n-1} and t_n:

$$f(\tau) = f(t_{n-1}) + [f(t_n) - f(t_{n-1})] \frac{\ln(\tau + 1)}{\ln(\Delta t + 1)} \tag{12-32}$$

This leads to

$$\alpha = 1 + \frac{1}{\Delta t} - \frac{1}{\ln(1 + \Delta t)} \tag{12-33}$$

Hwang et al.[14] used a similar but variable α, based on a dimensionless time τ:

$$\tau = \frac{t - t_{n-1}}{t_{n-1}} \qquad \tau_0 = \frac{\Delta t}{t_{n-1}} \tag{12-34}$$

Then

$$f(\tau) = f(t_{n-1}) + [f(t_n) - f(t_{n-1})] \frac{\ln(1 + \tau)}{\ln(1 + \tau_0)} \tag{12-35}$$

Integration leads to

$$\alpha = 1 + \frac{1}{\tau_0} - \frac{1}{\ln(1 + \tau_0)} \tag{12-36}$$

which is similar to Eq. (12-33) except that α varies with time.

It turns out that for either of the logarithmic cases $\alpha > \frac{1}{2}$ for all positive Δt and $\lim_{\Delta t \to 0} \alpha = \frac{1}{2}$. Therefore, these schemes are unconditionally stable.

Substitution of Eq. (12-31) into Eq. (12-29*b*) gives

$$\begin{aligned} [K_3]^T\{r_q(t_n)\} - \alpha\,\Delta t[K_2]\{r_p(t_n)\} = {} & [K_3]^T\{r_q(t_{n-1})\} + \Delta t(1-\alpha)[K_2]\{r_p(t_{n-1})\} \\ & + \alpha\,\Delta t\{M_3(t_n)\} + \Delta t(1-\alpha)\{M_3(t_{n-1})\} \\ & - \alpha\,\Delta t\{P_2(t_n)\} - \Delta t(1-\alpha)\{P_2(t_{n-1})\} \end{aligned} \tag{12-37}$$

When this is combined with Eq. (12-28), the entire system can be written as one set of matrix equations:

$$\begin{bmatrix} [K_1] & [K_3] \\ [K_3]^T & -\alpha\,\Delta t[K_2] \end{bmatrix} \begin{Bmatrix} r_q(t_n) \\ r_p(t_n) \end{Bmatrix} = \begin{Bmatrix} R_Q(t_n) \\ R_p(t_n) \end{Bmatrix} \tag{12-38}$$

Defining a general coefficient matrix $[K]$ and combined vectors of displacements and pore pressures and of loads and strains gives

$$[K]\{r\} = \{R\} \tag{12-39}$$

In these equations:

$$\begin{aligned} \{R_Q(t_n)\} &= -\{M_1(t_n)\} + \{M_2(t_n)\} - \{P_1(t_n)\} \\ \{R_p(t_n)\} &= [K_3]^T\{r_q(t_{n-1})\} + \Delta t(1-\alpha)[K_2]\{r_p(t_{n-1})\} \\ &\quad + \alpha\,\Delta t\{M_3(t_n)\} + \Delta t(1-\alpha)\{M_3(t_{n-1})\} - \alpha\,\Delta t\{P_2(t_n)\} \\ &\quad - \Delta t(1-\alpha)\{P_2(t_{n-1})\} \end{aligned} \tag{12-40}$$

All the terms in the coefficient matrix $[K]$ on the left side of Eqs. (12-38) and (12-39) can be assumed to be constant throughout the consolidation process or can be varied for different time intervals composing the entire time domain.

Choice of Elements

The results of the analysis in the preceding sections can only be applied when specific elements have been chosen. Several options have been employed by different researchers. Most applications have been limited to plane strain, and so further development will be limited to that case.

Sandhu and Wilson[11] introduced the composite element, illustrated in Fig. 12-3, consisting of a six-noded triangle for the displacement expansion, only three nodes being used for the pore pressures. The displacements vary quadratically over the element, while the stresses and strains obtained by differentiating the displacements vary linearly. The pore pressures vary linearly, too. Therefore, the element has the same degree of expansion for all stress components, both effective stresses and pore pressures.

The displacement expansion is required for displacements in two coordinate directions, u and v

$$\begin{Bmatrix} u \\ v \end{Bmatrix} = [N_u]\{q\} \tag{12-41a}$$

where $$\{q\}^T = [u_1 \quad u_2 \quad \cdots \quad u_6 \quad v_1 \quad v_2 \quad \cdots \quad v_6]$$

and $$[N_u] = \begin{bmatrix} [N_i] & [0] \\ [0] & [N_i] \end{bmatrix} = \begin{bmatrix} [N_1 \quad N_2 \quad \cdots \quad N_6] & [0] \\ [0] & [N_1 \quad N_2 \quad \cdots \quad N_6] \end{bmatrix} \tag{12-41b}$$

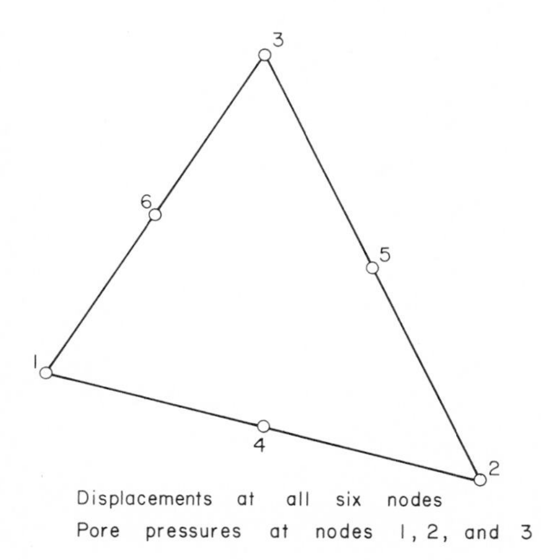

Figure 12-3 Linear strain and constant hydraulic gradient element.

The triangular coordinates (Fig. 12-4) are used to express $[N_i]$ as

$$[N_i] = [N_1 \quad N_2 \quad N_3 \quad N_4 \quad N_5 \quad N_6]$$
$$= [\zeta_1(2\zeta_1 - 1) \quad \zeta_2(2\zeta_2 - 1) \quad \zeta_3(2\zeta_3 - 1) \quad 4\zeta_1\zeta_2 \quad 4\zeta_2\zeta_3 \quad 4\zeta_1\zeta_3]$$
$$i = 1, 2, \ldots, 6 \qquad (12\text{-}41c)$$

ζ_i are defined in Fig. 12-4.

Similarly

$$[N_p] = [\zeta_1 \quad \zeta_2 \quad \zeta_3] = \{\zeta\}^T \qquad (12\text{-}42)$$

The $[B]$ matrices relating strains to displacements are obtained by differentiation. For example,

$$[N_u]_{,x} = \frac{1}{2A}[(4\zeta_1 - 1)b_1 \quad (4\zeta_2 - 1)b_2 \quad (4\zeta_3 - 1)b_3$$
$$4(\zeta_2 b_1 + \zeta_1 b_2) \quad 4(\zeta_2 b_3 + \zeta_3 b_2) \quad 4(\zeta_1 b_3 + \zeta_3 b_1)] \qquad (12\text{-}43a)$$

$$[N_u]_{,y} = \frac{1}{2A}[(4\zeta_1 - 1)a_1 \quad (4\zeta_2 - 1)a_2 \quad (4\zeta_3 - 1)a_3$$
$$4(\zeta_2 a_1 + \zeta_1 a_2) \quad 4(\zeta_2 a_3 + \zeta_3 a_2) \quad 4(\zeta_1 a_3 + \zeta_3 a_1)] \qquad (12\text{-}43b)$$

and hence

$$[B_e] = \frac{1}{2A}\begin{bmatrix} [N_u]_{,x} & [0] \\ [0] & [N_u]_{,y} \\ [N_u]_{,y} & [N_u]_{,x} \end{bmatrix} \qquad (12\text{-}43c)$$

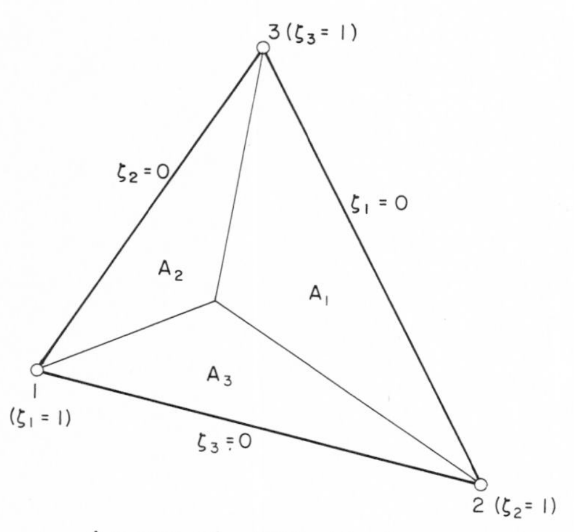

Figure 12-4 Triangular coordinates.

This can also be expressed

$$[B_e] = \frac{1}{2A}\begin{bmatrix} \{\zeta\}^T[U] & [0] \\ [0] & \{\zeta\}^T[V] \\ \{\zeta\}^T[V] & \{\zeta\}^T[U] \end{bmatrix} \tag{12-43d}$$

$$\text{where } [U] = \frac{1}{2A}\begin{bmatrix} 3b_1 & -b_2 & -b_3 & 4b_2 & 0 & 4b_3 \\ -b_1 & 3b_2 & -b_3 & 4b_1 & 4b_3 & 0 \\ -b_1 & -b_2 & 3b_3 & 0 & 4b_2 & 4b_1 \end{bmatrix} \tag{12-43e}$$

$$[V] = \frac{1}{2A}\begin{bmatrix} 3a_1 & -a_2 & -a_3 & 4a_2 & 0 & 4a_3 \\ -a & 3a_2 & -a_3 & 4a_1 & 4a_3 & 0 \\ -a_1 & -a_2 & 3a_3 & 0 & 4a_2 & 4a_1 \end{bmatrix} \tag{12-43f}$$

Also

$$[B_\Delta] = \frac{1}{2A}[\{\zeta\}^T[u] \quad \{\zeta\}^T[V]] \tag{12-44}$$

The pore pressure expansion involves

$$[B_q] = \frac{1}{2A}\begin{bmatrix} b_1 & b_2 & b_3 \\ a_1 & a_2 & a_3 \end{bmatrix} \tag{12-45}$$

These expressions can now be substituted into Eqs. (12-28) to obtain the matrices and vectors required for setting up the equations. The only variables will be ζ_1, ζ_1, or ζ_3, and any polynomial function of these can be integrated analytically over a triangle by using

$$\int_{\text{area}} \zeta_1^a \zeta_2^b \zeta_3^c \, dA = \frac{a!b!c!}{(a+b+c+2)!} 2A \tag{12-46}$$

Use of the foregoing formulation with the triangular element (Fig. 12-4) does not require numerical integration since the terms in the integrand can be integrated in closed form.

Yokoo et al.[13] used several different elements, all of which used the same expansion for the displacements and for the pore pressures. This makes $[N_p] = [N_u]$ for any choice of element, and the rest of the matrices are derived simply. The examples they presented include a two-noded bar element, a three-noded axially symmetric triangular ring element, and a four-noded rectangle.

Ghaboussi and Wilson[12] used an isoparametric element of four nodes in which the pore pressures were expanded according to the standard formula:

$$u = \sum_{i=1}^{4} h_i Q_{xi} \qquad v = \sum_{i=1}^{4} h_i q_{yi} \tag{12-47}$$

where the h_i are the same terms needed to describe the geometry of the element.

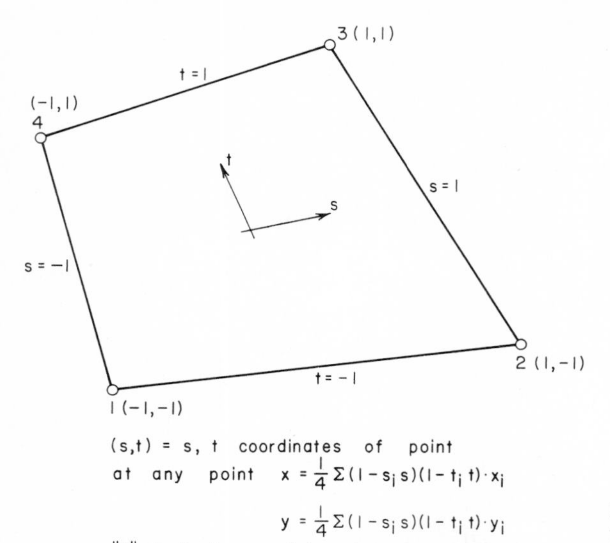

Figure 12-5 Isoparametric quadrilateral consolidation element.

Referring to Fig. 12-5, we can write

$$h_i = \tfrac{1}{4}(1 + s_i s)(1 + t_i t) \tag{12-48}$$

The displacement expansion used two additional nonconforming degrees of freedom:

$$u = \sum_{i=1}^{4} h_i q_{x_i} + (1 - s^2)u_5 + (1 - t^2)v_5 \tag{12-49}$$

$$v = \sum_{i=1}^{4} h_i q_{y_i} + (1 - s^2)u_5 + (1 - t^2)v_5 \tag{12-50}$$

where $q_{x_i} = u_1, u_2, u_3, u_4$ and $q_{y_i} = v_1, v_2, v_3, v_4$.

The two additional degrees of freedom are eliminated by static condensation after the element stiffness is completed. The derivation of the various matrices is straightforward, but numerical integration is necessary for nonrectangular elements.

A Hybrid Approach

Christian and Boehmer[2] used a somewhat different approach. They started from the principle of virtual work expressed in terms of effective stresses and pore pressures:

$$\int_V \{\bar{\sigma}\}^T \, \delta\{\epsilon\} \, dV + \int_V p\{L\}^T \, \delta\{\epsilon\} \, dV = \int_V \{\rho F_i\} \, \delta\{u\} + \int_S \{\bar{T}\}^T \, \delta\{u\} \, dS \tag{12-51}$$

where $\{L\} = \{1 \quad 1 \quad 0\}$ for plane strain and $\{1 \quad 1 \quad 1 \quad 0 \quad 0 \quad 0\}$ for a fully three-dimensional problem. Except for the term containing p, this is identical to the usual elastic finite element formulation, and the corresponding stiffness matrices and load vectors can be derived easily. They will turn out to be $[K_1]$, $\{M_1\}$, $\{M_2\}$, and $\{P_1\}$ in the notation used before. The contribution of the second term will lead to

$$\int_V [B]^T\{L\}p\, dV = \int_V [B_\Delta]^T p\, dV \tag{12-52}$$

and when $p = [N_p]\{p\}$, results are identical to Eq. (12-28); Christian and Boehmer used a quadrilateral element composed of four triangles, with the displacements of the center node eliminated by static condensation. A constant stress and pore pressure was assumed throughout the element. Thus $[N_p]$ was unity.

The change in volume of an element can be found by integrating the volumetric strain over the element:

$$\int_V \epsilon_{\text{vol}}\, dV = \Delta_{\text{vol}} \qquad \text{or} \qquad \int_V [B_\Delta]\{u_m\}\, dV = \Delta_{\text{vol}} \tag{12-53}$$

This gives nine equations per element, eight from Eqs. (12-51) and (12-52) expressing equilibrium and one from Eq. (12-53) expressing the constraint on volume change. The combined system for all equations becomes

$$\begin{bmatrix} [K_1] & [K_3] \\ [K_3]^T & [0] \end{bmatrix} \begin{Bmatrix} \{r_q\} \\ \{r_p\} \end{Bmatrix} = \begin{Bmatrix} \{R_q\} \\ \{R_p\} \end{Bmatrix} \tag{12-54}$$

In this system $\{R_p\}$ is the change in volume of each element; initially it will be zero. As consolidation proceeds, $\{R_p\}$ will have nonzero values. What is needed is a way of evaluating the change in volume with time.

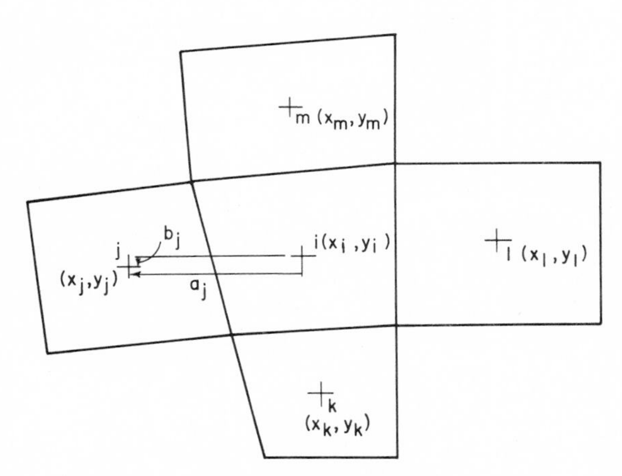

Figure 12-6 Elements used in evaluating rate of flow for simple quadrilateral.

The procedure adopted is illustrated in Fig. 12-6. The values of excess pore pressure at the centers of five elements are used to evaluate the quadratic variation of pore pressure

$$p = \alpha_1 + \alpha_2 x + \alpha_3 y + \alpha_y x^2 + \alpha_5 y^2 \tag{12-55}$$

The value of pore pressure at each node is then

$$\begin{Bmatrix} p_i \\ \vdots \\ p_m \end{Bmatrix} = [B] \begin{Bmatrix} \alpha_1 \\ \vdots \\ \alpha_5 \end{Bmatrix} \quad \text{and} \quad [B] = \begin{bmatrix} 1 & 0 & 0 & 0 & 0 \\ 1 & a_j & b_j & a_j^2 & b_j^2 \\ 1 & a_k & b_k & a_k^2 & b_k^2 \\ 1 & a_l & b_l & a_l^2 & b_l^2 \\ 1 & a_m & b_m & a_m^2 & b_m^2 \end{bmatrix} \tag{12-56}$$

where $a_j = x_j - x_i$, $b_j = y_j - y_i$, and so on for other subscripts.

The net rate of influx of fluid, which equals the rate of increase in volume, must be

$$\frac{k}{\gamma} \nabla^2 p = \frac{k}{\gamma} \begin{Bmatrix} \dfrac{\partial^2 p}{\partial x^2} \\ \dfrac{\partial^2 p}{\partial y^2} \end{Bmatrix} = \frac{1}{\gamma} \begin{bmatrix} 0 & 0 & 0 & 2k_x & 0 \\ 0 & 0 & 0 & 0 & 2k_y \end{bmatrix} [B]^{-1} \begin{Bmatrix} p_i \\ p_m \end{Bmatrix} \tag{12-57}$$

Then only the fourth and fifth rows of $[B]^{-1}$ need be evaluated; for details see Ref. 2.

The procedure is to solve initially for the displacements and excess pore pressures with no volume change. Next, from the excess pore pressures and from Eq. (12-57) the change in volume for each element during a small increment of time Δt is computed. This change in volume is then used in an updated $\{R_p\}$, and Eqs. (12-54) are solved for a new set of displacements and pore pressures, which are then used to evaluate new volumetric strains. The procedure continues as long as desired. It is an explicit marching scheme and subject to numerical instability, but it has been used successfully for solution of practical problems.

The Undrained Problem

If the loads are applied instantaneously, the initial condition will represent an undrained, incompressible material. The procedure by Christian and Boehmer[2] starts from such a case. The finite element techniques based on the Sandhu and Wilson functional can be used to solve for the undrained case by setting Δt equal to zero in Eq. (12-33). This leads to

$$\begin{bmatrix} [K_1] & [K_3] \\ [K_3]^T & [0] \end{bmatrix} \begin{Bmatrix} \{r_q\} \\ \{r_p\} \end{Bmatrix} = \begin{Bmatrix} \{R_Q\} \\ \{0\} \end{Bmatrix} \tag{12-58}$$

Yokoo et al.[23] have reported great difficulty in obtaining reasonable results for undrained analyses from the Sandhu and Wilson functional. They recommended changing the functional in such a way as to allow a node to have different values of pore pressure for each element to which it belongs, in other words, by relaxing the continuity of pore pressures. This is essentially a change in the element expansion rather than in the basic functional. An appropriate conclusion is that the pore pressure should not have the same expansion as the displacements. The procedures by Christian and Boehmer[2] and Sandhu and Wilson[11] use pore pressure expansions of the same order as the stress distributions resulting from the displacement expansions. This appears to be a reasonable and consistent procedure. The procedure used by Ghaboussi and Wilson[12] does not have such an agreement between pore pressure expansions and stress on strain expansions, but it uses a lower order expansion for pore pressures than for displacements.

The formulation in Ref. 2 has been used successfully to analyze several problems of stress distribution for both plane strain and axially symmetric conditions.[24,25] The conclusion is that one should not use the same expansion for both displacement and pore pressure if undrained cases are to be analyzed or if step loadings are to be used.

Even though a particular formulation can be made to solve the undrained problem, attempting to march out the time-dependent consolidation solution from the undrained solution may not work. This is because a change in the boundary condition on the pore pressure must occur at the drainage face in going from the undrained case to the consolidation condition. For elements with pore pressure nodes on that face a serious error is introduced.

2-3 PROGRAMMING CONSIDERATIONS

The programming of the consolidation analysis is reasonably simple for those familiar with finite element techniques. However, certain details are important if an efficient program is to be produced.

Equations (12-38) and (12-39) show the unknowns arranged with all displacements followed by the pore pressures. While this is convenient for derivation of expressions, it leads to a great waste of core storage. The square matrix $[K_1]$ has nonzero terms only within a relatively narrow band, provided the displacements are arranged in the order $u_1, v_1, u_2, v_2, \ldots$ instead of $u_1, u_2, \ldots, v_1, v_2, \ldots$. The rectangular matrices $[K_3]$ and $[K_3]^T$ extend the locations of the nonzero terms nearly to the edge of the matrix. This has the effect of creating a matrix with a very large bandwidth, and both the amount of required storage and the time for solution are increased greatly.

The difficulty is avoided by arranging all the unknowns for each node next to each other and rearranging the rows and columns of the coefficient matrix accordingly. For the Sandhu and Wilson procedure this will require some nodes with three unknowns and some with two, so that a bookkeeping array must be provided to keep track of what is represented by the terms in the vectors of "loads" and unknowns.

For the procedure by Christian and Boehmer the solution is to place the pore pressure unknowns and the volumetric strain "loads" for each element in their respective vectors immediately following the displacements or forces for the highest numbered node in that element. Bookkeeping is handled by two arrays, one keeping track of the location of the nodal terms and one keeping track of the location of element terms. Again, the coefficient matrix must be arranged accordingly. It will be noted that in this approach and in the undrained analysis for any element there are zeros on the diagonal of the stiffness matrix [see Eq. 12-58)]. These zeros become nonzeros during elimination because of the contribution of the terms in $[K_3]$. However, if the equations are arranged in such a way that the elimination process reaches one of these zero terms before it has been changed by the terms in $[K_3]$, the usual gaussian elimination routines without pivoting will come to a halt. It is therefore essential for the undrained problem that the nodes be numbered in such a way that nodes having all displacements fixed but pore pressures unspecified not be reached in the elimination process until some other nodes for the same element have been processed. In any case, an error message indicating that a zero has been encountered on the diagonal during elimination should be included in the program, and such a difficulty can be resolvable by simply renumbering the nodes.

Because of the many steps in the solution it is highly desirable that the entire coefficient matrix be stored in main core storage. Efficiency is enhanced for elastic problems with constant $\alpha\ \Delta t$ by performing the first step of gaussian elimination once and storing the resulting upper triangle. This is equivalent to factoring the coefficient matrix $[K]$ into an upper and a lower triangle:

$$[K] = [L][U] \tag{12-59}$$

If the diagonal terms in $[U]$ are unity, each term in $[L]$ is equal to the transposed term in $[U]$ multiplied by the corresponding diagonal term in $[L]$. In other words,

$$l_{ij} = l_{ii} U_{ji} \qquad \text{no summation} \tag{12-60}$$

The stored matrix can then consist of a half-band with the l_{ii} on the diagonal and the u_{ij} off the diagonal.

The solution of a system of simultaneous equations

$$[K]\{r\} = \{R\} \tag{12-61}$$

is equivalent to solving

$$[L]\{w\} = \{R\} \tag{12-62a}$$

and then

$$[U]\{r\} = \{w\} = [L]^{-1}\{R\} \tag{12-62b}$$

The terms in $\{w\}$ can be found from a frontal substitution

$$\begin{aligned} w_1 &= \frac{1}{l_{11}} R_1 \\ w_2 &= \frac{1}{l_{22}}(R_2 - l_{21} w_1) = \frac{R_2}{l_{22}} - u_{12} w_1 \\ w_j &= \frac{1}{l_{jj}}(R_j - l_{j,\,j-1} w_{j-1} - \cdots - l_{j,\,1} w_1) \\ &= \frac{R_j}{l_{jj}} - u_{j-1,\,j} w_{j-1} - \cdots - u_{1j} w_1) \end{aligned} \tag{12-63}$$

Then the terms in $\{q\}$ can be found from backsubstitution:

$$r_n = w_n \qquad r_{n-1} = w_{n-1} - u_{n-1,n} r_n$$

$$r_j = w_j - u_{j,j+1} r_{j+1} - \cdots - u_{jn} r_n \tag{12-64}$$

Solution for any time step consists of assembling the $\{R\}$ vector and performing a forward- and backsubstitution in turn. This is a very efficient procedure, and its efficiency argues against the variable $\alpha\ \Delta t$ approach.[14]

The choice of Δt will directly affect the accuracy of the solution and the time required to obtain it. For any procedure with $\alpha \geqslant \frac{1}{2}$ the use of a very large Δt will give stable solutions, and the final solution can be obtained directly by setting Δt to any extremely large value. However, the description of the details of the consolidation process requires a smaller Δt. The choice for a particular problem is best made by trying several values until the desired degree of accuracy is obtained. Details of such a quantitative study are given subsequently.

For an explicit scheme the choice of too large a Δt will lead to instability in the solution. A user of the Christian-Boehmer procedure must experiment to find the optimum value. Booker and Small[22] have shown that for the Sandhu functional[18] a stability criterion for constant loadings when $\alpha < \frac{1}{2}$ is

$$\Delta t \leqslant \frac{1}{\beta_N^2}\left(\frac{1}{2} - \alpha\right) \tag{12-65}$$

where the β_N^2 is the largest eigenvalue of

$$[K_2] - \beta_N^2 [K_3]^T [K_1]^{-1} [K_3] = \{0\}$$

which can also be expressed as the zeros of

$$\begin{vmatrix} [K_1] & [K_3] \\ [K_3]^T & \dfrac{1}{\beta_N^2}[K_2] \end{vmatrix} = 0 \tag{12-66}$$

Since evaluating this limit is quite cumbersome, in many practical cases a trial-and-error procedure with various Δt will give useful results faster. Better, $\alpha \geqslant \frac{1}{2}$ can be used.

12-4 APPLICATIONS

Most published results to date have concentrated on demonstrating that the results from the finite element method agree well with known analytic solutions. Christian et al.[10] analyzed the previously unsolved case of the consolidation of a layer with a rough rigid base under conditions of plane strain. They expressed the results for a flexible-strip load on the surface in terms of the ratio b/H between the half-width of the strip and the thickness of the consolidating layer. For $b/H > 1$ the time for 90 percent of consolidation is very close to that estimated from one-dimensional theory, and values of Poisson's ratio for the soil skeleton have little effect on the results. These results indicate that one-dimensional theory is

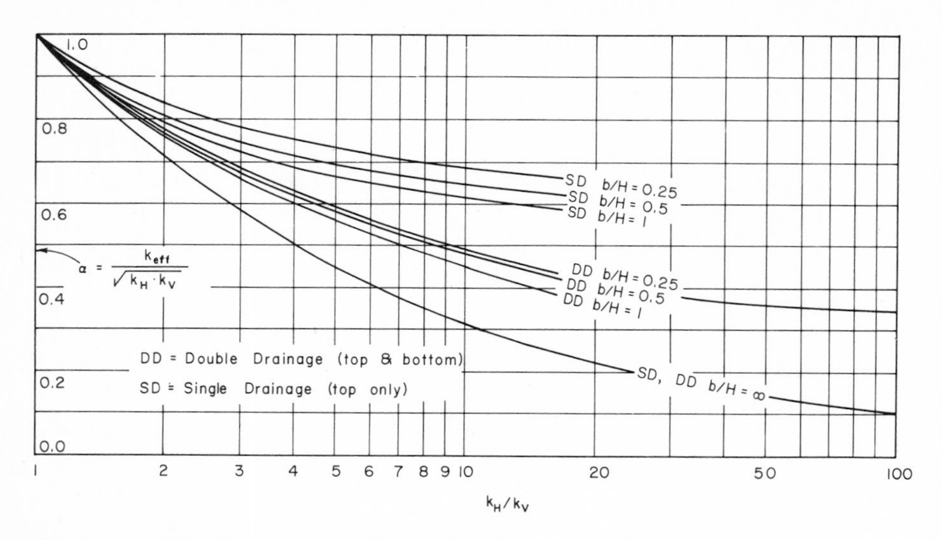

Figure 12-7 Effect of horizontal permeability at 85 percent consolidation.

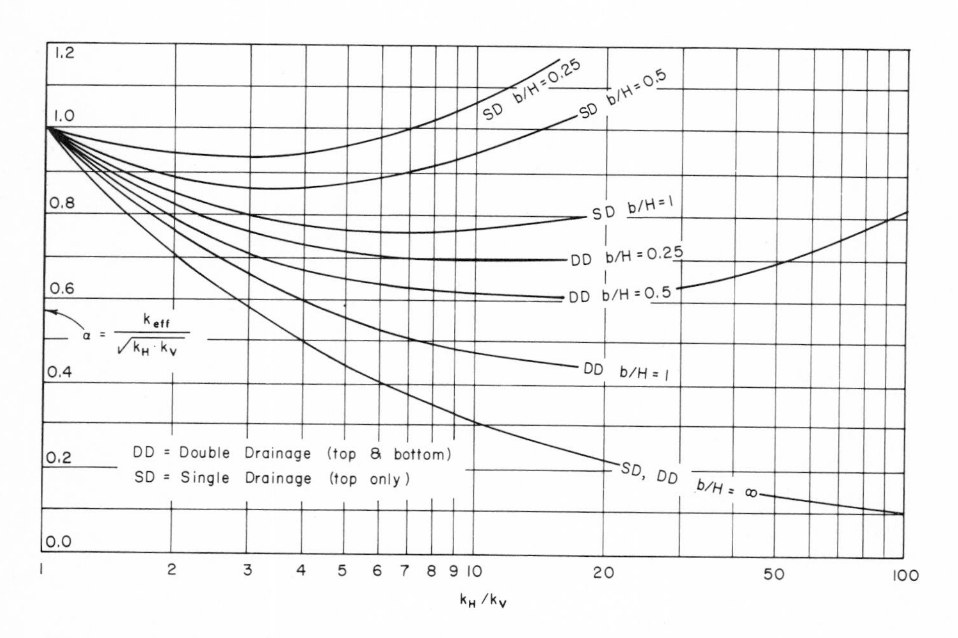

Figure 12-8 Effect of horizontal permeability at 50 percent consolidation.

adequate for settlement analysis when $b/H > 1$. For loads on thicker layers, settlement is accelerated and two-dimensional theory is necessary.

The effects of anisotropic permeability for the same problem are illustrated in Figs. 12-7 and 12-8. The abscissas are the ratios of horizontal to vertical permeability; the ordinate is the ratio between two permeabilities: the numerator is the isotropic permeability k_{eff} that would give the same amount of settlement of the centerline in the same time as is found from the anisotropic analysis, and the denominator is the equivalent permeability from steady-state flow $\sqrt{K_H K_V}$. These figures show that very large ratios of K_H/K_V can have relatively modest effects on the rate of consolidation settlement.

Example 12-1 As mentioned earlier, most previous results have considered verifications of finite element results by comparison with closed form solutions. Desai[15,16,26] has analyzed the formulation by Sandhu[18] from the viewpoint of practical applications. In these analyses, a comprehensive study was performed to examine the effects of a number of factors such as (large) variations in material properties, size of mesh (both spatial, Δx, Δy, and timewise, Δt), anisotrophy, and boundary conditions. In the following, we give a set of typical results from this study.

Figure 12-9 shows a three-layered soil system with a wide variation in the coefficient of permeability. In order to find the ranges of permeability k and elastic modulus E within which the numerical solution remains reliable, a number of values of $k_r = k_1/k_2$ and $E_r = E_1/E_2$, where the

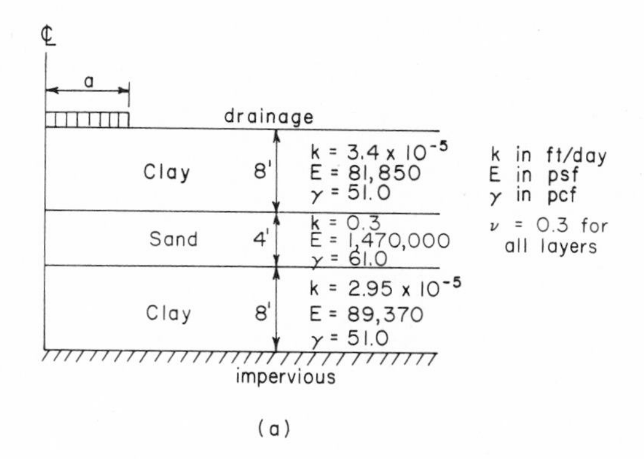

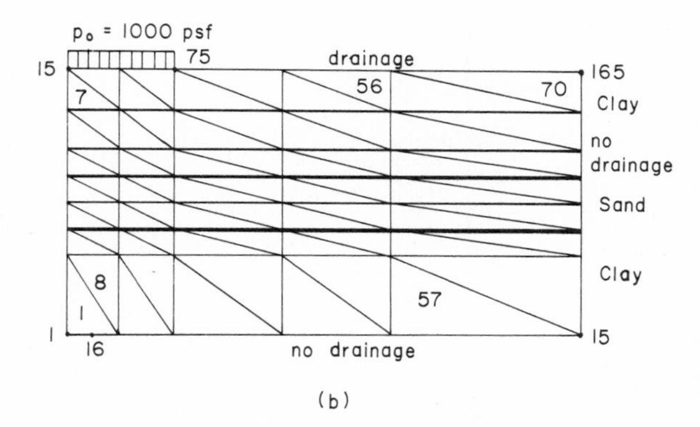

Figure 12-9 Analysis of three-layered system: (a) physical problem; (b) finite element mesh.[15,16]

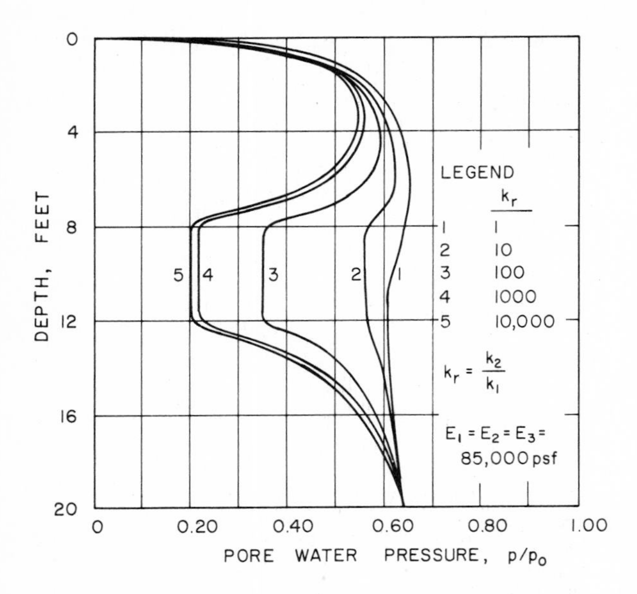

Figure 12-10 Distribution of pore water pressure at centerline for different k_r at time of 1.1 days.[15,16]

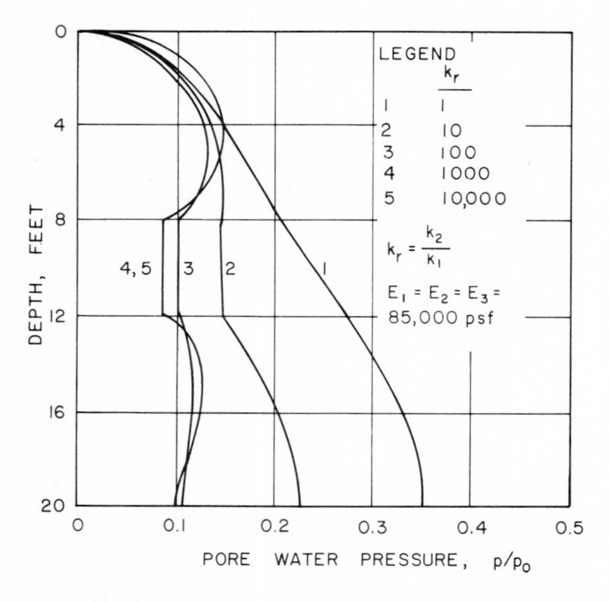

Figure 12-11 Distribution of pore water pressure at centerline for different k_r at time of 21.1 days.[15,16]

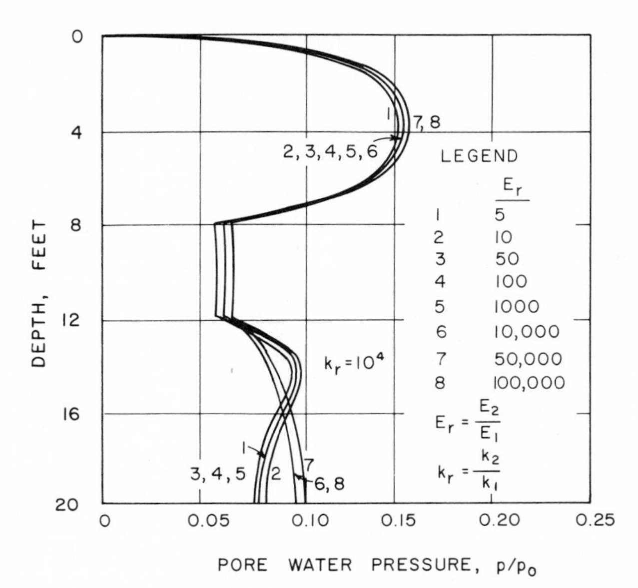

Figure 12-12 Distribution of pore water pressure at centerline for different E_r at time of 21.1 days and K_r of 10^4.[15,16]

subscripts 1 and 2 denote the top and the middle layers, respectively, were chosen, and the results at different spatial and temporal locations were plotted and scrutinized. Figures 12-10 to 12-12 show results for pore water pressures at typical time levels during consolidation. In another study, the average values of $\Delta\bar{x}$ and $\Delta\bar{y}$ and time increment $\Delta\bar{t}$ were varied. Typical results are shown in Fig. 12-13*a* and *b* in terms of a nondimensional factor λ. On the basis of the foregoing results, it was suggested that the numerical procedure would yield reliable results if the following criteria were observed:

$$k_r \leqslant 10^4 \qquad E_r \leqslant 10^4 \qquad \lambda = \frac{c_v \, \Delta\bar{t}}{\Delta\bar{x}^2 + \Delta\bar{y}^2} \leqslant 5 \times 10^{-3} \tag{12-67}$$

where c_v = coefficient of consolidation.

As stated previously, the procedure is stable mathematically;[22] however, the foregoing results indicate that the reliability and acceptable accuracy of the procedure can also depend on other physical factors.[27]

Figure 12-14*a* shows a practical problem involving a four-layered foundation in the Atlantic Ocean;[28] Fig. 12-14*b* shows the finite element mesh used for the problem. Typical finite element results are shown in Fig. 12-15, which includes a progressive history of deformations during consolidation under the foundation. The qualitative trends and quantative values when compared with results from one-dimensional consolidation[28] were found to be satisfactory.

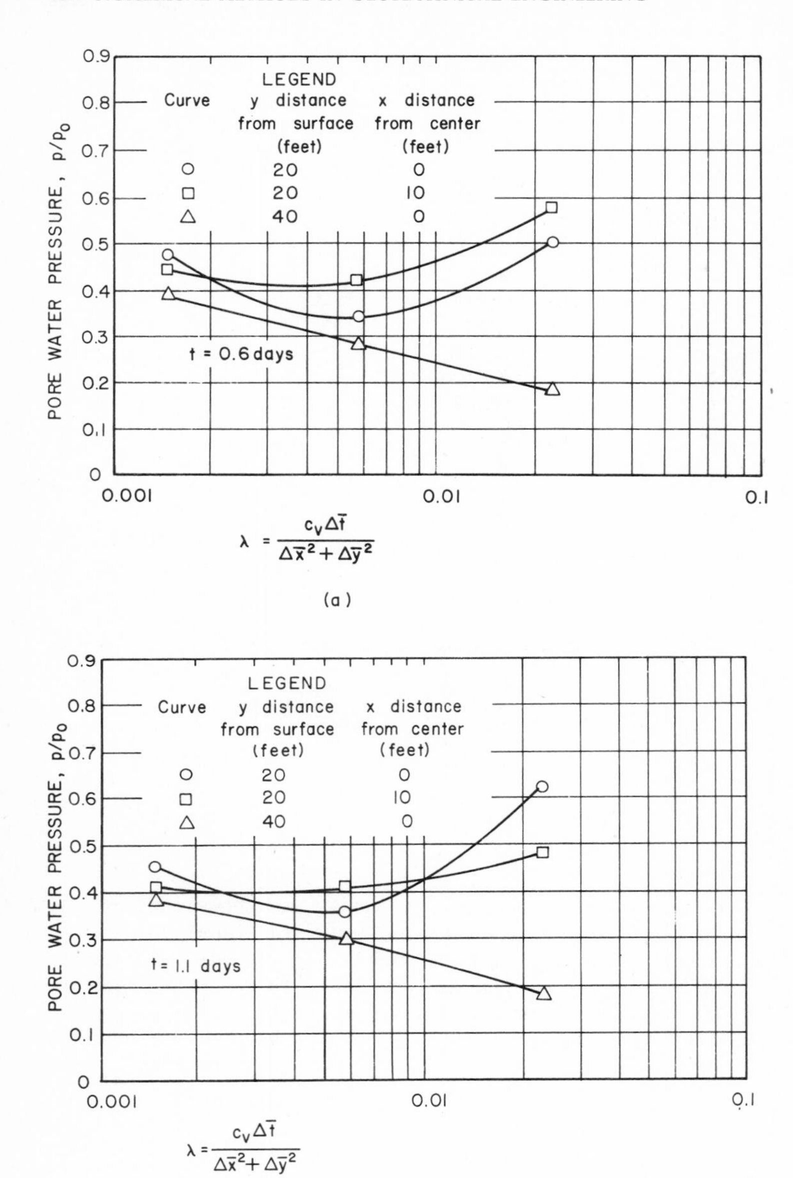

Figure 12-13 Effects of time step: pore water pressure at (*a*) $T = 2.5 \times 10^{-3}$; (*b*) $T = 4.5 \times 10^{-3}$.[15,16]

℄
196'
E in psf
k in ft/day
γ in pcf
106'
Embankment
6' Silty Clay Layer 1 $\bar{E}$ = 81,850 $\bar{k}$ = 3.4×10^{-5} γ = 51.0 $\bar{\nu}$ = 0.3
4' Sand Layer 2 $\bar{E}$ = 1,470,000 $\bar{k}$ = 0.3 γ = 61.0 $\bar{\nu}$ = 0.3
18' Plastic Clay Layer 3 $\bar{E}$ = 89,370 $\bar{k}$ = 2.95×10^{-5} γ = 51.0 $\bar{\nu}$ = 0.3
90' Sand Layer 4 $\bar{E}$ = 1,216,000 $\bar{k}$ = 0.3 γ = 66.0 $\bar{\nu}$ = 0.3
Impervious
(Not to scale)

(a)

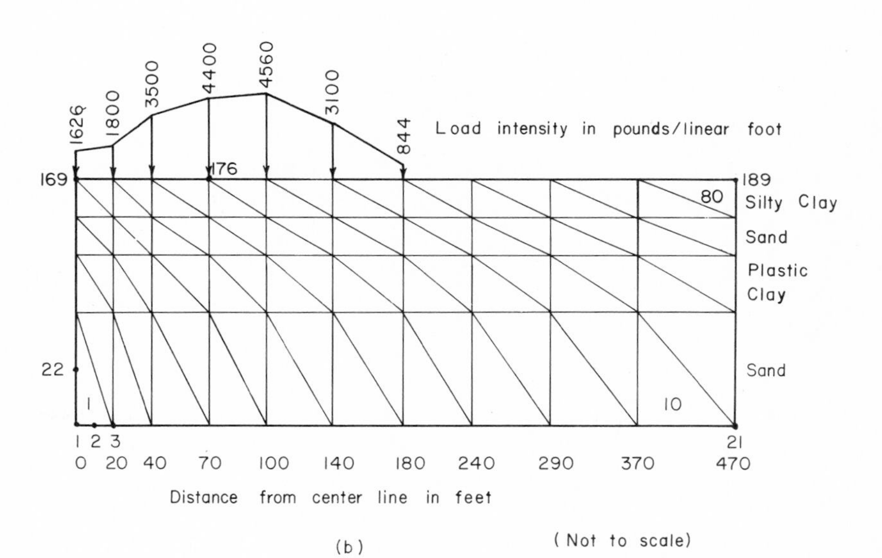

Figure 12-14 Field problem: (*a*) physical problem; (*b*) finite element mesh.[15, 16]

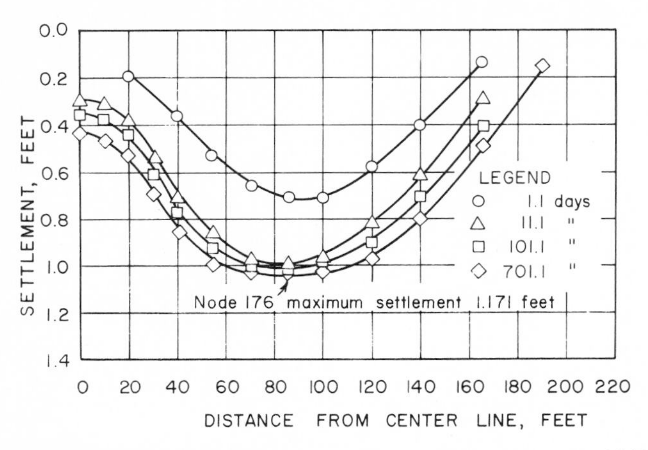

Figure 12-15 Time settlement results for problem of Fig. 12-14.[15,16]

12-5 EXTENSIONS AND FUTURE DEVELOPMENTS

This chapter has summarized the state of the art for numerical analysis of the classical Biot consolidation theory. Current research is active in several major directions.

Ghaboussi and Wilson[29] extended the functional formulation to include compressible pore fluid. The equation equivalent to Eq. (12-22) becomes

$$\Omega(u, p) = \int_V \left[\frac{1}{2} \bar{C}_{ijkl}\epsilon_{kl} * \epsilon_{ij} - \rho_w F_i * u_i + p * \theta u_{i,i} - \frac{1}{2} g * q_i * (p_{,i} + \rho_w F_i) - \frac{1}{2} p * \frac{1}{M} * p \right] dV - \int_{S_1} (\bar{T}_i * u_i)\, dS + \int_{S_2} (g * \bar{Q} * p)\, dS \tag{12-68}$$

where θ and M describe the compressibility of the fluid and solid particles, respectively. These are defined by the equations for total stress

$$\sigma_{ij} = \bar{C}_{ijkl}\epsilon_{kl} + \theta\, \delta_{ij} p \qquad p = \alpha M \epsilon_{ii} + M\zeta \tag{12-69}$$

where $\bar{C}_{ijkl}$ is the tensor of effective stress elastic constants and ζ is the volumetric strain component of the relative motion between the fluid and the soil. For solid particles and incompressible fluid, $\alpha = 1$ and $M = \infty$.

The terms for the compressible fluid are easily added to the derivation, giving a final set of equations similar to those for the incompressible fluid (12-37).

The major shortcoming of all the formulations described in this chapter is that they do not account for the nonlinearity of the stress-strain behavior of the soil, either plastic or viscous. These factors are likely to be far more important than the compressibility of the pore fluid in static problems. Work is now in progress on incorporating nonlinear stress-strain relations, and the results should provide useful insight into consolidation phenomena.

The functional formulation of Eqs. (12-22) and (12-68) is a special case of a general variational approach to coupled field problems developed by Sandhu and Pister;[19] the reader interested in the theoretical underpinnings of analysis of coupled problems should consult this paper.

REFERENCES

1. Biot, Maurice A.: General Theory of Three-dimensional Consolidation, *J. Appl. Phys.*, vol. 12, pp. 155–164, February 1941.
2. Christian, John T., and Jan Willem Boehmer: Plane Strain Consolidation by Finite Elements, *J. Soil Mech. Found. Div. ASCE*, vol. 96, no. SM4, pp. 1435–1457, July 1970.
3. Rendulic, L.: Porenziffer und Poren Wasserdruck in Tonen, *Bauingenieur*, vol. 17, pp. 559–564, 1936.
4. Davis, E. H., and H. G. Poulos: Rate of Settlement under Two- and Three-dimensional Conditions, *Geotechnique*, vol. 12, no. 1, pp. 95–114, March 1972.
5. Terzaghi, Karl: "Theoretical Soil Mechanics," John Wiley & Sons, Inc., New York, 1943.
6. Mandel, J.: Consolidation des sols, *Geotechnique*, vol. 3, pp. 287–299, 1957.
7. Cryer, C. W.: A Comparison of the Three-dimensional Theories of Biot and Terzaghi, *Q. J. Mech. Appl. Math.*, vol. 16, pp. 401–412, 1963.
8. Herrmann, L. R.: Elasticity Equations for Incompressible and Nearly Incompressible Materials by a Variational Theorem, *AIAA J.*, vol. 3, pp. 1896–1900, 1965.
9. Christian, John T.: Undrained Stress Distribution by Finite Elements, *J. Soil Mech. Found. Div. ASCE*, vol. 94, no. SM6, pp. 1333–1345, November 1968.
10. Christian, John T., Jan Willem Boehmer, and Phillipe P. Martin: Consolidation of a Layer under a Strip Load, *J. Soil Mech. Found. Div. ASCE*, vol. 98, no. SM7, pp. 693–707, July 1972.
11. Sandhu, Ranbir S., and Edward L. Wilson: Finite Element Analysis of Flow of Saturated Porous Elastic Media, *J. Eng. Mech. Div. ASCE*, vol. 95, no. EM3, pp. 641–652, June 1969.
12. Ghaboussi, Jamshid, and Edward L. Wilson: Flow of Compressible Fluid in Porous Elastic Media, *Int. J. Numer. Methods Eng.*, vol. 5, no. 3, pp. 419–442, 1973.
13. Yokoo, Yoshitsura, Kunio Yamagata, and Hiroaki Nagaoka: Finite Element Method Applied to Biot's Consolidation Theorem, *Soils Found.*, vol. 11, no. 1, pp. 29–46, March 1971.
14. Hwang, C. T., Norbert R. Morgenstern, and D. T. Murray: On Solutions of Plane Strain Consolidation Problems by Finite Element Methods, *Can. Geotech. J.*, vol. 8, no. 1, pp. 109–118, February 1971.
15. Desai, C. S.: Analysis of Consolidation by Numerical Methods, *Proc. Gen. Sess. Symp. Recent Dev. Anal. Soil Behavior Appl. Geotech. Struct., Univ. New South Wales, July 1975.*
16. Asproudas, S. A., and C. S. Desai: Analysis and Applications of a Finite Element Procedure for Consolidation, *VPI State Univ. Dept. Civ. Eng., Rep.* VPI-E 75.34, Blacksburg, Va., May 1975.
17. Gurtin, M. E.: Variational Principles for Elastodynamics, *Arch. Ration. Mech. Anal.*, vol. 16, no. 1, pp. 34–50, 1964.
18. Sandhu, Ranbir S.: Fluid Flow in Saturated Porous Elastic Media, Ph.D. thesis, Department of Civil Engineering, University of California, Berkeley, 1968.
19. Sandhu, Ranbir S., and K. S. Pister: A Variational Principle for Linear, Coupled Field Problems in Continuum Mechanics, *Int. J. Eng. Sci.*, vol. 8, pp. 989–999, 1970.
20. Sandhu, R. S.: *Proc. Symp. Appl. Finite Element Methods Geotech. Eng., Vicksburg, Miss., September 1972.*
21. Desai, C. S., and L. D. Johnson: Evaluation of Two FE Formulations for One-Dimensional Consolidation, *Int. J. Comput. Struct.*, vol. 2, pp. 469–486, 1972.
22. Booker, J. R., and J. C. Small: An Investigation of the Stability of Numerical Solutions of Biot's Equations of Consolidation, *Int. J. Solids Struct.*, 1975.
23. Yokoo, Yoshitsura, Kunio Yamagata, and Hiroaki Nagaoka: Finite Element Analysis of Consolidation Following Undrained Deformation, *Soils Found.*, vol. 11, no. 4, pp. 37–58, December 1971.

24. Carrier, W. D., III, and John T. Christian: Rigid Circular Plate Resting on a Non-homogeneous Elastic Half-Space, *Geotechnique*, vol. 21, no. 1, pp. 67–84, March 1973.
25. Carrier, W. D., III, and John T. Christian: Analysis of an Inhomogeneous Elastic Half-Space, *J. Soil Mech. Found. Div. ASCE*, vol. 99, no. SM3, pp. 299–306, March 1973.
26. Desai, C. S.: Evaluation of Numerical Procedures for Fluid Flow, *ASCE Nat. Conv. Denver, Colo., November 1975*, pap. 2609.
27. Desai, C. S.: Some Aspects of Theory and Applications, *Proc. 2d Int. Conf. Numer. Methods Geomech., Blacksburg, Va., June 1976.*
28. Saxena, S., private communications.
29. Ghaboussi, Jamshid, and Edward L. Wilson: Variational Formulation of Dynamics of Fluid-saturated Porous Elastic Solids, *J. Eng. Mech. Div. ASCE*, vol. 98, no. EM4, pp. 947–963, August 1972.

CHAPTER

THIRTEEN

FOUNDATIONS IN EXPANSIVE SOILS

R. L. Lytton

13-1 INTRODUCTION

Engineering design of foundations on expansive soil must be based upon some rational estimate of the amount of differential movement that the structure in contact with the soil will undergo. The predicted movement of the structure will depend, in turn, upon the expected movement of the mass of soil on which it rests. Numerical analyses of these soil movements can be divided into three classes: unsaturated moisture flow and estimation of equilibrium conditions, prediction of heave or shrinkage, and solution of coupled moisture-flow–elasticity boundary-value problems.

Once these movements of the soil mass have been predicted, they can be used in a soil-structure interaction analysis of overlying structures either approximately by superimposing the structure on the deformed soil or by solving the actual moisture-flow–soil-structure interaction problem step by step in time. The latter approach was followed for a beam resting on a drying soil[1] by using the finite element method and by assuming the soil to be nonlinearly elastic. These calculations require extended computer times and consequently such analyses may not be economical for design of individual foundations. However, they are valuable for predicting the observed behavior of carefully instrumented field tests and for guidance toward more simplified design calculations.

In addition to the mechanics of predicting the motion of an expansive clay mass, the analysis of two problems (1) pier and beam and (2) stiffened mat foundations will be discussed. Some of the design consequences of the viscoelasticity of the soil and concrete will be presented for stiffened mat foundations.

13-2 GEOTECHNICAL ASPECTS

Mechanics of Unsaturated Moisture Diffusion

The unsaturated flow of moisture in expansive clay is dominated by the cracking fabric and block structure of the clay, making it necessary to characterize the flow in two stages: (1) flow in the cracks, followed by (2) diffusion into the blocks, clods, and peds of the soil.

Numerical studies of flow in such cracked soil systems have been made[2] and have been roughly verified by trial-and-error calculations of field permeabilities.[3-5] The field permeabilities are on the order of 10^{-6} cm/s and have the same form of variation with soil suction as a laboratory sample. For instance,

$$k = \frac{k_0}{1 + a|h|^n} \tag{13-1}$$

where k_0 is the saturated permeability and a and n are material parameters. The form of Eq. (13-1) was first proposed by Gardner.[6] Typical values of k_0, a, and n are given in Table 13-1.

Many questions regarding the nature of the flow remain unresolved: e.g., whether the flow occurs mainly in the liquid, vapor, or adsorbed phase and whether solute and temperature gradients[7-9] and volume change[10,11] are important. For most practical purposes the mass velocity of flow v_i in the i direction can be expressed as

$$v_i = -k_{ij}\frac{\partial \psi}{\partial x_j} \tag{13-2}$$

where k_{ij} is the permeability tensor and the flow potential ψ is expressed as

$$\psi = -h + x_3 \pm \Omega \tag{13-3}$$

Table 13-1 Field values of expansive-clay permeability

Soil	k_0, cm/s	a	n	Ref.
Yazoo	4.5×10^{-7}	...	1.0	5
Lackland	2.7×10^{-6}	...	...	5
Horsham	2	...	...	2
West Laramie clay shale	2.7×10^{-6}	10^{-9}	3.0	4
Flagstaff Gully Dam	2.0×10^{-6}	...	...	3

The flow potential is made up of suction h, gravitational potential x_3, and the overburden potential Ω,[11–14] the last two of which nearly cancel each other while swelling is occurring and reinforce each other when shrinkage is taking place. In either case, they are usually so small in relation to the suction that we may consider the driving potential to be due to suction alone. Suction is defined as a free energy by an ideal gas law[15,16]

$$h = \frac{RT}{mg} \ln \frac{H}{100} \tag{13-4}$$

where R = universal gas constant
$= 8.314 \times 10^7$ ergs/mol·°C
T = absolute temperature
g = gravitational constant
m = mass of 1 mol of water
H = relative humidity, %

The suction is zero when the relative humidity is 100 percent and becomes a larger negative number as the humidity decreases. It is usually expressed as gram-centimeters per gram of water or simply as an energy head in centimeter units.

Equilibrium Condition

The equilibrium condition occurs when there is a steady flow of moisture and when neither swelling nor shrinking is taking place within a given profile. Under these conditions, Eq. (13-2) can be written

$$v_i = -k_{ij} \frac{\partial}{\partial x_j}(-h + x_3) \tag{13-5}$$

This equation can be integrated numerically in one dimension if there is a steady flux v_i and a known value of suction h somewhere in the profile. The following steps are used:

1. At nodal point m, where the value of suction h_m is known,

$$k_m = \frac{k_0}{1 + a|h_m|^n}$$

2. The change of suction is

$$\Delta h_{m+1} = \left(1 + \frac{v_3}{k_m}\right) \Delta x_3$$

3. The suction at nodal point $m + 1$ is

$$h_{m+1} = h_m + \Delta h_{m+1}$$

From this suction, a value of permeability at nodal point $m + 1$ can be calculated and the same pattern of calculations can be repeated until a complete suction profile is found. The results of these kinds of calculation are shown in Fig. 13-1,

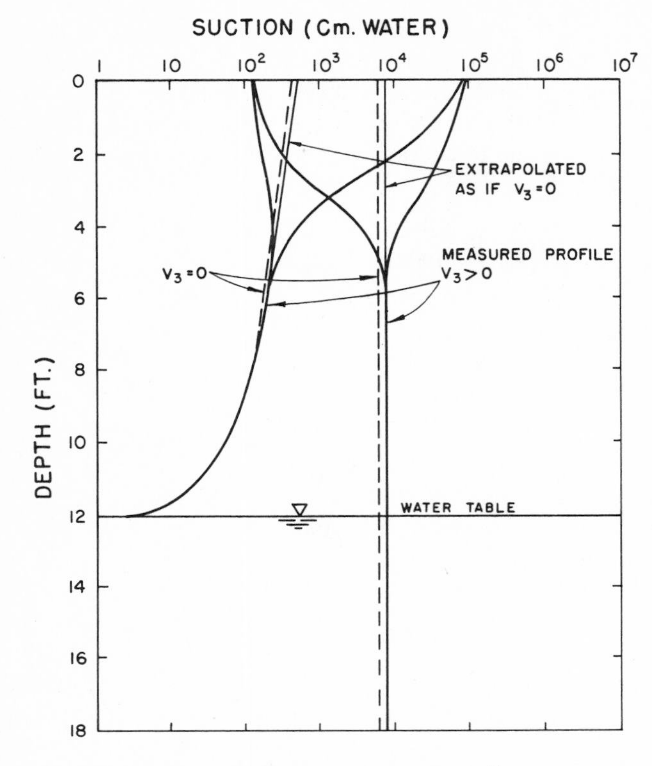

Figure 13-1 Typical profiles of suction vs. depth.[41]

which illustrates the different kinds of equilibrium profiles that can be expected when the suction is controlled by (1) high water tables and (2) a constant suction at depths that are remote from climatic influences. These relations have been verified by field observations.[17]

Moisture Diffusion

The process of moisture diffusion is derived by combining the flow equation (13-2), with the continuity equation

$$\frac{\partial \theta}{\partial t} = -\frac{\partial}{\partial x_i}(v_i) \tag{13-6}$$

to get the diffusion equation

$$\frac{\partial h}{\partial t} = \frac{\partial h}{\partial \theta}\frac{\partial}{\partial x_i}\left(k_{ij}\frac{\partial \psi}{\partial x_j}\right) \tag{13-7}$$

The term $\partial h/\partial \theta$ is the slope of the curve which relates suction to the volumetric water content θ. Some typical h-w curves for silt and clay are shown in Fig. 13-2.

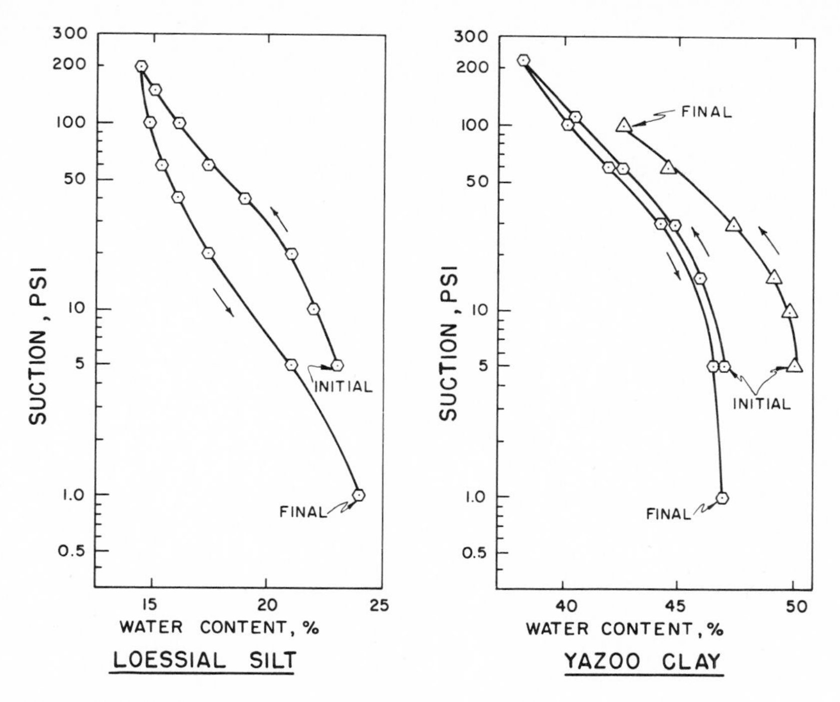

Figure 13-2 Typical suction-vs.-water content curves.[40]

The gravimetric water content w is related to the volumetric water content by

$$\theta = \frac{G}{1 + e} w$$

where G is the specific gravity of the solids and e is the void ratio.

Mechanics of Heave or Shrinkage

Ordinarily, the pressures applied to an expansive clay by a highway pavement or a light building are very small, rarely getting above 0.05 to 0.20 kg/cm^2 (0.7 to 2.8 lb/in^2). These pressures are insignificant compared with the kinds of swelling and shrinking stresses these soils can develop. Consequently, it is the usual practice to predict heaving or shrinkage based on the expected change of water volume[18,19] or void ratio[20] or expected strain due to a change of suction.[21] Some prediction methods use a combination of suction and water content change[5,13,22] to predict the vertical movement. In every case, the prediction is more accurate if the weight of the overburden, the stresses applied by the overlying structure, the cracking fabric of the soil, and the actual change of moisture or suction are used.

Figure 13-3 illustrates one of the major reasons for this; i.e., the amount of volume change that occurs depends not only upon how much the water content

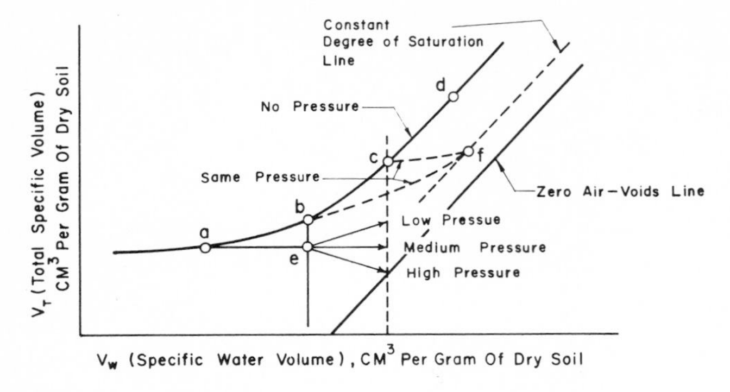

Figure 13-3 Pressure, volume, and water-volume relationships.[13]

changes but also upon the level of applied pressure and the initial moisture condition of the soil. The effect of pressure is seen by contrasting curve *abcd*, which represents the swell of a natural soil under no pressure, with the various curves originating at point *e*. Although the increase of water content from point *e* is the same as that between points *b* and *c*, there is a wide variety of possible volume changes depending upon the level of pressure applied to the soil. The effect of initial moisture condition is shown by comparing curve *bf* with curve *cf*. Both curves represent the swell of the natural soil under the same pressure. While both curves arrive at the same water content, the initially drier soil has swelled more.

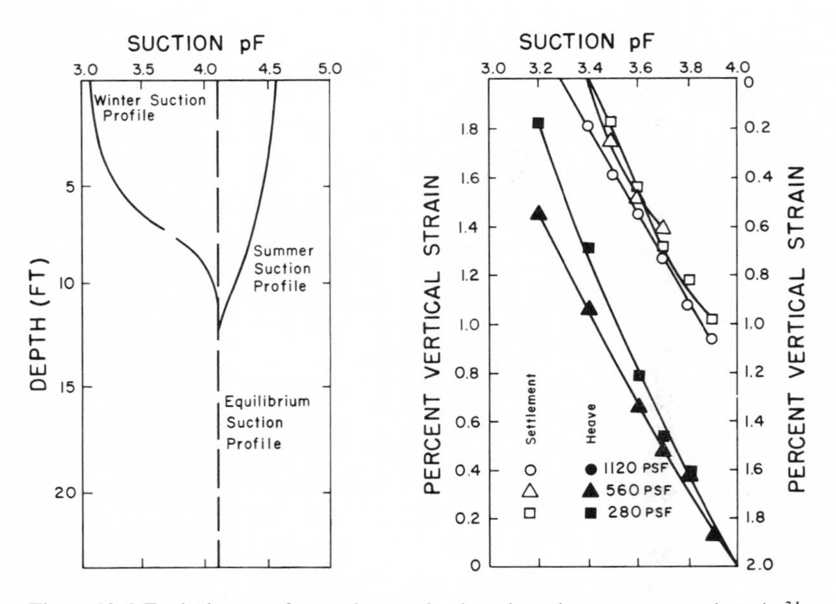

Figure 13-4 Typical curves for suction vs. depth and suction, pressure, and strain.[21]

A direct approach to predicting vertical swell uses the difference between the present suction condition and the expected equilibrium condition to calculate the total heave or shrinkage. Typical data used in this approach are illustrated in Fig. 13-4. As expected, the slopes of the heave and shrinkage curves are altered slightly by the applied pressure.

The numerical prediction of vertical movement ρ takes the form

$$\rho = \sum_{i=1}^{n} f_i \epsilon_i (\Delta x_3)_i \tag{13-8}$$

where f_i = cracking fabric factor (varies from $\frac{1}{3}$ for heavily cracked soil to 1 for tight soil with high lateral restraint)
$(\Delta x_3)_i$ = height of the ith layer of soil
ϵ_i = volume strain in that layer

Prediction of Heave: Uncoupled

The one-dimensional prediction of heave has several implied assumptions. One of the most important is that of uniform swelling or shrinking, which implies that any selected column of soil is typical of the entire soil mass. Another assumption is that there is no shear coupling between these soil columns. This assumption can result in significant errors in a predicted surface profile when the soil has a highly nonuniform moisture distribution.

When these nonuniform moisture conditions prevail, one of two uncoupled approaches may be adopted to predict the distortion of the surface of an expansive soil. One is based upon moisture diffusion and assumes no shear coupling between adjacent columns,[5,13] and the other is based upon elastic theory and assumes that the change of water content is known.[19] The moisture-diffusion approach uses Eq. (13-7) to calculate at every point the change of water content from some initial condition. This approach then uses pressure–water content–volume relations like those illustrated in Fig. 13-3 to calculate the vertical heave of all soil columns. The elastic-theory approach uses elastic moduli and expansion coefficients for the soil and assumes a water content change at every point. This approach then calculates stresses and displacements using an appropriate numerical representation of an elastic continuum. Example results of these approaches are given in Sec. 13-4.

Coupled Elasticity and Moisture Diffusion

The next obvious step is to couple the two approaches discussed above so that moisture diffusion is affected by changing stresses and the stresses and displacements, in turn, are influenced by the amount of moisture that diffuses. Richards[1] has made such calculations for the beam resting on an expansive clay that is drying out at the edges.

In general, there are six equations that must be satisfied simultaneously in the solution of such a problem, and they differ only slightly from those proposed by Biot.[23] They are specialized here for two dimensions.

Equilibrium equation:

$$\frac{\partial \sigma_{xx}}{\partial x} + \frac{\partial \tau_{xy}}{\partial y} + \rho F_x + \frac{\partial}{\partial x}(\chi h) = 0$$

$$\frac{\partial \tau_{xy}}{\partial x} + \frac{\partial \sigma_{xy}}{\partial y} + \rho F_y + \frac{\partial}{\partial y}(\chi h) = 0 \tag{13-9}$$

Constitutive equation:

$$\begin{Bmatrix} \sigma_{xx} \\ \sigma_{yy} \\ \tau_{xy} \end{Bmatrix} = \frac{E}{(1+\nu)(1-2\nu)} \begin{bmatrix} 1-\nu & \nu & 0 \\ \nu & 1-\nu & 0 \\ 0 & 0 & \dfrac{1-2\nu}{2} \end{bmatrix} \begin{Bmatrix} \epsilon_{xx} \\ \epsilon_{yy} \\ \nu_{xy} \end{Bmatrix} \tag{13-10}$$

Strain-displacement equation:

$$\begin{Bmatrix} \epsilon_{xx} \\ \epsilon_{yy} \\ \nu_{xy} \end{Bmatrix} = \begin{Bmatrix} \dfrac{\partial u}{\partial x} \\ \dfrac{\partial v}{\partial y} \\ \dfrac{1}{2}\left(\dfrac{\partial u}{\partial y} + \dfrac{\partial v}{\partial x}\right) \end{Bmatrix} \tag{13-11}$$

Darcy's law:

$$\begin{Bmatrix} v_x \\ v_y \end{Bmatrix} = - \begin{bmatrix} k_{xx} & k_{xy} \\ k_{xy} & k_{yy} \end{bmatrix} \begin{Bmatrix} \dfrac{\partial h}{\partial x} + \rho_w F_x \\ \dfrac{\partial h}{\partial y} + \rho_w F_y \end{Bmatrix} \tag{13-12}$$

Continuity equation:

$$\frac{\partial \theta}{\partial t} + \left(\frac{\partial v_x}{\partial x} + \frac{\partial v_y}{\partial y}\right) = 0 \tag{13-13}$$

Compressibility of unsaturated soil:

$$\frac{\partial h}{\partial t} = \frac{1}{\alpha}\frac{\partial h}{\partial \theta}\frac{\partial}{\partial t}\left(\frac{\partial u}{\partial x} + \frac{\partial v}{\partial y}\right) \tag{13-14}$$

where x, y are used in the place of x_1, x_2, a practice that is adopted frequently in the remainder of this chapter, and χ is a multiplying factor of the suction. If the suction is expressed as a pore pressure and is below about 5 kg/cm^2, it may be taken as 1.0; α is a multiplying factor which relates the change of volumetric water content θ to the change of total volume. It denotes the slope of the curve shown in Fig. 13-3. As is obvious from that figure, α varies with pressure; E is the modulus of elasticity; ν is Poisson's ratio; v_x, v_y are the average pore fluid velocities in the x and y directions; ϵ_{xx}, ϵ_{yy}, and γ_{xy} are the normal and shear strains; u and v are the

x and y displacements; F_x and F_y are the x and y body forces; ρ and ρ_w are the mass densities of the soil and the water; and σ_{xx}, σ_{yy}, and τ_{xy} are the normal and shear stresses.

One important soil property varies with suction and pressure level and must be determined experimentally; that is the amount of stress change equivalent to a given amount of suction change. In order to derive this relation, we first note that a change of isotropic stress $\Delta\sigma$ is related to volume strain by the bulk modulus K:

$$\Delta\sigma = K\frac{\Delta V}{V} \qquad (13\text{-}15)$$

where $\sigma = \frac{1}{3}(\sigma_{xx} + \sigma_{yy} + \sigma_{zz})$ in a three-dimensional test
ΔV = volume change
V = total volume

When the three-dimensional equation analogous to Eq. (13-14) is used, a change of suction Δh is related to volume strain as follows:

$$\Delta h = \frac{1}{\alpha}\frac{\partial h}{\partial \theta}\frac{\Delta V}{V} \qquad (13\text{-}16)$$

These two equations combine to give a relation between Δh and $\Delta\sigma$:

$$\Delta\sigma = \left[\frac{K\alpha}{\partial h/\partial \theta}\right]\Delta h \qquad (13\text{-}17)$$

The term in brackets is the χ factor in Eq. (13-9).

Methods of solving the coupled elasticity–moisture-diffusion equations (13.9) to (13.14) will be described subsequently in this chapter. Results of these calculations will give the stresses and displacements of the soil mass throughout its actively swelling zone.

13-3 DESIGN CONSIDERATIONS

Design Considerations for Drilled Pier Foundations

Drilled piers are commonly used to penetrate the actively swelling zone and to rest the foundation of a structure on a relatively inactive soil. Occasionally, underreamed footings are provided for these piers to give them resistance to uplift of the soil swelling around them. Only rarely is the extra bearing area of underreamed footings required to meet bearing capacity requirements in the stiff, overconsolidated soils that are typical of expansive clay. There are three major problems in designing drilled pier foundations: (1) predicting the vertical movement of each pier, (2) predicting the tensile stresses in each pier so that the sizes of the reinforcing steel can be selected, and (3) predicting the effect of the different amounts of movement of each pier upon the supported structure. The first two problems are solved by an analysis of a single pier; the third requires an analysis of the structure itself.

Vertical movement and tensile stresses The actual soil movement around a drilled pier can be complex. The soil pressures on the pier can be unsymmetrical due to several possible factors, e.g., a shrinkage crack on one side and soil in close contact on the other, and unsymmetrical moisture changes due to the fact that the pier is situated at the perimeter of the building. The lateral pressures and movments on these piers can be quite high[24] and can be neglected in design only if careful provisions have been made to prevent the development of shrinkage cracks or differential moisture around the pile. Superimposed upon this bending tendency is the tensile stress induced in the pier by swelling soil gripping the pier and pulling it upward and by anchoring the pier into the inactive zone with underreamed footings. Reinforcing steel must be placed in these piers to resist the tensile stresses caused by the swelling.

One-dimensional Differential Equation for a Drilled Pier

The stresses and displacements in a drilled pier are determined by solving the differential equation

$$E_c A \frac{d^2u}{dz^2} + PG(w - u) - \gamma_c A = 0 \tag{13-18}$$

where u, w = vertical displacements of pier and adjacent soil
E_c = elastic modulus of pier
A = cross-sectional area of pier
P = perimeter of pier
γ_c = unit weight of pier

G is called the *perimeter shear modulus*. Its units are typically pounds per cubic inch, or kilograms per cubic centimeter, and it relates the skin friction stress developed along the side of the pier to the relative movement between the soil and the pier.

Design Considerations for Stiffened Mat Foundations

As with drilled pier foundations, stiffened mats must also be designed to withstand the differential movement imposed by the supporting soil. The causes of the differential movement may be partly out of the designer's control: vegetation, slope, the condition of ground moisture at the time of construction, the naturally uneven nature of clay deposits, plumbing and utility defects, sun exposure, and drainage conditions around the perimeter of the building. Nevertheless, it is possible to predict, in a general way, the deformation patterns of the mat and their approximate scale. In analyzing a foundation to determine its design quantities, it is conservative and fairly realistic to assume the worst cases of soil support: center support, perimeter support, and corner support as shown in Fig. 13-5.

The last case, in which diagonally opposite corners support a rectangular mat, induces a very large twisting moment along the perimeter where liftoff occurs.

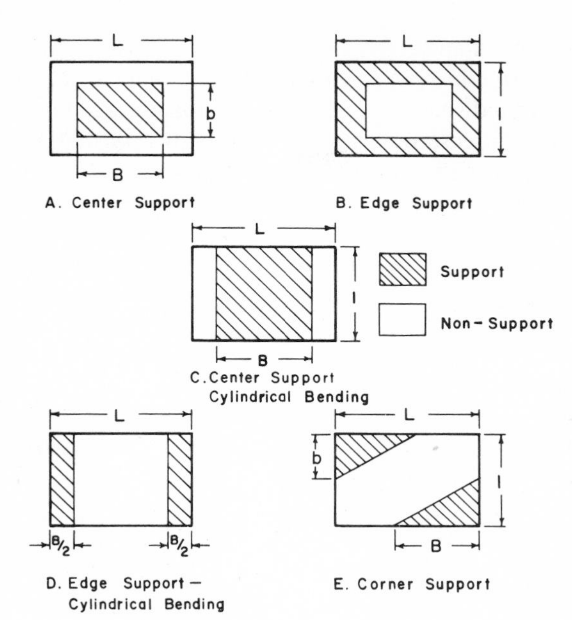

Figure 13-5 Soil-support patterns for maximum design quantities.

Economically, it is nearly impossible to reinforce these foundations against such high twisting moments. Consequently, it is a good general rule never to use foundation shapes that are subject to twisting moments, e.g., tee and ell shapes, and shapes with reentrant corners. If one must use such a shape, it is wise to provide some form of articulation between rectangular areas.

Despite the variety of influences which can cause differential movement, the amount of movement is constrained within fairly narrow limits in the range of 2 to 15 cm.

Shape of Soil Surface

Both analytical studies[1,19,21] and field measurements[25,26] have indicated that the surface of an expansive clay undergoing differential moisture change has a high point and a smooth transition to its low points. These shapes can be represented as exponential offsets y measured downward from the high point (as shown in Fig. 13-14)

$$y = cx^m \tag{13-19}$$

where x is the distance from the high point and m is an exponent that varies between about 2 and 8, increasing with the size of the building and the absence of a well developed cracking pattern in the soil.

The interaction of the soil with an overlying stiffened mat readjusts the shape of the surface to one that is flatter as the flexural stiffness of the mat increases.

There are two ways of analyzing these soil-structure interaction problems. One is termed *superposition* and the other a *true time-dependent interaction analysis*. The superposition method first predicts a final shape of the unloaded soil surface and then calculates the interaction of a superimposed flexural member with the distorted surface.[27] This method is simpler than the second and is useful in design calculations. The second method steps forward in time and includes the interaction effects at each time step.[1]

Beams and Plates on Distorted Surface

The differential equation of a beam on a distorted surface is

$$\frac{d^2}{dx^2}\left(EI\frac{d^2w}{dx^2}\right) - \frac{d}{dx}\left[GhB\frac{d}{dx}(w-y)\right] + kB(w-y) = q \qquad (13\text{-}20)$$

where w = deflection of beam
q = load per unit length
EI = flexural stiffness
Gh = coupled spring stiffness of foundation soil[27,28]
k = Winkler modulus
B = width of soil that cooperates in supporting beam

This equation applies wherever the beam is in contact with the soil. When the beam is not in contact with the soil, the soil-stiffness terms GhB and kB become zero. Finding where the beam lifts free of the mound requires a trial-and-error procedure regardless of whether the closed form or the finite difference form of the solution is used.

The governing differential equation for isotropic plates on a distorted surface is similar:

$$D\,\nabla^4 w - \nabla[Gh\,\nabla(w-y)] + k(w-y) = p \qquad (13\text{-}21)$$

where D is the flexural rigidity of the plate, p is the pressure on the plate, and the other terms are as previously defined. A more complicated formula for an orthotropic plate with stiffening beams can be found in Ref. 29.

13-4 NUMERICAL METHODS

The analysis of expansive clay foundations requires the solution of a variety of differential equations: moisture diffusion [Eq. (13-7)], coupled elasticity and moisture diffusion [Eqs. (13.9) to (13-14)], soil-structure interaction of drilled piers [Eq. (13-18)], and soil-structure interaction of stiffened beams and mats [Eqs. (13-20) and (13-21)]. The numerical methods used in solving these differential equations are discussed below.

Table 13-2 Moisture-diffusion computer programs

No. of dimensions	Type†	Time-stepping method	Coordinates‡	Ref.
1	FD	Explicit, implicit	R	30
1	FE	Forward difference	R	30
1	FD	Crank-Nicolson	R, A	4
2	FD	Crank-Nicolson	R	4
2	FD	Forward difference	R	31
2	FD	Alternating-direction explicit	R	5
2	FE	Crank-Nicolson	A	32

† FD = finite difference; FE = finite element.
‡ A = axisymmetric; R = rectangular.

Moisture Diffusion

Computer programs to solve Eq. (13-7) have been prepared in finite difference and finite element form with various methods for stepping forward in time. Some of them are listed in Table 13-2. Typical results of these computations are shown in Fig. 13-6 for the soil beneath a pavement in Horsham, Victoria, Australia.[31] The pF values recorded in this figure are the logarithm of the absolute value of suction:

$$\mathrm{p}F = \log |h|$$

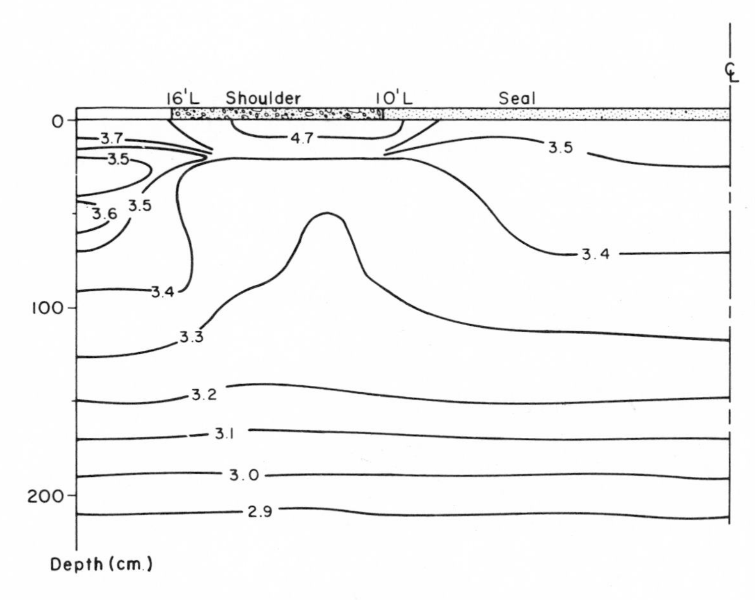

Figure 13-6 Typical suction contours beneath a pavement.[31]

Calculations of two-dimensional moisture diffusion are used to predict the final moisture condition beneath surfaces such as highway pavement and building foundations in order to estimate the change from the initial moisture condition and the surface distortion of the soil.

Finite difference formulation The finite difference methods as well as their stability and convergence characteristics were discussed in detail in Chap. 1. It must be noted carefully that even those methods which can be shown to be mathematically stable when the soil is homogeneous can give unstable results when the soil is inhomogeneous or when it is anisotropic and the directions of the principal permeabilities are more than 22.5° away from the directions of the coordinate axes.[12]

Finite element formulation The formulation of element equations and the assembly of the matrix of algebraic equation were described in Chap. 1. The element equation is repeated here for ready reference:

$$[k]\{h\} + [p]\{\dot{h}\} = \{Q(t)\} \tag{1-81}$$

In this equation $[k]$ is the element matrix of permeabilities, which represents the space derivative of suction h. The matrix $[p]$ is the collection of weights applied to each of the nodal points in representing the time derivative of suction $\dot{h}$. The $[p]$ matrix is multiplied by the average value of $\partial\theta/\partial h$ for that element. The vector $\{h\}$ is used here to represent the values of suction computed at the nodal points, and $\{Q(t)\}$ is the vector of flow input into the system.

Chapter 1 covers the formulation in rectangular coordinates; the formulation in axisymmetric coordinates is given here. The energy functional f, which must be minimized to satisfy the conditions of the anisotropic diffusion equation in axisymmetric cylindrical coordinates, is

$$f = \frac{1}{\gamma_w}\left[\frac{1}{2}k_{rr}\left(\frac{\partial h}{\partial r}\right)^2 + \frac{1}{2}k_{rz}\frac{\partial h}{\partial z}\frac{\partial h}{\partial r} + \frac{1}{2}k_{zr}\frac{\partial h}{\partial r}\frac{\partial h}{\partial z} + \frac{1}{2}k_{zz}\left(\frac{\partial h}{\partial z}\right)^2 + \frac{1}{2}\ln rk_{rr}\left(\frac{\partial h}{\partial r}\right)^2 + \ln rk_{rz}\frac{\partial h}{\partial z}\frac{\partial h}{\partial r} + \left(Q - c\frac{\partial h}{\partial t}\right)\right] \tag{13-22}$$

where

$$c = \frac{\partial \theta}{\partial h} \tag{13-23}$$

where k_{rr} = permeability in radial direction
k_{zz} = permeability in vertical direction
k_{rz}, k_{zr} = cross-permeability terms (usually equal)
r = radius to given point
Q = inflow rate to given element

The formulation based on Eq. (13-22) leads to a different $[k]$ matrix than that for rectangular coordinates

$$[k]_{rz} = \begin{bmatrix} k_{rr}(1 + \ln r) & k_{rz}(1 + \ln r) \\ k_{zr} & k_{zz} \end{bmatrix} \tag{13-24}$$

Details of this derivation are given by Lytton and Dunlap.[32]

Coupled Elasticity and Moisture Diffusion

The formulation of the coupled finite element equations for saturated flow has been given in Chap. 12 and will not be repeated here. As shown there, the element matrix is

$$\begin{bmatrix} [k]_1 & [c] \\ [c]^T & [k]_2 \end{bmatrix} \begin{Bmatrix} \{q\} \\ \{h\} \end{Bmatrix}_{t+1} = \begin{Bmatrix} \{Q\} \\ \{r_h\} \end{Bmatrix}_t \tag{13-25}$$

where $[k]_1$ = stiffness matrix
$[k]_2$ = permeability matrix
$[c]$ = coupling matrix which relates suction change to stresses
$\{q\}$ = displacement vector
$\{h\}$ = suction vector
$\{Q\}$ = load vector
$\{r_h\}$ = inflow vector

All that is involved in changing this form to one that can be used in solving the unsaturated flow equations [Eqs. (13-9) to (13-14)] is to multiply each term of the $[c]$ matrix by χ and each term of the matrix $[c]^T$ by $1/\chi$. With this simple change, existing computer programs for solving Biot equations can be modified to predict the movement and moisture diffusion within an expansive soil.

Various finite element computer programs have been written to solve these equations; Sandhu[33] uses a general variational principle, Hwang, Morgenstern, and Murray[34] use a Galerkin formulation, and Richards[1] uses a method which alternately solves the diffusion and elasticity portions of the problem separately and transfers the results between programs for each time step. While the first two methods solve simultaneously for displacements and pore pressures at each time step, Richards transfers suction changes to the elasticity solution and total stress changes back to the moisture-diffusion solution with each time step. With any of these schemes the geometry of the elements (as well as the elasticity, permeability, and other properties of the soil) can be changed between time steps to represent their dependence upon suction and pressure level.

Richards' method is basically an uncoupled approach; it has an advantage over existing coupled approaches since with his method it is possible to use more complex (and more realistic) forms of variation of volume strain with change of stress. For example, Henkel's law[35] can be used to obtain volume strain:

$$\frac{\Delta V}{V} = \frac{\Delta\sigma}{3} + a\,\Delta\tau_{\text{oct}} \tag{13-26}$$

from which the equivalent suction change is

$$\Delta h = \frac{1}{\alpha}\frac{\partial h}{\partial \theta}\left(\frac{\Delta\sigma}{3} + a\,\Delta\tau_{\text{oct}}\right) \tag{13-27}$$

In both these equations

$$\tau_{\text{oct}} = \tfrac{1}{3}[(\sigma_1 - \sigma_2)^2 + (\sigma_2 - \sigma_3)^2 + (\sigma_3 - \sigma_1)^2]^{1/2}$$

where σ_1, σ_2, and σ_3 are the principal stresses, and a is Henkel's pore pressure coefficient, which varies between 0.5 for normally consolidated clays and -0.1 for overconsolidated clays that are typical of expansive soils.

Soil-Structure Interaction of Drilled Piers

Two approaches are used in solving for the displacements and stresses induced in a drilled pier by the movement of expansive clay in contact with it. One is an axisymmetric elastic finite element analysis using elastic moduli and expansion coefficients. The other approach uses a finite difference solution of Eq. (13-18), which produces a tridiagonal matrix of equations as shown in Fig. 13-7. The matrix coefficients are

$$\begin{aligned} a_i &= \frac{2E_c A_i}{(\Delta z)^2} \\ b_i &= -\frac{P_i G_i}{2} - \frac{E_c(A_i + A_{i+1})}{(\Delta z)^2} \\ c_i &= -\frac{P_i G_{i+1}}{2} + \frac{E_c A_{i+1}}{(\Delta z)^2} \\ d_i &= \gamma_c A_{i+1} - \frac{P_i}{2}(G_i w_i + G_{i+1} w_{i+1}) \end{aligned} \tag{13-28}$$

where a_i, b_i, c_i, d_i are coefficients which appear in the matrix in Fig. 13-7. Δz is the vertical increment length, and the rest of the terms have been defined.

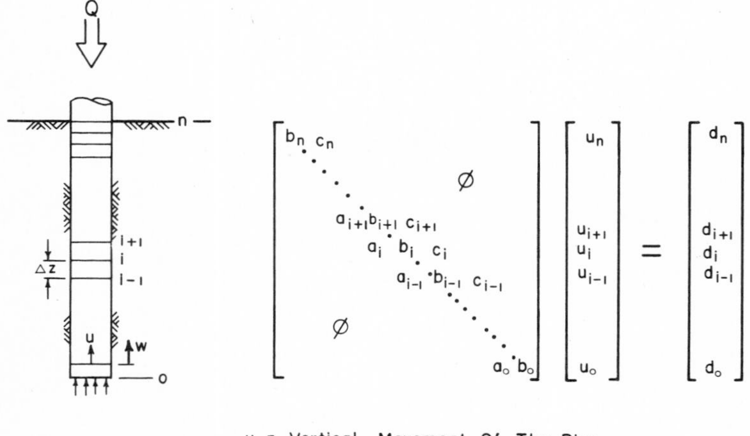

Figure 13-7 Finite difference analysis of drilled piers in expansive clay.

Boundary conditions The boundary conditions depend upon the sign of the difference between w and u. At the bottom of the pier, if $w - u$ is positive, the terms in brackets are deleted:

$$a_0 = -\frac{E_c A_1}{\Delta z} - \frac{P_0 G_1 \Delta z}{4}$$

$$b_0 = \frac{E_c A_1}{\Delta z} - \frac{P_0 G_1 \Delta z}{4} + [-KA_1] \tag{13-29}$$

$$d_0 = -\frac{\gamma_c A_1}{2} \Delta z - \frac{P_0 \Delta z}{4}(G_0 w_0 + G_1 w_1) + [-KA_1 w_0]$$

At the top of the pier, the boundary conditions are

$$b_n = -\frac{E_c A_n}{\Delta z} + \frac{P_n G_n \Delta z}{4}$$

$$c_n = \frac{E_c A_n}{\Delta z} + \frac{P_n G_{n-1} \Delta z}{4} \tag{13-30}$$

$$d_n = -Q + \frac{P_n \Delta z}{4}(G_{n-1} w_{n-1} + G_n w_n) - \frac{\gamma_c A_n}{2} \Delta z$$

where Q is a compressive load at the top of the pier. These equations allow the perimeter P_i, the perimeter shear modulus G_i, and the cross-sectional area of the pier A_i to be changed from one nodal point to the next within the computer program. They have the disadvantage that the elastic properties of the soil mass adjacent to the pier cannot be considered.

Soil-Structure Interaction of Beams and Plates

Equations (13-20) and (13-21) are the differential equations of a beam and a plate interacting with a coupled spring foundation which has some initial distortion. The Winkler or liquid subgrade foundation used in rigid pavement design is a special case of the coupled spring foundation. The foundation soil can also be represented numerically as an elastic continuum by using methods described in Chap. 5.

Several methods used to solve beam or plate-on-distorted-foundation problems are given in Table 13-3.

The finite difference formulation of the beam equation (13-20) produces a 5-wide banded matrix, and the plate equation (13-21) gives a 13-wide banded matrix. If the soil support is represented as a stiffness matrix of influence coefficients $[C]$, the finite difference formulation in the case of the beam or plate will take the form

$$[K]\{w\} - [C]\{w - y\} = \{Q\} \tag{13-31}$$

or

$$[[K] - [C]]\{w\} = -[C]\{y\} + \{Q\} \tag{13-32}$$

Table 13-3 Solutions of soil-structure interaction problems on expansive clay

Structural element	Flexible or rigid	Representation of soil	Numerical method used†	Analysis method used†	Ref.
Beam	Flexible	Winkler, coupled spring	FD	S	27
Beam	Flexible	Winkler, coupled spring	Closed form‡	S	27
Beam	Rigid	Winkler, coupled spring	Algebraic formula	S	27
Beam	Flexible	Elastic continuum	FE	TD	1
Plate	Flexible	Winkler	FD	S	36

† FD = finite difference; FE = finite element; S = superposition; TD = time-dependent solution

‡ The closed-form expressions are solutions to the differential equation of a beam on a mound. The solution must be found by trial and error.

where $[K]$ = beam or plate stiffness matrix

$\{w\}$ = downward deflection vector

$\{y\}$ = vector of offsets measured downward to soil surface from high point of initial distorted shape

$\{Q\}$ = load vector

The solution of this equation requires a trial-and-error procedure which zeros the rows and columns of the $[C]$ matrix wherever an individual value of y exceeds its corresponding value of w. Wherever this happens, it indicates that the structural element has lifted free of the soil.

The matrix formulation of this problem using a finite element representation of beams or plates in identical to Eqs. (13-31) and (13-32). Richards' time-dependent solution[1] noted in Table 13-3 treats the beam as a part of the elastic continuum. This solution also requires a trial-and-error process, but instead of zeroing elements of a $[C]$ matrix where the beam lifts free of the soil the modulus of the soil elements in the liftoff zones is reduced to a small value, on the order of 100 lb/in^2 (7.0 kg/cm^2), to simulate a vertical separation between the beam and the soil.

13-5 APPLICATIONS

The computation of expansive clay movements is used in predicting surface distortions of pavements and other overlying flexible structures and in calculating the stresses and displacements of beams, plates, drilled piers, and the structures built on them. This section presents typical results of these computations.

Surface Distortion

Two uncoupled methods of predicting surface distortion are used. One is based upon moisture diffusion, and the other is based upon elastic theory. Both have been discussed earlier in this chapter. The moisture-diffusion–volume-change approach has been used successfully in predicting the heave of a ponded surface of clay shale, as shown in Fig. 13-8. The change of suction was predicted using a two-dimensional finite difference computer program. The soil properties used in this problem are summarized in Table 13-4. However, there are discrepancies between the predicted and the measured heave due to the fact that the elastic properties of the soil are not taken into account. Each column of soil is assumed to swell independently of the ones next to it. Consequently, when there is a sharp change in moisture content, there is also reflected a sharp change in the predicted heave of the surface.

This difficulty was avoided by Livneh, Shklarsky, and Uzan[19] when they used a finite element program and an assumed pattern of moisture changes to predict heaving and induced stresses in an asphalt pavement. The assumed moisture changes were used to predict an initial strain in each element as in Chaps. 1 and 2. The element equations for this approach are:

Nodal forces:
$$\{F\} = \iiint_{\text{vol}} [B]^T[C]\{\epsilon_0\}\, d(\text{vol}) \tag{13-33}$$

Stresses:
$$\{\sigma\} = [C]\{\epsilon_0\} \tag{13-34}$$

Strains:
$$\{\epsilon_0\} = \Delta W \begin{Bmatrix} \alpha_1 \\ \alpha_2 \\ 0 \end{Bmatrix} \tag{13-35}$$

where $\{\epsilon_0\}$ = strain vector

α_1, α_2 = horizontal and vertical coefficients of expansion

Δw = change of gravimetric water content

and $[B]$ and $[C]$ are as defined in Chap. 1.

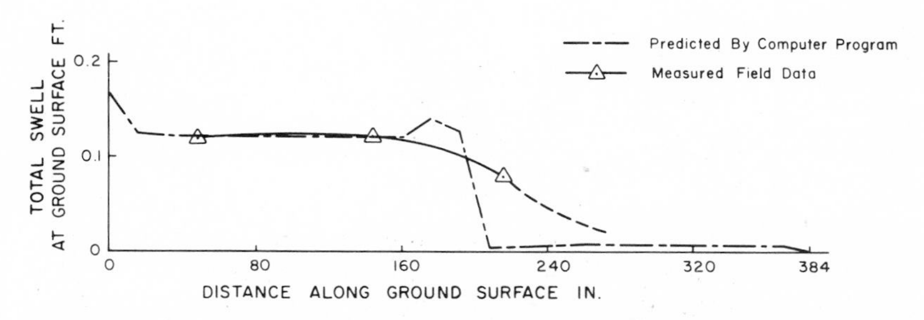

Figure 13-8 Surface heave of expansive clay shale predicted with the moisture-diffusion approach.[13]

Table 13-4 Expansive soil properties: diffusion and volume change

Permeability, cm/s	1.0×10^{-6}
Suction range, p*F* (wetting process)	1.5–3.5
Water content range, %	10–40
Swelling pressure (assumed), kg/cm²	6.3
Exponent relating pressure to total volume	1.2
Exponent relating water volume to total volume	2
Exponent relating suction to volumetric water content	4

The results of the computations are shown in Fig. 13-9. The soil properties used in the computations are shown in Table 13-5.

Comparison of the surface profiles in Figs. 13-8 and 13-9 shows that the moisture-diffusion approach results in abrupt changes in the heave profile whereas the elasticity approach, which has assumed no less abrupt a set of moisture changes, gives a smoother heave profile. Although these finite element calculations have not been verified by comparison with field experiment, it is reasonable to expect that if sufficiently accurate changes of water content are assumed, this approach will give results that are superior to the uncoupled moisture-diffusion approach.

Drilled Pier Foundations

Large bending moments and tensile stresses can be induced in drilled piers by the activity of surrounding expansive clay. An approximate method for calculating the bending moment is given by Komornik.[24] The tensile forces can be predicted

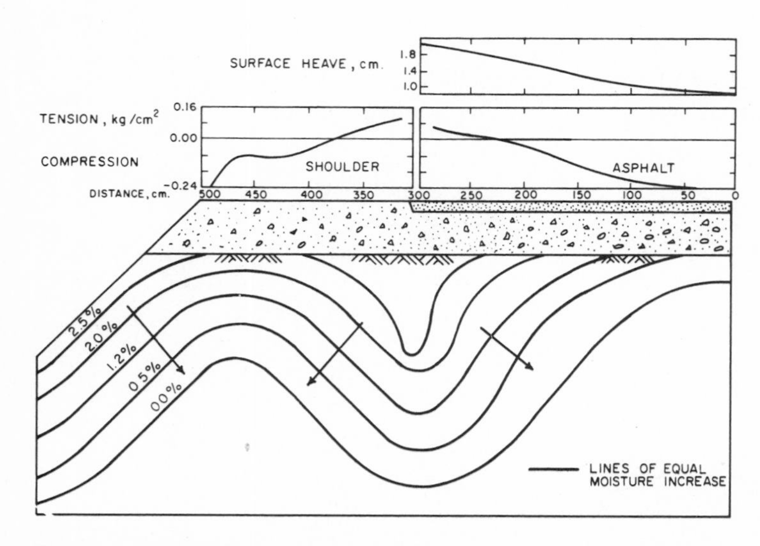

Figure 13-9 Surface heave of a pavement on expansive clay using the elasticity approach.[19]

Table 13-5 Expansive soil properties: elastic expansion

Direction	Modulus of elasticity E, kg/cm^2	Poisson's ratio ν	Coefficient of expansion α
Horizontal	20	0.25	0.50
Vertical	12	0.25	0.50

by finite difference or finite element techniques. The steps in the analysis are the same with either method.

1. Measure the initial moisture (or suction) profile and estimate the future moisture profile which represents the largest change from the initial condition. This will usually be a wetter profile.
2. Determine the soil properties consistent with the new profile. In the finite difference approach these are the relations between perimeter shear modulus and relative soil-pier movement. The finite element approach requires moisture-expansion coefficients and elastic moduli.
3. Perform the analysis. The finite element approach uses an axisymmetric computer program and the estimated moisture (or suction) change to determine the stresses in the pier and the total vertical movement of the pier at the ground surface. This method is similar to the approach used by Livneh, Shklarsky, and Uzan[19] illustrated in Fig. 13-9. The finite difference approach is one-dimensional and is based on Eqs. (13-18) and (13-28) to (13-30).

Results Typical results of the finite difference calculations are shown in Fig. 13-10. The maximum tensile stress usually occurs around the midheight of the pier and is used for determining the reinforcing-steel requirement. The tensile strength of concrete is around 350 to 400 lb/in^2 (25 to 38 kg/cm^2). Other material properties used in this problem are given in Table 13-6.

Several such analyses with different loads applied to the top of the pier result in a load-heave curve useful in the structural analysis of the grid system of beams which rests on the drilled piers and supports the structure. The load-heave curve shown in Fig. 13-11 was developed for the pier shown in Fig. 13-10.

Table 13-6 Material properties: drilled pier

Elastic modulus of concrete	3×10^6 lb/in^2	2.1×10^5 kg/cm^2
Perimeter shear modulus, range	70–2240 lb/in^3	1.9–62 kg/cm^3
Vertical heave at the surface	2.5 in	6.3 cm
Exponent relating heave to depth	3.0	
Soil subgrade modulus at tip of pier	16.0 lb/in^3	1.1 kg/cm^3

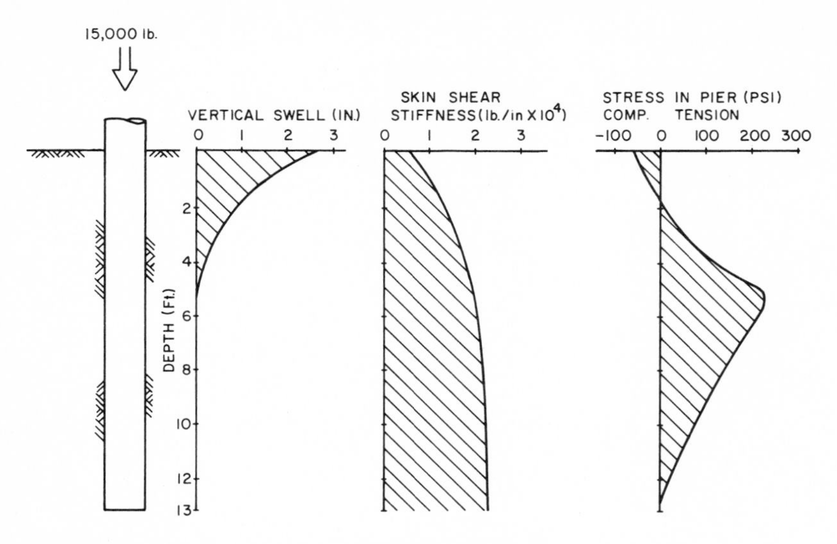

Figure 13-10 Typical results of calculation on a drilled pier.

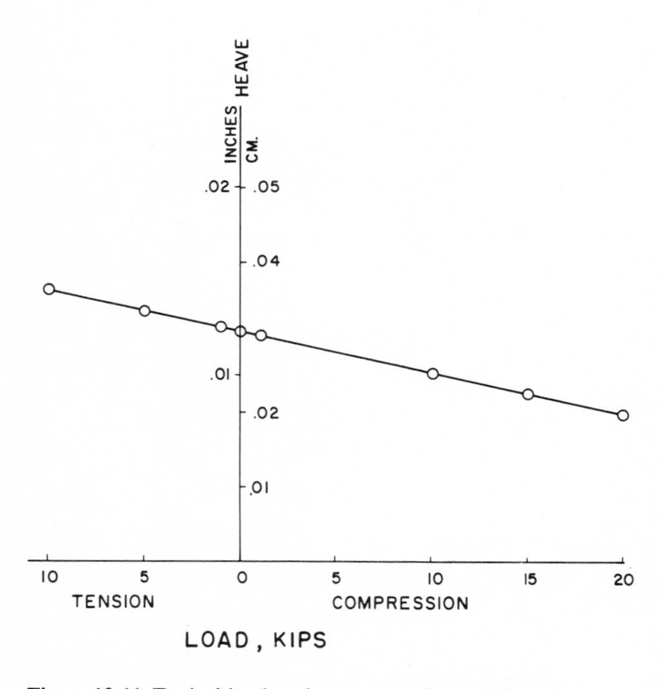

Figure 13-11 Typical load-vs.-heave curve for a drilled pier.

Analysis of Pier-and-Beam Foundations

The analysis of a pier-and-beam foundation can be made using either an elastic or a time-dependent approach, the latter being closer to the actual case but much more costly in computer time. The elastic analysis is simpler and may give results that are sufficiently accurate for design.

Elastic analysis With this approach, each pier is viewed as a spring support for the beam grid. The spring reaction is the load-heave curve illustrated in Fig. 13-11, and the initial elevation of each "spring" is its zero-load heave position. Standard structural-grid analysis programs can be used to determine the final position, deflections, moments, and shear in the structure. The major problem with this approach is in estimating the heave position of each pier at zero load. The best guides to this estimation are field observations and detailed analyses of typical soil movement using the expected moisture boundary conditions.

Time-dependent analysis The solution of the time-dependent soil-structure interaction problem for an entire structural foundation would be more complex. It would require interaction within each time step to make certain that the load on each pier is compatible with the amount of vertical movement that has taken place. The principles involved in all these calculations have been discussed previously, and further detailed discussion is beyond the scope of this chapter. An example of time-dependent analysis for beams on three supports has been described in Ref. 37.

Beams on a Distorted Foundation

The solution [Eq. (13-20)] requires a trial-and-error procedure to locate the points at which the beam lifts free of the soil. The equation no longer applies to the beam beyond the liftoff points. A schematic illustration of typical results of these calculations is shown in Fig. 13-12. The design quantities of moment, shear, and differential deflection are marked A through F. The shear stiffness term GhB would normally be so small as to be negligible, and the coefficient of subgrade reaction kB is analogous to the slope of the load-heave curve of the drilled pier illustrated in Fig. 13-11. Typical values of this long-term k vary between about 0.28 and 2.8 kg/cm^3. More detail on the determination of these soil constants is given in Refs. 27 and 28. The Winkler or coupled-spring foundation is convenient to use with either a finite element or finite difference method of computation since a foundation spring can be fit beneath each nodal point.

Rigid beam There are several advantages in considering the rigid beam. First, the differential equation (13-20) becomes an algebraic equation [Eq. (13-36)] since all derivatives of w with respect to x are zero:

$$\frac{d}{dx}\left(GhB\frac{dy}{dx}\right) + kB(w - y) = q \tag{13-36}$$

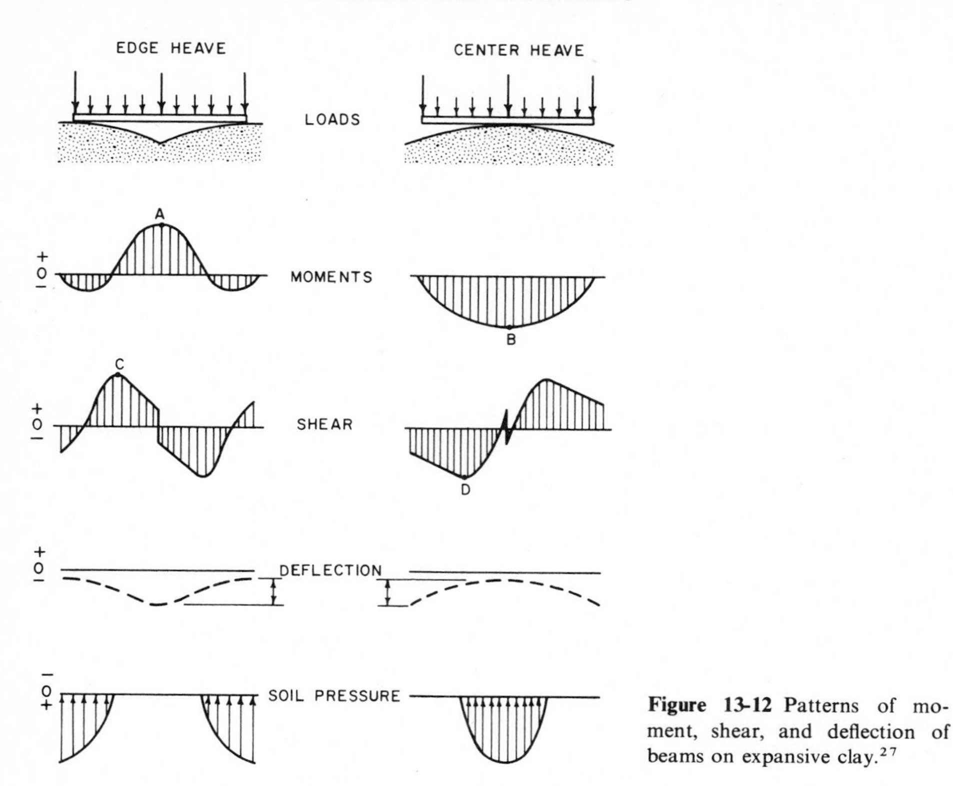

Figure 13-12 Patterns of moment, shear, and deflection of beams on expansive clay.[27]

Second, rigid-beam results have been shown[27] to represent upper-bound solutions for the moment and shear acting in a beam, making them suitable for design calculations. The steps in these calculations are as follows:[38]

1. Calculate a maximum moment M_0 assuming a rigid beam and rigid soil.
2. Calculate the one-dimensional moment M_1. Because the soil is not rigid, reduce the maximum moment M_0 by a moment correction term M_c, which accounts for the compressibility of the soil:

$$M_1 = M_0 - M_c \tag{13-37}$$

$$M_c = c\frac{TL}{8} \tag{13-38}$$

where T = total load acting on beam
L = length of beam
c = support index

$$c = \frac{m+1}{m+2}\left(\frac{m+1}{m}\frac{T}{A}\frac{1}{ky_m}\right)^{1/(m+1)} \tag{13-39}$$

where m = mound exponent
y_m = maximum expected differential movement of soil
T/A = average foundation pressure

Nomographs for the support index c have been presented in Ref. 38.

3. Calculate the design shear V_1. This is an empirical relation which is usually conservative:

$$V_1 = \frac{4M_1}{L} \tag{13-40}$$

4. Calculate the differential deflection Δ_1. Again, this is a conservative empirical relation:

$$\Delta_1 = \frac{M_1 L^2}{12E_c I} \tag{13-41}$$

where E_c is the elastic modulus of the concrete and I is the moment of inertia of the beam section.

5. Determine whether the selected beam section meets the design constraints for bending stress, shear stress, and differential deflection using an appropriate building code.

Plates on Distorted Foundation

Numerical studies with a finite difference computer program for an orthotropic plate with stiffening beams have shown that partial support produces patterns of moment, shear, and deflection like those shown in Fig. 13-13. These studies resulted in empirical rules for the relation of one-dimensional rigid-beam moments to two-dimensional maximum moments.[21,36] The procedure for a rectangular plate is as follows:

1. Calculate an M_1 in both the long and the short directions.
2. Calculate a design moment in each direction: M_L in the long direction and M_S in the short direction, according to the formulas

$$M_L = M_1\left(1.4 - 0.4\frac{L}{l}\right) \tag{13-42}$$

but not less than

$$M_L = M_1(1.5 - c) \tag{13-43}$$

$$M_S = M_1\left[1.0 + 0.9(1.2 - c)\left(\frac{L}{l} - 1\right)\right] \tag{13-44}$$

As the rectangle becomes more elongated, the moment in the long direction decreases and the moment in the short direction increases. Typical sizes of stiffening beams for residential construction are 24 to 36 in (61 to 92 cm) deep.

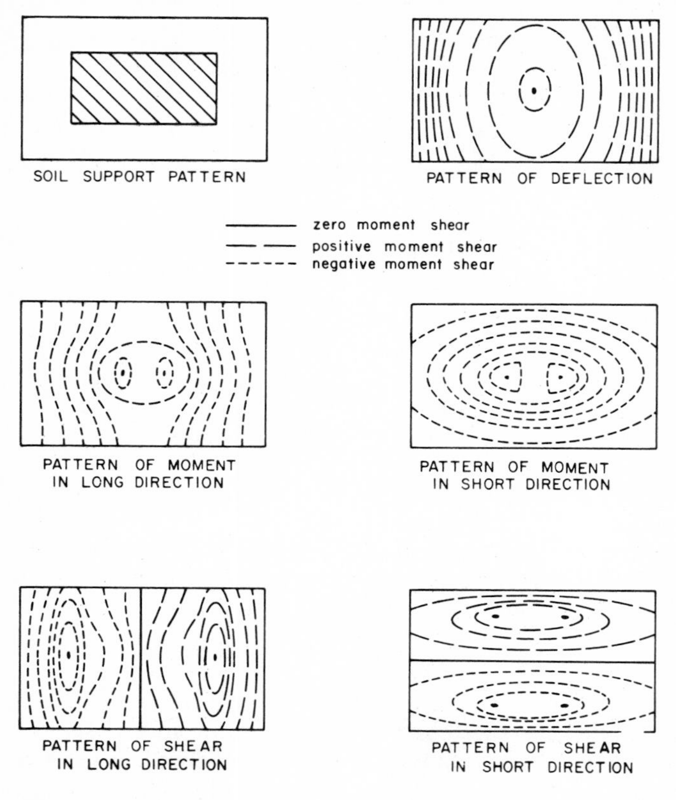

Figure 13-13 Patterns of soil support, moment, shear, and deflection for the center-support distortion mode.[36]

Viscoelasticity of the Soil

The method previously called superposition can be used to obtain an expression for the time-dependent maximum moment in the rigid beam shown in Fig. 13-14:

$$M = \frac{PL}{2} + \frac{pL^2}{8} - \int_0^l g\,dx \tag{13-45}$$

Letting

$$M_0 = \frac{PL}{2} + \frac{pL^2}{8} \tag{13-46}$$

and carrying out the integration in Eq. (13-45) gives the equation which relates the maximum moment of the support index $c(t)$:

$$M = M_0 - c(t)\frac{TL}{8} \tag{13-47}$$

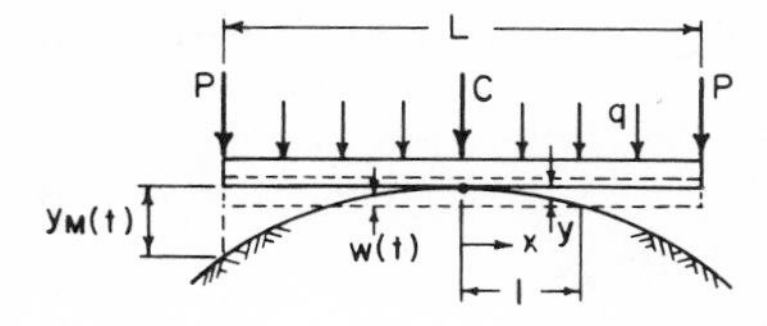

Figure 13-14 Rigid beam on a viscoelastic Winkler mound.

where

$$c(t) = Ct^{r/m}\left[r\beta(r, 1-n) - \frac{2q}{m+2}\beta(q, 1-n)t^{-q/m}\right] \tag{13-48}$$

and

$$C = \frac{KYL}{T}\left(\frac{2X^{1/m}}{L}\right)^{m+2} \tag{13-49}$$

$$r = \frac{m}{m+1}\left(n + \frac{q}{m}\right) \tag{13-50}$$

where $\beta(s, t)$ = beta function = $\Gamma(s)\Gamma(t)/\Gamma(s+t)$
$\Gamma(s)$ = gamma function
P = concentrated edge load
p = uniform load
g = subgrade reaction pressure

The remaining constants are defined below. The total load T is

$$T = 2P + R + pL \tag{13-51}$$

The downward offset from the high point of the mound is y, which is now a function of time and distance:

$$y(x, t) = y_m(t)\left(\frac{2x}{L}\right)^m \tag{13-52}$$

where y_m is the maximum differential movement and m is the mound exponent, which may range between 2 and 8. The beam penetration into the soil is w

$$w(x, t) = y_m(t)\left(\frac{2}{L}\right)^m (l^m - x^m) \tag{13-53}$$

where l is half the supported length at time t. The subgrade reaction pressure g is given by

$$g(x, t) = \int_0^t k(t-\tau)\frac{dw}{d\tau}\,d\tau \tag{13-54}$$

In this equation k, the subgrade modulus, is a function of time given by

$$k(t) = Kt^{-n} \tag{13-55}$$

where n is the exponential relaxation rate, which varies between about 0.25 and

1.0, and the constant K is the value of k at unit time. The maximum differential movement is

$$y_m(t) = Ytg \tag{13-56}$$

where q is the exponential rate of swelling, which varies between 0 and 1, and the constant Y is the value of y_m at unit time.

The term $c(t)$ is called the *support index* because it is a measure of the percentage of the beam that is supported by the soil. As $c(t)$ gets larger, the design moment M gets smaller. Consequently, as the exponents q, m, and n change, various conditions occur under which the design moment will either grow or diminish. A study of the soil support index $c(t)$ in Eq. (13-48) has shown that variations of the swelling rate constant q over its whole range has little effect on changing the supported length. The support index is changed most significantly (1) by variations in the rate of relaxation of the subgrade modulus n and (2) by the shape of the soil surface as described by the mound exponent m. This is shown in Fig. 13-15, where q is held constant and n and m are allowed to vary. The support index also changes with time, offering more support as the soil continues to swell and relax. The results of this study lead to several conclusions about design quantities.

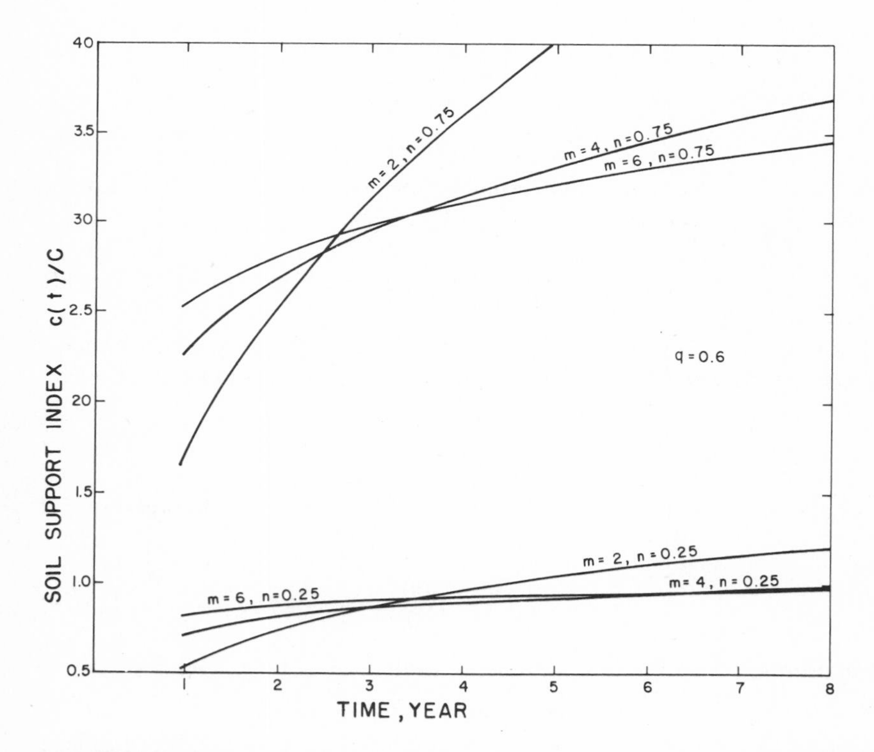

Figure 13-15 Variation of soil support index with mound shape (m) and relaxation of soil modulus (n).

1. The most crucial time in the life of a stiffened mat for developing critical bending stresses is in the first year or so.
2. The more a soil relaxes during the development of differential movement, the more support it will offer a stiffened mat and the less critical the stress condition becomes. If the soil does not relax but gets stiffer, the support and stress conditions can become more severe with time.

Further study into the effect of creep of concrete on deflections shows that under certain concrete creep rates the differential deflection can reach a maximum and then reverse itself. This result indicates that the time when stresses are critical may not coincide with the time when superstructure cracking, which is caused by deflection, is most severe.

With the viscoelastic equations (13-47) and (13-48), as well as others that apply to differential deflection and include the creep compliance of the concrete,[39] it is possible to determine simply the time-dependent response of structural elements, including the maximum values of moment, shear, and differential deflection which are useful in the design of stiffened mats on expansive clay.

REFERENCES

1. Richards, B. G.: Model for Slab Foundations on Expansive Clays, *Proc. 8th Int. Conf. Soil Mech. Found. Eng., Moscow, 1973*, vol. 2.2, pp. 185–191.
2. Richards, B. G.: A Mathematical Model for Moisture Flow in Horsham Clay, *Civ. Eng. Trans. Inst. Eng. Aust.*, vol. CE10. no. 2, pp. 220–224, October 1968.
3. Ingles, O. G., J. G. Lang, and B. G. Richards: Pre-equilibrium Observations on the Reconstructed Flagstaff Gully Dam, *Symp. Earth Rockfill Dams, Talwara, Punjab, India, November 1968*, vol. 1, pp. 162–170.
4. Lytton, R. L., and R. K. Kher: Prediction of Moisture Movement in Expansive Clays, *Cent. Highw. Res. Univ. Texas Res. Rep.* 118-3, Austin, May 1970.
5. Johnson, L. D., and C. S. Desai: A Numerical Procedure for Predicting Heave, *Proc. 2d Australia–New Zealand Conf. Geomech., Brisbane, July 1975.*
6. Gardner, W. R.: Laboratory Studies of Evaporation from Soil Columns in the Presence of Water Table, *Soil Sci.*, vol. 85, p. 244, 1958.
7. Philip, J. R.: The Physics of Water Movement in Porous Solids, Water and Its Conduction in Soils, *Highw. Res. Board Spec. Rep.* 40, p. 147, 1958.
8. Kemper, W. D.: Movement of Water as Affected by Free Energy and Pressure Gradients, part 1, *Proc. Soil Sci. Soc. Am.*, vol. 25, no. 4, pp. 255–260, 1961.
9. Aitchison, G. D., K. Russam, and B. G. Richards: Engineering Concepts of Moisture Equilibria and Moisture Changes in Soils, *Road Res. Lab. Rep.* 38, Crowthorne, 1966.
10. Smiles, D. E., and M. Rosenthal: The Movement of Water in Swelling Material, *Austr. J. Soil Res.*, vol. 6, 1968.
11. Philip, J. R.: Hydrostatics and Hydrodynamics in Swelling Soils, *Water Resour. Res.*, vol. 5, no. 5, 1969.
12. Lytton, R. L.: Theory of Moisture Movement in Expansive Clays, *Univ. Texas Cent. Highw. Res. Res. Rep.* 118-1, Austin, September 1969.
13. Lytton, R. L., and W. G. Watt: Prediction of Swelling in Expansive Clays, *Univ. Texas Cent. Highw. Res. Res. Rep.* 118-4, Austin, 1970.
14. Sokolov, M., and J. M. Amir: Moisture Distribution in Covered Clays, *Proc. 3d Int. Conf. Expansive Soils, Haifa, 1973*, vol. 1, pp. 129–136.

15. Edlefson, N. W., and A. B. C. Anderson: Thermodynamics of Soil Moisture, *Hilgardia*, vol. 15, pp. 31–298, 1943.
16. Statement of the Review Panel, "Moisture Equilibria and Moisture Changes in Soils Beneath Covered Areas: A Symposium in Print," pp. 7–21, Butterworths & Co. (Publishers), Ltd., London, 1965.
17. McQueen, I. S., and R. F. Miller: Soil Moisture and Energy Relationships Associated with Riporian Vegetation near San Carlos, Arizona, *Geol. Surv. Prof. Pap.* 655-E, 1972.
18. McDowell, C.: Interrelationship of Load, Volume Change, and Layer Thickness of Soils to the Behavior of Engineering Structures, *Proc. Highw. Res. Board*, vol. 35, p. 754, 1956.
19. Livneh, M., E. Shklarsky, and J. Uzan: Cracking of Flexible Pavements Based On Swelling Clay: Preliminary Theoretical Analysis, *Proc. 3d Int. Conf. Expansive Soils, Haifa, 1973*, vol. 1, pp. 257–266.
20. Jennings, J. E., and K. Knight: The Prediction of Total Heave from the Double Oedometer Test, *Symp. Expansive Clays, S. Afr. Inst. Civ. Eng. Trans.*, 1957.
21. Lytton, R. L., and J. A. Woodburn: Design and Performance of Mat Foundations on Expansive Clay, *Proc. 3d Int. Conf. Expansive Soils, Haifa, 1973*, vol. 1, pp. 301–308.
22. Richards, B. G.: Moisture Flow and Equilibria in Unsaturated Soils for Shallow Foundations, *Symp. Permeability Capillarity, ASTM, Spec. Tech. Pap.* 417.
23. Biot, M. A.: Theory of Elasticity and Consolidation for a Porous Anisotropic Solid, *J. Appl. Phys.*, vol. 26, pp. 182–185,
24. Komornik, A.: The Effect of Swelling Properties of Unsaturated Clay on Pile Foundations, D.Sc. thesis, Israel Institute of Technology, Haifa, May 1962.
25. De Bruijn, C. M. A.: Annual Redistribution of Soil Moisture Suction and Soil Moisture Density beneath Two Different Surface Covers and the Associated Heaves at the Onderstepoort Test Site near Pretoria, "Moisture Equilibria and Moisture Changes in Soils beneath Covered Areas: Symposium in Print," Butterworths & Co. (Publishers), Ltd., London, pp. 122–134, 1965.
26. Spotts, J. W.: A Mechanism of Gilgai Relief, Ph.D. thesis, Texas A & M University, 1974.
27. Lytton, R. L., and K. T. Meyer: Stiffened Mats on Expansive Clay, *J. Soil Mech. Found. Div. ASCE*, vol. 97, no. SM7, pp. 999–1019, July 1971.
28. Pasternak, P. L.: "On a New Method of an Elastic Foundation by Means of Two Foundation Constants" (in Russian), Gosuredarstvennoe Izadatelstvo Literaturi po Stroilelstvu i Arkhitekture, Moscow, 1954.
29. Panak, J. J., and H. Matlock: A Discrete-Element Method of Analysis for Orthogonal Slab and Arid Bridge Floor Systems, *Univ. Texas Cent. Highw. Res. Res. Rep.* 56–25, Austin, May 1972.
30. Desai, C. S., and L. D. Johnson: Some Numerical Procedures for Analysis of One-dimensional Consolidation, *Proc. Symp. Appl. Finite Element Method Geotech. Eng., Vicksburg, Miss., September 1972*, pp. 863–882.
31. Richards, B. G.: An Analysis of Subgrade Conditions at the Horsham Experimental Road Site Using the Two-dimensional Diffusion Equation on a High-Speed Digital Computer, "Moisture Equilibria and Moisture Changes in Soils Beneath Covered Areas: A Symposium in Print," pp. 243–258, Butterworths & Co. (Publishers), Ltd., London, 1965.
32. Lytton, R. L., and W. A. Dunlap: Deformation, Pore Pressure, and Stability Analysis of Large Storage Tank Foundations, *Proc. Symp. Appl. Finite Element Method Geotech. Eng., Vicksburg, Miss., September 1972*, pp. 767–798.
33. Sandhu, R. S.: Finite Element Analysis of Consolidation and Creep, *Symp. Appl. Finite Element Method Geotech. Eng., Vicksburg, Miss., September 1972*, pp. 697–738.
34. Hwang, C. T., N. R. Morgenstern, and D. W. Murray: Application of the Finite Element Method to Consolidation Problems, *Symp. Appl. Finite Element Method Geotech. Eng., Vicksburg, Miss., September 1972*, pp. 739–766.
35. Henkel, D. J.: The Shear Strength of Saturated Remolded Clays, *Res. Conf. Shear Strength Cohesive Soils, ASCE, Univ. Colorado, June 1960*, p. 533.
36. Lytton, R. L.: Design Criteria for Residential Slabs and Grillage Rafts on Reactive Clay, *CSIRO Div. Appl. Geomech.*, Melbourne, November 1970.
37. Mazurik, A., and A. Komornik: Interaction of Superstructure and Swelling Clay, *Proc. 3d Int. Conf. Expansive Soils, Haifa, 1973*, vol. 1, pp. 309–318.

38. Lytton, R. L.: Design Methods for Concrete Mats on Unstable Soils, *Proc. 3d Int. Conf. Materials Technol., Rio de Janiero, 1972*, pp. 170–177.
39. Lytton, R. L.: Stiffened Mat Design Considering Viscoelasticity, Site Conditions, and Geometry, *Proc. 3d Int. Conf. Expansive Soils, Haifa, 1974*, vol. 2.
40. Johnson, L. D.: Influence of Suction on Heave of Expansive Soils, *U.S. Army Eng. Waterw. Expt. Stn. Misc. Pap.* S-73-17, April 1973.
41. Aitchison, G. D., and B. G. Richards: The Fundamental Mechanisms Involved in Heave and Soil Moisture Movement and the Engineering Properties Which Are Important in Such Movement, *Proc. 2d Int. Res. Eng. Conf. Expansive Clay Soils, College Station, Tex., 1969*, pp. 66–84.

CHAPTER

FOURTEEN

FLOW THROUGH POROUS MEDIA

Chandrakant S. Desai

14-1 INTRODUCTION

The problem of flow or seepage of fluids through porous media becomes significant in such disciplines as soil and rock mechanics, groundwater resources, irrigation and drainage, hydraulics, hydrology, soil science, oil geology, and environmental engineering. Seepage through geologic materials usually involves flow through multiphase media: soil skeleton, liquid (water), and sometimes gases. For our present purposes we shall restrict attention to cases where the soil skeleton can be assumed to be incompressible or rigid. If the compressibility is considered, we deal with a coupled phenomenon; special cases of the coupled phenomenon are consolidation, swelling, electroosmosis, convection, and liquefaction. Some of these topics are covered elsewhere in this book.

A large number of procedures have been used for solution of fluid-flow problems. Analytical closed-form solutions to the governing differential equations and use of such laboratory devices as the electrical analogy, Hele-Shaw viscous flow, glass bead, blotting paper, centrifuge, and prototype models have been covered in various publications.[1-10] For certain situations, trial-and-error procedures, including sketching flow nets by hand, have also been used.[5,11,12]

The analytical solutions are generally possible only for cases which involve linear equations and in which geometry and boundary conditions can be described by simple functions. Analog devices, though applicable to complex problems, need changes in idealized parameters for each different problem. Construction of prototype models can be costly and time consuming, and it is difficult to evolve general solution procedures from laboratory or field model tests.

Many of the equations governing the flow problems are nonlinear, and most of the natural conditions are extremely complex. Analytical (closed-form) methods are usually not suitable for such problems, and recourse to the recently developed numerical methods becomes necessary.

14-2 GEOTECHNICAL ASPECTS

Since the composition of geologic media can exhibit significant spatial and temporal variations, such material parameters as permeability (conductivity, transmissivity) and effective porosity (specific yield, storage coefficient) relevant to analysis can vary widely. Selection of proper values of these parameters is highly essential for realistic answers. Their determination requires adequate field and laboratory tests.[3,5,11,12]

Nonhomogeneities in the materials usually pose no difficulties when numerical methods are employed. Special schemes may be necessary, however, to satisfy the continuity of flow conditions at the interfaces while using finite difference methods (Chap. 1). Numerical instabilities can arise if there is a very wide difference between the properties of adjacent layers. In the finite element method, where fluid heads are assumed as unknowns, this situation can perhaps be remedied by using higher order approximating models or finer meshes.[13] Alternatively, we can use formulations in terms of stream functions.[14,15]

Spatial and timewise variations of material properties such as permeability can be included conveniently in a numerical formulation. Material anisotropy is usually taken care of in the formulation of most of the numerical methods.

Constitutive Laws

Many numerical formulations in the past have been based on Darcy's law,[3,5,11] given by

$$v = -ki \tag{14-1}$$

where v = fluid velocity
k = coefficient of permeability
i = hydraulic gradient

Darcy's law may be valid only for situations involving fine materials and lower velocity gradients. Flow through coarser materials such as gravels and rock fills will be governed by nonlinear laws. Some non-Darcy laws employed in finite difference and finite element formulations are[16-19]

$$i = av + bv^2 \tag{14-2a}$$

$$i = cv^m \tag{14-2b}$$

in which a, b, c, and m are constants determined from experiments. Equation (14-2a) is often referred to as the *Forchheimer equation*, and Eq. (14-2b) is an exponential law.

14-3 NUMERICAL METHODS

The finite difference and the finite element methods seem to be the major techniques employed. Other methods such as combinations of closed form and numerical methods, method of characteristics, and integral equation methods have also been used. Appendix 14A lists references covering the theory and formulation procedures, and a review of applications of numerical methods to different types of problems. The applications have been classified as steady and unsteady (transient) problems and subclassified as confined and unconfined. A number of related topics such as dispersion and moisture movement are also included in the review in Appendix 14A. Additional information on the procedures employed and on the problems solved is provided by brief annotations with the references cited at the end of the chapter.

Governing Equations for Saturated Flow

The basic equation governing unsteady or transient flow of fluids through a saturated porous rigid medium can be expressed as[24–26]

$$\frac{\partial}{\partial x}\left(k_x \frac{\partial \phi}{\partial x}\right) + \frac{\partial}{\partial y}\left(k_y \frac{\partial \phi}{\partial y}\right) + \frac{\partial}{\partial z}\left(k_z \frac{\partial \phi}{\partial z}\right) + \bar{Q} = n \frac{\partial \phi}{\partial t} \qquad (14\text{-}3a)$$

where k_x, k_y, k_z = coefficients of permeability in x, y, z directions, respectively
t = time
$\phi = p/\gamma + z$ = fluid head or potential
p = pressure
γ = unit weight of water
z = elevation head
n = porosity or effective porosity
$\bar{Q}$ = specified fluid flux

The assumptions made in deriving Eq. (14-3a) are that the flow is continuous and irrotational, the fluid is homogeneous and incompressible, capillary and inertia effects are negligible, magnitudes of velocities are small, Darcy's law holds good, and x, y, and z are the principal directions of permeability.

If the conditions of the problem do not change with time, the right-hand side in Eq. (14-3a) vanishes, which results in an equation for steady-state flow. For steady-flow in isotropic media, the equation reduces to the well known Laplace equation

$$\nabla^2 \phi = 0 \qquad (14\text{-}3b)$$

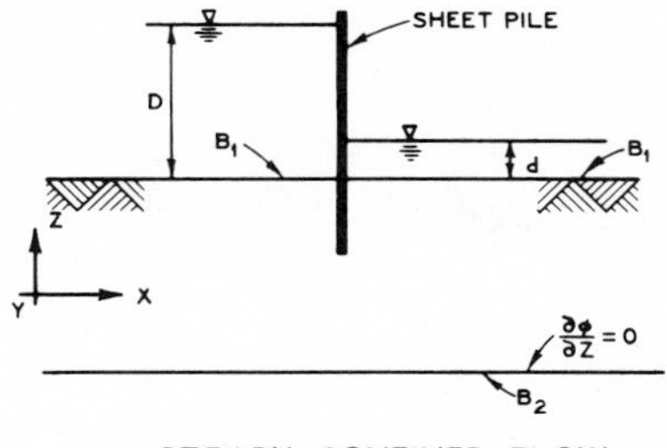

a. STEADY CONFINED FLOW

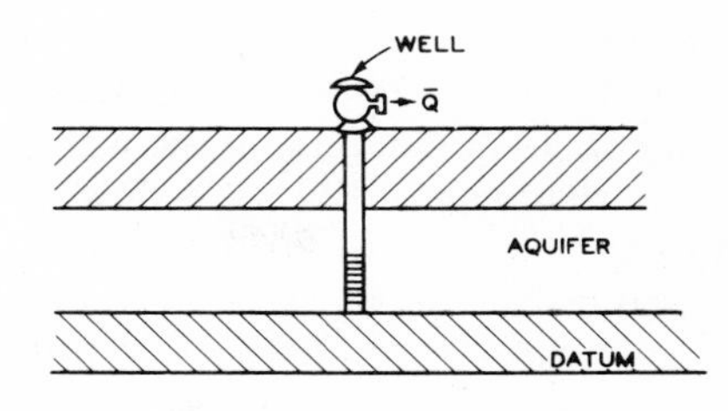

b. TRANSIENT CONFINED FLOW

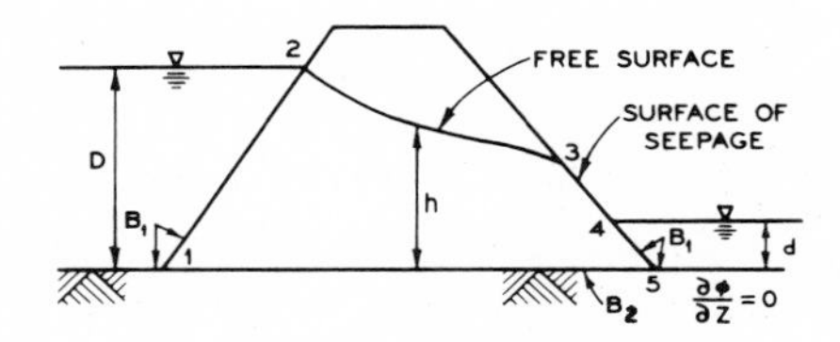

c. STEADY UNCONFINED FLOW

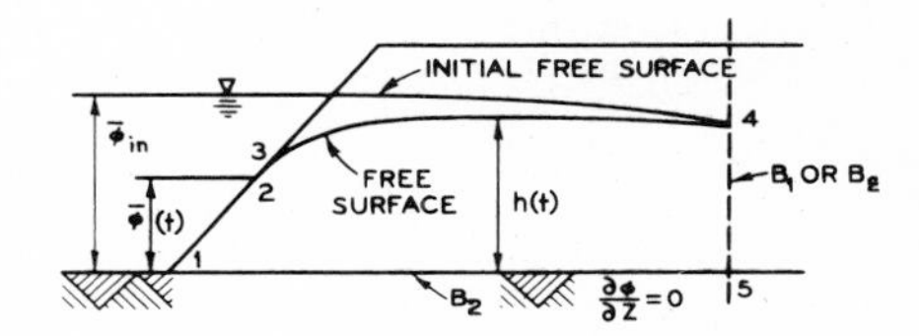

d. TRANSIENT UNCONFINED FLOW

Figure 14-1 Various categories of flow through porous media.

The case of steady flow in which fluid heads are known at all pervious external boundaries is called *steady confined* flow.[24] An example is the flow in the foundation of a sheet-pile wall (Fig. 14-1*a*). The flow toward a well in a saturated confined aquifer can be called *unsteady* or *transient* confined flow (Fig. 14-1*b*).

The category of flow involving a *free* or *phreatic* surface can be called *unconfined*. Seepage through an earth dam (Fig. 14-1*c*) under invariant upstream and downstream heads represents a *steady unconfined* condition, and flow through a riverbank under transient fluctuations of head in the river (Fig. 14-1*d*) can be termed *unsteady unconfined*.

Boundary Conditions

The following general boundary conditions can occur in flow problems (Fig. 14-1):

1. Head or potential boundary condition

$$\phi = \bar{\phi}(t) \text{ on part of the boundary } B_1 \tag{14-4a}$$

2. Flow boundary condition

$$k_x \frac{\partial \phi}{\partial x} l_x + k_y \frac{\partial \phi}{\partial y} l_y + k_z \frac{\partial \phi}{\partial z} l_z + \bar{q}(t) = 0 \tag{14-4b}$$

on part of the boundary B_2 where the intensity of flow $\bar{q}$ is prescribed; l_x, l_y, and l_z are direction cosines of the outward normal to the boundary.

A number of special cases of the foregoing boundary conditions are possible. An *equipotential* boundary can occur at the upstream and downstream faces of a dam (Fig. 14-1*c*) and along the walls of a well (Fig. 14-1*b*). An impervious boundary is characterized by zero flow condition, where the gradient of ϕ vanishes. This condition is also assumed at an axis of symmetry. A prescribed flow rate as the well discharges at different times is also a flow boundary condition.

Unconfined Flow

Existence of a free surface involves additional head and flow boundary conditions. For example, the head at a point on a free surface equals its elevation head, and flow across the surface vanishes. For steady unconfined flow, we must consider the following boundary conditions (Fig. 14-1*c*):

$$\phi = \begin{cases} D & \text{on 1-2} & (14\text{-}5a) \\ h & \text{on 2-3, } p = 0 & (14\text{-}5b) \\ z & \text{on 3-4, } p = 0 & (14\text{-}5c) \\ d & \text{on 4-5} & (14\text{-}5d) \end{cases}$$

and

$$\frac{\partial \phi}{\partial z} = 0 \qquad \text{on 1-5} \qquad (14\text{-}5e)$$

We notice that the conditions in Eqs. (14-5*b*) and (14-5*c*) are due to the free surface and that the region 3-4 constitutes the *surface of seepage*.

Unsteady unconfined seepage involves continuous movements of the free surface. The boundary conditions then are (Fig. 14-1*d*):

$$\phi = \begin{cases} \bar{\phi}(t) & \text{on 1-2} & (14\text{-}6a) \\ z(t) & \text{on 2-3} & (14\text{-}6b) \\ h(t) & \text{on 3-4} & (14\text{-}6c) \end{cases}$$

$$\frac{\partial h}{\partial t} = \frac{k}{n}\left(\frac{\partial \phi}{\partial z} - \frac{\partial \phi}{\partial x}\frac{\partial h}{\partial x}\right) \qquad \text{on 3-4} \qquad (14\text{-}6d)$$

$$\phi = d(t) \qquad \text{on 4-5 if heads are prescribed} \qquad (14\text{-}6e)$$

$$\frac{\partial \phi}{\partial x} = 0 \qquad \text{on 4-5 if no flow condition is assumed} \qquad (14\text{-}6f)$$

$$\frac{\partial \phi}{\partial z} = 0 \qquad \text{on 1-5} \qquad (14\text{-}6g)$$

Here Eqs. (14-6*b*) to (14-6*d*) represent conditions at the free surface, and the region 2-3 denotes the surface of seepage.

The foregoing boundary conditions exist as inherent natural properties of a problem. The process of numerical discretization may require assumptions regarding potential and flow conditions on certain discretized boundaries. We shall discuss these points subsequently.

Simplifications and Special Cases

Many problems can be idealized as one- and two-dimensional. These assumptions can provide more economical solutions compared with the three-dimensional idealization and can yield acceptable accuracies for certain problems from a practical viewpoint. In this section, we present some of the specializations that have been employed.

A simplified equation for determining the variations in a free surface based on the Dupuit assumption is expressed as[3–5,27]

$$k\left[\frac{\partial}{\partial x}\left(h\,\frac{\partial h}{\partial x}\right)+\frac{\partial}{\partial y}\left(h\,\frac{\partial h}{\partial y}\right)\right]=n\,\frac{\partial h}{\partial t} \tag{14-7a}$$

or

$$\frac{k}{2}\left(\frac{\partial^2 h^2}{\partial x^2}+\frac{\partial^2 h^2}{\partial y^2}\right)=n\,\frac{\partial h}{\partial t} \tag{14-7b}$$

where h is the height of the free surface. A one-dimensional form, often called the *Boussinesq equation*, is[4,27]

$$k\,\frac{\partial}{\partial x}\left(h\,\frac{\partial h}{\partial x}\right)=n\,\frac{\partial h}{\partial t} \tag{14-7c}$$

or

$$\frac{k}{2}\,\frac{\partial^2 h^2}{\partial x^2}=n\,\frac{\partial h}{\partial t} \tag{14-7d}$$

or

$$k\left[h\,\frac{\partial^2 h}{\partial x^2}+\left(\frac{\partial h}{\partial x}\right)^2\right]=n\,\frac{\partial h}{\partial t} \tag{14-7e}$$

A number of linearized versions of the above two- and one-dimensional equations are possible:[4,28,29]

$$k\bar{h}(x,t)\left(\frac{\partial^2 h}{\partial x^2}+\frac{\partial^2 h}{\partial y^2}\right)=n\,\frac{\partial h}{\partial t} \tag{14-8a}$$

and

$$k\bar{h}(x,t)\,\frac{\partial^2 h}{\partial x^2}=n\,\frac{\partial h}{\partial t} \tag{14-8b}$$

or

$$k\sqrt{u}\,\frac{\partial^2 u}{\partial x^2}=n\,\frac{\partial u}{\partial t} \tag{14-8c}$$

where $\bar{h}(x, t)$ and $\sqrt{u}$ are the mean heights of the external water level and $u = h^2$.

Unsaturated Flow

A general form of the equation governing unsteady flow in unsaturated media is[10,20,23]

$$\frac{\partial}{\partial x}\left[\rho k(\theta)\,\frac{\partial \phi}{\partial x}\right]+\frac{\partial}{\partial y}\left[\rho k(\theta)\,\frac{\partial \phi}{\partial y}\right]+\frac{\partial}{\partial z}\left[\rho k(\theta)\,\frac{\partial \phi}{\partial z}\right]=\frac{\partial(\rho\theta)}{\partial t} \tag{14-9}$$

where θ is the moisture content expressed as a volume fraction and ρ is the density of the fluid. For steady-state conditions, θ does not vary with time, and the right-hand side in Eq. (14-9) vanishes. The problem of unsaturated flow requires computation of the free surface that separates the saturated and unsaturated zones. The steady-state version of Eq. (14-9) can be considered valid for the saturated region below the free surface, and Eq. (14-9) can be assumed to govern the flow in the unsaturated zone.[23] The values of permeability k will be different for both unsaturated and saturated zones. In $\phi = p + z$, if $p \geqslant 0$, p is pressure head in the saturated zone, and if $p < 0$, it is soil moisture-tension head in the unsaturated zone. For $p \geqslant 0$, k can be assumed uniform in the region, whereas for $p < 0$, k can be a function[20,23] of p. Finite element schemes for unsaturated flow have been proposed in Refs. 27 and 30 and are reviewed in Ref. 26.

Finite Difference Method

In Chap. 1 we described various explicit and implicit schemes for different types of differential equations. Solutions for confined flow are usually straightforward. They can be obtained either by using relaxation schemes or by solving the final set of algebraic equations under prescribed boundary conditions. Solution for unconfined flow requires iterative schemes with Gauss-Seidel, successive over-relaxation, or Newton-Raphson methods (Chap. 1). Some of the iterative methods with the implicit finite difference method will be described in Examples 14-5 and 14-6. Often, it is possible to use finite difference alternating direction explicit (FD-ADE) methods. The free surface can be then determined by using an indirect method in which the ordinates of the free surface are located by searching for the points at which the total head equals the elevation head.[27,31] If the given boundary conditions permit, and if a proper choice of mesh is made, the ADE method can prove economical and also yield acceptable accuracy.

Finite Element Method

The finite element equations can be obtained by using a number of formulation procedures (Chap. 1). We shall consider here only some of the aspects of the flow problem.

Fluid head ϕ is usually adopted as the primary unknown parameter. We can use an alternative formulation in which the stream function ψ is assumed as the primary unknown.[14,15] The latter approach may become necessary for problems in which accurate satisfaction of continuity of flow is required.

The secondary quantities we often seek are the fluid velocity and the quantity of flow. If we use Darcy's law, these quantities can be expressed in matrix notation as[24]

$$\{v\} = \begin{Bmatrix} v_x \\ v_y \\ v_z \end{Bmatrix} = -[R] \begin{Bmatrix} \dfrac{\partial \phi}{\partial x} \\ \dfrac{\partial \phi}{\partial y} \\ \dfrac{\partial \phi}{\partial z} \end{Bmatrix} = -[R][B]\{\phi_n\} \tag{14-10a}$$

where $[R]$, the matrix of principal permeabilities, is

$$[R] = \begin{bmatrix} k_x & 0 & 0 \\ 0 & k_y & 0 \\ 0 & 0 & k_z \end{bmatrix} \tag{14-10b}$$

and matrices $[B]$ and $\{\phi_n\}$ have been defined in Chap. 1. Matrix $[R]$ need not be composed of principal permeabilities; arbitrary coordinate axes can be used with necessary transformations.

The quantity of flow Q can be evaluated from

$$Q = vA \tag{14-11a}$$

in which A is the cross-sectional area. One of the procedures to obtain Q is to compute it across a prescribed section line or plane.[32,33] For example, in a two-dimensional problem (Fig. 14-2) total flow across a section line per unit length can be expressed as

$$Q = \sum_{i=1}^{n} d_i\{T_i\}^T[R][B]\{\phi_n\} \tag{14-11b}$$

where d_i = mean height of element i on section line
n = number of elements on line
$\{T\}^T = [-\sin(\theta - \alpha) \quad \cos(\theta - \alpha)]$

θ and α are shown in Fig. 14-2.

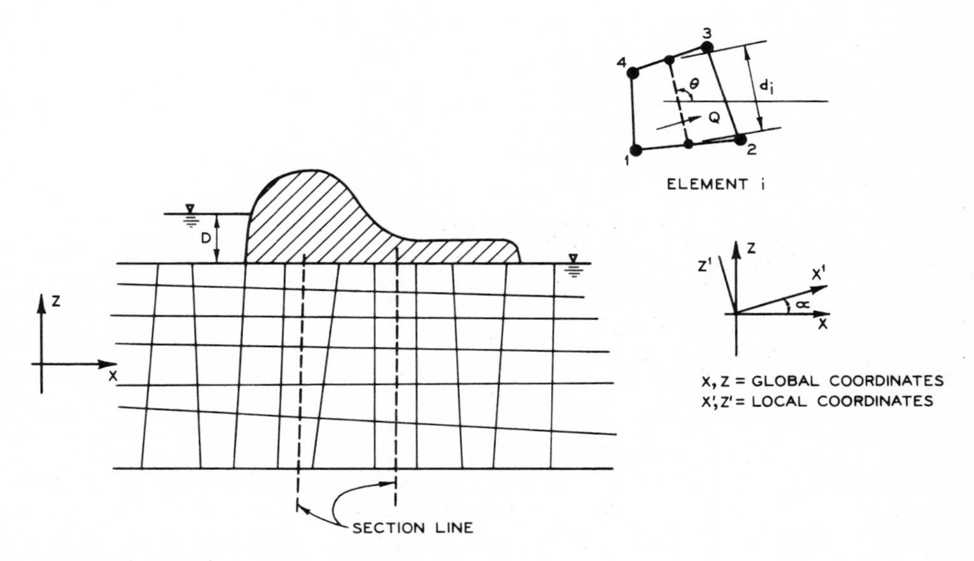

Figure 14-2 Computation of quantity of flow.

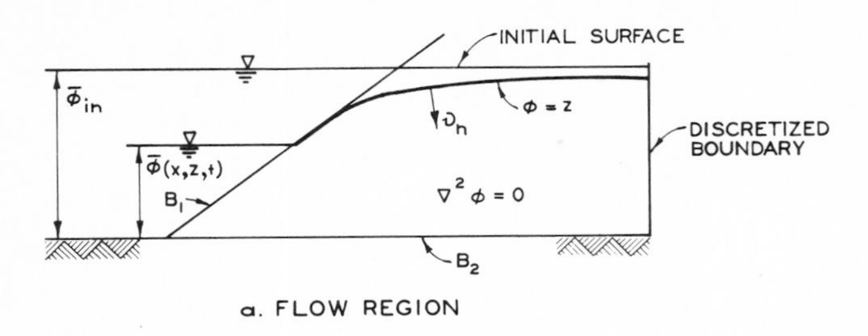

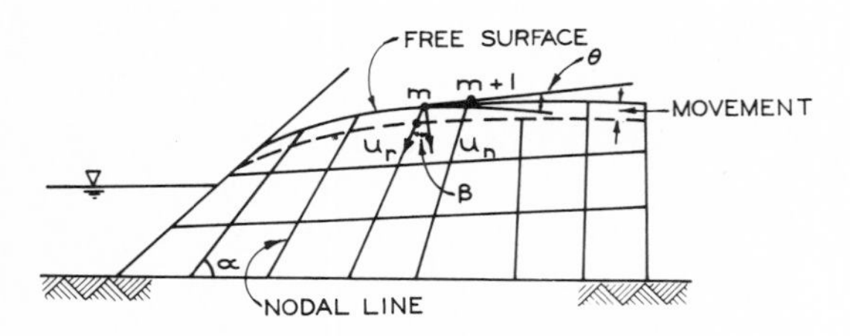

Figure 14-3 Location of transient free surface.

Procedures for Locating Free Surface

We shall first consider the case of transient seepage. Figure 14-3*a* depicts an earth bank for which we need to compute locations of the free surface as the consequence of sudden or gradual drawdown in the external head $\phi(x, y, t)$. The general procedure followed for determination of the transient locations of free surface can be expressed as[26]

$$X_i^{t+\Delta t} = X_i^t + \Delta t \dot{X}_i(t + \tau) \tag{14-12}$$

where X_i = coordinates of nodal points on free surface
N = number of free-surface nodes
Δt = time increment, $0 \leqslant \tau \leqslant \Delta t$

The dot denotes the rate of change of X. If $\tau = 0$, we essentially obtain the forward-difference integration in time. For that case, it is usually necessary to preselect a small value of Δt to assure convergence and accuracy.

The procedure usually starts from an initially known position for the free surface (Fig. 14-3*a*). The nonzero values of normal velocities at the nodes are computed, and the subsequent locations (Fig. 14-3*b*) of the nodes are computed by using Eq. (14-12). Details of one of the special forms of Eq. (14-12) are given by Desai[25,34] and France et al.[35] Other forms of Eq. (14-12) have been used by various investigators.[36–38] The concept of lagrangian coordinates for free surface movements governed by non-Darcy law has been used by McCorquodale.[19]

The scheme in Eq. (14-12) was modified[39,40] to include an iterative procedure in which $\dot{X}_i$ was computed at locations other than at $\tau = 0$. Based on the Lipschitz condition,[41] the size of Δt can be decreased or increased automatically such that convergence and stability are assured at each time step.

Steady Unconfined Flow

Taylor and Brown,[42] Finn,[43] and Neuman and Witherspoon[44] solved the steady free-surface flow by using an iterative procedure in which the free surface and the mesh were updated by using empirical laws. The procedure first assumed one of the conditions at the free surface, say $p = 0$, and then during iterations satisfied the other condition, $v_n = 0$ across the surface.

The foregoing procedure for the transient free-surface case can be specialized for steady flow with a free surface.[25,45] On the basis of an initial estimate of the free surface, we can obtain velocities at the nodes on the free surface by performing a cycle of the finite element method. By adopting a small Δt, the free surface and the mesh can be progressively updated until the boundary conditions in Eqs. (14-5*b*) and (14-5*c*) are approximately satisfied. Subsequently, we shall describe some numerical properties of this procedure and discuss the selection of Δt.

An Alternative New Scheme

Most of the investigations in the past have included the flow domain below the free surface in the finite element (or the finite difference) mesh. The mesh is then collapsed or expanded progressively on the basis of the changing locations of the free surface with time (Fig. 14-4*a*). This usually involves revision of at least a portion of the mesh and of the element matrices in that portion at each level of time or iteration.

It is possible to devise an alternative procedure in which the mesh for the entire zone (Fig. 14-4*b*) is used at all levels and the free surface is located by using some additional techniques. Here, the necessity of revising the mesh is avoided, which can result in considerable savings in computational effort. A scheme based on this idea was used in conjunction with the finite difference method.[27] A modified form of the scheme proposed by Desai[46] for use with the finite element method has been called the *residual flow* or *potential method*.

Residual Flow or Potential Method

At any instant, the finite element method is used to compute nodal potentials in the entire mesh (Fig. 14-4*b*) for given boundary conditions. An approximate location of the free surface is obtained by searching along each nodal line for the point at which Eq. (14-6*a*) is fulfilled.

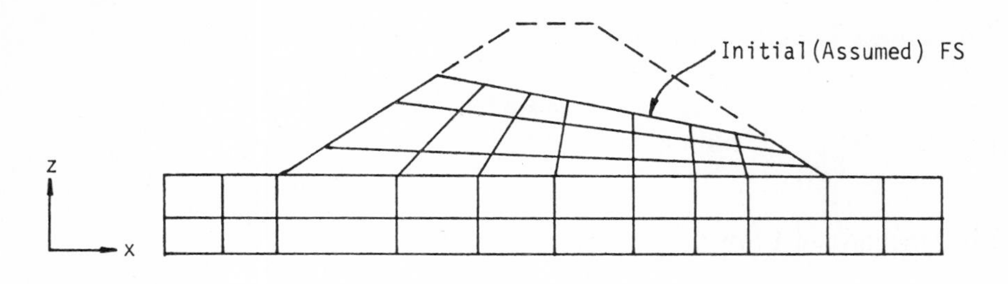

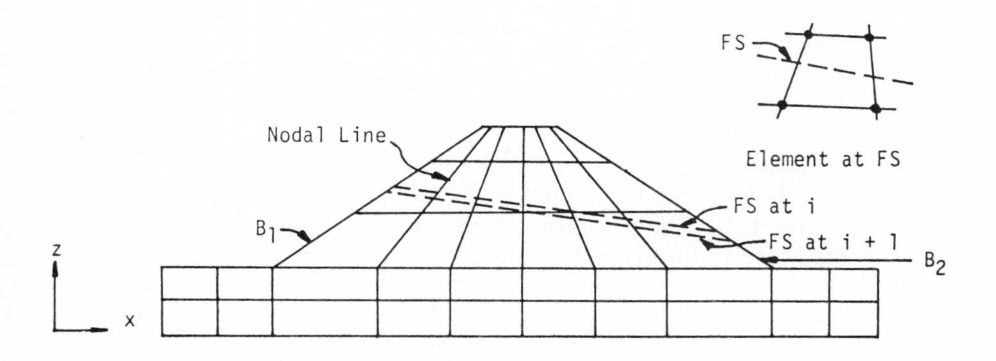

Figure 14-4 Discretization for conventional and proposed procedures.

The approximate location is then adjusted or corrected by using the residual approach. The residual vector is computed as

$$\{\bar{Q}_0\} = \iiint_V \{N\}^T\{\bar{Q}\}\, dV \tag{14-13a}$$

where $\{\bar{Q}\} = \{V_n\}A \qquad \{V_n\} = -[\bar{R}][B(s_i, t_i)]\{\phi_n\}$

The residual vector is computed for all elements at the free surface and a finite element cycle is performed to evaluate nodal potentials due to $\{Q_0\}$, which are added to the nodal potentials due to external boundary conditions as

$$\{\phi_{nj}\} = \{\phi_{no}\} + \sum_{i=1}^{j}\{\Delta\phi_{nj}\} \tag{14-13b}$$

where j is the stage of iteration and $\{\phi_{no}\}$ the nodal potentials with the given boundary conditions. The approximate location of the free surface is determined when the normal velocity is $\leqslant \epsilon$, where ϵ is a small number.

Limitations of Numerical Procedures

Certain limitations and difficulties can be encountered with the numerical methods. Abrupt changes in geometries and material properties can cause computational difficulties and instabilities. For instance, location of the free surface at the intersection of the core and shell of a dam requires special schemes to avoid numerical instabilities.

Although it is possible to use schemes that are unconditionally stable from the mathematical viewpoint, solutions of acceptable accuracy can be obtained only if spatial and time meshes of proper magnitudes are selected. Some criteria for selection of the meshes will be described subsequently.

The numerical formulations described herein and used commonly may not be suitable for all problems. For instance, the concept of a free surface may not be valid for less permeable cohesive soils in which substantial capillary fringes exist and the pore spaces can remain saturated under large negative pore pressures.[26,30,47]

For coarser materials, Darcy's law may not be valid, the flow can become turbulent, and the procedures need modifications (Example 14-10). Darcy's law can be considered valid for grain sizes less than approximately 0.05 cm, that is, medium to coarse sand.[18,26,47]

The assumption of a rigid soil-water system can limit the height of the zone of flow that can be analyzed. The rate of drawdown can have significant influence on the numerical procedure, and the rate may have to be limited so that the pressure lag caused by the compressibility of the system is small enough to justify the assumption. Isaacs and Mills[47] estimated that for the assumption to be valid, the time for drawdown should be greater than $10^{-3}H^2/k$, where H is the total drawdown.

Selection of Meshes

Numerical convergence and stability A numerical method can be adopted as a general procedure only when it can yield consistent answers for major variations in the parameters influencing the class of problems. Numerical convergence and stability of a procedure can provide a good index for reliability of a procedure. Moreover, knowledge of these factors aids the user in the proper choice of spatial and timewise subdivisions to assure adequate accuracy with optimum economy.

In some of the explicit and implicit finite difference schemes for linear equations governing confined flow, criteria for convergence and stability have been obtained (Chap. 1). Only limited studies have considered these factors for flows governed by nonlinear equations involving a free surface. This is truer in the case of the finite element procedures. We shall describe briefly some of the available data and recent developments that can be useful in the selection of finite difference and finite element meshes.

In the finite difference successive overrelaxation procedure for partially saturated flow, Reisenauer[20] limited the value of the relaxation factor ω to 1.2 to reduce instability. If instability occurred during computations, ω was reduced to a value less than unity. He also used special schemes, e.g., arranging the equations in an optimum manner and the concept of movable initial point, to speed convergence and to reduce instability. Freeze and Witherspoon[48] obtained a value of $\omega = 1.85$ by a trial-and-error procedure so as to obtain speedy convergence. An optimum value for ω of 1.9 was reported by Todsen.[49]

Taylor and Luthin[23] used a variable Δt to reduce computer time and to reduce instability. For instance, an initial value of $t = 10^{-3} k_0$ was used, where k_0 is the initial value of conductivity. As the procedure progressed, the value of Δt was increased by a factor of 1.05 to 1.10 till it reached $0.55 k_0$. The upper limit of Δt was based on the maximum permissible change in ϕ and k. For the particular problem solved (Example 14-6) the maximum change in ϕ was 0.04 in. If Δt went beyond the permissible value, it was reduced by a factor of 0.95. Pinder and Bredehoeft[50] allowed the time increment to vary in a geometric progression

$$\Delta t_{k+1} = \Delta t_k + 0.25\ \Delta t_k \tag{14-14}$$

in which k is the time step and initial $\Delta t = 1$ ms.

Terzidis[51] derived stability criteria for a number of finite difference schemes for solution of dimensional flow governed by the Boussinesq equation (14-7). The criteria are

$$\left.\begin{aligned} &\lambda < \frac{1}{4\bar{h}_k} && \text{simple explicit scheme} \\ &\lambda > \frac{1}{-4\bar{h}_{k+1}} && \text{simple implicit scheme} \\ &\lambda < \frac{1}{2(\bar{h}_k - \bar{h}_{k+1})} && \text{Crank-Nicholson implicit scheme} \\ &\lambda < \frac{1 - \frac{1}{3}(\bar{h}_k + \bar{h}_{k+1})}{2(\bar{h}_k - \bar{h}_{k+1})} && \text{higher order implicit scheme} \end{aligned}\right. \tag{14-15a}$$

where $\lambda = \Delta t/(\Delta x)^2$
$\bar{h}_k$ = average value of h at time level k
$h = h/h_0$
h = depth of free surface
h_0 = characteristic depth

From a number of numerical solutions, Verma and Brutsaert[52] found that the simple explicit scheme can be stable for Δt derived from the criterion

$$\lambda < \frac{1}{2\bar{h}_k} \tag{14-15b}$$

For transient free-surface non-Darcy flow in two dimensions, McCorquodale[19] presented a criterion

$$\Delta t \lesssim \frac{\Delta \bar{x}}{\sqrt{gh}} \tag{14-16}$$

where g = gravitational constant
$\Delta\bar{x}$ = average horizontal distance between nodes on free surface
h = depth to free surface

For rapid drawdown (Example 14-10) a value of $t = 0.1$ s was used. Isaacs and Mills[47] used an empirical law in which Δt was determined from a specified maximum nodal displacement for each adjustment of the free surface.

With the explicit finite difference method for unsteady flow in wells, Ruston and Tomlinson[53] found the following criterion:

$$\frac{\Delta t T}{\Delta x^2 S} < 0.25 \tag{14-17a}$$

where T is the transmissibility and S the storage coefficient. They conducted analyses for the ADI procedure and found that the accuracy of the solution is influenced by such factors as changes in head on outside boundaries, changes in head at inside boundaries (at wells), and changes in discharge at wells. On the basis of numerical results for various values $(\Delta t T)/(\Delta x^2 S)$, a criterion for achieving acceptable accuracy was obtained as

$$\frac{\phi_1 - \phi_2}{\phi_1 - \phi_0} < D \tag{14-17b}$$

where ϕ_1 = value of head at node after one complete iteration of solution 1 with time step t
ϕ_2 = value of head at same node after two complete iterations of solution 2 with time step $\Delta t/2$
ϕ_0 = starting head at node

The values of D for constant discharge and a sudden change in head at a well were found to be 0.3 and 0.05, respectively. It was found that Δt can be increased, after the occurrence of a change in head or discharge, by a factor of 1.25 after each time step.

Desai[45] performed a quantitative study to investigate convergence and stability behavior of steady unconfined flow through berms (Fig. 14-8) around cofferdams (Example 14-2). Convergence behavior of typical nodes on the free surface obtained by using the forward difference form of Eq. (14-12) is shown in Fig. 14-5 for different values of $\Delta t = 0.001$, 0.01, 0.1, and 1.0 day. The solution became unreliable for values of Δt beyond 0.1 day. For this type of problem, it appeared that the following criteria can be used for selection of Δt:

$$\Delta t \lesssim 10^{-5}\bar{k} \qquad \text{or} \qquad \frac{\bar{k}\,\Delta t}{\Delta R} \lesssim 1 \tag{14-18}$$

where

$$\bar{k} = (k_h^2 + k_v^2)^{1/2} \qquad \Delta R = \frac{1}{n}\sum_i^n \frac{\Delta x_i + \Delta y_i}{2}$$

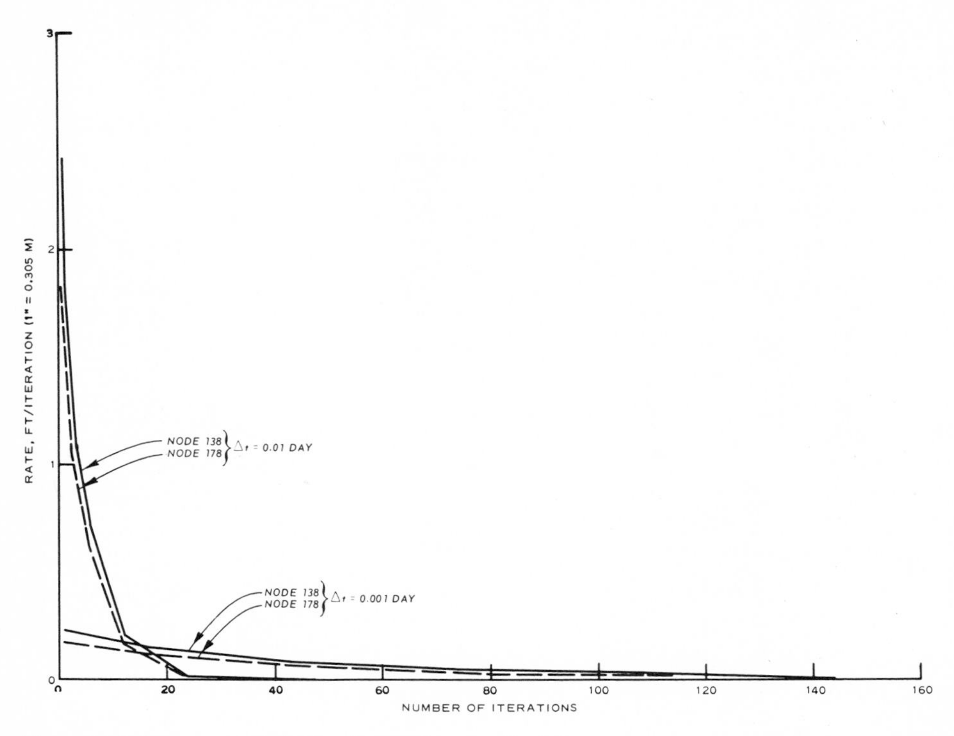

Figure 14-5 Convergence behavior for steady unconfined flow.

where n = number of elements in zone, i.e., allowed to move, adjacent to free surface
Δx, Δy = dimensions of element
h = horizontal direction
v = vertical direction

In a quantitative study, Desai[26,54] examined a number of finite element solutions for transient unconfined flow through a viscous flow model with an entrance slope angle[27] of 45° subjected to drawdown conditions. Three different values were used for each of the parameters, mesh = 18, 44, and 147 nodes, $k = 2.36$, 0.236, and 0.0236 cm/s, $\Delta t = 10$, 50, and 100 s, and rates of drawdown $R = 0.027$, 0.054, and 2.5 cm/s; the last value of $R = 2.5$ cm/s yielded almost sudden drawdown. A value of $n = 1$ was used. On the basis of these numerical exercises, it was found that reliable solutions can be obtained if the following criterion is satisfied:

$$k\,\Delta t\left(\frac{1}{\Delta\bar{x}^2} + \frac{1}{\Delta\bar{y}^2}\right)R \leqslant 0.05 \tag{14-19}$$

where $\Delta\bar{x}$ and $\Delta\bar{y}$ are mean element dimensions.

A somewhat rigorous procedure, proposed by Sandhu, Rai, and Desai[39,40] for transient free-surface seepage, provides a self generating mesh that assures convergence at every time step. It involves a number of iterations at each time step, based on average or mean rates of movements of the free surface. In addition, a mathematical check is enforced to satisfy the Lipschitz condition[41]

$$\lambda = \left| \frac{x^{i+1} - x^i}{x^i - x^{i-1}} \right| < 1 \tag{14-20}$$

where x is the solution and i the step of iteration. If at any stage and at any node the condition in Eq. (14-20) is violated, the time step is reduced by half, and if λ is too small, say less than 0.25, Δt is increased by a factor of 2.0.

Comments Most of the foregoing criteria are based on trial-and-error procedures and are empirical in nature. General criteria from mathematical considerations have not yet been developed.

From the user's viewpoint, a good strategy would be to adopt a reasonably small mesh that is consistent with economy. One or more of the criteria summarized above can be used in selecting Δt. If time and resources permit, it may be advisable to perform a trial-and-error analysis to obtain an optimum magnitude of Δt for the problem on hand. It will be economical to increase the magnitude of Δt after a change, e.g., in head or in discharge, has taken place.

A mathematically stable procedure may not necessarily mean that one can use any (large) size of spatial and temporal increments. Required accuracy should be a guiding consideration in the choice of the spatial and time steps.

Computational Time

Published data on the computer times required for different problems have been scanty. Table 14-1 gives computer times reported by a number of investigators.

Discretization of Infinite Media

Figure 14-1 shows problems in which the geometry and the potential and flow conditions on part(s) of the boundaries are known uniquely. Introduction of such conditions in a numerical method can be straightforward. In many practical situations, however, we may encounter geometries and boundary conditions that cannot be defined. For instance, in the case of flow in or out of riverbanks, tidal beaches, and extensive aquifers, we have to deal with infinite extents of the media. It is then necessary to include only significant finite zones in an analysis, and we have to make proper assumptions concerning potential and flow conditions on the discretized boundaries (Fig. 14-6*a*). Proper choice of these conditions will depend upon the geological properties and conditions of groundwater flow and will require engineering judgment.

On the basis of a number of numerical results from finite element analysis, and laboratory data on a large viscous flow model simulating long riverbanks,

Table 14-1 Computational times for flow problems

Method	Ref.	Details of problem and computer	Time
FE	73	Isotropic heterogeneous system with 121 nodes, 40 time steps; CDC 6400	12.3 s or 0.0025 s/node/time step
		Anisotropic heterogeneous system with 580 nodes, 100 time steps; CDC 6400	513 s or 0.009 s/node/time step
FD	23	See Example 14-6	4.6 min
FD	52	One-dimensional flow, 51 grid points, double precision, total time $t = 10$ min reached in 2000 time intervals; IBM 360/65	1 min
FD	86	Flow through a rectangular dam 16×24 ft with unit grid size, 15 to 23 adjustment of free surface; IBM 7090	47–63 s
FE	25, 28, 34, 55	Steady confined flow through foundation of a dam, 100 nodes; GE 430	12 s
		Unconfined flow through earthen banks, 100 nodes (Example 14-8); GE 430	3.5 s/cycle
		One-dimensional flow, 7 nodes; GE 430 time-sharing	1.25 s/time step
FE	79	One-dimensional flow, 20 nodes 4 time-level steps; IBM 360/75	1.63 s/time step

Desai[34,55] proposed criteria to determine extents of discretized zones for free surface flow through earth banks. It was found that if the end boundary is placed beyond a distance of about $8H$ to $12H$, measured from the final point of drawdown (Fig. 14-6*a*), the behavior of the free surface near the sloping face of the bank will not be influenced significantly. The assumption of an "impervious" base in an infinite medium will be approximately valid if the bottom boundary is placed beyond a distance of about $3H$ to $6H$ (Fig. 14-6*a*).

Three possible boundary conditions were assumed to occur at the discretized end boundary, namely impervious, constant level, and equipotential (Fig. 14-6*b*). Both the impervious and constant-level conditions yielded about the same results, which compared well with observations, whereas the equipotential condition gave

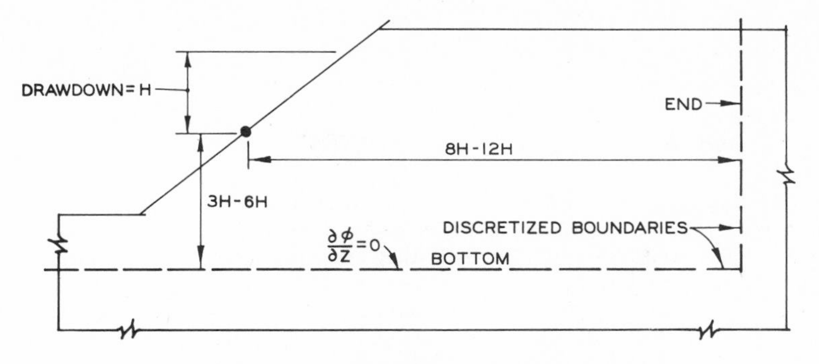

a. DISCRETIZED END AND BOTTOM BOUNDARIES

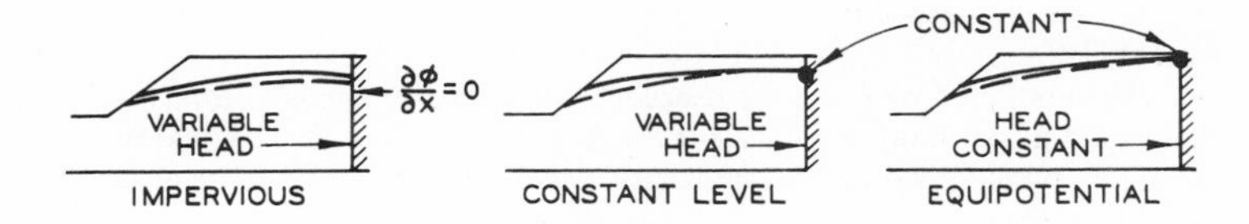

b. DIFFERENT BOUNDARY CONDITIONS

Figure 14-6 Location of discretized boundaries for "infinite" media.

results that differed from the other two and from the observations. Hence, it was concluded that for long homogeneous banks, the boundary condition at large distances can be assumed to be impervious or constant level.

For the problem of steady unconfined seepage in long embankments for tailing ponds, Kealy and Williams[56] found that realistic solutions can be obtained only if the end boundary is treated as a flow line. Assumption of an equipotential boundary did not yield satisfactory comparisons with observations. Fang and Wang[38] have discussed the problem of proper boundary conditions on discretized regions for transient unconfined flow in sandy tidal beaches.

14-4 APPLICATIONS AND DESIGN ANALYSIS

In view of the significant amount of reported data on correlations between numerical results and observations, we can conclude, with a sense of confidence, that the numerical methods can provide reliable basis for design analysis. In this section, we shall present some typical applications, most of which involve correlations between numerical solutions and laboratory and/or field observations. In order to obtain some idea of comparisons, we shall also describe two problems for which both finite element and finite difference methods have been used.

The aim of any solution procedure is to help the engineer obtain design guidelines by using numerical procedures. As a step toward this goal, a set of curves for stability analyses of earth banks subjected to transient unconfined flow conditions is included as the last example.

Example 14-1 Seepage through foundation of a dam[25] A dam 80 ft long resting on a two-layered foundation 40 ft deep is shown in Fig. 14-7*a*. Within each layer the soils are assumed homogeneous. The finite element method (Sec. 14-3) using an isoparametric element (Chap. 1) was used, and the foundation was discretized as shown in Fig. 14-7*b* with 105 nodes and 80 elements. The bottom boundary was assumed impervious, and the end boundaries were placed 60 ft from the upstream and downstream faces of the dam. A differential head of 100 ft was applied.

The equipotentials computed from nodal heads obtained from the finite element method are plotted on the left half in Fig. 14-7*b*. Figure 14-7*c* and *d* shows computed uplift pressures on the dam foundation and the flow net, respectively. The computed values compared well with those from graphical solutions.[11,25]

The quantity of flow computed by using the procedure described previously [Eq. (14-11)] was found to be about 0.785 ft^3 per unit time for a 1-ft length of the foundation. The quantity of flow from a flow net drawn by hand with a trial-and-error procedure was found to be equal to about 0.81, showing a discrepancy of about 3 percent between the two values.

An example for steady free-surface flow is given in Ref. 25.

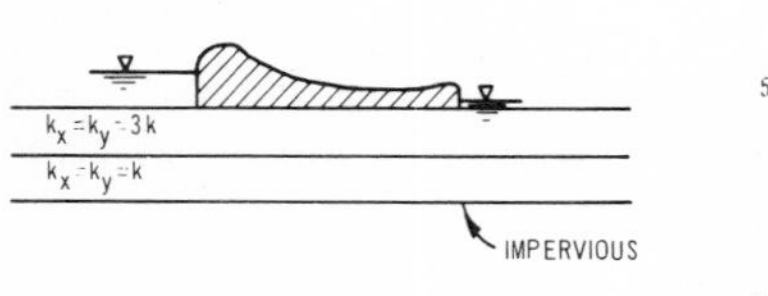

a. DAM ON LAYERED FOUNDATION

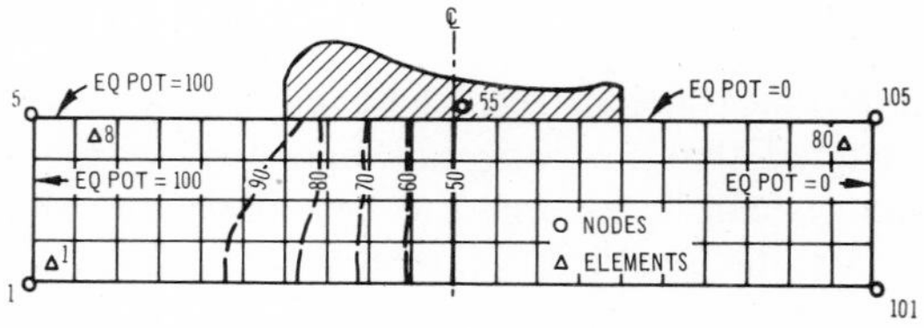

b. FE MESH AND EQUIPOTENTIALS

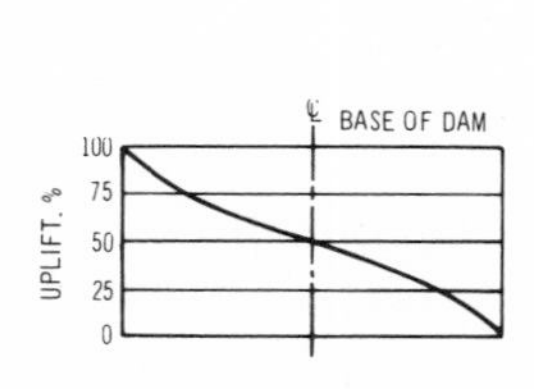

c. UPLIFT PRESSURE ON BASE OF DAM

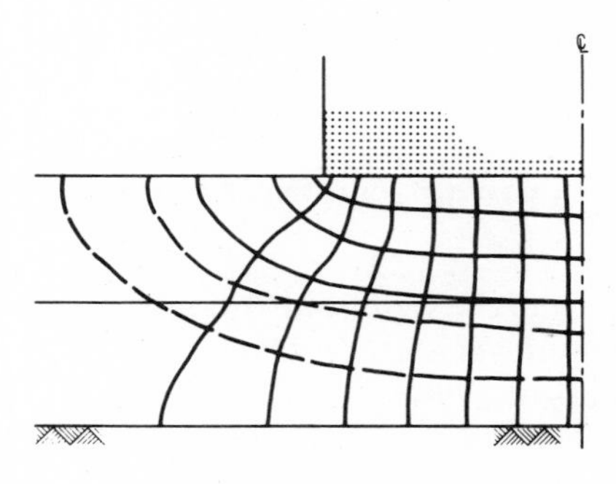

d. FLOW NET

Figure 14-7 Seepage through foundation of a dam.

Example 14-2 Flow through foundation of, and berm around, a cofferdam[45] Earthen berms are often provided at the inner side of cofferdams to increase their stability. Figure 14-8*a* shows such a cofferdam constructed in a layered foundation system with anisotropic permeabilities given as $k_h = 4k_v$, where k_h and k_v are the horizontal and vertical permeabilities, respectively. Under the differential head of 79 ft ($H = 428 - 295 = 79$ ft) the flow takes place through the foundation and then enters the berm. In order to compute seepage gradients and forces for design of the berm, it is necessary to compute the location of the free surface.

The finite element method code for steady unconfined flow[34] was used to solve the problem. The cofferdam was assumed to be impervious, and the permeability of the berm was assumed to be equal to that of the foundation soils immediately below. The discretized region and end boundaries and applied potentials are shown in Fig. 14-8*b*.

An initial estimate for the location of the free surface was required to start the iterative procedure. It was found that a consistent and accurate initial estimate was needed to get a properly converging solution. Since the water rises up through the berm, a more direct procedure was to start from the ground surface, elevation 349, as the initial estimate. Accordingly, the mesh region, shown as the dashed zone in Fig. 14-8*b*, was allowed to move during the iterative scheme. Figure 14-8*b* shows the free surface at equilibrium after about 30 iterations. A value of $\Delta t = 0.01$ day was used; here Δt can be essentially a parameter. Equipotentials as percentages of the differential head are also shown in Fig. 14-8*b*. We have already discussed in Sec. 14-3 the criterion for selection of Δt for this problem.

Three-dimensional analysis For certain situations such as arbitrary layering and near junctions of different structures (dam and abutment) it becomes necessary to perform a three-dimensional analysis. An example of the use of three-dimensional analysis is shown in Fig. 14-8*c*, in which the cofferdam problem was analyzed using a three-dimensional formulation and code described by Desai.[33] The results

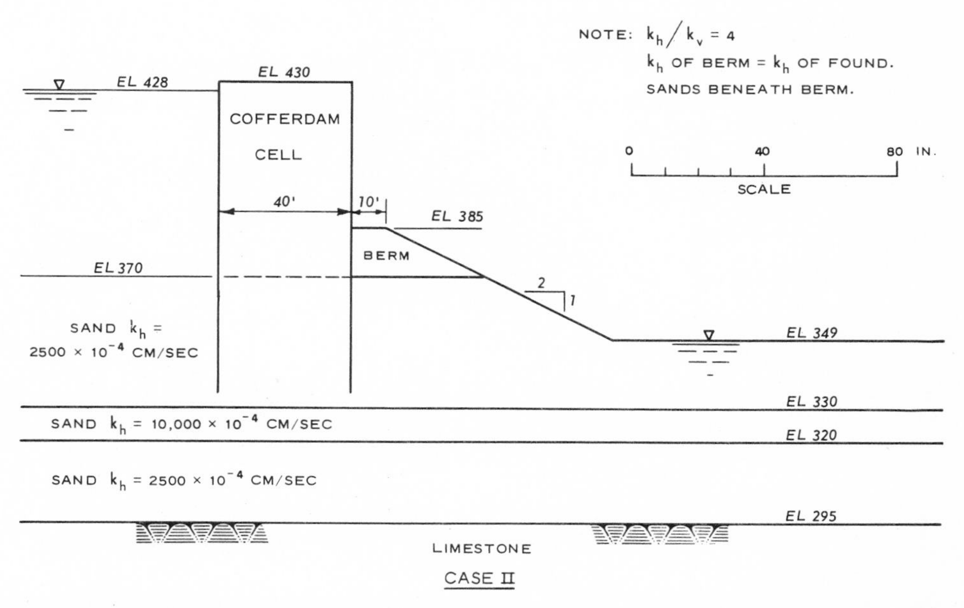

Figure 14-8 Seepage through layered foundation and berm around a cofferdam: (*a*) details of cofferdam foundation.

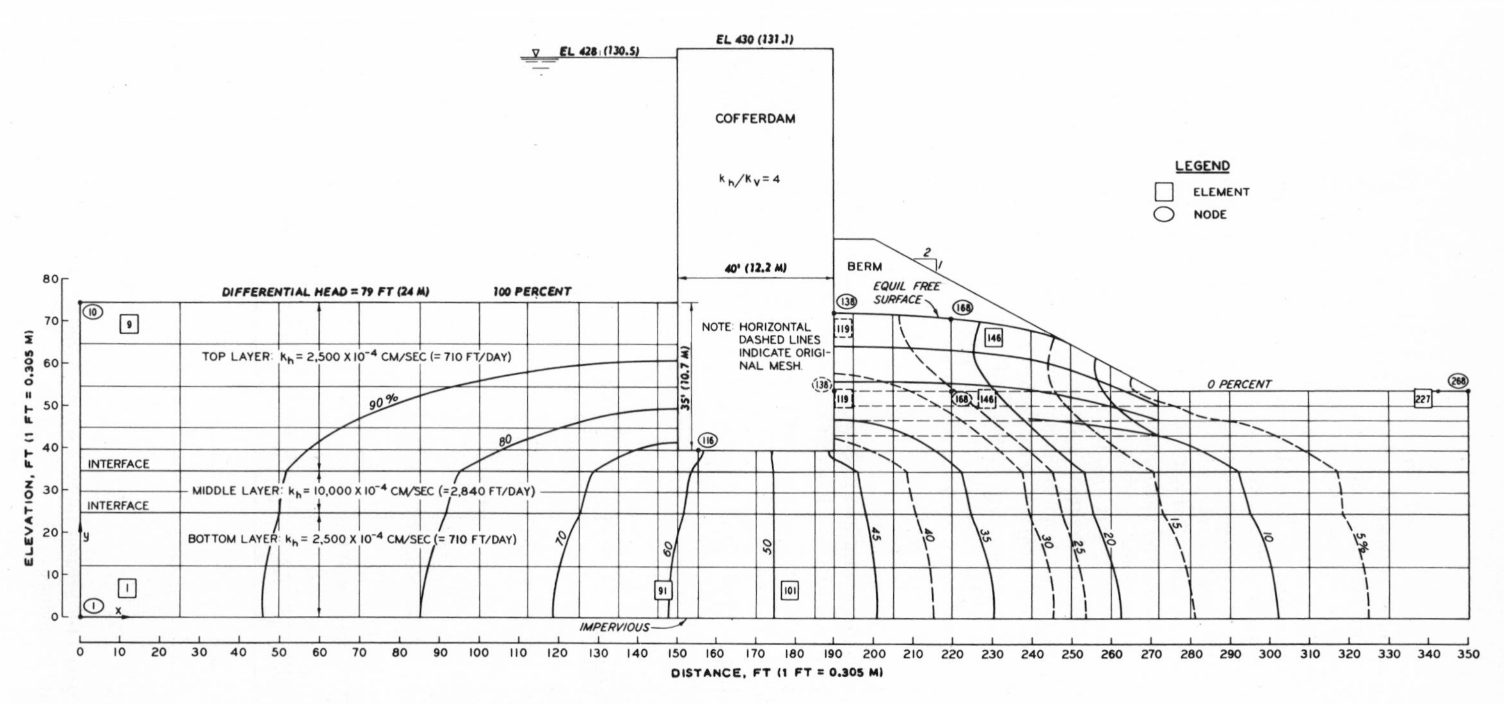

Figure 14-8 (continued) (*b*) Finite element results with computed equipotentials.

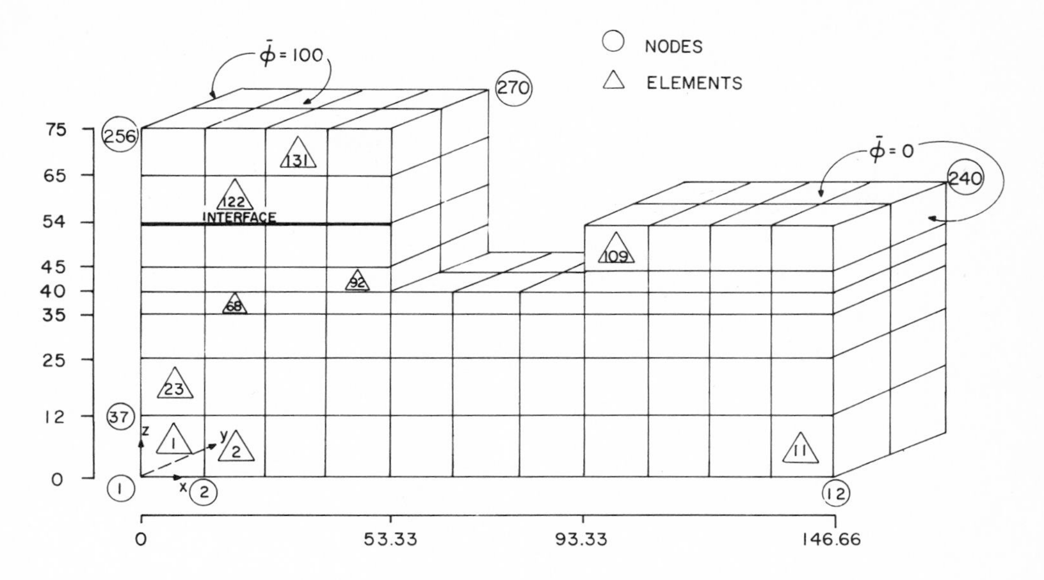

Figure 14-8 (continued) (*c*) Three-dimensional mesh and analysis for cofferdam problem.

are not significantly different from the foregoing two-dimensional analysis since the layering is uniform and the system is homogeneous; moreover, the boundary conditions in the three-dimensional analysis involve essentially a plane flow simulation. However, for problems involving arbitrary changes in material properties and geometry (in the third direction) and existence of such structures as galleries, pipes, and junctions, the three-dimensional analysis will yield different and more realistic results than the two-dimensional idealization.

Example 14-3 Evapotranspiration from an aquifer[57] If the water table is near the ground surface, water can escape by evaporation from capillary fringes above the water table and by transpiration from plants having their roots in the water table or in the capillary fringe (Fig. 14-9*a*). The solid curve in Fig. 14-9*b* shows measured variations in depth to water below the land surface with time in the Punjab region of Pakistan. After introduction of intensive irrigation in the year 1902, the water level in this region rose due to leakage at an approximately uniform rate of 2.3 ft/day until about 1920. At later times, the hydrograph flattened, showing increased evapotranspiration losses as the water table approached the land surface.

The governing equation for this problem is similar to Eq. (14-3*a*) and was solved by Prickett and Lonnquist[57] using the FD-ADI method. As shown in Fig. 14-9*c*, the rate of evapotranspiration Q_{et} was assumed to be a linear function of the difference between the elevation of the land surface and the elevation of the water table. The quantities of recharge due to leakage and of evapotranspiration were converted into inputs as fluid flux at the nodes in the finite difference grid. The rate of 2.3 ft/yr yields leakage of about 9.45×10^{-3} gal/day·ft^2 with a storage coefficient value of 0.20. For 1000-ft grid intervals, the rate comes to 9.45×10^3 gal/day per node. The slope of the line $R_{i,j}$ in Fig. 14-9*c* was computed as $(9.45 \times 10^3)/(640 - 613) = 350$ gal/day·ft because at elevation 640 (Fig. 13-9*b*) the evapotranspiration rate was equal to the recharge rate and the

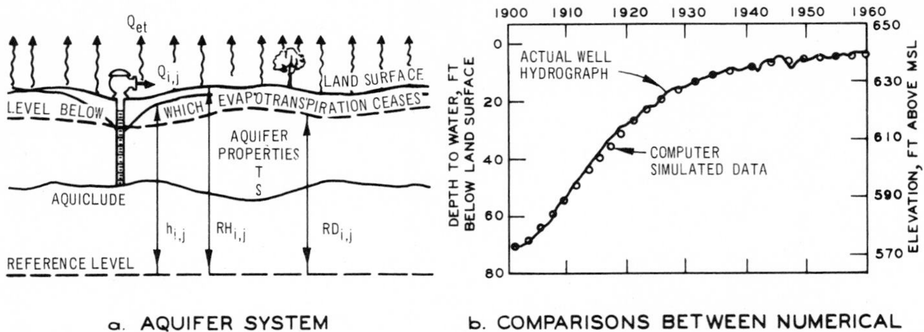

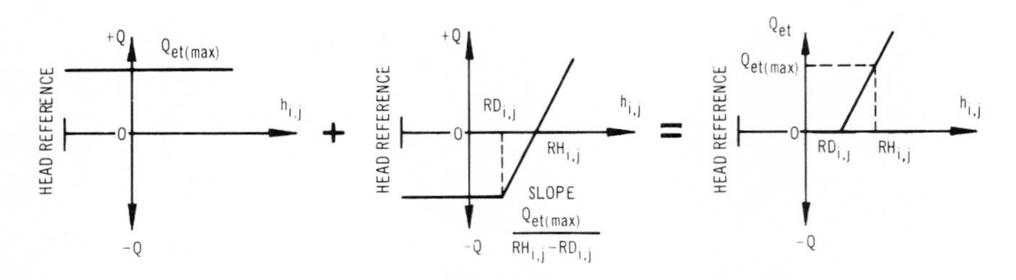

Figure 14-9 Finite difference solution for aquifer analysis.[57]

evapotranspiration rate was assumed to be zero at elevation 613. The maximum value of Q_{et} was computed from

$$R_{i,j} = \frac{Q_{et,\,max}}{RH_{i,j} - RD_{i,j}} \tag{14-21}$$

to be equal to 1.05×10^4 gal/day. The net value of applied nodal flux was $1.05 \times 10^4 - 9.45 \times 10^3 = 1.05 \times 10^3$ gal/day.

Initial heads in the aquifer were assumed to be at elevation 572 ft and $\Delta t = 365$ days was used. The wells in the area were assumed to be fully penetrating, and the magnitude of drawdown was small compared with the saturated thickness of the aquifer. Figure 14-9*b* shows a satisfactory correlation between the computed and measured depths.

Example 14-4 Analysis for Cambrian-Ordovician aquifer of northeastern Illinois[57] The pumpage of water from this aquifer has increased significantly in recent years, and in some areas near Chicago dewatering of the aquifer has occurred. This can be due to conversion from artesian to water table conditions, reduction in saturated thickness of the aquifer, and change of flow from two to three dimensions. With these complexities, the simplified models for the flow are no longer valid. The FD-ADI for Eq. (14-3) was used by Prickett and Lonnquist[57] to include a number of factors affecting the flow.

Figure 14-10*a* shows the main locations of pumping and the rates of pumpage. Figure 14-10*b* shows the finite difference mesh for the region. The aquifer, which is about 1000 ft deep,

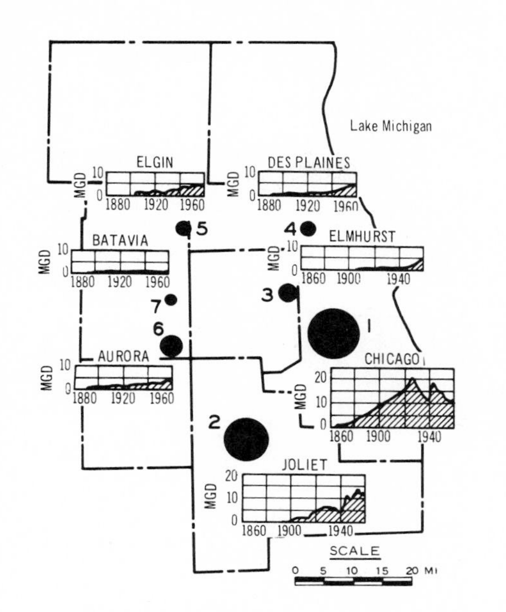

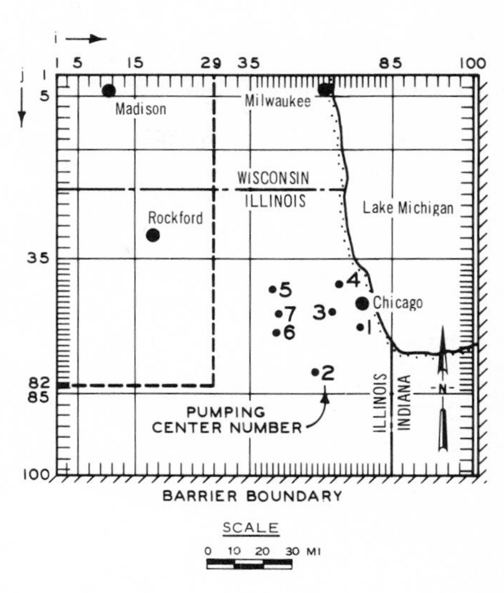

a. LOCATIONS OF PUMPING AND PUMPAGE RECORDS

b. FINITE DIFFERENCE MESH

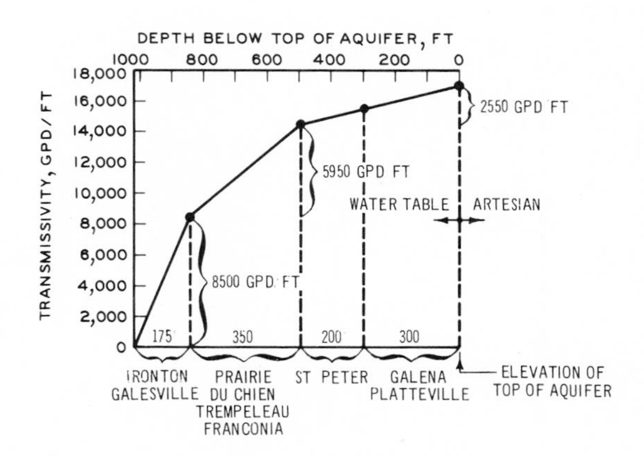

c. VARIATIONS OF TRANSMISSIVITIES FOR VARIOUS COMPONENTS

Figure 14-10 Finite difference prediction for an aquifer behavior.[57]

consists of a combination of dolomites and sandstones over a shale of low permeability. Leaky artesian conditions exist in the aquifer, in which the major recharge occurs in the western and northwestern zone of Chicago region. The leakage coefficient (ratio of vertical conductivity to the thickness of the confining bed) was about 2.5×10^{-7} gal/day·ft^3. Transmissivities of various components of the aquifer are shown in Fig. 14-10*c*. This figure was used to reduce the transmissivity if the water level fell below the aquifer. A storage coefficient of 0.05 was used.

Figure 14-11*a* and *b* shows typical comparisons between the measured and computed values of declines in water levels and plots of declines in water levels vs. time in certain deep wells, respectively. The agreement between the predictions and the observation is good.

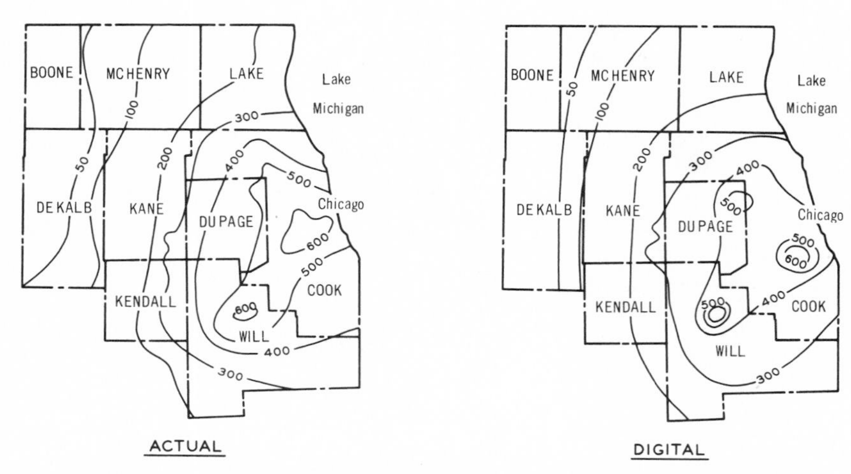

a. COMPARISONS FOR DECLINES IN PIEZOMETRIC SURFACE DURING 1864 – 1958

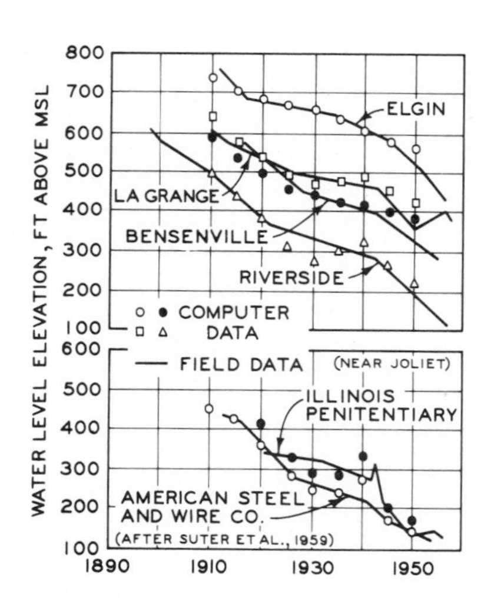

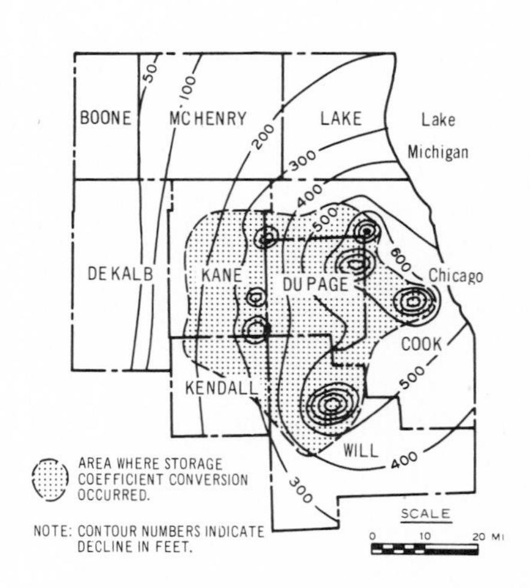

b. COMPARISONS BETWEEN ARTESIAN PRESSURE IN WELLS

c. PROJECTED DECLINES IN PIEZOMETRIC SURFACE 1864 – 1995

Figure 14-11 Finite difference prediction for an aquifer; further details of problem in Fig. 14-10.[57]

With the increase in the development of groundwater, the storage coefficient may experience conversion from an artesian to a water table condition. Assuming that in the year 1995 there will be increased pumping up to 145 Mgal/day, the predictions for water table declines computed from the finite difference method are shown in Fig. 14-11*c*. Here, the region that can be expected to experience conversion of the storage coefficient is shown dotted.

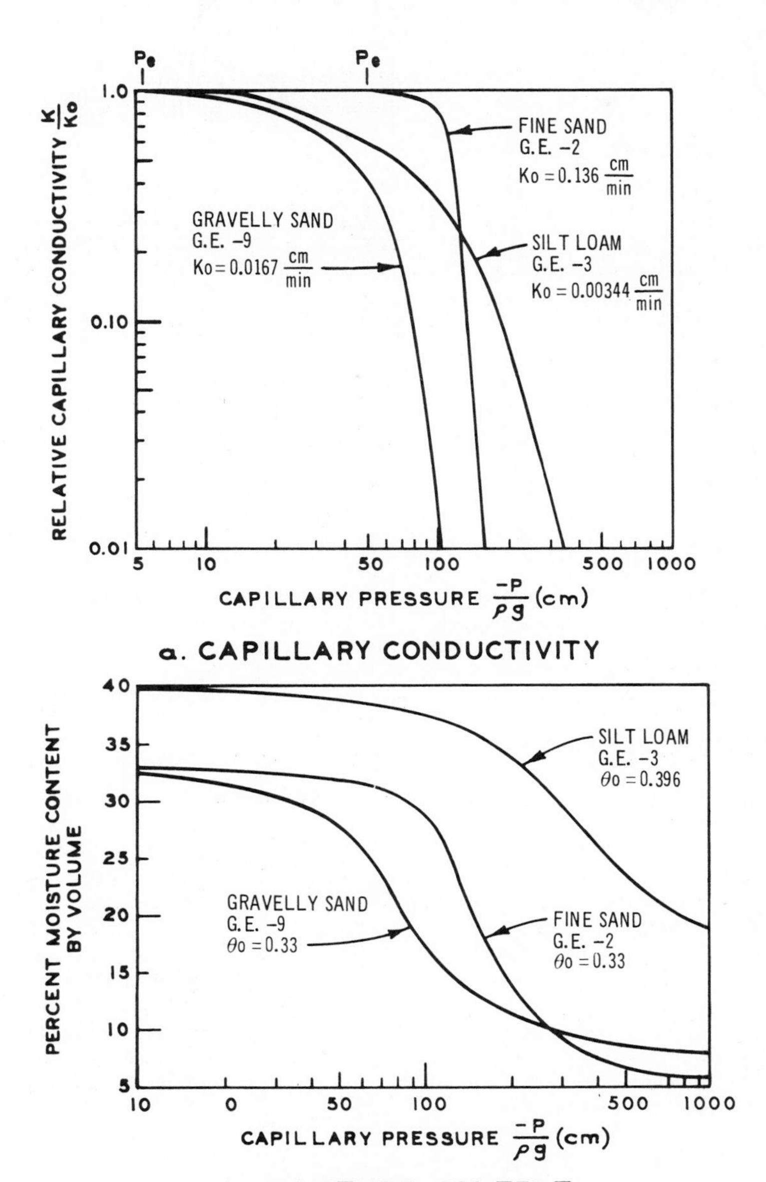

Figure 14-12 Relationships between conductivity, moisture content, and pressure.[20]

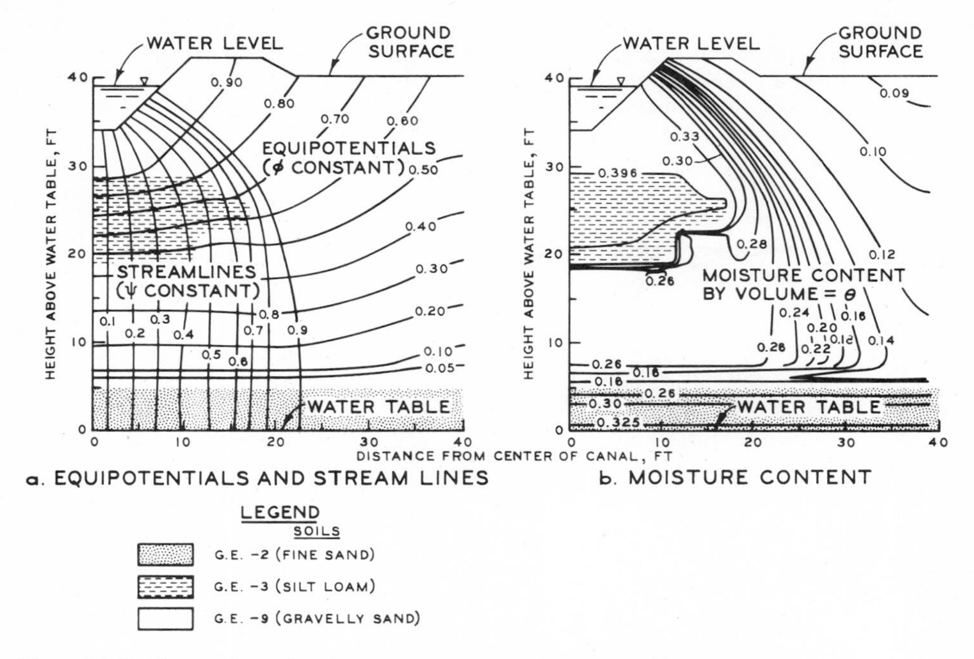

Figure 14-13 Distributions of moisture content, potentials, and streamlines.

Example 14-5 Steady flow from a canal in nonhomogeneous soils By using the steady-state version of Eq. (14-9) Reisenauer[20] solved the problem of *partially saturated* flow in heterogeneous soils. The following relations were used between the capillary conductivity k, capillary pressure head p_c, hydraulic pressure p, and moisture content:

$$\theta = F(p_c) \qquad k = f(p_c) \qquad p_c = -\frac{p}{\rho g} = z - \phi \tag{14-22}$$

In the finite difference method, central differences were used for approximating the first derivative, and a combination of forward and backward differences was used for the second derivatives. The resulting nonlinear equations were solved by using a modified Gauss-Seidel iterative scheme, in which the overrelaxation factor was computed internally for saturated zones. In the partially saturated zones, an upper limit of 1.2 was placed on the overrelaxation factor to assure stability (Sec. 14-3).

Figure 14-12*a* and *b* shows the relationship [Eq. (14-22)] between k and θ and p_c for three soils that occur in the nonhomogeneous soils below a 15-ft canal. The solution procedure starts with an initial estimate of ϕ, for which values of k are obtained from Eq. (14-22). Improved values of ϕ are now computed using the finite difference equations; this permits computation of k and θ from Eq. (14-22). The procedure is continued till convergence is obtained. Computed values of equipotentials and streamlines are shown in Fig. 14-13*a*, and Fig. 14-13*b* shows the distributions of moisture contents in the heterogeneous regions.

Example 14-6 Unsteady flow in gravity wells We now consider free-surface transient and unsaturated flow in a gravity well (Fig. 14-14*a*) solved by Taylor and Luthin.[23] It was assumed

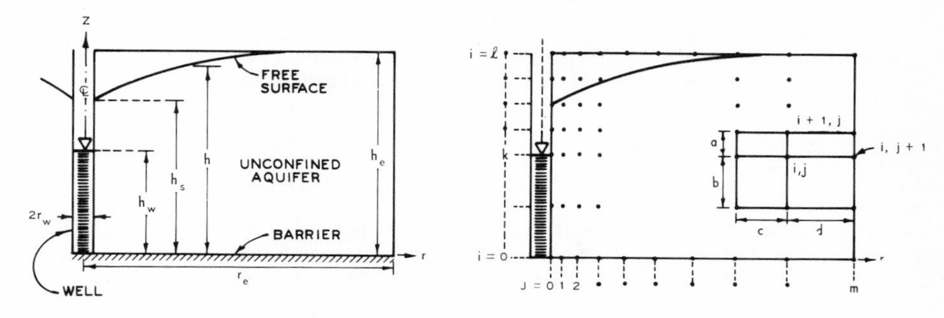

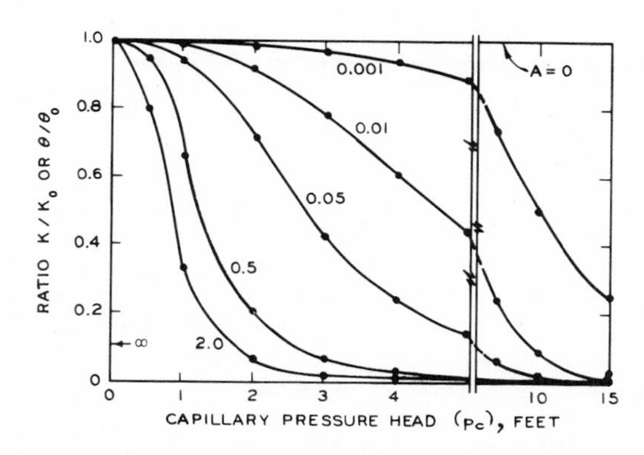

Figure 14-14 Finite difference solutions for unsteady flow in a gravity well.[23]

that the flow in the saturated and unsaturated regions was respectively governed by

$$\frac{k_0}{r}\frac{\partial \phi}{\partial r} + \frac{\partial}{\partial r}\left(k_0 \frac{\partial \phi}{\partial r}\right) + \frac{\partial}{\partial z}\left(k_0 \frac{\partial \phi}{\partial z}\right) = 0 \tag{14-23a}$$

$$\frac{k}{r}\left(\frac{\partial \phi}{\partial r}\right) + \frac{\partial}{\partial r}\left(k \frac{\partial \phi}{\partial r}\right) + \frac{\partial}{\partial z}\left(k \frac{\partial \phi}{\partial z}\right) = \frac{\partial \theta}{\partial t} \tag{14-23b}$$

where k_0 and k are conductivities in the saturated and unsaturated zones, respectively, and r and z are coordinate directions. Equations (14-23) are special cases of Eq. (14-9); Eq. (14-23*a*) was used by Reisenauer[20] (Example 14-5) for steady flow in unsaturated media.

The finite difference procedure employed here was similar to that used by Reisenauer.[20] Unique relations were assumed between the water content, conductivity, and capillary pressure head as

$$k = \frac{k_0}{Ap_c^3 + 1} \qquad \theta = \frac{\theta_0}{Ap_c^3 + 1} \tag{14-24}$$

where θ_0 is the constant value of θ for positive capillary pressure (assumed to be equal to 0.4 for this example) and A is a constant. Variations of and k with p_c are depicted in Fig. 14-14*c*.

Figure 14-14*a* shows a fully penetrating well in an uniform aquifer. The water is assumed to be pumped at a steady rate Q, causing lowering of the free surface. It was assumed that at a large distance r_e the free surface was at the ground level. The boundary conditions assumed at the well below the free surface were

$$\phi = \begin{cases} h_w & 0 \leqslant z \leqslant h_w \\ z & h_w \leqslant z \leqslant h_s, \text{ surface of seepage} \end{cases} \tag{14-25}$$

At the well above the free surface and at the lower and upper boundaries, no flow conditions were assumed. The initial conditions were (1) water level in well at h_w and free surface $p = 0$ at ground level, (2) ϕ = arbitrary estimates at grid points, (3) $p = h - z$ at r_e, (4) $k = 1.0$, (5) $p = 0.0$, and (6) $\theta = 0.4$.

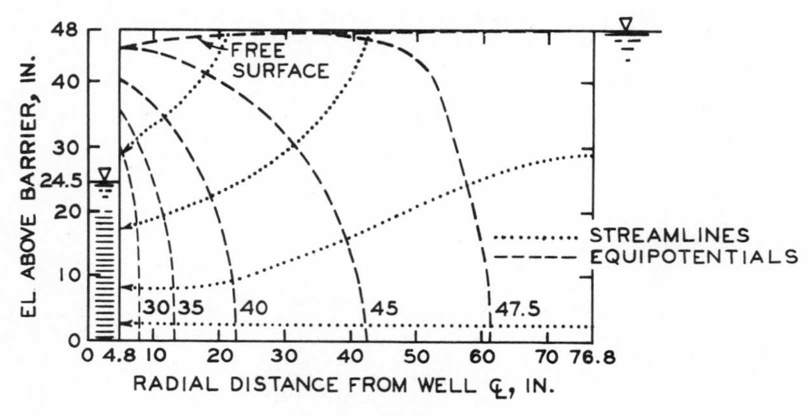

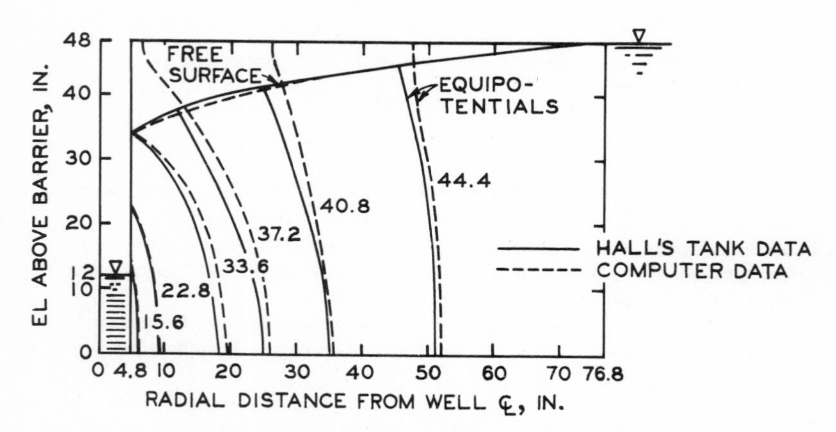

Figure 14-15 Details of results for problem in Fig. 14-14.[23]

With these initial conditions, the finite difference equations for Eq. (14-23a) were solved and the values of k, ϕ, p, and θ were adjusted. The finite difference form of Eq. (14-23b) was solved next for all points where $p \leqslant 0$, except along the periphery of the well, to compute changes in θ for a time increment Δt. Values of k, ϕ, p, and θ were updated, using Eq. (14-24) wherever applicable. Now for all points where $p > 0$, except those at the well, finite difference equations for the steady-state equation were solved to evaluate ϕ and p. Next, the water level h_w was modified and the new elevations for the free surface were computed. This was done by comparing the calculated inflow with that prescribed. Elevation of the free surface was found by locating grid points at which p changes sign and by extrapolation.

The foregoing procedure of computing changes in θ in the unsaturated zone, steady-state analysis in the saturated zone, and modifications in boundary conditions was repeated until desired conditions were reached. We have already covered (Sec. 14-3) the criterion used for varying the time increment Δt.

Computed equipotentials and streamlines at two typical time levels, 0.006 day and equilibrium, are shown in Fig. 14-15a and b. In Fig. 14-15b, the computed values are compared with laboratory test results with a sand-tank model in which the prescribed pumping rate was $Q = 2450k_0$ and for which variation of parameters were assumed to be governed by Eq. (14-24). Good agreement can be seen between the two results. It was found that no significant changes occurred in the computed potentials and locations of the free surface for variations in A (Fig. 14-14c) from 2 to 20 ft^{-3}. This may be due to the fact that only a small extent of unsaturated zone is involved in this problem.

Example 14-7 Unsteady free surface flow under gradual drawdown[34] Knowledge of seepage forces in earthen banks and dams caused by flow due to external fluctuations in the water head becomes essential for design of stable banks. Seepage forces in turn depend upon the location of the phreatic surface. Usually, the critical location occurs during drawdown or fall in the water levels. Rise and drawdown of water level in the Mississippi River often cause failure of the silty and sandy banks. Figure 14-16a shows a typical cross section of the river bank, and Fig. 14-16b shows the history of river levels and corresponding measured heads in two piezometers installed in the bank.

The finite element procedure (Sec. 14-3) was used to compute the changing free surface and heads in the flow region under the drawdown in the river that took place between Apr. 30 and May 30, 1965. The rate of drawdown R was about 0.67 ft/day. The permeability and the porosity for the soils obtained from laboratory and field measurements are shown in Fig. 14-17a. Criteria used for locating the discretized boundaries and potential and flow conditions were described in Sec. 14-3.

The numerical values of ϕ and the measured values of heads in the piezometers are compared in Fig. 14-17a. Good agreement is seen between the computed and observed values. Both the finite difference and finite element methods were also used for computation of heads in a viscous flow model that simulates the rise and flow conditions in the river.[27,34] Figure 14-17c shows typical flow nets for rise and drawdown conditions in the model as evaluated from the finite difference analysis.[27] Numerical stability criteria for these problems were discussed in Sec. 14-3.

Example 14-8 Flow in parallel ditches and drains[28] A simplified finite element formulation for flow that can be idealized as one-dimensional was obtained by using Galerkin's method.[28] One-dimensional elements (Chap. 1) were used, and the surface of seepage was determined by using an iterative procedure with the method of fragments.[5,27,28]

Figure 14-18a shows the problem of drawdown in parallel drains, and Fig. 14-18b shows results from the finite element procedure using one-dimensional elements compared with those obtained by using higher order isoparametric elements[35] and those from experiments with an analog model;[58] values of $k = 0.0674$ in/s (0.171 cm/s) and $n = 0.886$ were used.

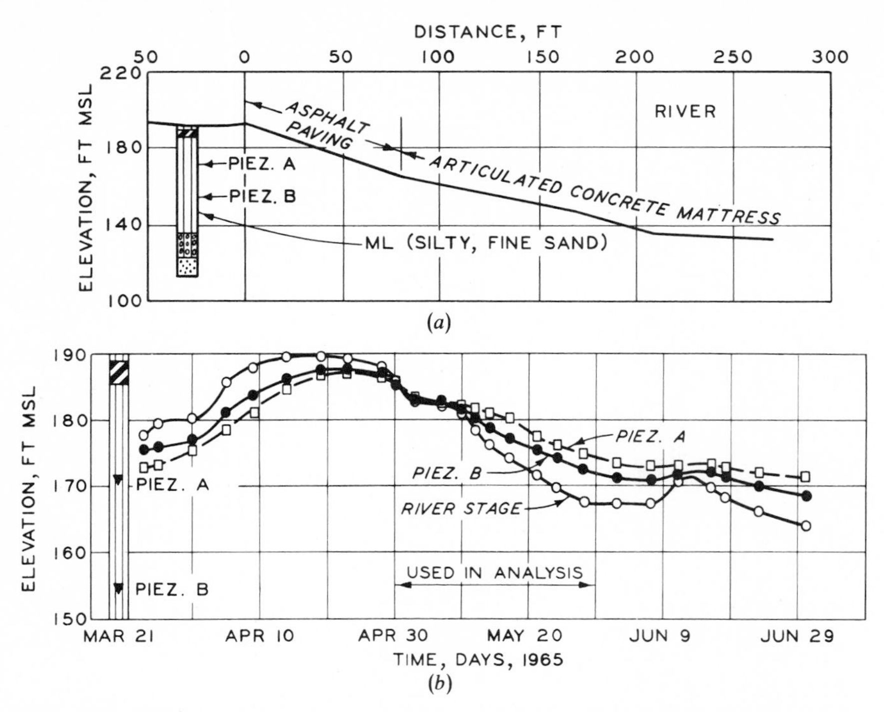

Figure 14-16 Details of section and history of head variations:[34] (*a*) cross section and locations of piezometers; (*b*) history of river levels and head variations during 1965.[34]

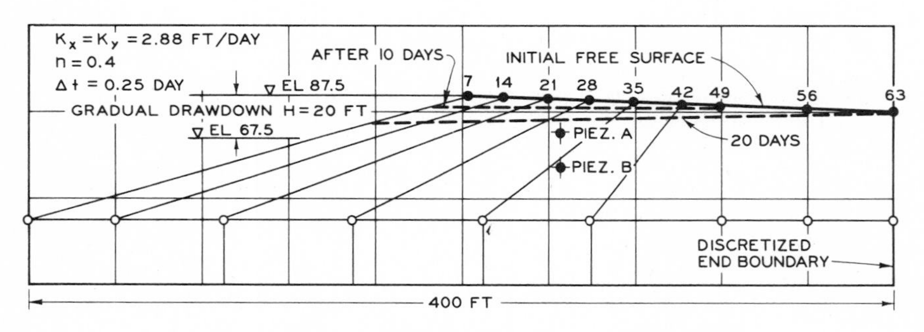

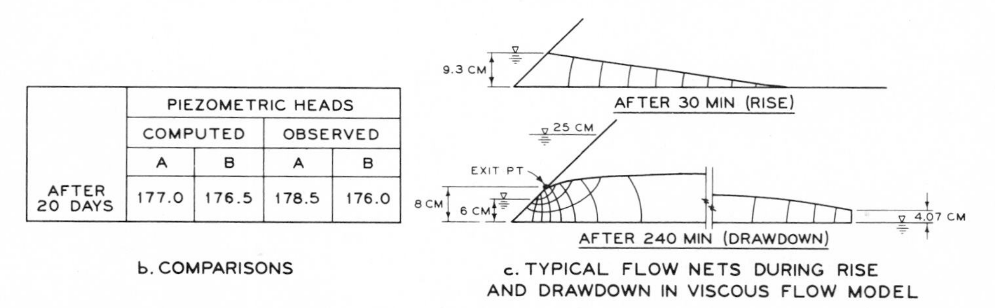

	PIEZOMETRIC HEADS			
	COMPUTED		OBSERVED	
	A	B	A	B
AFTER 20 DAYS	177.0	176.5	178.5	176.0

Figure 14-17 Finite element results for transient free-surface flow.[34]

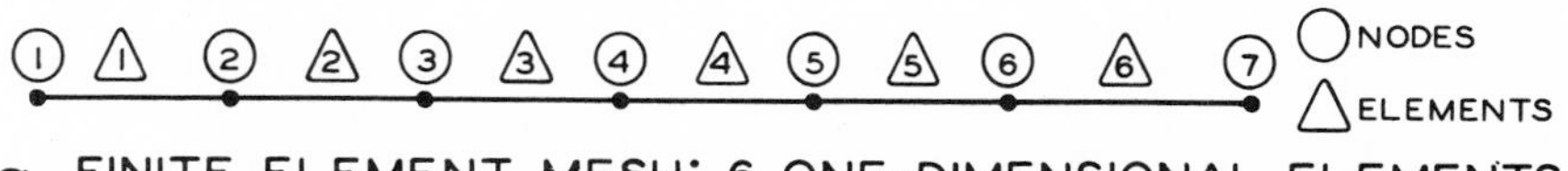

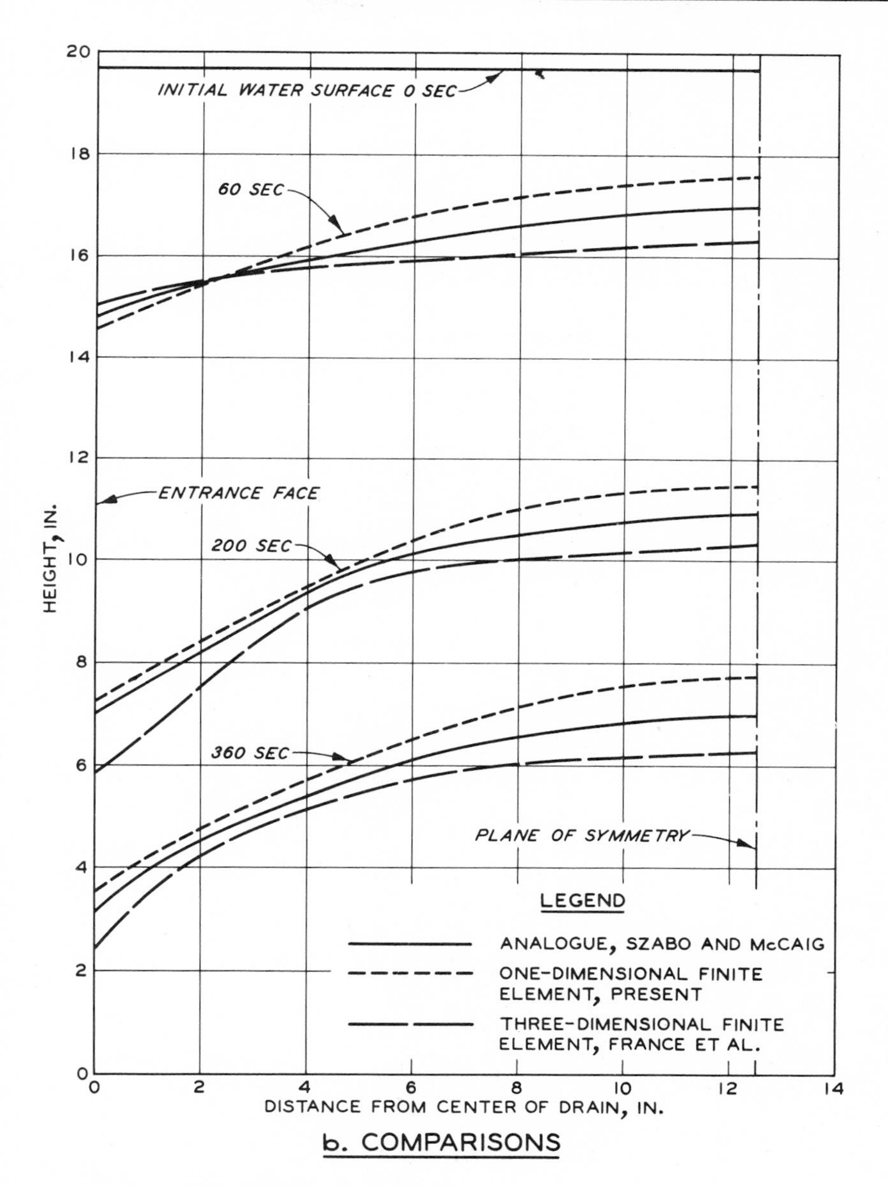

Figure 14-18 Flow in parallel ditches and drains.[28]

Example 14-9 Non-Darcy flow through dams and banks Darcy's law [Eq. (14-1)] may not be valid for flow through such materials as rock fills because the fluid velocity in these situations may not be a linear function of hydraulic gradients. Formulations and applications for flow governed by non-Darcy laws [Eqs. (14-2*a*) and (14-2*b*)] using the finite difference and the finite element methods have been reported by various investigators.[16–19]

A variational functional for non-Darcy flow used to derive the finite element equations was[19]

$$A = \iint \left[k|\nabla\phi|(\phi_x^2 + \phi_y^2) + \frac{a}{6b\alpha}(2 - \alpha|\nabla\phi|)\sqrt{1 + \alpha|\nabla\phi|} \right] dx\, dy \qquad (14\text{-}26)$$

in which $\alpha = 4b/a^2$, $k|\nabla\phi|$ is the hydraulic conductivity, and the subscripts in ϕ_x and ϕ_y denote the partial derivative with respect to x and y, respectively. Minimization of A yielded a set of nonlinear equations that were solved by using an iterative scheme with successive overrelaxation.

Steady unconfined flow Volker[17] used triangular finite elements with linear variations for ϕ to obtain comparisons between the finite element method predictions and laboratory test results. The laboratory tests were conducted in a flume 2 ft wide and 2 ft deep with a clear perspex side for observations. Piezometric heads were measured at a number of locations. Figure 14-19*a* shows a model bank that consisted of a crushed aggregate of $\frac{3}{4}$ in nominal size with a mean geometric diameter of 0.54 in. The values of the parameters in Eq. (14-2) obtained from laboratory pemeameter tests were

$$a = 0.319 \text{ s/ft} \qquad b = 11.821 \text{ s}^2/\text{ft}^2 \qquad c = 8.893 \qquad m = 1.745 \qquad (14\text{-}27)$$

Figure 14-19*a* shows comparisons between laboratory data, results based on Darcy's law, and those based on two nonlinear relations, Eqs. (14-2*a*) and (14-2*b*). The two nonlinear equations yielded almost identical results and compared well with laboratory results. The measured and computed values of heads at various locations in the model are tabulated in Fig. 14-19*b*. The measured value of the quantity of discharge was 0.083 ft^3/s·ft, which compared well with the computed value of 0.090 ft^3/s·ft.

Unsteady unconfined flow McCorquodale[19] used Eq. (14-2*a*) in a finite element formulation and performed a number of laboratory tests involving transient free-surface conditions. One of the test setups used is shown in Fig. 14-20*a*. The measured material parameters were $a = 0.01$ s/cm and $b = 0.01$ s^2/cm^2, and the value of drainable porosity $n = 0.4$ was used.

A sudden drawdown of about 51 cm was allowed, and the locations of the free surface were recorded photographically. Correlations between the computed and observed movements of the free surface at different times are shown in Fig. 14-20*b*.

Example 14-10 Comparisons between finite difference and finite element methods *Aquifer analysis* In this example, we shall discuss analyses of an aquifer performed by using both the finite difference and the finite element methods.[59] The FD-ADI was used for Eq. (14-3*a*). The finite element equations were obtained by using Galerkin's method (Chap. 1).

The field problem considered was the aquifer adjacent to Musquodoboit River in Nova Scotia. The aquifer, which is about 4800 ft wide, 5700 ft long, and 62 ft thick, consists of glaciofluvial deposits of coarse sand, gravel, cobbles, and boulders deposited in an U-shaped glacial valley (Fig. 14-21*a*), cut into the slates and quartzites of the Meguma group and the granite intrusives of the Devonian period.[59] The values of transmissivity and storage coefficient were evaluated from pumping tests; further details of the numerical procedures and geological properties can be found in Ref. 59.

Figure 14-21*b* and *c* shows the finite difference and finite element meshes for the aquifer region. Comparisons of the history of drawdown in the three wells are shown in Fig. 14-21*d*. In addition to the foregoing aquifer problem, Pinder and Frind[59] employed both methods for the problem of a well in an infinite aquifer, and (as discussed subsequently) derived a number of useful conclusions regarding the finite element and finite difference methods.

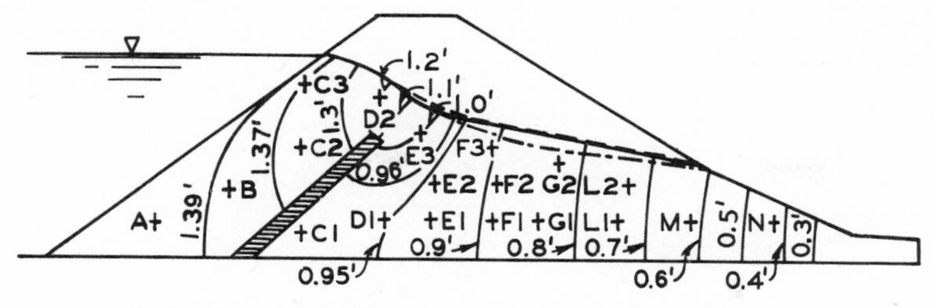

——— NONLINEAR FINITE ELEMENT SOLUTION
—·—·— EXPERIMENTAL FREE SURFACE
– – – – DARCY FREE SURFACE

a. COMPARISONS BETWEEN COMPUTED AND OBSERVED FREE SURFACES

POINT (1)	PIEZOMETRIC HEAD VALUE			
	EXPERIMENTAL (2)	FORCHHEIMER (3)	EXPONENTIAL (4)	DARCY (5)
A	1.383	1.382	1.382	1.380
B	1.379	1.379	1.379	1.371
C1	0.850	0.950	0.950	0.960
C2	1.358	1.356	1.356	1.331
C3	1.379	1.374	1.374	1.371
D1	0.850	0.950	0.950	0.980
D2	1.190	1.170	1.170	1.170
E1	0.830	0.930	0.930	0.960
E2	0.840	0.950	0.950	0.980
E3	0.890	0.980	0.980	1.000
F1	0.800	0.890	0.890	0.900
F2	0.810	0.900	0.900	0.910
F3	0.830	0.910	0.910	0.930
G1	0.760	0.840	0.840	0.840
G2	0.750	0.850	0.850	0.840
L1	0.680	0.750	0.750	0.730
L2	0.690	0.750	0.750	0.730
M	0.570	0.620	0.620	0.600
N	0.410	0.450	0.450	0.440

b. PIEZOMETRIC HEADS AT DIFFERENT POINTS IN DAM

Figure 14-19 Non-Darcy flow through a model dam.[17]

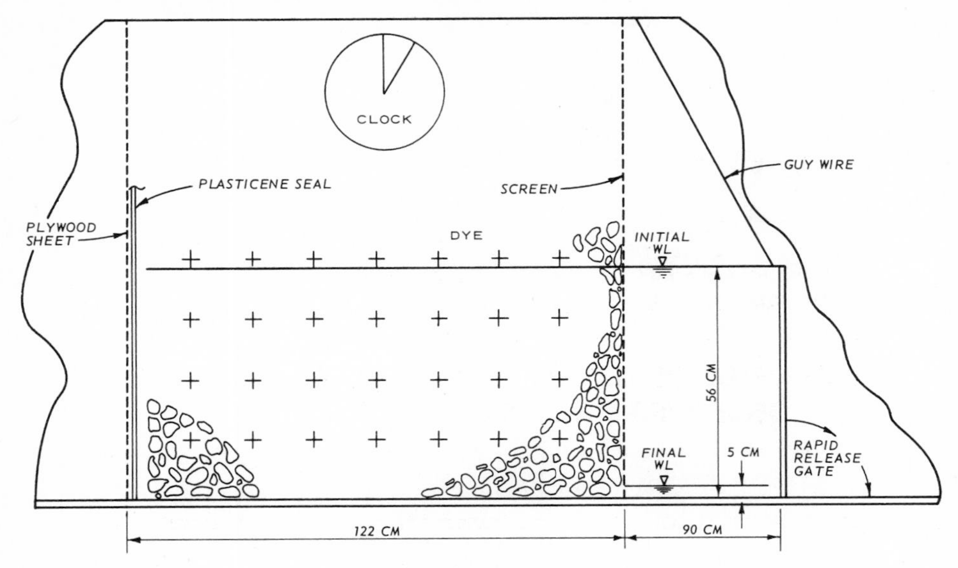

a. TYPICAL MODEL FOR RECTANGULAR BANK

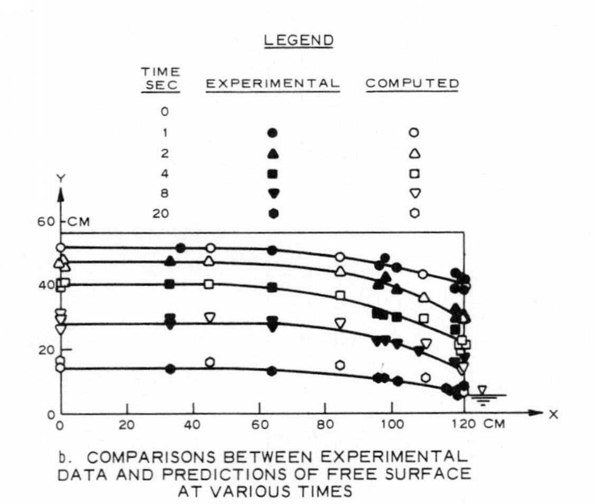

b. COMPARISONS BETWEEN EXPERIMENTAL DATA AND PREDICTIONS OF FREE SURFACE AT VARIOUS TIMES

Figure 14-20 Unsteady non-Darcy flow.[19]

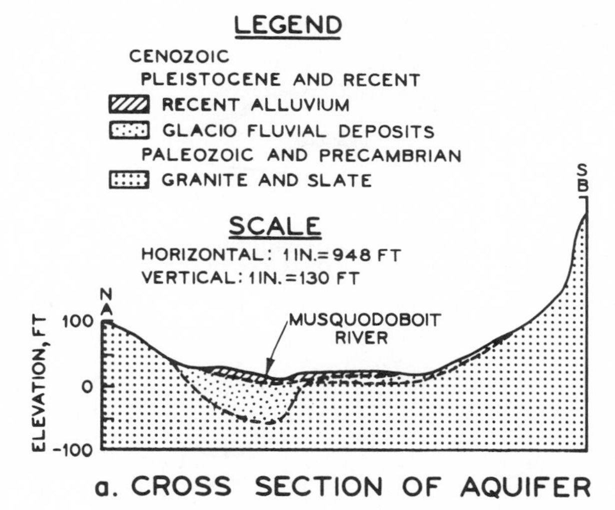

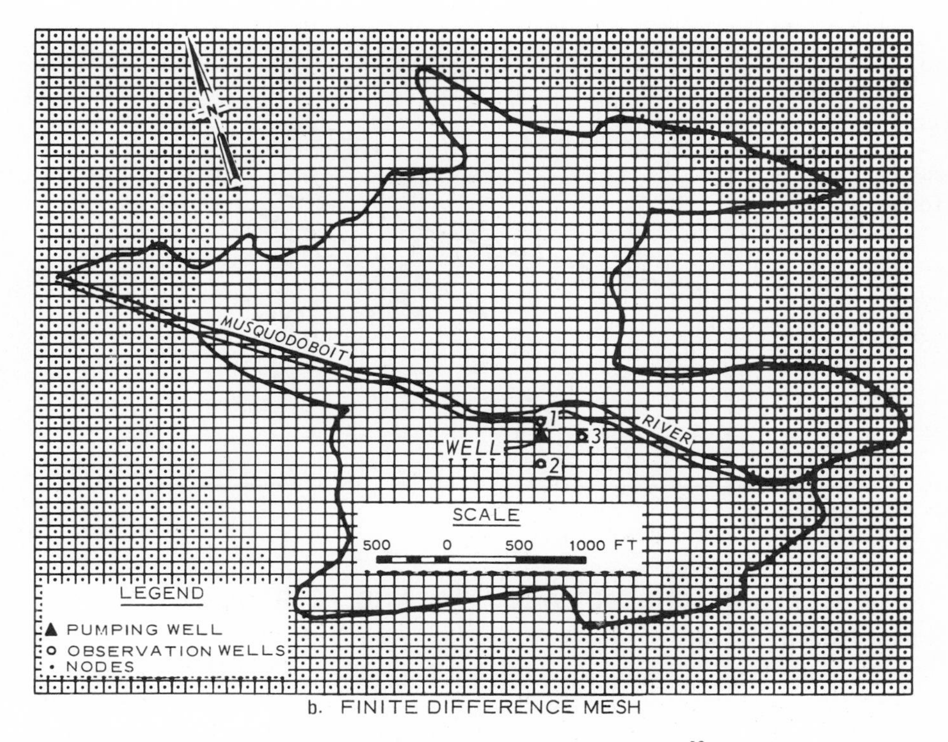

Figure 14-21 Finite element and finite difference analysis of an aquifer.[59]

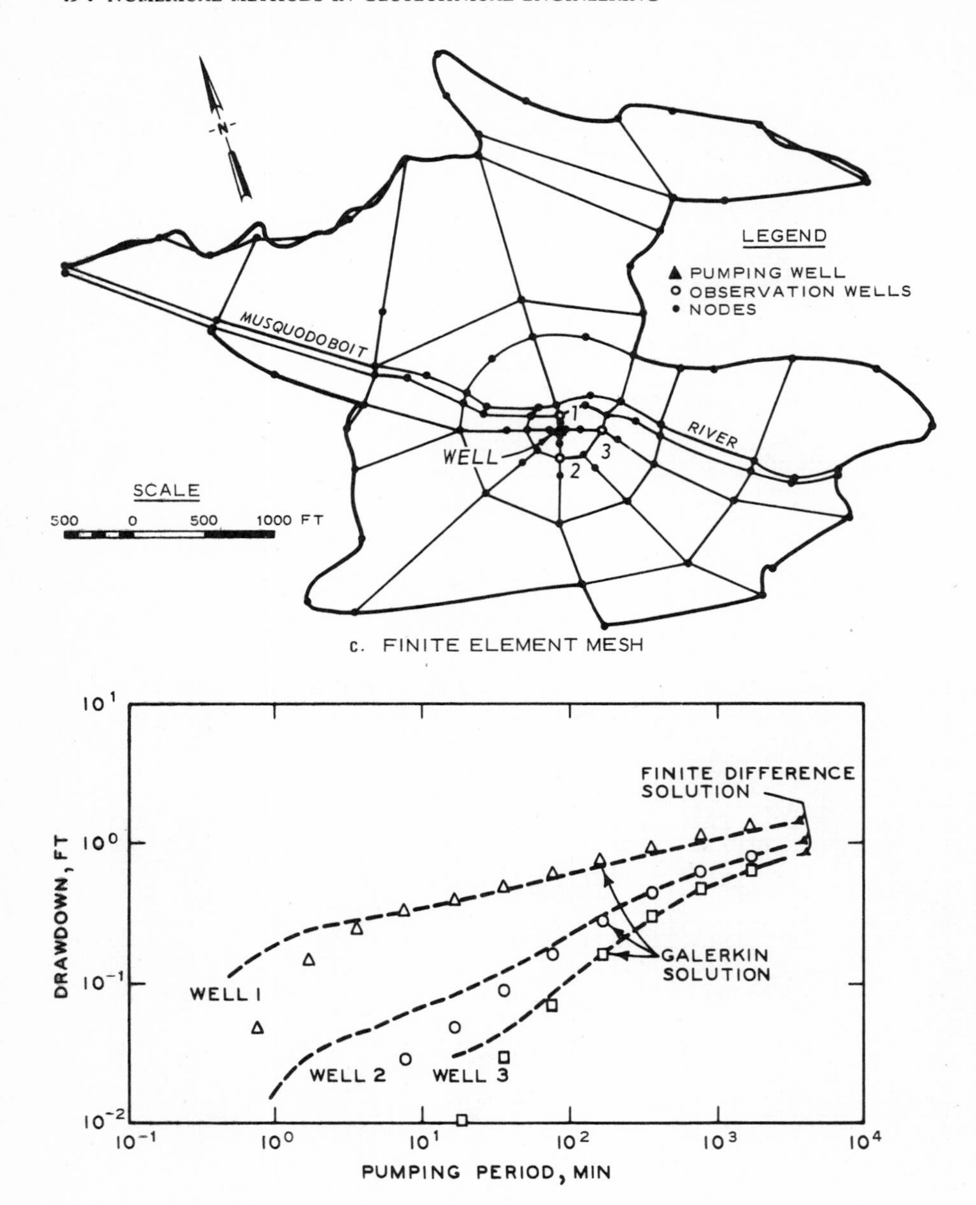

Figure 14-21 (continued).

Sudden drawdown in the viscous flow model Desai[55] compared solutions from the FD-ADE[27,31] and the finite element method (Sec. 14-3) with experimental results from a viscous flow parallel-plate model.[31] Sudden drawdown conditions were allowed, and the locations of the free surface were obtained by taking photographs at various time levels. Value of k for the model was about 0.846 in/s, and a value of unity was used for n.

Figure 14-22 shows comparisons between the numerical solutions and observations. Both finite difference and finite element methods yielded satisfactory agreement with the laboratory data. The ADE procedure yielded adequate accuracy and can prove more economical than the ADI and finite element procedures for problems in which the temporal history of heads on a boundary, e.g., the sloping face in Fig. 14-22, is known.

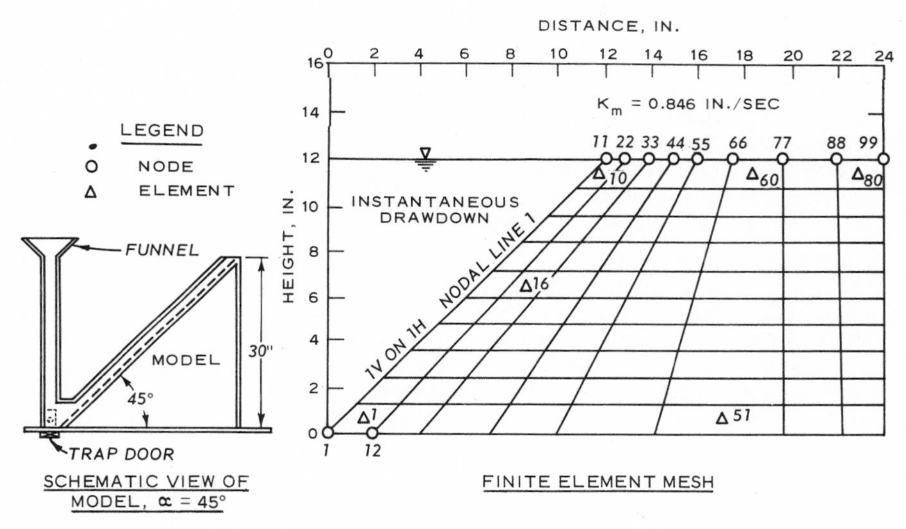

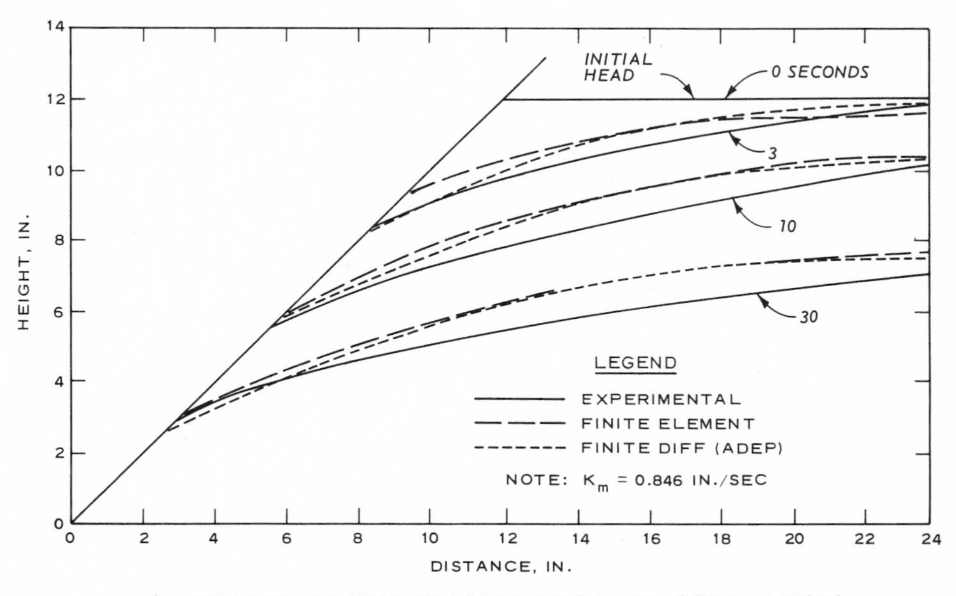

Figure 14-22 Finite element and finite difference results for sudden drawdown in viscous flow model.[55]

Comparison between finite element and finite difference methods On the basis of the foregoing two studies involving both the finite element and finite difference methods, the following observations are offered concerning a comparison between the two procedures:

1. Both methods have advantages, and their choice will depend upon the particular problem on hand.
2. For the problems considered, both methods yielded almost the same accuracy. However, for many problems, with a carefully designed mesh and a higher order isoparametric element, the finite element may yield the same accuracy for fewer nodes, compared with the finite difference method.
3. With an elimination scheme such as the Cholesky method involving essentially only the forward and backward substitutions, the computational effort in the finite element method can be competitive with the efficient FD-ADI.[59]
4. Formulation and development of a computer code for the finite element method can be much more involved than the finite difference schemes. Care is always necessary in generating error free input data for the finite element method.[24]

Example 14-11 Design charts The finite element and finite difference methods[25,27,29,46,54] were used to evolve design charts for computation of the factor of safety of slopes subjected to transient unconfined-flow conditions. Two phases were involved: (1) computation of a time-dependent free surface and (2) use of a limit equilibrium (wedge, slices or Swedish) (Fig. 14-23*b*) for computation of the factor of safety for the free-surface locations at different times. The following nondimensional parameters (Fig. 14-23*a*) were defined:

$$\frac{R}{k_h} \quad \frac{H_d}{H} \quad \frac{H_e}{H} \quad \frac{k_h}{k_v} \quad \frac{c'}{\gamma H} \quad \phi' \tag{14-28}$$

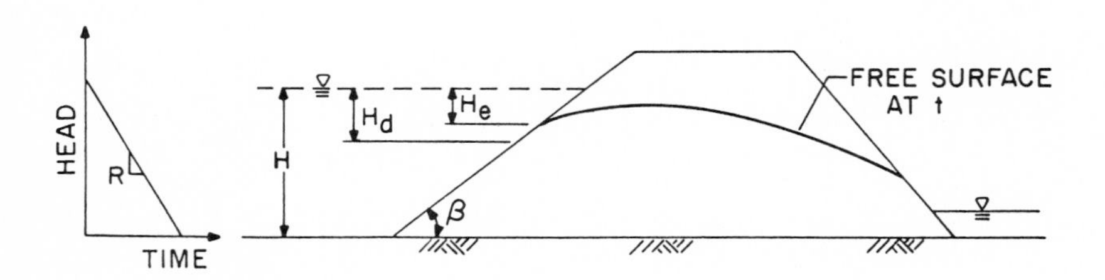

(a) STEP 1: LOCATION OF FREE SURFACE

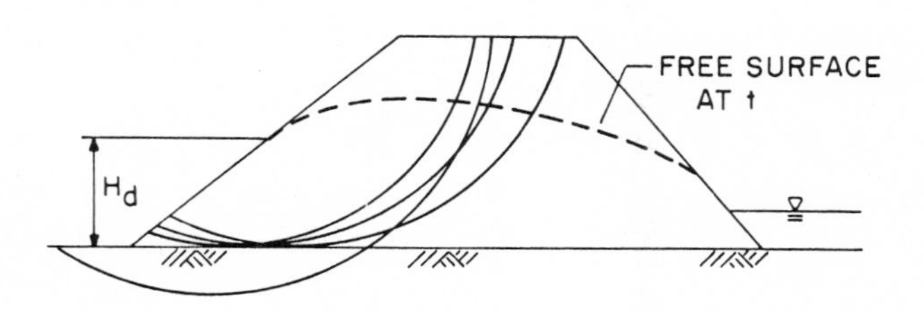

(b) STEP 2: COMPUTATION OF FS AT TIME t

Figure 14-23 Steps in development of design charts.[60]

where R = rate of drawdown
H_d/H = drawdown ratio
H = height of slope
H_e/H = ratio of exit point
α = entrance angle of slope
c', ϕ' = strength parameters for soil
γ = density

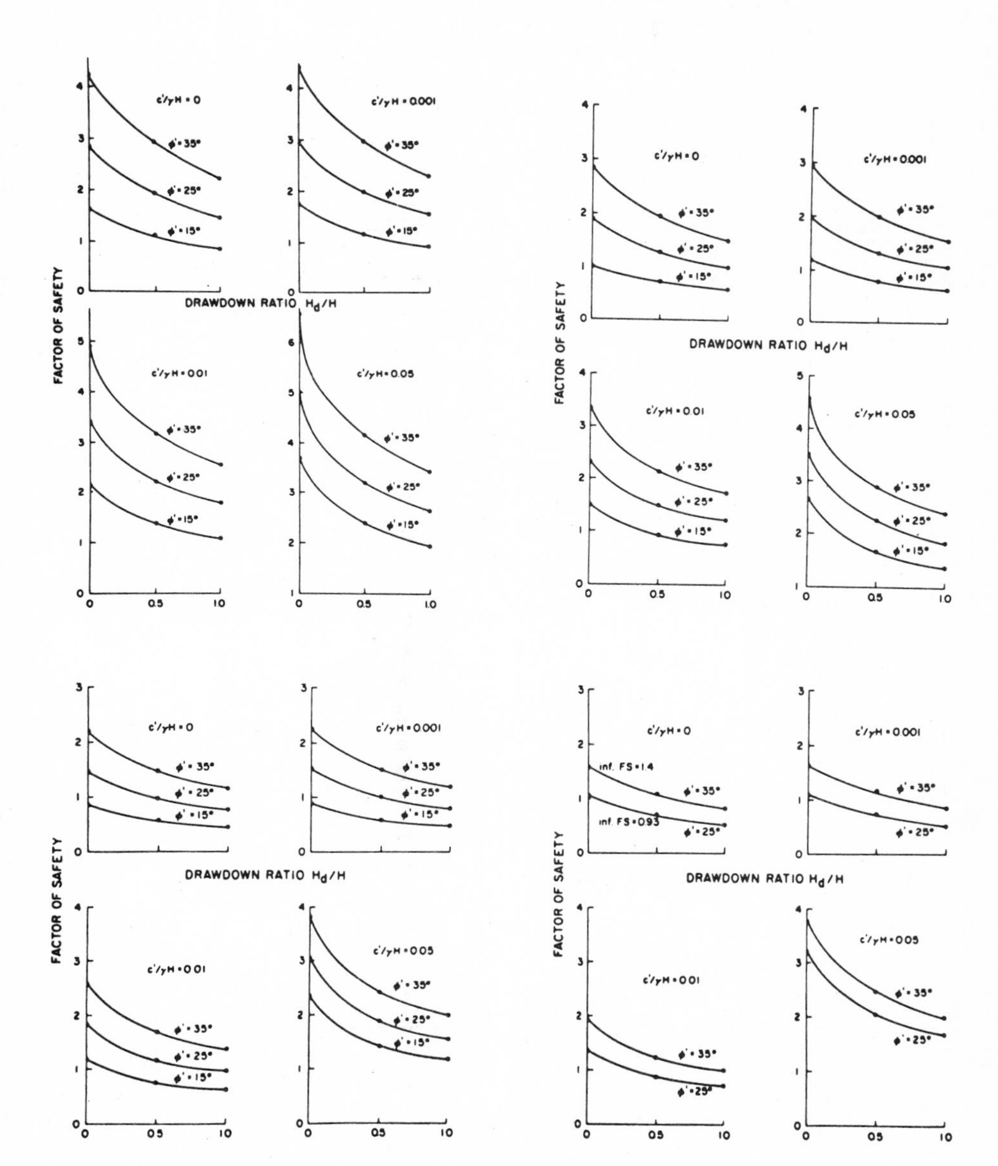

Figure 14-24 Typical design charts for factor-of-safety-vs.-drawdown ratio.[60]

A large number of computer solutions were obtained, and relations between H_e/H and H_d/H for different R/k_h, H_e/H, and R/k_h for $H_d/H = 1$ (end of drawdown) and factor of safety vs. time were established. The final results were expressed as design charts giving relations between the factor of safety, drawdown ratio H_d/H, and values of ϕ' and $c'/\gamma H$ for different slope angles.[60] A typical design chart for different slope angles is shown in Fig. 14-24; the results are plotted in a form[61] familiar to designers. The plots can be used to design safe slopes for critical drawdowns.[60]

It was found that the critical factor of safety during the drawdown was about 10 to 15 percent higher than those obtained from the assumptions of sudden drawdown employed in conventional designs. Thus use of numerical schemes that yield a history of seepage forces during drawdown can permit more economical designs.

14-5 FLOW IN JOINTED MEDIA

The problem of flow is formulated by constructing *deterministic* and/or *probalistic* models.[18,26] In the deterministic approach, two concepts, *particulate* and *continuum*, have been used. All the foregoing works have been based essentially on the continuum concept, in which both the solid medium and liquid are assumed to be continuous.

For discontinuous or jointed media it will be necessary to incorporate theories that can permit discontinuities in the geologic media and in flow. Exact mathematical theories have not yet been established. Most of the past studies using numerical methods have relied essentially on the continuum approach, with use of special (empirical) models for the discontinuities.[62–65] Detailed consideration of this topic is beyond the scope of this book; we shall briefly review some of the models used. For details the reader may consult Ref. 65.

Models for Joints

The discontinuous or jointed (rock) is assumed to be incompressible. It is often assumed that the flow occurs only through joints. Although Darcy's law has often been used, it has been found that the constitutive law is essentially nonlinear; one such law reported by Maini et al.[64] is

$$C \frac{\partial p}{\partial r} = (v_m)^n \tag{14-29}$$

where v_m = mean velocity

$\partial p/\partial r$ = hydraulic gradient

C = constant, dependent on geometry of joints or fractures and viscosity of fluid

n = degree of nonlinearity

The permeability of joints is usually derived from special forms of Navier-Stokes equations as[27,29,63]

$$v_{xj} = \frac{-g(2a_j)^2}{12v}\frac{\partial \phi}{\partial x} = -k_j \frac{\partial \phi}{\partial x} \qquad v_{yj} = \frac{-g(2a_j)^2}{12v}\frac{\partial \phi}{\partial y} = -k_j \frac{\partial \phi}{\partial y} \tag{14-30}$$

where a_j = half-width of joint
ν = kinematic viscosity
k_j = permeability of joint

A number of schemes can be used for simulating flow through joints in a finite element procedure. An open joint can be considered as a discrete conduit with permeability computed as in Eq. (14-30) and porosity equal to unity. Special joint elements for open and filled joints can be used with their permeability matrices established on the basis of the concept similar to that used for simulation of deformations in joints (see Chap. 4).

Numerical solutions for flow through jointed media are of recent origin, and the research activity in this direction is expected to increase. An important topic will be determination of realistic material models both for deformations of the medium and for flow. It will require use of adequate laboratory and field techniques. Moreover, simulation of the spatial distribution of joints will require field mapping.

Perhaps a combination of deterministic and probabilistic approaches will prove successful for the problem of flow through discontinuous media.

APPENDIX 14A REVIEW

Theory and Formulation Procedures

Method†	Ref.
FD	10, 20–23, 27, 29, 48, 50, 66–69
FE	15, 18, 24–26, 28, 33, 34, 36, 42, 43, 44, 55, 59, 70–79

Application	Steady-flow problems			Unsteady-flow problems		
	Sub-class†	Numerical method†	Ref.	Sub-class†	Numerical method†	Ref.
Aquifer analysis, performance evaluation; groundwater development; flow in wells; saturated and unsaturated flows	C	FD	80, 81	C	FD	48, 50, 53, 57, 66, 73, 88–91
		FE	82	C	FE	36, 82, 59
				U	R-K	92, 93
					CFD	94
					FD	19, 52, 95, 96
Flow through foundations of dams, sheet piles, coffer-dams	C	FD	83			
		FE	25, 32, 70, 84			
Flow through earth-rock fill dams, banks, berms	U	FD	58, 85, 86	U	FD	27, 29, 31, 55, 58
					FE	19, 25, 29, 34–36, 55, 79, 97, 98,
		FE	16, 17, 25, 42–45, 56, 84			

Application	Steady-flow problems			Unsteady-flow problems		
	Sub-class†	Numerical method†	Ref.	Sub-class†	Numerical method†	Ref.
Flow to and from canals,	U	FD	20–22, 87	U	FD	49, 69, 99, 100
ditches, drains		FE	82		FE	25, 35
Dispersion of contaminants,					FD	101
salt water intrusion					FE	102–107
Miscellaneous:						
Flow in sandy tidal beaches	...	...	...	U	FE	38, 108
Moisture movement in soils	...	...	...		FD	67, 109–112
Disposal of liquid wastes through						
underground chimneys	...	...	...		FD	113

† C = confined, U = unconfined, FD = finite difference, FE = finite element, R-K = Runge-Kutta, CFD = closed/finite difference.

REFERENCES

1. Theis, C. V.: The Relation between the Lowering of the Piezometer Surface and the Rate and Duration of Discharge of a Well Using Ground Water Storage, *Trans. Am. Geophys. Union*, vol. 16, pp. 519–524, 1935.
2. Muskat, M.: "The Flow of Homogeneous Fluids through Porous Media," McGraw-Hill Book Company, New York, 1937.
3. Todd, D. K.: "Ground Water Hydrology," John Wiley & Sons, Inc., New York, 1960.
4. Polubarinova-Kochina, P. Ya.: "Theory of Ground Water Movement," R. J. M. De Wiest, trans., Princeton University Press, Princeton, N.J., 1962.
5. Harr, M. E.: "Groundwater and Seepage," McGraw-Hill Book Company, New York, 1962.
6. *Proc. Symp. Transient Ground Water Hydraul., Colorado State University, Fort Collins, Colo.*, July 1963.
7. De Wiest, R. J. M. (ed.): "Flow through Porous Media," Academic Press, Inc., New York, 1963.
8. Davis, S. N., and R. J. M. De Wiest: "Hydrology," John Wiley & Sons, Inc., New York, 1966.
9. Hubbert, M. K.: "The Theory of Ground-Water Motion and Related Papers," Hafner Publishing Company, Inc., New York, 1969.
10. Remson, I., G. M. Hornberger, and F. J. Molz: "Numerical Methods in Subsurface Hydrology," Wiley-Interscience, New York, 1971.
11. Terzaghi, K., and R. B. Peck: "Soil Mechanics in Engineering Practice," John Wiley & Sons, Inc., New York, 1967.
12. Cedergren, H. R.: "Seepage, Drainage, and Flow Nets," John Wiley & Sons, Inc., New York, 1967.
13. Desai, C. S., and L. D. Johnson: Evaluation of Two Finite Element Formulations for One-dimensional Consolidation, *Int. J. Comput. Struct.*, vol. 2, no. 4, September 1972.
14. Ponter, A. R. S.: The Application of Dual Minimum Theorems to the Finite Element Solution of Potential Problems with Special Reference to Seepage, *Int. J. Numer. Methods Eng.*, vol. 4, no. 1, January-February 1972.
15. Desai, C. S.: Overview, Trends and Projections: Theory and Applications of the Finite Element Method in Geotechnical Engineering, *Proc. Symp. Appl. Finite Element Methods Geotech. Eng., U.S. Army Eng. Waterw. Expt. Stn., Vicksburg, Miss., September 1972.*
16. Fenton, J. D.: Hydraulic and Stability Analyses of Rockfill Dams, *Univ. Melbourne, Dept. Civ. Eng.*, DR 15, July 1968. (Finite difference and finite element method, successive overrelaxation; non-Darcy flow through dams and overtopped dams and stability of such structures.)

17. Volker, R. E.: Nonlinear Flow in Porous Media by Finite Elements, *J. Hydrol. Div. ASCE*, vol. 95, no. HY6, November 1969.
18. Trollope, D. H., K. P. Stark, and R. E. Volker: Complex Flow through Porous Media, *Aust. Geomech. J.*, vol. G1, no. 1, 1971.
19. McCorquodale, J. A.: Variational Approach to Non-Darcy Flow, *J. Hydrol. Div. ASCE*, vol. 96, no. HY11, November 1970.
20. Reisenauer, A. E.: Methods for Solving Problems of Multidimensional Partially Saturated Steady Flow in Soils, *J. Geophys. Res.*, vol. 68, no. 20, October 1963.
21. Jeppson, R. W.: Axisymmetric Seepage through Homogeneous and Nonhomogeneous Porous Mediums, *Water Resour. Res.*, vol. 4, no. 6, December 1968. (Finite difference; applies to flow from ponds, with permeability as a function of depth.)
22. Jeppson, R. W.: Free-Surface Flow through Heterogeneous Porous Media, *J. Hydrol. Div. ASCE*, vol. 95, no. HY1, January 1969. (Finite difference; Newton-Raphson; flow from canals and permeability as a function of depth.)
23. Taylor, G. S., and J. N. Luthin, Computer Methods for Transient Analysis of Water Table Aquifer, *Water Resour. Res.*, vol. 15, February 1969.
24. Desai, C. S., and J. F. Abel: "Introduction to the Finite Element Method: A Numerical Method for Engineering Analysis," Van Nostrand Reinhold Company, New York, 1972.
25. Desai, C. S.: Finite Element Procedures for Seepage Analysis Using an Isoparametric Element, *Proc. Symp. Appl. Finite Element Method Geotech. Eng., U.S. Army Eng. Waterw. Expt. Stn., Vicksburg, Miss.*, September 1972. (Applications to flow through dam foundations, dams, riverbanks, and in parallel drains.)
26. Desai, C. S.: Finite Element Methods for Flow in Porous Media, chap. 8 in Gallagher, R. H., *et al.* (eds.), "Finite Elements in Fluids," John Wiley & Sons, Inc., London, 1975; also *Int. Symp. Finite Element Methods Flow Prob., Univ. Wales, Swansea, January 1974.*
27. Desai, C. S., and W. C. Sherman: Unconfined Transient Seepage in Sloping Banks, *J. Soil Mech. Found. Div. ASCE*, vol. 97, no. SM2, February 1971. (Finite difference, ADE; correlations with data from tests with a large parallel-plate model.)
28. Desai, C. S.: An Approximate Solution for Unconfined Seepage, *J. Irrig. Drain Div. ASCE*, vol. 99, no. IR1, March 1973.
29. Desai, C. S.: Seepage in Mississippi River Banks: Analysis of Transient Seepage Using Viscous Flow Model and Numerical Methods, *U.S. Army Eng. Waterw. Expt. Stn.*, Misc. Pap. S-70-3, Vicksburg, Miss., February 1970. (Finite difference, ADE, finite element method; free-surface flow through parallel-plate models.)
30. Neuman, S. P.: Galerkin Approach to Unsaturated Flow in Soils, *Proc. Int. Symp. Finite Element Methods Flow Prob., Univ. Wales, Swansea, January 1974.*
31. Dvinoff, A. H., and M. E. Harr: Phreatic Surface Location after Drawdown, *J. Soil Mech. Found. Div. ASCE*, vol. 97, no. SM1, January 1971. (Finite difference flow through viscous flow models and dams; curves for drawdown.)
32. King, G. J. W., and R. N. Chowdhury: Finite Element Solution for Quantity of Steady Seepage, *Civ. Eng. Public Works Rev.*, vol. 66, no. 785, December 1971. (Procedure for computing quantity of flow through foundations of dams.)
33. Desai, C. S.: 3-D Seepage through Porous Media and CODE SEEP-3DFE, *VPI State Univ. Dept. Civ. Eng. Rep.* VPI-E-75.29, Blacksburg, Va., May 1975.
34. Desai, C. S.: Seepage Analysis of Earth Banks under Drawdown, *J. Soil Mech. Found. Div. ASCE*, vol. 98, no. SM1, November 1972.
35. France, P. W., C. J. Parekh, J. C. Peters, and C. Taylor: Numerical Analysis of Free Surface Seepage Problems, *J. Irrig. Drain Div. ASCE*, vol. 97, no. IR1, March 1971. (Finite element method, time-integration scheme; flow through dams and in parallel drains.)
36. Neuman, S. P., and P. A. Witherspoon: "Analysis of Unsteady Flow with a Free Surface Using the Finite Element Method," *Water Resour. Res.*, vol. 7, no. 3, June 1971. (Iterative time-integration scheme; flow through zoned dams and toward wells.)
37. Cheng, R. T., and C. Y. Li: On the Solution of Transient Free-Surface Flow Problems in Porous Media by Finite Element Method, *J. Hydrol.*, vol. 20, 1973.

38. Fang, C. S., and S. N. Wang: Groundwater Flow in a Sandy Tidal Beach, 2: Two-dimensional Finite Element Analysis, *Water Resour. Res.*, vol. 8, no. 1, February 1973.
39. Sandhu, R. S., I. S. Rai, and C. S. Desai: Variable Time Step Analysis of Unconfined Seepage, *Proc. Int. Symp. Finite Element Methods Flow Prob., Univ. Wales, Swansea, January 1974.*
40. Desai, C. S., and R. S. Sandhu: Finite Element Analysis of Unconfined Flow with a Variable Time-Step Procedure, *Proc. Army Sci. Conf., West Point, N.Y., May 1974.*
41. Henrici, P.: "Elements of Numerical Analysis," John Wiley & Sons, Inc., New York, 1964.
42. Taylor, R. L., and C. B. Brown: Darcy Flow Solutions with a Free Surface, *J. Hydrol. Div. ASCE*, vol. 93, no. HY2, March 1967. (Finite element; iterative procedure for locating steady free surface through dams and for flow toward wells.)
43. Finn, W. D. L.: Finite Element Analysis of Seepage through Dams, *J. Soil Mech. Found. Div. ASCE*, vol. 93, no. SM6, November 1967. (Iterative procedure for steady free-surface flow through dams.)
44. Neuman, S. P., and P. A. Witherspoon: Finite Element Method for Analyzing Steady Seepage with a Free Surface, *Water Resour. Res.*, vol. 6, no. 3, June 1970.
45. Desai, C. S.: Free Surface Seepage through Foundation and Berm of Cofferdams, *Ind. Geotech. J.*, vol. 5, no. 1, January-March 1975.
46. Desai, C. S.: Finite Element Residual Schemes for Unconfined Flow, *Int. J. Numer. Methods Eng.*, vol. 10, pp. 1415–1418.
47. Isaacs, L. T., and K. G. Mills: discussion of Numerical Analysis of Free Surface Seepage Problems, *J. Irrig. Drain. Div. ASCE*, vol. 98, no. IR1, March 1972.
48. Freeze, R. A., and P. A. Witherspoon: Theoretical Analysis of Regional Groundwater Flow, I: Analytical and Numerical Solutions to the Mathematical Model, *Water Resour. Res.*, vol. 2, 1966, pp. 641–656.
49. Todsen, M.: On the Solution of Transient Free-Surface Flow Problems in Porous Media by Finite Difference Methods, *J. Hydrol.*, vol. 12, no. 3, February 1971. (Successive overrelaxation; applies to ditch drainage and rectangular dams; error analysis.)
50. Pinder, G. F., and J. D. Bredehoeft: Application of the Digital Computer for Aquifer Evaluation, *Water Resour. Res.*, vol. 4, no. 5, October 1968.
51. Terzidis, G.: Computational Schemes for the Boussinesq Equation, *J. Irrig. Drain. Div. ASCE*, vol. 94, no. IR4, 1968.
52. Verma, R. D., and W. Brutsaert: Unsteady Free Surface Ground Water Seepage, *J. Hydrol. Div. ASCE*, vol. 97, no. HY8, August 1971. (Finite difference, successive overrelaxation, explicit; solves for flow through rectangular regions.)
53. Ruston, K. R., and L. M. Tomlinson: Digital Computer Solutions of Ground-Water Flow, *J. Hydrol.*, vol. 12, no. 4, March 1971. (Finite difference, ADI; examines effects of magnitude of Δt; compares implicit, explicit, and resistance-network methods.)
54. Desai, C. S.: Evaluation of Numerical Procedures for Fluid Flow, *ASCE Natl. Conv., Denver, Colo., November 1975*, pap. 2609.
55. Desai, C. S.: Seepage in Mississippi River Banks; Analysis of Transient Seepage Using Viscous Flow Model, and Finite Difference and Finite Element Methods, *U.S. Army Eng. Waterw. Expt. Stn. Tech. Rep.* 1, Vicksburg, Miss., May 1973. (FD-ADE, finite element; flow through banks and dams, viscous flow model, and design aspects.)
56. Kealy, C. D., and R. E. Williams: Flow through a Tailing Pond Embankment, *Water Resourc. Res.*, vol. 7, no. 4, February 1971.
57. Prickett, T. A., and C. G. Lonnquist: Selected Digital Computer Techniques for Groundwater Resource Evaluation, *Ill. State Water Supply*, Bull. 55, Urbana, Ill., 1971.
58. Szabo, B. A., and I. W. McCaig: A Mathematical Model for Transient Free Surface Flow in Nonhomogeneous or Anisotropic Porous Media, *Water Resour. Bull.*, vol. 4, no. 3, September 1968. (Finite difference; compares numerical results with data from analog model for flow through rectangular banks and drains.)
59. Pinder, G. F., and E. O. Frind: Application of Galerkin's Procedure to Aquifer Analysis, *Water Resour. Res.*, vol. 8, no. 1, February 1972. (Finite difference and finite element; evaluates different isoparametric elements.)

60. Desai, C. S.: Analysis of Slopes under Drawdown by Numerical Method, *Geotech. Eng. Div. ASCE*, vol. 103, no. GT7, July 1977.
61. Morgernstern, N. R.: Stability Charts for Earth Slopes during Rapid Drawdown, *Geotechnique*, vol. 13, pp. 121–131, 1963.
62. Noorishad, T., P. A. Witherspoon, and T. L. Brekke: A Method for Coupled Stress and Flow Analysis of Fractured Rock Masses, *Univ. Calif., Geotech. Eng. Rep.* 71-6, Berkeley, 1971.
63. Wittke, W., P. Rissler, and S. Semprich: Räumliche, laminare, und turbulente Strömung in klüftigen Fels nach zwei verschiedenen Rochenmodellen, *Proc. Symp. Percolation Fissured Rock, Stuttgart, 1972.*
64. Maini, Y. N. T., J. Noorishad, and J. Sharp: Theoretical and Field Considerations on the Determination of *in Situ* Hydraulic Parameters in Fractured Rock, *Proc. Symp. Percolation Fissured Rock, Stuttgart, 1972.*
65. *Proc. Symp. Percolation Fissured Rock, Stuttgart*, 1972.
66. Klute, A.: A Numerical Method for Solving the Flow Equation for Water in Unsaturated Materials, *Soil Sci.*, vol. 73, pp. 105–116, 1952.
67. Hanks, R. J., and S. A. Bowers: Numerical Solution of the Moisture Flow Equation into Layered Soils, *Proc. Soil Sci. Soc. Am.*, vol. 26, no. 6, November–December 1962. (Finite difference, Crank-Nicholson; applies to moisture movement through soil masses.)
68. Brakensiek, D. L.: Finite Differencing Methods, *Water Resour. Res.*, vol. 3, no. 3, 3d quart. 1967. (Reviews calculus of finite differences.)
69. Rubin, J.: Theoretical Analysis of Two-dimensional Transient Flow of Water in Unsaturated and Partly Unsaturated Soils, *Proc. Soil Sci. Soc. Am.*, vol. 32, September-October 1968. (Finite difference, ADI; solves problems of drainage in ditches and soil slabs.)
70. Zienkiewicz, O. C., and Y. K. Cheung: Finite Elements in the Solution of Field Problems, *Engineer*, Sept. 24, 1965. (Perhaps the earliest paper showing the use of finite element methods for flow problems.)
71. Zienkiewicz, O. C., P. Mayer, and Y. K. Cheung: Solution of Anisotropic Seepage by Finite Elements, *J. Eng. Mech. Div. ASCE*, vol. 92, no. EM1, February 1966. (Solves for flow through foundations of sheet piles and dams.)
72. Zienkiewicz, O. C., A. K. Bahrani, and P. L. Arlett: Solution of Three-dimensional Field Problems by the Finite Element Method, *Engineer*, Oct. 27, 1967.
73. Javandel, I., and P. A. Witherspoon: Application of the Finite Element Method to Transient Flow in Porous Media, *Trans. Soc. Petrol. Eng.*, vol. 243, pp. 241–251, September 1968. (Finite element methods; applies to flow toward wells in aquifers.)
74. Neuman, S. P., and P. A. Witherspoon: Variational Principles for Confined and Unconfined Flow of Ground Water, *Water Resour. Res.*, vol. 6, no. 5, October 1970.
75. Zienkiewicz, O. C., and C. J. Parekh: Transient Field Problems: Two-dimensional and Three-dimensional Analysis by Isoparametric Finite Elements, *Int. J. Numer. Methods Eng.*, vol. 2, no. 1, 1970.
76. Zienkiewicz, O. C.: "The Finite Element Method in Engineering Science," McGraw-Hill Publishing Company, Ltd., London, 1971.
77. Desai, C. S., J. T. Oden, and L. D. Johnson: Numerical Analysis of Some Time Dependent Problems, *U.S. Army Eng. Waterw. Expt. Stn.*, Vicksburg, Miss. (Evaluates a number of finite difference and finite element schemes and applies concept of space-time elements.)
78. Verruijt, A.: Solution of Transient Groundwater Flow Problems by the Finite Element Method, *Water Resour. Res.*, vol. 8, no. 3, June 1972. (Approximation to time derivative made before introduction of a variational principle.)
79. Bruch, J. C.: Nonlinear Equation of Unsteady Ground-Water Flow, *J. Hydrol. Div. ASCE*, vol. 99, no. HY3, March 1973. (Finite element methods; implicit, Newton-Raphson; one-dimensional free-surface flow through rectangular banks.)
80. Remson, I., C. A. Appel, and R. A. Webster: Groundwater Models Solved by Digital Computer, *J. Hydrol. Div. ASCE*, vol. 91, no. HY3, May 1965. (Finite difference, Gauss-Seidel; analyzes performance of aquifers and examines effects of new reservoirs on them.)
81. Freeze, R. A., and P. A. Witherspoon: Theoretical Analysis of Regional Groundwater Flow, 1:

Analytical and Numerical Solutions to the Mathematical Model, *Water Resour. Res.*, vol. 2, no. 4, 4th quart. 1966. (Finite difference, successive overrelaxation.)

82. Finite Element Solution of Steady State Potential Flow Problems, *U.S. Army Eng. Dist. Hydrolog. Eng. Center* 723-G2-L2440, Sacramento, Calif., November 1970. (Solves problems of groundwater flow, radial flow to well in leaky aquifer, and seepage from ditch, considering effects of capillarity.)
83. Tomlin, G. R.: Seepage Analysis through Zoned Anisotropic Soils by Computer, *Geotechnique*, vol. 16, no. 3, September 1966. (Finite difference, relaxation; applies to flow through zoned foundations.)
84. Guellec, P.: Calculation of Flows in Porous Media by the Finite Element Method, *Lab. Ponts Chaussees Rapp. Rech.* 11, November 1970. (Solves for flow toward wells, in foundation of sheet piles, and through rectangular dams.)
85. Curtis, R. P., and J. D. Lawson: Flow over and through Rock-fill Banks, *J. Hydrol. Div.* ASCE, vol. 93, no. HY5, September 1967. (Finite difference, Gauss-Seidel; solves for turbulent flow through rock-fill banks.)
86. Finnmore, E. J., and B. Perry: Seepage through an Earth Dam Computed by the Relaxation Method, *Water Resour. Res.*, vol. 4, no. 5, October 1968. (Finite difference, relaxation; applies to flow through rectangular dams.)
87. Jeppson, R. W.: Seepage from Channels through Layered Porous Mediums, *Water Resour. Res.*, vol. 4, no. 2, April 1968. (Finite difference; considers both potential and stream functions as independent variables.)
88. Vacher, J. P., and V. Cazabat: Écoulement des fluides dans les milieux poreux stratifiés résultat obtenus sur le modèle du bicouche avec communication, *Rev. IFP*, vol. 16, no. 11, 1961. (Finite difference, ADI; considers flow toward wells in layered aquifers.)
89. Freeze, R. A.: Three-dimensional, Transient, Saturated-Unsaturated Flow in a Groundwater Basin, *Water Resour. Res.*, vol. 7, no. 2, April 1971. (Finite difference, successive overrelaxation; considers flow in small geologic basins including infiltration, recharge, and stream base flow.)
90. Ruston, K. R., and L. A. Wedderburn: Aquifers Changing between the Confined and Unconfined State, *Ground Water*, vol. 9, no. 5, September-October 1971. (Electrical analogy and finite difference.)
91. Huang, Y. H.: Computer Analysis of Flow Toward Artesian Wells, *J. Hydrol. Div. ASCE*, vol. 97, no. HY9, September 1971. (Finite difference; solves for flow toward fully and partially penetrating wells.)
92. Singh, R., and J. B. Franzini: Unsteady Flow in Unsaturated Soils from a Cylindrical Source of Finite Radius, *J. Geophys. Res.*, vol. 72, no. 4, February 1967. (Numerical integration by fourth-order Runge-Kutta formulas.)
93. Marino, M. A., and W. W. G. Yeh: Nonsteady Flow in a Recharge Well–Unconfined Aquifer System, *J. Hydrol.*, vol. 16, no. 2, June 1972. (Coupling of fourth-order Runge-Kutta scheme with lagrangian interpolation.)
94. Karadi, G., R. J. Krizek, and H. Elnaggar: Unsteady Seepage Flow between Fully Penetrating Trenches, *J. Hydrol.*, vol. 6, pp. 417–430, 1968. (Combined closed form and finite difference; considers infiltration and evaporation.)
95. Vachaud, P. G., and P. Guelin: Équations et modèles mathématiques pour le calcul des transferts d'eau dans la zone de sol non saturée, *Houille*, no. 8, 1969. (Finite difference; considers movement of water between ground surface and water table.)
96. Hornberger, G. M., J. Ebert, and I. Remson: Numerical Solution of the Boussinesq Equation for Aquifer-Stream Interaction, *Water Resour. Res.*, vol. 6, no. 2, April 1970. (Finite difference, Douglas-Jones predictor-corrector; considers recession and flow due to changes in stream stages.)
97. Mills, K. G.: Computation of Post-drawdown Seepage in Earth Dams by Finite Elements, M.S. thesis, University of Queensland, 1971. (Time-integration scheme; solves for flow through zoned dams.)
98. Tulk, J. D., and G. P. Raymond: Drainage of Granular Soils, *Proc., Spec. Conf. Finite Element Methods in Civil Eng.*, McGill Univ., Montreal, *June 1972*. (Finite element methods; applies to flow through granular fills.)

99. Karadi, G. M., R. J. Krizek, and M. Rechea: Critical Evaluation of Certain Methods of Unsteady GroundWater Hydraulics, *Water Resour. Bull.*, vol. 6, no. 3, May-June 1970. (Finite difference, semidiscretization; considers tile drainage.)
100. Herbert, R., and M. Zytynski: A New Technique for Time-variant Ground Water Flow Analysis, *J. Hydrol.*, vol. 16, no. 2, June 1972. (A special finite difference method in which zone of saturation is divided into small increments in horizontal direction and the full depth of aquifer is included in vertical direction; applies to flow toward trenches.)
101. Shamir, U. Y., and D. R. F. Harleman: Numerical Solutions for Dispersion in Porous Mediums, *Water Resour. Res.*, vol. 3, no. 2, 2d quart. 1967.
102. Guymon, G. L.: A Finite Element Solution of the One-dimensional Diffusion-Convection Equation, *Water Resour. Res.*, vol. 6, no. 1, February 1970.
103. Guymon, G. L., V. H. Scott, and L. R. Herrman: A General Numerical Solution of the Two-dimensional Diffusion-Convection Equation by the Finite Element Method, *Water Resour. Res.*, vol. 6, no. 6, 1970.
104. Guymon, G. L.: Note on the Finite Element Solution of the Diffusion-Convection Equation, *Water Resour. Res.*, vol. 8, no. 5, October 1972.
105. Nalluswami, M., R. A. Longenbaugh, and D. K. Sunada: Finite Element Method for the Hydrodynamic Dispersion Equation with Mixed Partial Derivatives, *Water Resour. Res.*, vol. 8, no. 5, October 1972.
106. Amend, J. H., D. N. Contractor, and C. S. Desai: Oxygen Depletion and Sulfate Production in Strip Mine Spoil Dams, *Proc. 2nd Int. Conf. on Num. Methods in Geomech., Blacksburg, Va., June 1976.*
107. Desai, C. S., and D. N. Contractor: Finite Element Analysis of Flow, Diffusion and Salt Water Intrusion in Porous Media, *Proc. U. S. Germany Symp. Formul. Comput. Alg. Fin. Elem. Anal., MIT, Cambridge, Mass., August 1976.*
108. Harrison, W., C. S. Fang, and S. N. Wang: Groundwater Flow in a Sandy Tidal Beach, 1: One-dimensional Finite Analysis, *Water Resour. Res.*, vol. 7, no. 5, October 1971. (Compares numerical results with field observations.)
109. Ashcroft, G., D. D. Marsh, D. D. Evans, and L. Boerma: Numerical Method for Solving the Diffusion Equation, I: Horizontal Flow in Semi-infinite Media, *Proc. Soil Sci. Soc. Am.*, vol. 26, no. 6, November-December 1962. (Finite difference, implicit; considers moisture movement through soils.)
110. Remson, I., R. L. Drake, S. S. McNeary, and E. M. Wallo: Vertical Drainage of Unsaturated Soil, *J. Hydrol. Div. ASCE*, vol. 91, no. HY1, 1965. (Finite difference, implicit, Gauss-Seidel; solves for flow in unsaturated soil slabs.)
111. Jensen, M. E., and R. J. Hanks: Nonsteady State Drainage from Porous Media, *J. Irrig. Drain. Div. ASCE*, vol. 93, no. IR3, September 1967. (Finite difference, implicit.)
112. Wang, F. C., and V. Laxminarayan: Mathematical Simulation of Water Movement through Unsaturated Nonhomogeneous Soils, *Proc. Soil Sci. Soc. Am.*, vol. 32, May-June 1968. (Finite difference, explicit-implicit.)
113. Korver, J. A.: Fluid Flow from Nuclear Chimneys, *Water Resour. Res.*, vol. 2, no. 2, 2d quart. 1966.

CHAPTER

FIFTEEN

STATIC ANALYSIS OF EARTH RETAINING STRUCTURES

G. Wayne Clough and Yuet Tsui

15-1 INTRODUCTION

The application of numerical methods to analysis of retaining structures begins with some of the earliest published works in geotechnical engineering, those of Coulomb[1] and Rankine.[2] The methods of Coulomb and Rankine were the forerunners of the limit approach, which is the basis of several modern analysis techniques. Limit analysis has been used with success to predict collapse loads for earth retaining structures, but it cannot predict deformations associated with the limit loads and cannot yield information before the limiting state.

Newer numerical methods are oriented toward predicting not only earth pressures but also deformations of the soil mass and retaining structures. This is in recognition of the fact that in many earth retaining structure problems it is more important to form a reasonable prediction of the deformations that occur in the soil behind the structure than of the earth pressures acting on the structure.

This chapter will be oriented toward methods which presently have the strongest capabilities in the analysis of soil-structure interaction. The most obvious example of such a method is the finite element technique, and its use will be covered in some detail. However, other useful methods will also be reviewed, with some emphasis on techniques which have been developed recently and show promise for future extension.

Table 15-1 Numerical methods used in soil-structure interaction analysis

Method	Ref.	Form of solution
Upper limit analysis	11	Finite difference
Lower limit analysis	12–14	Finite difference, optimization
Beam or slab on elastic subgrade	15–18	Finite difference, discrete element
Associated stress and velocity fields	19–22	Finite difference
Finite element	23–45	Finite element

15-2 REVIEW OF NUMERICAL METHODS FOR EARTH RETAINING STRUCTURES

Most numerical analysis tools for retaining structures are of recent origin; however, the early work of many contributors in the field of geotechnical engineering provided information useful in modern analytical tools. Developments in theoretical analysis were provided by Coulomb,[1] Rankine,[2] Kotter,[3] Reissner,[4] Westergaard,[5] and Terzaghi,[6,7] and substantial improvements were made in our understanding of soil-structure interaction processes by the experimental and field work of Meem,[8] Terzaghi,[9] and Peck.[10] More recently, many others, too numerous to mention, have also contributed.

From these works our present numerical tools have come, at least in part. Some prominent modern methods and their solution forms are shown in Table 15-1. Examples of each of these methods are discussed in the subsequent paragraphs.

Limit Analysis

Upper limit methods Adaptation of upper limit analysis methods for computer solution has extended their usefulness. Upper limit approaches based on slope-stability methods† have been the easiest to program and have been applied to a wide range of earth pressure problems. Janbu[11] developed the first generalized procedure of slices for earth pressure analysis.

In this approach, the soil behind a retaining structure is assumed to fail along some arbitrary slip surface. The soil mass above the slip surface is broken into finite slices, and the retaining structure is assumed to serve as a reaction force, holding the soil mass in equilibrium. By following the procedure proposed by Janbu,[11] two force-equilibrium equations for each slice and the moment-

† These methods may not be rigorous upper limit procedures.

equilibrium equation for the entire mass are solved to yield the necessary reaction force to be exerted by the structure upon the soil mass. The equilibrium equations must be solved by iteration because the factor of safety is included on both sides of the equations.

Methods of slices like that of Janbu[11] have advantages in that arbitrarily stratified soil deposits can be analyzed, drained and undrained soil strengths can easily be accommodated, and the mechanics of the methods are known to most geotechnical engineers. The disadvantages, which are common to all limit procedures, are (1) that no information on deflections is obtained; (2) effects of interaction of materials with disparate stress-strain characteristics are not considered; (3) the influence of construction sequence or unusual initial stress conditions are not accounted for; and (4) a trial procedure must be employed to locate the most critical failure surface.

Lower limit methods Numerical analysis of earth retaining structures has also been employed using lower limit procedures, most commonly following a finite difference scheme like that proposed by Sokolovski.[12] More recently Lysmer[14] has developed a lower limit analysis, somewhat similar to the finite element method, which shows promise of being more flexible than the Sokolovski technique.

In the Sokolovski approach, the equations of equilibrium in two dimensions are combined with the Mohr-Coulomb failure criterion to give a pair of differential equations for the stresses along the *stress characteristics*, lines along which the Mohr-Coulomb criterion is satisfied. The resulting differential equations can be solved by using finite difference techniques for stresses in the soil mass and along the soil-structure interface. The development of the finite difference equations is given in detail by Sokolovski[12] and in a more readily readable form by Scott[13] and will not be repeated here.

The solution yields a prediction for collapse load, the earth pressure distribution on the structure, and the stress distribution in the soil mass involved in the assumed collapse mechanism. While this method suffers from many of the same disadvantages as the upper limit procedures, it does provide information about distribution of stress on the earth retaining structure as well as the resultant load. However, the upper limit approaches have been more readily adapted to design, and applications of the Sokolovski method have been made for only relatively simple problems.[12,13]

A new form of lower-limit analysis suited for computer solution has been proposed by Lysmer.[14] In this technique, the soil mass adjacent to the structure is broken into a series of elements connected at nodal points at corners of the elements. Stress boundary conditions are specified at each node on a boundary, yield constraints are assumed, and by means of linear programming techniques, optimum lower bound values of stresses within the elements are obtained, which are in equilibrium everywhere, satisfy the boundary conditions, and do not exceed the yield constraint.

Because the method generates its own stress field, it would appear to be ideally suited for use in analysis of more complex problems; however, Lysmer notes that the cost of the analysis increases very rapidly with the size of the mesh configuration, and its application has currently been limited to simple problems. Difficulties may also be anticipated in determining appropriate boundary conditions for advanced problems, and because deflections are not predicted by this technique, its role in design work will probably be small.

Beam on Elastic Subgrade

The approach using a beam or slab on an elastic subgrade has found application in numerical analysis of sheet piles[15] and braced excavations.[16] Both finite difference and discrete element methods have been employed for the solution of the governing equations. In either case, the elastic subgrade is assumed to generate a reactive pressure proportional to the deflection (Winkler's hypothesis) and, in essence, consists of a bed of springs. The soil response is characterized by a spring-constant parameter k.

Finite difference methods Finite difference equations for the differential equations describing the bending of a beam on a Winkler subgrade have been presented by Gleser[47] and Palmer and Thompson[48] and are relatively straightforward to develop. These equations have been extended to the special problem of the sheet pile by Turabi and Balla;[15] solution of the equations is usually of the iterative type.

The finite difference solution requires a knowledge of the stiffness parameters of the structure k of the soil and boundary conditions for the extremities of the sheet pile. The solution yields information about the moments, shears, and deflections of the structure as well as soil pressures on the structure. It is important that the soil parameter k need not be a constant and instead can be a function of depth or pressure.

Discrete element methods The problem of a beam or slab on a Winkler subgrade can also be formulated through a stiffness or discrete element approach. Matlock and Haliburton[18] have done so for beams, and Hudson and Matlock[17] have reported a solution for slabs. The discrete element form of solution gives more flexibility in simulating cracking of the slab or beam and in modeling nonhomogeneity and nonlinearity in the subgrade.

Haliburton[16] has demonstrated the use of the discrete element beam on a Winkler subgrade for analysis of problems including braced excavations and sheet piles. Forms for the nonlinear response of a soil in the sheet-pile and braced-excavation problems, which can be incorporated into the solution, have been suggested by Haliburton. Difficulties are encountered in simulating effects of construction sequence and wall friction; however, Haliburton suggests approximate techniques for these purposes.

The advantages of either of the solutions using the beam on an elastic foundation over a solution by limit analysis lie in their ability to account for structure flexibility and soil stiffness. Thus the effects of stress redistribution as a result of differential structural deflections are accommodated. The limitations of a beam or elastic foundation approach involve (1) the inherent difficulty of determining a spring constant for the soil and giving it a physical meaning; (2) inability to simulate construction sequence directly, effects of unusual initial soil stresses, and development of wall friction; and (3) lack of information about surface movements behind the retaining structure.

Associated Stress and Velocity Fields

In the previous discussion of lower limit methods, the Sokolovski[12] approach for calculation of limit loads on earth retaining structures was reviewed. It was noted that this approach, as originally proposed, could only be used for limiting states of stress and did not yield information about deformations. However, a new method, proposed by Roscoe[21] and extended by James et al.[20] and Serrano,[22] utilizes the Sokolovski technique in predicting both stress and strain distribution in soil-structure interaction problems for conditions before and at the limit. The method is in the early stages of development and has been only applied to simple problems.

The method involves development of consistent stress and strain fields for fixed increments of structural deflection or load. To begin, a load increment is specified, and an initial stress field is predicted by the method of Sokolovski.[12] Next, an initial strain field is developed, as described by James et al.[20] and Serrano,[22] using assumed soil parameters. After determination of the strain field it will generally be found that the assumed soil parameters are not consistent with the calculated stresses and strains. An iterative procedure is then adopted to generate stress and strain fields which are consistent with all parametric assumptions. At convergence a complete distribution for the stress and strain in the soil is obtained, as well as earth pressures on the structure. The results show excellent comparison with detailed observations made in model tests by James and Bransby.[46] However, the method is not fully developed, and it is difficult to evaluate its potential as a design tool.

Fundamental questions remain concerning the appropriateness of the assumed stress-strain laws and the effects on the predictions caused by noncoincidence of principal stress and strain directions.[19] Practical questions about how to assume wall friction distributions and to account for construction sequences, seepage loadings, and structural flexibility must also be answered. Under any circumstances, the method does provide the geotechnical engineer with a powerful tool to enhance his understanding of the processes involved in soil-structure interaction.

Finite Element Method

Few of the early applications of the finite element method in geotechnical engineering involved soil-structure interaction because of extra complications in the analysis. The complications included such factors as the need to simulate relative

Table 15-2 Soil-structure interaction problems to which the finite element method has been applied

Type of problem	Ref.
Simple retaining structures	24, 27, 34, 36, 41
Navigation locks	26, 31, 32
Braced walls	23, 30, 38, 39, 40, 42
Anchored walls	28, 33, 37, 39, 43–45
Bulkheads	23
Bridge piers	35
Cut and cover tunnels	29

movements between the soil and the structure and construction sequences and numerical problems caused by the disparate stress-strain characteristics of the soil and the structure. Fortunately, solutions to these problems have been obtained, and documented applications of the finite element method are expanding. Descriptions of recent applications are shown in Table 15-2. The problems range from simple retaining structures to such complex problems as anchored and braced excavations, navigation locks, and cut and cover tunnels.

The theoretical basis for the finite element method has been well established (Chap. 1) and need not be repeated here. Special problems related to its application to earth retaining structures, however, are covered in some detail in Sec. 15-4. The advantages of the finite element method in analysis of earth retaining structures lie in its ability to predict both earth pressures and deformations with a minimum of simplifying assumptions. Both structure and soil are considered interactively so that the effects of structural flexibility are taken into account. Limitations in using the method primarily derive from our inability to prescribe appropriate constitutive behavior for soil and to determine the parameters needed for the constitutive models.

15-3 ROLE OF NUMERICAL ANALYSIS

The descriptions given so far have shown that numerical analysis can play an important role as a research tool in studying earth retaining structures. Its use as a design tool, however, has been somewhat limited; some of the reasons for this minimal design role may be (1) satisfaction with conventional methods; (2) difficulties with older numerical methods; (3) cost; and (4) character of the geotechnical engineer. The last reason can be explained by the fact that most geotechnical engineers enjoy their work precisely because they often design using considerable "engineering judgment" and less reliance on theoretical tools.

Satisfaction with the present state of the art of design methods for earth-structure interaction is to a degree substantiated by studies of failures of earth retaining systems such as that reported by Sowers and Sowers.[49] They note that few of the failures included in their study could be attributed to inadequacies in conventional earth pressure theory but were largely due to environmental factors. These facts may be accepted at face value; however, they do nothing to assure the designer that his conventional design is economical, nor do they ensure that designs for problems beyond the conventional will perform satisfactorily. Unconventional problems and development of more economical designs are clearly areas where numerical analysis tools can play a role, assuming that the analyses can be performed for a reasonable cost and with a minimum of oversimplifying assumptions. The growing number of successful, documented numerical analyses of practical problems suggests that these possibilities can be realized. This is particularly true for the finite element method.

Given the advantages of newer numerical methods and proof that they can be used with confidence, under appropriate circumstances the geotechnical engineer should find them attractive accessories to his conventional design and his engineering judgment. It remains clear, however, that the use of numerical methods should never preclude the generous use of engineering judgment, because while the results obtained may be more detailed and comprehensive, simulations of the capricious elements of nature and man's work are always highly idealized.

15-4 USE OF THE FINITE ELEMENT METHOD

Because of its clear advantages over other methods, only the finite element method will be considered in detail, i.e., how it is applied to retaining-structure analysis. The application of this method involves a number of special problems not encountered in other applications. Although solutions to these problems are not complete, in most cases reasonable and practical means are available to deal with them. These techniques are discussed in the following paragraphs.

The Finite Element Mesh

Construction of a finite element mesh is the first step in the idealization of a problem. Hard and fast rules cannot be established for drawing the mesh; it is in fact more of an art than a science, and the details will depend to some extent on such subjective factors as how accurate the results need to be and the zones from which information obtained will be most valuable. It is well known, however, that in areas of expected stress concentrations the mesh must be refined.

In the soil, stress concentration problems become particularly acute where sudden changes in structure geometry occur, e.g., the corners of an underground culvert. Refinement in such areas will improve the solution; however, it must be remembered that the soil has an infinite number of degrees of freedom and may often deform in such a manner that it is not feasible to have enough elements to simulate the actual behavior accurately. Thus a compromise must be accepted between economics and solution accuracy.

Adequate numbers of elements for the structure must also be ensured, since otherwise poorly conditioned structural response can easily result. This may create some problems for the geotechnical engineer since as a user he often wishes to employ as few elements as possible for the structure. The necessary minimum number of structural elements is sensitive to the type of element being used. Studies of this problem have been made and are reported by Clough and Ducan[26] for simple elements and Felippa[50] and Doherty et al.[51] for higher order elements. The element proposed by Doherty et al.[51] is particularly useful in that an accurate representation of structural bending can be obtained with fewer elements while requiring less computer time to obtain a solution than with most element types used for the problem.

Numerical Problems

When both the structure and the soil are represented in a finite element mesh, the wide disparity between the stress-strain properties can create numerical difficulties in elements near the boundary between the materials. These problems seem to occur if the number of significant digits being used to represent numbers in the computer is seven or less. Use of seven significant digits is a standard representation for single-precision accuracy on many models of computers, e.g., the IBM 360 or 370 series.

The effects can be seen in Fig. 15-1, which is a plot of the percentage difference obtained between a vertical force applied to a flexible slab and the resultant of the vertical stresses calculated from soil elements beneath the slab. The difference for seven-digit accuracy can be seen to diverge from the correct result as soon as the ratio of the soil modulus to the slab modulus reaches 1:300, a low ratio for most soil-structure interaction problems. For example, if the slab is concrete and the soil a medium clay, this ratio can easily reach 1:1000.

The numerical difficulties can be resolved by using a 16-digit representation of numbers. This representation is available in single precision on some computer systems but on most others can only be obtained using double precision. As shown in Fig. 15-1, vertical equilibrium is obtained under these circumstances with modulus ratios even up to 1 : 1,000,000, which is characteristic of the ratio of the modulus of very soft clay to that of steel.

In using programs developed on machines with single precision with 16-digit accuracy, care should be taken to convert the program to double precision if the user's system has only seven-digit single-precision accuracy.

Boundary Conditions

Boundary conditions in a soil-structure interaction analysis involve locations of the mesh boundaries, assumptions of symmetry, plane strain or plane stress conditions, and interface between the structure and the soil. Each of these aspects can be significant.

Mesh boundaries Locations of most of the mesh boundaries are dictated by the physical characteristics of the problem. At least one lateral boundary, however, is

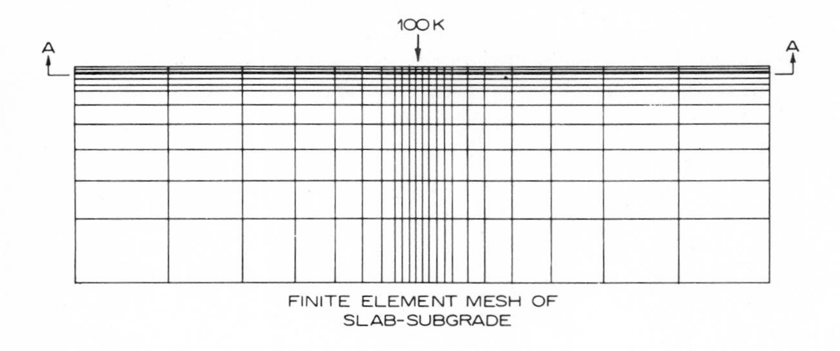

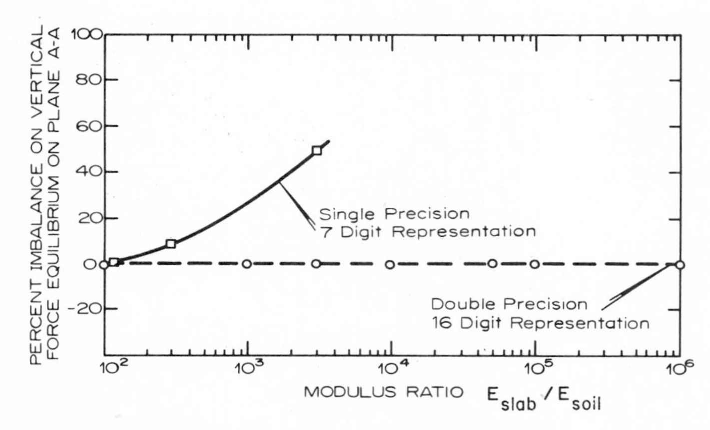

Figure 15-1 Effect of modulus disparity and digit representation on the accuracy of computed results.

usually positioned so as to have negligible influence on the problem and to simulate the effects of a semi-infinite medium. For most problems the boundary should be located by trial.

Assumptions of symmetry and plane strain or plane stress conditions If possible, it is desirable to analyze problems assuming axisymmetric, plane strain or plane stress conditions since all these conditions require only two degrees of freedom per node and the cost of such an analysis is greatly reduced from the three degrees of freedom per node of a three-dimensional analysis. Many problems are accurately modeled by two-degree-of-freedom systems, e.g., a storage tank idealized as axisymmetric about its central axis or a long cantilever wall which obviously fits the plane strain case. Other problems, e.g., the gate bays of lock structures, should be modeled by a three-degree-of-freedom system. There are a large class of intermediate geotechnical problems which are not plane strain cases but which characteristically have a repetitious load pattern, e.g., braced or anchored walls. Traditionally these problems are assumed to be plane strain, and recent finite element analyses of these systems have also employed this assumption[23,28,30,33,37–40,42,43] by analyzing only a vertical section through the wall.

The question of the accuracy of assuming plane strain conditions for anchored walls has been studied by Tsui and Clough,[52] where it was demonstrated that most anchored walls have a tie back spacing that is close enough to justify the assumption of plane strain conditions. An example is shown in Fig. 15-2, where the three-dimensional pressures acting on a 1-ft-thick concrete wall (the stiffness of which is equivalent to heavy sheet piling or a heavy soldier-pile wall) resting against a medium clay and acted upon by tie back loads spaced at 10-ft centers are compared with those assumed in equivalent plane strain conditions. The earth pressure is defined as a dimensionless parameter I_p, which equals $l_0^2\, p/P$, where l_0 is the characteristic length of the slab, p the earth pressure, and P the prestress load. The maximum deviation of the plane strain pressure from the

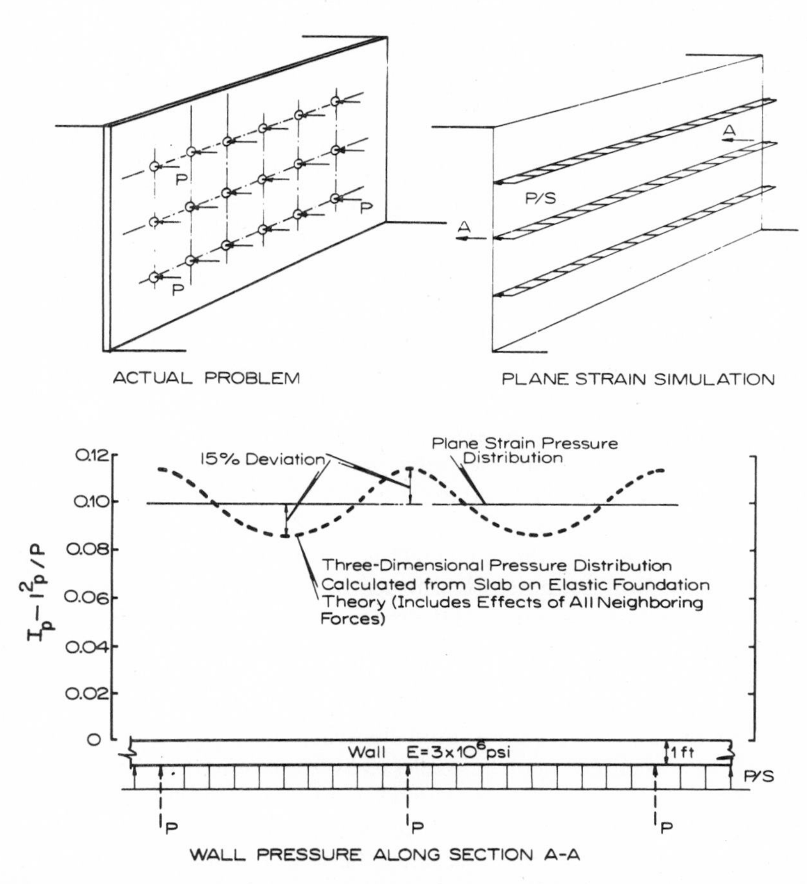

Figure 15-2 Comparison of three-dimensional and plane strain earth pressures behind a tied back slurry wall.

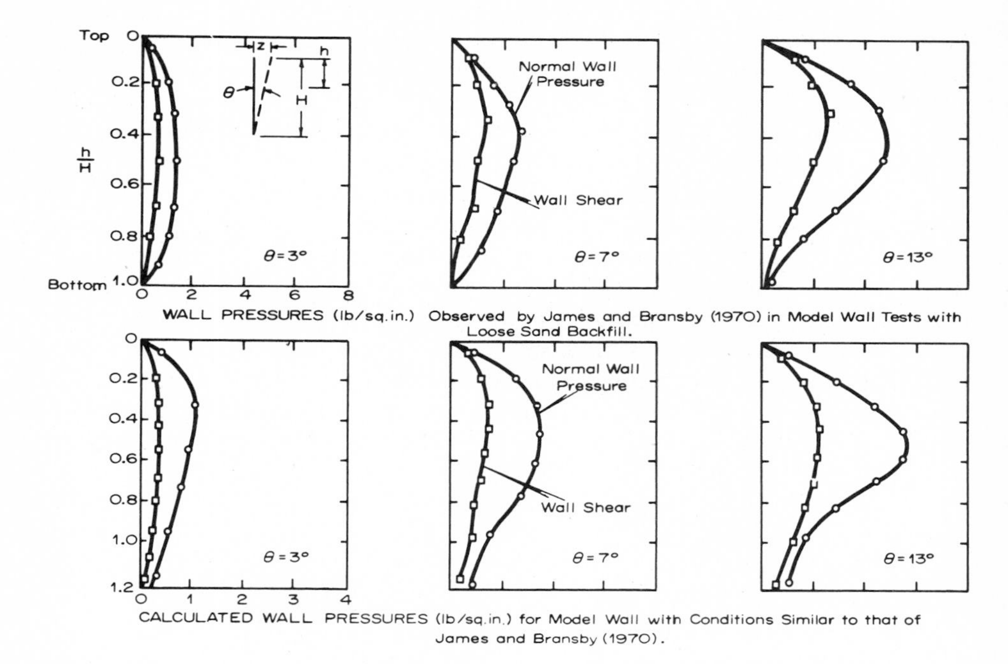

Figure 15-3 Comparisons of observed and calculated interface shear and normal stresses for a model retaining wall.

three-dimensional pressures is only 15 percent. In the general case the deviation of three-dimensional pressures from a plane strain distribution was defined in terms of the soil stiffness, the wall stiffness, and the tie back spacing. The results showed that the stiffer the wall, the closer the tie back spacing, and the softer the soil, the more accurate the assumption of plane strain conditions.

Another important aspect of the representation of braced or anchored walls is how to simulate discontinuous wall elements such as soldier piles, braces, or tie backs in a plane strain analysis. In the study by Tsui and Clough[52] it was shown that the most accurate representation is one in which the bending stiffness of the soldier piles in the wall and the axial stiffners of the braces or tie backs are put on a per foot basis. For example, for a brace its axial stiffness in the plane strain analysis is the actual axial stiffness of the brace divided by the brace spacing. The analysis using this representation yields results characteristic of the average condition between the braces and the midpoint between the braces.

Interface representation In conventional finite element theory, the displacements along the boundary between adjacent finite elements are required to be compatible; i.e., no gaps may open or relative displacements occur between adjacent elements. Relative displacements do occur at the interface between soil and structure and play an important role in the interaction between soil and structure. The nature of the interface behavior depends upon the roughness of the structural material or the angularity of the soil.

The interface behavior can be represented adequately in finite element analyses using the one-dimensional slip element defined in Chap. 4. Models for the behavior of the slip element can be determined from interface (direct shear) tests between the soil and the structure. The element is placed along the boundaries between the two-dimensional elements representing the soil and the structure wherever it is anticipated that relative movements may occur.

The results of an example demonstrating behavior of the one-dimensional element are shown in Fig. 15-3. The shear and normal stresses calculated in the one-dimensional elements between a rigid wall and a loose-sand backfill for various degrees of wall rotation about the wall toe in the passive sense are shown in this figure. The interface elements were assigned properties simulating a rough interface. Also shown in Fig. 15-3 are measured values of shear and normal stresses reported by James and Bransby[46] from tests with conditions similar to those assumed in the analysis. The observed and calculated behavior show good qualitative agreement in relation to the development of normal and shear stresses. Quantitative agreement was not obtained, at least in part, because the tests of James and Bransby [46] were conducted on a small scale model and the scale effects were significant; e.g., the friction angle of the sand was reported as high as 49° at the very low pressures involved in the test.

Construction-Sequence Simulation

As first pointed out by Terzaghi,[6] the sequence of construction can have a significant influence on earth pressures and structural behavior. Embankment fill and backfill settlements have been shown by Clough and Woodward[53] to be calculated rationally only if the analytical technique can simulate the construction sequence. Lambe[54] demonstrated the disastrous effects variations in water levels caused by excavations can have on settlements behind braced excavations. Thus it is essential that a reliable analytical technique be capable of simulating the anticipated construction processes as realistically as possible.

Simulation of construction procedures is usually best accomplished in finite element analyses by dividing the loading sequence into small increments, by analyzing for the effects of each increment in sequence, and by superimposing the results of each increment on the preceding results to obtain the cumulative stress and displacement conditions. If nonlinear material behavior is assumed, it may be necessary to iterate on each increment to ensure convergence.

The effects of many of the construction processes, e.g., excavation, backfill placement, dewatering, and placement and prestressing of braces or anchors, are well defined and have been included explicitly in finite element analyses. Other processes such as excavation of trenches with slurry replacement and tremieing and setup of concrete in slurry filled trenches are not so well defined, but their simulation can be accomplished. Finally, there are processes which are by nature random, and no method of accounting for them can ever be precise, e.g., wedging between braces and wales of braced excavations or overexcavation for tunnel liners and wall lagging.

General techniques for simulation of excavation, backfill placement, and dewatering have been described in Chaps. 1 and 16. Specific applications of these procedures to analysis of earth retaining structures are given in Refs. 26, 36, 37, 40, and 53.

Problems related to modeling of braces and tie backs are discussed by Palmer and Kenney,[38] Wong,[42] Clough et al.,[28] and Clough and Tsui.[39] Generally braces and tie rods can be represented by one-dimensional spring or bar elements. Stiffness of the elements should be defined on a per foot basis, as described earlier in this section. Further consideration can be given to reduction of the brace stiffness if poor wedging procedures are to be employed between the braces and the wales. This problem is discussed by Palmer and Kenney[38] and Wong.[42]

Simulation of the construction of cast-in-place foundation walls in slurry trenches presents some special difficulties, which are not fully resolved. Clough[29] has noted that the lateral pressure exerted by the concrete triemied into the slurry trench can be very high, up to three times that of the original lateral pressure in the ground. Unfortunately it is not clear at present how much the pressures could be reduced by partial setup in the concrete during the tremie or how much the pressures are subsequently relieved by concrete shrinkage. Preliminary indications from field studies[55,56] are that the pressures are relieved to some extent but still remain above those originally in the soil. Future field studies should yield information which will allow for more accurate simulation of the slurry-wall problem.

Unfortunately, regardless of the accuracy of the techniques for simulation of basic construction processes there are many details in construction which possibly cannot be simulated rationally but which can strongly influence the performance of the system. Knowledge of the effects of these details is essential to keep a reasonable perspective about possible perturbations due to parameters not included in the analysis. For example, Lambe[57] has shown that variations in wedging practices on a bracing job resulted in differences of up to 100 percent in measured bracing deflections, and Cording[58] demonstrated variations of surface settlement of 400 percent due to differences in overexcavation practices in shield tunneling.

Modeling for Media

Constitutive models must be established for both the structure and the soil. Of these two materials, the choice of a model for the soil represents a more formidable task; however, because of the significance of structural deflections care must be exercised in selecting suitable parameters to characterize the structural material.

Structural materials The structural material in most soil-structure applications may be assumed to be linear elastic, but in the case of reinforced concrete, this assumption must be carefully considered in the light of the possibility of long-term creep or cracking of the concrete. An analytical representation of creep and cracking in reinforced concrete is still in the research stages, but an empirical approach like that recommended by the ACI Committee Report[59] on deflections in reinforced concrete flexural members can serve as a practical substitute. A modulus reduced in accordance to data obtained from creep tests and past experience is employed to allow for greater structural deflections, while a reduced section modulus is employed to account for cracking. For some problems, the effects of creep and cracking may lead to doubling or tripling of the structural deflections.

Soils The question of constitutive models for soils has been discussed in Chaps. 2 and 3. In finite element analyses of earth retaining structures, elastoplastic,[24,41] linear elastic,[36] and nonlinear elastic models[23,26,31,34,37–39,42–45] have been employed. Reasonable comparisons with observed behavior have been made with most of the models, although nonlinear elastic models remain the most useful for the present because of their accessibility, their proved success in analyses of practical problems, the simple tests needed for the parameters, and the fact that they can be applied to soils in drained and undrained conditions.

However, no analytical soil model can currently cope with all the aspects of soil behavior involved in earth-structure interaction. Thus it is important to examine carefully the stress paths to which the soils will be subjected and to conduct a correspondingly careful laboratory test program which accounts for the actual stress paths. The observed behavior can be compared to the analytical capabilities of the soil model to determine whether the model is adequate.

The stress path in an earth-structure interaction problem is often complex because the loading sequence usually involves excavation, backfilling, dewatering, and possibly concentrated loading by surcharge or prestress loads—all of which may be applied in different sequences. As an example of a relatively simple case the construction of a prestressed anchored wall is shown in Fig. 15-4. In this problem, the soil element immediately behind the wall, element A in Fig. 15-4, is first subjected to an extension type of loading during the initial excavation step. This loading is shown in the upper right-hand part of Fig. 15-4 in the form of a stress-path plot obtained from a finite element analysis. After the first excavation, an anchor is installed, and a prestress load is applied near element A. As seen from the stress path, this loading results in a lowering of the shear-stress level at A since it is compressive, as opposed to the initial extension loading. Subsequently, element A undergoes a series of unload-reload cycles as further excavation and prestress occurs. Element B, on the front side of the wall and adjacent to A, undergoes a different stress path (see lower left side), again with pronounced cycles of behavior under the different types of loading.

In this problem the soil model should account for the differences in soil behavior for loading and unloading response. This behavior rules out the use of a simple nonlinear model and requires instead a model which is capable of shifting from loading to unloading behavior on each loading step. Note also that the use of such a model does not guarantee simulation of the actual behavior. In addition,

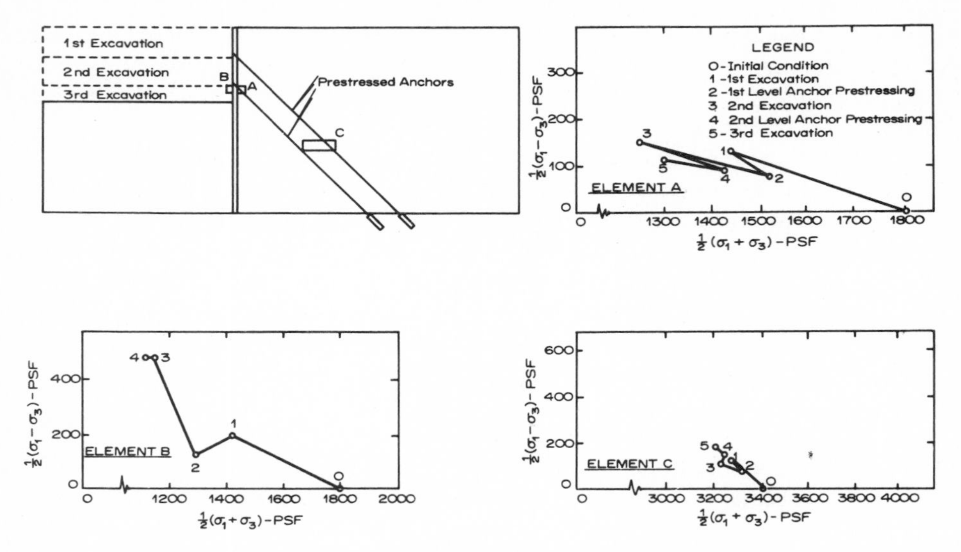

Figure 15-4 Total stress and stress paths for soil elements during construction of a tied back wall.

iteration is required at each stage of loading in order to develop information for deciding whether the state of stress is in a loading or unloading zone.

Because many soil-structure interaction problems involve dewatering effects, a more complex stress path even than that shown for the idealized anchored wall problem is involved. No soil model has been advanced which can consider all the potential effects of these very complex stress paths; and the blind application of even a sophisticated load-unload model can lead to erroneous predictions. The real response of the soil should be ascertained in tests which reasonably simulate some of the important stress paths involved in the problem. Lambe[60] and Ladd[61] have discussed this concept thoroughly, and it has particular relevance for those employing the finite element method.

15-5 ANALYSIS OF A TIED BACK WALL: A CASE STUDY

As documented in Table 15-2, the finite element method has been applied to a relatively wide variety of earth retaining-structure problems. A broad review of these applications can be found in Ref. 25. In this section an application will be described for a particular problem, the case of a tied back wall used to support a 30-ft excavation into soft varved clay in Hartford, Conn. The excavation was completed in 1973.

The finite element analyses were undertaken in the design stage before construction. Of primary interest was the anticipated wall and soil deformations and the degree of mobilized shearing resistance. A detailed description of the studies is given by Murphy, Clough, and Woolworth.[37]

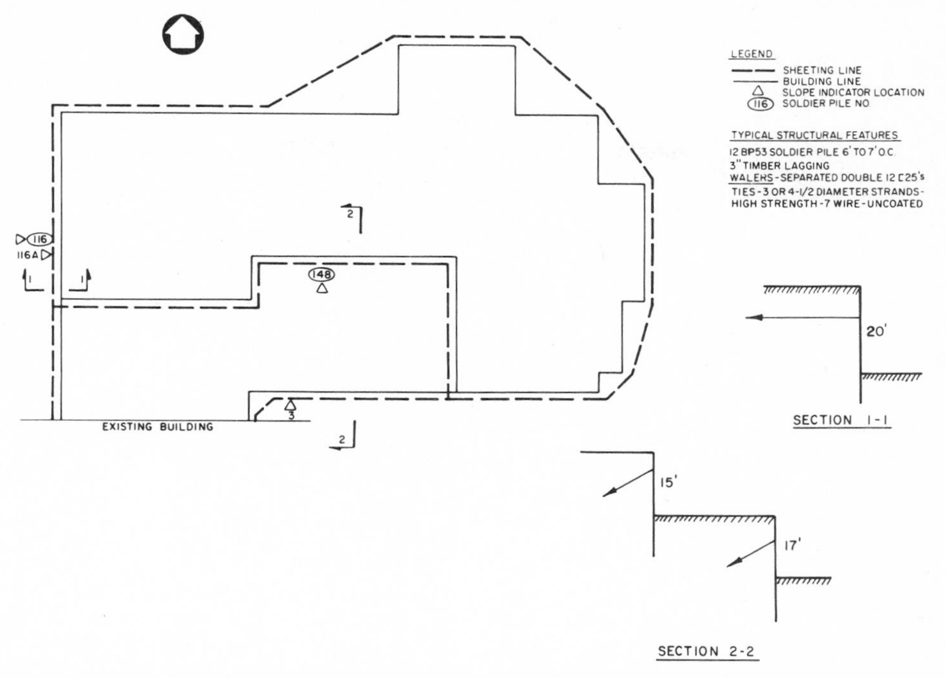

Figure 15-5 Plan view, sections, and structural details of the Hartford tied back wall.[37]

Project Description

The Hartford excavation was supported by a Berlin-type steel H beam and timber lagging wall supported by prestressed tendon anchors. The wall H beams and anchors were set into a very dense till deposit underlying the varved clay. A plan view of and sections through the excavation is shown in Fig. 15-5. The excavation varied from a simple one-tier cut on the north, east, and west sides of the excavation to a two-tiered cut on the south side. The underlying till was exposed in the bottom of the excavation on the east side of the cut, while on the west side some 20 feet of varved clay lay between the excavation bottom and the top of the till.

The geometry and various structural features of the wall were selected by the contractor. The system, described in Fig. 15-5, was designed to withstand a conventional trapezoidal earth pressure distribution, the magnitude of which was computed in terms of an undrained shear strength of 400 lb/ft^2 and a unit weight of 120 lb/ft^2. The actual range of measured laboratory undrained strength values was 400 to 1200 lb/ft^2.

Finite Element Modeling

The finite element analyses were performed assuming plane strain conditions to exist. Two meshes were used in the analyses, one for the one-tier areas of the cut

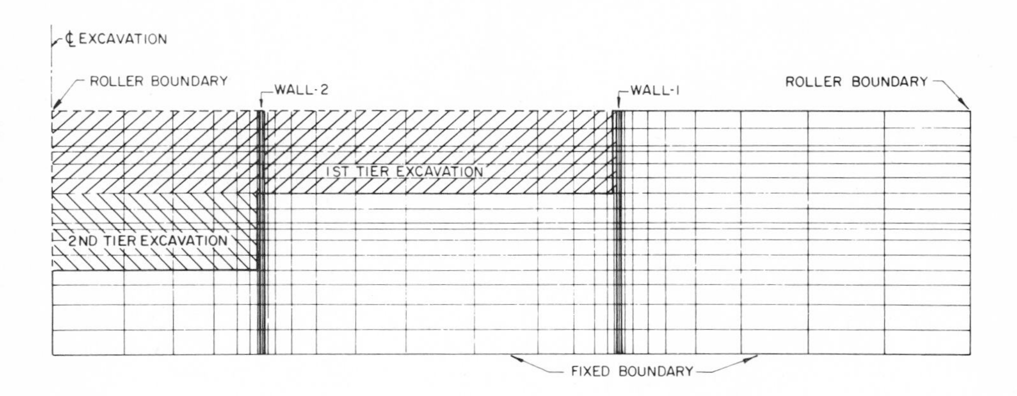

Figure 15-6 Finite element mesh for analysis of two-tiered excavation area, Hartford wall.[37]

and one for the two-tier area; the 523-element mesh for the two-tier area is shown in Fig. 15-6. The mesh allowed for use of a six-step construction sequence, which closely followed that actually used in the field, consisting of alternating steps of excavation and prestressing. The rigid base at the bottom of the mesh was assumed to represent the top of the till layer.

The configuration of the mesh can be seen to be regular with no inclined elements. No attempt was made to model the inclined tie backs and anchors in the mesh schematically since it was not necessary in this problem. The anchors were set into the "rigid" till layer and would be unmoving; the tie backs simply served as spring connections between the wall and the unmoving anchors; thus tie backs were represented by inclined spring supports at the face of the wall which resisted outward movement of the wall. Each spring was installed at the appropriate time in the simulation, immediately after a prestressing force was applied to the wall and before the subsequent excavation step. The axial stiffnesses of the springs were set equal to the actual tie back stiffnesses divided by the tie back spacing.

One-dimensional slip elements were employed on either side of the wall to allow for possible relative movements. Adhesion on the interface was assumed equal to one-half the soil shear strength. Two-dimensional (isoparametric quadrilateral elements[51]) were employed for the wall and soil. The flexural stiffness of the wall elements was set equal to the flexural stiffness of a soldier pile in the actual wall divided by the soldier-pile spacing.

The soil model employed was a nonlinear elastic model described by Clough and Duncan.[27] The model assumes that as long as shear stresses increase, the soil behavior is nonlinear but that upon a decrease in shear stresses the soil behavior is linear elastic. For saturated clays in undrained loading the model requires as input an initial tangent modulus, an unload-reload modulus, and an undrained shear strength.

Soil Material Parameters

Selection of material parameters was necessary for only the varved clay since the till was assumed to be rigid, an assumption later borne out by the observational data. Tests performed on the varved clay consisted of unconfined compression and drained and undrained triaxial and field vane tests. These results were supplemented by more extensive data available on similar varved clays reported by Ladd and Wissa[62] and from other projects.

The test results indicated that the varved clay is a very complex material with a depositionally induced anisotropy and a scatter in material properties caused by variations in sizes of clay and silt varves. It was also apparent from the site that the upper 10 ft of the clay was overconsolidated. Undrained shear strengths in the upper zone were on the order of 1200 lb/ft^2, while below this zone the shear strength dropped to about 400 lb/ft^2.

Under the pressure of time and economics allowable for the design studies it was apparent that consideration of all the complexities of the material behavior at the site would not be feasible. Instead a decision was made to use a simplified but conservative soil representation in the analyses and to monitor the excavation behavior closely to check on the analysis results. Early confirmation could be obtained of the analysis predictions by comparing predicted and observed behavior for the first stages of construction.

For the analyses the soil was assumed to be homogeneous with a constant strength with depth. Because undrained conditions appeared to give more critical behavior than drained, only the undrained was considered. An undrained shear strength of 400 lb/ft^2 was assigned to the varved clay, a value on the lower side of all the strength-test results. The initial tangent and unload-reload modulus values for the nonlinear soil model were assumed to be 600 times the undrained shear strength based primarily on recent experiences in predicting field performance in soft clays.[63,64] Initial stresses were assigned to the soil elements based on an at-rest soil condition with lateral stresses assumed equal to the vertical. The lateral stresses assumed were selected to represent an average condition in the soil since available data would suggest that the lateral stresses in the upper overconsolidated zone should be greater than 1 while in the lower zone they should be less than 1.[63]

Results

The predicted wall and soil deflections and zones of high shear stresses for the two-tiered area are shown in Fig. 15-7. Wall movements on the order of 3 in were predicted, and a sizeable zone in which the shear stress would exceed 0.6 times the shear strength was indicated. The zone of high shear stress is in the form of the critical failure zone that would be predicted by a classical wedge-stability analysis. Similar results were found for the one-tiered cuts.

The data indicated that the system would perform satisfactorily although the movements would be relatively large for such a small excavation. This indicated

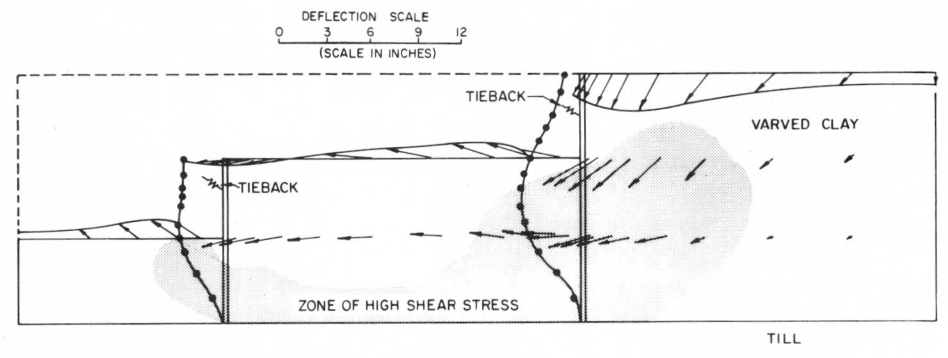

Figure 15-7 Predicted movement patterns and zone of high shear stress, Hartford wall.[37]

that the system should perform satisfactorily particularly in view of the fact that conservative soil parameters had been selected.

Instrumentation consisting of survey control and inclinometers were activated as construction was undertaken. As the instrumentation data became available after the excavation was taken to the first tier level, it became apparent that, as anticipated, the actual behavior of the system would be better than predicted. The analyses had been conservative, and observed deflections were on the order of one-half those predicted. Subsequently, the original analyses were repeated in an effort to improve the predictions for the final stages of construction. In the new analyses the varved clay was assigned a shear strength of 800 lb/ft^2, a value double that originally assumed and a value much more in line with the average shear strength of the deposit. The construction sequence simulated in the

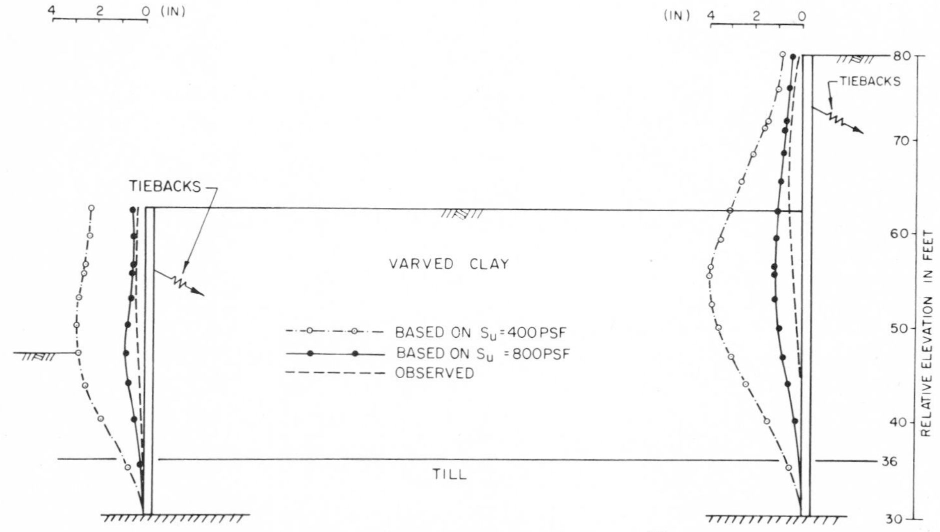

Figure 15-8 Predicted and observed wall movements, Hartford wall.[37]

analyses was also modified to better model that actually used in the field. This primarily involved addition of a surcharge loading behind the wall, which resulted in additional system movements.

Wall deformations from the new analyses are shown in Fig. 15-8 along with those predicted by the original design analyses and those observed with the wall inclinometers. The displacements computed based on an undrained shear strength of 800 lb/ft^2 were quite close to the observed and much closer than those originally computed in the design stages. The design predictions were conservative; however, the results were acceptably accurate to serve as a useful guide in gaging the system acceptability and in establishing a base from which to judge the observed behavior.

REFERENCES

1. Coulomb, C. A.: "Éssai sun une application des règles des maximis et minimis à quelques problèmes de statique," *Mem. Acad. R. Sci.*, vol. 7, Paris, 1776.
2. Rankine, W. J. M.: On the Stability of Loose Earth, *Phil. Trans. R. Soc.*, vol. 147, p. 9, 1857.
3. Kotter, F.: "Die Bestimmung des Druckes an gekrummten Gleitflachen, *Sitzungsb., K. Preuss. Akad. Wiss.*, Berlin, 1903.
4. Reissner, H.: Zum Erddruch Problem, *Sitzungsber. Berlin Math. Geo.*, vol. 23, p. 14, 1924.
5. Westergaard, H. M.: Stresses in Concrete Pavement Computed by Theoretical Analysis, *Public Roads*, vol. 7, no. 2, pp. 25–35, 1926.
6. Terzaghi, K.: A Fundamental Fallacy in Earth Pressure Computations, *J. Boston Soc. Civ. Eng.*, vol. 23, p. 71, 1936.
7. Terzaghi, K.: General Wedge Theory of Earth Pressures, *Trans. ASCE*, vol. 106, 1941, p. 68.
8. Meem, J. C.: The Bracing of Trenches and Tunnels, with Practical Formulas for Earth Pressures, *Trans. ASCE*, vol. 60, pp. 1–23, 1908.
9. Terzaghi, K.: Large Retaining Wall Tests, I: Pressure of Dry Sand, *Eng. News-Rec.*, vol. 3, pp. 136–140, February, 1934.
10. Peck, R. B.: Earth Pressure Measurements in Open Cuts, Chicago Subway, *Trans. ASCE*, vol. 108, pp. 1008–1036, 1943.
11. Janbu, N.: Earth Pressures and Bearing Capacity Calculations by Generalized Procedure of Slices, *Proc. 4th Int. Conf. Soil Mech. Found. Eng., London, 1957*, vol. 2, pp. 207–212.
12. Sokolovski, V. V.: "Statics of Granular Media," Pergamon Press, New York, 1965.
13. Scott, R. F.: "Principles of Soil Mechanics," Addison-Wesley Publishing Company, Inc., Reading, Mass., 1963.
14. Lysmer, J.: Limit Analysis of Plane Problems in Soil Mechanics, *J. Soil Mech. Found. Div. ASCE*, vol. 96, no. SM4, pp. 1311–1334, July 1970.
15. Turabi, D. A., and A. Balla: Distribution of Earth Pressure on Sheet-Pile Walls, *J. Soil. Mech. Found. Div. ASCE*, vol. 94, no. SM6, pp. 1271–1301, November 1968.
16. Haliburton, T. A.: Numerical Analysis of Flexible Retaining Structures, *Proc. J. Soil Mech. Found. Div. ASCE*, proc. pap. 6221, November 1968.
17. Hudson, W. R., and H. Matlock: Cracked Pavement Slabs with Nonuniform Support, *Proc. J. Highw. Div. ASCE*, vol. 93, no. HW1, pp. 19–42, 1967.
18. Matlock, H., and T. A. Haliburton: A Finite Element Method of Solution for Linearally Elastic Beam-Columns, *Univ. Texas Cent. Highw. Res., Res. Rep.* 56-1, Austin, February 1965.
19. James, R. G., and J. A. Lord: An Experimental and Theoretical Study of an Active Earth Pressure Problem Relevant to Braced Cuts, *Proc. 5th Eur. Conf. Soil Mech. Found. Eng., Madrid, 1972*, vol. 1, pp. 29–38.

20. James, R. G., I. A. A. Smith, and P. L. Bransby: The Prediction of Stress and Deformations in a Sand Mass Adjacent to a Retaining Wall, *Proc. 5th Eur. Conf. Soil Mech. Found. Eng., Madrid, 1972*, vol. 1, pp. 39–46.
21. Roscoe, K. H.: The Influence of Strains in So.1 Mechanics, *Geotechnique*, vol. 20, no. 2, pp. 129–170, 1970.
22. Serrano, A. A.: The Method of Associated Field of Stress and Velocity and Its Application to Earth Pressure Problems, *Proc. 5th Eur. Conf. Soil Mech. Found. Eng., Madrid, 1972*, vol. 1, pp. 77–84.
23. Bjerrum, L., C. J. F. Clausen, and J. M. Duncan: Earth Pressures on Flexible Structures: A State-of-the-Art Report, *Proc. 5th Eur. Conf. Soil Mech. Found. Eng., Madrid, 1972*, vol. 2, pp. 169–196.
24. Canizo, L., and C. Sagaseta: Earth Pressure of an Elastoplastic Soil upon a Moving Rigid Wall, *Proc. 5th Eur. Conf. Soil Mech. Found. Eng., Madrid, 1972*, vol. 1, pp. 125–133.
25. Clough, G. W.: Application of the Finite Element Method to Earth-Structure Interaction, *Proc. Conf. Appl. Finite Element Method Geotech. Eng., U.S. Army Eng. Waterw. Expt. Stn., Vicksburg, Miss., May 1972.*
26. Clough, G. W., and J. M. Duncan: Finite Element Analyses of Port Allen and Old River Locks, *U.S. Army Eng. Waterw. Expt. Stn. Contract Rep.* S-69-6, Vicksburg, Miss., September 1969.
27. Clough, G. W., and J. M. Duncan: Finite Element Analyses of Retaining Wall Behavior, *J. Soil Mech. Found. Div. ASCE*, vol. 97, no. SM12, December 1971.
28. Clough, G. W., P. R. Weber, and J. Lamont: Design and Observation of a Tied-back Wall, *Proc. Spec. Conf. Perform. Earth Earth-Supported Struct., Purdue Univ., 1972*, vol. 1, pt. 2, June, pp. 1367–1390.
29. Clough, G. W.: Analytical Problems in Modeling Slurry Wall Construction, *U.S. Dept. Trans. Res. Rep.*, September 1973.
30. Cole, K. W., and J. B. Burland: Observation of Retaining Wall Movements Associated with Large Excavation, *Proc. 5th Eur. Conf. Soil Mech. Found. Eng., Madrid, 1972*, vol. 1, pp. 445–454.
31. Desai, C. S., L. D. Johnson, and C. M. Hargett: Analysis of Pile-Supported Gravity Lock, *J. Geotech. Eng. Div., ASCE*, vol. 100, no. GT9, September 1974.
32. Duncan, J. M., and G. W. Clough: Finite Element Analyses of Port Allen Lock, *J. Soil Mech. Found. Div. ASCE*, vol. 97, no. SM8, pp. 1053–1068, August 1971.
33. Egger, P.: Influence of Wall Stiffness and Anchor Prestressing on Earth Pressure Distributions, *Proc. 5th Eur. Conf. Soil Mech. Found. Eng., Madrid, 1972*, vol. 1, pp. 259–264.
34. Girijavallabhan, C. V., and L. C. Reese: Finite Element Method for Problems in Soil Mechanics, *J. Soil Mech. Found. Div. ASCE*, vol. 94, no. SM2, pp. 473–496, March 1968.
35. Moore, H. E.: Finite Element Analyses of the Earth Pressures against a Bridge Pier, *Proc. 9th Eng. Geol. Soils Eng. Symp., Boise, Idaho, April 1971.*
36. Morgenstern, N. R., and Z. Eisenstein: Methods of Estimating Lateral Loads and Deformations, *Proc. 1970 Spec. Conf. Lateral Stresses Ground Des. Earth-Retaining Struct., Cornell Univ., June 1970*, pp. 51–102.
37. Murphy, D. J., G. W. Clough, and R. S. Woolworth: Temporary Excavations in Varved Clay, *J. Geotech. Div. ASCE*, vol. 101, March 1975.
38. Palmer, J. H. L., and T. C. Kenney: Analytical Study of a Braced Excavation in Weak Clay, *Can. Geotech. J.*, vol. 9, pp. 145–164, May 1972.
39. Clough, G. W. and Y. Tsui: Performance of Tied-back Walls in Clay, *J. Geotech. Div. ASCE*, vol. 100, no. GT12, pp. 1259–1274, December 1974.
40. Ward, W. H.: Remarks on Performance of Braced Excavations in London Clay, *Proc. Conf. Perform. Earth Earth-Supported Struct., Purdue Univ., 1972*, vol. 3.
41. Simpson, B., and C. P. Wroth: Finite Element Computations for a Model Retaining Wall in Sand, *Proc. 5th Eur. Conf. Soil Mech. Found. Eng., Madrid, 1972*, vol. 1, pp. 85–94.
42. Wong, I. H.: Analysis of Braced Excavations, D.Sc. thesis, Massachusetts Institute of Technology, Cambridge, Mass., 1971.
43. Stroh, D.: Berechnung tiefer Baugruben nach der finite Element Methode und Vergleich mit Messungen, *Univ. Darmstadt, Inst. Bodenmech. Grundbau Res. Rep.*, 1973.
44. Clough, G. W.: Deep Excavations and Retaining Structures, *Conf. Found. Tall Build., Lehigh Univ., August 1975.*

45. Denby, G. M.: Temporary Berms for Minimizing Settlements behind Braced Excavations, M.S. thesis, Duke University, Durham, N.C., 1975.
46. James, R. G., and P. L. Bransby: Experimental and Theoretical Investigations of a Passive Earth Pressure Problem, *Geotechnique*, vol. 20, no. 1, pp. 17–37, 1970.
47. Gleser, S. M.: Lateral Load Tests on Vertical Fixed Head and Free-Head Piles, *Proc. ASTM Tech. Publ.* 154, 1954.
48. Palmer, L. A., and J. B. Thompson: The Earth Pressure and Deflection along the Embedded Lengths of Piles Subjected to Lateral Thrust, *Proc. 2d Int. Conf. Soil Mech. Found. Eng.*, vol. 5, pp. 156–161, 1948.
49. Sowers, G. B., and G. F. Sowers: Failures of Bulkhead and Excavation Bracing, *Civ. Eng.*, vol. 37, no. 1, pp. 72–77, 1967.
50. Felippa, C. A.: Refined Finite Element Analysis of Linear and Non-linear Two-dimensional Structures, *Univ. California*, SESM Rep. 66-22, Berkeley, 1966.
51. Doherty, W. P., E. L. Wilson, and R. L. Taylor: Stress Analysis of Axisymmetric Solids Utilizing Higher-Order Quadrilateral Finite Elements, *Univ. California Struc. Eng. Lab. Rep.* SESM 69-3, Berkeley, January 1969.
52. Tsui, Y., and G. W. Clough: Plane Strain Approximations in Finite Element Analyses of Temporary Walls, *Proc. Bi-annu. ASCE Geotech. Conf., Austin, Tex., June, 1974.*
53. Clough, R. W., and R. J. Woodward: Analysis of Embankment Stress and Deformations, *J. Soil Mech. Found. Div. ASCE*, vol. 93, no. SM4, proc. pap. 5329, pp. 529–549, July 1967.
54. Lambe, T. W.: Braced Excavations, *Prepr. Proc. 1970 Spec. Conf. Lateral Stresses Ground Des. Earth-Retaining Struct., Cornell Univ., June 1970*, pp. 149–218.
55. DiBiagio, E., and F. Myrvoll: Full Scale Field Tests of a Slurry Trench Excavation in Soft Clay, *Proc. 5th Eur. Conf. Soil Mech. Found. Eng.*, vol. 1, pp. 461–472.
56. DiBiagio, E., and J. A. Roti: Earth Pressure Measurements on a Braced Slurry-Trench Wall in Soft Clay, *Proc. 5th Eur. Conf. Soil Mech. Found. Eng., Madrid, 1972*, vol. 1, pp. 473–484.
57. Lambe, T. W.: *Discuss. Biannu. ASCE Conf. Perform. Earth Struct., Purdue Univ., June 1972.*
58. Cording, E. J.: Movements around a Soil Tunnel on the Washington Metro, *Tunneling Technol. Newsl.*, no. 1, April 1973.
59. American Concrete Institute Committee 435: Deflections of Reinforced Concrete Flexural Members, *Proc. Am. Concr. Inst.*, vol. 63, pp. 637–667, June 1966.
60. Lambe, T. W.: Methods of Estimating Settlement, *Proc. ASCE Settlement Conf., Northwestern Univ., June 1964.*
61. Ladd, C. C.: Stress-Strain Modulus of Clay from Undrained Triaxial Tests, *J. Soil Mech. Found. Div. ASCE*, vol. 90, no. SM5, September 1964.
62. Ladd, C. C., and E. Z. Wissa: Geology and Engineering Properties of Connecticut Valley Varved Clays with Special Reference to Embankment Construction, *MIT Res. Rep.* R70-56, September 1970.
63. Ladd, C. C.: Test Embankment on Sensitive Clay, *Proc. ASCE Spec. Conf. Perfor. Earth Earth-supported Struct.*, vol. 1, pt. 1, *Purdue Univ., 1972.*
64. D'Appolonia, D. J., and T. W. Lambe: Method for Predicting Initial Settlement, *J. Soil Mech. Found. Div. ASCE*, vol. 96, no. SM2, March 1970.

CHAPTER

SIXTEEN

EMBANKMENTS AND EXCAVATIONS

Fred H. Kulhawy

16-1 INTRODUCTION

Before the 1950s, engineers had to rely upon highly idealized elasticity and plasticity solutions for an analysis of embankment and excavation problems. Many of these solutions have been discussed by Sherard et al.[1] who noted their inadequacy for analysis or design because they are unable to model the field geometry, loading conditions, and material behavior effectively. However, for simplified field cases, solutions like those compiled by Poulos and Davis[2] may be useful for preliminary evaluation.

It was not until 1952 that a realistic beginning was made into the evaluation of the stresses within embankments. Bishop[3] then conducted a relaxation solution of the Airy stress function for the case of a homogeneous symmetrical triangular elastic embankment and foundation with 3:1 side slopes. This solution provided a great deal of insight into the internal stresses within an embankment-foundation system, but it was realized that these and related techniques were not sufficiently flexible to accommodate complex geometries, loading conditions, and material behavior.

The development of the finite element method (Chap. 1) provided a suitable alternative which was flexible enough to accommodate a wide range of geotechnical problems. Following its formalization in the early 1960s, applications of this method were initiated for embankment and excavation studies, which can be divided into four main stages as follows:

1. *Construction stage* During this stage the embankment is constructed or the excavation is opened.
2. *Adjustment stage* The final geometry has been reached, but excess pore pressure dissipation and consolidation are occurring, the reservoir loads are being imposed for a dam embankment, and steady-state seepage patterns are developing.
3. *Long-term stage* The final geometry and loading patterns are established, but variations in loading and time-dependent variations of material properties are occurring.
4. *Transient stage* At any time during the above three stages an embankment or excavation may be subject to earthquake, blast, seepage, or other transient loading.

In the following sections an attempt will be made to summarize the techniques presently available for modeling embankment and excavation behavior during these stages and the results obtained to date. Since transient phenomena such as earthquakes are discussed in Chaps. 19 and 20 and supported excavations are discussed in Chap. 15, neither of these topics will be covered here.

16-2 PHYSICAL MODELING

Because the finite element method itself is an approximate representation of the true physical system, considerable care must be exercised in the establishment of the analytical model to ensure that no significant errors are introduced. Details of these requirements are discussed in the following sections.

Element Type

Of the great variety of elements available it has been shown that elements with linear strain capabilities are perhaps the most appropriate for problems such as embankments and excavations. The higher order elements may not be necessary and the lower order (constraint strain) element does not provide sufficient accuracy in any but very simple problems. For embankment and excavation problems, elements of arbitrary quadrilateral shape are found to be most appropriate because of the efficiency of modeling placement or excavation in essentially horizontal layers.

Discretization of Embankments and Excavations

Discretization of a system such as an embankment on a "rigid" foundation is straightforward because the boundaries are defined, but in problems involving "infinite" media, e.g., embankment foundations and excavations in deep geologic masses, finite boundaries must be established. It is helpful if a relatively hard

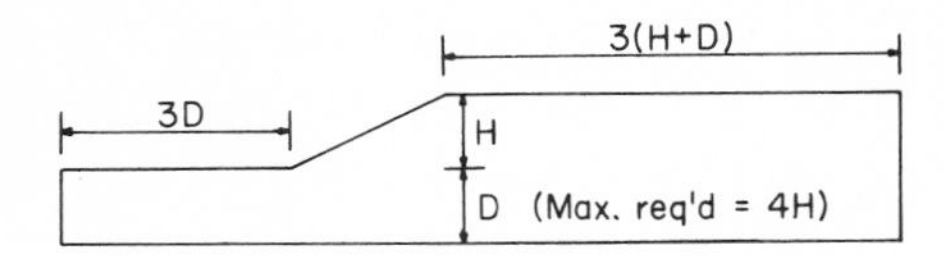

A. Excavation (Adapted from Dunlop and Duncan, 4; Dunlop, et. al., 5; Constantopoulos, et. al., 6)

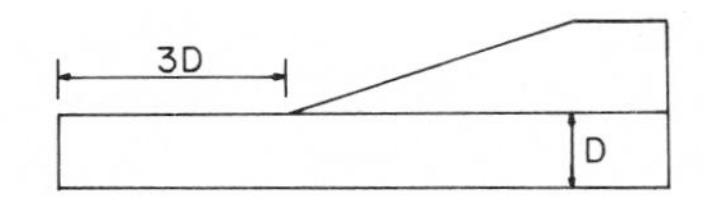

B. Embankment and Foundation (Estimated from above references)

Figure 16-1 Discretization of excavations and embankments.

material is encountered at shallow depth, in which case the boundary for the finite element mesh is established at the interface of the relatively hard material. It has been the writer's experience that a relatively hard material can be defined as one which has a modulus equal to 500 to 1000 times that of the overlying material.

When a relatively hard material is not present, it is necessary to establish finite boundaries within which the significant influence of the construction operations occurs. Figure 16-1 summarizes the results of several studies to determine the minimum extents required to model infinite media. For excavations, the boundaries depend upon the slope height and the depth to a firm stratum if it is relatively shallow. For embankments, no thorough boundary studies appear to have been conducted, but from excavation boundary studies the dimensions given in Fig. 16-1*b* appear to be reasonable; however, they are subject to change.

Simulation of Construction Operations

Embankments and excavations are constructed sequentially, and for realistic solutions to these problems the construction sequences should be simulated as carefully as possible. Also, it has been demonstrated analytically[7,8] that the final computed results are different for a solution based upon sequential construction and one based upon single stage construction. The only exception is for an excavation in homogeneous linearly elastic material; in this case the final stresses and displacements will be the same regardless of the number of excavation steps. It should also be noted that the final stresses in a homogeneous linear elastic embankment do not differ appreciably with number of placement lifts but the final displacements are greatly affected.

Figure 1-22 shows schematically the processes of excavation and embankment placement which include (1) establishing the initial stresses in the geologic mass and (2) incrementally adding or removing layers of material. At each increment the computed displacements, strains, and stresses are summed to provide a complete construction history. Details of these procedures are given below.

Initial stresses The initial stresses $\{\sigma_0\}$ are introduced into the discretized system in either of two ways. First a finite element analysis could be conducted, in which gravity loading is applied to all elements to establish the initial state of stress, after which excavation or placement would begin. The second approach, by far the most commonly used with horizontal ground surfaces, is to compute the initial stresses directly as a function of depth below ground surface. In other words

$$\sigma_y \approx \gamma H \qquad \sigma_x \approx K_0 \sigma_y$$
$$\tau_{xy} \approx 0 \qquad \text{for horizontal ground surfaces} \tag{16-1}$$

where σ_y = initial vertical stress
σ_x = initial horizontal stress
τ_{xy} = initial shear stress
γ = unit weight of soil or rock
H = depth below ground surface for point in question
K_0 = lateral earth pressure coefficient

The values of γ and K_0 are introduced in either a total or effective stress capacity, depending upon whether a total or effective stress analysis is to be conducted.

Excavation The excavation surface shown in Fig. 1-22*b* is considered to be a stress free surface. This condition is satisfied by applying equivalent nodal point forces which are equal but opposite in direction to the existing stresses along the surface at a given stage of excavation. At the same time the moduli of the "removed" elements are reduced to very small values.

Numerous procedures have been proposed for implementing these operations.[5,8,10–15] For quadrilateral linear strain elements the generalized procedure developed by Clough and Duncan,[12] with minor modifications,[16] is effective. In this approach, the nodal point forces to be applied along the excavation surface are computed from the nodal point stresses, which are interpolated from the center stresses of the adjacent elements.

The basic interpolation formula is

$$\sigma = a_1 + a_2 x + a_3 y + a_4 xy \tag{16-2}$$

where σ = (known) element stress
x, y = coordinates at which stress is known
$a_{1,2,3,4}$ = interpolation coefficients

For the four elements surrounding a given nodal point and the three stresses in each element, Eq. (16-2) becomes

$$\{\sigma_e\} = [m]\{a\} \tag{16-3}$$

where $\{\sigma_e\}$ = known element stress vector
$[m]$ = known stress coordinate matrix
$\{a\}$ = unknown interpolation coefficient vector

The nodal point stresses $\{\sigma_n\}$ of the element to be excavated are then

$$\{\sigma_n\} = [n]\{a\} = [n][m]^{-1}\{\sigma_e\} \tag{16-4}$$

in which $[n]$ is the known coordinate matrix for the I, J, K, L nodal points, Fig. 16-2.

Using the principle of virtual work and the linear boundary stress distribution shown in Fig. 16-2, the equivalent horizontal and vertical nodal point forces can be established for each nodal point. For example, the vertical force at J is

$$F_y^J = \tfrac{1}{6}[(XJI)\sigma_{y_I} + 2(XJI + XKJ)\sigma_{y_J} + (XKJ)\sigma_{y_K} + (YIJ)\tau_{xy_I} + 2(YIJ + YJK)\tau_{xy_J} + (YJK)\tau_{xy_K}] \tag{16-5}$$

This operation is repeated for all eight nodal point forces, resulting in

$$\{F_n\} = [H]\{\sigma_n\} = [H][n][m]^{-1}\{\sigma_e\} = [Q]\{\sigma_e\} \tag{16-6}$$

where $\{F_n\}$ = 8 × 1 nodal force vector
$[H]$ = 8 × 12 boundary geometry matrix
$\{\sigma_n\}$ = 12 × 1 nodal stress vector
$[Q]$ = 8 × 12 resultant matrix relating unknown nodal forces and known element center stresses

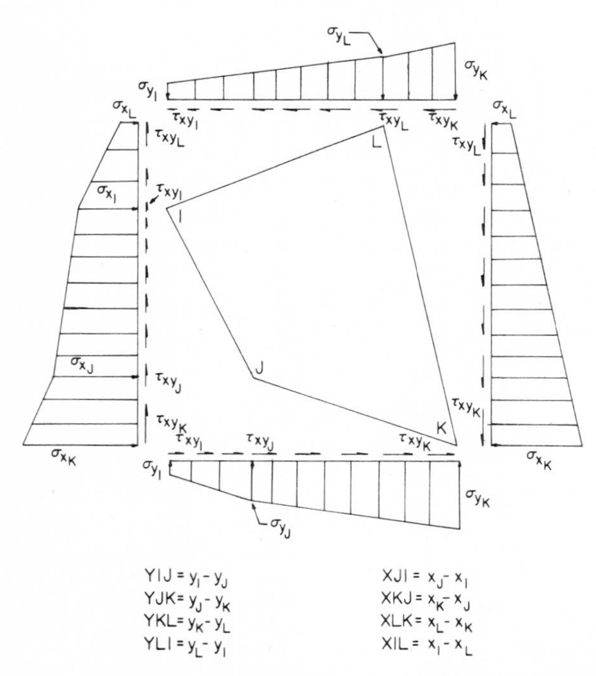

Figure 16-2 Quadrilateral element and boundary stress distribution.[12]

This approach is implemented by specifying the four elements surrounding each nodal point along the excavation boundary. Usually two of these elements are to be excavated and the other two remain as a portion of the geologic mass. A general procedure for computing forces at the excavated surface is described in Chap. 1.

Placement The embankment placement procedure is outlined in Fig. 1.22*a*, which shows that the final foundation stresses will result from the initial foundation stresses plus the effect of successive placement of all of the embankment layers, but that the stresses within any layer will be from the effect of placement of the layer itself plus the placement of subsequent layers only.

Two main procedures have been proposed for implementing these placement operations. The first, known as the *dense liquid procedure*, was proposed by King[17] and has been used effectively with subsequent modifications by several authors.[18–21] In this procedure, the effect of an element being placed is evaluated by multiplying the unit weight and volume of the element to give the total applied force. This force is then divided equally between the element nodal points and applied to the system. During this application, the modulus of the newly placed elements is reduced, usually by a factor of 1000, to simulate placement of loose material with weight but no stiffness. The modulus is then returned to its normal value before placement of a subsequent layer.

In the second approach, known as the *surface-pressure procedure*,[22–24] placement of a new layer is simulated by applying a pressure, converted to equivalent nodal forces, to the surface of the previously placed layer with a magnitude equal to the overburden pressure at the base of the new layer. Lefebvre and Duncan[22] noted that the results from the two procedures were virtually identical.

With both procedures, the stresses in a newly placed layer are assigned values of vertical stress commensurate with the overburden pressure at the element centroid and of horizontal stress equal to the vertical stress multiplied by $\nu/(1 - \nu)$, in which ν is Poisson's ratio. In addition, the displacements at the top of a newly placed layer are set equal to zero. The advantages of these modifications and the subsequent improved accuracy of the computed results are discussed in detail by Kulhawy et al.[20]

Required number of layers Excavation or fill placement is commonly accomplished by successive removal or placement of thin layers of soil or rock. For simulation accuracy, the actual construction sequence should be followed as closely as possible, but, as in the case with large earthworks involving possibly hundreds of layers, this becomes impractical from computer storage and cost standpoints. It is therefore necessary to establish the minimum number of layers required to provide optimum accuracy and economy. As discussed previously, the computed stresses are relatively insensitive to number of layers employed, but the computed displacements are quite sensitive to the number of layers; therefore the displacements can be used to establish criteria.

Clough and Duncan[12] and Kulhawy et al.[20] established these criteria by using a one-dimensional laterally restrained column built up of, or excavated in, finite layers and compared the computed results with a closed-form solution for an infinite number of layers of infinitesimal thickness. In addition they assumed that the modulus of the column material was stress dependent in the form

$$E = Kp_a\left(\frac{\sigma_3}{p_a}\right)^n \tag{16-7}$$

where E = modulus
K = modulus number
σ_3 = confining pressure
n = exponent expressing stress dependency
p_a = atmospheric pressure in consistent stress units

Since the modulus is dependent upon the confining pressure, another variable is introduced; this is the state of construction at which the modulus is evaluated. Three cases are possible: (1) the initial stresses before layer placement or excavation, (2) the final stresses after layer placement or excavation, and (3) the average of the past and present stresses. For both the final and average stresses, one cycle of iteration will be necessary for each layer placement, so that the final stresses will be known for evaluation of the modulus. The studies mentioned previously showed that the average stress approach is more accurate than the final-stress approach, and since both require twice as much computer time as the initial-stress approach, only the comparison between the initial- and average stress solutions will be considered.

Figure 16-3 shows the displacement errors for placement as a percentage difference between the finite and infinitesimal layer solutions. It can be seen that for a given value of n and a selected allowable error in the displacements, the

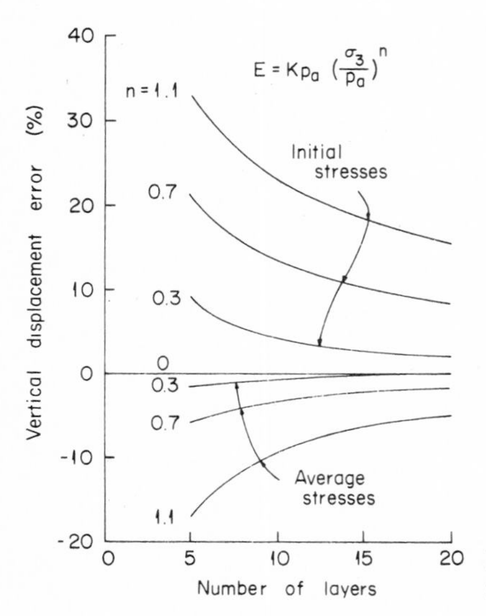

Figure 16-3 Error in simulation of placement for column model with stress-dependent modulus.[12,20]

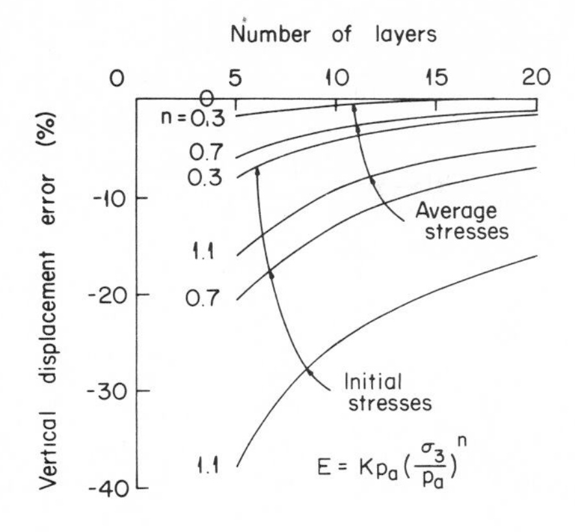

Figure 16-4 Error in simulation of excavation for column model with stress-dependent modulus.[12]

initial-stress solution will overestimate the displacements and require substantially more layers than the average stress solution, which will underestimate the displacements. Even though the solution time per layer is twice as large for the average stress solution, the use of fewer layers indicates that the average stress approach is a more efficient method of simulating fill placement for nearly all cases. Similar conclusions were recently reached by others.[25] Figure 16-4 shows the displacement errors for excavation, from which it can be seen that (except that both types of solutions underestimate the displacements) the same conclusions can be reached as for the placement simulation.

It must be noted, however, that the modulus value of soil and rock not only increases with increasing confining pressure, as assumed in these column models, but also decreases with increasing stress level. Since the stress level usually is increased during placement or excavation, it can be expected that the actual numerical inaccuracy for the displacements will be no more than that shown in Figs. 16-3 and 16-4 and possibly less. Although these figures are based upon a somewhat simplified representation of all the variables involved, it is felt that these curves provide a simple and useful guideline for determining the number of layers required for embankment or excavation analyses.

Discontinuities and Interfaces

Natural discontinuities such as joints, faults, etc., occur in virtually all natural rock masses and in many overconsolidated soil deposits. Similarly, structural interfaces can occur in embankment or excavation problems where dissimilar materials are placed adjacent to each other, e.g., adjacent zones in a zoned embankment, an embankment and its foundation, or a backfill and retaining wall system. Conventional finite elements cannot model these features adequately because the elements are positively joined at the nodal points and subsequently prevent the relative movement between adjacent materials. A suitable analytical model must be able to allow this relative movement for realistic solutions to be obtained.

Descriptions and uses of two- and three-dimensional interface elements are given in Chaps. 4 and 7. Applications to embankment and excavation problems including nonlinearity and stress dependency are given in Refs. 12, 16, 26–28. This type of element has proved to be an effective approach for analyzing the effect of discontinuities and interfaces and can be used in situations where relative movements are possible.

Two-dimensional vs. Three-dimensional Analyses

Commonly finite element analyses of embankments and excavations are conducted as two-dimensional representations of the true three-dimensional problem, but very little work has been done in determining the relative applicability of these solutions until recently, when three-dimensional programs became available.[29] For embankments and excavations which are very long, plane strain analyses of the transverse sections would be appropriate, except where discontinuities or other geological features intersect the transverse section obliquely. If this occurs, a plane strain analysis could only provide a very rough approximation because the problem is truly three-dimensional and should be analyzed as such.

For dam embankments within constrained valley walls, the behavior is particularly complex because there are many modes of potential load transfer which can occur and affect the results.[30] Limited analyses of homogeneous linear elastic dams in V-shaped valleys with 1:1, 3:1, and 6:1 slopes have recently been conducted to investigate these phenomena.[22,23] The main results of this study indicate that (1) plane strain analyses of the maximum transverse section are reasonably accurate with valley walls at slopes of 3:1 or flatter but are not very accurate with valley walls at slopes of 1:1, (2) plane stress analyses of the maximum longitudinal section do not provide accurate results, and (3) plane strain analyses of the maximum longitudinal section provide reasonably accurate results for all valley wall slopes investigated. Similar conclusions have also been reached for a symmetrical central-core dam.[24,31]

Finite Element Mesh Design

The final design of the finite element mesh is arrived at as a compromise, in which the user attempts to get the most accurate results for the least time and effort. The first step is to establish the initial and final external geometry of the slope or embankment to be analyzed, taking into account the boundary requirements and the possible necessity of a three-dimensional analysis. The internal geometry, e.g., zoning, layering, discontinuities, and interfaces, is then introduced, and planes of symmetry are sought. It is helpful if planes of symmetry are present because then it is not necessary to analyze the entire system. Once the required geometry is established, the minimum number of layers to be included in the embankment or excavation is determined from the procedures discussed previously.

At this point the system is ready to be divided into finite elements, but it must be remembered that in general the smaller the element, the more accurate and

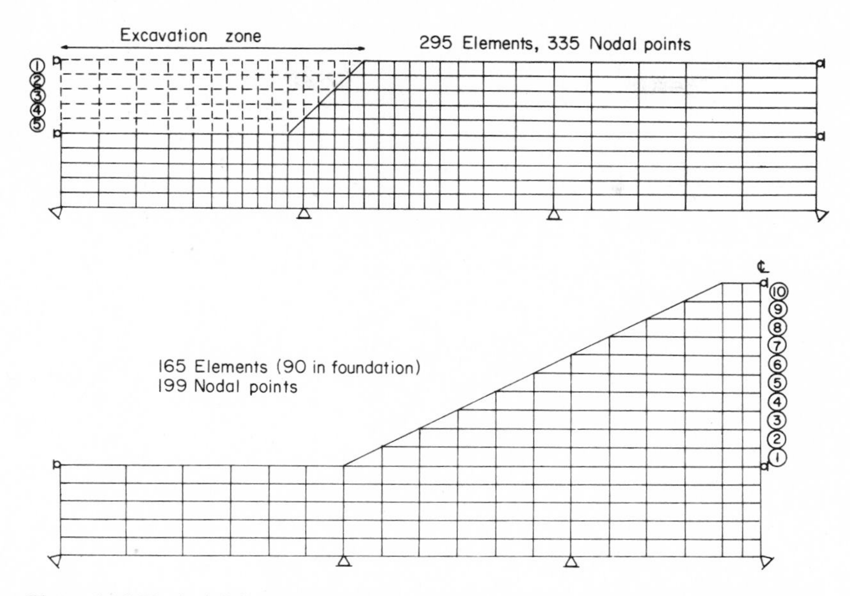

Figure 16-5 Typical finite element meshes for a slope excavation and an embankment.

expensive the solution. However, it is necessary to use smaller elements in regions of stress concentration, away from which the elements can be made progressively larger without significant sacrifice of accuracy. There are no definitive criteria with regard to this increasing spacing and subsequent element sizes, but the writer and others[5] have found that a gradual increase in element size away from the main zones of interest is suitable. In addition, it was found that for embankment and excavation problems, element aspect ratios (base to height) up to about 5 do not seriously influence the results and that aspect ratios of about 10 away from the main region of interest would be allowable.

Figure 16-5 shows typical finite element meshes for a simple excavation and for a simple embankment and foundation. For the excavation example it can be seen that the number of layers and slope angle control the element size at the slope but with increasing distance from the slope the element size is increased. Quadrilateral elements are used throughout the mesh even though it appears that triangular elements are used at the slope face. In reality these are quadrilaterals with the fourth nodal point at the midpoint along the slope. The above comments are also true for the embankment, but it can be seen that with a flatter slope angle elements with a larger aspect ratio are used. In both these cases, if relatively hard material were not close to the ground surface, as implied in this figure, the vertical spacing of elements with depth could increase.

Figure 16-6 shows an example finite element mesh for a central core dam in a V-shaped valley with rigid abutments; for this case two planes of symmetry were present so that only one-fourth of the dam had to be analyzed. Otherwise all other criteria as discussed above are applicable as extended to the third dimension.

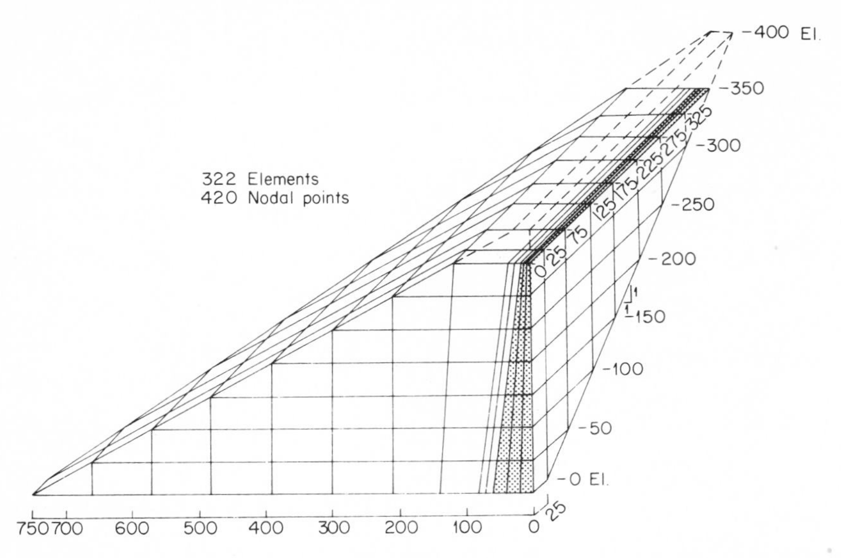

Figure 16-6 Finite element mesh for a three-dimensional analysis of a symmetrical center, core dam.[24,31]

16-3 MODELING OF MATERIAL BEHAVIOR

Stress–Strain–Volume Change Behavior

The development of approaches for modeling the stress–strain–volume change behavior of soil and rock materials has been and will continue to be the subject of much research and controversy for years to come. A number of models have been proposed including linear, bilinear and nonlinear elastic, elastoplastic, and visco-plastic models. Various other empirical and semiempirical models based upon either secant or tangent methods of evaluating the required stress-strain parameters have also been used. These theories for stress-strain behavior of soil and rock materials are discussed in detail in Chaps. 2 and 3, and the theories for discontinuities are discussed in Chap. 4. The pertinent aspects of these approaches for embankment and excavation analyses are discussed below.

Excavation and placement involve incremental changes in stress during construction. In addition the behavior of soil and rock materials, as well as discontinuities and interfaces, is both nonlinear and stress dependent. These two facts require a model which can accommodate these forms of behavior, but most of the approaches presently available presume an idealized model which does not take into account all aspects of the observed behavior. To date (1974) the empirical and semiempirical proposals[18,20,32–41] have been most successful in representing the actual behavior. Of these proposals, the commonly used is the hyperbolic representation for the tangent modulus, as proposed by Duncan and Chang,[34] and for the tangent Poisson's ratio, as proposed by Kulhawy et al.;[20]

both are based upon the earlier representations proposed by Kondner for non-linearity[42] and Janbu for stress dependency.[43]

These hyperbolic tangent relationships, while popular and effective at the present time, have certain drawbacks because they do not have a sound theoretical base and they cannot presently accommodate failure and/or local yielding in a straightforward manner. A number of recent approaches may prove to be quite useful for strain-softening behavior (Chap. 2), but as yet they have not been widely used. This topic of postfailure behavior is widely debated at present and needs careful consideration.

Consolidation, Secondary Compression, and Creep

During and after construction, soil and rock materials exhibit time-dependent deformations under constant total stress; these deformations develop because of consolidation, secondary compression, and creep. A number of procedures for incorporating these phenomena in finite element analyses have been proposed (Chaps. 2 and 12) but only limited work has been done to apply these procedures to embankment and excavation problems. It should be noted, however, that these phenomena introduce a time dependency to the deformations which may play an important role in the overall evaluation of embankments and excavations.

16-4 DESIGN USE

The use of finite element methods in the analysis of numerous types of embankment and excavation problems has been extensive in recent years, and a paper by Duncan[44] has summarized many of them. The following sections attempt to show the areas in which solutions or numerical approaches are available and summarize some of the key points which have been established from these analyses.

End of Construction Analyses

The behavior of embankments and excavations during and up to the end of construction has been the subject of greatest study. All the solutions discussed below have assumed total stress conditions during which fine-grained materials behave in an undrained manner while coarse-grained materials behave in a drained manner.

Embankments Among the generalized studies conducted are the following: two-dimensional analyses of homogeneous embankments or fills on rigid foundations[18,20,45] or on deformable foundations;[5,18,46–51] two-dimensional analyses of zoned embankments on rigid foundations;[20,52] and three-dimensional analyses of homogeneous embankments[22,23] or zoned embankments[24,31] on rigid foundations. From these studies the following trends have been noted:

1. The stresses vary in direct proportion to the unit weight and height but are not affected by the modulus.
2. The displacements vary in direct proportion to the unit weight and the square of the height and vary inversely with the modulus.
3. Poisson's ratio effects are more complicated, but in general a decrease in Poisson's ratio causes decreases in the horizontal and vertical stresses and the horizontal displacements, increases in the vertical displacement, and essentially no change in the shear stresses.
4. The effect of side slope changes are also complicated, but in general an increase in the side slope causes decreases in the horizontal and vertical stresses and increases in the shear stresses and in the horizontal and vertical displacements.
5. The vertical stresses at the base of a dam are not affected by foundation flexibility, but both the horizontal and shear stresses decrease with increasing flexibility. Both the horizontal and vertical displacements at the base of the dam increase with increasing flexibility.
6. In zoned embankments, increasing difference in properties between adjacent zones leads to greater load transfer and susceptibility to local failure and cracking.
7. Three-dimensional effects are not very pronounced at valley-wall slopes of 3:1 or flatter but are significant with valley walls at 1:1.
8. Anisotropy of the embankment fill, assuming anisotropy planes normal and parallel to a given layer, causes changes in all the stresses and displacements except for the vertical stresses, which remain constant.
9. Pore pressures in embankments and foundations under undrained conditions can be reasonably well predicted with the computed total stress changes and the conventional pore pressure parameters, and the values are not very sensitive to differences in material models assumed.

The case histories analyzed by two-dimensional methods have included instrumented test fills or embankments on soft foundations[37,53–59] and homogeneous or zoned dams on rigid foundations, such as Otter Brook,[18–20,41,60] El Infiernillo,[61,62] Llyn Brianne,[39,40] Scammonden,[38] Wilmot,[32] and Oroville.[63,64] The only three-dimensional case history to date has been Duncan Dam.[36] All these analyses have shown that the behavior of embankments during construction can be reasonably well predicted if an appropriate analytical model is used.

The study conducted for the 770-ft-high Oroville Dam, shown in Fig. 16-7, provides a good example of the applicability of the criteria presented in Secs. 16-2 and 16-3 for embankment analyses. The properties of the shell and core materials were modeled with hyperbolic relationships and were determined, respectively, from drained triaxial tests and unconsolidated-undrained triaxial tests with volume change measurements conducted on laboratory samples modeling the gradation and compaction conditions of the materials as placed. The properties of the other zones were estimated from these tests and related studies. The incremental analyses were conducted in three stages to simulate the field construction

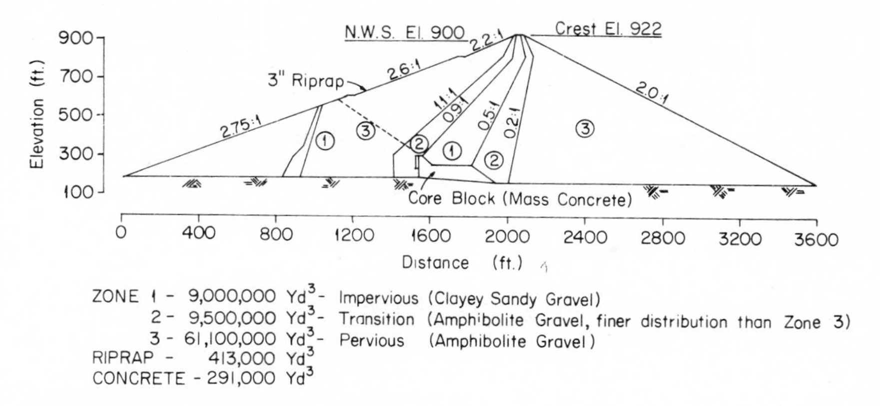

Figure 16-7 Oroville Dam, maximum section.[63,64]

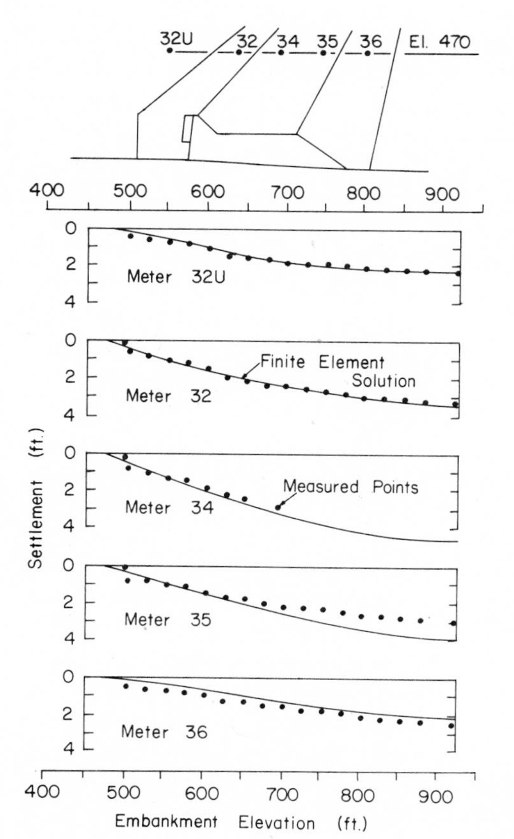

Figure 16-8 Settlements at elevation 470 in Oroville Dam.[63,64]

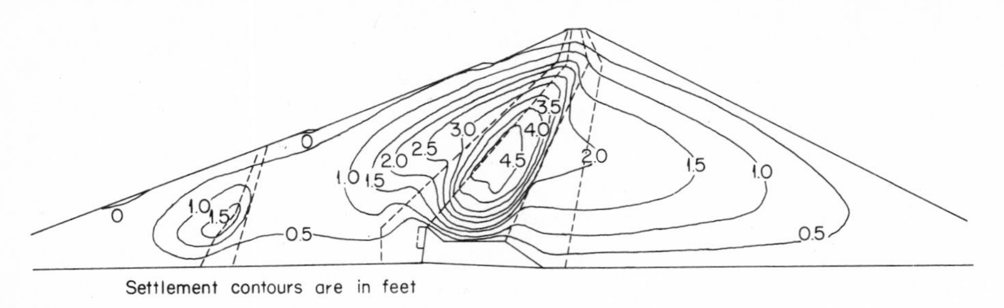

Figure 16-9 Contours of calculated settlements in Oroville Dam.[63,64]

sequence as closely as possible. The first stage simulated construction of the core block in 9 lifts with 89 elements, the second stage simulated construction of the 400-ft-high cofferdam upstream from the core block in 9 lifts with 111 elements, and the third stage simulated construction of the main embankment over the core block and against the cofferdam in 12 lifts with 249 elements.

The results of this analysis were presented in the form of principal stress and displacement contours at the end of construction and as variations of the displacements with increasing embankment elevation for the locations within the embankment where settlement and horizontal displacement devices were installed. Results such as the settlements presented in Figs. 16-8 and 16-9 showed good agreement with the measured field values. A complete presentation of the results of this analysis is given by Kulhawy and Duncan.[63,64]

Excavations Among the studies conducted have been the following: generalized two-dimensional analyses of excavations in homogeneous soil or rock[4–6,9,13–15,60,65–69] and in jointed or discontinuous rock;[13,67,70] two-dimensional analyses of explosive excavation slopes;[71] three-dimensional analyses of excavations in jointed or discontinuous rock;[72,73] and two-dimensional case history studies of large excavations.[11,74] From these generalized studies the following trends have been noted:

1. The stress changes vary in direct proportion with the unit weight and height but are not affected by the modulus.
2. The displacements vary in direct proportion with the unit weight and the square of the height and inversely with the modulus.
3. The effects of Poisson's ratio have not really been investigated because nearly all the studies considered either a single Poisson's ratio value throughout (in the case of rock excavation analyses) or completely undrained analyses in soil, during which Poisson's ratio was adopted to be approximately equal to 0.5.
4. The initial horizontal stress coefficient K_0 is perhaps the most important variable in excavation analyses in homogeneous material because increases in this coefficient cause increased slope displacements and increased zones of yield or failure within the excavated slope.

5. For normally consolidated deposits, the zone of failure begins beneath the slope and progresses up toward the crest and down toward the toe. For overconsolidated deposits, the zone of failure begins beneath the base of the excavation and progresses up toward the base and down toward a firm layer beneath.
6. Linear and bilinear excavation analyses overestimate the extent of the failure zone.
7. Tension zones develop in the top of the excavated slope.
8. Increases in the slope angle generally cause increased displacements, shear stresses, and extent of failure zone.
9. Anisotropic material properties do not affect the results markedly.
10. Pore pressures in excavations can be reasonably well predicted with the computed total stress changes and the conventional pore pressure parameters, but the values are sensitive to differences in material models assumed.
11. The effect of a single prominent joint or discontinuity in the slope, intersecting the base of the excavation, increases the displacements and extents of the tensile zones as the initial horizontal stresses increase and the joint becomes more flexible.

From limited case history analyses it can be said that the behavior of excavations during construction may be reasonably well predicted if appropriate physical and material models are employed.

A note should be made at this point regarding tension zones. Virtually all the excavation solutions and some of the embankment solutions have shown tensile stresses in a portion of the slope. Since neither soil nor rock can sustain any appreciable tension, the solutions should be evaluated in the light of this fact. Zienkiewicz et al.[75] have proposed an approach to this problem. When tension greater than the tensile strength develops, an iterative process is performed in which the excess tensile stresses are relieved and redistributed to the adjacent elements. Details of this procedure are given in Chaps. 1 and 18. This approach is not often used with surface excavation or embankment problems because the computed tensile stresses are commonly small.

Adjustment and Long-Term Analyses

After construction of an embankment or an excavation (and to a certain degree during construction as well) the embankment or excavation undergoes an adjustment period, during which excess pore pressure dissipation and consolidation occurs, steady-state seepage patterns develop, and for a dam embankment the reservoir loads are imposed. After the adjustment period, the long term state is reached during which variations in the loads, as well as creep or secondary compression, can occur. All these conditions can be superimposed upon the results of the end of construction analysis.

Consolidation Excess pore pressure dissipation and consolidation involve time-dependent phenomena, as discussed in Sec. 16-3 and Chap. 12, but as of this writing, they have yet to be applied directly to embankment or excavation problems, even though the basic analytical approaches have been developed. If the time dependency of displacements is not of great concern but the end of construction and after consolidation displacements are, an approach like that employed by Raymond[57] for embankments on clay foundations and by Chang and Duncan[11] for deep excavations can be used. With this approach two analyses are conducted: the first uses undrained material properties to simulate rapid construction, and the second uses drained material properties to simulate very slow construction. In this way bounds are established on the displacements even though the time dependency is not included. Figure 16-10 shows typical results obtained using this approach for the 160-ft-deep Buena Vista Pumping Plant excavation;[11] it can be seen that this approach provides satisfactory bounds on the displacements.

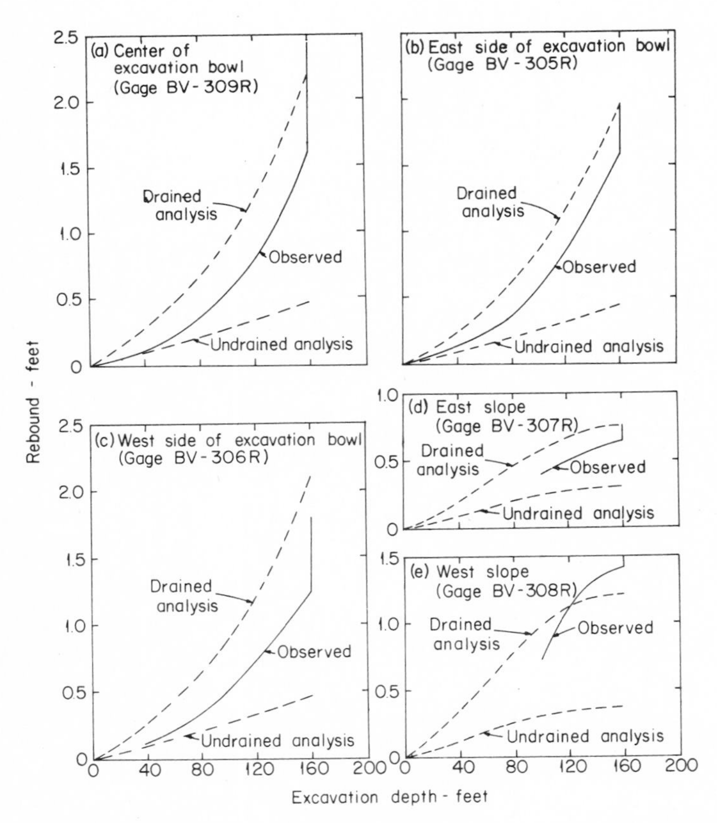

Figure 16-10 Variations of calculated and observed rebounds with excavation depth.[11]

Reservoir filling During and after construction of a dam embankment, the reservoir level begins to rise behind the embankment, subjecting it to a complex loading pattern. This pattern can be divided into four main components: (1) water load on the core, which causes downstream and downward movements; (2) water load on the upstream foundation, which causes upstream and downward movements; (3) bouyant uplift in the upstream shell, which causes upward movements in the upstream shell; and (4) softening from wetting of the upstream shell, which causes settlements in the upstream shell.

Nobari and Duncan[76,77] have investigated these components and have proposed the following analytical model. The water loads on the core and upstream foundation can be introduced as boundary pressures, and the bouyant uplift in the upstream shell can be modeled by applying an upward pressure to the submerged elements, corresponding to the depth of submergence. The effect of softening from wetting can be established by first conducting shear tests on the shell material in both the wet and dry states. From these test results the stress decrease from the dry to wet states can be established as a function of stress level and confining pressure. These stress decreases are then applied to the shell elements during submergence. The loading from the water and the unloading from the bouyancy and the softening, which are all functions of reservoir level, are all applied at the same time as equivalent nodal point forces.

The applicability of this approach was demonstrated by a study of the movements in Oroville Dam during reservoir filling.[76,77] The results of this analysis showed good agreements for the horizontal movements but relatively poor and underestimated agreements for the settlements. Two major reasons can be cited for these differences: (1) no consolidation, secondary compression, creep, or seepage was accounted for in the analyses, and the subsequent movements from these effects were not included; (2) the starting point for these analyses was taken as the end of construction results by Kulhawy and Duncan,[63,64] which show good but

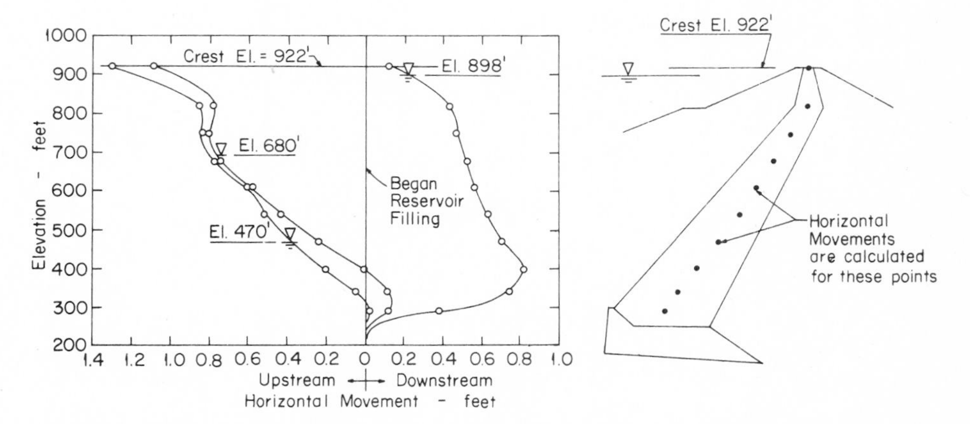

Figure 16-11 Calculated horizontal movements at three stages of reservoir level, Oroville Dam, shell compacted to 90 percent relative density.[76,77]

not exact comparisons of the stresses and displacements in Oroville Dam. Both these factors could have affected the results, but the approach is very reasonable, and the pattern of the results is similar to observed field behavior. For example, it is known that during reservoir filling the initial embankment movements are upstream and then, as the final pool elevation is reached, the movement is downstream. Figure 16-11 shows the results of the analysis for Oroville Dam in which this type of behavior was computed.

Seepage After excavation of a slope below the groundwater table or after reservoir filling as discussed above, the flow pattern through soil or rock gradually adjusts itself until a steady-state flow pattern develops. The resulting pore pressures change the effective stresses, and subsequent movements can be expected to occur. Details on evaluating seepage patterns are discussed in Chap. 14, but it should be noted that seepage effects on subsequent movements can be treated numerically in the same manner as stress changes caused by wetting.

Creep or secondary compression After consolidation and excess pore pressure dissipation has occurred, the continuing time-dependent deformations caused by creep or secondary compression become fully effective. To accommodate for this, approximate time laws can be employed, as suggested by Palmerton,[78] Finn and Emery,[79] and others, as discussed in detail in Chaps. 2 and 3. In the numerical analysis, the stress-displacement behavior is computed for successively increasing time periods with time as the only variable. The results of Finn and Emery[79] for rock slopes and of Palmerton[78] for levee foundations show that observed field behavior can be estimated qualitatively, but much work is still needed in this area to develop a definitive approach.

Stability Evaluation

In addition to computing the stresses, strains, and displacements, the results of the finite element solutions can be used effectively to (1) evaluate the possibility of development of tension zones and subsequent cracking within an embankment, (2) assess the possibility of hydraulic fracturing, (3) isolate zones of local yielding, and (4) compute the overall degree of safety of an embankment or excavation.

Tension zones and cracking A number of studies have been conducted to evaluate the tension zones and cracking in embankments. The earliest of these[80–82] were based upon idealized linear elastic material properties or moduli which increased with confining pressure and upon the embankment being constructed in one stage. These highly idealized conditions exaggerate the tension zones but nevertheless could provide a rough qualitative assessment of cracking potential. The results of these studies showed that in the longitudinal section the dimensions of the tension zone increase with embankment height, the dimensions of the tension zone and the magnitude of tension vary inversely with Poisson's ratio, greater tension

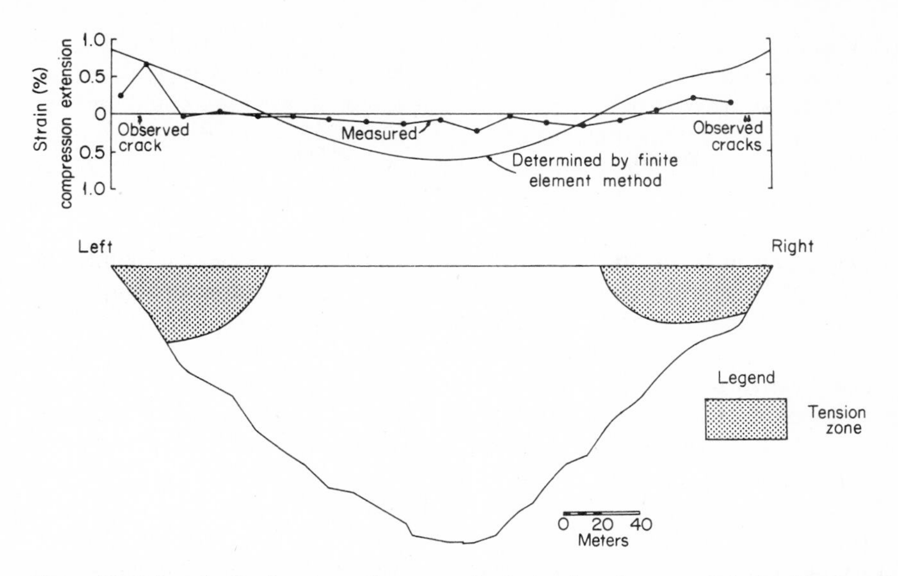

Figure 16-12 Longitudinal strains at the crest and calculated tension zones in Infiernillo Dam (looking downstream).[80]

develops with steeper abutments, interior tension zones develop along the interface of a dam and compressible foundation, and tension zones are reduced with stiffer embankments on compressible foundations. An example showing the measured and computed tension zones and longitudinal strains in the crest of Infiernillo Dam is shown in Fig. 16-12.

In the transverse section, these studies showed that when the shell is more compressible than the core, tension zones can develop in the upper parts of the shell and core but when the core is more compressible or the Poisson's ratio of the core is low, tension zones can develop in the upper half of the core. When compared with observed tension zones in several other embankments, the results of these studies showed a good qualitative agreement.

Further two-dimensional[81] and three-dimensional[35] studies were conducted to investigate these cracking phenomena. These studies showed that when incremental analyses and nonlinear material properties are used, the tension zones are greatly reduced or eliminated, suggesting that the tension zones really develop as a function of foundation compressibility, time-dependent material properties, and loading after construction. It is also quite possible that significant mismatch of material properties between the shell and core may lead to tension zones. Eisenstein et al.[36] further showed that with an incremental nonlinear three-dimensional analysis, the computed zones of tensile stress compared favorably with the observed crack locations in Duncan Dam.

Conduits and cutoff walls Since conduits are frequently required at the base of embankments, studies were conducted[48,49,80,84,85] to evaluate the effect of embankment and foundation geometry and material properties on the spreading of conduits and on the development of tension zones. The results of these linear elastic analyses showed that the observed spreading of conduits could be reasonably well predicted and that the presence of a conduit could lead to small tension zones in the crest above the conduit. It was also found that tension zones would develop around rectangular conduits but not around semielliptical conduits.

Similarly, cutoff walls are frequently required in embankment foundations. Several authors[80,84] investigated this problem by using the same techniques as above and found that the presence of a cutoff wall induces tension in the crest above the cutoff wall and in the foundation adjacent to the upper part of the wall. It was also noted that a large negative skin friction could be transferred to the wall, which could exceed the strength of the wall in its lower section. All these undesirable results became worse as the thickness of the cutoff wall increased.

Hydraulic fracturing Hydraulic fracturing or the formation of cracks by excessive water pressure has been responsible for the development of leaks in the cores of several dams. Nobari et al.[52] have investigated this phenomenon using incremental nonlinear analyses for zoned embankments based upon the procedures discussed in Secs. 16-2 and 16-3. To evaluate the possibility of hydraulic fracturing, they compared the maximum principal stress σ_1 to the reservoir water pressure u and located the most critical depth at the upstream face, where the ratio of σ_1/u would be a minimum. If σ_1/u was less than unity, hydraulic fracturing would be possible for a core with no tensile strength. The results of this study showed that the likelihood of hydraulic fracturing is greater for higher dams and can be reduced by (1) making the core wider, (2) compacting the core at a lower water content, (3) using a transition zone which is less stiff than the shell, and (4) using a sloping core instead of a vertical central core. A typical example showing potential hydraulic fracturing is shown in Fig. 16-13.

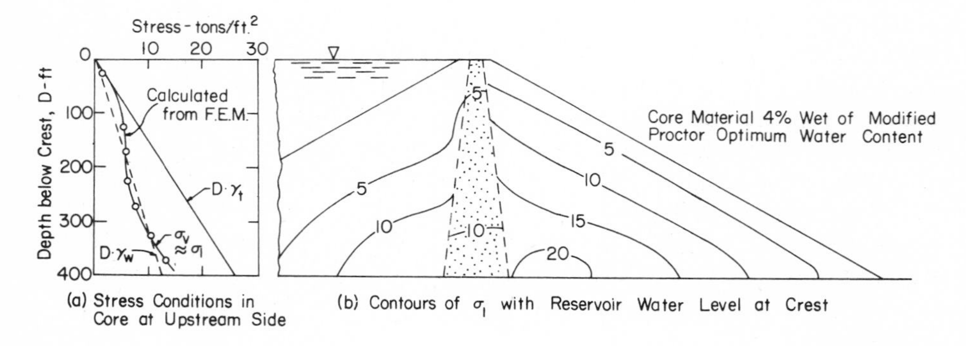

Figure 16-13 Stress conditions at upstream side of core and contours for hydraulic-fracturing example.[52]

Local yielding and overall stability The results obtained from finite element analyses can also be used to assess the stability of embankments and excavation slopes, both in terms of overall stability and in terms of local yielding. As discussed previously with excavation slopes,[4,6,9,68] the progression of the failure zone can be followed by utilizing an appropriate failure criterion and allowing yielding during excavation. The overall stability can be evaluated, as shown by Dunlop et al.[5] and others, by evaluating the computed mobilized shear stresses along a critical shear surface and comparing these stresses with the strength along the surface. The overall stability computed in this manner agrees favorably with that computed by conventional limit equilibrium methods. However, the main advantage of the finite element approach is that by comparing the largest mobilized shear stress within the excavated slope with the average shear stress needed for limit equilibrium, an assessment can be made of the required factor of safety from limit equilibrium to prevent local overstress or yielding. Table 16-1 shows typical results obtained for four different excavation slopes and for initial horizontal total stress coefficients representative of normally consolidated and overconsolidated deposits. It can be seen that values of the factor of safety from 2 to 3 are required to prevent local overstress in normally consolidated deposits but that values from 4 to 5 are required for overconsolidated deposits. Results like these are very helpful in assessing the potential for progressive failure of an excavation slope.

A similar approach can be used to evaluate the stability of embankment slopes from the results of finite element analyses. Wright et al.[86] and Wang and

Table 16-1 Values of factor of safety required to prevent local overstress along critical shear surface for excavated slopes[9]

Slope	K_0	$c/\gamma H$ Largest shear stress	$c/\gamma H$ Average shear stress	Factor of safety
3:1	0.81	0.31	0.16	1.94
	1.60	0.70	0.16	4.37
1.5:1	0.81	0.36	0.175	2.06
	1.60	0.78	0.175	4.45
Vertical	0.81	0.57	0.26	2.19
	1.60	1.01	0.26	3.89
Vertical cut to rigid base	0.81	0.77	0.26	2.96
	1.60	1.32	0.26	5.07

Table 16-2 Values of factor of safety required to prevent local overstress along critical shear surface in embankment slopes[86]

$\lambda_{c\phi} = \dfrac{\gamma H \tan \phi}{c}$	Slope ratio		
	1.5:1	2.5:1	3.5:1
0	1.46	1.44	1.49
2	1.32	1.23	1.23
5	1.34	1.18	1.14
20	2.27	1.21	1.10
50	4.36	1.37	1.12

Sun[87] have shown that the factor of safety, as evaluated from the computed stresses along a critical shear surface, is essentially the same or slightly larger than the factor of safety computed from the modified Bishop method and that the factor of safety increases slightly with increasing values of Poisson's ratio. Again the main advantage of the finite element method is the assessment of local yield or the required limit equilibrium factor of safety to prevent local overstress. Table 16-2 shows the computed factors of safety required to prevent local overstress in embankment slopes. It can be seen that (except for steep slopes in highly frictional materials) local overstress is not a significant factor, but it should be noted that these values may be very different in zoned embankments.

16-5 EFFORT, COST, AND VALUE OF ANALYSES

Finite element analysis for embankments and excavations has achieved the stature of a so-called standard method of analysis but, unlike many other standard methods, it is strictly a computer method and is therefore dependent upon computer systems and the reliability of the program being used. A program or code should never be treated as a black box method unless the user is familiar with the internal workings of the program and its reliability. Any organization which contemplates conducting its own analyses should be prepared to invest the equivalent of several months of engineer time so that the person will be able to learn the internal workings of the available computer programs and to be sufficiently versed in the programs to make minor changes and adaptations as needed.

It has been the writer's experience that simple two-dimensional embankment and excavation problems can be analyzed for several days of experienced engineer time and several hundred dollars of computer time if an appropriate preexisting program is available and the problem and material properties can be easily evaluated from existing data. If extensive modifications must be made or new features incorporated, one must expect that several months to a year or more may have to be invested, along with several thousand dollars of computer time, before an appropriate program can be built.

It must also be remembered that the larger the number of variables and different conditions to be considered, the larger the required effort will be. Even though the unit solution time may be minimal because of efficient solution routines, very complex incremental time-dependent three-dimensional solution involving asymmetrical geometries may cost several thousand dollars of computer time and perhaps a couple weeks of experienced engineer time if an available program can be utilized directly. These time and cost limitations, in addition to computer capabilities, must always be kept in mind.

However, reliable results can be obtained if one approaches the problem carefully and intelligently; guidelines have been proposed in the previous sections. If a sound physical model is established, if the material properties are properly evaluated, and if all loadings are correctly imposed, there is no reason why good results cannot be obtained.

REFERENCES

1. Sherard, J. L., R. J. Woodward, S. F. Gizienski, and W. A. Clevenger: "Earth and Earth-Rock Dams, pp. 377–383, John Wiley & Sons, Inc., New York, 1963.
2. Poulos, H. G., and E. H. Davis: "Elastic Solutions for Soil and Rock Mechanics," John Wiley & Sons, Inc., New York, 1974.
3. Bishop, A. W.: The Stability of Earth Dams, Ph.D. dissertation, University of London, 1952.
4. Dunlop, P., and J. M. Duncan: Development of Failure around Excavated Slopes, *J. Soil Mech. Found. Div. ASCE*, vol. 96, no. SM2, pp. 471–493, March 1970.
5. Dunlop, P., J. M. Duncan, and H. B. Seed: Finite Element Analyses of Slopes in Soil, *Univ. California Dept. Civ. Eng. Soil Mech. Bitum. Mat. Rep.* TE-68-3, Berkeley, Calif., May 1968; also *U.S. Army Eng. Waterw. Exp. Stn., Contr. Rep.* S-68-6, Vicksburg, Miss.,
6. Constantopoulos, I. V., J. T. Christian, and R. V. Whitman: Initiation of Failure in Slopes in Overconsolidated Clays and Clay Shales, *U.S. Army Eng. Nucl. Cratering Group, NCG Tech. Rep.* 29, Livermore, Calif., November 1970.
7. Brown, C. B., and L. E. Goodman: Gravitational Stresses in Accreted Bodies, *Proc. R. Soc. Lond.*, Ser. A, vol. 276, no. 1367, December 17, 1963, pp. 571–576.
8. Goodman, L. E., and C. B. Brown: Dead Load Stresses and the Instability of Slopes, *J. Soil Mech. Found. Div. ASCE*, vol. 89, no. SM3, May, 1963, pp. 103–134.
9. Duncan, J. M., and P. Dunlop: Slopes in Stiff Fissured Clays and Shales, *J. Soil Mech. Found. Div. ASCE*, vol. 95, no. SM2, pp. 467–492, March 1969.
10. Brown, C. B., and I. P. King: Automatic Embankment Analysis: Equilibrium and Instability Conditions, *Geotechnique*, vol. 16, no. 3, September 1966, pp. 209–219.
11. Chang, C.-Y., and J. M. Duncan: Analysis of Soil Movement around a Deep Excavation, *J. Soil Mech. Found. Div. ASCE*, vol. 96, no. SM5, pp. 1655–1681, September 1970.
12. Clough, G. W., and J. M. Duncan: Finite Element Analyses of Port Allen and Old River Locks, *Univ. California Dept. Civ. Eng. Geotech. Eng. Rep.* TE-69-3, Berkeley, Calif., September 1969; also *U.S. Army Eng. Waterw. Exp. Stn. Contr. Rep.* S-69-3, Vicksburg, Miss.,
13. Duncan, J. M., and R. E. Goodman: Finite Element Analyses of Slopes in Jointed Rock, *U.S. Army Eng. Waterw. Exp. Stn. Cont. Rep.* S-68-3, Vicksburg, Miss., February 1968.
14. Finn, W. D. L.: Static and Seismic Analysis of Slopes, *Rock Mech. Eng. Geol.*, vol. 4, no. 3, pp. 268–277, 1966.
15. Finn, W. D. L.: Static and Dynamic Stresses in Slopes, *Proc. 1st Congr. Int. Soc. Rock Mech., Lisbon, 1966*, vol. 2, pp. 167–169.
16. Kulhawy, F. H.: Analysis of Underground Openings in Rock by Finite Element Methods, *Syracuse Univ. Contr. Rep. to U.S. Bur. Mines*, Syracuse, N.Y., April 1973.

17. King, I. P.: Finite Element Analysis of Two-dimensional Time-dependent Stress Problems, *Univ. California Dept. Civ. Eng. Struct. Mater. Res. Rep.* 65-1, Berkeley, Calif., January 1965.
18. Clough, R. W., and R. J. Woodward III: Analysis of Embankment Stresses and Deformations, *J. Soil Mech. Found. Div. ASCE*, vol. 93, no. SM4, pp. 529–549, July 1967.
19. Finn, W. D. L.: Static and Seismic Behavior of an Earth Dam, *Can. Geotech. J.*, vol. 4, no. 1, pp. 28–37, February 1967.
20. Kulhawy, F. H., J. M. Duncan, and H. B. Seed: Finite Element Analyses of Stresses and Movements in Embankments during Construction, *Geotech. Eng. Rep.* TE-69-4, Berkeley, November 1969; also *U.S. Army Eng. Waterw. Exp. Stn. Contr. Rep.* S-69-8, Vicksburg, Miss.,
21. Raphael, J. M., and R. W. Clough: Construction Stresses in Dworshak Dam, *Univ. California Dept. Civ. Eng. Struct. Mater. Res. Rep.* 65-3, Berkeley, Calif., April 1965.
22. Lefebvre, G., and J. M. Duncan: Three Dimensional Finite Element Analyses of Dams, *Univ. California Dept. Civ. Eng. Geotech. Eng. Rep.* TE-71-5, Berkeley, Calif., May 1971; also *U.S. Army Eng. Waterw. Exp. Stn. Contr. Rep.* S-71-6, Vicksburg, Miss.,
23. Lefebvre, G., J. M. Duncan, and E. L. Wilson: Three-dimensional Finite Element Analyses of Dams, *J. Soil Mech. Found. Div. ASCE*, vol. 99, no. SM7, pp. 495–507, July 1973.
24. Palmerton, J. B., and G. Lefebvre: Three-dimensional Behavior of a Central Core Dam, *U.S. Army Eng. Waterw. Exp. Stn. Res. Rep.* S-72-1, Vicksburg, Miss., December 1972.
25. Naylor, D. J., and D. B. Jones: The Prediction of Settlement within Broad Layered Fills, *Geotechnique*, vol. 23, no. 4, pp. 589–594, December 1973.
26. Clough, G. W., and J. M. Duncan: Finite Element Analyses of Retaining Wall Behavior, *J. Soil Mech. Found. Div. ASCE*, vol. 97, no. SM12, pp. 1657–1673, December 1971.
27. Duncan, J. M., and G. W. Clough: Finite Element Analyses of Port Allen Lock, *J. Soil Mech. Found. Div. ASCE*, vol. 97, no. SM8, pp. 1053–1068, August 1971.
28. Zienkiewicz, O. C., B. Best, C. Dullage, and K. G. Stagg: Analysis of Nonlinear Problems in Rock Mechanics with Particular Reference to Jointed Systems, *Proc. 2d Congr. Int. Soc. Rock Mech., Belgrade, 1970*, vol. 3, pp. 501–509.
29. Wilson, E. L.: SOLID SAP: A Static Analysis Program for Three Dimensional Solid Structures, *Univ. California Dept. Civ. Eng. Struct. Mater. Res. Rep.* SESM 71–19, Berkeley, Calif., September 1971.
30. Squier, L. R.: Load Transfer in Earth and Rockfill Dams, *J. Soil Mech. Found. Div. ASCE*, vol. 96, no. SM1, pp. 213–233, January 1970.
31. Palmerton, J. B.: Application of Three-dimensional Finite Element Analysis, *Proc. Symp. Appl. Finite Element Method Geotech. Eng., U.S. Army Eng. Waterw. Exp. Stn., Vicksburg, Miss., 1972*, pp. 155–214.
32. Boughton, N. O.: Elastic Analysis for Behavior of Rockfill, *J. Soil Mech. Found. Div. ASCE*, vol. 96, no. SM5, pp. 1715–1733, September 1970.
33. Desai, C. S.: Nonlinear Analyses Using Spline Functions, *J. Soil Mech. Found. Div. ASCE*, vol. 97, no. SM10, pp. 1461–1480, October 1971.
34. Duncan, J. M., and C.-Y. Chang: Nonlinear Analysis of Stress and Strain in Soils, *J. Soil Mech. Found. Div. ASCE*, vol. 96, no. SM5, pp. 1629–1653, September 1970.
35. Eisenstein, Z., A. V. G. Krishnayya, and N. R. Morgenstern: An Analysis of Cracking in Earth Dams, *Proc. Symp. Appl. Finite Element Method Geotech. Eng. U.S. Army Eng. Waterw. Exp. Stn., Vicksburg, Miss., 1972*, pp. 431–455.
36. Eisenstein, Z., A. V. G. Krishnayya, and N. R. Morgenstern: An Analysis of the Cracking at Duncan Dam, *Proc. ASCE Spec. Conf. Perform. Earth Earth-supported Struct., Purdue Univ., June 1972*, vol. 1, pt. 1, pp. 765–778.
37. Foott, R., and C. C. Ladd: The Behavior of Atchafalaya Test Embankments During Construction, *MIT Dept. Civ. Eng. Soils Publication* 322, Cambridge, Mass., May 1973.
38. Penman, A. D. M., J. B. Burland, and J. A. Charles: Observed and Predicted Deformations in a Large Embankment Dam during Construction, *Proc. Inst. Civ. Eng.*, vol. 49, pp. 1–21, May 1971.
39. Penman, A., and J. A. Charles:Construction Deformations in Rockfill Dam, *J. Soil Mech. Found. Div. ASCE*, vol. 99, no. SM2, pp. 139–163, February 1973.

40. Penman, A. D. M., and J. A. Charles: Construction Deformations in a Rockfill Dam, *Ground Eng.*, vol. 6, no. 2, pp. 24–36, March 1973.
41. Resendiz, D., and M. P. Romo: Analysis of Embankment Deformations, *Proc. ASCE Spec. Conf. Perform. Earth Earth-supported Struct., Purdue Univ., June 1972*, vol. 1, pt 1, pp. 817–836.
42. Kondner, R. L.: Hyperbolic Stress-Strain Response: Cohesive Soils, *J. Soil Mech. Found. Div. ASCE*, vol. 89, no. SM1, pp. 115–143, February 1963.
43. Janbu, N.: Soil Compressibility as Determined by Oedometer and Triaxial Tests, *Proc. Eur. Conf. Soil Mech. Found. Eng., Weisbaden, 1963*, vol. 1, pp. 19–25.
44. Duncan, J. M.: Finite Element Analyses of Stresses and Movements in Dams, Excavations and Slopes, *Proc. Symp. Appl. Finite Element Method Geotech. Eng. U.S. Army Eng. Waterw. Exp. Stn., Vicksburg, Miss., 1972*, pp. 267–326.
45. Chowdhury, R. N.: Deformation Problems in Anisotropic Soil, *Proc. Spec. Conf. Finite Element Methods Civ. Eng., McGill Univ., Montreal, 1972*, pp. 653–672.
46. D'Appolonia, D. J., T. W. Lambe, and H. G. Poulos: Evaluation of Pore Pressures beneath an Embankment, *J. Soil Mech. Found. Div. ASCE*, vol. 97, no. SM6, pp. 881–897, June 1971.
47. Mitchell, J. K., and W. S. Gardner: Analysis of Load-bearing Fills over Soft Subsoils, *J. Soil Mech. Found. Div. ASCE*, vol. 97, no. SM11, pp. 1549–1571, November 1971.
48. Mueser, Rutledge, Wentworth & Johnston: Report on Study of Movements of Articulated Conduits under Earth Dams on Compressible Foundations, *Contr. Rep. to USDA Soil Conserv. Serv.*, June 1968.
49. Rutledge, P. C., and J. P. Gould: Movements of Articulated Conduits under Earth Dams on Compressible Foundations, chap. 7 in R. C. Hirschfeld and S. J. Poulos (eds.), "Embankment Dam Engineering," pp. 209–237, John Wiley & Sons, Inc., New York, 1973.
50. Thoms, R. L., and A. Arman: Analysis of Stress Distribution under Embankments on Soft Foundations, *Louisiana State Univ. Eng. Res. Bull.* 99, Baton Rouge, 1969.
51. Thoms, R. L., and A. Arman: Photoelastic and Finite Element Analysis of Embankments Constructed over Soft Soils, *Highw. Res. Board, Rec.* 323, Washington, D.C., 1970, pp. 71–86.
52. Nobari, E. S., K. L. Lee, and J. M. Duncan: Hydraulic Fracturing in Zoned Earth and Rockfill Dams, *Geotech. Eng. Rep.* TE-73-1, Berkeley, January 1973; also *U.S. Army Eng. Waterw. Exp. Stn. Contr. Rep.* S-73-2, Vicksburg, Miss.,
53. Bozozuk, M., and G. A. Leonards: The Gloucester Test Fill, *Proc. ASCE Spec. Conf. Perform. Earth Earth-supported Struct., Purdue Univ., June 1972*, vol. 1, pt. 1, pp. 299–317.
54. Hollingshead, G. W., and G. P. Raymond: Prediction of Undrained Movements Caused by Embankments on Muskeg, *Can. Geotech. J.*, vol. 8, no. 1, pp. 23–35, February 1971.
55. Ladd, C. C.: Test Embankment on Sensitive Clay, *Proc. ASCE Spec. Conf. Perform. Earth Earth-supported Struct., Purdue Univ., June 1972*, vol. 1, pt. 1, pp. 101–128.
56. Lambe, T. W., D. J. D'Appolonia, K. Karlsrud, and R. C. Kirby: The Performance of the Foundation under a High Embankment, *J. Boston Soc. Civ. Eng.*, vol. 59, no. 2, pp. 71–94, April 1972.
57. Raymond, G. P.: The Kars (Ontario) Embankment Foundation, *Proc. ASCE Spec. Conf. Perform. Earth Earth-supported Struct., Purdue Univ., June 1972*, vol. 1, pt. 1, pp. 319–340.
58. Raymond, G. P.: Prediction of Undrained Deformations and Pore Pressures in Weak Clay under Two Embankments, *Geotechnique*, vol. 22, no. 3, pp. 381–401, September 1972.
59. Wroth, C. P., and B. Simpson: An Induced Failure at a Trail Embankment, II: Finite Element Computations, *Proc. ASCE Spec. Conf. Perform. Earth Earth-supported Struct., Purdue Univ., June 1972*, vol. 1, pt. 1, pp. 65–79.
60. Finn, W. D. L., and A. P. Troitskii: Computation of Stresses and Strains in Dams Made of Local Materials, Earth Slopes, and Their Foundations, by the Finite Element Method, *Hydrotech. Construc.*, no. 6, pp. 492–499, June 1968.
61. Alberro, J.: Stress-Strain Analysis of El Infiernillo Dam, *Proc. ASCE Spec. Conf. Perform. Earth Earth-supported Struct., Purdue Univ., June 1972*, vol. 1, pt. 1, pp. 837–852.
62. Skermer, N. A.: Finite Element Analysis of El Infiernillo Dam, *Can. Geotech. J.*, vol. 10, no. 2, pp. 129–144, May 1973.

63. Kulhawy, F. H., and J. M. Duncan: Nonlinear Finite Element Analysis of Stresses and Movements in Oroville Dam, *Univ. California Dept. Civ. Eng. Geotech. Eng. Rep.* TE-70-2, Berkeley, Calif., January 1970.
64. Kulhawy, F. H., and J. M. Duncan: Stresses and Movements in Oroville Dam, *J. Soil Mech. Found. Div. ASCE*, vol. 98, no. SM7, pp. 653–665, July 1972.
65. Bhattacharyya, K. K., and S. H. Boshkov: Determination of the Stresses and the Displacements in Slopes by the Finite Element Method, *Proc. 2d Cong. Int. Soc. Rock Mech., Belgrade, 1970*, vol. 3, pp. 339–344.
66. Dodd, J. S., and H. W. Anderson: Tectonic Stresses and Rock Slope Stability, *Proc. 13th Symp. Rock Mech. Stability Rock Slopes, Univ. Illinois, August–September 1971*, pp. 171–182.
67. Goodman, R. E., and J. M. Duncan: The Role of Structure and Solid Mechanics in the Design of Surface and Underground Excavations in Rock, chap. 105 in M. Te'eni (ed.), "Structure, Solid Mechanics and Engineering Design," pp. 1379–1404, John Wiley & Sons, Inc., New York, 1971.
68. Lo, K. Y., and C. F. Lee: Stress Analysis and Slope Stability in Strain-softening Materials, *Geotechnique*, vol. 23, no. 1, pp. 1–11, March 1973.
69. Pariseau, W. G., B. Voight, and H. D. Dahl: Finite Element Analyses of Elastic-Plastic Problems in the Mechanics of Geologic Media: An Overview, *Proc. 2d Congr. Int. Soc. Rock Mech., Belgrade, 1970*, vol. 2, pp. 311–323.
70. Ko, K. C.: Discrete Element Technique for Pit Slope Analysis, *Proc. 13th Symp. Rock Mech. Stability Rock Slopes, Univ. Illinois, August–September 1971*, pp. 183–199.
71. Gates, R. H.: Slope Analysis for Explosive Excavations, *Proc. 13th Symp. Rock Mech. Stability Rock Slopes, Univ. Illinois, August–September 1971*, pp. 243–268.
72. Mahtab, M. A.: Three Dimensional Finite Element Analysis of Jointed Rock Slopes, Ph.D. dissertation, Department of Civil Engineering, University of California, Berkeley, 1970.
73. Mahtab, M. A., and R. E. Goodman: Three Dimensional Finite Element Analysis of Jointed Rock Slopes, *Proc. 2d Congr. Int. Soc. Rock Mech., Belgrade, 1970*, vol. 3, pp. 353–360.
74. Pariseau, W. G., and K. Stout: Open Pit Mine Slope Stability: The Berkeley Pit, *Proc. 13th Symp. Rock Mech. Stability Rock Slopes, Univ. Illinois, August–September 1971*, pp. 367–395.
75. Zienkiewicz, O. C., S. Valliappan, and I. P. King: Stress Analysis of Rock as a "No Tension" Material, *Geotechnique*, vol. 18, no. 1, pp. 56–66, March 1968.
76. Nobari, E. S., and J. M. Duncan: Effect of Reservoir Filling on Stresses and Movements in Earth and Rockfill Dams, *Geotech. Eng. Rep.* TE-72-1, Berkeley, January 1972; also *U.S. Army Eng. Waterw. Exp. Stn., Contr. Rep.* S-72-1, Vicksburg, Miss.,
77. Nobari, E. S., and J. M. Duncan: Movements in Dams Due to Reservoir Filling, *Proc. ASCE Spec. Conf. Perform. Earth Earth-supported Struct., Purdue Univ., June 1972*, vol. 1, pt. 1, pp. 797–815.
78. Palmerton, J. B.: Creep Analysis of Atchafalaya Levee Foundation, *Proc. Symp. Appl. Finite Element Methof Geotech. Eng., U.S. Army Eng. Waterw. Exp. Stn., Vicksburg, Miss., 1972*, pp. 843–862.
79. Finn, W. D. L., and J. J. Emery: Stresses and Deformations in Creeping Rock Slopes, *Proc. 2d Congr. Int. Soc. Rock Mech., Belgrade, 1970*, vol. 4, pp. 235–240.
80. Covarrubias, S. W.: Cracking of Earth and Rockfill Dams, *Harvard Univ., Soil Mech. Ser.* 82, Cambridge, Mass., April 1969; also *U.S. Army Eng. Waterw. Exp. Stn. Cont. Rep.* S-69-5, Vicksburg, Miss.,
81. Covarrubias, S. W.: Comparison of Observed and Theoretical Tensile Strains in the Crests of Two Earth and Rockfill Dams, *U.S. Army Eng. Waterw. Exp. Stn. Contr. Rep.* S-71-11, Vicksburg, Miss., April 1971.
82. Lee, K. L., and C. K. Shen: Horizontal Movements Related to Subsidence, *J. Soil Mech. Found. Div. ASCE*, vol. 95, no. SM1, pp. 139–166, January 1969.
83. Strohm, W. E., Jr., and S. J. Johnson: The Influence of Construction Step Sequence and Nonlinear Material Behavior on Cracking of Earth and Rock-Fill Dams, *U.S. Army Eng. Waterw. Exp. Stn. Misc. Pap.* S-71-10, Vicksburg, Miss., May 1971.

84. Casagrande, A., and S. W. Covarrubias: Tension Zones in Embankments Caused by Conduits and Cutoff Walls, *Harvard Univ. Soil Mech. Ser.* 85, July 1970; also *U.S. Army Eng. Waterw. Exp. Stn. Contr. Rep.* S-70-7, Vicksburg, Miss.,
85. Hughes, J. M. O.: Culvert Elongations in Fills Founded on Soft Clays, *Can. Geotech. J.*, vol. 6, no. 2, pp. 111–117, May 1969.
86. Wright, S. G., F. H. Kulhawy, and J. M. Duncan: Accuracy of Equilibrium Slope Stability Analyses, *J. Soil Mech. Found. Div. ASCE*, vol. 99, no. SM10, pp. 783–791, October 1973.
87. Wang, F.-D., and M.-C. Sun: Slope Stability Analysis by the Finite Element Stress Analysis and Limiting Equilibrium Method, *U.S. Bur. Mines, Rep. Invest.* 7341, Washington, D.C., January 1970.

CHAPTER

SEVENTEEN

NUMERICAL AND PHYSICAL MODELING

Ian M. Smith

17-1 INTRODUCTION

Modeling has not traditionally played a significant role in geotechnical engineering design processes. Whereas few major steel or concrete structures or hydraulic works are built without an assessment first being made of their likely performance by means of scaled physical models, construction of large earth dams, excavations, and foundations of heavy structures has been embarked upon without previous model evaluations of likely performance. Indeed predictive power has been assumed to be so uncertain that the dominant design philosophy in geotechnical engineering has been the "observational method," summed up in the phrase "learn as you go."[32] The designer is relieved of the need to make good predictions, because he can, and ought to, modify poorer ones in the light of experience gained by observation as the works proceed. Following an initial site exploration "sufficient to establish at least the general nature, pattern and properties of the deposits, but not necessarily in detail" the designer must have in mind "a plan of action for every unfavourable situation that might be disclosed by the observations."[32]

The practitioners of the method are aware that there are situations in which it cannot be successful. For example, "if the character of the project is such that the design cannot be altered during construction, the method is inapplicable."[32] One could add that if the critical event in the life of a structure does *not* occur during construction or is not analogous to an event occurring during construction, the method cannot furnish a successful design. This is the case in many critical geotechnical designs at present under way. Nuclear power plant design to resist earthquakes and oil production platform design for deepwater locations are obvious examples.

In these situations there is no escape from the need to *predict* how the structure will behave under the critical loading conditions. Lambe[23] has referred to the need to improve predictive capability in geotechnics, which is usually based on modeling the design situation. It is the purpose of the present chapter to discuss the relative significance of two types of modeling, numerical and physical, with particular reference to four types of geotechnical design problems for which correlations between numerical predictions and field or laboratory observations have been possible. Particular emphasis is given to recent developments in numerical modeling (finite elements) and physical modeling (large centrifuges).

Description of Problems

Four areas of numerical and physical modeling are covered in this chapter. Some of these areas are the subjects of other chapters, as listed below, but the emphasis differs. The problems are:

1. Deformation and failure of undrained slopes of soil, both built-up and excavated (see also Chap. 16)
2. Rate of consolidation of soft foundations under earth embankments during and after construction (see also Chap. 12)
3. Deformation and stability of flexible sheetpiling retaining sands and clays
4. Static and dynamic behavior of gravity ocean-bed oil-production platforms (see also Chap. 20)

In all cases the relative merits of physical and numerical modeling are discussed and correlations between the two methods (and where possible with field performance) are given.

Principles of Modeling in Geotechnical Engineering

In 1957, Rocha[35] discussed the possibility of solving soil mechanics problems by means of models. The conclusions he reached are of interest in the light of subsequent developments in numerical and physical modeling. The main drawback of physical models was stated to be that "when the weight of the mass [of soil] itself is taken into account the materials of the prototype cannot be used for the construction of the models." Taylor,[60] discussing seepage through earth-dam models, went further, saying that "the use of the soil from the site offers no real advantages and often has the disadvantage of excessive capillary effects." These two references, however, are among the few in which a model of a field situation was contemplated. It is this type of modeling with which this chapter is concerned rather than the modeling of idealized situations at small scale to check theoretical earth pressure or bearing capacity theories, e.g., the classic experiments of Terzaghi.[62]

Rocha[35] listed the following shortcomings of analytical modeling at that time:

1. In spite of soil's being a two- or three-phase material, this fact was seldom recognized. Only undrained or fully drained conditions could be analyzed.
2. Only unidirectional drainage was considered in the theory of consolidation, whereas in practice drainage in two and three dimensions nearly always occurs.
3. The mathematical analysis becomes insuperably difficult when nonlinear relations are considered.
4. Difficulties are met when dealing with masses with heterogeneity or anisotropy.
5. Problems are often wrongly idealized as two-dimensional.
6. Great difficulties must be faced whenever the boundary conditions, i.e., the shapes and loadings applied to the surface, are complex.

The extent to which the above shortcomings still apply is examined in the following sections on recent developments in physical and numerical modeling.

Centrifugal Modeling

The heterogeneity of soil deposits and the difficulties of artificially creating fabrics with natural properties make use of the prototype soil imperative in model construction. Previously mentioned objections to doing so[35,60] can largely be overcome by subjecting the model to a high acceleration field produced centrifugally. Work of this nature was first described in geological contexts in the United States in about 1930 and by Pokrovsky[33] in soil mechanics.

Consider an element of a soil model rotating at constant angular velocity ω at radius r from the axis of a centrifuge. The centrifuge axis is coaxial with earth's gravitational acceleration. The element of soil is subject to the accelerations $d^2r/dt^2 - r\omega^2$ radially, $2(dr/dt)\omega$ tangentially, and g axially, where g is typically 9.81 m/s^2 but varies somewhat with location on the earth's surface. In "static" problems or problems involving laminar fluid flow, dr/dt is negligible, so that the acceleration field is $r\omega^2$ radially, zero tangentially, and g axially. In order that the centrifugal field should predominate over the earth's field, $r\omega^2$ is commonly of the order of $100g$. This could be achieved by spinning a model at 3 m radius and 170 r/min or at 1 m radius and 300 r/min. A detailed discussion of centrifuge technology is beyond the scope of this work, but the following points should be noted:

1. Since the centrifugal field varies directly with r, it varies across a model, typically by ± 10 percent.
2. In dynamic problems, dr/dt may not be negligible and may destroy similarity between model and prototype.

3. While the model is being accelerated to the required speed, similarity does not hold.

Similarity and scaling laws Let the centrifugal acceleration exceed the earth's gravitational acceleration by a factor N. That is,

$$\frac{g_m}{g_p} = N \tag{17-1}$$

Further, in a model geometrically similar to a prototype, let the "gravitational" stresses be similar; i.e.,

$$\rho_m g_m L_m = \rho_p g_p L_p \tag{17-2}$$

Since the prototype soil is used in the model,

$$\frac{L_m}{L_p} = \frac{g_p}{g_m} = \frac{1}{N} \tag{17-3}$$

Thus stress similarity is assured in a 1:N scale centrifuge model. A consequence of this is that 1:100 scale models of earth slopes can be caused to fail due to their own self weight, so that Rocha's argument[35] ceases to hold. In the same way, capillary rise is given by

$$h_c = \frac{2T_s}{\rho_w g R} \tag{17-4}$$

so that for the prototype soil and pore fluid in a model, capillary rise scales as N and similarity is maintained. Thus it *is* advantageous to use the prototype soil in seepage models.[60]

Primary consolidation is governed by the H^2 scaling law, so that a given degree of consolidation in a model will occur N^2 times faster than in the prototype. Thus at 100g, 1 h in a model represents about 1 yr in the prototype.

In vibration problems, if the soil can be assumed to be elastic, dynamic similarity of elastic and inertial forces follows the Cauchy or Mach dimensionless-model law:

$$\left(\frac{Et^2}{\rho L^2}\right)_m = \left(\frac{Et^2}{\rho L^2}\right)_p \tag{17-5}$$

Thus time scales as N, and periods of vibration in a model would be typically 1/100-times those in the prototype. Note that the centrifugal force field enables sandlike elastic models to be scaled as N rather than $\sqrt{N}$, the Froude scaling, which is necessary in vibration studies using scaled models under normal gravity.[24]

The above considerations of scaling laws for vibration are true only for elastic materials. Viscous effects such as creep and damping cannot readily be modeled, and the interpretation of model test results in vibration problems is obviously much more complex than in static cases.

In addition, it is evident that different physical processes obey different time-scaling laws, so that if different effects are occurring simultaneously, faithful modeling is impossible. For example, consider a scaled model of an ocean-bed oil-production platform bearing on its soil foundations and subjected to wave-induced oscillations. In order to model correctly the frequency of the structure-soil interaction system, waves would have to be generated to scale in the model, with a period of $1/N$ times that of ocean waves. For $N = 100$ this would mean a period of some $\frac{1}{10}$ s. However the drag effects of the waves on the structure may follow Froude's law, which would require a period of 1 s. Similarly if drainage were occurring in the foundation soils by means of primary consolidation, the necessary period would be 1 ms. If the oscillatory motion is relatively short-lived, as in an earthquake, it would be legitimate to scale time as N for the duration of the shaking and as N^2 for the subsequent dissipation of excess pore water pressure (assuming that ground accelerations of the required magnitude with a period of some $\frac{1}{300}$ s could be produced in the centrifugal environment). In the study of response of ocean-bed structures to storms the solution again appears to be limited to the simulation of the storm at N followed by a dissipation phase at N^2. If the two phenomena are inextricably linked, modeling in the centrifuge is inappropriate. In such conditions, however, it should be noted that the governing mathematical equations would be complicated.

Importance of stress path As well as scaling difficulties in time-dependent problems, centrifugal modeling, as conducted at the present time, suffers from the difficulties inherent in adding material to, or removing material from, a model at full speed. Fluids or powders can relatively easily be introduced and water can be removed to simulate rapid drawdown in an earth dam,[1] but the commonest method of testing is to make a model with the *final* desired geometry, e.g., of an earth embankment or cutting, and then subject that model to a gradually increasing centrifugal field. Thus the model is only in similarity with its prototype when the desired acceleration is finally reached. The stress path to which the model is subjected is one of gradual gravity turn-on. In the field, however, the construction of embankments or cuttings involves a continual change in geometry at constant gravity. Thus in many instances the centrifuged model is not a true model of the field situation.

Conclusions Centrifugal modeling removes many of the objections previously raised to modeling in soil mechanics using prototype soils. (The alternative of using artificially weak soils has never been seriously pursued due to the critical influence of thixotropy and so on.[37,48,63]) Nevertheless models will generally fail to be precisely analogous to prototypes for various reasons, e.g., time-scaling and viscous effects in dynamic problems and stress-path considerations. Therefore the thesis put forward in this chapter is that a *combination* of physical and numerical modeling is necessary to improve predictive capability in geotechnical engineering. Some difficulties inherent in numerical modeling are now discussed.

Numerical Modeling Using Finite Elements

Development of finite element techniques has removed some, but not all, of Rocha's previously mentioned analytical difficulties in soil mechanics.[35] Nonlinear material laws, heterogeneous materials, and complex boundary conditions are nowadays fairly readily dealt with in analysis. However, although the true nonlinear, contractive-dilatant elastoplastic behavior of soils can in principle be analyzed,[52,53] this has not yet proved to be economically feasible in practical geotechnical problems. Further, it is still not possible to take proper account in analysis of the solid, fluid, and gaseous phases of the soil, and most analyses are essentially total-stress analyses. The exception to this rule is the use of Biot's theory for consolidation analysis, and while there are situations in which this is acceptable, there are many others in which the lack of coupling between shear-stress changes and changes in excess pore water pressure render the results meaningless.[51] Finally, even with the ever-increasing power of computers, three-dimensional analyses are costly, and idealizations of real problems into two-dimensional equivalents are often imperative.

Role of Physical and Numerical Modeling

In the previous sections it has been suggested that, for different reasons, centrifuged and finite element models of real constructions will generally tend to be unfaithful models. There are however further considerations which, if valid, limit the applicability *both* of physical and numerical modeling.[10] Many authors[7,11,22,31] have referred to the crucial influence of discontinuities on the behavior of geotechnical constructions. If these discontinuities cannot be scaled, as suggested by Roscoe[36] and Palmer and Rice[31] in the case of propagation of progressive failure surfaces, the consequences for physical modeling are clearly unfavorable. Nor are present finite element models capable of sustaining the propagation of a thin rupture zone in a previously undetermined direction. Nevertheless many cases will be cited, in the four problem areas considered, in which known prototype behavior has been predicted rather well from models. The roles of physical and numerical models will be seen to be complementary. Centrifuge-model tests which follow incorrect stress paths can be analyzed, followed by an analysis of the prototype's stress path. Conversely, since realistic numerical models for partial drainage are not yet available, data from the physical model may be the only means of interpolating between the initial (undrained) and terminal (fully drained) states. Due to the difficulties of scaling in dynamic problems, model and prototype vibrations can be studied separately with representative damping characteristics. Physical models are also an excellent means, under relative closely controlled laboratory conditions, of evaluating the applicability of analytical assumptions, e.g., the assumptions inherent in Biot's consolidation theory, in practical situations. All these points are amplified in a later section on applications, after the geotechnical aspects of constructing adequate physical models have been considered and the special numerical techniques appropriate in the finite element calculations have been outlined.

17-2 GEOTECHNICAL ASPECTS OF THE PROBLEMS

In contrast to the established *observational method*, which requires an initial site investigation only to establish the general nature of the deposits at a site, a predictive method involving modeling depends upon a thorough site investigation which will enable an adequate model of the site to be constructed. The scale of typical significant features on site will also determine the scale at which a centrifuge or finite element model can be representative of that site.

Modeling a Site

Many of the attitudes which led to the development of the observational method can be traced to the quality of site investigations and sampling techniques. Reliance on drillers' descriptions of severely disturbed material and on undrained shear strengths of "undisturbed" samples taken from small diameter drill holes in which water pressures were not balanced resulted in pictorial models of subsoil conditions of almost bewildering complexity (Fig. 17-1). By adopting many of the suggestions of Hvorslev,[18] Rowe[38,41] has been able to show that when improved methods of site investigation, sampling, and laboratory testing are employed, typical sites are far less inhomogeneous than might previously have been supposed. In constructing a model of a site, the salient properties to be modeled are the stiffnesses, strengths, and drainage characteristics of the various strata or regions. Rowe's procedure for obtaining the relevant information can be briefly outlined; it usually results in the division of a subsoil into about three or four regions. The procedure is as follows:

1. Relatively closely spaced small diameter (say 50-mm) *continuous* samples are taken (Fig. 17-2*a*), which on being split open enable the different strata to be classified according to their fabrics.
2. A small number of "large" diameter (260-mm) holes are drilled from which 250-mm-diameter specimens representative of the various strata can be prepared. (In addition, if the opportunity arises, block samples usually 1 m square are taken from test pits.)
3. Unconsolidated undrained triaxial tests on the large specimens give measures of the undrained stiffnesses E and strengths C_u of the strata. Variations in undrained strength with depth so obtained usually exhibit markedly less variation than conventional $U4$ strengths.[38]
4. Oedometer tests in which the direction of predominant drainage is governed by the fabric give measures of the drainage properties (k or C_v) and drained stiffnesses (m_v or E') of the ground. Permeabilities often change markedly with changing effective stress, e.g., in soft alluvial deposits.

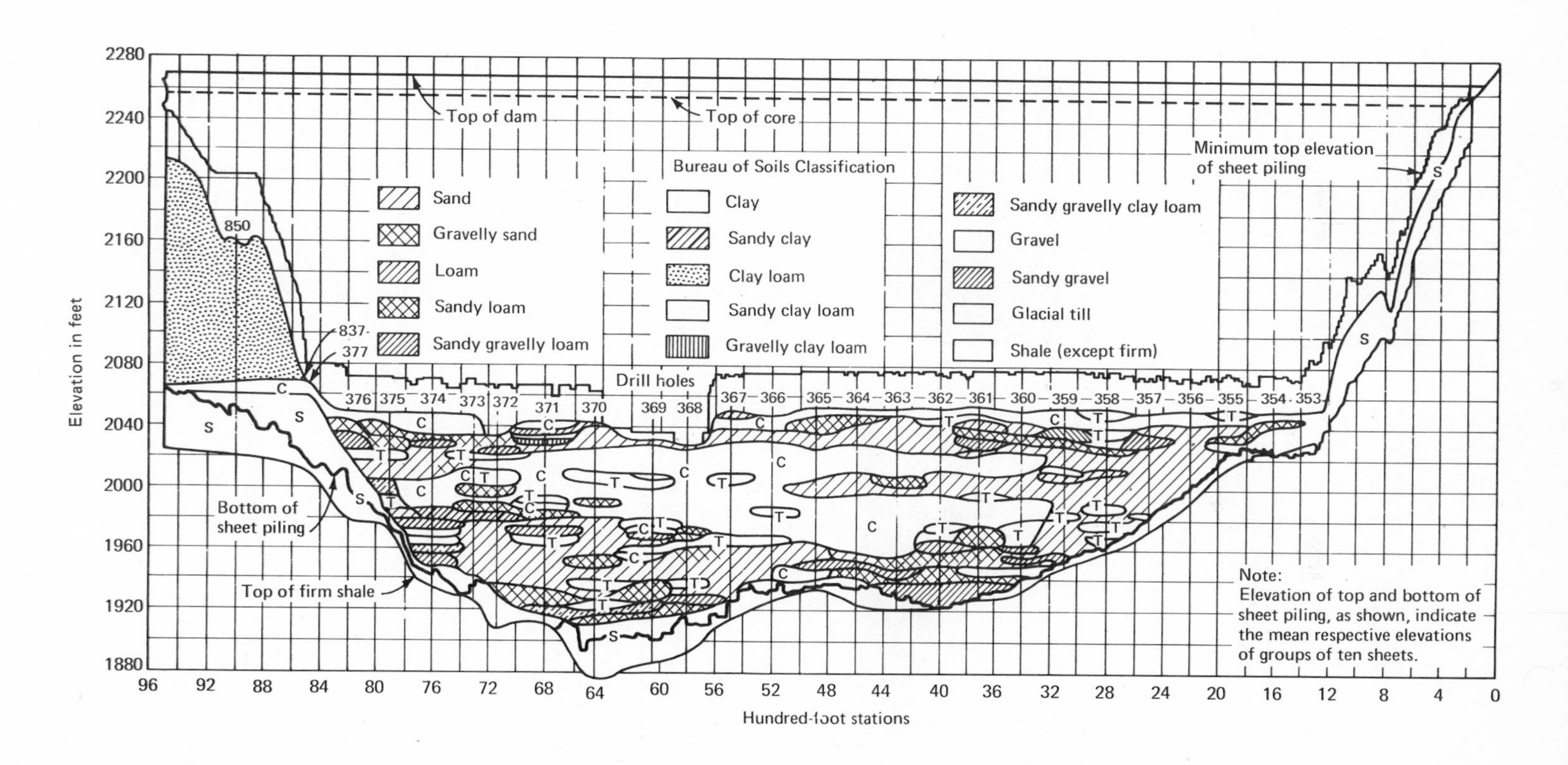

Figure 17-1 Soil profile along cutoff wall at Fort Peck Dam.[60]

(*a*) (*b*)

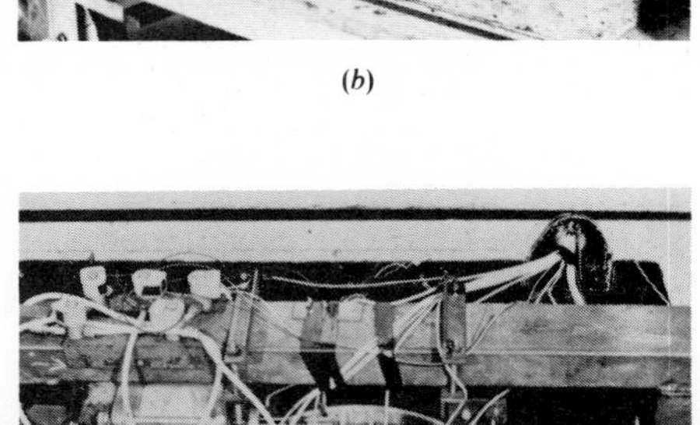

(*c*) (*d*)

Figure 17-2 (*a*) 50-mm-diameter continuous cores partially dried and split; (*b*) $1\frac{1}{2}$-t centrifuge model of earth dam; (*c*) failure of model centrifuged sheet-pile wall in clay; (*d*) model offshore oil production platform.

5. The laboratory information is supplemented where possible by in situ tests.
6. Blocks of undisturbed or compacted material from the site are used in the construction of models to be tested in a centrifuge.

Examples of applications of this procedure to numerical and physical models are given in a later section.

Centrifuge design is heavily dependent upon the types of problem to be modeled and not least upon financial criteria, since to a rough approximation machine cost increases with the square of the scale. Thus although the heterogeneity of soils leads to a demand for large models, the maximum mass of soil carried in the present generation of machines in the Soviet Union and in England is some 2 to 3 t at a mean radius of some 3 to 4 m. Acceleration fields created in these machines are in the range $50g$ to $300g$ but if consolidation phenomena are being studied, clearly one would not wish to go beyond $100g$. At this level, the thinnest layer of soil which can practically be constructed would be about 3 mm thick, which would model a 1-ft layer in the prototype. Similarly on the basis of *grain size*, medium silt in the model would represent coarse gravel. Since prototype soils are generally used in models, crushing of granular media can be expected to be more severe since the load per contact is much greater,[46] especially during bedding down.

Experience of large oedometer tests as simple one-dimensional consolidation models[39] indicates that in order to include sufficient fabric of a major stratum for the model to be representative, some 50 to 100 mm in thickness must be used. Since, as has been stated, sites commonly comprise up to six such strata, typical centrifuge models should be up to 0.6 m in depth. Plan areas are governed by the need to minimize interference from the container walls. As in the case of numerical models, "long" structures such as embankments are usually tested in section only. In this case it is important that the width-to-height ratio be about 2:3 and that the side walls be lubricated. The above considerations have led[38] to the design of a machine at the University of Manchester capable of testing 2.0 by 1.0 by 0.6 m models at up to 140*g* at mean radius 2.9 m, with the principal objective of studying plane deformation of earth dams. It must be emphasized that a need to study different classes of problems could lead to quite different machine specifications.

Constitutive Laws

Since centrifuge models are built from the prototype soil and brought into similarity with the prototype, the prototype constitutive behavior should be exactly reproduced under a subsequently imposed stress path. There are, however, practical difficulties in model construction which can lead to deviations from the desired behavior. These factors will be considered next in a separate section. In the present section, numerical constitutive assumptions are discussed for the four application problems in turn.

Problem 1: Deformation and failure of undrained slopes Recent reviews of finite element analyses of soil slopes have shown that the vast majority of contributions on the topic have treated the soil as an elastic material either linear or nonlinear (see Chaps. 2 and 16). Some work on elastoplastic analysis of excavations has been reported, and there seems to be no doubt that the elastoplastic constitutive behavior of soil should be recognized in calculations. While the elastoplastic behavior of drained and partially drained soils is complex,[28,50–53] it has been suggested[54,55] that the particular case of undrained deformation and failure in saturated clays is amenable to fairly standard analytical treatment, which can be recommended for immediate practical implementation. The clay is treated as being a linearly elastic–perfectly plastic material, as shown in Fig. 17-3. The elastic moduli have usually to be selected initially on the basis of triaxial test data but can subsequently be modified after some centrifuge tests have been analyzed. Piecewise linear elastic behavior could also be assumed if the desire for accuracy warrants it, but the cost of analysis is then increased. Yield can be assumed to take place according to the von Mises or Tresca criteria, and there are arguments in favor of each. After yielding, no-volume-change plastic flow occurs.

Problem 2: Rate of consolidation of soft foundations The only constitutive law currently available for the analysis of consolidation problems involving two- and three-dimensional deformations is Biot's poroelastic theory.[6] This constitutive

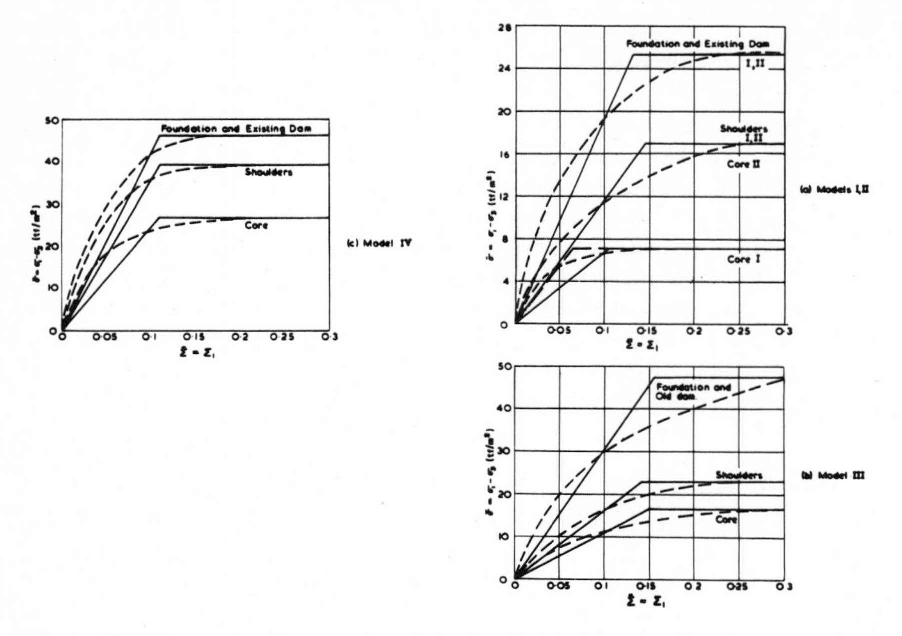

Figure 17-3 Measured and assumed undrained stress-strain curves for clays; data from triaxial tests.[54]

law does recognize the two- or three-phase nature of soil and involves the coupling of a linearly elastic solid soil skeleton and its pore fluid. Some idealized problems have been considered in the literature (see Chaps. 2 and 12), and it has been shown that despite the formidable objections to the theory on the grounds that shear-stress changes do not influence excess pore water pressure,[51] results of very great practical importance can be achieved in the particular case of fairly extensively loaded foundations.[56] The constitutive parameters required are the coefficients of permeability of the soil k_x, k_y, and k_z and the stiffness and Poisson's ratio of the skeleton (assumed isotropic) in terms of effective stresses, E' and ν'. The permeability should be obtained, preferably by direct measurement or alternatively by deduction from rates of excess pore water pressure dissipation, in oedometer tests on large specimens (250 mm diameter, 100 mm high). These will ordinarily be supplemented by in situ permeability measurements, but the laboratory tests are necessary in order to *predict* the change in permeability with increasing effective stress, which will usually be very marked in soft foundation deposits. Stiffness of the skeleton can also be measured in the oedometer, where in plane strain

$$\frac{1}{m_v} = \frac{E'(1 + \nu')}{(1 - \nu')(1 - 2\nu')} \tag{17-6}$$

Since m_v is the parameter measured, E' and ν' are not independently measureable.

To avoid numerical complications ν' can be chosen to be 0.3 so that m_v is a direct measure of E'. Again this will ordinarily change with increasing effective stress in soft soils. The constitutive law proposed is therefore a piecewise linear effective-stress-dependent Biot law.

Problem 3: Deformation and stability of flexible sheet piling For piles in clay the difficulties of handling other than the fully undrained or fully drained conditions are the same as those encountered in problem 1. Undrained conditions have been analyzed using the elastoplastic type of constitutive law shown in Fig. 17-3. For piles in sand, a nonlinear elastic constitutive law has been used to represent the sand's behavior,[8,57] as shown in Fig. 17-4*b*, and somewhat modified to introduce a measure of yield and no-volume-change plastic flow,[57] as shown in Fig. 17-4*c*. The methods by which the elastic parameters are obtained have been clearly stated in Chap. 2.

Problem 4: Static and dynamic behavior of heavy ocean-bed structures Due to the difficulties of site investigation and sampling in deep waters (200 m and more) the constitutive properties of stiffness, strength, and permeability cannot be obtained

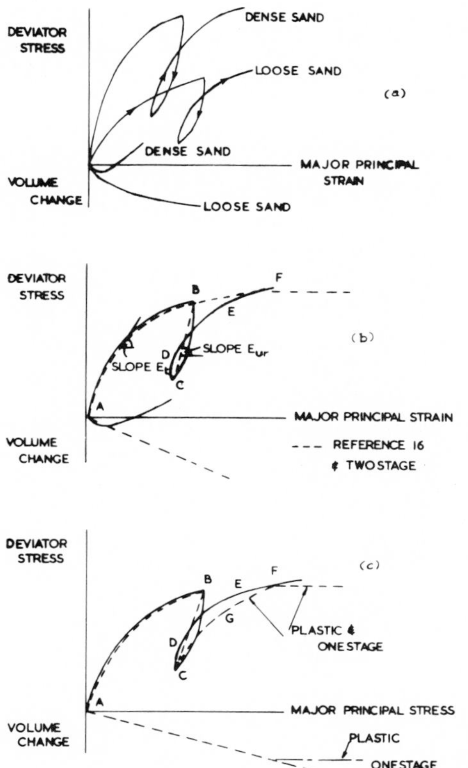

Figure 17-4 Stress-strain relationships for drained sand.[57]

with the precision which has become common on land. At the present stage of development, elastic stress-strain laws are used, and among problems of interest are the eigenvalues of the structure-soil system and the variation of these eigenvalues with local softening. Damping characteristics are extremely uncertain, and such experience as exists in other fields is borrowed.[47] At best, Rayleigh damping with strain-dependent damping ratios is assumed.

Special Factors: Softening, Interfaces, etc.

Whereas in reality there is often a gradual transition from one soil horizon to another, discrete element numerical models contain abrupt changes between elements, at best across special interfaces. However, due to scaling difficulties, abrupt changes in soil type involving an interface are also common in centrifugal models. For example, if it is desired to model a thin, soft, saturated clay layer sandwiched between two drier layers, suction of water from the saturated layer is likely to occur to an unacceptable degree during model preparation. Thus polyethylene sheeting is sometimes used to encapsulate layers and maintain their undrained state. Similarly, greased tracing paper has been used[14] to simulate joints in centrifuged models of rock slopes. To this extent all models tend to be more comparable with one another than with prototypes. Nondilatant joint elements[15] are used in the finite element models of the wall-soil interface in application 3. If softening is really related to particle size,[36] modeling of any sort is subject to error.

17-3 NUMERICAL METHODS

Since numerical techniques are considered in detail in Chap. 1, only a brief mention is necessary here of the methods used in the application problems. Three facets of numerical analysis will be seen to be involved in the applications as follows:

1. Solution of linear and nonlinear sets of simultaneous equations generated by elastic and elastoplastic finite element formulations
2. Integration of first-order time-dependent problems arising in analysis of primary consolidation
3. Isolation of (lowest) eigenvalues in dynamic problems

The methods are now considered in turn.

Elastoplastic Analyses

The method used successfully for the simple constitutive assumptions made is the *initial-stress method*.[69,53] For more complex constitutive laws it may be preferable to use hybrid methods.[52] The method can incorporate a *failure criterion*, which, although arbitrary, has been shown to lead to close correlations with good upper- and lower-bound plasticity solutions[54,55] and with $\phi_u = 0$ slip path analyses commonly used in practice. The criterion used is that the change in any term in the incremental load vector between successive iterations should not exceed one-thousandth of the maximum term in that load vector after 30 iterations. If this criterion is not met, the structure is said to have failed. While it may be that this criterion is dependent on the types of problem under consideration and on the types of element used (linear isoparametric quadrilaterals in this case), it is extremely important that such a link between finite element computations and established geotechnical engineering practice be forged.

First-Order Time-dependent Problems

The typical problem, which arises for example in the analyses of consolidation problems, is the first-order ordinary differential equation in time:

$$[A] \frac{d\{u\}}{dt} = -[B]\{u\} + \{S\} \tag{17-7}$$

subject to some starting condition on $\{u\}$,

$$\{u\}_{t=0} = \{u_0\} \tag{17-8}$$

Instead of proceeding to generate the Crank-Nicolson and other methods by finite differencing Eq. (17-7), one could integrate the equation explicitly to give

$$\{u\}_{t+\Delta t} = \exp(-\Delta t[A]^{-1}[B])(\{u\}_t - [B]^{-1}\{S\}) + [B]^{-1}\{S\} \tag{17-9}$$

While direct evaluation of the matrix

$$\exp(-\Delta t[A]^{-1}[B]) = \exp(-p[C]) \tag{17-10}$$

is impractical, it is of interest to examine *rational approximations* to $\exp(-p[C])$. For example, Padé approximations[12,17] can be used.

Alternatively a class of approximations to e^{-z} with denominators of the form $(1 + \gamma z)^n$ is given by the Nørsett approximations.[29]

A third class of approximations is obtained using iterated methods. For example, the fully implicit and Crank-Nicolson methods are expanded to read

$$e^{-z} = \frac{1}{(1 + z/n)^n} \quad \text{and} \quad e^{-z} = \frac{(1 - z/2n)^n}{(1 + z/2n)^n} \tag{17-11}$$

respectively. Thus it becomes a matter of whether n iterations with time step $n\,\Delta t$ yield more stable or accurate answers than n time steps with time step Δt.

Fuller details of all these methods are given in Ref. 49. The great advantage of

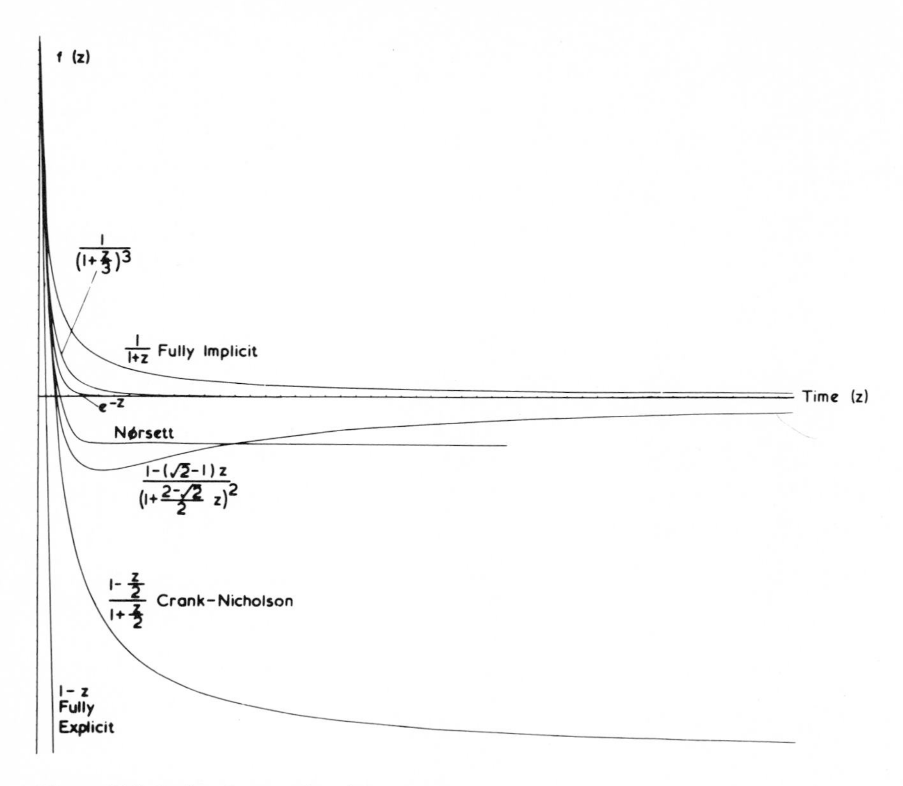

Figure 17-5 Rational approximations to e^{-z}.

these rational approximations to e^{-z} from an engineering point of view is that the approximations can be plotted as a graph and compared with the true curve. This is done in Fig. 17-5 for various Padé, Nørsett, and iterated methods. The reasons for instabilities and oscillations in these methods[58] thus becomes clear in that the difference between the true solution and the approximate one, i.e., the error, tends to be unbounded in the explicit method with increasing time step, one sign in the fully implicit methods leading to monotonic but excessively damped behavior and both signs in methods in which the approximation crosses the abscissa, such as the Crank-Nicolson method, leading to an oscillating result.

This is demonstrated in a consolidation problem governed by the simple Terzaghi theory by the results shown in Fig. 17-6. The explicit method is unstable in this case for $\Delta t \geqslant 0.25$. However all the implicit methods give good results up to $\Delta t \approx 1.0$. Thereafter the oscillatory behavior of the Crank-Nicolson method is apparent, as is the excessive damping in the fully implicit case. At $\Delta t = 5$ the Nørsett approximation is still excellent, although even this becomes oscillatory at $\Delta t = 10$. In this instance the iterated methods are less accurate than multistep equivalents.

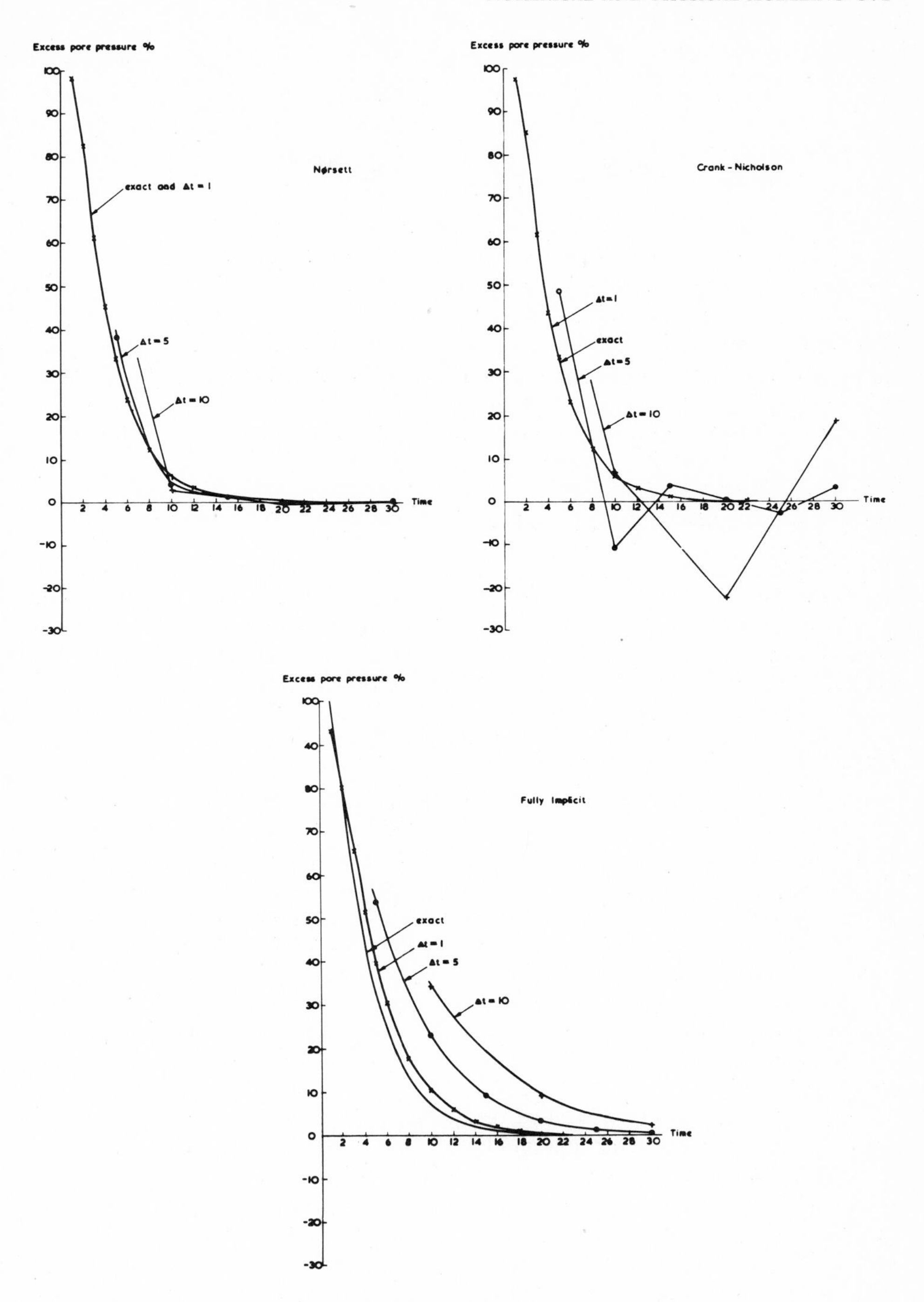

Figure 17-6 Comparison of numerical procedures for integrating in time.

Eigenvalue Problems

Wilkinson[67] studied these problems extensively, but there are few efficient algorithms available for dealing with the sparse banded matrices generated by finite element idealizations.[21] Out-of-core[3] and in-core[4] methods have recently been outlined which are promising in this regard. Algorithms based on Wilkinson's work have been implemented in the application described.

17-4 APPLICATIONS AND DESIGN ANALYSIS

The four application problems are considered in turn. Relevant studies using physical and numerical models are briefly reviewed, and comparisons of results from the two methods are given where possible with correlations with field work. In all cases the relative merits of finite element and centrifugal model studies are assessed.

Deformation and Failure of Undrained Slopes

Many finite element analyses of slope deformations have been published both for built-up slopes and for excavations (see Chap. 15). Few[54,55] involve direct computation of a failure condition for the slope, the usual method being to perform subsequent slip-path limiting equilibrium calculations where the slip paths pass through the highly sheared zones. As mentioned previously, elastic soil material is usually assumed.

Slope problems have also been extensively studied using centrifugal models since the pioneering work of Pokrovsky in the early 1930s using small model earth slopes some 550 by 60 mm in plan and 200 mm deep.[33] Work has continued in the Soviet Union[5,61,64] on deformation and failure of clay slopes and rock-fill dams under steadily increasing centrifugal fields, and failures due to rapid drawdowns have also been studied.[1,19] In Japan, models of a 90-m-high rock-fill dam have been tested[26] as well as slopes in undisturbed and reconsolidated clays.[27] Simulation of cuttings made in stiff fissured clay has been attempted,[25] and the difficulties of sufficiently accurate measurement of ground displacements using photogrammetry discussed.[9] Interaction between two different modes of failure of model flood-protection embankments, namely by increasing centrifugal force and by increased artesian foundation pressure at constant centrifugal force, has also been considered.[10]

The present example studied in detail is a zoned earth embankment dam.[38,42,54] A typical centrifuge model of such a dam, having a mass of $1\frac{1}{2}$ t is shown in Fig. 17-2*b*. Considerations leading to the adoption of the height-to-width ratio and model scale and the difficulties of measuring pore water pressures in models have already been published.[38,42] Typical stress-strain curves for model clays, obtained from unconsolidated-undrained triaxial tests on undisturbed and compacted materials, and the approximations to those used in finite element calculations are shown in Fig. 17-3.

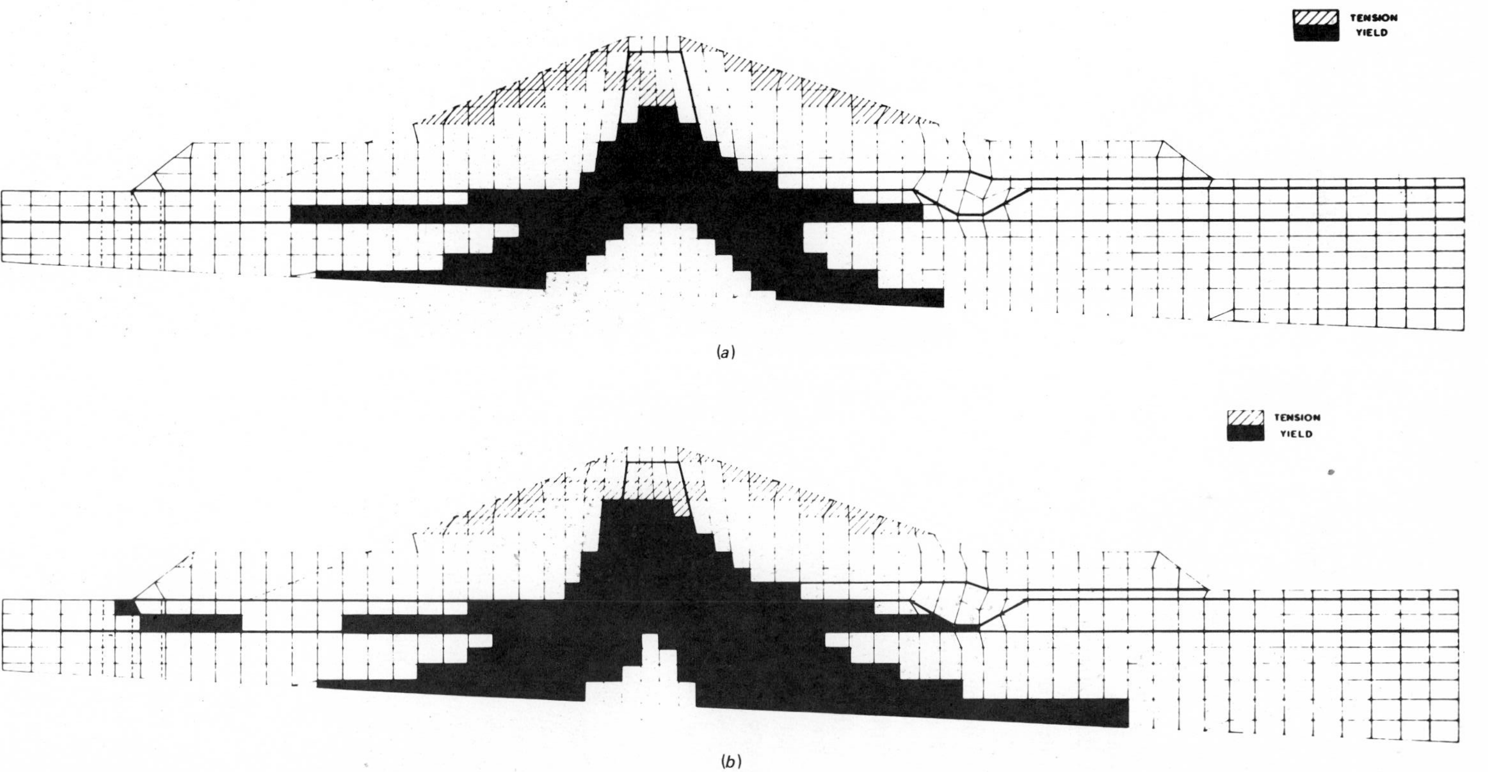

Figure 17-7 Computed zones of yield and tension: dam under centrifugal loading at (*a*) $N = 155$ (no downstream berm); (*b*) $N = 184.7$ (downstream berm added).[54]

The dam shown in Fig. 17-7 is founded on weathered clay overlying unweathered clay and sands. Its core is made of the weathered foundation clay compacted wet, while imported siltstone, compacted dry, forms the shoulders. Several centrifuge models were built at a scale of 1:134, but even at this scale it was necessary to test the upstream and downstream sections separately (see also Fig. 17-2*b*). Finite element checks on the validity of this truncation were useful. The model foundations were constructed using 1 by 1 m block undisturbed samples from the site in order to preserve the natural fabric. Water content and compactive efforts representative of site conditions were used for the dam materials.

A comparison of the results obtained from physical and numerical models of this type has shown that:[54]

1. The effects of truncating models in the manner shown are slight.
2. The numerical method accurately predicts the zones of yielding and of tension observed in centrifugal models.
3. The centrifuge applies a gravity turn-on type of loading to a model which is unrepresentative of the layered construction sequence in the field.
4. This results in excessive tensile zones behind the crests of slopes and approximately halfway down the slopes.
5. Nevertheless, computation shows that yielding and failure in model and prototype should occur at about the same load.
6. Although finite element failure loads agree well with $\phi_u = 0$ calculations, models fail at higher loads because the clays do not remain undrained.

The power of the elastoplastic finite element calculation incorporating a failure criterion is shown by the results in Fig. 17-7. Various configurations for the dam cross section were investigated at the design stage, with differing provisions for berms, etc. When a trial slip-path method is used for the analysis of the undrained condition, many thousands of paths have therefore to be investigated.

The calculations show that without the downstream berm, the factor of safety is 1.15 associated with a downstream slip, while the addition of the berm raises the factor of safety to 1.4 associated this time with upstream failure. This information, together with likely zones of yielding in which piezometers should be installed and the complete deformation history, was obtained in two computer runs at a cost similar to that involved in repeated slip-path trials.

Relative merits of physical and numerical models of undrained slopes The essential feature that emerges from the above remarks is that neither physical nor numerical models faithfully represent field conditions but that they tend to be inadequate in different ways. Therefore a combination of both types of study can lead to predictions of field behavior with a tolerable level of confidence. The disadvantages of physical models are:

1. They are unable to model prototype stress paths of embanking and excavation due to the difficulties of moving material in the high acceleration environment.

2. The time taken to accelerate a model to achieve similarity may be long in comparison with field construction times, leading to unrepresentative drainage.
3. Pore suctions during model construction sometimes lead to a need to isolate wet regions from suction and lead to difficulties in pore water pressure measurement.

These are offset by the following advantages:

1. Use of real soil at prototype stress levels assures correct modeling of complex constitutive behavior, involving stress-strain response and drainage.
2. Collapse and postcollapse mechanisms, often involving large displacements, are clearly demonstrated.
3. Excess pore water pressures can be recorded and used in effective stress designs.
4. Three-dimensional effects are readily modeled.
5. Many real soils tend to drain somewhat during "undrained" deformation, and this effect is correctly modeled.

In contrast, the deficiencies of numerical modeling are:

1. Idealized and probably unrepresentative constitutive behavior has to be assumed.
2. Total stress analyses must be conducted, with no direct prediction of excess pore water pressures.
3. Analysis of three-dimensional effects is at present costly.
4. Rupture propagation and postfailure mechanisms are not predicted.

Compensating advantages are:

1. Finite element models, certainly in two dimensions, can be more extensive than typical centrifuge models. Similarly the boundary conditions in the centrifuge may not be precisely those in the prototype. The centrifuge model can therefore be modeled numerically and compared with the numerical model of the prototype.
2. In the same way, the stress path and construction sequence in centrifuge model and prototype can be independently modeled numerically.
3. The cost of minor alterations to finite element models (geometry, constitutive behavior, etc.) is negligible compared with the cost of building a new centrifuge model.

Rate of Consolidation of Soft Embankment Foundations

Many numerical models of settlement of embankment foundations have not incorporated the embankment stiffness in the analysis.[23] The only available methods of doing so are based on the use of Biot's consolidation theory, which has recently been applied to practical field problems.[56]

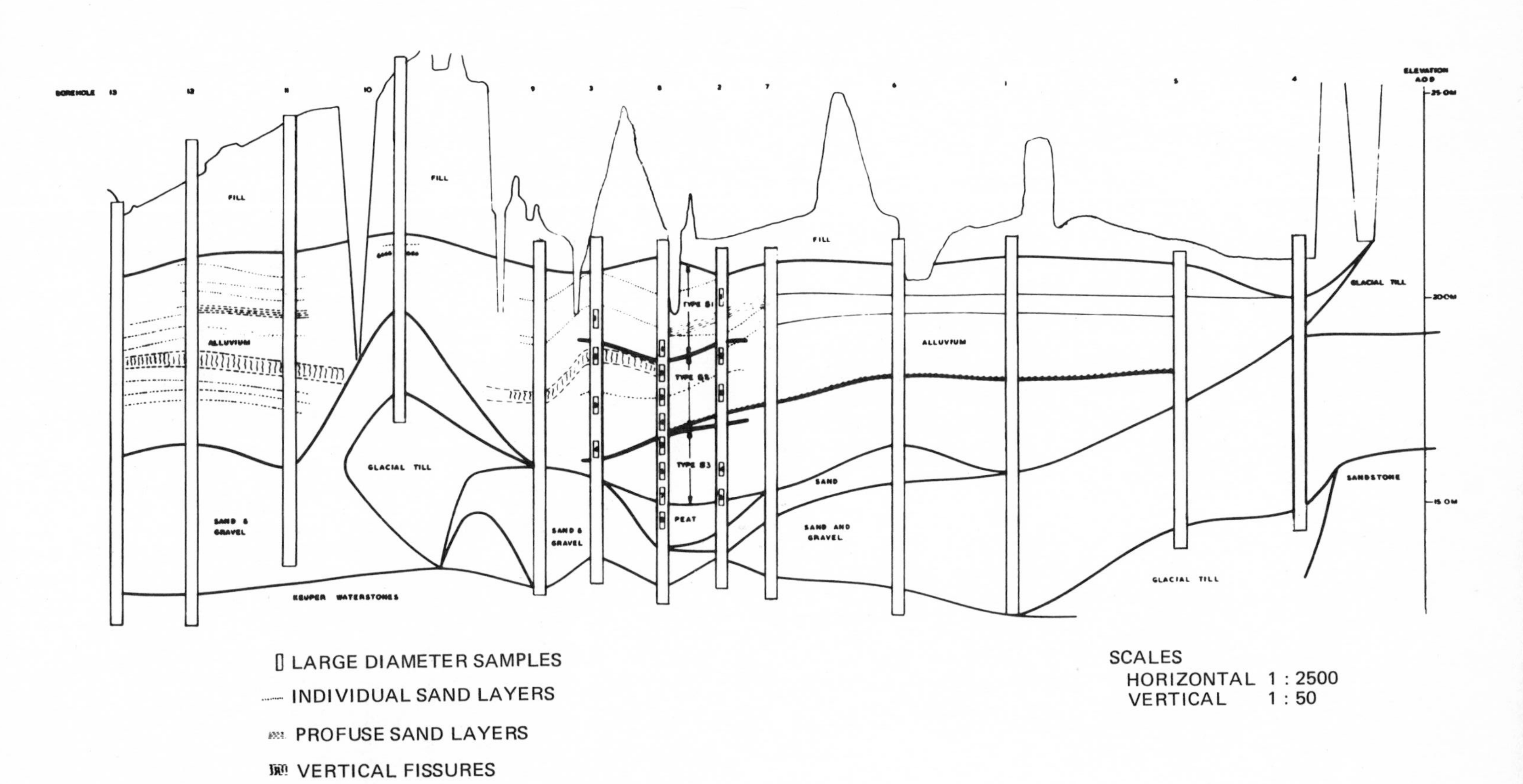

Figure 17-8 Site stratigraphy

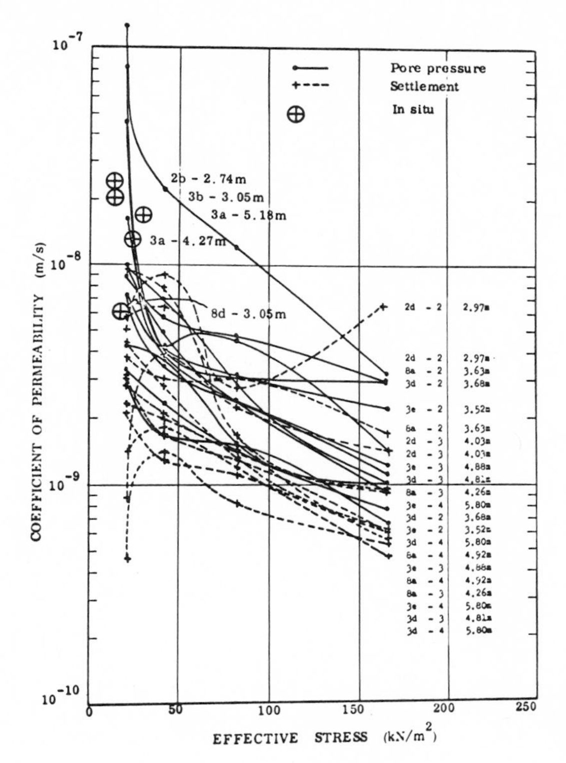

Figure 17-9 Vertical flow oedometer test, permeability vs. effective stress.

Centrifugal testing of soft foundations has been carried out in the Soviet Union,[34] and measurements of excess pore water pressures induced beneath model embankments in the centrifuge have been published.[38,43] Other work on similar lines concentrated on measurement of displacement fields at a boundary of a glass-contained model embankment and foundation.[2]

The present example is concerned with a road embankment constructed on soft alluvium. The site investigation with continuous sampling (Fig. 17-2*a*) led to the modeling of the site by three dominant soil types (Fig. 17-8). Large-diameter samples representative of the three soil types were extracted and their compressibilities and permeabilities determined in oedometer tests, with results shown in Fig. 17-9. [The predicted change in permeability with effective stress was subsequently confirmed by in situ permeability tests (Fig. 17-10).]

The program of construction on site involved embanking to the first berm level in some 70 days followed by a rest period of some 60 days and completion in 210 days. This was simulated in the finite element model by adding layers gradually to the model, thus building up its stiffness with time. The effect on the foundation pore pressures and a comparison with field piezometer measurements are shown in Fig. 17-11. The embankment was believed to be about 10 times as

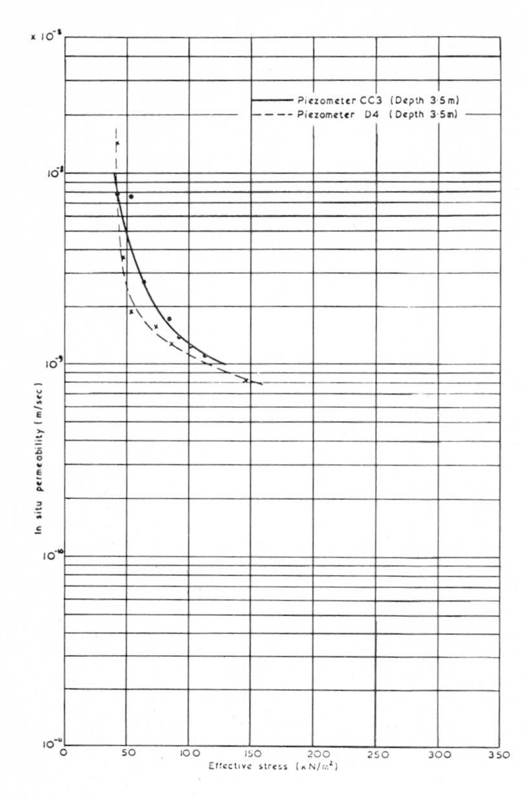

Figure 17-10 Insitu permeability-effective stress relationship.

stiff as the foundation, and the influence of relative stiffness on the distribution of excess pore pressure across the foundation section is shown in Fig. 17-12. Poisson's ratio for the bank also has a significant influence.

The same problem has been analyzed assuming gravity turn-on gradually applied to the complete bank with time. Apart from the transients during pore pressure build-up, the pore pressure response has been found to be similar to that in the incrementally constructed model and hence in the prototype. Thus it appears that the centrifuge can be successfully used to model such problems.

This particular road embankment has not been modeled centrifugally, but Fig. 17-13 shows the distribution of excess pore pressure across a centrifuged model foundation compared with finite element predictions. The form of results, indicating load transfer through the stiffer shoulders, is predicted and contrasts with the previous case of a homogeneous embankment.

The advantages and disadvantages of the two types of modeling are again as stated for the first problem. The Biot constitutive law cannot be expected to be valid if there is significant shearing without increase in mean pressure[51] or significant plastic flow but appears to work rather well in the present instance, at least for extensive embankments.

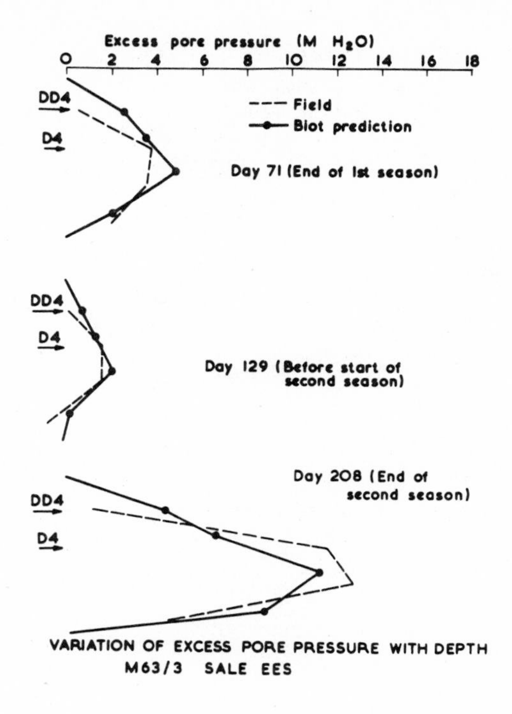

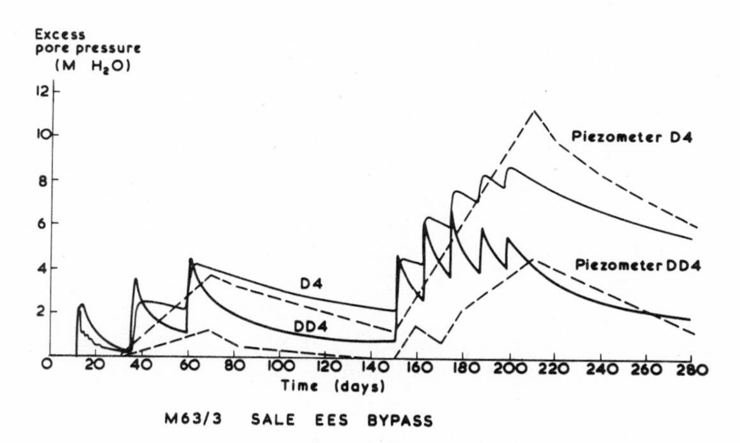

Figure 17-11 Comparison of computed foundation pore pressures with field piezometer measurements.[56]

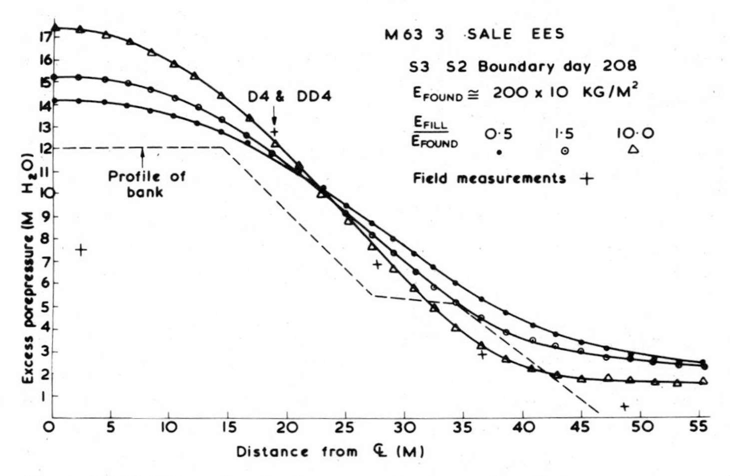

Figure 17-12 Effect of embankment properties on foundation pore pressures.[56]

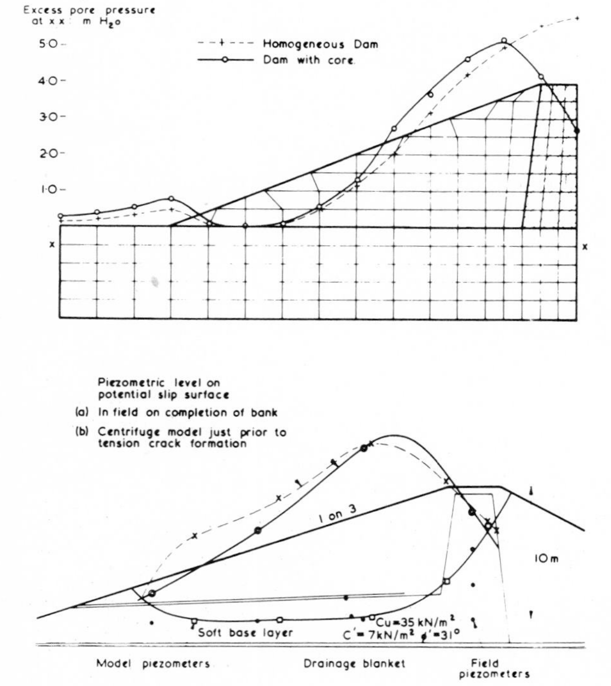

Figure 17-13 Effect of embankment inhomogeneity on foundation pore pressures.

Deformation and Stability of Flexible Sheet Piling

Rowe[44,45] conducted many laboratory tests at normal gravity on 0.76- to 1.07-m-high model sheet-pile walls retaining cohesionless materials. As he has pointed out,[42] his results were in agreement with those of Tschebotarioff,[64] who worked with walls 2.7 m high (a scale factor of about 3) and a further factor of about 6 would achieve field scale. It is therefore unlikely that the small-scale results are significantly in error for this particular case, which is a deformation rather than a stability problem, no field study yet having been reported of failure of a sheet-pile wall in bending.[42]

A comprehensive series of numerical comparisons with Rowe's physical-model test data have been presented,[57] which indicate (Fig. 17-14) that the measured bending moments in flexible walls can be predicted to within about 5 percent by a finite element method using an appropriate constitutive assumption (Fig. 17-4*c*). Since the computations assumed full-scale dimensions, there does not seem to be any need for centrifugal model studies of problems of this class. Rowe only tested uniform cohesionless materials (sands, silts, ashes), but the computations can readily be extended to nonuniform soils. It may even be impossible to conduct relevant centrifugal model studies because of the critical effects of very small yield in the anchors, which would have to be controlled in the centrifuge to less than $\frac{1}{10}$ mm. Stress-path considerations are also likely to be very significant in such a structure-soil interaction problem. The relevant physical tests appear to be those previously conducted by Rowe and by Tschebotarioff. However, since Rowe's data from some 1000 tests occupying 10 years were reproduced in a matter of weeks on the computer, given normal turnaround times, the conclusion must be reached that, if computer solutions had been available, so many physical tests would not have been necessary. Conversely, though, if no physical test data had been available, it would not have been possible to isolate a relevant numerical method and constitutive assumption for the computations.

Flexible walls in clay For walls driven into clays which remain undrained, one design problem is that of safeguarding the immediate stability of the wall. Centrifuge studies[27] have shown overall collapse of walls, and a typical failure of a centrifuged model wall in clay is shown in Fig. 17-2*c*. This problem is therefore essentially analogous to the slope-stability problems considered previously, and finite element computations based on elastoplastic constitutive laws can again be used to predict the progress of yielding culminating in a collapse load. The centrifuge, however, is useful for modeling the progress of drainage with time for an initially stable wall, leading to a long-term working load condition which ought to be similar to that previously measured and computed for sands. Nevertheless some clays will be more compressible even than the loose silt already tested.[44] Centrifugal studies of this kind are progressing.

Relative merits of physical and numerical models For walls in cohesionless materials, previous physical model studies appear to be more appropriate than centrifugal studies, which suffer from difficulties in simulating field stress paths and in

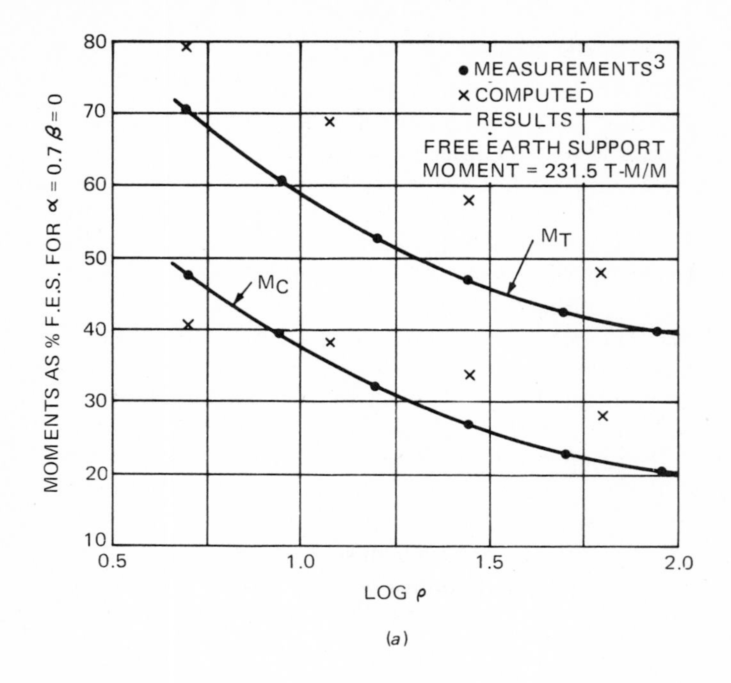

(a)

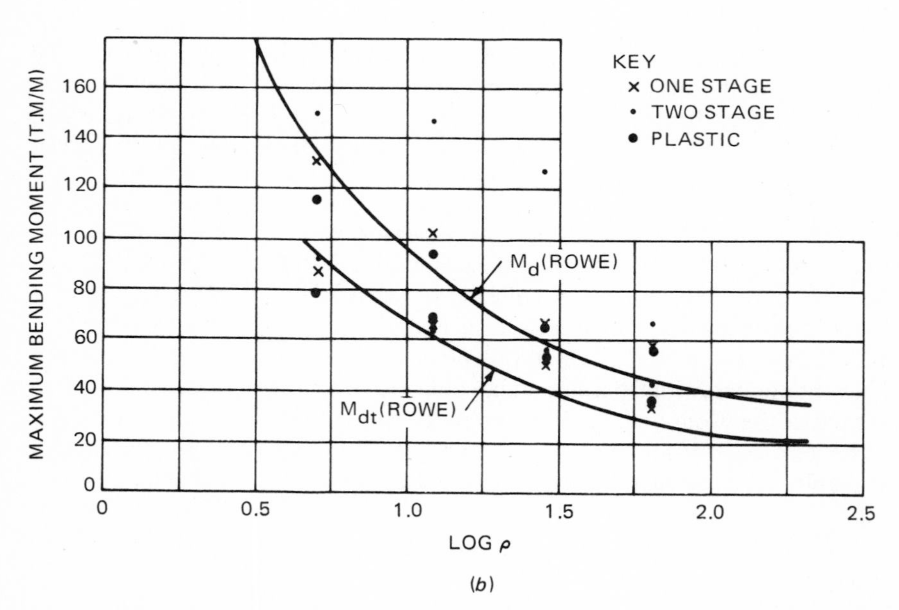

(b)

Figure 17-14 (*a*) Computed and measured moments in encastré walls; (*b*) computed and measured moment reductions in anchored walls.[57]

controlling anchor yields. Currently available numerical models appear to be adequate but may need to be supplemented by 1:10 to 1:20 scale physical models at normal gravity in which field stress paths can be modeled and inhomogeneities or divergences from standard conditions, e.g., inclined dredge lines, considered.

For walls in clay the centrifuge is useful for investigating "immediate" stability and for studying the process in deformation of initially stable walls with time. Numerical models provide checks on the initial stability and on the "long-term" behavior of walls if this should prove to be similar to that of walls in sand.

Static and Dynamic Behavior of Heavy Ocean Bed Structures

When novel forms of construction are proposed for which there are no precedents, models are the only means of assessing likely or possible field behavior. As was stated in Sec. 17-1, the traditional observational method is no longer helpful, since these structures are built in relatively shallow waters, towed to their ocean location in perhaps 100 m depth of water, and sunk to the bed. The design problems involve likely oscillations of the structure-foundation system under normal working conditions and under maximum storm loads, which may not occur for some months or years after construction. There is a similarity with the design of structures to resist earthquakes, but the modes of deformation are quite different.

Models of the interaction between oil production platforms and their foundations under oscillating loads have been tested in a centrifuge.[59] As yet the studies have been limited to true modeling of the foundation soils and to representing the structure by a flat plate rocked by means of hydraulic jacks (Fig. 17-2*d*).

Since the modeling law for vibrations indicates that frequencies of oscillation should probably follow the model scale, frequencies of the order of 10 Hz are required to model storm loading on a 1:100 scale model. Such frequencies have not yet been achieved in the centrifuge, and present work has been limited to 1 Hz. However, valuable information has been obtained regarding the mechanisms of deformation of such structure-soil systems under load.[59] Local yielding at the edges of structures has produced a progressive settlement with increasing wave loading, which agrees well with prototype observations at Ekofisk.[13] It should be noted, however, that the magnitude of the Ekofisk settlements is of the order of 250 mm, so that crushing effects[46] are likely to mean that 1:100 scale deformations will not occur in the models. An extremely important aspect of the models is that they allow an estimate of the damping properties of the foundation soils to be made.

Numerical models of ocean-bed structures are being intensively studied by designers and by classification societies. A typical structure and finite element model are shown in Fig. 17-15. A three-dimensional finite element model shows local yielding at the edges of a structure as experienced in the physical models (Fig. 17-16), and two-dimensional dynamic analyses indicate the influence of local yielding on the eigenvalues of the system (Fig. 17-17).

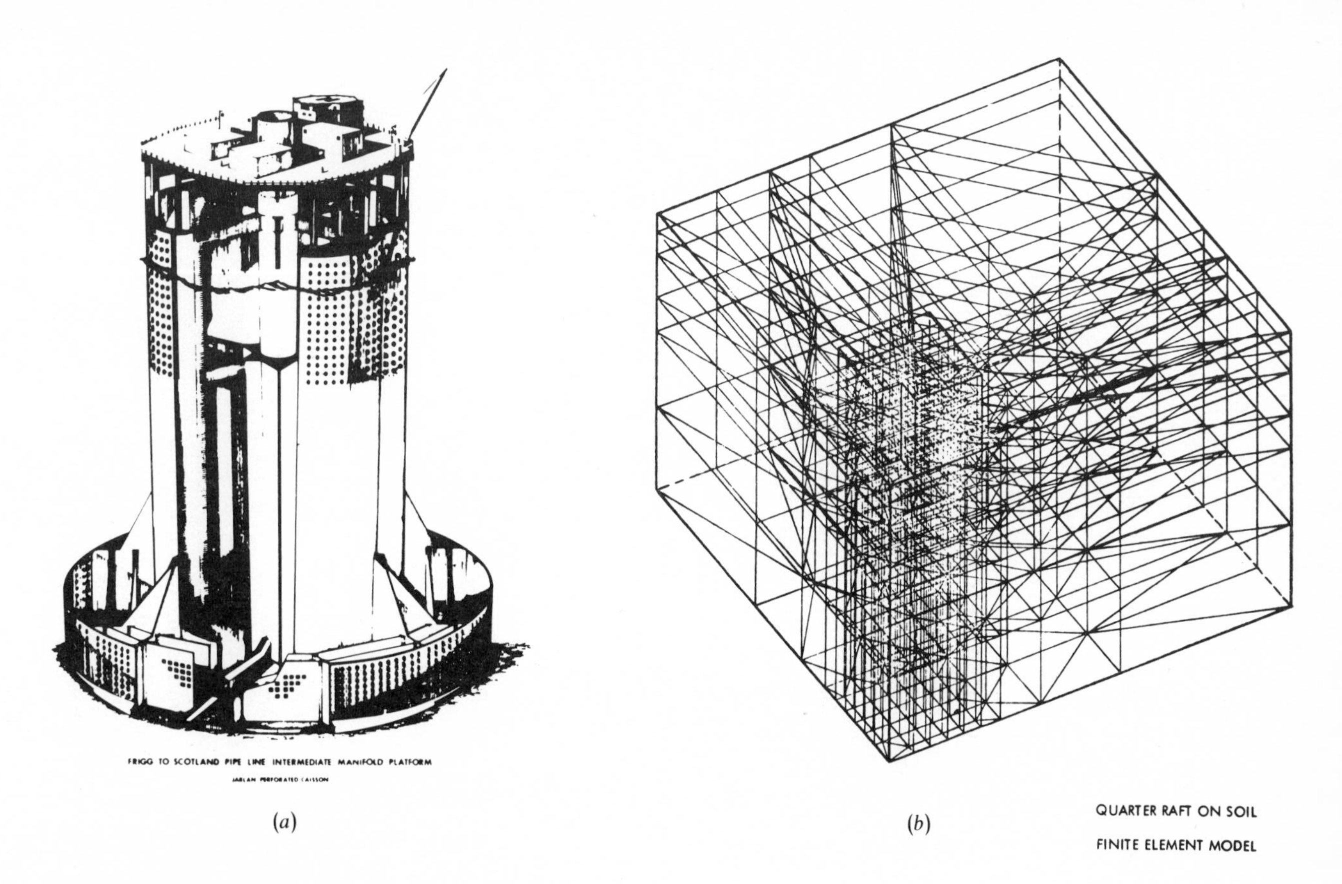

Figure 17-15 (*a*) Typical structure; (*b*) typical finite element model.

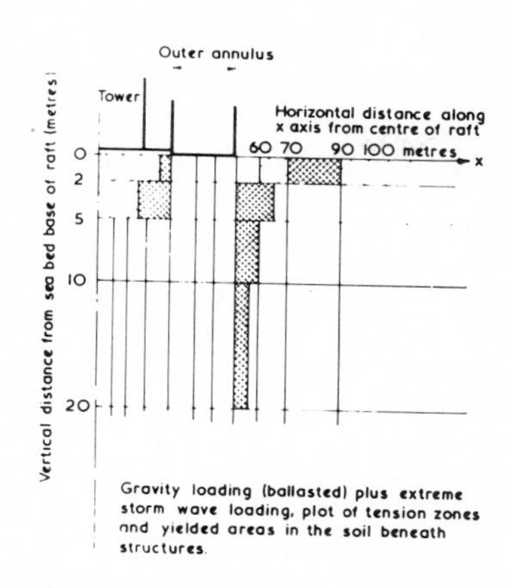

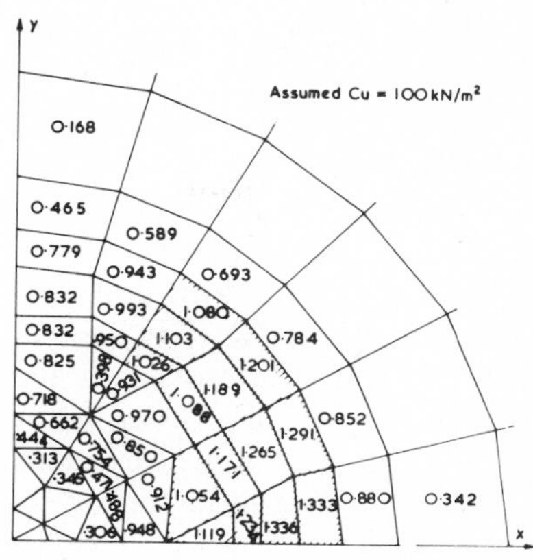

Figure 17-16 Local foundation yielding.

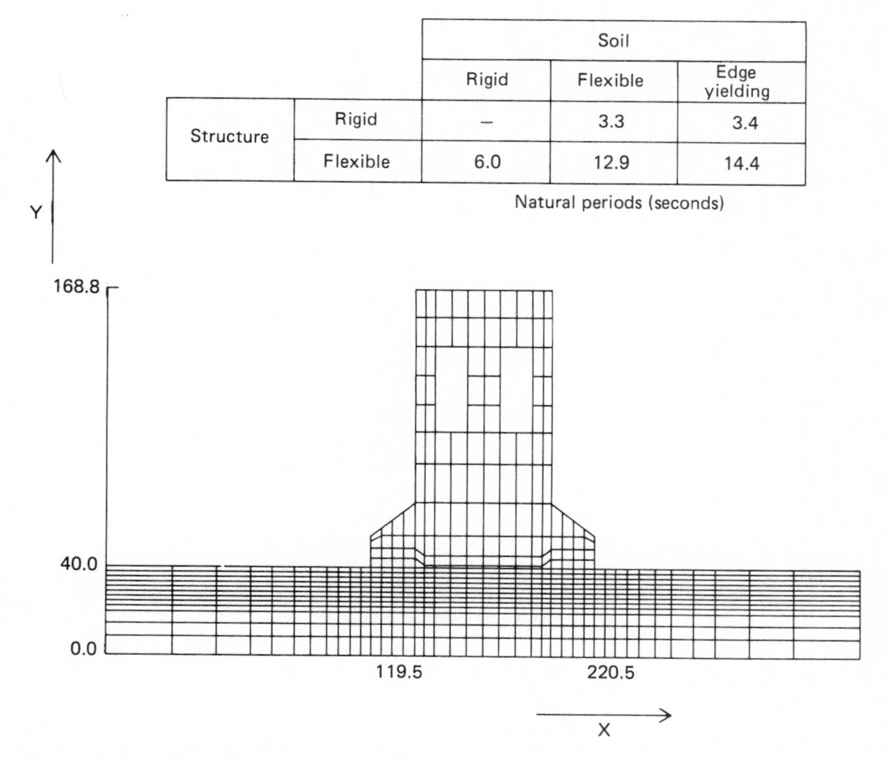

		Soil		
		Rigid	Flexible	Edge yielding
Structure	Rigid	–	3.3	3.4
	Flexible	6.0	12.9	14.4

Natural periods (seconds)

Figure 17-17 Effect of foundation weakening on natural periods of structure-foundation system.

Relative merits of physical and numerical models In centrifuge models, the three-dimensional nature of problems can be investigated and a measure of damping effects in the real soil can be achieved. However, the models are at present incomplete and will probably never purport to be true models of the freely vibrating field situation. Numerical models of the full three-dimensional problems are expensive and suffer from the usual uncertainties about constitutive laws. However eigenvalues can be obtained and the equations of motion integrated. As usual, maximum benefits are likely to accrue from a combined study, employing both physical and numerical models in a complementary way. It has been shown by both methods that local yielding rather than general shear is likely to occur under platforms, in contrast to the earthquake situation. For this reason it is important to obtain the stress paths undergone by the soil from model studies. The "index" type of softening tests on soil elements used in earthquake studies are not likely to be relevant.

REFERENCES

1. Avgherinos, P. J., and A. N. Schofield: Drawdown Failures of Centrifuged Models, *Proc. 7th Int. Conf. Soil Mech. Found. Eng., Mexico, 1969*, vol. 2.
2. Bassett, R. H.: Centrifugal Model Tests of Embankments on Soft Alluvial Foundations, *Proc. 8th Int. Conf. Soil Mech. Found. Eng., USSR, 1973*, vol. 2.2.
3. Bathe, K. J., and E. L. Wilson: Large Eigenvalue Problems in Dynamic Analysis, *J. Eng. Mech. Div. ASCE*, vol. 98, no. EM6, December 1972.
4. Bathe, K. J., and E. L. Wilson: Eigensolution of Large Structural Systems with Small Bandwidth, *J. Eng. Mech. Div. ASCE*, vol. 99, no. EM3, June 1973.
5. Belfer, S. M.: The Influence of Soil Condition on Deformation of the Base of an Embankment, *Soil Mech. Found. Eng. USSR*, November–December 1964.
6. Biot, M. A.: General Theory of Three-dimensional Consolidation, *J. Appl. Phys.*, vol. 12, 1941.
7. Bjerrum, L.: Problems of Soil Mechanics and Construction on Soft Clays and Structurally Unstable Soils, *Proc. 8th Int. Conf. Soil Mech. Found. Eng., USSR, 1973*, vol. 3.
8. Bjerrum, L., et al.: Stability of Flexible Structures, *Proc. 5th Eur. Conf. Soil Mech. Found. Eng., Madrid, 1972*, vol. 2.
9. Bolton, M. D., et al.: Ground Displacements in Centrifuged Models, *Proc. 8th Int. Conf. Soil Mech. Found. Eng., USSR, 1973*, vol. 1.1.
10. Bolton, M. D., et al.: Modelling, *Proc. Symp. Plasticity Soil Mech., Cambridge, 1973.*
11. De Beer, E. E.: The Scale Effect on the Phenomenon of Progressive Rupture in Cohesionless Soils, *Proc. 6th Int. Conf. Soil Mech. Found. Eng., Montreal, 1965*, vol. 2.
12. Donnelly, J. D. P.: The Padé Table in Handscombe (ed.), "Methods of Numerical Approximation," Pergamon Press, New York, 1965.
13. Foss, I.: Ekofisk Settlements and the Steady Sea Lab, *Ground Eng.*, vol. 7, no. 4, July 1974.
14. Goldstein, M., et al.: Stability Investigation of Fissured Rock Slopes, *Proc. 1st Int. Congr. Rock Mech., Lisbon, 1966*, vol. 2.
15. Goodman, R. E., et al.: A Model for the Mechanics of Jointed Rock, *J. Soil Mech. Div. ASCE*, vol. 94, no. SM3, May 1968.
16. Graney, L., I. M. Smith, and J. Walsh: Response of Unconfined Aquifers to Pumping, *Proc. Symp. Numer. Methods Fluid Flow Prob., Swansea, 1974.*
17. Graves-Morris, P. R. (ed.): "Padé Approximants and Their Applications," Academic Press Inc., New York, 1973.
18. Hvorslev, M. J.: "Subsurface Exploration and Sampling of Soils for Engineering Purposes," The Engineering Foundation, New York, 1949.

19. Hydroprojekt: "Research Studies for Designing and Erection of Earth and Rockfill Dams," All-Union Designing, Surveying and Scientific Research Institute, USSR, 1973.
20. Idriss, I. M., et al.: Seismic Response by Variable Damping Finite Elements, *J. Geotech. Eng. Div. ASCE*, vol. 100, no. GT1, January 1974.
21. Jennings, A.: A Direct Iteration Method of Obtaining Latent Roots and Vectors of a Symmetric Matrix, *Proc. Camb. Phil. Soc.*, vol. 63, 1967.
22. Kerisel, J.: Scaling Laws in Soil Mechanics, *Proc., 3d Pan-Am. Conf. Soil Mech. Found. Eng., Venezuela, 1967*, vol. 3.
23. Lambe, T. W.: Predictions in Soil Engineering, *Geotechnique*, vol. 23, no. 2, 1973.
24. Lundgren, H.: Dimensional Analysis in Soil Mechanics, *Acta Polytechn. Scand.*, vol. 237, 1957.
25. Lyndon, A., and A. N. Schofield: Centrifuged Model Test of a Short Term Failure in London Clay, *Geotechnique*, vol. 20, no. 4, 1970.
26. Mikasa, M., et al.: Centrifugal Model Test of a Rockfill Dam, *Proc. 7th Int. Conf. Soil Mech. Found. Eng., Mexico, 1969*, vol. 2.
27. Mikasa, M., and N. Takada: Significance of Centrifugal Model Tests in Soil Mechanics, *Proc. 8th Int. Conf. Soil Mech. Found. Eng., Moscow, 1973*, vol. 1.2.
28. Naylor, D. J., and O. C. Zienkiewicz: The Adaptation of Critical State Soil Mechanics for Use in Finite Elements, *Proc. Roscoe Mem. Symp., Cambridge, 1971*.
29. Nørsett, S. P.: One Step Methods of Hermite Type for Numerical Integration of Stiff Systems, *Bit*, vol. 14, 1974.
30. Palmer, J. H., and T. C. Kenney: Analytical Study of a Braced Excavation in Weak Clay, *Can. Geotech. J.*, vol. 9, no. 2, 1972.
31. Palmer, A. C., and J. R. Rice: The Growth of Slip Surfaces in the Progressive Failure of Overconsolidated Clay, *Proc. R. Soc., Ser. A*, vol. 332, 1973.
32. Peck, R. B.: The Observational Method in Applied Soil Mechanics, *Geotechnique*, vol. 19, no. 2, 1969.
33. Pokrovsky, G. I., and I. Fedorov: Studies of Soil Pressures and Soil Deformations by Means of a Centrifuge, *Proc. 1st Int. Conf. Soil Mech. Found. Eng., Cambridge, 1936*, vol. 1.
34. Polshin, D. E., et al.: Centrifugal Model Testing of Foundation Soils of Building Structures, *Proc. 8th Int. Conf. Soil Mech. Found. Eng., Moscow, 1973*, vol. 1.3.
35. Rocha, M.: The Possibility of Solving Soil Mechanics Problems by the Use of Models, *Proc. 4th Int. Conf. on Soil Mech. Found. Eng., London, 1957*, vol. 1.
36. Roscoe, K. H.: Soils and Model Tests, *J. Strain Anal.*, vol. 3, 1968.
37. Rowe, P. W.: Sheet Pile Walls in Clay, *Proc. Inst. Civ. Eng.*, vol. 7, July 1957.
38. Rowe, P. W.: The Relevance of Soil Fabric to Site Investigation Practice, *Geotechnique*, vol. 22, no. 2, 1972.
39. Rowe, P. W.: Failure of Foundations and Slopes on Layered Deposits in Relation to Site Investigation Practice, *Proc. Inst. Civ. Eng.*, suppl. pap. 7057S, 1968.
40. Rowe, P. W.: The Influence of Geological Features of Clay Deposits on the Design and Performance of Sand Drains. *Proc. Inst. Civ. Eng.*, suppl. pap. 7058S, 1968.
41. Rowe, P. W.: Representative Sampling in Location Quality and Size, *ASTM Symp. Sampling Soil Rock, Toronto, 1970*, Spec. Tech. Pap. 483.
42. Rowe, P. W.: Large Scale Laboratory Model Retaining Wall Apparatus, *Proc. Roscoe Mem. Symp., Cambridge, 1971*.
43. Rowe, P. W.: Embankments on Soft Alluvial Ground, *Q. J. Eng. Geol.*, vol. 5, 1972.
44. Rowe, P. W.: Anchored Sheet-Pile Walls, *Proc. Inst. Civ. Eng.*, vol. 1, pt. 1, 1952.
45. Rowe, P. W.: Limit Design of Flexible Walls, *Proc. Midland Soil Mech. Soc.*, vol. 1, 1957.
46. Rudnitski, N. Y.: Centrifugal Modelling of the Settlements of a Sand Base under a Rigid Foundation Slab with Allowance for Its Unit Weight, *Soil Mech. Found. Eng. USSR*, vol. 9, 1972.
47. Seed, H. B., and I. M. Idriss: Soil Moduli and Damping Factors for Dynamic Response Analysis, *Univ. California Earthquake Eng. Res. Cent. Rep.* EERC 70-10, Berkeley, 1970.
48. Seed, H. B., and H. A. Sultan: Stability of Sloping Core Earth Dams, *J. Soil Mech. Div. ASCE*, vol. 93, no. SM4, 1967.
49. Siemieniuch, J. L., and I. Gladwell: On Time Discretisations for Linear Time-dependent Partial

Differential Equations, *Manchester Univ. Dep. Math. Numer. Anal. Rep.* 5, September 1974.
50. Simpson, B., and C. P. Wroth: Finite Element Computations for a Model Retaining Wall in Sand, *Proc. 5th Eur. Conf. Soil Mech. Found. Eng., Madrid, 1972*, vol. 1.
51. Smith, I. M.: Plane Plastic Deformation of Soil, *Proc. Roscoe Mem. Symp., Cambridge, 1971*.
52. Smith, I. M.: Incremental Numerical Solution of a Simple Deformation Problem in Soil Mechanics, *Geotechnique*, vol. 20, no. 4, 1970.
53. Smith, I. M., and S. Kay: Stress Analysis of Contractive or Dilative Soil, *J. Soil Mech. Found. Div. ASCE*, vol. 97, no. SM7, 1971.
54. Smith, I. M., and R. Hobbs: Finite Element Analysis of Centrifuged and Built-up Embankments, *Geotechnique*, vol. 24, no. 4, 1974.
55. Smith, I. M.: Numerical Analysis of Plasticity in Soils, *Proc. Symp. Plasticity Soil Mech., Cambridge, 1973*.
56. Smith, I. M., and R. Hobbs: Biot Analysis of Consolidation beneath Embankments, *Geotechnique*, vol. 26, no. 1, 1976.
57. Smith, I. M., and R. Boorman: The Analysis of Flexible Bulkheads in Sands. *Proc. Ins. Civ. Eng.*, vol. 57, 1974.
58. Smith, I. M., et al.: Rayleigh-Ritz and Galerkin; Finite Elements for Diffusion-Convection Problems, *Water Resourc. Res.*, vol. 9, no. 3, 1973.
59. Stubbs, S. B.: Seabed Foundation Considerations for Gravity Structures, *Proc. Conf. Offshore Struct., London, October, 1974*.
60. Taylor, D. W.: Fundamentals of Soil Mechanics, John Wiley & Sons, Inc., New York, 1948.
61. Ter-Stepanian, G. I., and M. N. Goldstein: Multistoreyed Landslides and the Strength of Soft Clays, *Proc. 7th Int. Conf. Soil Mech. Found. Eng., Mexico, 1969*, vol. 2.
62. Terzaghi, K.: Old Earth Pressure Theories and New Test Results, *Eng. News-Rec.*, Sept. 30, 1920.
63. Tripathi, S. N., and V. J. Patel: Model Testing Techniques for Earth Dams, *Sols—Soils*, vol. 5, no. 16, 1966.
64. Tschebotarioff, G. P.: Final Report on Large Scale Earth Pressure Tests with Model Flexible Bulkheads, Princeton University, 1949.
65. Vaughan, P. R., M. L. G. Werneck, and M. M. A. F. Hainza: discussion at sess. 3 and 4, *Symp. Field Instrum., London, 1973*.
66. Vutsel, V. I., et al.: Use of the Method of Centrifugal Modelling in Investigations of Hydraulic Structures, *Hydrotech. Construc. USSR*, no. 8, 1973.
67. Wilkinson, J. H.: "The Algebraic Eigenvalue Problem," Clarendon Press, Oxford, 1965.
68. Wilson, E. L., and K. J. Bathe: Linear and Nonlinear Analysis of Complex Structures, *Proc. 5th World Conf. Earthquake Eng., Rome, 1973*, sess. 5.
69. Zienkiewicz, O. C., et al.: Elasto-plastic Solutions of Engineering Problems: "Initial Stress" Finite Element Approach, *Int. J. Numer. Methods Eng.*, vol. 1, 1969.

CHAPTER

EIGHTEEN

STATIC ANALYSIS FOR UNDERGROUND OPENINGS IN JOINTED ROCK

W. Wittke

18-1 INTRODUCTION

In Germany construction of underground openings in jointed rock is of considerable importance, along with planning of traffic routes, water power, irrigation projects, underground storage, and mining operations. Evaluating the stability and design of lining and safety measures for structures of this type is therefore one of the main tasks of rock mechanics engineers. Safe and economical solutions can be based only on a comprehensive knowledge of the stress-strain characteristics of jointed rock masses. The following introductory general remarks are therefore concerned with stress changes and displacements in rock masses due to excavation of openings.

The so-called *primary state of stress*, which exists before excavation of a tunnel, depends on the load of the overburden and consequently on the topography of the surface as well. Since these stresses due to gravity are very often superimposed by internal stresses of tectonic or other origin, the principal stresses determining the primary state of stress may have any magnitude and orientation and can be evaluated only by in situ measurements.

When the support of the rock mass adjacent to the tunnel is removed by excavation, the primary state of stress is changed. Assuming that the original

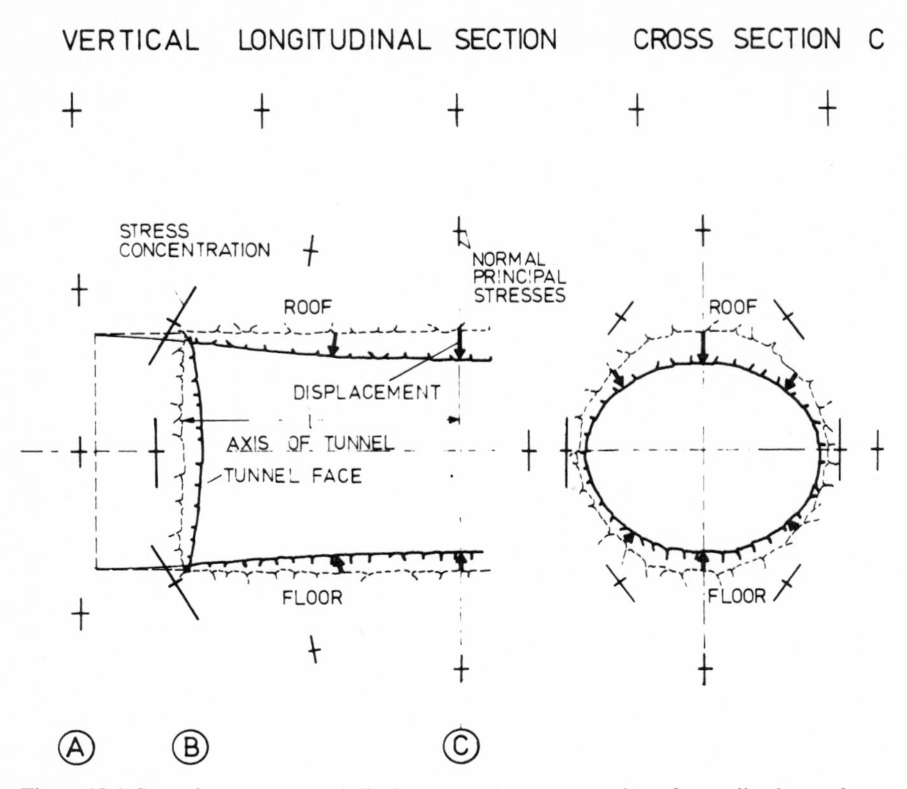

Figure 18-1 Secondary stresses and displacements due to excavation of an unlined tunnel.

maximum principal stress is vertical, that the rock mass with regard to its stress strain characteristics is homogeneous, isotropic, and time independent, and tha the tunnel is stable without an artificial support, the stress redistribution can bε illustrated as shown in Fig. 18-1. From cross section *C-C* it can be seen thai excavation causes a rearrangement of the stresses from the excavated to the adjacent rock mass, resulting in the so-called *secondary state of stress* and the *secondary displacements*.

In the area of cross section *B*, however, the secondary stress and displacements are not yet completely developed because in front of the tunnel face the unexcavated rock still gives considerable support to the overburden. Due to the high stress concentrations in this zone, however, the tunnel face is deformed, and first displacements of the adjacent tunnel walls toward the tunnel occur. The magnitude of the displacements, however, decreases with increasing distance from the tunnel face, and at a certain distance the primary stresses are not changed (see section *A*, Fig. 18-1). In conclusion it can be stated that the stresses and displacements due to tunnel excavation can be evaluated in a satisfactory manner only by a three-dimensional analysis.

The state of stresses and displacements represented in Fig. 18-1 advances with tunnel excavation, and the secondary displacements occur immediately after each step of excavation, but if the stress-strain behavior of the rock mass in question is time dependent, it is possible that the final resulting displacements occur not at a constant distance from the tunnel face, e.g., in section *C*, Fig. 18-1, but at a certain time, which depends on the rate of advance of tunnel excavation and the time dependency of the stress-strain characteristics of the rock mass. The problem is even more complex if the mechanical properties of the rock mass are inhomogeneous and anisotropic and if the primary state of stress does not correspond to the assumptions adopted for the example illustrated in Fig. 18-1. The primary and secondary stresses and displacements are no longer symmetrical to the vertical section through the tunnel axis.

Let us return to our illustrative example (Fig. 18-1) and consider the influence of a lining. Such a measure is actually required only if the strength of the rock mass adjacent to the tunnel will be exceeded and the unlined tunnel will be unstable. If this lining is installed a short distance from the tunnel face, for example, $a \geq l$ (Figs. 18-1 and 18-2), it will not be subjected to any loading because, as mentioned above, the total displacements resulting from rock excavation have already occurred (in our example at section *C*). If, however, the lining is installed directly at the tunnel face, then depending on its stiffness it will serve partly as a substitute for the excavated rock and consequently the secondary displacements in cross section *C* will be diminished in comparison with those of the unlined tunnel (Figs. 18-1 and 18-2). The stabilizing effect of a lining as a safety measure therefore depends considerably on how far from the tunnel face it is installed. In rock with time-dependent stress-strain characteristics the time of installation is also of importance.

It has already been pointed out that the stability of underground openings is decisively influenced by the mechanical properties of the rock mass in question. In this context the discontinuities occur in the rock mass as throughgoing faults, master joints, and as the result of sedimentation, schistosity, and tectonic jointing. Thanks to their genesis in certain homogeneous regions the last three types of discontinuity are usually arranged in sets of series of nearly parallel planes, the spacing of which is small in comparison with common dimensions of underground openings. An idealized example of a cubic section from a rock mass traversed by three sets of discontinuities (K_1 to K_3) is shown in Fig. 18-3, where the parameters describing the geometry of the discontinuities of the various series are compiled and explained for the typical discontinuity K_{2j} of series K_2:

1. *Orientation* Described by the angles of strike β_{2j} and dip γ_{2j}:
   Strike β_{2j} = angle between contour line of K_{2j} and north
   Dip γ_{2j} = angle between line of dip of K_{2j} and the horizontal
2. *Spacing* b_{2j} = mutual distance of two adjacent discontinuities of series K_2
3. *Plane degree of separation* K_{2j} = ratio of separated sections to total area covered by discontinuity K_{2j}
4. *Opening* $2a_{2j}$ = spacing of walls of discontinuity K_{2j}

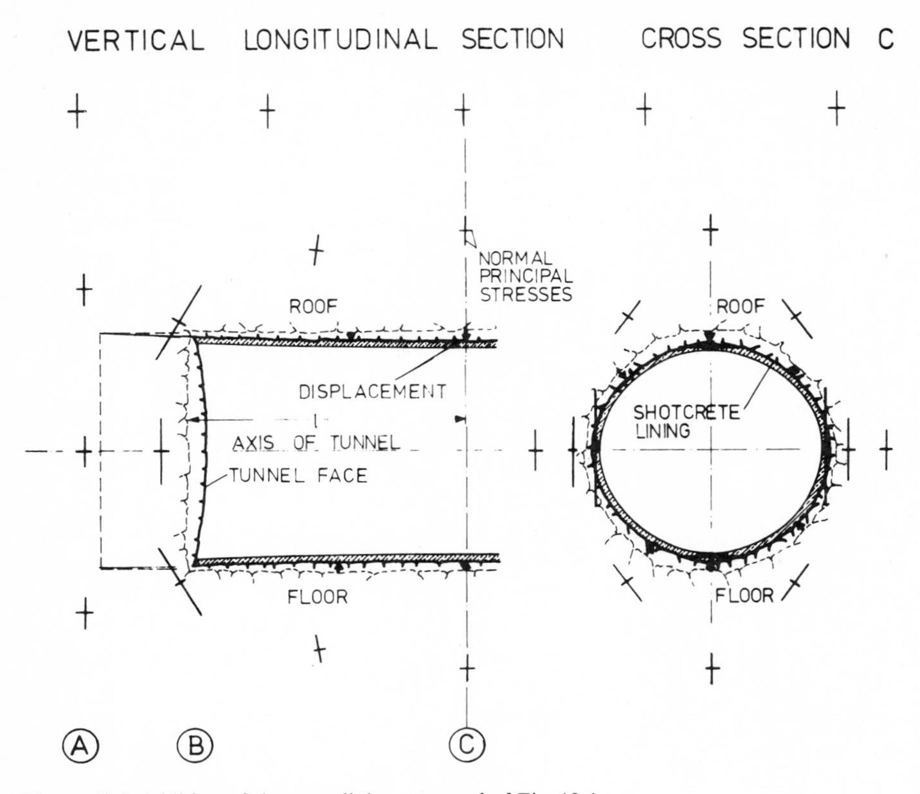

Figure 18-2 Addition of shotcrete lining to tunnel of Fig. 18-1.

5. *Roughness or unevenness* Irregularities of walls of discontinuity K_{2j}
6. *Coatings and fillings* Thin layers of, say, soil along the walls of the discontinuities and fillings of soil or minerals of open discontinuities

These parameters can be measured and documented during explorations, e.g., in outcrops and boreholes. The measured values, however, usually reveal considerable scatter and must be evaluated statistically.

Unlike the series of discontinuities, faults and master joints in the area of an underground opening usually are isolated. On the other hand, discontinuities of this type usually traverse the rock mass over long distances. The geometry of these discontinuities can be described by the parameters outlined under 1, 4, 5 and 6.

In the terminology of rock mechanics intact rock, as distinguished from rock mass, is the unjointed material the rock is composed of. In plutonic, volcanic, and sedimentary rock the grains or crystals of the intact rock are often randomly oriented. The orientation of grains or crystals parallel to bedding, or schistosity, is responsible for a parallel texture in metamorphic and various sedimentary rocks.

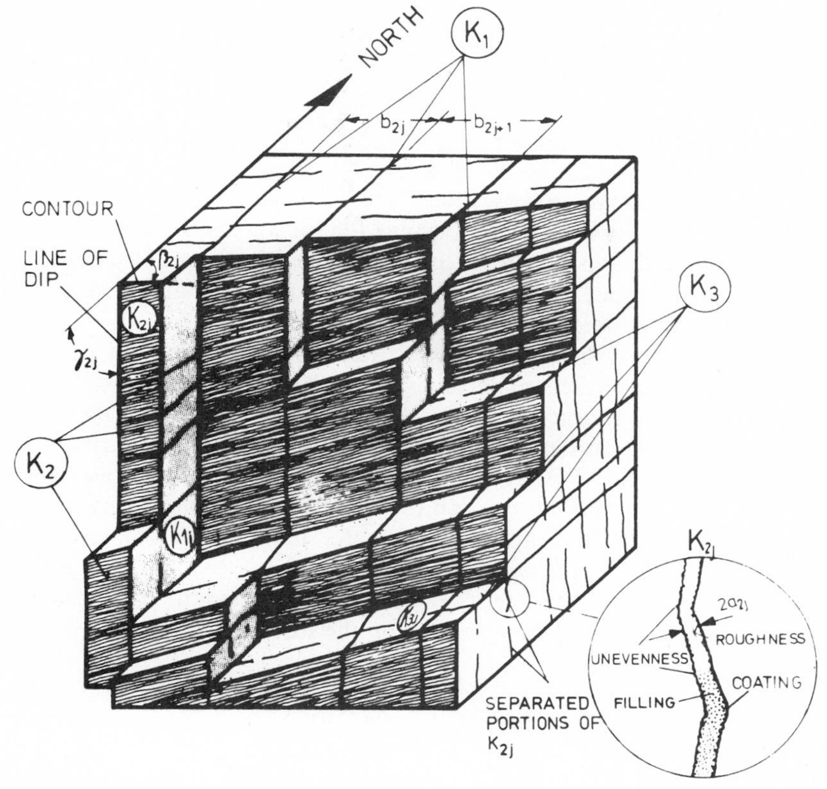

Figure 18-3 Model of jointed rock mass.

Orthogonal to a parallel texture of this type the intact rock often has a distinct cleavability.

Evidently the mechanical characteristics of a rock mass are influenced by a stress-strain relationship valid for both the intact rock and the discontinuities. Because of the many and various factors influencing the mechanical properties of rock masses, a stress-strain relationship of common applicability cannot be evaluated at present.

An extensive literature exists on the stress-strain characteristics of the various types of intact rock, from which it can be concluded that for a random texture when stresses well below the corresponding strength are applied, a linear elastic isotropic stress-strain law can be assumed. For a parallel texture, however, the intact rock is likely to follow an anisotropic stress-strain law. A nonlinear stress-strain law can be meaningful particularly for stresses of the order of magnitude of the strength. The strength of intact rock types with a random texture usually is isotropic and can be described by a curved Mohr envelope. With a distinct parallel texture the shear strength parallel to the texture is usually less than in other orientations and consequently the strength of these types of intact rock is very anisotropic. Since postfailure strains for certain intact rock types are accompanied by loosening of the structure, volume changes occur. A distinct time dependency occurs in saline rock types, e.g., potash; for other types of intact rock it usually can be observed only if the applied stresses are of the order of magnitude of the strength.

Swelling due to chemical processes, e.g., the alteration of anhydrite into gypsum due to water intrusion, is of importance for the stress-strain characteristics of a number of rock types. Alterations of the rock type, as in case of a sedimentary rock with different layers, mean that inhomogeneities also have to be considered in stability calculations of underground openings.

The stress-strain relationships of the discontinuities are usually very dependent on the parameters described under 3 to 6, Fig. 18-3. How and to what magnitude the stress-strain characteristics of a rock mass are influenced by these parameters depends mainly on the orientation and spacing of the discontinuities of the various series (1, 2, Fig. 18-3). In this context the decrease of the shear and tensile strength along the discontinuities and the corresponding anisotropy of the rock mass strength is of major importance for the stability of underground openings.

Depending on the case in question, the stress-strain law for the rock mass is more strongly influenced by the intact rock or by the discontinuities.

18-2 APPLIED ROCK MECHANICAL MODEL

A rock mechanical model oriented toward application must simplify the real stress-strain relationship in a manner which, on one hand, permits the solution of practical problems at a reasonable cost and, on the other hand, still gives results which are sufficiently close to reality. A model of this type applicable in all cases is not yet available, and therefore the model must be evaluated from case to case. For each model it must be determined which parameters have to be accounted for and which can be neglected. Furthermore in every case it must be determined whether the parameters introduced into the model can be determined by the techniques available in rock mechanics technology and by what means.

A number of authors have introduced various rock mechanical models into their finite element programs.[3,4,6,8] The author of this chapter, in connection with his consulting activities, has evaluated a rock mechanical model which has already been applied to a number of underground projects.[13,14,16,17] The satisfactory experience resulting from these applications has led to the conclusion that this model gives useful results in many practical cases. This model will subsequently be described, and it will be applied in Secs. 18-3 and 18-4.

The model is based on the idealized representation of a regularly jointed rock, described in Sec. 18-1 (Fig. 18-3). One of the series of discontinuities shown in Fig. 18-1 can also represent a direction of latent cleavability due to bedding or schistosity.

The stress-strain law for this idealized rock mass is assumed to be linear elastic as long as the applied stresses do not exceed the strength, described subsequently in more detail. In order to account for a case which very often occurs when rock with a parallel texture has to be considered, a transversely anisotropic elastic stress-strain relationship is also adopted. This stress-strain law can be described in a right-hand cartesian coordinate system (x, y, z) as follows:

$$\begin{aligned} \{\sigma\} &= [D]\{\epsilon\} \\ \{\sigma\}^T &= \{\sigma_x \quad \sigma_y \quad \sigma_z \quad \tau_{xy} \quad \tau_{yz} \quad \tau_{zx}\} \\ \{\epsilon\}^T &= \{\epsilon_x \quad \epsilon_y \quad \epsilon_z \quad \gamma_{xy} \quad \gamma_{yz} \quad \gamma_{zx}\} \end{aligned} \tag{18-1}$$

$$[D] = \begin{bmatrix} \dfrac{E_1(1 - nv_2^2)}{(1 + v_1)(1 - v_1 - 2nv_2^2)} & \dfrac{E_1 v_2}{1 - v_1 - 2nv_2^2} & \dfrac{E_1(v_1 + nv_2^2)}{(1 + v_1)(1 - v_1 - 2nv_2^2)} & 0 & 0 & 0 \\ & \dfrac{E_2(1 - v_1)}{1 - v_1 - 2nv_2^2} & \dfrac{E_1 v_2}{1 - v_1 - 2nv_2^2} & 0 & 0 & 0 \\ & & \dfrac{E_1(1 - nv_2^2)}{(1 + v_1)(1 - v_1 - 2nv_2^2)} & 0 & 0 & 0 \\ & \text{sym} & & 2G_2 & 0 & 0 \\ & & & & 2G_2 & 0 \\ & & & & & \dfrac{E_1}{1 + v_1} \end{bmatrix}$$

with $n = E_1/E_2$. The x and z axes are oriented parallel and the y axis orthogonal to the texture. The Young's moduli E_1 and E_2 are valid for compression normal and parallel to the texture, respectively. Poisson's ratio v_2 describes the strain parallel to the texture in orthogonal compression, and v_1 stands for strain parallel to texture in parallel compression, which is also perpendicular to the strain. Finally G_2 is the corresponding shear modulus. The special isotropic linearly elastic case results from Eq. (18-1) by accounting for the relationships given by

$$E_1 = E_2 \qquad v_1 = v_2 \qquad G_1 = G_2 = \frac{E}{2(1 + v)} \tag{18-2}$$

The strength is described separately for the intact rock and the discontinuities by the Mohr-Coulomb criterion of failure. The peak strength of isotropic intact rock is represented by the envelope 1-2-3, which is given by the strength parameters σ_t, c, and ϕ (Fig. 18-4*a*). The residual strength is assumed to be given by an angle of friction only (strength line 4-5, Fig. 18-4*a*). If the intact rock reveals a parallel texture, it is possible to account for reduced tensile and shear strength by introducing the failure criteria represented in Fig. 18-4*b*. Then the peak and residual strengths shown in Fig. 18-4*a* are valid only for orientations deviating from those given by the texture.

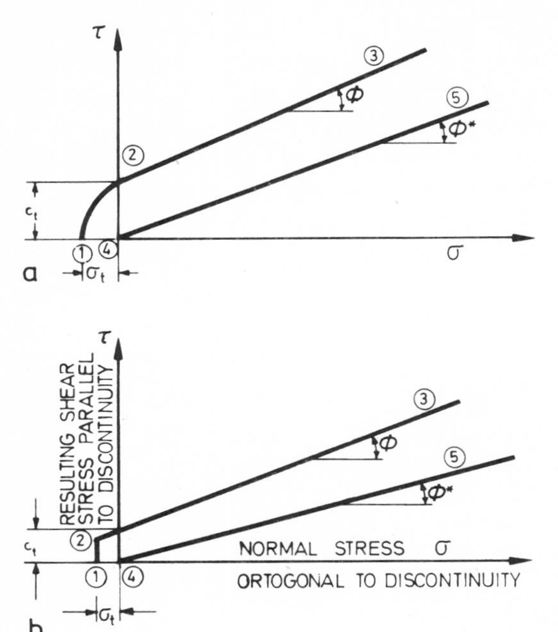

Figure 18-4 Mohr-Coulomb failure criterion.

As with a parallel texture, the rock mass strength can also be decreased by one or more discontinuities. The described model in such cases is based on the assumption that the tensile strength σ_t orthogonal and the shear strength (c, ϕ) parallel to such discontinuities are given by the straight line 1-2-3 in Fig. 18-4*b*. The residual strength here is also assumed to be due to friction only (ϕ, straight line 4-5, Fig. 18-4*b*).

If the failure criterion represented in Fig. 18-4*b* is adopted, then in a rock with a parallel texture the shear strength parallel to the texture and the tensile strength orthogonal to it are diminished in comparison with all other directions in every point of the rock mass. The corresponding assumption is adopted for a series of discontinuities if the spacing of the discontinuities of one series is small in comparison with the dimensions of the underground structure in question. Consequently with regard to a violation of the failure criterion for such discontinuities the rock mass is considered as a quasi-homogeneous continuum. Deviating from this assumption, single master joints and faults within the calculations are represented by special rows of elements, the strength of which is assumed according to Fig. 18-4*b*. In summary it can be stated that the influence of the discontinuities on the stress-strain behavior of a rock mass within the described model is considered mainly with regard to the corresponding strength anisotropy of rock masses.

A stress-strain law for the postfailure region has not been formulated. For the assumed model it follows from the iterative calculation procedure applied in case of local failure within the rock mass and explained in Sec. 18-3.

The simplifying assumptions described above do not allow for a distinct nonlinear stress-strain law for stresses well below the strength. Nor can an eventual time-dependent stress-strain law or the occurrence of volume changes as a consequence of local failure or swelling characteristics be accounted for by the described rock mechanical model. The practical examples in Secs. 18-4 to 18-6 will show how far reaching these simplifying assumptions are.

18-3 PROGRAM

Since the fundamentals of the finite element method were discussed in Chap. 1, only some special points concerning the finite element program to be dealt with in this chapter and developed for the investigation of three-dimensional stress-strain problems in connection with underground openings in rock will be discussed. This finite element program is based on a subdivision of the continuum into tetrahedra. Consequently the displacements within each element are assumed to be linearly dependent on the coordinates, and the stresses and strains are assumed to be constant. When the selected elements in zones of high stress gradients are sufficiently small, this assumption gives results which are sufficiently accurate for practical application and the selection of improved elements and corresponding displacement functions can therefore be abandoned. The subdivision of a given section of a continuum into finite elements usually begins with a subdivision into three-dimensional eight-cornered elements, which in the next step are subdivided into tetrahedra. These subdivisions, however, cannot be uniquely performed; e.g., two possibilities of subdivision are shown for the example of a cube in Fig. 18-5*a* and *b*. In these two cases the orientation and consequently the stiffness of the tetrahedra are different. Calculations have shown that even in cases of

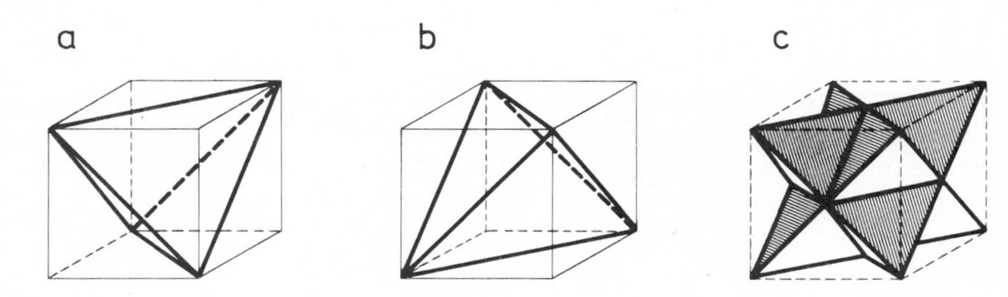

Figure 18-5 Subdivision of eight-noded element into tetrahedra: (*a*), (*b*) two arrangements of five tetrahedra; (*c*) two tetrahedra penetrating each other, increasing the stiffness by a factor of 1.5.

equal displacements of the nodes of an eight-cornered element the stresses resulting for the tetrahedra vary with the selected subdivision and deviate from the correct solution. To eliminate this cause of error, a special subdivision of the eight-cornered elements into two tetrahedra which penetrate each other was selected (Fig. 18-5*c*). The stiffness of these two elements is multiplied by a factor of 1.5 for agreement with the stiffness of the original eight-cornered element. At the end of the calculation the mean value of the stresses resulting for both tetrahedra is evaluated and assumed to be representative for the original eight-cornered element. By means of numerous examples it can be proved that, with regard to the overall assumptions the finite element method is based on, this procedure gives results which are better than those of a subdivision into five tetrahedra and which are sufficiently close to the correct solution.

The calculation of an underground opening begins with the evaluation of a representative section, including the opening and the adjacent rock mass, to such an extent that zones in which stress changes due to the excavation occur are included. For this purpose a three-dimensional section which is bounded by vertical and horizontal planes can usually be defined. Along the boundaries of this section given nodal forces and displacements must be introduced as boundary conditions.

The calculation usually begins with the evaluation of the primary state of stress, and the evaluation of the corresponding displacements, which have already occurred in nature. If the primary stresses depend only on the weight of the overburden, they can be approximated in a first calculation step by accounting for the elastic constants and the specific weight of the rock mass as well as the topography at the surface. In other cases corresponding measurements of the primary stresses should be the starting point of the calculations and should be accounted for by corresponding boundary conditions, e.g., nodal forces and displacements.

In a second calculation step, the so-called secondary case, the stress changes and the displacements due to the planned opening are investigated. For the excavation this can be done by eliminating the elements located within the planned opening. Various stages of excavation, such as those occurring in the excavation of a cavern in benches, are accounted for in separate calculation steps, in which the state of stress of a certain stage of excavation is considered the primary state of stress for the following excavation stage. For the evaluation of the influence of a shotcrete, cast-in-place concrete, or other types of lining special rows of elements are provided. In calculating the primary state of stresses the mechanical properties of the rock mass are attached to these elements, whereas for the corresponding secondary-stress state the stiffness and strength of the material of the lining are introduced. Since the rows of elements representing the lining in the calculation of the secondary case with adjusted stiffness are again subjected to the primary displacements, the secondary stresses in the lining, which in reality is subjected solely to the secondary displacements due to rock excavation, cannot be evaluated simply by subtracting the primary stresses of the corresponding elements. Obviously an additional adjustment of the evaluated secondary stresses by specially determined initial stresses is required. By these means it can be arranged that the elements representing the lining are subjected only to the secondary stresses. By the same procedure it is also possible to assign to the lining a certain loading which corresponds to only part of the secondary displacements. The influence of pre-stressed anchors and other loads applied in the secondary case is represented by corresponding nodal forces, which for an anchor are applied at its head, and adhesive stretch.

When the strength is exceeded locally, an iterative procedure based on the so-called *initial stress approach* is applied. For an explanation of the procedure applied in the program described, in Fig. 18-6*a* an idealized, isotropic, weightless system consisting of nine quadratic elements is shown. This system is loaded by the maximum and minimum normal principal stresses σ_1^0 and σ_3^0, the corresponding state of stress being denoted by $\{\sigma^0\}$. Further, plane strain conditions are assumed. For the middle element a linearly elastic behavior is adopted until the criterion of failure represented in Fig. 18-6*b* is violated, whereas the adjacent eight elements, the so-called *surrounding U*, are assumed to be linearly elastic at any stress level. The graphic representation in Fig. 18-6*b* shows that the state of stress due to loading $\{\sigma^0\}$ violates the failure criterion for element *e*. The admissible state of stresses is given by $\{\sigma_{Br}\}_e$, the corresponding stress circle touching the Mohr-Coulomb envelope (Fig. 18-6*b*). The differential stress $\{\Delta\sigma^1\}_e = \{\sigma^0\}_e - \{\sigma_{Br}\}_e$, that is, the difference between the calculated and the admissible residual state of

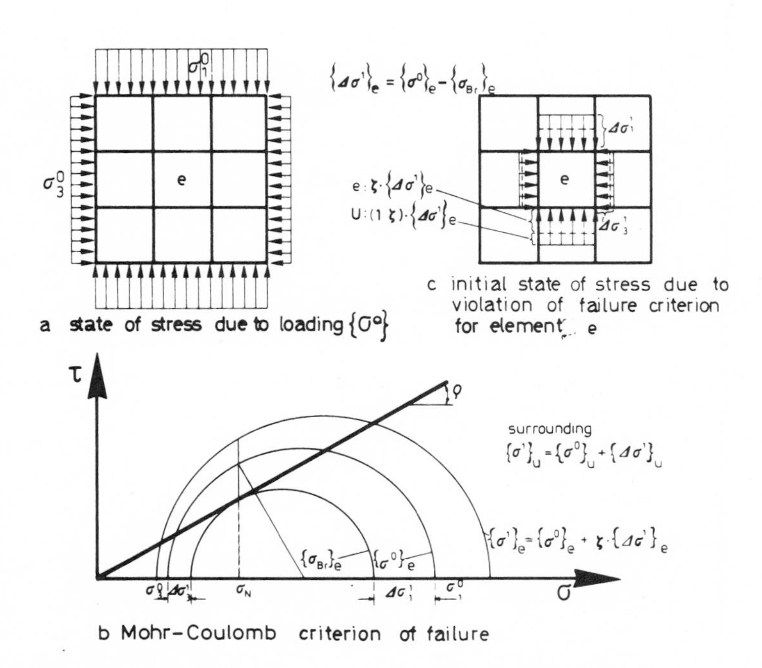

Figure 18-6 Iterative calculation of stress redistribution in case of failure.

stress, must be redistributed to the surrounding U. The admissible state of stress, or state of stress at failure $\{\sigma_{Br}\}_e$, in the described procedure is evaluated from the state of stress $\{\sigma^0\}_e$, as shown in Fig. 18-6*b*, where σ_N is the normal stress on the failure plane. This assumption ($\sigma_N = \text{const}$) is only possible, however, if the element considered has boundary conditions permitting it to sustain a certain stress σ_3. If this is not the case, e.g., at the walls of an underground opening, then it is assumed that the corresponding elements for which the failure criterion is violated do not sustain any stresses and $\{\sigma^0\}_e$, the total state of stress of element e, is redistributed to the surrounding U.

The redistribution of stresses from element e to the surrounding U is simulated by means of initial stresses in an iterative calculation. In the first iterative step the system is loaded by the initial state of stress $\{\Delta\sigma^1\}_e$, represented in Fig. 18-6*c*. This stress is distributed according to the stiffness of the system (1) to element e with the fractional part of $\zeta\{\Delta\sigma^1\}_e$ and (2) to the surrounding U with the fractional part of $(1-\zeta)\{\Delta\sigma^1\}_e$, where it creates the state of stress $\{\Delta\sigma^1\}_u$. In these equations the factor ζ, which depends on the stiffness of the system and varies from $0 < \zeta < 1$,† describes the fraction of the initial stresses distributed to element e. If this initial state of stress is superimposed on the state of stress due to loading, the following stresses result (Fig. 18-6*c*):

For the surrounding U: $\quad \{\sigma^1\}_u = \{\sigma^0\}_u + \{\Delta\sigma^1\}_u$

For element e: $\quad \{\sigma^1\}_e = \{\sigma^0\}_e + \zeta\{\Delta\sigma^1\}_e$

This iterative calculation is continued to step n (Fig. 18-7*a* to *c*). Then the initial state of stress results in

$$\{\Delta\sigma^n\}_e = \{\Delta\sigma^1\}_e[1 + \zeta + \zeta^2 + \cdots + \zeta^n]$$

and for $n \to \infty$ the initial state of stress converges toward

$$\{\Delta\sigma^n\}_e = \{\Delta\sigma^1\}_e \frac{1}{1-\zeta}$$

This again is distributed according to the stiffness of the system as follows: (1) to element e with the fractional part of

$$\zeta\{\Delta\sigma^n\}_e = \frac{1}{1-\zeta}\{\Delta\sigma^1\}_e$$

and (2) to the surrounding U with the fractional part of $(1-\zeta)\{\Delta\sigma^n\}_e = \{\Delta\sigma^1\}_e$, where it creates the state of stress $\{\Delta\sigma^n\}_u$. If this final initial state of stress is

† For $\zeta \geqslant 1$ equilibrium within the system is not possible.

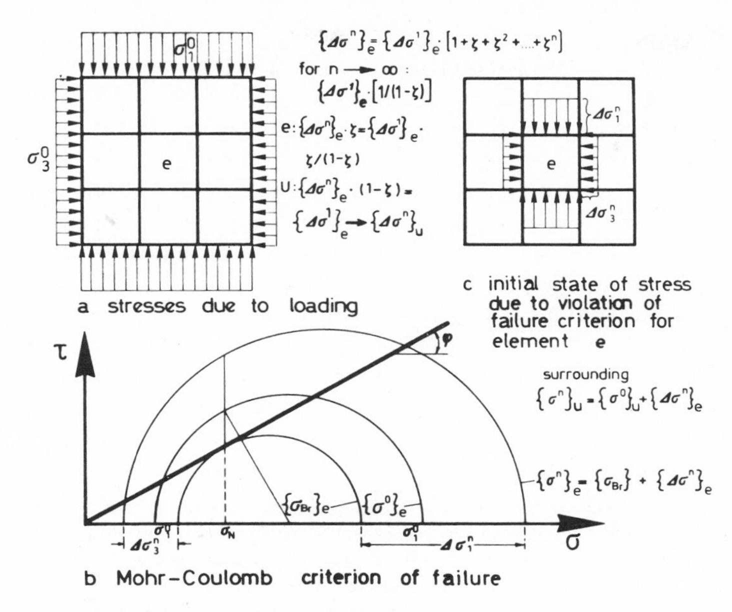

Figure 18-7 Step n of iteration of Fig. 18-6.

superimposed on the state of stress due to loading, the following stresses result (Fig. 18-7*b*). For the surrounding U

$$\{\sigma^n\}_u = \{\sigma^0\}_u + \{\Delta\sigma^n\}_u$$

where $\{\Delta\sigma^n\}_u$ just corresponds to the stress portion $\{\Delta\sigma^1\}_e$ which element e could not sustain. For element e

$$\{\sigma^n\}_e = \{\sigma^0\}_e + \zeta\{\Delta\sigma^n\}_e = \{\sigma^0\}_e + \frac{\zeta}{1-\zeta}\{\Delta\sigma^1\}_e = \{\sigma_{Br}\} + \{\Delta\sigma^1\}_e + \frac{\zeta}{1-\zeta}\{\Delta\sigma^1\}_e$$

$$= \{\sigma_{Br}\} + \frac{\zeta}{1-\zeta}\{\Delta\sigma^1\}_e = \{\sigma_{Br}\} + \{\Delta\sigma^n\}_e$$

(see Fig. 18-7*b*), where the stress portion exceeding the state of stress at failure $\{\Delta\sigma^n\}_e$ has served to deform element e to just such an extent that the nonadmissible stress fraction $\{\Delta\sigma^1\}_e$ can be redistributed to the surrounding U. By these means the required plastic deformation of element e has been simulated, and for evaluation of the stresses resulting for element e it is assumed that this deformation does not create stresses. Therefore in the last iterative step the stress $\{\Delta\sigma^n)_e$ is subtracted from the stress $\{\sigma^n\}_e$ resulting in the state of stress at failure $\{\sigma_{Br}\}_e$ for element e, and consequently equilibrium for the total system is restored. The iterative calculation is correspondingly performed for a general case; then, however, the convergence ($n \to \infty$) cannot be proved as described above.

18-4 INVESTIGATIONS OF ROCK MECHANICS FOR THE UNDERGROUND POWERHOUSE AT WEHR†

Structure

The Hornberg pumped storage scheme of the Hotzenwald plant is at present under construction in the southern Black Forest, Germany. It consists of an upper and lower reservoir, with a capacity of approximately 4×10^6 m^3 and a net head of approximately 600 m. Another major structure is the underground powerhouse at Wehr, in which four units (pumps, turbines, and generators) with a capacity of 1160 MVA are to be installed. The powerhouse cavern is 19 m wide, 33 m high, and 219 m long. The rock overburden amounts to approximately 350 m (Figs. 18-8 and 18-9).

† It is a great honor to have been asked to serve as a consultant for the projects described in Secs. 18-4 to 18-6. I should like to convey my thanks to the owners of these projects, Prof. Pfisterer, of the Schluchseewerk AG, Freiburg (Sec. 18-4), Baudirektor L. Carl, of the Talsperrenneubauamt, Nürnberg (Sec. 18-5), and Dipl.-Ing. W. Lenssen, chief of the water-power section of Rheinisch Westfälische Elektrizitätswerke AG, Essen (Sec. 18-6) for the considerable support they provided.

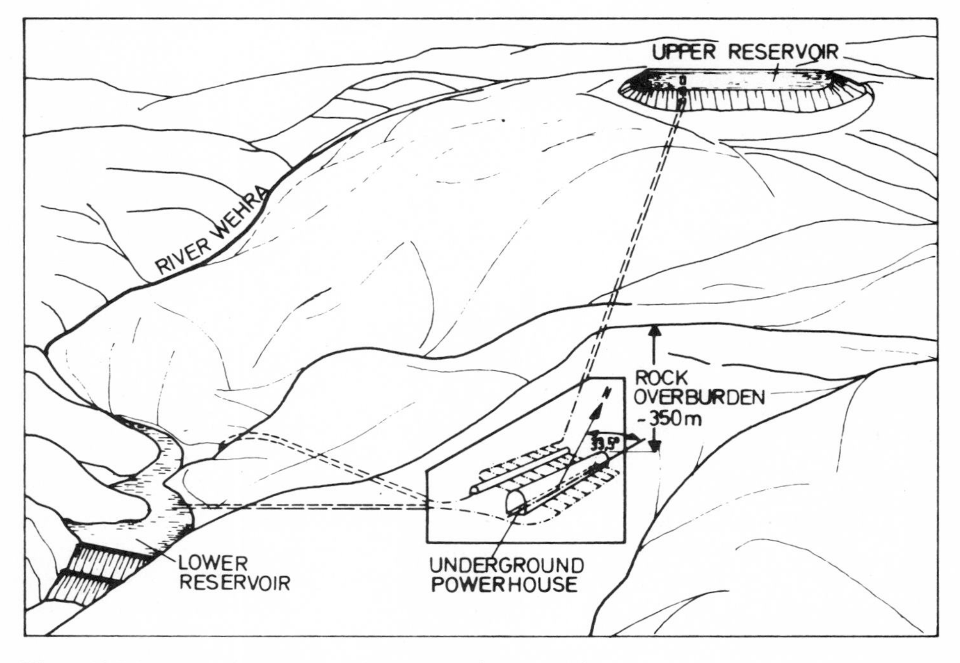

Figure 18-8 Hotzenwald-Hornberg plant, pumped storage scheme.

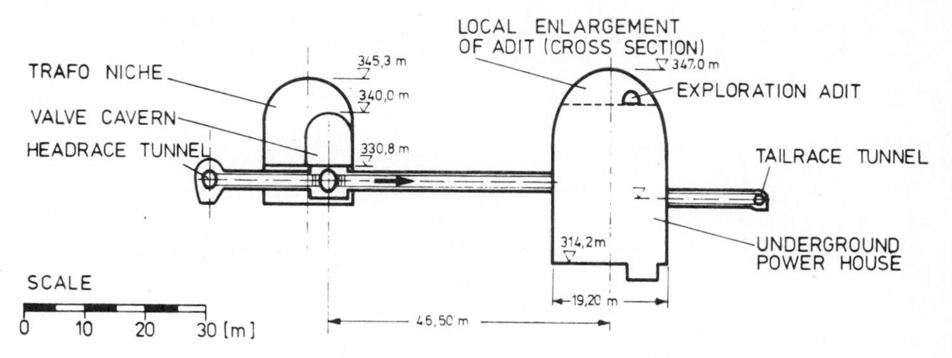

Figure 18-9 Cross section of underground powerhouse.

Engineering Geological Investigations

Approximately 10 years before construction began, an exploratory adit was excavated. In the area of the cavern it bifurcates into two parallel adits approximately 25 m above the final cavern roof. From the resulting geological conditions the direction of the cavern axis was determined as N33.5°E (Fig. 18-8).

In a later phase of the investigations an additional horizontal exploration adit was excavated in the area of the vault of the planned cavern. This so-called *vault adit*, which runs approximately parallel to the cavern axis, was extended beyond both end slabs of the cavern. To execute in situ tests of rock mechanics and to permit installation of extensometers before construction the cross section of the vault adit was widened to match that of the cavern vault at four places over a length of 3 m each (Fig. 18-9). Further information on the geology resulted from a number of core drillings, which were performed from the vault adit. From the explorations performed it followed that the planned cavern would be located in gneiss of two different forms. In the northeast section of the cavern the gneiss has a granitelike texture similar to that of the *Albtal granite* occurring in the vicinity of the project area. Consequently this rock was named *Albtalgranitelike gneiss* (AG). In the southwest area of the cavern the gneiss has a distinct parallel texture. For this rock type the abbreviation LG is used. The boundary between rock types AG and LG approximately follows a vertical plane, which strikes at N160°E and intersects the vault adit at a distance of 140 m from the northeast end of the cavern. Along this plane no discontinuity was developed within the rock mass.

As a result of a detailed engineering geological mapping of the roof adit and the other explorations, it was found that five master joints (1, 3, 4, 5, and *PK* in Fig. 18-10) and one fault (2 in Fig. 18-10) exist in the area of the powerhouse. Except for the so-called parallel master joint *PK*, all these discontinuities strike at N150°E to N170°E and dip steeply to northeast and southwest, respectively. Depending on their strike, they intersect the cavern axis at angles between 45 and 65°. The parallel joint *PK*, on the other hand, strikes at acute angles to the cavern axis. In the strike and dip of this discontinuity a considerable scatter could be

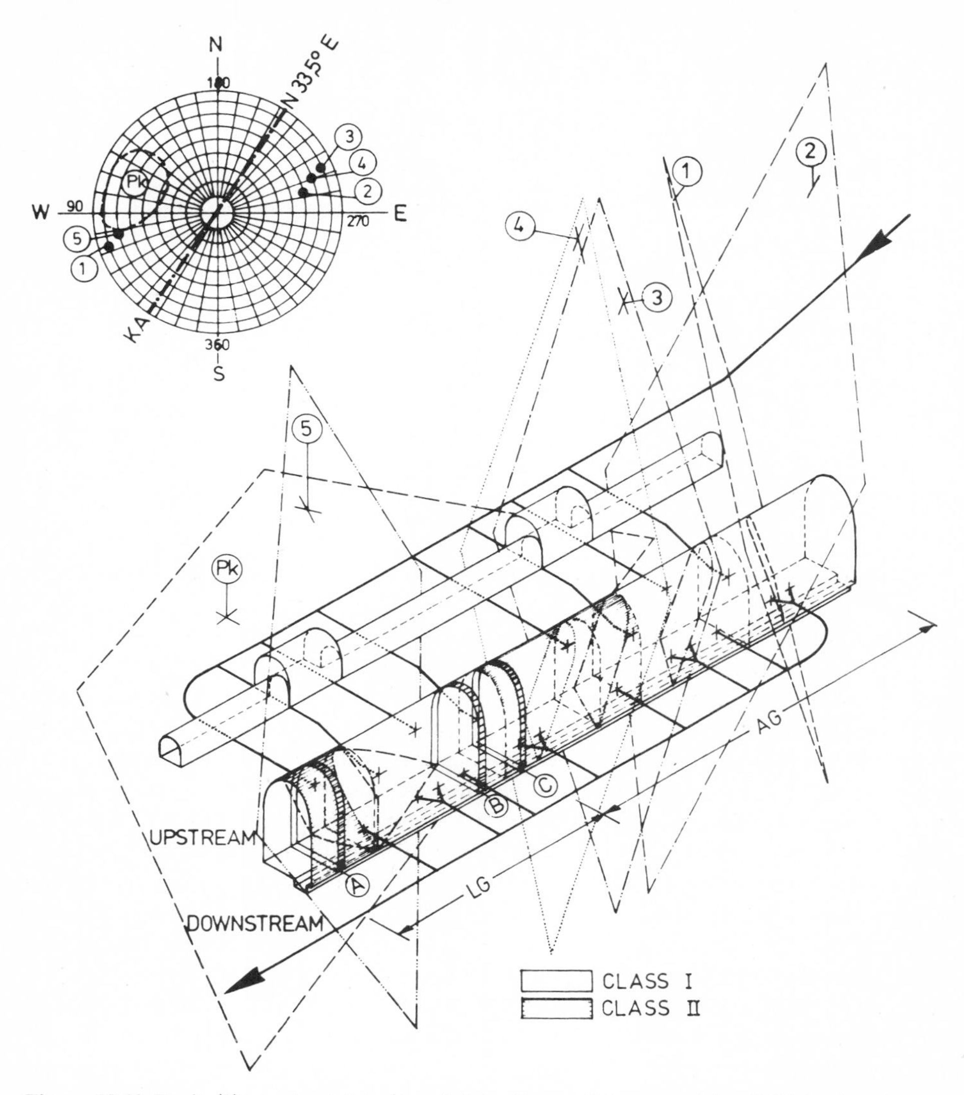

Figure 18-10 Fault (2), master joints (1 and 3 to 5), parallel master joint (*PK*), and rock-mass classification.

observed, and consequently its importance for the stability of the cavern could be realized only after excavation of the vault adit. The parallel master joint intersects the upstream wall of the powerhouse at the southwest end at the elevation of the roof and at station +150 m (Figs. 18-10 and 18-14) at the elevation of the floor. The parallel master joint and the upstream wall of the cavern intersect at an acute angle and form a large rock mass wedge, which opens upward and has a tendency to slide into the cavern.

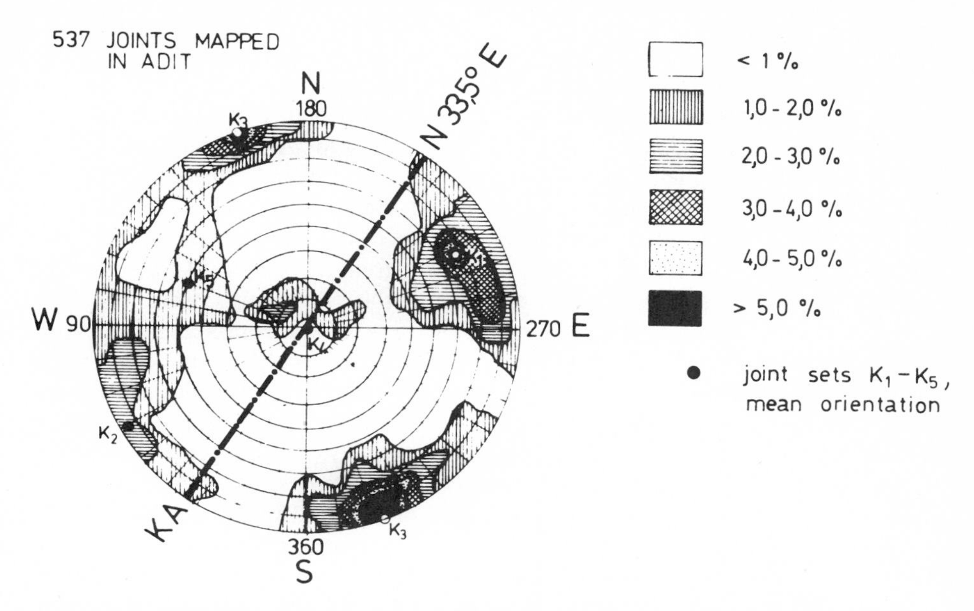

Figure 18-11 Representation of orientation of joints by polar projection of lower hemisphere.

The appearance of all master joints and of the fault zone differs from place to place although on the average it is similar. Over large extents these discontinuities are centimeters to decimeters wide and reveal coatings and fillings of barite, calcite, and cohesive soils. Seepage water exists along all throughgoing discontinuities, whereas the intermediate rock mass is practically impermeable.

Besides these master joints and the fault zone all joints intersecting the walls or the roof of the vault adit over $\gtrsim 1$ to 2 m were also mapped. For these 537 minor joints an evaluation of the measurements with regard to the orientation was performed. The resulting representation in a polar projection of the lower hemisphere reveals a scatter which is typical for the rock type in question (Fig. 18-11). With certain idealizations, however, the evaluation results in five joint sets (K_1 to K_5), including approximately 80 percent of the mapped minor joints. The mean orientation of these five joint sets is also represented in Fig. 18-11. Only a minority of the joints reveal fillings of calcite, barite, kaolin, and hematite iron, but the fillings do not cover any greater area. Small quantities of seepage were observed for approximately 4 percent of all mapped joints. From a statistical evaluation of the mapping with regard to frequency and extent of the joints it could be concluded that the intensity of jointing is not regular along the axis of the vault adit. When certain idealizations were applied, two rock classes I and II with equal jointing could be evaluated, the corresponding zones where these classes occur being represented in Fig. 18-10. Of course during exploration these zones could only be localized within the vault adit and had to be verified parallel to excavation.

Rock Mechanical Investigations

For the evaluation of the strength and deformability characteristics of the rock mass a number of rock mechanical tests were performed. First the intact rock strength was determined in the laboratory separately for Albtalgranitelike gneiss and parallel textured gneiss by means of unconfined compression tests, Brazilian tests, and specially developed one-dimensional shear tests.[16] These tests for the Albtalgranitelike gneiss resulted in an average unconfined compressive strength of $\sigma_c = 860\ \mathrm{kg}_f/\mathrm{cm}^2$, a uniaxial tensile and shear strength of $\sigma_t = 73\ \mathrm{kg}_f/\mathrm{cm}^2$ and $c_t = 300\ \mathrm{kg}_f/\mathrm{cm}^2$, respectively. For the gneiss the test results were distinctly dependent on the orientation of the parallel texture. The unconfined compression tests for loading parallel and orthogonal to the texture resulted in mean strengths of $\sigma_c = 645$ to $760\ \mathrm{kg}_f/\mathrm{cm}^2$; where the loading was applied at angles around 45° to the texture it was $\sigma_c = 460\ \mathrm{kg}_f/\mathrm{cm}^2$. The mean tensile strength σ_t and shear strength c_t were measured at $\sigma_t = 88\ \mathrm{kg}_f/\mathrm{cm}^2$ and $c_t = 80\ \mathrm{kg}_f/\mathrm{cm}^2$ for loading parallel or diagonally to texture and at $\sigma_t = 43\ \mathrm{kg}_f/\mathrm{cm}^2$ and $c_t = 247\ \mathrm{kg}_f/\mathrm{cm}^2$ where the corresponding loading was applied orthogonal to the texture.

To determine the deformability characteristics of the rock mass flat-jack tests (LFJ),[9] ultrasonic tests, and borehole deformation tests were performed in the vault adit. For comparison prisms of intact rock from the sites of the LFJ tests tested in the laboratory showed a relatively low deformability of the rock mass. A detailed evaluation of all tests, especially accounting for the relationship between static and dynamic moduli, resulted in the following rock mass deformation moduli:

$$V = \begin{cases} 600 \text{ to } 900 \times 10^3\ \mathrm{kg}_f/\mathrm{cm}^2 & \text{for rock class I} \\ 500 \text{ to } 800 \times 10^3\ \mathrm{kg}_f/\mathrm{cm}^2 & \text{for rock class II} \end{cases}$$

Immediately adjacent to the fault zone and the master joints the deformability of the rock mass is higher.

From a long time LFJ test it was found that the time dependency of the stress-strain characteristics of the rock mass is small. Since only 20 percent of the total displacements occurred with a certain time delay, the time dependency could be neglected in the stability analyses with fair approximation.

The in situ stresses were evaluated from tests with the so-called small flat jack (SFJ).[10] The principle consists in cutting a slot and using a flat jack to apply to its side walls a uniform pressure with a magnitude that compensates for the deformation adjacent to the slot due to slot cutting. This pressure is equal to the normal stress acting perpendicular to the jack before slot cutting. The measurements resulted in a vertical normal stress approximately equal to the weight of the overburden ($\sigma_1 \approx 100\ \mathrm{kg}_f/\mathrm{cm}^2$) and in horizontal stresses of approximately $\sigma_2 \approx \sigma_3 = 12\ \mathrm{kg}_f/\mathrm{cm}^2$.

The strength parameters of the minor jointing were estimated from the intact rock strength parameters and the corresponding degrees of separation. Similarly, the normal stiffness and the shear parameters of the parallel master joint were evaluated since the varying appearance of this discontinuity made it impossible to determine these properties by in situ tests.

Rock Mass Properties for Stability Analysis

From the results of the geological and rock mechanical investigations the representative mechanical parameters were derived and attached to the corresponding cross sections *A*, *B*, and *C*, analyzed in the finite element calculations (Fig. 18-10). For the Young's modulus a value was selected which can be applied to both rock classes I and II thanks to the observed scatter of the test results. For comparison, calculations were performed with a considerably lower Young's modulus (see Table 18-1).

Table 18-1 gives selected parameters describing the peak and residual shear strength, c_t, ϕ_t and c_r, ϕ_r, respectively, and the tensile strength σ_t along the discontinuities of the minor joint net, which, as mentioned, were derived from the intact-rock strength and the degrees of separation. Since the parallel master joint considerably influences the stability of the cavern, a major part of the analyses was concentrated on this problem. Because of the variable opening and appearance of this discontinuity and because the success of the planned improvement by grouting could not have been foreseen, the determination of the normal stiffness was very difficult. The analyses were carried out on an average thickness of $d = 0.1$ m and a Young's modulus of $E = 500\ \text{kg}_f/\text{cm}^2$ and revealed the same normal stiffness $d = 1.0$ m and $E = 5000\ \text{kg}_f/\text{cm}^2$, respectively, for the unimproved discontinuity. For the zones of the parallel master joint improved by grouting, the parameters $d = 1.0$ m and $E = 100{,}000\ \text{kg}_f/\text{cm}^2$ were adopted in the corresponding calculations. The angle of friction, on the other hand, could be estimated relatively well because the parallel master joint is a throughgoing discontinuity with silty coatings on the walls and locally even silty fillings, which could not be grouted by cement. Consequently the same angle of friction had to be adopted for the grouted and the ungrouted joint. For the analysis of cross section *A*, $\phi = 10°$ was assumed; for alternative calculations $\phi = 25°$ and for cross section *B*, $\phi = 25°$. Cohesion and tensile strength according to the above statements were assumed to be zero (Table 18-1).

The remaining master joints and the fault intersect the cavern axis approximately orthogonally, which with regard to stability is relatively favorable. Therefore no detailed analysis of the influence of these discontinuities was performed. By reducing the shear and tensile strength for the joint sets K_1 and K_2 parallel to these throughgoing discontinuities their influence was roughly estimated during the analysis of section *C*.

Stability Analysis

Since the length of the powerhouse is large compared with its width and height, a comprehensive three-dimensional analysis including the total structure did not promise to be very useful. Three rock mass slabs *A*, *B*, and *C*, 1 m thick and vertically oriented to the cavern axis, were selected for the stability analyses (Fig. 18-10). It could be assumed that these slabs were representative with respect to location and appearance of the discontinuities.

Table 18-1 Parameters selected for finite element calculations

	Cross section (Fig. 18-10)		
	A	*B*	*C*
Rock mass	LG†	LG†	AG†
Rock overburden H_u, m	350	350	350
Unit weight, Mg_f/cm^2	2.5	2.5	2.5
Young's modulus *E*, Mg_f/cm^2	700, 400‡	700, 400‡	700, 400‡
Poisson's ratio v	0.18	0.18	0.18
Intact rock:			
Unconfined compressive strength σ_c, kg_f/cm^2	645	645	860
Uniaxial tensile strength σ_t, kg_f/cm^2	43	43	73
Uniaxial shear strength c_t, kg_f/cm^2	80	80	300
Values for minor joints, kg_f/cm^2:			
K_1, c_t	14.7	0	0
σ_t	8.6	0	0
K_2, c_t	0	14.7	0
σ_t	0	8.6	0
K_3, c_t	0	0	0
σ_t	0	0	0
K_4, c_t	...	0	0
σ_t	...	0	0
K_5, c_t	14.7	55.6	0
σ_t	8.6	30.0	0
Angle of friction of joints ϕ_t, deg	25	25	25
Residual shear strength, kg_f/cm^2:			
c_r	0	0	0
ϕ_r	40	40	0
Parallel master joint, kg_f/cm^2:			
σ_t	0	0	
c_t	0	0	
Angle of friction ϕ, deg	10 25§	25	
Young's modulus *E*, Mg_f/cm^2:			
Not grouted	0.5	5.0	
Grouted	...	100	
Assumed thickness *d*, m	0.1	1.0	

† LG = gneiss, parallel texture; AG = gneiss, granitelike texture.

‡ The calculations were carried out with $E = 700\ Mg_f/cm^2$ for the majority of the investigated cases; the others used $E = 400\ Mg_f/cm^2$.

§ For alternative calculations.

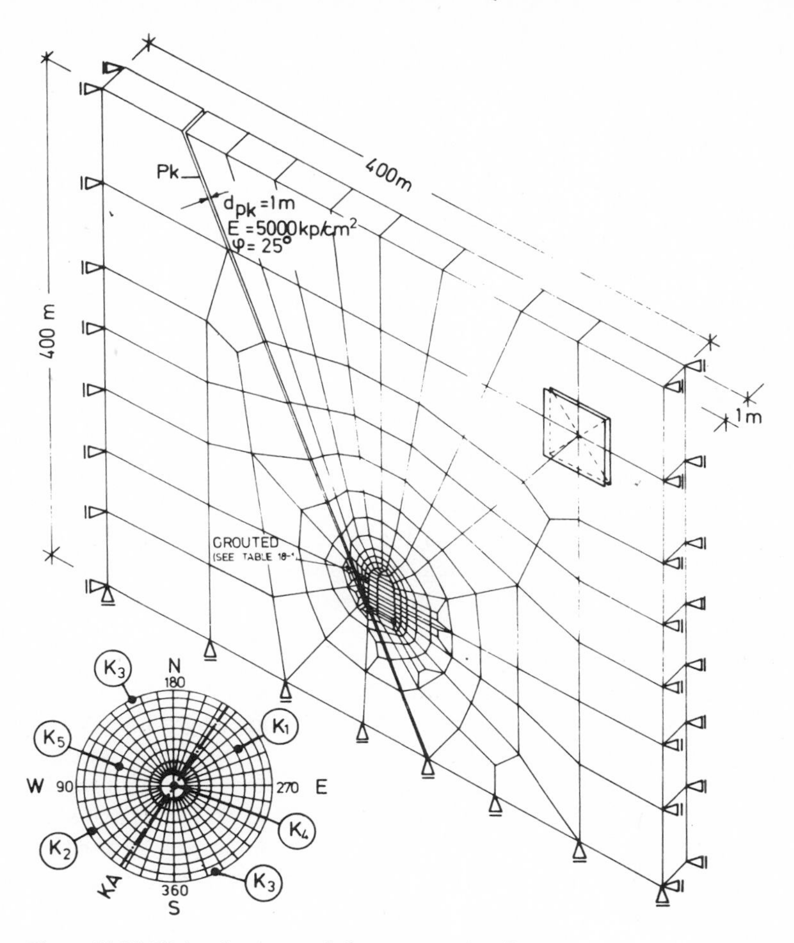

Figure 18-12 Finite element mesh for cross section *B*.

Cross section *A* represents zones where the parallel master joint *PK* intersects the upstream wall at the elevation of the cavern roof (Fig. 18-10). The analysis of cross section *B* on the other hand permits conclusions with regard to the stability of the cavern in zones where the master joint intersects the upstream wall near to floor. Cross section *C*, which is located in more distinctly jointed rock of class II, served for the stability analysis of areas not influenced by master joints or the fault (Fig. 18-10). As described above, at least a rough estimate of the influence of the latter discontinuities (1 to 5 in Fig. 18-10) on the stability of the cavern could be derived from the analysis of cross section *C* by reducing the strength parameters of joint sets K_1 and K_2.

The structure of the finite element networks can be seen from the example selected for cross section *B* (Fig. 18-12). It represents a rock mass slab 400 m wide, 400 m high, and 1 m thick, including the cavern. The master joint *PK* was represented by a special row of elements with low normal stiffness and shear strength (Table 18-1). The strike was assumed to be approximately parallel to the cavern axis. According to the assumption of the method and because of the higher stress gradients expected adjacent to the cavern, a finer element mesh was assumed for this area.

Various stages of excavation were also investigated. This was of special importance for cross section *B*, since (as mentioned below) the stability of the rock wedge bounded by master joint *PK* and the upstream wall of the cavern had to be continuously observed during construction. Consequently the stages of excavation at elevation +326 m and +320 m were specially analyzed.

For cross section *C* the result of the analysis was that stability of the powerhouse could be achieved without any lining or safety measures and for cross section *A* the stability of the cavern could be proved. It was found, however, that the stresses and displacements within the rock mass were strongly dependent on the normal stiffness and shear strength of the master joint *PK*. For small normal stiffness and shear strength remarkable relative shear displacements of the opposite walls of master joint *PK* occurred, and the stress distribution around the cavern was asymmetric. Consequently, as excavation progressed, relative displacements of the cavern wall adjacent to the line of intersection with master joint *PK* took place.

The results of the analyses of cross section *B* were of special importance for the stability of the cavern. With progressive excavation, stress concentrations occurred within the rock mass wedge (described above) bounded by joint *PK* and the upstream cavern wall. These stress concentrations resulted in plastic deformations along the joints of the various sets. The corresponding shear failures are marked by circles in the plots of the principal stresses in Fig. 18-13*a*, *c*, and *e*. As soon as the foot of this wedge was excavated, the whole wedge slid toward the cavern and consequently was unloaded, the corresponding stresses due to overburden being transmitted to the rock mass downstream of the downstream cavern wall (Fig. 18-13*e* and *f*). As a consequence of this process the upstream wall moved toward the cavern and was stretched. The corresponding displacements increased from the roof toward the floor of the cavern. This differed from the other cross sections investigated in that the stability of the cavern without additional measures could not be proved. Even after 40 iterative steps the analysis still resulted in increasing displacements of the tip of the wedge, from which it could be concluded that the stability of the cavern was endangered.

Additional calculations showed, however, that installation of prestressed anchors would limit the displacements to ~1 cm and the rock mass wedge could be stabilized.

Rock Mass Excavation, Lining, and Safety Measures

For the cross sections not influenced by master joints and faults a light lining accounting for the minor jointing was planned despite the favorable results of the analyses. The lining consisted of approximately 15-cm-thick square wire

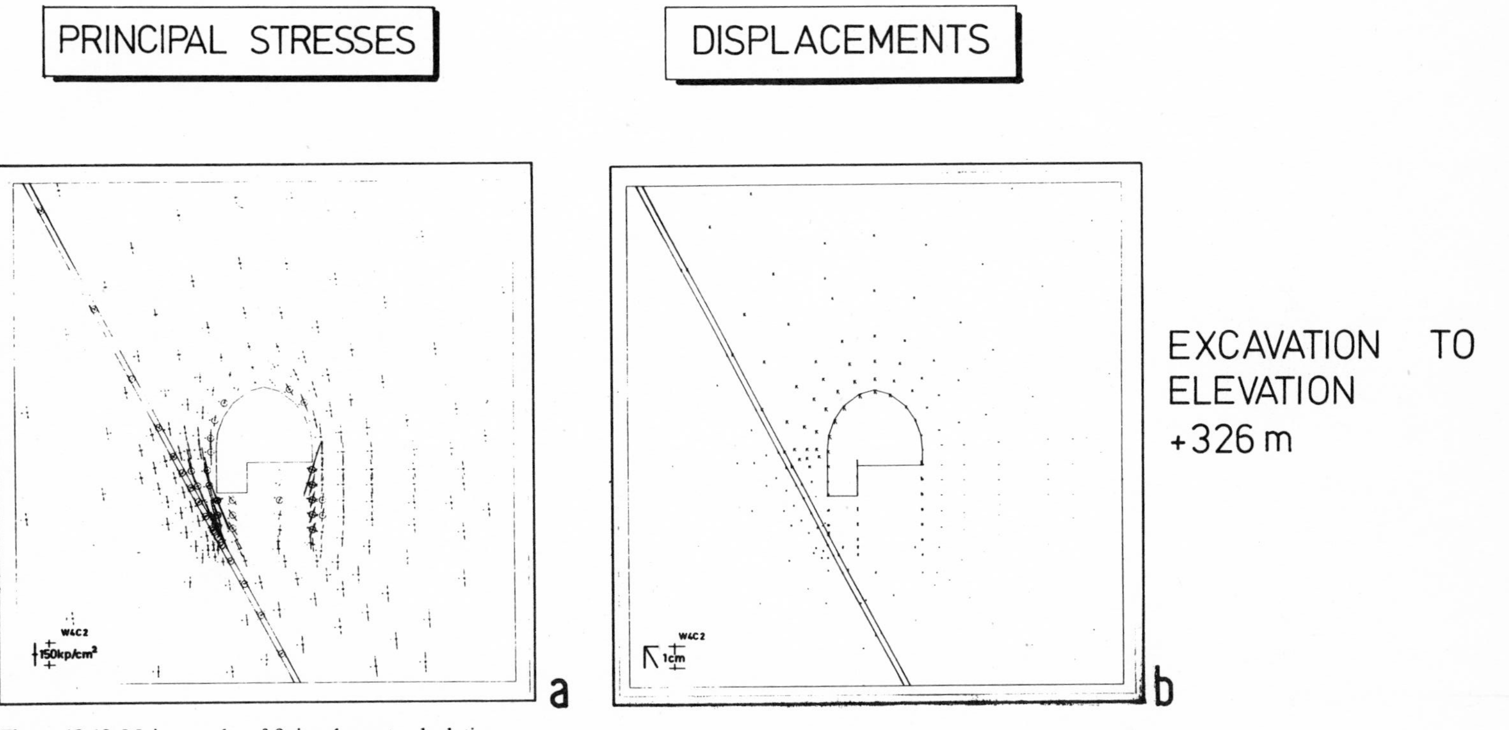

Figure 18-13 Major results of finite element calculation.

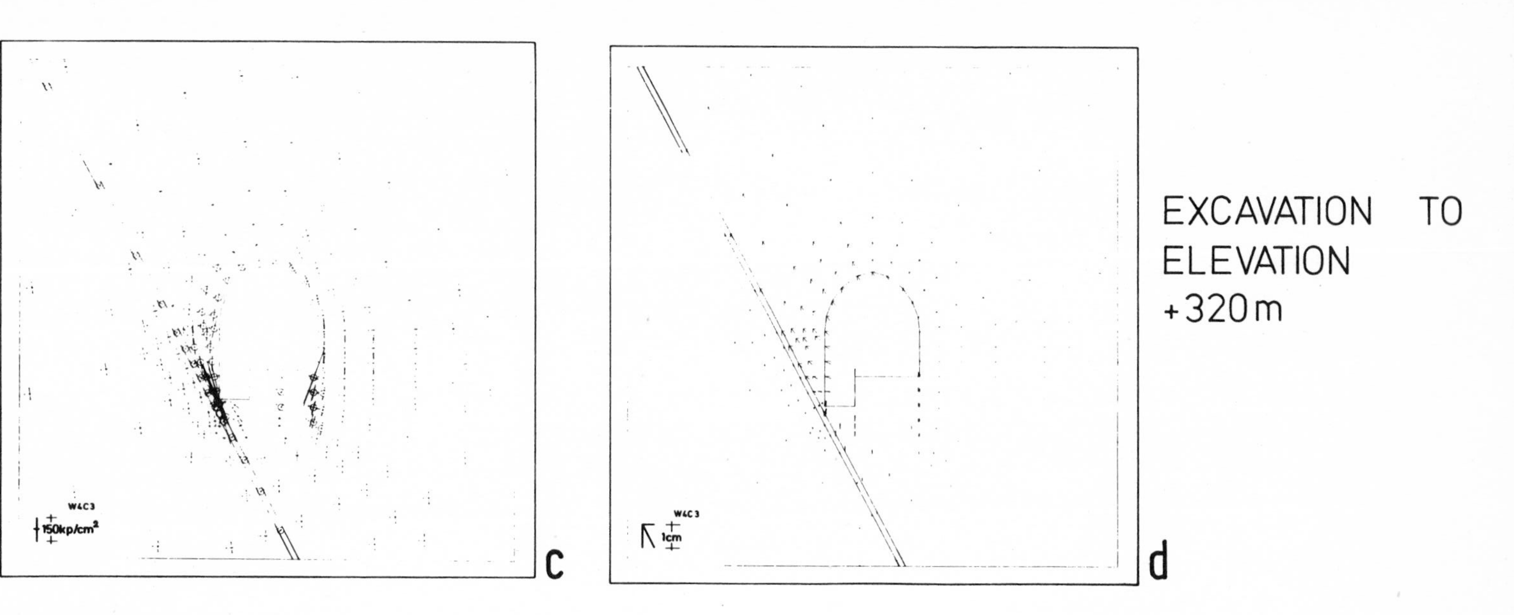

Figure 18-13 (continued)

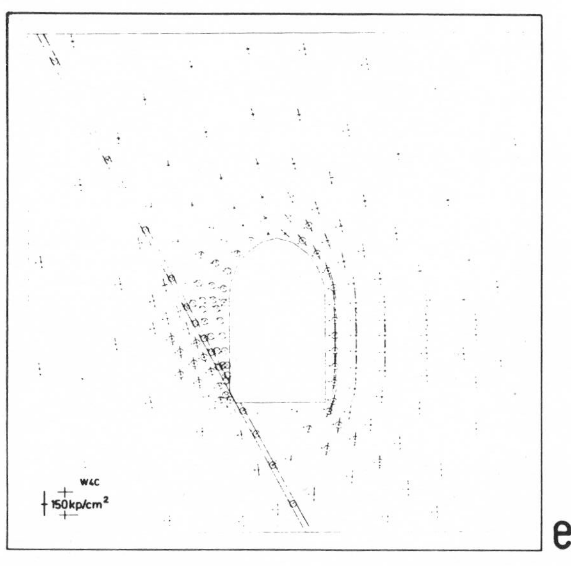

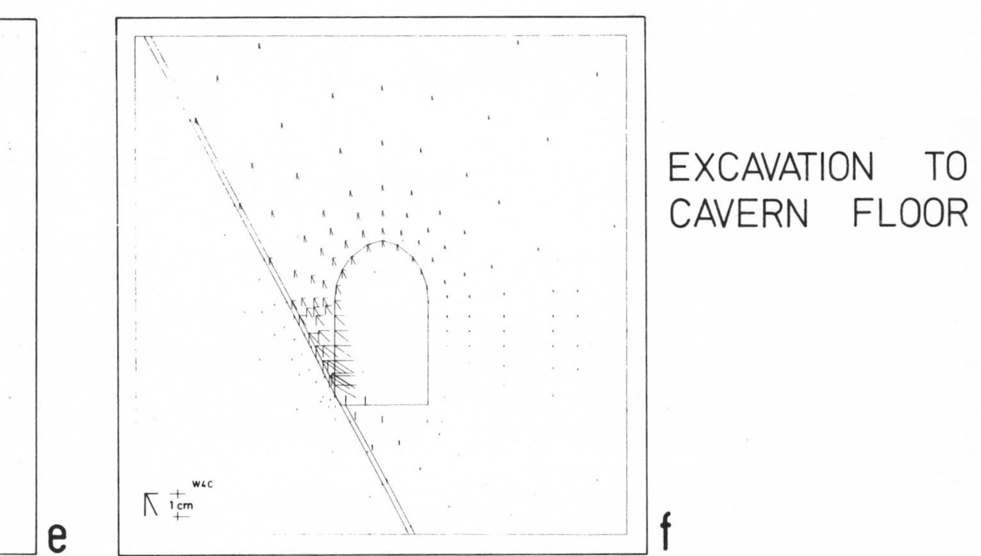

Figure 18-13 (continued)

mesh and additional Perfo bolts with a spacing of 1.5 to 2.5 m, diameter of 22 to 24 mm, and length of 4 m.

To increase the normal stiffness and to decrease permeability the master joints and the fault zone in the vault area were grouted. In the area of the cavern walls this was done as the excavation of the corresponding benches advanced. Farther along the intersections of the master joints, the fault, and the cavern wall, the amount of bolting and reinforcement of the shotcrete was increased.

For improvement of the stability of the rock mass wedge between the master joint *PK* and the upstream cavern wall a comprehensive grouting program was started before excavation was begun, proceeding from the vault adit and extending over the total height of the cavern. Altogether 76 t of cement were used. Dividing the corresponding grout volume by the grouted area gives an average thickness of the grouted master joint of 3 to 4 cm. According to the result of the analyses, 82 prestressed anchors, with a capacity of 170 t each, lengths from 13 to

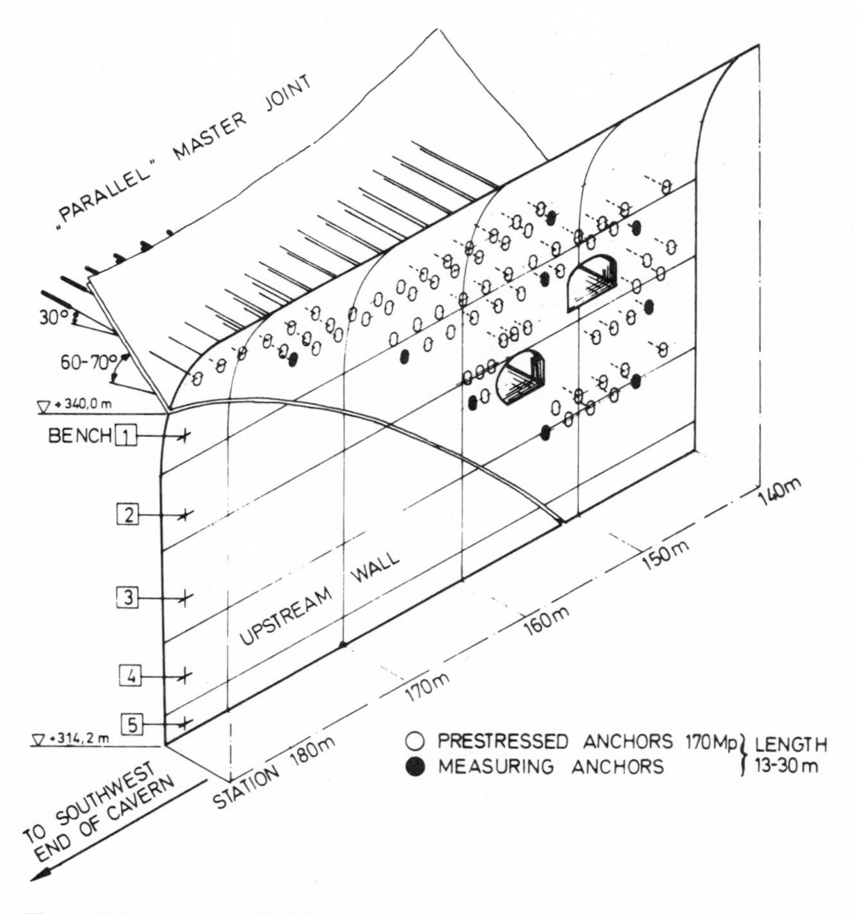

Figure 18-14 Arrangement of 82 prestressed anchors.

30 m, and an upward inclination of 30°, were installed from the upstream cavern wall (Fig. 18-14). They penetrated the parallel master joint *PK* and fixed the rock mass wedge into the rock beyond. To prevent rock loosening as far as possible the installation of the anchors followed excavation of the benches. Since the analyses showed that it could be assumed that the wedge would slide off only toward the end of the excavation (Fig. 18-13), and since displacements due to this sliding could only be limited and not prevented, an increase in the prestressing load of the anchors already installed was to be expected. Assuming a low Young's modulus for the rock mass, this stress increase was evaluated from the results of the analyses at approximately 400 to 1900 kg_f/cm^2. Consequently for all anchors an initial prestressing load of only 140 t was applied, and 9 of the 82 anchors were installed as measuring anchors. Since all anchors remained ungrouted during the construction period, from the results of the measurements it would have been possible to release the prestressing load at any time. The shotcrete of the vault zone and both cavern walls of this area was reinforced by an additional layer of square wire mesh to compensate for the stress redistribution resulting from sliding of the rock mass wedge.

Since the drainage of the parallel master joint also required special attention, at elevation 330 m and at the cavern floor horizontal weep holes penetrating this discontinuity at a certain distance from the cavern were installed. Corresponding drainage holes also were drilled in the area of the remaining master joints and the fault.

Control Measurements during Construction

To control the rock mass displacements during and after construction 13 multiple extensometers were installed in the area of the master joints before the excavation started. Further one- and two-point extensometers served as a control measure for the measuring anchors and eventual displacements of the cavern walls. Consequently by continuous comparison of the results of these measurements with those of the analyses it was possible to evaluate the stability of the cavern at any time, and if necessary it would have been possible to increase the amount of lining and safety measures.

Of special importance in this context were the results of extensometers *S*II (Fig. 18-15) and *S*III (Fig. 18-16), by which the rock mass wedge between the upstream cavern wall and master joint *PK* could be monitored. Extensometer *S*II was installed at station $KA + 150$ (see Figs. 18-14 and 18-15) before the excavation of the cavern started from the valve cavern, already excavated at that time. Fix points 32 and 34 of this extensometer were located close to the tip of the wedge. In Fig. 18-15 the displacements measured after excavation of the vault and the five benches are plotted and compared with the corresponding results of the analyses for the full excavation. The results of the measurements show that during and after excavation of the vault and the first two benches practically no displacements occurred. Excavation of the third bench initiated the sliding of the wedge, which increased with further excavation.

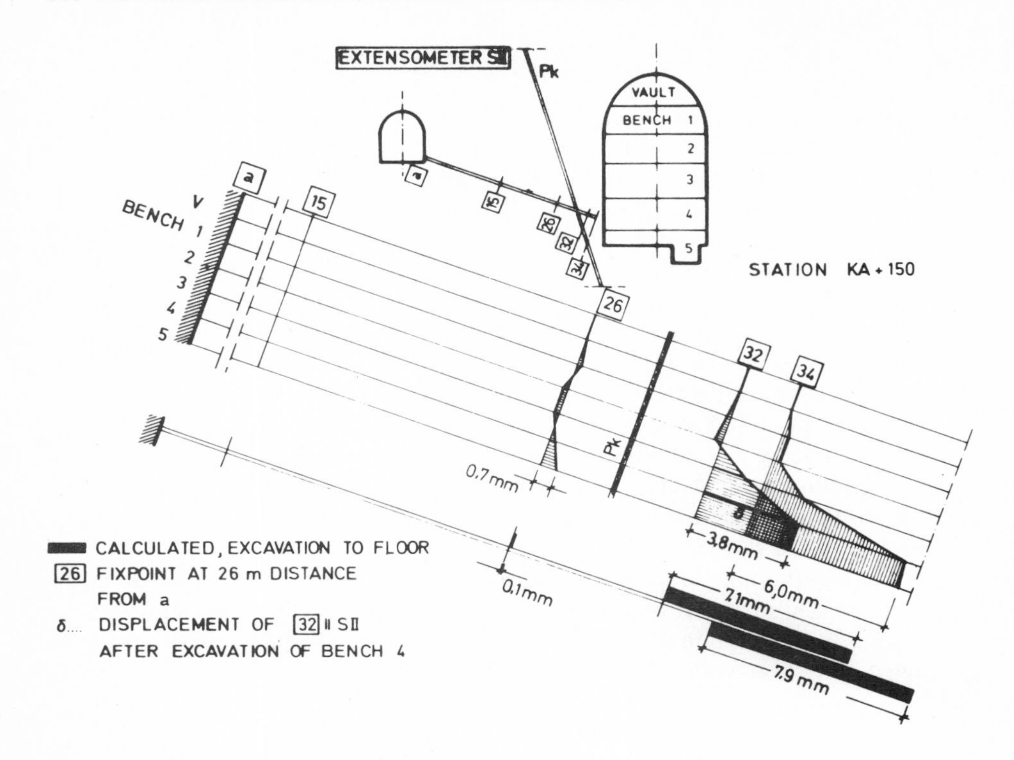

Figure 18-15 Results from extensometer *S*II.

Extensometer *S*III was installed at station $KA + 156$ m (Figs. 18-14 and 18-16) from a gallery (excavated at an early stage) which connects the pressure shaft with a turbine. Here for the fix points 12, 20, and 30 (Fig. 18-16) located above the parallel master joint, displacements toward point *a* were measured beginning after excavation of bench 2 and increasing with the progress of excavation. These displacements are also due to sliding off of the wedge. After excavation of benches 4 and 5 for fix points 20 and 30, a decrease of the displacements was recorded. This result is in accordance with the expected redistribution of stresses to the rock mass beyond the downstream cavern wall due to unloading of the rock mass wedge.

The magnitude of the displacements measured for both extensometers *S*II and *S*III is in satisfactory agreement with the forecast based on the results of the analyses (Fig. 18-17) and permits the conclusion that the decisive mechanical parameters of the rock mass and the parallel master joint *PK* were evaluated from the results of the investigations with satisfactory accuracy. A summary of the most remarkable measured and calculated displacements from the area adjacent to cross section *B* (see Fig. 18-12) proves that the rock mass wedge between the upstream cavern wall and parallel master joint *PK* deformed as expected.

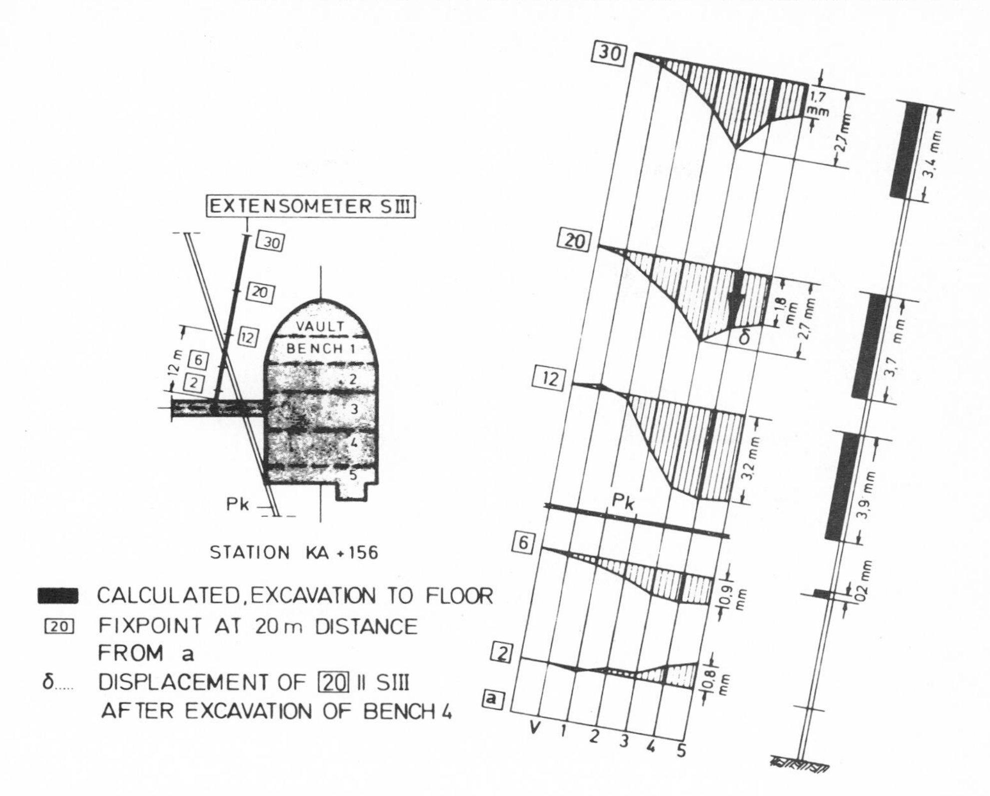

Figure 18-16 Results from extensometer *S*III.

Furthermore, from these results and those of other measurements, particularly the measuring anchors, it can be concluded that the stress increase of the prestressed anchors amounted to $\sigma_{\max} \approx 1000\ \mathrm{kg}_f/\mathrm{cm}^2$ and thus did not exceed the allowable limits. For the other installed multiple extensometers not mentioned here, the tendency and in most cases the absolute magnitude of the measurements are in good agreement with the forecast based on the analyses.

In Fig. 18-18 the displacements measured for the fix points of extensometers *S*II and III3 are plotted vs. time. The corresponding diagrams also show the days on which the excavation of benches 3 and 4 adjacent to the master joint *PK* was performed. It can be seen that on Feb. 14, 1973, when bench 3 was excavated, and on Feb. 28 and March 8, 1973, when bench 4 was excavated, sudden displacements were measured for the fix points of extensometer *S*II located close to the tip of the wedge. Further for extensometer III3, beyond the downstream cavern wall, a sudden increase of the displacements occurred on Feb. 14, 1973 for most of the fix points. Thus these results prove that the sliding off of the rock mass wedge and the redistribution of stresses beyond the downstream wall of the cavern occurred immediately after the corresponding excavation, as predicted. Furthermore, the

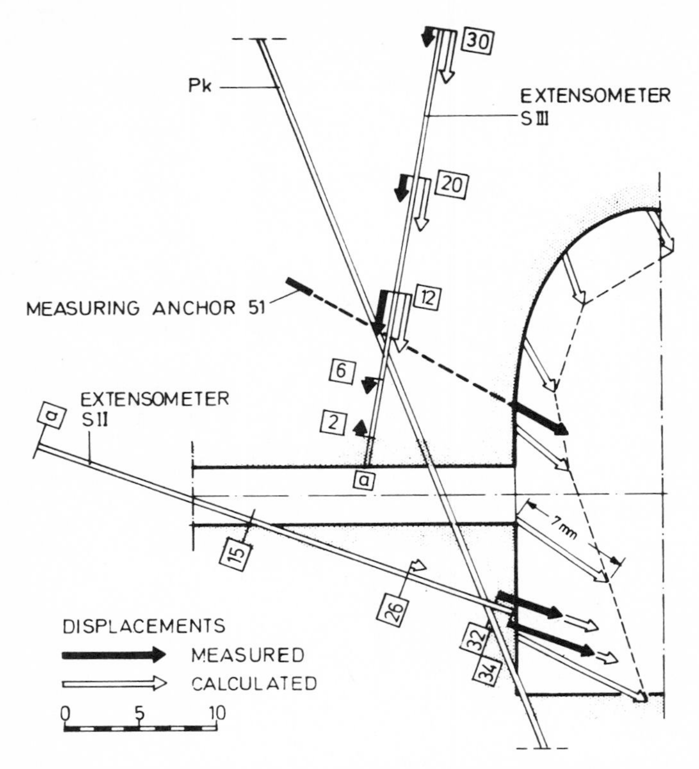

Figure 18-17 Comparison between field measurements and calculations.

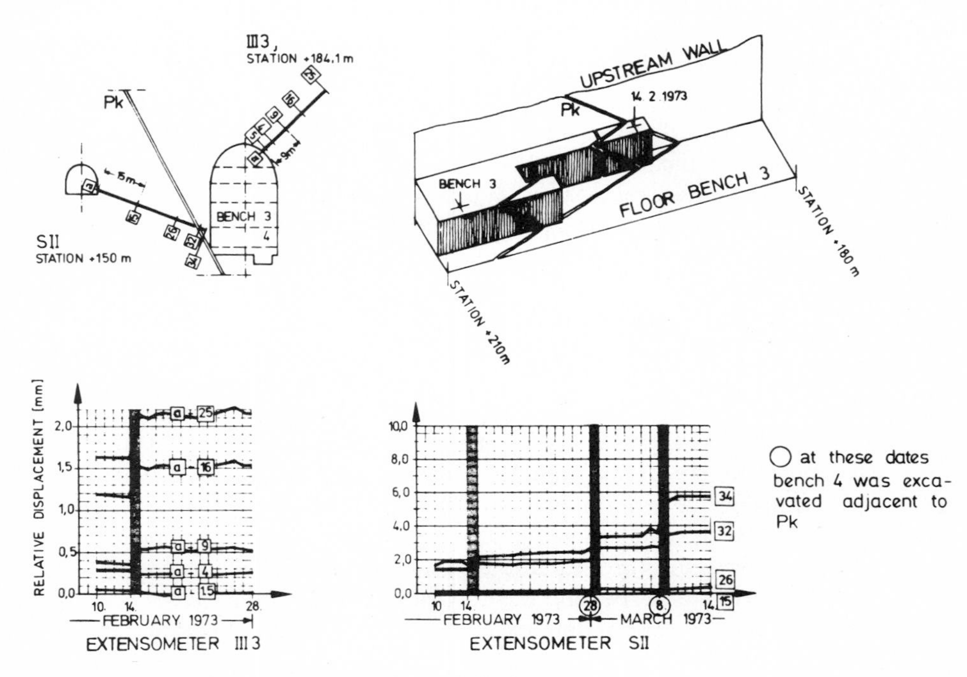

Figure 18-18 Relation between excavation and displacements.

results show that the stress-strain behavior of the rock mass is time independent (Fig. 18-18). A more detailed description of the results of the various measurements can be found in the literature.[7,16]

Summary

In summary the described example proves that comprehensive geological and rock mechanics investigations, a finite element analysis, and permanent measurements during construction make it possible to evaluate safe and economic solutions for complicated rock mechanics projects. For example, it was possible to determine the lining and the safety measures well in advance, and no modifications were required during construction. Investigations of this kind are therefore expected to prove of considerable importance. A photograph of the powerhouse is shown in Fig. 18-19.

Figure 18-19 Underground powerhouse at Wehr during construction, Spring, 1973.

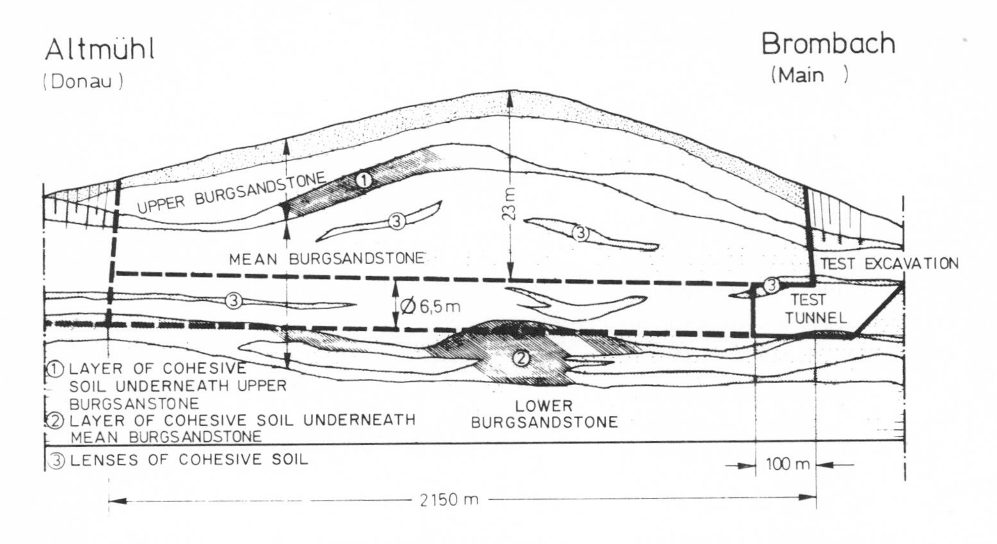

Figure 18-20 Geologic section of Altmühl Tunnel between the Rhine and Danube Rivers.

18-5 ALTMÜHLTUNNEL

Structure

To increase the low-water discharge in the upper catchment area of the Main it is planned to divert high-water discharges from the Altmühl River, a tributary of the Danube, to the Brombach River, a tributary of the Main. For this project a tunnel approximately 2.2 km long, the so-called Altmühltunnel, crossing the main watershed between the Rhine and the Danube is designed for a maximum water level at the elevation of the roof.

In 1973 a 100-m-long test tunnel, which later will be part of the planned tunnel, was excavated (Fig. 18-20). The internal diameter of the test tunnel is approximately 6.5 m, and the gradient is 1.25 percent. The average rock overburden amounts to ~ 13 m. Adjacent to the portal, which is located in a test excavation, the overburden is only 3 m. The results of the test tunnel were expected to provide the information required for the design of the main tunnel, e.g., for the stability analyses and the planning of the excavation procedure as well as the lining.

Geological Conditions

From the results of the engineering geological explorations, e.g., core drillings and surface mappings, it could be concluded that the substratum consists of medium- to coarse-grained Keuper sandstone with horizontal stratification. Two through-going layers of cohesive soil separate this sandstone into the upper, mean, and lower

Table 18-2 Idealized underground profile

Layer	Depth, m	Description
a	0–3.0	Organic soil and sand
b	3.0–4.5	Upper Burg sandstone
c	4.5–7.6	Layer of cohesive soil layer underneath upper Burg sandstone
d	7.6–11.2	Mean Burg sandstone
e	11.2–12.4	Layer of cohesive soil at roof of test tunnel
f	12.4–17.6	Mean Burg sandstone
g	17.6–17.9	Layer of cohesive soil at test-tunnel floor
h	< 17.9	Mean Burg sandstone

Burg sandstone (Fig. 18-20). In each of these sandstone layers, cohesive soil lenses of limited extent are intercalated. These two layers and the lenses consist mainly of clayey, sandy silt and clayey, silty sand. The Altmühltunnel along a major part of its line will be located within the mean Burg sandstone and at the elevation of its floor will intersect the underlying layer of cohesive soil (Fig. 18-20).

For the area of the test tunnel from the explorations the idealized underground profile shown in Table 18-2 resulted (Fig. 18-21). Parallel bedding discontinuities with 0.8 to 1.0 m spacing occur in the sandstone. Further occasional discontinuities with a mean strike of N80°E and 80° dip to the south were explored in the sandstone. The joints intersect the axis of the tunnel at an acute angle of 15°. The estimated mean extent of these discontinuities amounts to 35 m^2, that is, approximately the area of the tunnel cross section. The mean spacing is 2.5 m. Because of the low permeability of the intermediate clay layer for the upper and mean Burg sandstone two separate groundwater tables were explored.

Mechanical Properties of the Rock Mass

For the evaluation of the stress-strain properties and the strength of the sandstone and the layers of cohesive soil laboratory rock mechanics and soil mechanics tests on samples taken from core borings were performed. The corresponding test results revealed a considerable scatter, and only a limited number of tests could be performed within the scope of the explorations made before the test tunnel was excavated. Nevertheless it was possible to derive the mechanical parameters of the rock mass representation for the stability analyses of the test tunnel from the test results and an additional appraisal of the influence of the bedding and the other discontinuities. It was intended that the corresponding mechanical parameters for the final tunnel would be evaluated from the results of measurements performed during excavation of the test tunnel.

Stability Analyses

A finite element analyses for a representative cross section and the portal was performed to investigate the stability of the test tunnel. Because of the uncertainties with regard to the mechanical parameters, the latter were varied in the analysis; i.e., a so-called parametric study was performed. Since the influence of the steeply dipping discontinuities on the stability of the tunnel was practically negligible, as could be evaluated from preliminary calculations, the problem was symmetrical to the tunnel axis and consequently the analysis could be limited to one-half of the tunnel and the adjacent rock mass. Most of the calculations were based on a standard cross section of 20 m width and 30 m height (Fig. 18-21). The tunnel within this cross section is represented with a diameter of $D = 6.5$ m; the overburden amounts to 13.5 m (Fig. 18-21). The thickness of the analyzed rock mass slab measured parallel to the tunnel axis is 1 m. The calculations are based on the underground profile described above (Fig. 18-21). The thickness of the proposed shotcrete lining was assumed to be 16 cm. At the tunnel roof slightly prestressed 10-t rock bolts were provided, the average prestressing load amounting to 1 kg_f/cm^2. The adhesive stretch of these bolts is located in the sandstone above the layers of cohesive soil at the tunnel roof (Fig. 18-21). The in situ stresses have been assumed according to the weight of the overburden and the corresponding Poisson's ratio ν ($\sigma_v = \gamma h$, $\sigma_h = [\nu(1 - \nu)]\sigma_v$) resulting in the stress ratios represented for loading cases *B*5 and *G*1 of Table 18-3. In the calculations the un-

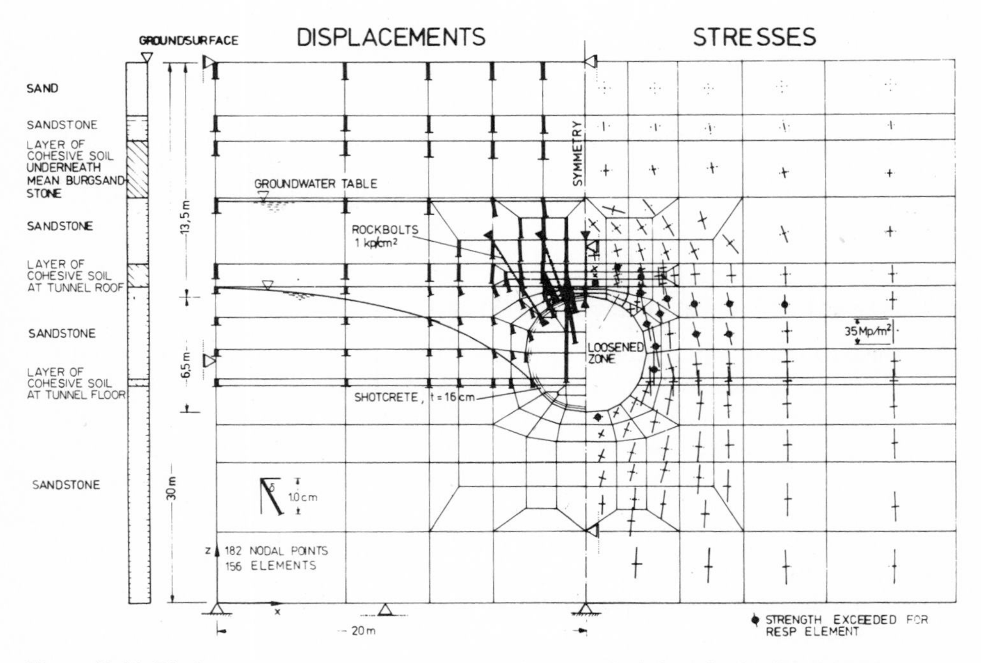

Figure 18-21 Displacements and stresses from finite element calculations for loading case *B*5.

Table 18-3 Parameters of rock mass, soil, and shotcrete introduced into the finite element analysis

	E, kg $/\text{cm}_f^2$						Unit weight,† Mg_f^2/m^3		In situ stresses‡ $K_0 = \frac{\sigma_h}{\sigma_v}$	
	Loading cases $B5$, $G1$	Loading case $G2$	ν	Cohesion c_t, kg_f/cm^2	Angle of friction ϕ, deg	Tensile strength σ_t, kg_f/cm^2	γ_d	γ'	Loading cases $B5$, $G1$	Loading case $G2$
Sandstone	30,000	15,000	0.15	20	30	5	2.1	1.2	0.2	2.0
Sand	500		0.3	0	35	0	1.9	—	0.43	
Layer of cohesive soil under mean Burg sandstone and at tunnel floor	50	25	0.4	0	30	0	2.2	1.2	0.67	
Layer of cohesive soil at tunnel roof	100	50	0.4	0	30	0	2.2	—	0.67	
Reinforced shotcrete	200,000		0.18	70	20	15	2.4	—		

	Strike	Dip	Degree of separation κ, %	Cohesion c_t, kg_f/cm^2	Angle of friction ϕ, deg	Tensile strength σ_t, kg_f/cm^2
Parallel bedding discontinuities in sandstone§	0°	0°	90	2.0	30	0.5

† γ_d = dry, γ' = submerged.
‡ σ_h = horizontal normal stress, σ_v = vertical normal stress.
§ One set of steeply dipping discontinuities has not been accounted for in the analyses presented in this chapter.

disturbed groundwater table and a depression due to change of drainage conditions after tunnel excavation were also accounted for (Fig. 18-21).

As an example for the various analyzed loading cases resulting from a variation of the mechanical parameters and the amount of applied safety measures, the results for loading case *B*5 are represented in Fig. 18-21. Loading case *B*5 is characterized by the Young's moduli and the in situ stresses compiled in the corresponding columns (loading cases *B*5 and *G*1) of Table 18-3. To account for the expected loosening due to blasting a very low Young's modulus was assumed for the zone of the sandstone layer above the tunnel roof, marked in Fig. 18-21. As a consequence this zone sustains practically no stresses, and in the calculations it was assumed to be fixed by slightly prestressed rock bolts, as outlined above. A shotcrete lining was not foreseen in the analysis of loading case *B*5. The representation of the evaluated principal normal stresses (Fig. 18-21) shows that the assumed loosened zone leads to the result that the stress redistribution due to tunnel excavation occurs in the sandstone layer above the layer of cohesive soil at the tunnel roof. The representation of the evaluated displacements results in a uniform settlement of approximately 5 mm at the surface due to tunnel excavation. This settlement mainly results from a compression of the layers of cohesive soil at the tunnel roof and floor due to the stress redistribution caused by tunnel excavation. Because of its comparatively low compressibility the sandstone does not contribute significantly to the evaluated settlements.

In spite of local failures in the sandstone and clay adjacent to the tunnel, the analysis of loading case *B*5 resulted in a convergency of the evaluated displacements and consequent stability of the tunnel. The considerable displacements toward the tunnel resulting for zones adjacent to the tunnel roof led to the conclusion that in addition to rock bolting a shotcrete lining should be installed immediately after excavation. This was of special importance at the portal of the test tunnel, where the layer of cohesive soil was expected to occur immediately at the tunnel roof.

For the stability analysis of the tunnel portal, a 45-m-long, 20-m-wide, and 30-m-high three-dimensional section at the bottom of the large test excavation was determined (Fig. 18-22). Thanks to symmetry here as well, only one-half of the tunnel and the adjacent rock mass had to be analyzed. The selected section includes the first 39 m of the tunnel and a 6-m-long section of the test excavation. The underground conditions right at the portal deviate slightly from those evaluated for the standard cross section. In particular the layers of cohesive soil discussed previously are located here immediately at the tunnel roof and floor, respectively; however, the mechanical parameters could be assumed according to those of loading case *B*5 (Table 18-3).

Figure 18-22 shows a perspective representation of the displacements for a loading case in which the roof and both walls of the tunnel are lined by shotcrete from the portal to the end of the section considered. The settlement of the roof was evaluated as 3.4 mm at the portal and as 5 mm at the end of the considered section. At the surface a depression curve results from the analysis, whereas the usual heaving of the tunnel floor is negligibly small. In another loading case the

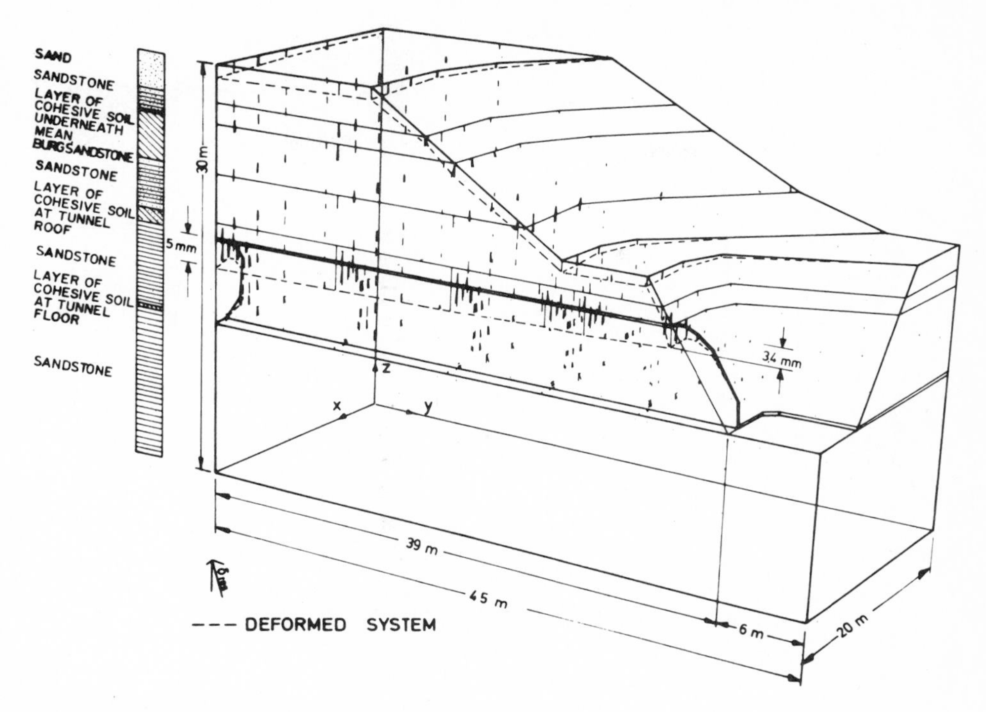

Figure 18-22 Computed deformations, isometric view.

tunnel was assumed to be excavated to a 20-m length, the first 6 m at the portal being lined by shotcrete. These calculations showed that the roof of the unlined part of the tunnel is considerably deformed toward the tunnel. Consequently the unlined section of the tunnel was to be kept as short as possible. From the various cases similarly analyzed it was concluded that the portal including the first 39 m of the tunnel would be stable if a shotcrete lining installed immediately after excavation was planned. Such a shotcrete lining transmits a major part of the stresses originally transmitted by the excavated rock.

As a result of these three-dimensional analyses and the calculations for the standard cross section, a 16-cm-thick lining of shotcrete with an unconfined compressive strength of 250 kg_f/cm^2 reinforced by an inside and outside square wire mesh of 1.58 cm^2/m was recommended as a preliminary lining for the test tunnel. Further rock bolting at the roof applying a slight prestressing load of 1 kg_f/cm^2, as described above, was recommended; finally, according to the analyses, it was suggested that the shotcrete and the anchors be installed immediately after each blast or at least every other blast.

Measurements during Construction

To monitor the stability of the test tunnel during construction and to evaluate the mechanical parameters applicable for the design of the main tunnel the recommended measuring program consisted of surface leveling, extensometers, conver-

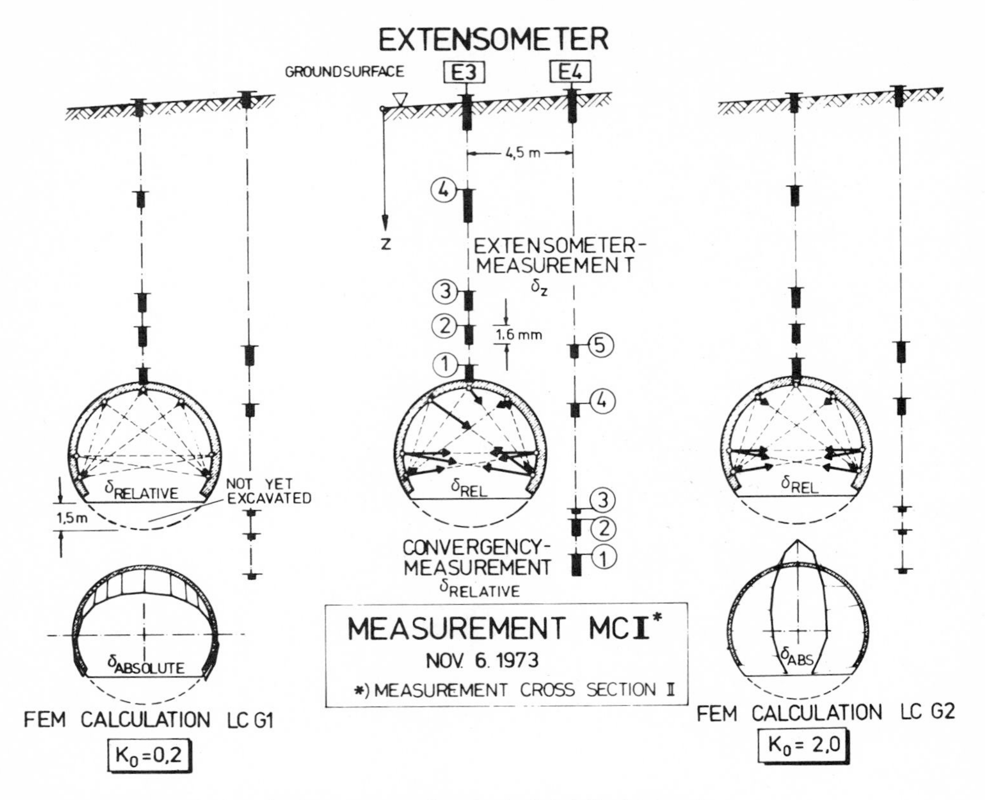

Figure 18-23 Comparison between measured and calculated displacements.

gency measurements along the tunnel circumference, and stress measurements in the shotcrete lining. In cross section I at the tunnel portal extensometers were installed from the test excavation before tunnel excavation started. Standard cross sections II and IV (29 and 68 m from the portal, respectively) were equipped with two extensometers each (Fig. 18-23), seven convergency bolts (Fig. 18-23), and eight Gloetzl pressure cells in the lining, four of the cells being radially oriented for measurement of the tangential stresses and (Fig. 18-24) four being tangentially installed for measurement of the contact pressure between rock and lining. The extensometers of sections II and IV were installed from the surface before tunnel excavation started. Further seepage measurements and geological mapping of the tunnel walls were performed during excavation.

As a consequence of the tunnel excavation an approximately homogeneous heaving of the rock above the tunnel of 6 to 8 mm was measured for cross section I. A further homogeneous heaving was recorded during excavation of a 1.5-m-high part of the tunnel cross section at the floor, which was excavated only at the end of the construction period beginning at the end of the test tunnel and ending at the portal.

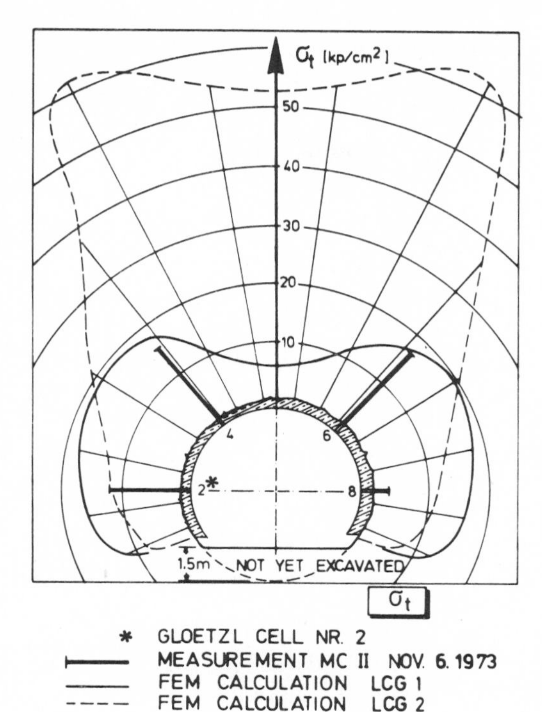

Figure 18-24 Comparison between measured and calculated tangential stresses.

In measurement cross section II a settlement of 1 to 2 mm and a heaving of approximately 2 mm at the floor were recorded (Fig. 18-23) for the stage of excavation 1.5 m above the final tunnel floor at the roof. The displacements slowly increased with excavation and reached the final value after the tunnel excavation was approximately 30 m ahead of the measurement section. Immediately after the excavation passed the measurement section, approximately 20 to 30 percent of the resulting displacements occurred (see also Figs. 18-1 and 18-2). The corresponding convergency measurements reveal a relative convergency of $2\delta_h = 6$ to 9 mm for the tunnel walls and $2\delta_v = 1$ to 3 mm (Fig. 18-23) between the tunnel roof and the lower part of the walls. However, from the results of these convergency measurements the absolute displacements of the convergency bolts cannot be derived. It must be realized that the convergency bolts could be installed only after the corresponding cross section was excavated and consequently only the portion of the displacements occurring later could be measured. The same conclusion applies for the stresses measured by the pressure cells. Tangential stresses in the lining of $\sigma = 12$ to 17 kg_f/cm^2 (Fig. 18-24) and radial normal stresses between the lining and rock of $\sigma_r = 0.3$ to 0.5 kg_f/cm^2 resulted from the measurements.

The results from measurement cross section IV were in good agreement with those of cross section II, due probably to equal rock mass composition.

Comparison between Results of Measurements and Calculations

The results of the measurements and the results of the analyses confirmed the stability of the test tunnel. For a verification of the assumptions a recomputation of measurement cross sections II and IV, based on the method described above, was performed. For these computations according to the network already described (Fig. 18-21) a finite element mesh was selected accounting for the stage of excavation represented in Figs. 18-23 and 18-24; that is, the excavation reaches only 1.5 m above the tunnel floor, and the lining has not yet been installed down to the bottom of the tunnel.

For loading case *G*1 the same parameters as in loading case *B*5 were applied (Table 18-3). A comparison of the resulting displacements with the measured values for loading case *G*1 reveals a good agreement for the fix points of the extensometers, whereas for the convergency a considerable deviation results (Fig. 18-23). In particular, the horizontal convergency of the tunnel walls could not be reproduced by the calculations (Fig. 18-23). For the tangential stresses in the shotcrete lining a reasonable agreement was found (Fig. 18-24). The reasons for differences between results of measurements and calculations probably are (1) that the assumed horizontal in situ stresses are too low and (2) that the Young's moduli assumed for the sandstone and the layers of cohesive soil are too high. Consequently in a further loading case *G*2 for the sandstone and the layers of cohesive soil Young's moduli of the order of magnitude of the lower boundary of the corresponding laboratory tests were assumed (see loading case *G*2, Table 18-3). As a consequence of information about the tectonic conditions in the project area orthogonal to the tunnel axis, a horizontal in situ stress of $\sigma_h = 2\sigma_v$ was assumed for the sandstone layers (see loading case *G*2, Table 18-3), the corresponding vertical normal stress σ_v being assumed according to the weight of the overburden and the horizontal normal stress parallel to the tunnel axis according to the corresponding Poisson's ratio. A comparison of the measured and calculated displacements for this case (loading case *G*2) reveals a good match for the extensometers and the horizontal convergency of the tunnel wall. When it is remembered that the unavoidable delay in installing the pressure cells meant that the measured tangential stresses in the lining would necessarily be smaller than calculated, these stresses are also reproduced satisfactorily by the calculations (Fig. 18-24).

As a consequence, even at a small depth remarkably high horizontal primary stresses of $\sigma_h = 2\sigma_v$ certainly exist in the sandstone layers. A verification of this result also follows from the heavings measured in measurement cross section I for the rock above the portal and from the convergency measurements in cross sections II and IV, which revealed additional horizontal displacements of the tunnel walls resulting from excavation of the 1.5-m-thick part of the cross section at the tunnel floor at the end of construction. During construction, when the installed shotcrete lining at the tunnel roof was only ~ 8 cm thick and rock bolts had not yet been installed, an audible shear failure of the lining approximately 11 m long occurred suddenly along the centerline of the tunnel roof. If the comparatively high normal stresses in the lining at the tunnel roof as evaluated from loading case

*G*2 are considered (Fig. 18-24), this observation is also probably due to the high horizontal in situ stresses.

Summary and Conclusions

The good agreement of the analyses (loading case *G*2) and the corresponding in situ measurements for the test tunnel made it possible to apply the resulting mechanical parameters (Table 18-3) to the design of the main tunnel. This example shows that by application of the finite element method the in situ stresses and mechanical parameters of the rock mass can be evaluated from measurements of the displacements in a testing tunnel if it is adequately equipped.

Furthermore, from an extrapolation of the results of the core drillings performed along the axis of the main tunnel, idealized characteristic cross sections can be derived (*a* to *f*, Fig. 18-25). These cross sections can be expected to be representative for certain sections and percentages of the tunnel length and differ in the varying number, thickness, and elevation of silty clay (Fig. 18-25).

From these cross sections and the parameters described above (Table 18-3) the final design of the main tunnel could be evaluated with the help of extensive finite element calculations. In addition the test tunnel provided comprehensive information concerning the construction procedure envisaged. In view of the fact that the test tunnel can later serve as part of the main tunnel, the costs of excavation and lining the test tunnel are completely justified.

The construction of the main tunnel began in April 1975. A photograph of the testing tunnel is shown in Fig. 18-26. A more detailed description of the project and the analyses can be found elsewhere.[17]

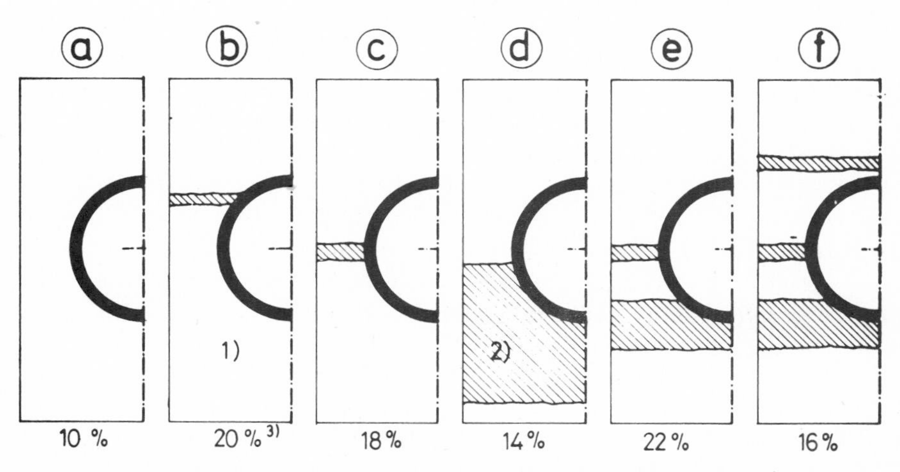

Figure 18-25 Six characteristic profiles.

Figure 18-26 Tunnel during construction.

18-6 BREMM TEST CAVERN

Introduction

During the feasibility study for the Bremm pumped-storage plant of the Rheinisch Westfälische Electricity Supply Board (RWE) in Germany, a part of the vault at the planned underground powerhouse was excavated for testing purposes. The Bremm test cavern is located in the Eifel mountains adjacent to Bremm, a small village on the Mosel River. This large-scale in situ test was performed in connection with other geological explorations in order to gain experience with regard to adequate construction procedures and to determine the mechanical properties of the rock mass so that an economic and safe design of the extensive underground works could be planned for this project.

Structure

The test cavern has a semicircular cross section, a width of 24 m, a length of 30 m, and a height of 9 m, as can be seen in the perspective representation of Fig. 18-27. The rock overburden amounts to $H_0 = 240$ m. The angle between north and the horizontal cavern axis is N10°E (Fig. 18-27, Table 18-4).

Geology

The geological mappings of numerous core drillings and some exploration adits excavated in the project area were performed by the geologist of the Lahmeyer Consulting Engineers, Frankfurt, and analyzed in cooperation with the author.

Table 18-4 Mechanical parameters of rock mass and lining introduced into finite element calculations

Orientation	Strike	Dip
Axis of cavern (horizontal)	N10°E	
Bedding K_3	N62°E	51°NW
Joint set K_1	N140°E	80°NE
K_2	N5°E	90°

Elastic constants for rock mass			
E_1	300,000 kg_f/cm^2	E_2	60,000 kg_f/cm^2
ν_1	0.25	ν_2	0.05
G_2	28,400 kg_f/cm^2		

Strengths				Angle of friction	
	Compressive strength σ_c, kg_f/cm^2	Tensile strength σ_t, kg_f/cm^2	Cohesion c_t, kg_f/cm^2	Peak shear strength ϕ, deg	Residual shear strength ϕ^*, deg
Intact rock	470	47	140	28	28
Bedding K_3	...	0.5	20	25	15
K_1, K_2	...	10	42	25	15
Shotcrete†	300	15			

Anchors
Mono rock anchors, 16 t each, 6–10 m long, 0.6 anchor/m^2

† Shotcrete: $E_B = 350{,}000\ kg_f/cm^2$, $\nu_B = 0.18$.

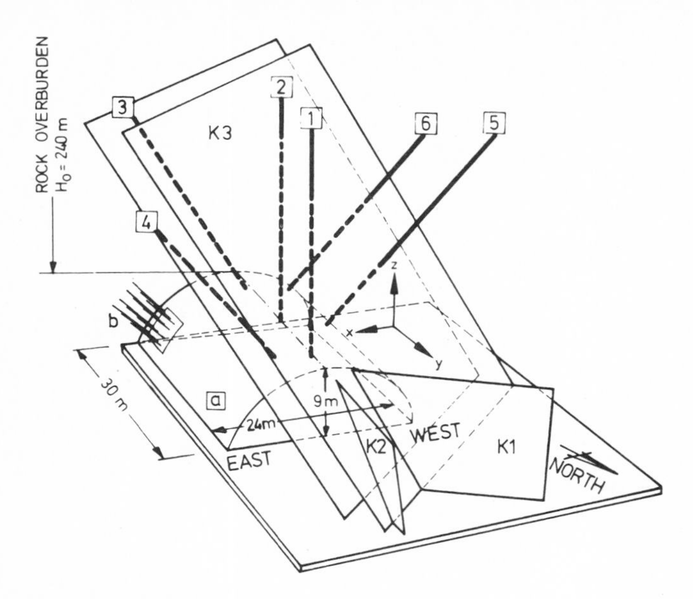

Figure 18-27 Model of test cavern.

From these investigations it was concluded that the rock in the project area consists of a clayey schist with quartzitic intercalations. The orientation of the schistosity in the cavern area is practically constant and approximately parallel to the bedding. The mean angle of strike of the bedding planes, denoted K_3 in Fig. 18-27, is N62°E, and the mean angle of dip is 51° northwest (Table 18-4). Consequently the bedding planes intersect the cavern axis diagonally (Fig. 18-27, Table 18-4). In addition two sets of approximately vertically dipping minor joints K_1 and K_2 were evaluated from the geological mappings. The joints of set K_1 are approximately obliquely oriented to the bedding, and the joints of set K_2 strike approximately parallel to the cavern axis. The spacings and extent of the joints K_1 and K_2 were statistically evaluated from the results of the geological mappings. Single throughgoing master joints parallel to bedding K_3 and joints sets K_1 and K_2 are not dealt with in this chapter.

Mechanical Parameters

From a statistical evaluation of the geological mapping with regard to structure the parameters describing the strength and deformability of the intact rock were determined from unconfined compression and cylinder-splitting tests on cores taken from the drillings. In these laboratory tests the orientation of the bedding and the schistosity K_3 were varied with regard to the axis of the investigated core samples.

The test results revealed a considerable strength anisotropy of the intact rock. For shear parallel to bedding the resulting mean shear parameters are

$c_t = 20\ kg_f/cm^2$ (cohesion) and $\phi = 25°$ (angle of friction), whereas the residual shear strength is determined by an angle of friction $\phi^* = 15°$. For the tensile strength orthogonal to K_3 a value of $\sigma_t = 0.5\ kg_f/cm^2$ resulted. These data were equated (in analyses discussed below) with the average rock mass strength along K_3. For directions deviating from that of the bedding the mean unconfined compressive strength of the intact rock resulted in $\sigma_c = 470\ kg_f/cm^2$, which corresponds to a cohesion of $c_t = 140\ kg_f/cm^2$ and an angle of friction of $\phi_t = 28°$, the residual angle of friction also being assumed to be $\phi^* = 28°$ (Table 18-4).

From these test results and the statistically evaluated degree of separation of joints K_1 and K_2 a homogeneous reduction of the intact rock strength along joint sets K_1 and K_2 was evaluated (see row 3 of the strength parameters in Table 18-4).

Following the same principle, the parameters describing the deformability of the rock mass were derived from the laboratory test results and those of the geological mapping. Here too a high degree of anisotropy was found, resulting in a transversely anisotropic stress-strain behavior, as described in Sec. 18-2. The mean Young's moduli for compression parallel and orthogonal to bedding were evaluated at $E_1 = 300{,}000$ and $E_2 = 60{,}000\ kg_f/cm^2$, respectively. The corresponding Poisson's numbers were $\nu_1 = 0.25$ and $\nu_2 = 0.05$ and the shear modulus $G_2 = 28{,}400\ kg_f/cm^2$ (Table 18-4).

Because of the limited extent of the rock mechanics tests performed and the scatter of the test results, the mechanical parameters given in Table 18-4 were varied within the subsequently described analyses.

Since measurements of the in situ stresses were not available before the excavation of the test cavern began, the primary state of stress in the cavern area was evaluated from a separate analysis based on the same rock mechanics parameters (Table 18-4) and applying the weight of the rock overburden. The result of the analysis will be discussed subsequently.

The assumed parameters describing the strength and the deformability of the shotcrete lining subsequently recommended are also given in Table 18-4.

Stability Analysis

The stability of this underground structure was evaluated by a finite element analysis. A representative section was evaluated, neglecting the stabilizing effect of the end slabs of the cavern (Fig. 18-27). This section is confined by two bedding planes with a spacing of 5 m and by two vertical and two horizontal planes mutually parallel to the cavern axis and having a spacing of 75 m (Fig. 18-28).

By applying the mechanical rock mass parameters compiled in Table 18-4 in a first calculation step the expected in situ stresses were evaluated. This resulted in a vertical maximum principal stress equal to the weight of the overburden. The minimum and intermediate principal stresses were oriented orthogonal and parallel to the strike of the bedding, respectively.

From this result the stresses and displacements due to excavation of the test cavern were evaluated. For the loading case presented here a lining consisting of

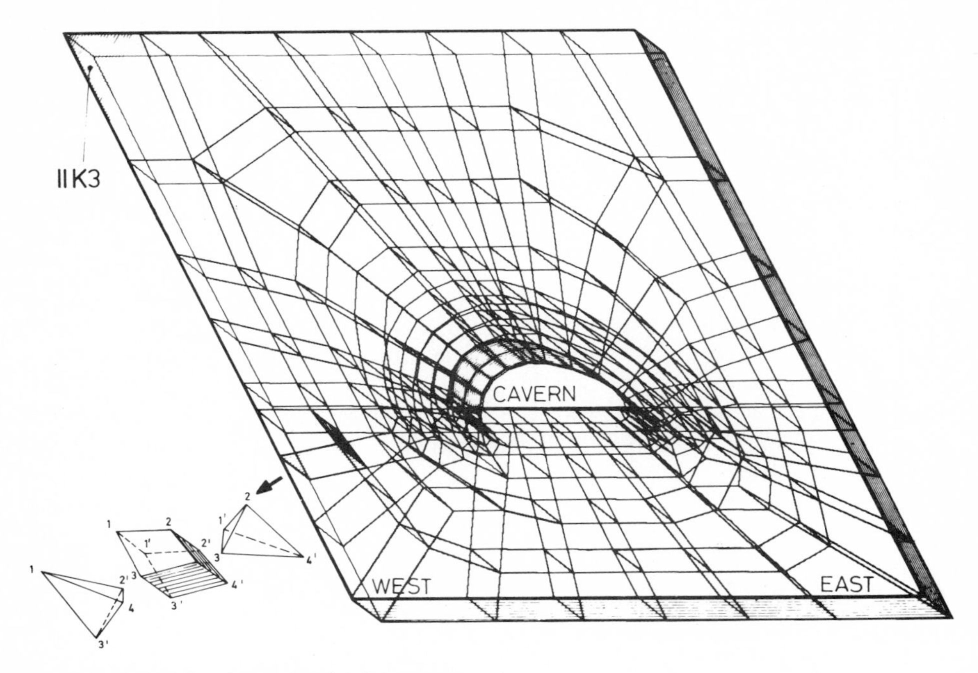

Figure 18-28 Finite element mesh of test cavern.

20-cm-thick reinforced shotcrete was recommended. The influence of the recommended anchoring, consisting of slightly prestressed 16-t anchors 6 to 10 m long, was evaluated for this loading case. For the applied mechanical parameters see Table 18-4. From the most important results presented in Fig. 18-29 it can be seen that the resulting displacements are approximately orthogonal to the bedding (Fig. 18-29*a*), that the western part of the vault deforms toward the cavern more intensely than the eastern part, and that the heaving of the cavern floor is smaller than the settlement of the roof (Fig. 18-29*b*). These results are due to the orientation of the bedding with regard to the cavern axis and the anisotropic deformability of the rock mass. As a further consequence of this anisotropy above the western part of the vault and underneath the eastern part of the floor of the cavern, a certain loosening orthogonal to the bedding occurs (Fig. 18-29*a* and *c*). Consequently the tensile strength orthogonal to the bedding (due to the unfavorable angle between the maximum principal stresses and the bedding) and the shear strength parallel to the bedding are exceeded within these zones (Fig. 18-29*c*). Since the extent of these so-called *plastic zones* was limited, however, and since convergence of the calculated displacements was found, the analysis resulted in stability of the cavern if the lining and anchoring are applied.

The results of comparative calculations applying reduced shear and tensile strength along the bedding and variations of the ratio of the moduli E_1 and E_2 will not be discussed, and a more extensive anchoring applied in a faulted zone will not be reported on.

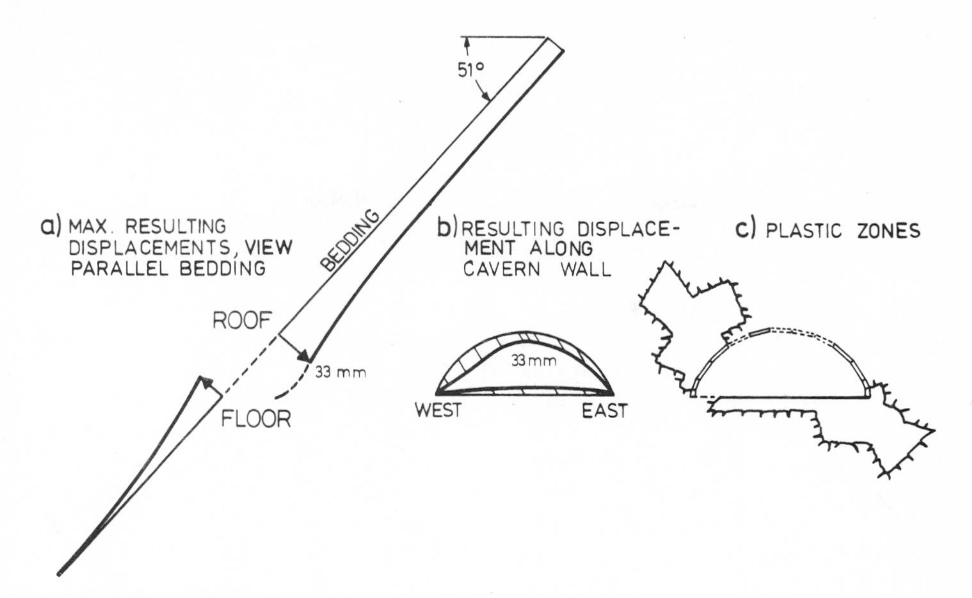

Figure 18-29 Results for loading case 4.

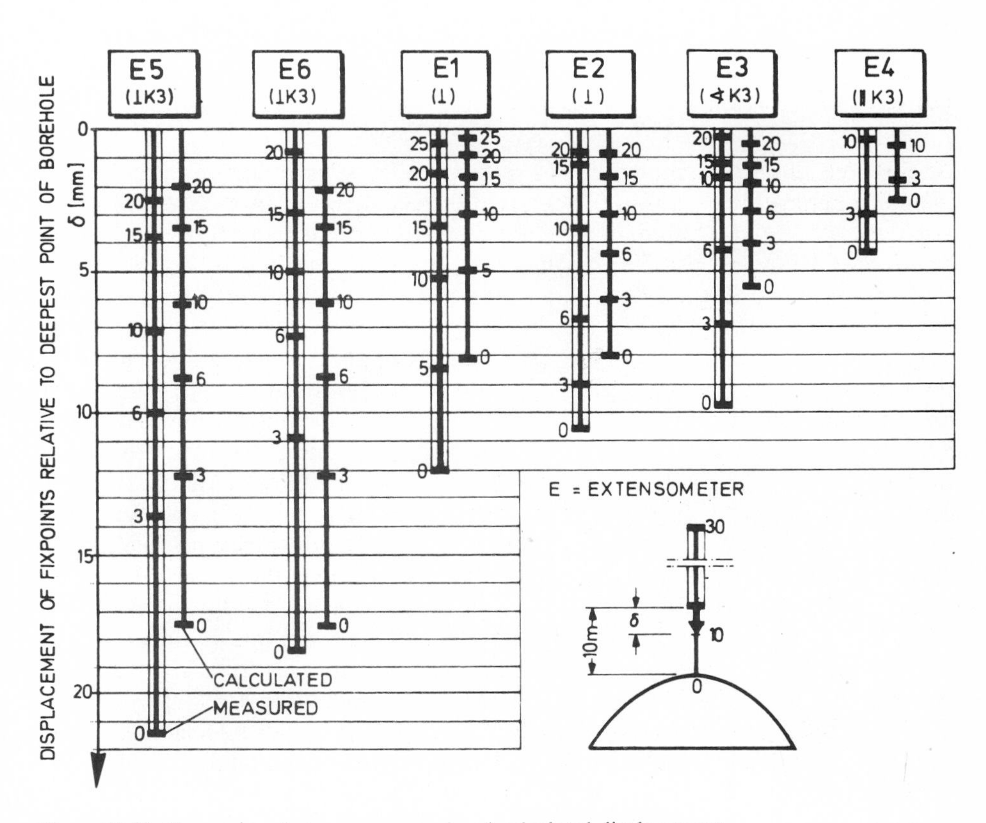

Figure 18-30 Comparison between measured and calculated displacements.

Measurements during Construction

For monitoring the stability of the cavern during and after construction five 30-m-long extensometers with five fix points (*E*1 to *E*3, *E*5, *E*6) and one 20-m-long extensometer *E*4 with four fix points were installed as soon as the corresponding extensometer head at the cavern wall was accessible (Fig. 18-27). Extensometers *E*1 and *E*2 are vertical and monitor the deformations of the rock mass above the vault of the cavern. Extensometers *E*3 and *E*4 were oriented parallel and *E*5 and *E*6 orthogonal to the bedding because in these directions the minimum (*E*3, *E*4) and maximum (*E*5, *E*6) displacements were to be expected (see Fig. 18-29).

The measured displacements increased according to the stage of excavation and reached the final value immediately after excavation was completed. Consequently it could be concluded that the time dependency of the stress-strain behavior of the rock mass, as in Secs. 18-4 and 18-5, can be neglected.

The measured resulting displacements are represented in Fig. 18-30. To make comparison with the calculated values possible it must be remembered that a certain fraction of the total displacements had already occurred when the extensometers were installed. Consequently the stages of excavation prevailing when the extensometers were installed have been analyzed, applying the parameters compiled in Table 18-4. The displacements resulting from the latter calculations

Figure 18-31 Bremm test cavern during construction.

were deducted from those calculated for the stage of full excavation (Fig. 18-29), and the resulting differences are plotted in Fig. 18-30. A comparison of these values with the results of the measurements reveals a remarkable agreement with regard to the tendency and the absolute magnitude. The displacements parallel to bedding *K*3 (*E*3 and *E*4), as expected, are considerably smaller than those orthogonal to bedding (*E*5 and *E*6, Fig. 18-30). A photograph of the test cavern during construction is given in Fig. 18-31. Further information on this project can be derived from the literature.[13]

Conclusions

The results of these investigations are expected to provide the basis for the final design of the underground structures of the planned power plant. Consequently the analyses for the Bremm test cavern represent another example of the applicability and usefulness of the evaluated rock mechanical model and the corresponding finite element program.

REFERENCES

1. Desai, C. S., and J. F. Abel: "Introduction to the Finite Element Method: A Numerical Method for Engineering Analysis," Van Nostrand Reinhold Company, New York, 1972.
2. Goodman, R. E.: On the Distribution of Stresses around Circular Tunnels in Nonhomogeneous Rocks, *Proc. 1st Congr. Int. Soc. Rock Mech., Lisbon, 1966.*
3. Goodman, R. E., R. L. Taylor, and T. L. Brekke: A Model for the Mechanics of Jointed Rock, *J. Soil Mech. Found. Div. ASCE*, vol. 94, no. SM3, pp. 637–659.
4. Grob, H., K. Kovari, F. Vannotti, und C. Amstad: Praktische Anwendung der Finite Element Methode in der Tunnelstatik, *Proc. 2d Congr. Int. Soc. Rock Mech., Belgrad 1970*, vol. 2, pp. 897–903.
5. Lekhnitskii, S. G.: Theory of Elasticity of an Anisotropic Elastic Body, Holden-Day, Inc., Publisher, San Francisco, 1963.
6. Malina, H.: Berechnung von Spannungsumlagerungen in Fels und Boden mit Hilfe der Elementenmethode, *Veroff. Inst. Boden- Felsmech. Karlsruhe* Heft 40, 1969.
7. Pfisterer, E., W. Wittke, and P. Rissler: Untersuchungen, Berechnungen und Messungen beim Bau der Maschinenkaverne Wehr, *Proc. 3d Congr. Int. Soc. Rock Mech., Denver 1974*, vol. IIB, pp. 1308–1317.
8. Reyes, S. F., and D. K. Deere: Elastic Plastic Analysis of Underground Openings by the Finite Element Method, *Proc. 1st Congr. Int. Soc. Rock Mech., Lisbon 1966*, pp. 477–486.
9. Rocha, M., and I. N. Da Silva: A New Method for the Determination of Deformability in Rock Mechanics, *Proc. 2d Congr. Int. Soc. Rock Mech., Belgrade, 1970*, vol. 1, p. 2-21.
10. Rocha, M., J. B. Lopes, and J. N. Silva: A New Technique for Applying the Method of the Flat Jack in the Determination of Stresses inside Rock Masses, *Proc. 1st Congr. Int. Soc. Rock Mech., Lisbon, 1966*, vol. 2, p. 57.
11. Rodatz, W., and M. Wallner: Untersuchung des räumlichen Spannungs- und Verformungszustandes an der Ortsbrust eines Tunnels nach der Finite Element Methode, pt. 1, *Dtsch. Ges. Erd-Grundbau Natl. Tag. Felshohlraumbau, Essen, March 1974*, pp. 107–118.
12. Wilson, E. L.: Finite Element Analysis of Two-dimensional Structures, Ph.D. thesis, University of California, Berkeley, 1963.

13. Wittke, W., W. Rodatz, and M. Wallner: Three-dimensional Calculation of the Stability of Caverns, Tunnels, Slopes and Foundations in Anisotropic Jointed Rock by Means of the Finite Element Method, *Schriftenr. Dtsch. Ges. Erd- Grundbau Dtsch. Beitr. Geotech.*, no. 1, 1972.
14. Wittke, W., and E. Pfisterer: Geotechnical Investigations for the Underground Powerhouse Wehr of the Hotzenwald Pumped Storage Schemes, West Germany, *Symp. Hydro-Electr. Pumped Storage Schemes, Athens, November 1972.*
15. Wittke, W., and S. Semprich: 3-D Finite Elements for Foundations in Soil, *Proc. 8th Int. Conf. Soil Mech. Found. Eng. Moscow, August 1973.*
16. Wittke, W., E. Pfisterer, and P. Rissler: Bemessung der Auskleidung der Maschinenkaverne Wehr nach der Methode finiter Elemente, pt. 1, *Dtsch. Ges. Erd- Grundbau Natl. Tag. Felshohlraumbau, Essen, March 1974*, pp. 123–148.
17. Wittke, W., L. Carl, and S. Semprich: Felsmessungen als Grundlage für den Entwurf einer Tunnelauskleidung, Strasse Brücke Tunnel 27, 1975, Heft 1, pp. 1–8.
18. Zienkiewicz, O. C.: "The Finite Element Method in Engineering Science," McGraw-Hill Publishing Company, Ltd., London, 1971.
19. Zienkiewicz, O. C., B. Best, C. Dullage, and K. G. Stagg: Analysis of Non-linear Problems in Rock Mechanics with Particular Reference to Jointed Rock Systems, *Proc. 2d Congr. Int. Soc. Rock Mech., Belgrade, 1970*, p. 8-14.

CHAPTER

NINETEEN

SOIL AMPLIFICATION OF EARTHQUAKES

José M. Roësset

The effect of local soil conditions on the amplitude and frequency content of earthquake motions has been the subject of considerable interest and research in recent years. Physically the problem is to predict the characteristics of the seismic motions that can be expected at the free surface (or at any depth) of a soil stratum. Mathematically the problem is one of wave propagation in a continuous medium.

If the medium is linearly elastic and the geometry is relatively simple, analytical solutions can be obtained for any kind of waves using a method presented[1] as early as 1950. In practice, since the wave content of a potential earthquake is hard to predict, solutions are often limited to the simple case of shear waves propagating vertically. Many results for this case have been presented,[2–5] and a discrete model[6,7] with lumped masses and springs, based on a finite difference formulation, has enjoyed great popularity among practicing engineers. The continuous and the discrete formulations are equivalent.[8] In addition, continuous solutions for waves with arbitrary angles of incidence and combinations of shear and dilatational plane waves have been obtained.[9,10] This chapter describes the formulation and solution of the one-dimensional amplification problem.

19-1 THREE-DIMENSIONAL WAVE EQUATIONS FOR A LINEAR MATERIAL

The dynamic equilibrium equations for the three-dimensional case are

$$\begin{aligned}
\frac{\partial \sigma_{xx}}{\partial x} + \frac{\partial \sigma_{xy}}{\partial y} + \frac{\partial \sigma_{xz}}{\partial z} &= \rho \frac{\partial^2 u_x}{\partial t^2} = \rho \ddot{u}_x \\
\frac{\partial \sigma_{xy}}{\partial x} + \frac{\partial \sigma_{yy}}{\partial y} + \frac{\partial \sigma_{yz}}{\partial z} &= \rho \frac{\partial^2 u_y}{\partial t^2} = \rho \ddot{u}_y \qquad (19\text{-}1) \\
\frac{\partial \sigma_{xz}}{\partial x} + \frac{\partial \sigma_{yz}}{\partial y} + \frac{\partial \sigma_{zz}}{\partial z} &= \rho \frac{\partial^2 u_z}{\partial t^2} = \rho \ddot{u}_z
\end{aligned}$$

where σ = component of stress
u_x, u_y, u_z = components of displacement vector $\{v\}$
ρ = mass density of material

Defining a rotation vector $\{\Omega\}$ with components Ω_x, Ω_y, Ω_z gives

$$\Omega_x = \frac{1}{2}\left(\frac{\partial u_z}{\partial y} - \frac{\partial u_y}{\partial z}\right) \qquad \Omega_y = \frac{1}{2}\left(\frac{\partial u_x}{\partial z} - \frac{\partial u_z}{\partial x}\right) \qquad \Omega_z = \frac{1}{2}\left(\frac{\partial u_y}{\partial x} - \frac{\partial u_x}{\partial y}\right) \quad (19\text{-}2)$$

and substituting the equations of linear elasticity from Chap. 2, one can rewrite Eq. (19-1) as

$$\begin{aligned}
(\lambda + 2G)\frac{\partial \epsilon_{\text{vol}}}{\partial x} + 2G\left(\frac{\partial \Omega_y}{\partial z} - \frac{\partial \Omega_z}{\partial y}\right) &= \rho \ddot{u}_x \\
(\lambda + 2G)\frac{\partial \epsilon_{\text{vol}}}{\partial y} + 2G\left(\frac{\partial \Omega_z}{\partial x} - \frac{\partial \Omega_x}{\partial z}\right) &= \rho \ddot{u}_y \qquad (19\text{-}3) \\
(\lambda + 2G)\frac{\partial \epsilon_{\text{vol}}}{\partial z} + 2G\left(\frac{\partial \Omega_x}{\partial y} - \frac{\partial \Omega_y}{\partial x}\right) &= \rho \ddot{u}_z
\end{aligned}$$

Obtaining the derivative of the first equation of (19-3) with respect to x, the derivative of the second with respect to y, and the derivative of the third with respect to z, and adding the three expressions gives

$$(\lambda + 2G)\left(\frac{\partial \epsilon_{\text{vol}}}{\partial x^2} + \frac{\partial \epsilon_{\text{vol}}}{\partial y^2} + \frac{\partial \epsilon_{\text{vol}}}{\partial z^2}\right) = (\lambda + 2G)\, \nabla^2 \epsilon_{\text{vol}} = \rho \frac{\partial^2 \epsilon_{\text{vol}}}{\partial t^2} \qquad (19\text{-}4)$$

If, on the other hand, one obtains the derivative of the third equation with respect to y and subtracts from this expression the derivative of the second equation with respect to z, taking into account that

$$\frac{\partial^2 \Omega_x}{\partial x^2} + \frac{\partial^2 \Omega_y}{\partial x\, \partial y} + \frac{\partial^2 \Omega_z}{\partial x\, \partial z} = 0$$

one has

$$G\left(\frac{\partial^2 \Omega_x}{\partial x^2} + \frac{\partial^2 \Omega_x}{\partial y^2} + \frac{\partial^2 \Omega_x}{\partial z^2}\right) = G\, \nabla^2 \Omega_x = \rho \frac{\partial^2 \Omega_x}{\partial t^2} \qquad (19\text{-}5)$$

Similarly
$$G\,\nabla^2\Omega_y = \rho\frac{\partial^2\Omega_y}{\partial t^2} \tag{19-6}$$

and
$$G\,\nabla^2\Omega_z = \rho\frac{\partial^2\Omega_z}{\partial t^2} \tag{19-7}$$

The equations of motion can thus be written as

$$(\lambda + 2G)\,\nabla^2\epsilon_{\text{vol}} = \rho\frac{\partial^2\epsilon_{\text{vol}}}{\partial t^2} \qquad G\,\nabla^2\Omega_x = \rho\frac{\partial^2\Omega_x}{\partial t^2} \tag{19-8}$$

$$G\,\nabla^2\Omega_y = \rho\frac{\partial^2\Omega_y}{\partial t^2} \qquad G\,\nabla^2\Omega_z = \rho\frac{\partial^2\Omega_z}{\partial t^2}$$

with the additional condition

$$\frac{\partial\Omega_x}{\partial x} + \frac{\partial\Omega_y}{\partial y} + \frac{\partial\Omega_z}{\partial z} = 0$$

Calling
$$\frac{\lambda + 2G}{\rho} = v_p^2 \tag{19-9}$$

$$\frac{G}{\rho} = v_s^2$$

then
$$\nabla^2\epsilon_{\text{vol}} = \frac{1}{v_p^2}\frac{\nabla^2\epsilon_{\text{vol}}}{\partial t^2} \qquad \nabla^2\{\Omega\} = \frac{1}{v_s^2}\frac{\partial^2}{\partial t^2}\{\Omega\} \tag{19-10}$$

For a steady state harmonic motion with frequency ω, a general solution of these equations is given by

$$\epsilon_{\text{vol}} = A \exp\left[\frac{i\omega}{v_p}(v_p t - l_x x - l_y y - l_z z)\right]$$

$$\{\Omega\} = \{B\} \exp\left[\frac{i\omega}{v_s}(v_s t - l_x x - l_y y - l_z z)\right] \tag{19-11}$$

with $\quad l_x^2 + l_y^2 + l_z^2 = 1 \qquad l_x B_x + l_y B_y + l_z B_z = 0 \qquad i^2 = -1$

If all three components of $\{l\}$ are equal to or less than 1, they can be interpreted as direction cosines, and $\{l\}$ is then a unit vector indicating a direction. This is the direction of propagation of *body waves*.

Considering first the scalar expression for ϵ_{vol} in (19-11) and defining

$$f_p = \exp\left[\frac{i\omega}{v_p}(v_p t - l_x x - l_y y - l_z z)\right] \qquad A_p = \frac{iv_p}{\omega}A \tag{19-12}$$

one finds the corresponding displacements

$$\begin{aligned} u_{xp} &= A_p f_p l_x \\ u_{yp} &= A_p f_p l_y \qquad \text{or} \qquad \{U\}_p = A_p f_p\{l\} \\ u_{zp} &= A_p f_p l_z \end{aligned} \tag{19-13}$$

This indicates that the motion takes place entirely along the direction of propagation $\{l\}$, with an amplitude A_P and a velocity of propagation v_p. The solution is then defined as a dilatational, or P, wave. Similarly, from

$$f_s = \exp\left[\frac{i\omega}{v_s}(v_s t - l_x x - l_y y - l_z z)\right] \tag{19-14}$$

the displacements corresponding to the vector expression for $\{\Omega\}$ are

$$u_{xs} = 2i\frac{v_s}{\omega}(l_z B_y - l_y B_z)f_s$$

$$u_{ys} = 2i\frac{v_s}{\omega}(l_x B_z - l_z B_x)f_s \qquad \text{or} \qquad \{U\}_s = 2i\frac{v_s}{\omega}f_s\{B\} \times \{l\} \tag{19-15}$$

$$u_{zs} = 2i\frac{v_s}{\omega}(l_y B_x - l_x B_y)f_s$$

This indicates that the motion has no component along the direction of propagation; in other words, it is entirely in a plane perpendicular to that direction.

Except for the particular case $l_x = l_y = 0$, $l_z = \pm 1$, where the direction of propagation coincides with the global z axis, it is possible to find the components of the motion along two orthogonal directions in a plane perpendicular to $\{l\}$: one in a horizontal plane (perpendicular to the global z axis) u_{SH}, the other in a vertical plane (containing the global z axis) u_{SV}.

Let us define

$$A_{\text{SH}} = 2i\frac{v_s}{\omega}\frac{B_z}{\sqrt{l_x^2 + l_y^2}} \qquad A_{\text{SV}} = 2i\frac{v_s}{\omega}\frac{l_x B_y - l_y B_x}{\sqrt{l_x^2 + l_y^2}} \tag{19-16}$$

Then

$$u_{\text{SH}} = A_{\text{SH}} f_s \qquad u_{\text{SV}} = A_{\text{SV}} f_s \tag{19-17}$$

The solution corresponds then to two families of shear waves propagating along the direction of $\{l\}$ with a velocity of propagation v_s and amplitudes of motion A_{SH}, A_{SV}. These are called SH and SV waves, respectively.

The displacements in terms of these amplitudes can be rewritten as

$$u_{xs} = \frac{l_x l_z A_{\text{SV}} - l_y A_{\text{SH}}}{\sqrt{l_x^2 + l_y^2}} f_s$$

$$u_{ys} = \frac{l_y l_z A_{\text{SV}} + l_x A_{\text{SH}}}{\sqrt{l_x^2 + l_y^2}} f_s \tag{19-18}$$

$$u_{zs} = -\sqrt{l_x^2 + l_y^2}\, A_{\text{SV}} f_s$$

In the particular case $l_x = l_y = 0$, $l_z = \pm 1$, no proper distinction can be made between SH and SV waves; however, one can arbitrarily define in this case

$$u_{xs} = A_{\text{SV}} f_s \qquad u_{ys} = A_{\text{SH}} f_s \qquad u_{zs} = 0 \tag{19-19}$$

or perhaps, more consistently, assuming that this is a limit condition attained when l_x and l_y tend to zero at the same rate so that $\lim l_x/\sqrt{l_x^2 + l_y^2} = \lim l_y/\sqrt{l_x^2 + l_y^2} = 1/\sqrt{2}$

$$u_{xs} = \frac{1}{\sqrt{2}}(A_{SV} - A_{SH})f_s \qquad u_{ys} = \frac{1}{\sqrt{2}}(A_{SV} + A_{SH})f_s \qquad u_{zs} = 0 \quad (19\text{-}20)$$

Allowing thus for this particular case, a general solution of the equations of motion can be written as

$$\begin{aligned} u_x &= u_{xp} + u_{xs} \\ u_y &= u_{yp} + u_{ys} \qquad \text{or} \qquad \{U\} = \{U\}_p + \{U\}_s \qquad (19\text{-}21) \\ u_z &= u_{zp} + u_{zs} \end{aligned}$$

It is worth noting that this general solution is still valid when one of the three components l_x, l_y, l_z is not real, although the physical interpretation of a direction of propagation must then be reconsidered. If, for instance, two of the components (say l_x and l_y) are real and the third (say l_z) is imaginary, the solutions represent waves which propagate along a direction in the xy plane and whose amplitude increases or decreases (depending on the sign of l_z) exponentially with z. These are referred to as *generalized surface waves* (*generalized Love waves* when there is only shear distortion, *generalized Rayleigh waves* if there are both volumetric change and shear distortion). In an infinite medium the solution can always be expressed in terms of only body waves (l_x, l_y, and l_z are all real for each individual wave). The existence of generalized surface waves (complex values of l_x, l_y, or l_z) depends on the boundary conditions of the problem (free boundary, surfaces of discontinuity in material properties, etc.).

In practice there will be an internal dissipation of energy in the material. It is generally accepted that most of this dissipation takes place in soils through internal friction rather than through linear viscoelastic properties. Constant viscous damping produces an energy loss per cycle that increases linearly with frequency. A hysteretic damping, on the other hand, produces an energy loss per cycle that is frequency independent but depends on the amplitude of the strains. Experimental evidence indicates that the second case is closer to the true behavior of soils. In order to maintain linearity of the solution, the amplitude dependence is dropped, using what is normally called a *linear hysteretic damping*. This type of damping is reproduced in the previous formulation by assuming complex values of the elastic parameters λ', G' of the form $\lambda(1 + 2iD)$ and $G(1 + 2iD)$, where D is the damping ratio. The wave propagation velocities v'_p and v'_s are then complex too, of the form $v_p\sqrt{1 + 2iD}$ and $v_s\sqrt{1 + 2iD}$.

19-2 PLANE WAVES

The overall problem of following an earthquake as it propagates from its source is basically a three-dimensional wave propagation problem. By assuming a line source of relatively large length, or by considering only the effects at some distance from the source, the problem can be reduced to a two-dimensional one, where all the waves propagate in directions parallel to a plane (say the xz plane), and the motion is therefore independent of the third coordinate (y in this case).

If all derivatives in the y direction are zero, the equations of motion (19-8) are simply

$$(\lambda + 2G)\,\nabla^2 e = \rho \frac{\partial^2 e}{\partial t^2} \qquad G\,\nabla^2 u_y = \rho \ddot{u}_y \qquad G\,\nabla^2 \Omega_y = \rho \frac{\partial^2 \Omega_y}{\partial t^2} \tag{19-22}$$

where now $\nabla^2 = \partial^2/\partial x^2 + \partial^2/\partial z^2$ and the general solution for a steady state harmonic motion becomes (since $l_y = 0$)

$$u_x = A_P l_x f_p + A_{SV} l_z f_s \qquad u_y = A_{SH} f_s \qquad u_z = A_P l_z f_p - A_{SV} l_x f_s \tag{19-23}$$

where

$$f_p = \exp\left[\frac{i}{v_p}(v_p t - l_x x - l_z z)\right] \tag{19-24}$$

$$f_s = \exp\left[\frac{i\omega}{v_s}(v_s t - l_x x - l_z z)\right] \tag{19-25}$$

If l_x and l_z are both real (with absolute value less than 1), they can be interpreted again as direction cosines, or $l_x = \sin\alpha$, $l_z = \cos\alpha$, where α is the angle between the direction of propagation of the wave and the z axis.

This shows that in the case of plane waves, the displacement u_y in the y direction is uncoupled from the displacements u_x and u_z in the x and z directions. The first one results only from the propagation of SH waves, while the other two are functions of both SV and P waves. Each problem can thus be studied independently.

A further simplification is introduced if the direction of propagation of the waves is assumed to be vertical (parallel to the z axis). The results will then be also independent of x ($l_x = 0$, $l_z = \pm 1$) and the problem becomes a one-dimensional case. Each one of the components of motion is then uncoupled.

$$u_x = A_{SV} f_s \qquad u_y = A_{SH} f_s \qquad u_z = A_P f_p \tag{19-26}$$

with

$$f_p = \exp\left[\frac{i\omega}{v_p}(v_p t \pm z)\right] \tag{19-27}$$

$$f_s = \exp\left[\frac{i\omega}{v_s}(v_s t \pm z)\right] \tag{19-28}$$

19-3 AMPLIFICATION OF SH WAVES

Consider first a homogeneous half-space, as shown in Fig. 19-1, with a free surface at $z = 0$ and two trains of SH waves, one traveling upward with amplitude A_{SH} (incoming waves), the other traveling down with amplitude A'_{SH} (reflected waves).

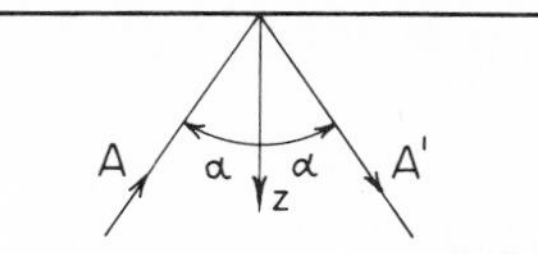

Figure 19-1 Reflection of SH wave at free surface.

The corresponding displacements u_x and u_z are zero, and

$$u_y = \left[A_{SH} \exp\left(\frac{i\omega}{v_s} z \cos\alpha\right) + A'_{SH} \exp\left(-\frac{i\omega}{v_s} z \cos\alpha\right)\right] f_s(x, t) \quad (19\text{-}29)$$

where
$$f_s(x, t) = \exp\left[\frac{i\omega}{v_s}(v_s t - x \sin\alpha)\right]$$

The boundary condition at $z = 0$ requires that the shear stress σ_{yz} vanish for any x and t or that $\partial u_y/\partial z = 0$. Thus

$$A_{SH} = A'_{SH} \quad (19\text{-}30)$$

The motion, out of the plane, in the half-space due to an incoming train of SH waves with amplitude A_{SH} and angle α is therefore

$$u_y = A_{SH}\left[\exp\left(\frac{i\omega}{v_s} z \cos\alpha\right) + \exp\left(-\frac{i\omega}{v_s} z \cos\alpha\right)\right] f_s(x, t) \quad (19\text{-}31)$$

and at the free surface $u_y = 2A_{SH} f_s(x, t)$.

If there is no internal dissipation of energy in the material, so that v_s is real, the two exponentials can be combined and expression (19-31) becomes

$$u_y = 2A_{SH} \cos\left(\frac{z}{v_s} \cos\alpha\right) f_s(x, t) \quad (19\text{-}32)$$

Homogeneous Layer of Finite Depth

Consider now a homogeneous layer of soil of finite and constant thickness h resting on a half-space (elastic rock), as indicated in Fig. 19-2. Representing the parameters and coefficients for the soil and the rock with the same letters and a subscript n ($n = 1$ for the soil, 2 for the rock), measuring the depth z for each of them independently from its top and calling

A = amplitude of waves traveling up
A' = amplitude of waves traveling down
$p = \omega/v_s \cos\alpha$

leads to

$$u_{yn} = (A_n e^{ip_n z_n} + A'_n e^{-ip_n z_n}) f_{sn}(x, t) \quad (19\text{-}33)$$

with
$$f_{sn}(x, t) = \exp(i\omega t) \exp\left(-\frac{i\omega}{v_{sn}} x \sin\alpha_n\right) \quad (19\text{-}34)$$

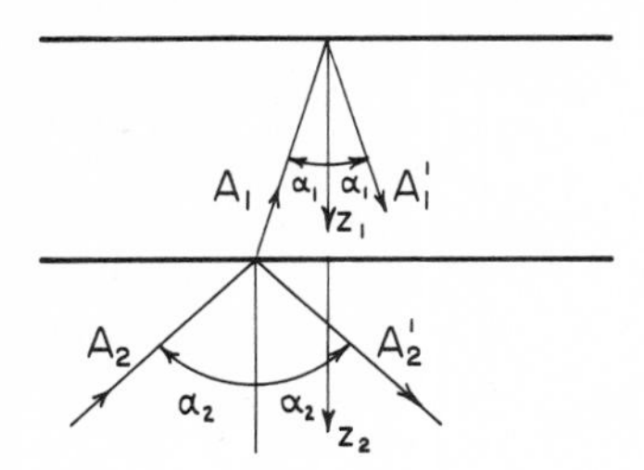

Figure 19-2 Reflection and refractions of SH wave for layer on half-space.

The free boundary condition at the top of the soil stratum yields again

$$A_1 = A'_1 \tag{19-35}$$

At the interface between the soil and the rock, the continuity of displacements and shear stresses requires that

$$u_{y1}(h) = u_{y2}(0) \tag{19-36a}$$

$$\sigma_{yz1}(h) = \sigma_{yz2}(0) \tag{19-36b}$$

Condition (19-36*a*) yields

$$A_1(e^{ip_1h} + e^{-ip_1h})f_{s1}(x, t) = (A_2 + A'_2)f_{s2}(x, t) \tag{19-37}$$

and since this identity must be satisfied for any x, t,

$$\frac{i\omega}{v_{s1}} \sin \alpha_1 = \frac{i\omega}{v_{s2}} \sin \alpha_2 \tag{19-38a}$$

and

$$A_1(e^{ip_1h} + e^{-ip_1h}) = A_2 + A'_2 \tag{19-38b}$$

It should be noticed that these two conditions impose also continuity of shear strains γ_{xy}. Expression (19-38*a*) yields

$$\frac{\sin \alpha_1}{v_{s1}} = \frac{\sin \alpha_2}{v_{s2}} \tag{19-39}$$

which can be identified as Snell's law of refraction and which provides the angle of the waves in the soil, given the angle of incidence in the rock. Condition (19-36*b*) becomes

$$A_1 iG_1 p_1(e^{ip_1h} - e^{-ip_1h}) = iG_2 p_2(A_2 - A'_2) \tag{19-40}$$

Combining (19-36*b*) and (19-40) gives

$$A_2 + A'_2 = A_1(e^{ip_1h} + e^{-ip_1h}) \qquad A_2 - A'_2 = A_1(e^{ip_1h} - e^{-ip_1h})\frac{G_1 p_1}{G_2 p_2} \tag{19-41}$$

The ratio

$$\frac{G_1 p_1}{G_2 p_2} = \frac{G_1 \cos \alpha_1 \, v_{s2}}{G_2 \cos \alpha_2 \, v_{s1}} = \sqrt{\frac{\rho_1 G_1}{\rho_2 G_2}} \frac{\cos \alpha_1}{\cos \alpha_2} \tag{19-42}$$

called the admittance ratio between the soil and the rock, will be represented by q. Its inverse is the impedance ratio.

Then

$$A_2 = \tfrac{1}{2}A_1[(1+q)e^{ip_1h} + (1-q)e^{-ip_1h}]$$
$$A'_2 = \tfrac{1}{2}A_1[(1-q)e^{ip_1h} + (1+q)e^{-ip_1h}] \qquad (19\text{-}43)$$

For a train of waves with amplitude A_{SH} traveling up through the underlying rock, the resulting displacements are thus given by

$$u_{y1} = A_1(e^{ip_1z_1} + e^{-ip_1z_1})f_s(x, t)$$
$$u_{y2} = (A_{SH}e^{ip_2z_2} + A'_{SH}e^{-ip_2z_2})f_s(x, t)$$

with
$$f_s(x, t) = \exp(i\omega t)\exp\left(-\frac{i\omega}{v_{s2}}x \sin\alpha_2\right)$$

$$p_1 = \frac{\omega}{v_{s1}}\cos\alpha_1 \qquad p_2 = \frac{\omega}{v_{s2}}\cos\alpha_2 \qquad (19\text{-}44)$$

$$\frac{\sin\alpha_1}{v_{s1}} = \frac{\sin\alpha_2}{v_{s2}}$$

and
$$A_1 = \frac{2A_{SH}}{(1+q)e^{ip_1h} + (1-q)e^{-ip_1h}}$$

$$A'_{SH} = A_{SH}\frac{(1-q)e^{ip_1h} + (1+q)e^{-ip_1h}}{(1+q)e^{ip_1h} + (1-q)e^{-ip_1h}} \qquad q = \sqrt{\frac{\rho_1 G_1 \cos\alpha_1}{\rho_2 G_2 \cos\alpha_2}} \qquad (19\text{-}45)$$

Amplification Functions

Equations (19-44) and (19-45) provide the complete solution for an incoming train of SH waves. It is possible now from these expressions to define several amplification functions. Using Fig. 19-3 as reference, one can define three amplification functions AF_1, AF_2, and AF_3, as follows.

The amplification function $AF_1(\omega)$ is the ratio of the amplitude of motion at point A (free surface of the soil) to the amplitude of motion at point B (interface between soil and rock). Then

$$AF_1(\omega) = \frac{u_{y1}(0)}{u_{y1}(h)} = \frac{u_{y1}(0)}{u_{y2}(0)} = \frac{2A_1}{A_2 + A'_2} \qquad \text{or} \qquad AF_1(\omega) = \frac{2}{e^{ip_1h} + e^{-ip_1h}} \qquad (19\text{-}46)$$

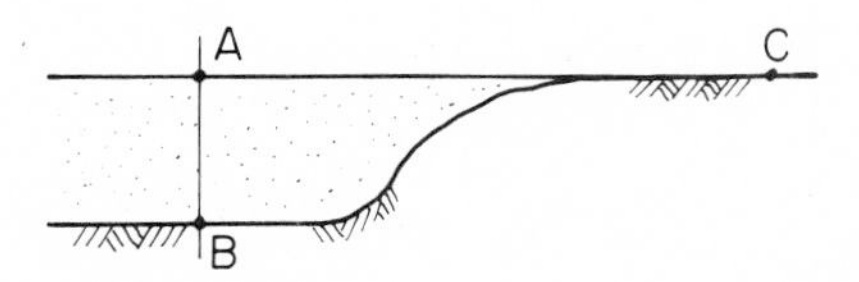

Figure 19-3 Soil amplification geometry.

If p_1 is real (there is no internal damping in the soil),

$$\mathrm{AF}_1(\omega) = \frac{2}{2 \cos p_1 h} = \frac{1}{\cos p_1 h} \tag{19-47}$$

and it can be seen that this expression becomes infinite (resonance condition) for

$$\cos p_1 h = 0 \qquad \text{or } \omega_n = \frac{(2n-1)\pi}{2} \frac{v_{s1}}{h \cos \alpha_1} \qquad n = 1, 2, \ldots \tag{19-48}$$

These are the natural frequencies of the system. In particular if $\alpha_2 = 0$, $\alpha_1 = 0$ and the problem reduces to the one-dimensional case (waves travelling vertically). Then

$$\omega_n = \frac{(2n-1)\pi}{2} \frac{v_{s1}}{h}$$

are the typical natural frequencies of the soil stratum in shear.

It should be noted that this amplification function is independent of the properties of the rock. It corresponds to the assumption of a rigid base, where the displacement is specified, e.g., a shaking table. It is also called the *rigid rock amplification.*

The amplification function $\mathrm{AF}_2(\omega)$ is the ratio of the amplitude of motion at point A (free surface of the soil) to the amplitude of motion which would occur at B if there were no soil on top.

The corresponding displacement at B would be then $2A_{\mathrm{SH}}$, and

$$\mathrm{AF}_2(\omega) = \frac{A_1}{A_{\mathrm{SH}}} = \frac{2}{(1+q)e^{ip_1h} + (1-q)e^{-ip_1h}} \tag{19-49}$$

Again if there is no damping in the soil, the expression can be rewritten as

$$\mathrm{AF}_2(\omega) = \frac{2}{2 \cos p_1 h + 2iq \sin p_1 h} = \frac{1}{\cos p_1 h + iq \sin p_1 h} \tag{19-50}$$

This is now a complex number. The magnitude of the amplification is given by

$$|\mathrm{AF}_2| = \frac{1}{\sqrt{\cos^2 p_1 h + q^2 \sin^2 p_1 h}} \tag{19-51}$$

and it can be seen that the denominator will never become zero (there is no resonance even if there is no damping in the soil).

Stationary values of $|\mathrm{AF}_2|$ are obtained at

$$(1-q^2) \sin 2p_1 h = 0$$

$$\omega_n = \frac{n\pi v_{s1}}{2h \cos \alpha_1} \tag{19-52}$$

The values

$$\omega_n = \frac{(2n-1)\pi}{2} \frac{v_{s1}}{h \cos \alpha_1}$$

correspond to maxima of $|\mathrm{AF}_2|$ and the values

$$\frac{2n\pi}{2} \frac{v_{s1}}{h \cos \alpha_1}$$

to minima. The maxima occur at the same frequencies for which AF_1 showed resonance. Their value now is

$$|\mathrm{AF}_2|_{\max} = \frac{1}{q} = \sqrt{\frac{\rho_2 G_2 \cos \alpha_2}{\rho_1 G_1 \cos \alpha_1}} = \frac{\gamma_2 v_{s2} \cos \alpha_2}{\gamma_1 v_{s1} \cos \alpha_1} \tag{19-53}$$

where γ is the unit weight of the material.

This second definition is usually known as *elastic rock amplification.* It will always yield values of the amplification smaller than those corresponding to AF_1 since it allows for dissipation of part of the energy of the incoming waves (after reflection at the free surface of the soil) through the waves with amplitude A'_2 that travel downward into the rock. This effect is often referred to as *radiation* or *geometric damping.* To visualize its importance better, one can write for small values of the internal damping in the soil D

$$|\mathrm{AF}_1|_{\max} \approx |\mathrm{AF}_1(\omega_n)| = \frac{4}{(2n-1)\pi} \frac{1}{2D}$$

$$|\mathrm{AF}_2|_{\max} \approx |\mathrm{AF}_2(\omega_n)| = \frac{1}{q + 2D[(2n-1)\pi]/4} \tag{19-54}$$

or

$$\frac{1}{|\mathrm{AF}_2|_{\max}} = q + \frac{1}{|\mathrm{AF}_1|_{\max}}$$

In particular it is possible to think of the reduction in the maxima of AF_2 with respect to those of AF_1 as the result of an additional, equivalent damping (the radiation damping) of the form

$$D_{\mathrm{rad}} = \frac{2}{\pi} q \frac{\omega_1}{\omega} \tag{19-55}$$

The radiation damping thus has an effect which decreases with increasing frequency and which is a function of the factor q. Its relative importance will also depend on the value of the internal damping in the soil. For very light excitations (low levels of strain and small values of internal soil damping) and properties of the rock not very different from those of the soil, the effect of the radiation damping will be very important. On the other hand, when there is a marked difference in elastic properties between the soil and the underlying rock and moderate to large motions are considered (high levels of strain and a substantial amount of internal damping in the soil), the effect of the radiation damping will be small and AF_1 and AF_2 may be very similar.

A third amplification function $\mathrm{AF}_3(\omega)$ could be defined as the ratio of the amplitude of motion at A (free surface of the soil) to the amplitude of motion at C, a point on a hypothetical outcropping of rock at the same elevation of point A. It is assumed in this definition that point C is at a distance from A which is a multiple of the wavelength, so that $(\omega/v_{s1})(x_c - x_A) \sin \alpha_1 = 2n\pi$ and the corresponding exponential is unity.

The magnitude of AF_3 is equal to that of AF_2 if there is no damping in the rock. Otherwise there would be a small difference due to the attenuation over the thickness h. The distinction between AF_2 and AF_3, however, is a refinement of relatively little practical interest.

Effect of the Angle of Incidence

The above formulas have been derived for the general case of a train of SH waves traveling through the rock at an angle α_2 with respect to the vertical (z) axis. Making α_2 equal to zero, one obtains the more usual solution for the one-dimensional amplification. Figures 19-4 and 19-5 show the magnitude of the amplification functions AF_1 and AF_2 for a soil layer with $h = 100$ ft, $v_s = 750$ ft/s, $\gamma = 125$ lb/ft^3, and a linear hysteretic damping of 0, 0.05, and 0.10. The underlying rock (for AF_2) is assumed to have $v_s = 4500$ ft/s and $\gamma = 140$ lb/ft^3 with $D = 0$.

Table 19-1 shows the possible effect of α_2 on the amplifications. Since the soil is much softer than the rock, the angle of the waves in the soil stratum is always very small (nearly normal incidence at the surface). The effect on the natural frequencies of the soil layer is also negligible in this case for all practical purposes. The effect on the rigid rock amplification AF_1 (not shown) would be correspondingly meaningless. The effect on the elastic rock amplification AF_2,

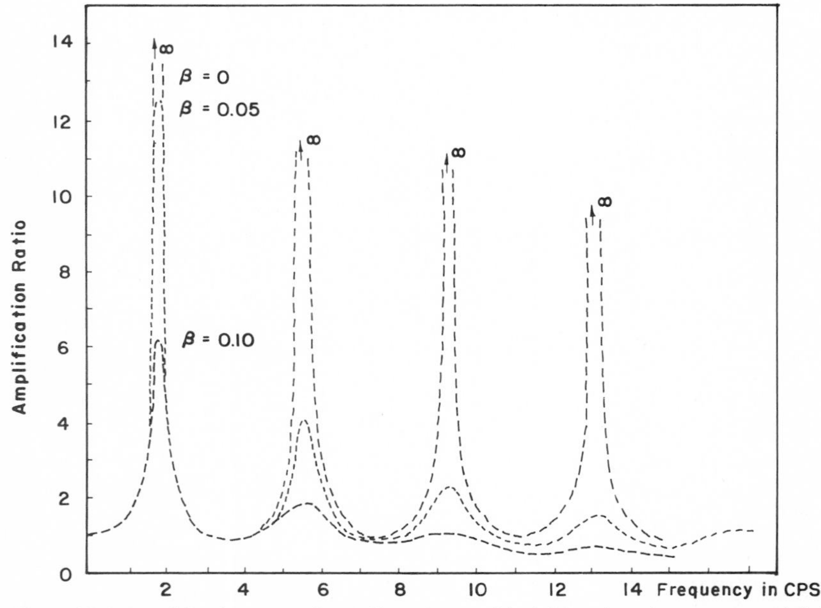

Figure 19-4 Amplification curve for uniform layer with rigid rock and constant modal damping.

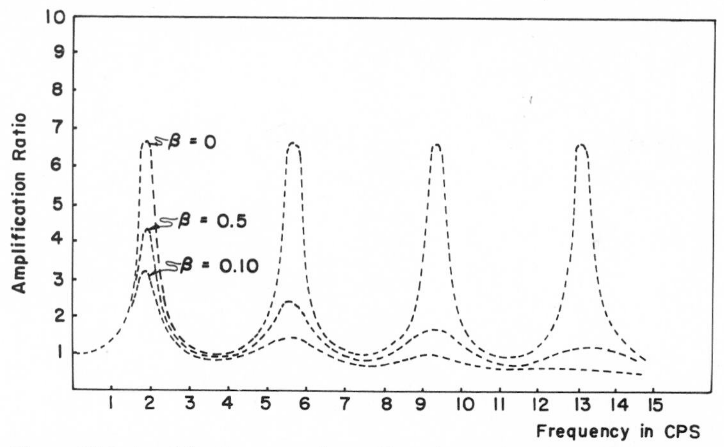

Figure 19-5 Amplification curve for uniform layer with elastic rock and constant modal damping.

however, is noticeable and very significant for moderately large angles of incidence in the rock in spite of the fact that the incidence in the soil is almost normal. Furthermore the effect is in all cases a reduction of the amplification values. Figure 19-6 shows the magnitude of AF_2 as a function of frequency for different values of α_2, the same profile, and an internal damping in the soil $D = 0.05$.

Soil Stratum with Multiple Layers

When the soil deposit is made of several layers with different properties (Fig. 19-7), it can again be assumed that for each layer the solution is of the form

Table 19-1 Effect of the angle of incidence of SH waves

Angle in rock α_2	Angle in soil α_1	Ratio of natural frequencies $\dfrac{\omega_n(\alpha_2)}{\omega_n(0)} = \dfrac{1}{\cos \alpha_1}$	Ratio of peak amplifications $\dfrac{AF_2(\alpha_2)}{AF_2(0)} = \dfrac{\cos \alpha_2}{\cos \alpha_1}$
0°	0	1	1
10°	1°40′	1.0004	0.98
20°	3°	1.002	0.94
30°	4°45′	1.004	0.87
40°	6°10′	1.006	0.77
45°	6°45′	1.007	0.71
50°	7°18′	1.008	0.64
60°	8°17′	1.011	0.50
70°	9°	1.013	0.35
80°	9°23′	1.013	0.17
90°	9°37′	1.020	0

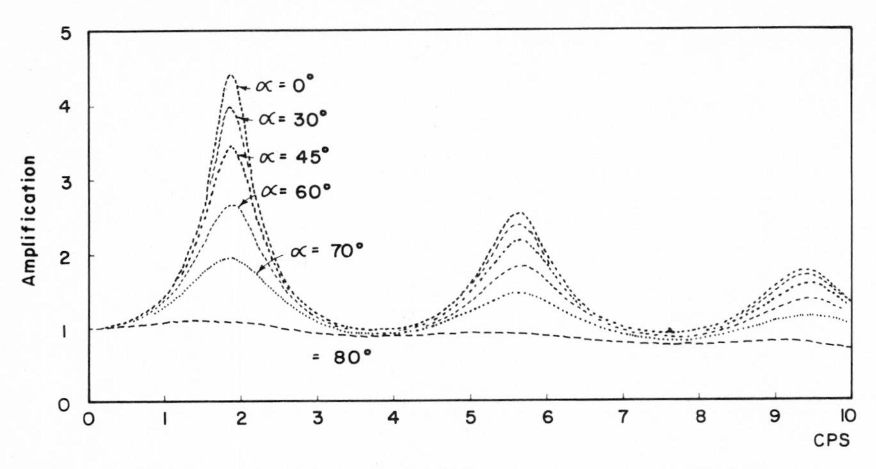

Figure 19-6 Effect of angle of incidence on amplification for layer on elastic rock.

of Eq. (19-33). The compatibility conditions at the interface of two layers force again

$$f_{s1}(x, t) = f_{s2}(x, t) = \cdots = f_{sm}(x, t) = f_{s,\,m+1}(x, t)$$

or

$$\frac{\sin \alpha_1}{v_{s1}} = \frac{\sin \alpha_2}{v_{s2}} = \cdots = \frac{\sin \alpha_m}{v_{sm}} = \frac{\sin \alpha_{m+1}}{v_{sm+1}} \tag{19-56}$$

and

$$\begin{aligned} A_{n+1} &= \tfrac{1}{2}[A_n(1 + q_n)e^{ip_nh_n} + A'_n(1 - q_n)e^{-ip_nh_n}] \\ A'_{n+1} &= \tfrac{1}{2}[A_n(1 - q_n)e^{ip_nh_n} - A'_n(1 + q_n)e^{-ip_nh_n}] \end{aligned} \tag{19-57}$$

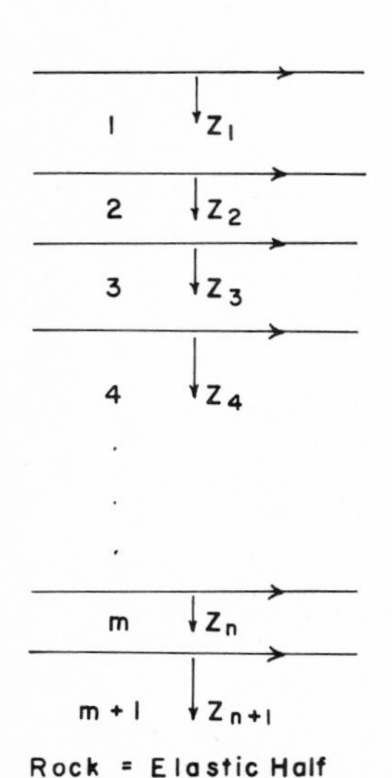

Figure 19-7 Multiple strata over elastic rock.

where h_n is the thickness of the nth layer, $p_n = \omega/v_{sn} \cos \alpha_n$, and

$$q_n = \frac{\gamma_n v_{sn}}{\gamma_{n+1} v_{sn+1}} \frac{\cos \alpha_n}{\cos \alpha_{n+1}} \qquad n = 1 \text{ to } m \tag{19-58}$$

Since in the top layer (due to the free boundary condition) $A_1 = A'_1$, it is possible in a recursive way to compute A_2 and A'_2 in terms of A_1, then A_3 and A'_3 in terms of A_2 and A'_2, and so on, obtaining finally an expression for $A_{m+1} = A_{\text{SH}}$, the amplitude of the incoming wave in the rock, in terms of A_1.

The inverse of this expression provides the value of A_1 in terms of A_{SH}, and all other amplitudes can then be computed. Since all values of A_n and A'_n are known, the displacements, stresses, and strains are defined at any point of the soil stratum as a function of the frequency ω for a steady state harmonic motion.

The amplification function with rigid rock is then given by

$$\text{AF}_1(\omega) = \frac{2A_1}{A_{m+1} + A'_{m+1}} \tag{19-59}$$

and the amplification with elastic rock

$$\text{AF}_2(\omega) = \frac{A_1}{A_{\text{SH}}} = \frac{A_1}{A_{m+1}} \tag{19-60}$$

The explicit expression for these amplification functions becomes too long even for two layers. However, the numerical computation proceeding from layer to layer is simple and adapts itself very well to a digital computer.[8,11]

The effect of an elastic rock at the bottom of the soil stratum is again to introduce radiation damping, decreasing the peaks of the amplification function at the resonant frequencies. While it is not possible in this case to find a simple formula to estimate the magnitude of the radiation damping, a proper understanding of its nature is important, particularly in the selection of the rigid base for a deep soil profile if the amplification with rigid rock AF_1 is to be used. If, for instance, there is a shallow layer of soft soil over a deep layer of stiff soil and a base rock with properties not too different from those of the layer above it, it may be wiser to cut at the level of the first interface (neglecting thus the amplification at the first resonant frequency, which involves the deep layer but where there may be substantial radiation) than to go down all the way to the rock, introducing very large but fictitious amplifications in the low frequency range. The importance of this effect will depend again on the magnitude of the motion, which through the levels of strain influences the expected amount of internal damping one should use for the soil.

The effect of the angle of incidence of the incoming train of waves in the rock is also similar to that described for a homogeneous layer. It can be expected in most practical situations to have incidences in the top layers, near the surface, very nearly normal. The change in the natural frequencies of the stratum with the angle of incidence will be in most cases negligible, but the reduction in the peak values of the elastic rock amplification may be important for moderate to large angles in the rock.

An analogous treatment can be applied to the amplification of SV and P waves[10] but is outside the scope of the present chapter. For vertical propagation P and SV effects are uncoupled, and the solution for the SV waves is identical to that for the SH waves. The solution for the P waves is identical in form with $\lambda + 2G$ substituted for G in all expressions. When the angle of incidence of either P or SV waves is not zero, there is a coupling between them and the more complicated analysis of Ref. 10 is required.

19-4 AMPLIFICATION OF SEISMIC MOTIONS

Fourier Analyses

Given a certain accelerogram representing an earthquake record at the outcropping of rock or at the interface between soil and rock, and assuming that this motion is the result of a specific train of waves, the corresponding accelerogram at the free surface of the soil can be obtained by:

1. Obtaining the Fourier transform of the input earthquake
2. Multiplying this Fourier transform by the amplification or transfer function of the soil
3. Obtaining the inverse Fourier transform of the product

The Fourier transform of a function of time $f(t)$ can be visualized as a limiting case of a Fourier series expansion. It is given by

$$F(\omega) = \int_0^{\infty} f(t)e^{-i\omega t}\, dt \qquad \text{if } f(t) = 0 \text{ for } t \leqslant 0 \tag{19-61}$$

$f(t)$ is then said to be the inverse Fourier transform of $F(\omega)$

$$f(t) = \frac{1}{2\pi}\int_{-\infty}^{+\infty} F(\omega)e^{i\omega t}\, d\omega \tag{19-62}$$

$F(\omega)$ is a complex function; when it is written

$$F(\omega) = C(\omega) - iS(\omega) \tag{19-63}$$

$C(\omega) = \int_0^{\infty} f(t) \cos \omega t\, dt$ is the cosine transform, and $S(\omega) = \int_0^{\infty} f(t) \sin \omega t\, dt$ is the sine transform. If on the other hand, $F(\omega)$ is expressed as

$$F(\omega) = E(\omega)e^{-i\phi(\omega)t}$$

then

$$E(\omega) = \sqrt{C(\omega)^2 + S(\omega)^2} \qquad \phi(\omega) = \tan^{-1}\frac{S(\omega)}{C(\omega)} \tag{19-64}$$

$E(\omega)$ represents the amplitude Fourier spectrum and $\phi(\omega)$ the phase spectrum. The amplitude spectrum has an important physical meaning. Given two values of frequency ω_1, ω_2, the area under the curve $E(\omega)$ from ω_1 to ω_2 gives the

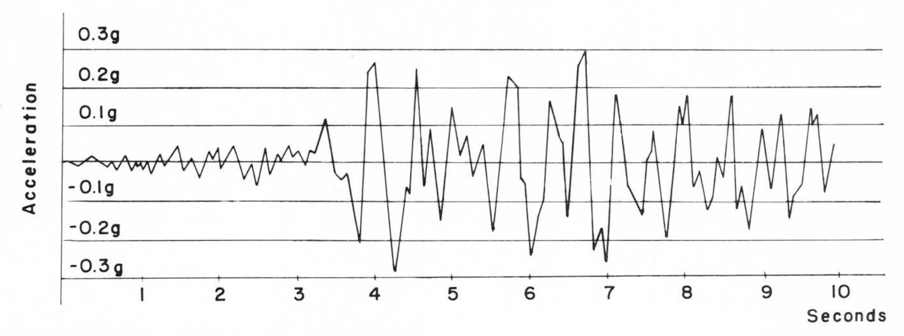

Figure 19-8 Accelerogram of Taft earthquake.

amplitude of motion in this range. A simple look at the Fourier amplitude spectrum provides a qualitative picture of where most of the amplitude of the motion is.

The determination of the Fourier transform is extremely simple in a digital computer by means of the fast Fourier transform techniques (Cooley-Tuckey algorithm). While numerically one does not really obtain the Fourier transform but a Fourier series, the accuracy is good if there is some damping in the system and one takes care to provide a sufficient duration of zero motion after $f(t)$ to allow for the damping out of free-vibration terms in the response.

The transfer function of the soil, or the amplification function as described above, is a complex function. Since most modern compilers permit the use of complex variables, direct multiplication of these two functions, normally evaluated at equal frequency intervals, offers no problem.

The inverse Fourier transform of this product will again be a real function of time, representing the time history of acceleration on top of the soil. This inverse transformation is normally performed with the same subroutine used to obtain the direct Fourier transform.

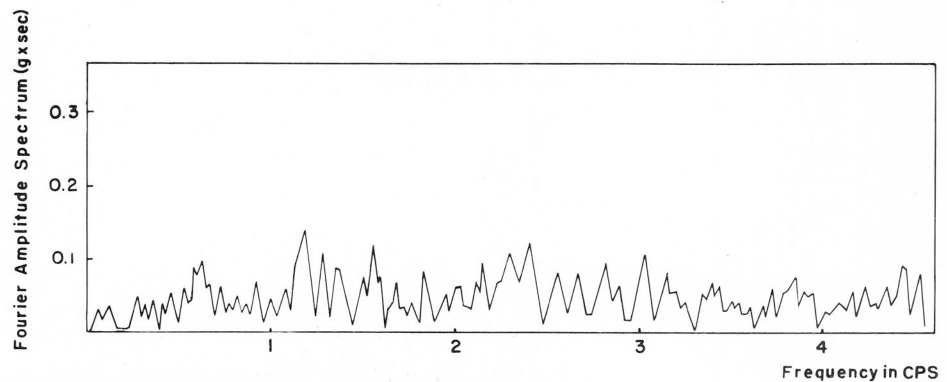

Figure 19-9 Fourier amplitude spectrum for Taft earthquake.

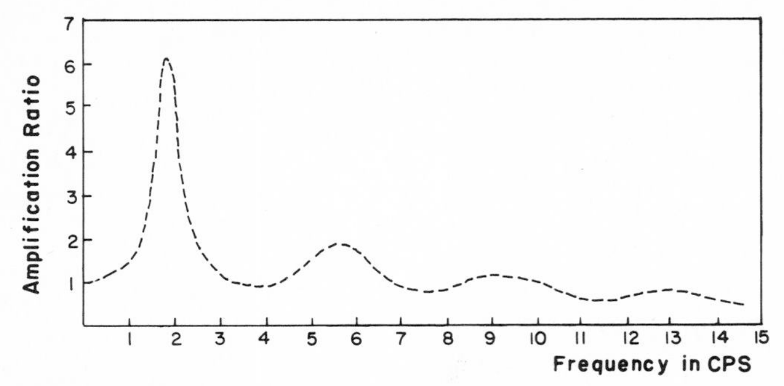

Figure 19-10 Amplification function for soil.

Figures 19-8 to 19-12 summarize the procedure (the phase spectra are not plotted). Figure 19-8 shows a record of the 1952 Taft earthquake, and Fig. 19-9 its amplitude Fourier spectrum. Figure 19-10 shows the amplification curve for the soil profile used before as an example (SH waves with normal incidence and rigid rock). The product of the two amplitude spectra appears in Fig. 19-11, and the time history of the motion at the free surface of the soil in Fig. 19-12. It can be seen that in this case the peak acceleration is amplified by a factor of nearly 2.

While the amplification of the peak acceleration is an incomplete description of soil amplification effects, it has been widely used. Most commonly it is expressed as the ratio of peak accelerations at the surface and at bedrock.

Figure 19-13 shows curves of acceleration ratio computed approximately by Madera[12] for a uniform soil profile with 0.02, 0.05, and 0.10 internal damping, one-dimensional shear wave amplification for elastic rock, and two different values of the impedance ratio, $\gamma_2 v_{s2}/\gamma_1 v_{s1}$, as a function of the fundamental period

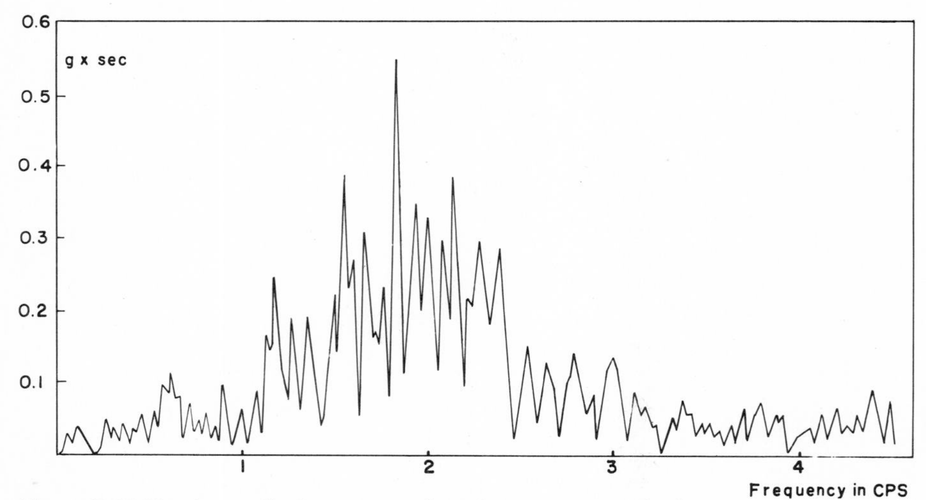

Figure 19-11 Fourier amplitude spectrum of accelerogram at top of soil.

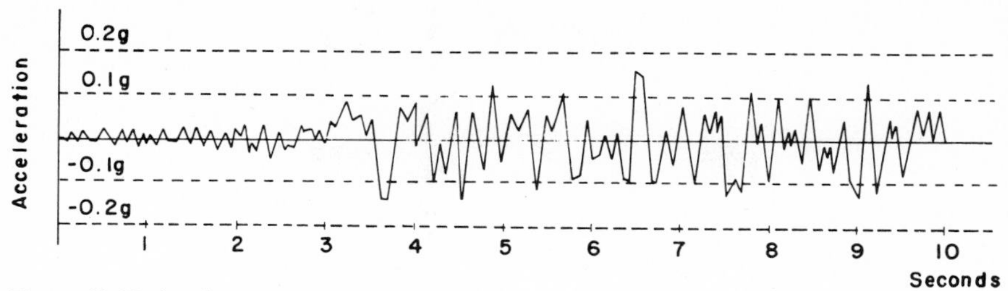

Figure 19-12 Accelerogram at top of soil.

of the stratum. For soft or deep layers of soil, with relatively long periods and appreciable internal damping, the maximum acceleration may actually be reduced. It should also be pointed out that the effect of the impedance ratio (and therefore of the elastic rock) becomes less important with increasing internal damping.

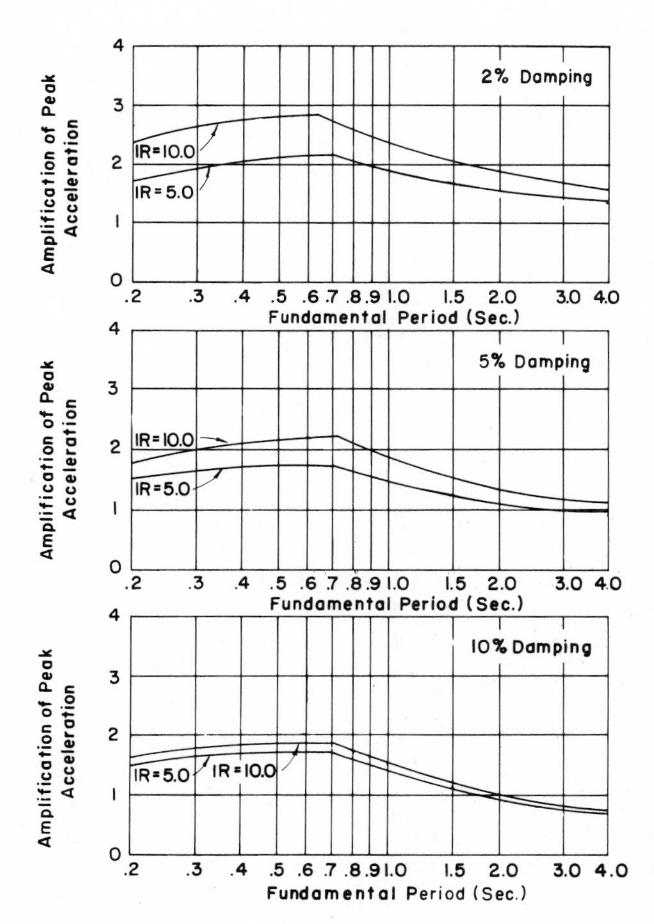

Figure 19-13 Approximate acceleration ratios for uniform soil layer.

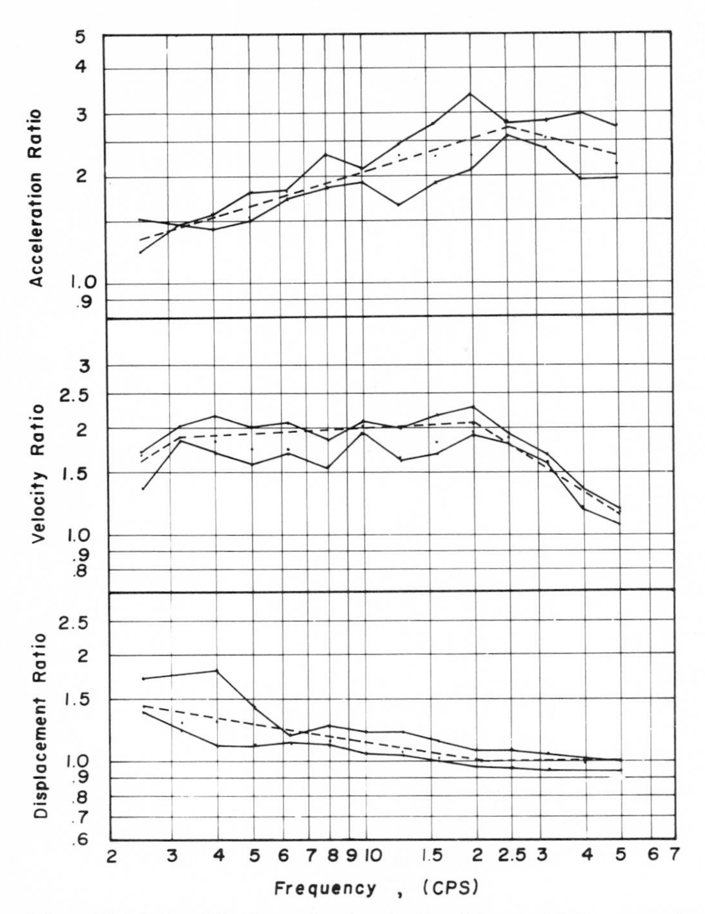

Figure 19-14 Amplification ratios for single soil layer with 2 percent damping.

Figures 19-14 to 19-16 show the results of a more accurate study by Stern[13] using five different earthquakes as input, a soil deposit with the characteristics of the example already used, one-dimensional SH wave amplification with elastic rock, and values of internal damping of 0.02, 0.05, and 0.10. Bounds and average values for the five earthquakes are given for the amplification of maximum ground acceleration, velocity, and displacement, three parameters which provide a better representation of earthquake characteristics than the peak acceleration alone.[14] The natural frequency of the layer was varied by changing the thickness of the soil stratum, but the impedance ratio remained constant.

Ratio of Response Spectra

The best way to represent the possible effects of an earthquake on a structure or a piece of equipment is through the use of design response spectra. The displace-

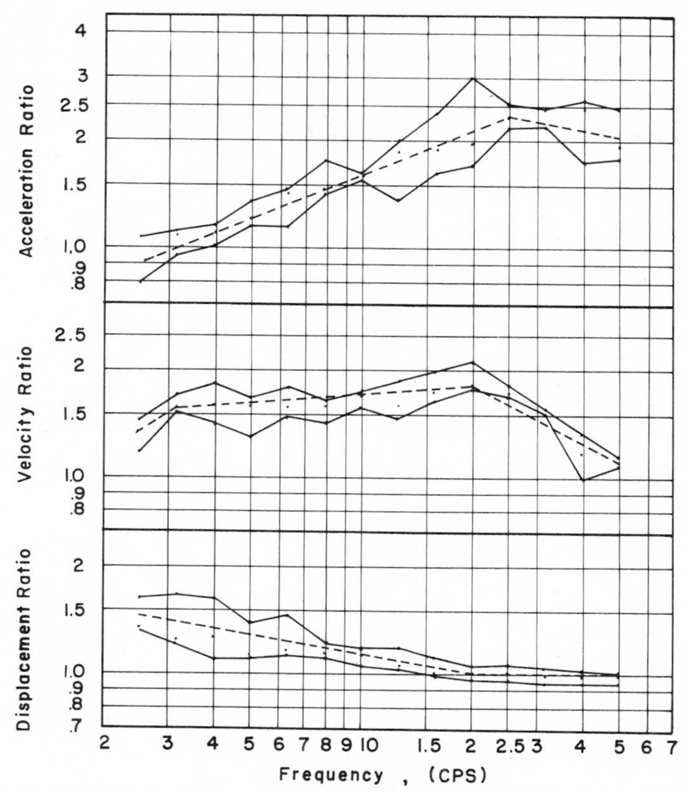

Figure 19-15 Amplification ratios for single soil layer with 5 percent damping.

ment response spectrum gives the maximum relative displacement of a one-degree-of-freedom system with a specified ratio of critical damping D, as a function of its frequency, when the system is subjected to a particular earthquake. In the same way one could define a velocity response spectrum (maximum relative velocity) or an acceleration response spectrum (maximum absolute acceleration). For undamped systems the maximum absolute acceleration is equal to ω^2 times the maximum relative displacement, where ω is the natural frequency of the one-degree-of-freedom system. While this relationship does not hold when the system is damped, for values of D of interest in structural problems the error introduced by accepting it is negligible. The maximum relative velocity of a one-degree-of-freedom system with natural frequency ω is also very close to ω times the maximum relative displacement, except for small values of ω. It is thus more common to define a pseudo-velocity (fictitious velocity) equal to ω times the maximum relative displacement and a pseudo-acceleration equal to ω^2 times this quantity. In this way the relative displacement response spectrum, the pseudo-velocity response spectrum, and the pseudo-acceleration response spec-

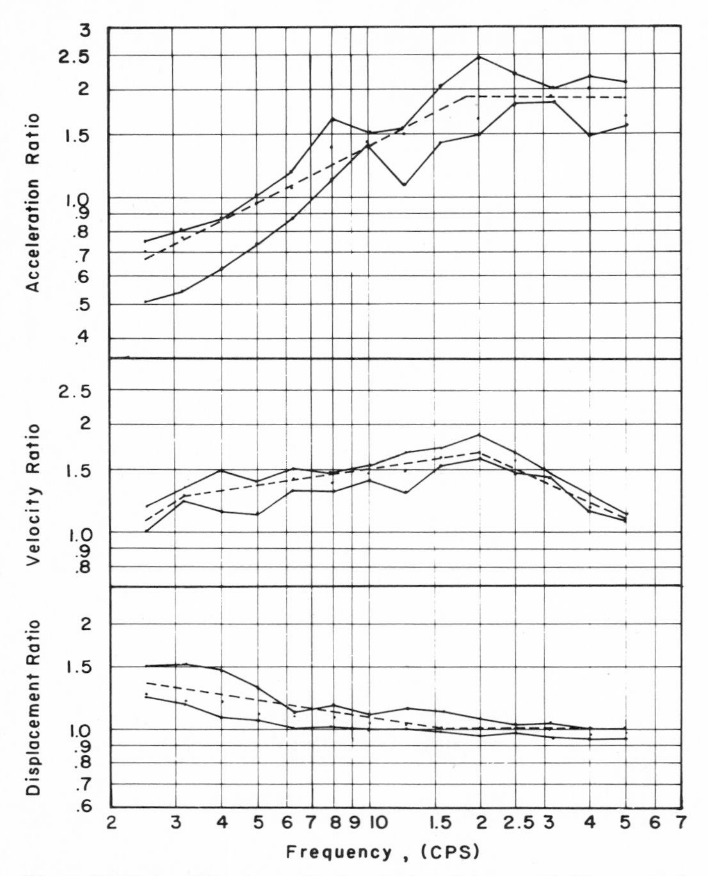

Figure 19-16 Amplification ratios for single soil layer with 10 percent damping.

trum are uniquely related to one another, and it is enough to compute one of them.

Given an accelerogram representing an earthquake record at the free surface of a soil deposit, the interface between soil and rock, or a hypothetical outcropping of rock, the corresponding response spectrum can be computed in two different ways:

1. From the Fourier transform of the accelerogram, multiplying it by the transfer function of a one-degree-of-freedom linear oscillator, inverting the result, and finding the maximum value over time.
2. By integrating numerically through a step-by-step procedure the equation of motion of a one-degree-of-freedom system.

The first procedure is normally referred to as integration in the frequency domain, whereas in the second case the solution is said to be carried out in the

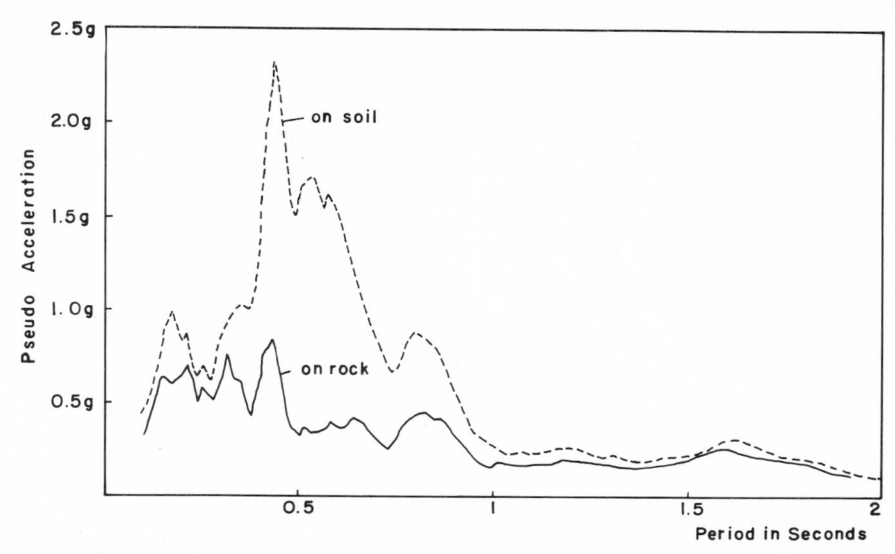

Figure 19-17 Pseudo-acceleration response spectra on rock and on top of soil, 2 percent structural damping.

time domain. While the first method represents a consistent continuation of the procedure followed to determine the amplified motion, the second is normally more economical of computer time.

Figure 19-17 shows the pseudo-acceleration response spectrum of the Taft earthquake, selected as input at the soil-rock interface and the same spectrum for the resulting motion at the free surface of the soil deposit. Figure 19-18 shows the ratio of both response spectra for 0.02 and 0.05 structural damping (obtained by dividing the corresponding values at each frequency). Shown in this figure also, with a dashed line, is the magnitude of the amplification curve. The similarity between these curves is apparent but there are also noticeable differences. In particular:

1. The amplification curve tends to zero as the frequency increases. The ratio of response spectra, on the other hand, tends to a finite value, which is the ratio of the maximum acceleration on top of the soil to the maximum acceleration of the input earthquake.
2. For a given set of soil properties the amplification curve is a function only of these properties. The ratio of response spectra depends not only on the soil properties but also on the amount of structural damping of the one-degree-of-freedom oscillators and on the earthquake.
3. The ratio of response spectra is smoother than the amplification curve, with lower peaks and less pronounced valleys. It becomes smoother as the structural damping increases and for damping values of 20 or 25 percent it is practically

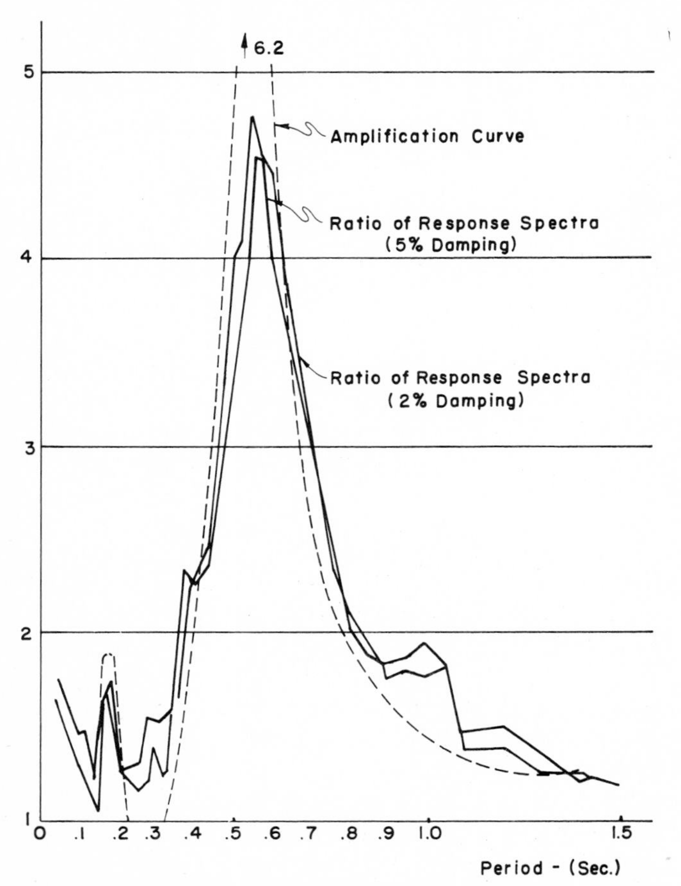

Figure 19-18 Ratio of response spectra between top of soil and rock.

constant over a long range of frequencies. On the other hand, for very small values of structural damping, the ratio of response spectra would be closer to the amplification curve. This implies that the ratio of response spectra, particularly for moderate to high values of structural damping, is not as good an indicator of the soil characteristics as the amplification function.

4. It should be expected that the agreement between the amplification curve and the ratio of response spectra would be better for high values of internal soil damping, since this would tend to eliminate the transients (free vibration terms) and to furnish a motion which is more nearly periodic.

The methods previously outlined are mainly intended to consider an earthquake at the base of the soil or the outcropping of rock, to filter it through the soil, and to obtain the resulting accelerogram at the free surface and the corresponding response spectra. If, on the other hand, it is desired to use the ratio of response spectra as a representative characteristic of the soil profile for a given intensity of motion, instead of the amplification curve, it would be necessary to perform the analysis for several inputs, representing samples of earthquakes with the same statistical properties and intensity of motion. The resulting ratios of response spectra should then be treated statistically, obtaining a smooth average and probabilistic bounds. The present practice of using a single artificial earthquake with a smooth response spectrum which envelops a specified design spectrum over the whole range of frequencies is somewhat arbitrary and leads to inconsistencies, since the resulting motion at the free surface of the soil will no longer exhibit these characteristics.

The use of a family of similar earthquakes (real and artificial) in order to be able to interpret the results statistically and to obtain smooth ratios of response spectra, as suggested above, may be too cumbersome and costly for many practical applications. As a result, attempts have been made to derive simple, approximate rules through which a smooth ratio of response spectra could be obtained directly from the amplification function.[15] These rules consist in general in the application of reduction factors at the resonant peaks of the amplification function and magnification factors at the valleys.

Additional Considerations

Implementation of a computer program to obtain the amplification functions for a horizontally stratified soil deposit and a specified train of body waves, determining the motion at the free surface of the soil for a given input earthquake at bedrock or at the outcropping of rock and its response spectra, is extremely simple. Proper application of such a program and the underlying theory in practical situations, however, requires engineering judgment in order to:

1. Decide on the type of body waves which should be assumed, their angle of incidence, and their relative amplitudes when there is more than one type of incident wave
2. Decide on the type of amplification function, rigid rock or elastic rock, which should be used
3. Select an appropriate set of soil properties consistent with the intensity of motion and the expected level of strains in order to reproduce, at least approximately, the true nonlinear behavior of the material

These three factors have been listed in what is believed to be increasing order of importance. They clearly represent limitations of the theory of soil amplification. They do not detract, however, from its practical usefulness; they just point out the need to realize that the results are always engineering estimates and that sound judgment is required to make them valid. This situation is common to most practical civil engineering problems and is one which geotechnical engineers are particularly used to.

The Inverse Problem

The preceding formulations have been primarily oriented toward the determination of the expected motion at the free surface of a soil deposit given the motion at bedrock or at the outcropping of rock. There are practical situations where the inverse problem is of interest; i.e., given a motion specified as the design earthquake at the surface of the soil, determine the motion or strains at any point within the soil and particularly a compatible earthquake at bedrock or at the outcropping of rock. Its justification is clear when the soil deposit can be classified as firm ground, since most of the earthquakes which have been used to derive design response spectra from statistical considerations have been recorded on similar soils. Use of amplification theory in this case, taking as input a firm ground motion, would lead to a double amplification.

As long as the material can be considered linearly elastic, the solution of the inverse problem, often referred to as a *deconvolution process*, offers little difficulty. For a specified train of waves the existence and uniqueness of the solution can be guaranteed, and the procedures and formulas presented in Sec. 19-3 can be used to obtain displacements at the bottom of the soil or the amplitudes of the waves incident in the rock from the displacements at the free surface. The transfer function from the top of the soil to the bottom (or to the outcropping of rock) is just the inverse of the transfer function from the bottom (or the outcropping of rock) to the top.

Problems may arise, however, and often do, when an iterative procedure is used to match moduli and damping values to the levels of strain in order to simulate nonlinear behavior. Convergence of this trial-and-error scheme for the inverse problem is not guaranteed, particularly for deep soil strata and severe shaking.

Problems may also arise when it is desired to specify directly the motion expected at a certain depth within the soil stratum, e.g., at the foundation level.

19-5 FINITE DIFFERENCE MODELS

Given the equations of motion of the wave propagation problem, the formulation and solution using any of the numerical methods described in Chap. 1 are extremely simple.

Basic Models

A discrete model suggested by Seed and Idriss[6,7] for the one-dimensional amplification case (waves with normal incidence and horizontal layers of soil of constant thickness) became particularly popular among practicing engineers. As presented in the original derivation, the model was intended only for the rigid rock case (motion specified at the soil-rock interface), and the discretization was achieved by physically lumping masses and connecting them with springs and dashpots (Fig. 19-19). Calling h_j the thickness of the jth sublayer with modulus G_j

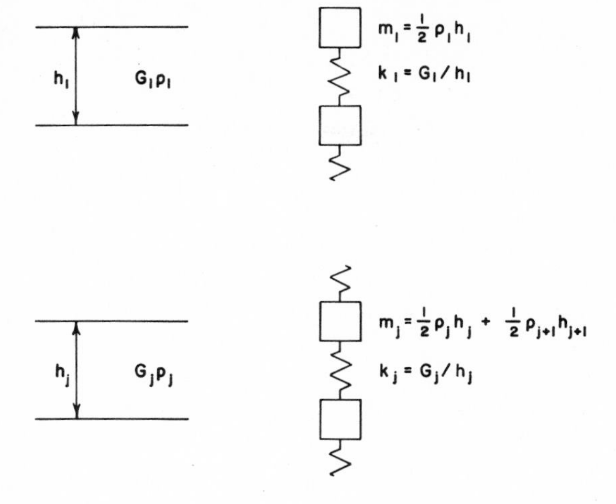

Figure 19-19 Lumped mass and spring model.

and density ρ_j, and $m_j = \frac{1}{2}\rho_j h_j + \frac{1}{2}\rho_{j-1} h_{j-1}$ for $j = 2$ to n, $m_1 = \frac{1}{2}\rho_1 h_1$; $k_j = G_j/h_j$, $j = 1$ to n, the corresponding system of linear differential equations (for no damping) can be written

$$[M]\{\ddot{U}\} + [K]\{U\} = \{J\}\frac{G_n}{h_n}u_s = \{J\}k_n u_s \tag{19-65}$$

where

$$[M] = \begin{bmatrix} m_1 & & & \\ & m_2 & & \\ & & \ddots & \\ & & & m_n \end{bmatrix} \tag{19-66}$$

is the diagonal mass matrix

$$[K] = \begin{bmatrix} k_1 & -k_1 & & & \\ -k_1 & k_1 + k_2 & -k_2 & & \\ & -k_2 & k_2 + k_3 & & -k_3 \\ & & & \ddots & \\ & & & & k_{n-1} + k_n \end{bmatrix} \tag{19-67}$$

is a tridiagonal stiffness matrix

$$\{J\} = \begin{Bmatrix} 0 \\ 0 \\ \vdots \\ 0 \\ 1 \end{Bmatrix} \tag{19-68}$$

and u_s is the specified displacement at the base.

Equation (19-65) can be written in an alternate form by defining the relative displacement of each mass with respect to the base $y_i = u_i - u_s$. Then

$$[M]\{\ddot{Y}\} + [K]\{Y\} = -[M]\{I\}\ddot{u}_s \tag{19-69}$$

where $\{I\}$ is a vector with all components unity and $\ddot{u}_s$ represents the specified input accelerogram. This discrete model can also be derived from a central finite difference formulation for the case of a uniform layer and rigid rock.

For elastic rock, the analytical solution of the equation of motion, presented in Sec. 19-3, gives in the rock

$$\begin{aligned} u_{y2} &= (A_2 e^{ip_2 z_2} + B_2 e^{-ip_2 z_2})e^{i\omega t} \\ \sigma_{yz2} &= ip_2 G_2(A_2 e^{ip_2 z_2} - B_2 e^{-ip_2 z_2})e^{i\omega t} \end{aligned} \tag{19-70}$$

In particular, since $u_{n+1} = u_{y2}(0)$ and $\tau_{n+1} = \sigma_{yz2}(0)$, if u_s is the displacement that would occur on top of the rock, without any soil,

$$u_{n+1} = (A_2 + B_2)e^{i\omega t} \qquad \tau_{n+1} = ip_2 G_2(A_2 - B_2)e^{i\omega t} \qquad u_s = 2A_2 e^{i\omega t} \tag{19-71}$$

Thus
$$\tau_{n+1} = ip_2 G_2(u_s - u_{n+1}) = i\rho_2 v_{s2}\omega(u_s - u_{n+1})$$
or
$$\tau_{n+1} = \rho_2 v_{s2}(\dot{u}_s - \dot{u}_{n+1}) \tag{19-72}$$

The equation of motion at point $n + 1$ can be written

$$\rho_1 \ddot{u}_{n+1} = \frac{\partial}{\partial z}(\tau_{n+1}) \tag{19-73}$$

Defining the stress at midpoint of the nth sublayer as

$$\tau_{n+1/2} = \frac{G_1}{h}(u_{n+1} - u_n) \qquad \text{and} \qquad \frac{\partial}{\partial z}(\tau_{n+1}) = \frac{2}{h}(\tau_{n+1} - \tau_{n+1/2})$$

Eq. (19-73) becomes

$$\rho_1 \ddot{u}_{n+1} + \rho_2 v_{s2}\frac{2}{h}(\dot{u}_{n+1} - \dot{u}_s) + \frac{2}{h}\frac{G_1}{h}(u_{n+1} - u_n) = 0 \tag{19-74}$$

or
$$\left(\rho_1 \frac{h}{2}\right)\ddot{u}_{n+1} + \rho_2 v_{s2}\dot{u}_{n+1} + \frac{G_1}{h}(u_{n+1} - u_n) = \rho_2 v_{s2}\dot{u}_s \tag{19-75}$$

This, combined with the previous equations for $i = 2$ to n, represents the dynamic-equilibrium equations of a discrete system, as shown in Fig. 19-20*c*. It can be seen that the system now has $n + 1$ masses and a viscous dashpot under the bottom mass. It is not subjected to a base motion but to a force, applied at the bottom mass, equal to $\rho_2 v_{s2} \dot{u}_s$, where $\dot{u}_s$ is the velocity of the motion expected at the outcropping of rock. This model was suggested by Tsai.[16]

The finite difference formulation becomes harder when the soil deposit has several layers with different properties; the existence of discontinuities, as represented by the abrupt changes in material properties at the interfaces between

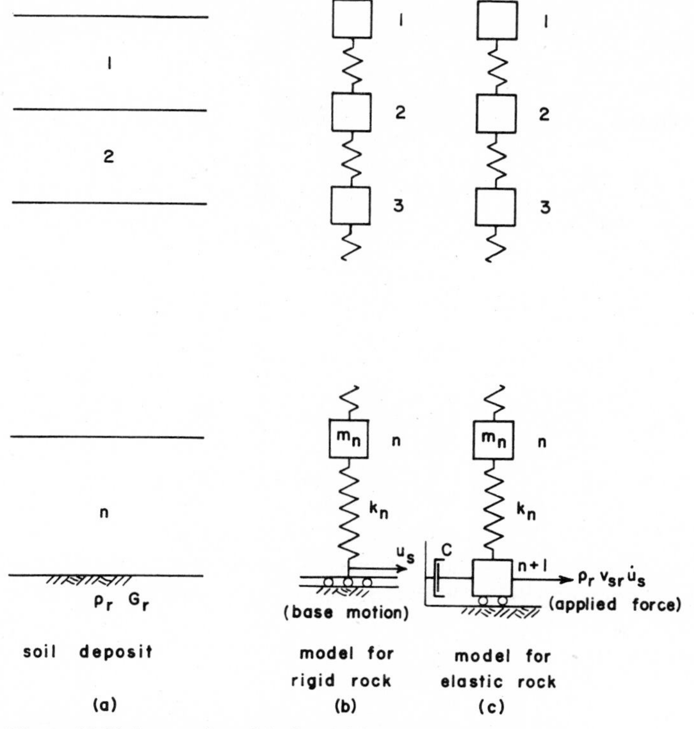

Figure 19-20 Lumped models for rigid rock and elastic rock.

layers, does not lend itself to a proper treatment by this method. Assuming instead a continuous variation of properties, where G_j represents the value of the shear modulus at middepth of each sublayer and ρ_j represents the density at the top (jth interface, counting the free surface as 1), and following the approach of Ang and Newmark,[17] it is possible to define stress points and mass points (Fig. 19-21).

At each stress point (middepth of the corresponding layer)

$$\tau_j = G_j \frac{1}{h_j}(u_{j+1} - u_j) \tag{19-76}$$

At each mass point

$$\rho_j u_j = \frac{2}{h_j + h_{j-1}}(\tau_j - \tau_{j-1}) \tag{19-77}$$

The equations for the interior points of the mesh can thus be written as

$$\tfrac{1}{2}\rho_j(h_{j-1} + h_j) + \left[-\frac{G_{j-1}}{h_{j-1}}u_{j-1} + \left(\frac{G_{j-1}}{h_{j-1}} + \frac{G_j}{h_j}\right)u_j - \frac{G_j}{h_j}u_{j+1}\right] = 0 \tag{19-78}$$

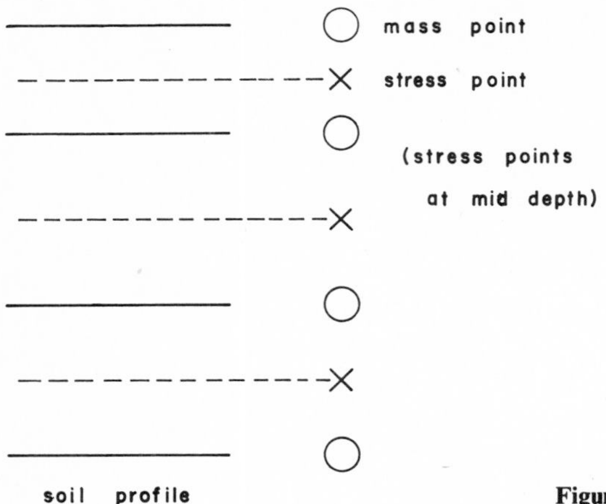

Figure 19-21 Mass-point and stress-point model.

A comparison with the original physical model shows that the only difference between the models is in the mass term. Since in most cases the variation of ρ is not large, this factor is not important. If, on the other hand, a sudden and important change in mass density occurs at the interface between two sublayers, the term ρ_j would have to be replaced in (19-78) by an average value at that point. The physical model would suggest a value

$$\rho_{j,\,\mathrm{av}} = \frac{\rho_{j-1} h_{j-1} + \rho_j h_j}{h_{j-1} + h_j} \tag{19-79}$$

which is just a weighted average of the densities of the two sublayers with the thicknesses as weighing factors.

Internal Soil Damping

To reproduce the internal dissipation of energy in the soil, in the discrete model, one could select, in principle, a set of arbitrary dashpots and place one between each two masses, in parallel with the spring k. If c_j is the constant of the dashpot between masses j and $j + 1$, the equations of motion would just be modified by the addition of a term $[C]\{\dot{U}\}$ or $[C]\{\dot{Y}\}$, where C is a damping matrix entirely similar to $[K]$ (the stiffness matrix of the system) but with c_j instead of k_j.

For the finite difference formulation the same result would be achieved by assuming that each soil layer (or sublayer) had a viscosity η_j. The equation of motion for a linear viscoelastic material is then

$$G \frac{\partial^2 u}{\partial z^2} + \eta \frac{\partial^3 u}{\partial z^2\, \partial t} = \rho \frac{\partial^2 u}{\partial t^2} \tag{19-80}$$

The discrete dashpots resulting from application of finite differences are

$$c_j = \frac{\eta_j}{h_j} \tag{19-81}$$

While this procedure is mathematically straightforward, some difficulties arise in the selection of the values of c or η for practical applications and in the physical interpretation of the system's behavior.

A constant c, or a constant η, over each sublayer implies a dissipation of energy of a viscous type (increasing with frequency). Experimental evidence seems to indicate, however, that the internal dissipation of energy in the soil is mostly of a hysteretic nature. Ideally one would like to have for each layer of soil curves relating modulus and damping to level of strain. Then, for an expected or assumed strain, one would pick the appropriate value of G_j and an appropriate amount of linear hysteretic damping D_j.

If the problem is to be solved in the frequency domain, following a procedure parallel to that described in Sec. 19-4 for the continuous formulation, the simplest option is to assume a complex modulus of the form $G_j(1 + 2iD_j)$. The terms corresponding to internal soil damping are then included in the complex stiffness matrix. Alternatively, for a steady state harmonic motion with frequency ω, $\{\dot{U}\} = i\omega\{U\}$, and the complex part of the stiffness matrix multiplied by $\{U\}$ can be interpreted as a term $[C]\{\dot{U}\}$, where $[C]$ is the internal damping matrix.

If the problem is to be solved in the time domain, it is not possible to have terms that are frequency dependent. An approximate way to simulate hysteretic behavior is to estimate first the amount of damping d_i in mode i from the damping ratios of the different layers D_j, using weighted modal-damping rules. If $\{\phi_i\}$ and ω_i are the mode shapes and natural frequencies of the undamped system, and if $[Q]$ is the modal matrix which has the mode shape $\{\phi_i\}$ as its ith column, a matrix $[C]$ can be defined

$$[C] = [M][Q][B][Q^T][M] \tag{19-82}$$

where $[B]$ is a diagonal matrix with $b_i = 2d_i\omega_i$.

If the solution is to be carried out through a modal analysis, determination of the values d_i is all that is necessary and the matrix $[C]$ need not be formed.

Other procedures have often been used to find a damping matrix $[C]$, defining it as proportional to the stiffness matrix, the mass matrix, or a combination of both. One must be careful, in using these methods, to avoid having a damping which increases with frequency, filtering out all high frequency components of the motion.

For elastic rock the evaluation of $[C]$ becomes more cumbersome and problematic. In general, an additional term must be added to account for radiation effects.[8]

Methods of Solution

Three different approaches can be followed to solve the equations of motion of the discrete system: a solution in the frequency domain following the same steps described in Sec. 19-4; a direct solution in the time domain, integrating the set of differential equations numerically; and a modal analysis.

Solution in the frequency domain Given the equations of motion in the form

$$[M]\{\ddot{U}\} + [C]\{\dot{U}\} + [K]\{U\} = \{J\}f \tag{19-83}$$

where $\{J\}$ is a vector with all components 0, except the last one, which is unity, $f = k_n u_s$ for rigid rock and $\rho_r v_{sr} \dot{u}_s$ for elastic rock, and the terms of $[K]$ and $[C]$ can be complex and frequency dependent; assuming a steady state periodic motion and excitation with frequency ω of the form $e^{i\omega t}$ leads to

$$\dot{u}_s = i\omega u_s \qquad \ddot{u}_s = -\omega^2 u_s \qquad \{U\} = i\omega\{U\} \qquad \{\ddot{U}\} = -\omega^2\{U\}$$

Equation (19-83) can thus be rewritten as

$$([K] + i\omega[C] - \omega^2[M])\{U\} = \{J\}gu_s \tag{19-84}$$

where $g = k_n$ for rigid rock and $i\omega\rho_r v_{sr}$ for elastic rock.

For a given frequency ω, this represents a set of linear algebraic equations with complex coefficients which can be solved by any appropriate numerical scheme, e.g., a Gauss elimination. The displacement of any mass is then

$$\{U\} = ([K] + i\omega[C] - \omega^2[M])^{-1}\{J\}gu_s \tag{19-85}$$

and accelerations, strains, or stresses can be computed. The amplification function can then be obtained at each frequency as u_1/u_s. A proper definition of this function will require repeated solutions of Eq. (19-84) for different values of ω. If the fast Fourier transform is going to be used for the following steps, the amplification function must be tabulated at equal frequency intervals.

Determination of the design motion at the top of the soil deposit is then performed as described in Sec. 19-4, obtaining the Fourier transform of the input earthquake, multiplying it by the amplification function of the soil, and finding the inverse Fourier transform of the product. This approach is restricted to linear behavior (at least in each cycle of an iterative scheme), as is the continuous solution.

Direct solution in the time domain If the terms of the damping matrix $[C]$ and the stiffness matrix $[K]$ are restricted to be real and frequency independent, expression (19-83) represents a set of ordinary differential equations with constant coefficients. A variety of methods (some based on physical approximations, others based on mathematical approximations) are available for the step-by-step integration of this system of equations.

It is generally preferred in this case to work with relative displacements $\{Y\} = \{U\} - \{I\}u_s$. Equation (19-83) then becomes

$$[M]\{\ddot{Y}\} + [C]\{Y\} + [K]\{Y\} = -[M]\{I\}\ddot{u}_s \tag{19-86}$$

and the forcing function is given by the accelerogram of the input earthquake (assumed to occur at the base of the stratum or at the outcropping of rock, depending on the formulation), instead of the time history of displacements or velocity.

When this approach is followed, the time history of acceleration at the free surface of the soil $\ddot{u}_1 = \ddot{y}_1 + \ddot{u}_s$ can be obtained directly, bypassing entirely the determination of the amplification function. The procedure is particularly interesting if one wishes to perform a true nonlinear analysis, instead of the iterative linear scheme needed by the other methods. It is then necessary to define the nonlinear characteristics of each spring k_j. Hysteretic dissipation of energy will be accounted for through the nonlinear behavior, and the matrix $[C]$ should therefore include only whatever damping may exist in the soil for very low levels of strain (in addition to the bottom dashpot for the elastic rock case). This damping will be very small (it is often neglected) and may be viscous.

Many methods may be still used for the numerical integration of the nonlinear equations of motion. One must be careful, however, with methods which may introduce fictitious damping into the system to make the procedure stable and with methods which are reported as unconditionally stable when used for linear systems but which lose this property for nonlinear systems.

Modal analysis Most engineers working in problems of structural dynamics are thoroughly familiar with the use of modal analysis. Not only must the matrices $[K]$ and $[C]$ have terms that are real and frequency independent in this case, but in addition the matrix $[C]$ must satisfy the orthogonality condition $\{\phi_i\}^T[C]\{\phi_j\} = 0$ for $i \neq j$, where $\{\phi\}$ are the eigenvectors (mode shapes) of the free vibration problem.

If we call $b_i = 2d_i\omega_i = \{\phi_i\}^T[C]\{\phi_i\}$ with the $\{\phi_i\}$ normalized with respect to the mass matrix ($\{\phi_i\}^T[M]\{\phi_i\} = 1$) and define the participation factor of the ith mode for Eq. (19-86) as

$$\Gamma_i = \{\phi_i\}^T[M]\{I\} \tag{19-87}$$

the original system of n (or $n + 1$) linear differential equations can be uncoupled into n (or $n + 1$) independent equations of the form

$$\ddot{a}_i(t) + b_i \dot{a}_i(t) + \omega_i^2 a_i(t) = -\ddot{u}_s \tag{19-88}$$

The solution can then be expressed as

$$\{Y\} = \sum_i a_i(t)\Gamma_i\{\phi_i\} \qquad \{\ddot{Y}\} = \sum_i \ddot{a}_i(t)\Gamma_i\{\phi_i\}$$

or

$$\{\ddot{U}\} = \{I\}\ddot{u}_s + \sum_i \ddot{a}_i(t)\Gamma_i\{\phi_i\} \tag{19-89}$$

Each term of the summation represents the contribution of a mode to the total solution. Identification of each one of these terms often provides valuable insight into the dynamic behavior of the system and its response. In addition the summation does not have to extend in general to all the modes.

Modal analysis requires a linear system. The expected earthquake at the free surface or time histories of strains can be obtained directly, bypassing the amplification function (solution in the time domain). It is possible, however, to find the amplification function by defining a steady state harmonic motion and excitation. Then

$$\mathrm{AF}(\omega) = 1 + \sum_j \frac{\omega^2 \Gamma_j \phi_{j1}}{\omega_j^2 - \omega^2 + 2id_j \omega \omega_j} \tag{19-90}$$

which is clearly a complex function if there is damping in the system.

Accuracy of the Discrete Model

An important step in the use of a discrete model is the selection of an appropriate mesh size, i.e., a sufficient number of masses, to obtain a reasonable accuracy. For a uniform layer of soil, the natural frequencies are given by

$$\omega_{it} = \frac{(2n-1)\pi}{2} \frac{v_s}{H} \tag{19-91}$$

Table 19-2 shows the values of ω_i / ω_{1t} resulting from a discrete model using 1, 2, 4, 8, and 16 masses. It can be seen from this table that the highest modes computed from the discrete system are grossly in error in comparison with the exact values; however, the first $n/2$ frequencies (if n is the number of masses) are within 10 percent of the true solution.

In order to decide on the necessary number of masses, one must consider not only the geometry and physical characteristics of the soil deposit but also the range of frequencies of interest and the type of results desired. Some formulas presented in the past[8] have been primarily intended as a guideline to reproduce a

Table 19-2 Natural frequencies from a discrete model[8] ω_i / ω_{it}

		Number of masses				
ω_i	Theoretical value	1	2	4	8	16
1	1	0.8989	0.9732	0.9935	0.9985	0.9995
2	3	...	2.3529	2.8289	2.9586	2.9895
3	5	...	...	4.2373	4.8019	4.9505
4	7	...	...	5.0000	6.4620	6.8611
5	9	...	...	...	7.8740	8.7146
6	11	...	...	...	8.9888	10.4712
7	13	...	...	...	9.7561	12.1212
8	15	...	...	...	10.1266	13.6986
9	17	...	...	...	...	15.0943
10	19	...	...	...	...	16.3934
11	21	...	...	...	...	17.4672
12	23	...	...	...	...	18.4336
13	25	...	...	...	...	19.2308
14	27	...	...	...	...	19.8020
15	29	...	...	...	...	20.2020
16	31	...	...	...	...	20.4082

natural frequency ω_i with great accuracy (within 1 or 2 percent). It is clear, however, from simple considerations of dynamics, that such accuracy may not be necessary if i is moderately large, there is damping in the system, and one is interested in the amplification function, response spectra, or such measures of the motion at the free surface of the soil as the maximum ground acceleration.

If there is some damping, the amplification function at a frequency ω_i is not only affected by the ith mode but also by the previous ones (1 to $i - 1$), which are reproduced with higher accuracy; a small error in the value of ω_i produces mainly a shift in the corresponding peak, which is less and less marked with increasing i. This effect would be even less noticeable in the ratio of response spectra (as defined in Sec. 19-4), since this function is smoother (with wider and less pronounced peaks) than the amplification function, and the participation of high modes in such response parameters as the maximum acceleration at the top may be almost negligible. It must finally be realized that an excessive accuracy in the mathematical determination of the high frequencies is hard to justify when there is a substantial uncertainty in the physical properties of the soil and the characteristics of the expected earthquake.

Taking all these factors into account, it would seem that for most practical cases reproducing the maximum frequency of interest ω_{max} within 10 percent is probably more than adequate. Additional limitations should be imposed to guarantee that each layer with different properties is accounted for and to ensure a minimum total number of masses (so as to have at least a few modes with reasonable approximation). This would lead to

$$n - 1 \geqslant \frac{2H}{\pi v_s} \omega_{max} \tag{19-92}$$

or

$$h = \frac{H}{n} \leqslant \frac{\pi v_s}{2\omega_{max}} \left(1 - \frac{1}{n}\right) \approx \frac{\pi v_s}{2\omega_{max}} \qquad \text{with } n \geqslant 8 \tag{19-93}$$

For the general case of multiple layers with different properties,

$$h_j \leqslant \frac{\pi v_{sj}}{2\omega_{max}} \text{ and not larger than layer thickness} \tag{19-94}$$

At the frequency ω_{max} the corresponding wavelength would be

$$l_j = v_{sj} T_{max} = \frac{2\pi v_{sj}}{\omega_{max}}$$

Expression (19-94) suggests therefore a maximum distance between masses of one-quarter of the shortest wavelength of interest ($h_j \leqslant l_j/4$). This implies at least one point between nodes of the wave (Fig. 19-22). A general rule of thumb in the application of the finite difference method to the solution of eigenvalue problems recommends the existence of two or three points between nodes of the wave, which would yield $h_j \leqslant l_j/6$ or $l_j/8$, but this is the requirement of having the corresponding mode reproduced with great accuracy. Considering again

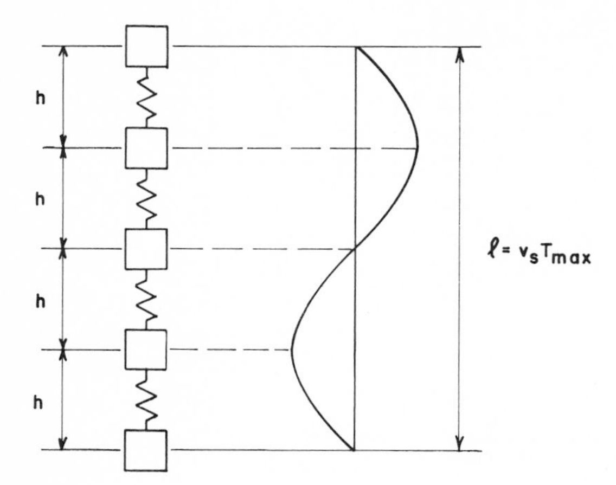

Figure 19-22 Required masses to simulate mode.

the decreasing participation factor of the higher modes in the amplification problem, formula (19-94) seems justified. Some judgment, however, is necessary in each individual case.

19-6 FINITE ELEMENT MODELS

The finite element method has been used extensively in the solution of problems of soil dynamics, particularly when dealing with soil amplification and a geometry which is not one-dimensional, when studying soil-structure interaction effects (vibrations of foundations), and when it is desired to solve in a single step the combined dynamic system formed by the soil and the structure, under a base-motion excitation.[18-22] All these cases are of interest in the seismic analysis of nuclear power plants. Some of them are further discussed in detail in Chap. 20.

One-dimensional Amplification

While the use of the finite element method is particularly useful for two-dimensional situations, it is interesting to look at the results of this formulation in the simple one-dimensional case. Again, for a soil deposit formed by n parallel layers or sublayers of soil, as shown in Fig. 19-19, and assuming a linear variation of the displacement u (u_y or u_x, depending on the type of shear wave), over each sublayer

$$u = u_j\left(1 - \frac{z_j}{h_j}\right) + u_{j+1}\frac{z_j}{h_j}$$

$$\gamma = \frac{\partial}{\partial z} = \frac{1}{h_j}(u_{j+1} - u_j) \tag{19-95}$$

$$\tau = G\gamma = \frac{G_j}{h_j}(u_{j+1} - u_j)$$

The expression for the strain energy is then

$$\frac{1}{2}\{U\}^T[k]\{U\} = \int_0^{h_j} \frac{1}{2}\frac{G_j}{h_j^2}(u_{j+1} - u_j)^2\, dz_j = \frac{1}{2}\frac{G_i}{h_j}(u_{j+1} - u_j)^2 \qquad (19\text{-}96)$$

and the stiffness matrix of the element (segment) $j, j + 1$ is

$$[k]_j = \begin{bmatrix} \dfrac{G_j}{h_j} & -\dfrac{G_j}{h_j} \\ -\dfrac{G_j}{h_j} & \dfrac{G_j}{h_j} \end{bmatrix} \qquad (19\text{-}97)$$

In a similar way writing the expression for the kinetic energy or the work done by the inertia forces, one can derive a consistent mass matrix for the element

$$[m]_j = \begin{bmatrix} \dfrac{\rho_j h_j}{3} & \dfrac{\rho_j h_j}{6} \\ \dfrac{\rho_j h_j}{6} & \dfrac{\rho_j h_j}{3} \end{bmatrix} \qquad (19\text{-}98)$$

From the individual element matrices $[k]_j$ and $[m]_j$, one can assemble the total stiffness matrix $[K]$ of the system and the total mass matrix $[M]$. The stiffness matrix $[K]$ is exactly the same as for the finite difference formulation. The mass matrix, on the other hand, is no longer a diagonal matrix (corresponding to a lumped mass model) but has the same tridiagonal structure as the stiffness matrix.

Table 19-3 shows the natural frequencies (in the form of the ratio ω_i/ω_{1t}) for a uniform layer of soil, with the above formulation. Table 19-2 had the corresponding values for the finite difference model; these values would apply also to a finite element formulation with lumped masses instead of consistent mass matrices. A comparison of the two tables indicates that while the results with lumped masses are always smaller than the true frequencies, consistent mass matrices overestimate them. For the first modes of any column, the error in the two formulations is about the same. For higher modes results obtained with consistent masses are better than those of the lumped mass model, although still not very good. This comparison suggests that a refinement of the finite element formulation is possible by considering a fraction of the total mass (30 or 40 percent) as lumped masses and the remaining as a consistent mass matrix. Kausel has recommended a mass matrix of the form

$$[m]_j = \begin{bmatrix} \frac{1}{2}\alpha\rho_j h_j + \frac{1}{3}(1-\alpha)\rho_j h_j & \frac{1}{6}(1-\alpha)\rho_j h_j \\ \frac{1}{6}(1-\alpha)\rho_j h_j & \frac{1}{2}\alpha\rho_j h_j + \frac{1}{3}(1-\alpha)\rho_j h_j \end{bmatrix} \qquad (19\text{-}99)$$

with α of the order of 0.3 to 0.4, which corresponds to the above suggestion.

This model can, of course, be extended to the solution of P waves with normal incidence by simply replacing G_j with $\lambda_j + 2G_j$. The introduction of the appropriate boundary conditions at the bottom of the soil deposit for rigid rock or elastic rock, the different ways of including the effect of internal damping in the soil, and the possible methods of solution are all as described in Sec. 19-5 for the finite difference model.

Table 19-3 Natural frequencies from finite element model ω_i/ω_{it}

i	Theoretical value	Number of elements: 1	2	4	8	16
1	1	1.1027	1.0259	1.0064	1.0016	1.0004
2	3	...	3.5837	3.1749	3.0435	3.0109
3	5	...	...	5.7674	5.2024	5.0505
4	7	...	...	8.3402	7.5543	7.1384
5	9	...	...	...	10.1512	9.2949
6	11	...	...	...	12.9462	11.5390
7	13	...	...	...	15.6180	13.8891
8	15	...	...	...	17.3913	16.8595
9	17	...	...	...	...	18.9573
10	19	...	...	...	...	21.6749
11	21	...	...	...	...	24.4790
12	23	...	...	...	...	27.2955
13	25	...	...	...	...	29.9924
14	27	...	...	...	...	32.3698
15	29	...	...	...	...	34.1671
16	31	...	...	...	...	35.1580

Amplification of SH Waves

Consider again a soil deposit with one-dimensional geometry (parallel layers of soil) but the more general case of an incident SH wave in the rock with angle of incidence different from zero. The analytical solution indicates that for this case the displacement u_y can be expressed as the product of a function of z, variable from layer to layer, and a function of x which is the same for all points (and, of course, a function of t). Taking advantage of this fact, the displacement expansion for the finite element formulation can be taken as

$$u_y = \left[u_j\left(1 - \frac{z_j}{h_j}\right) + u_{j+1}\frac{z_j}{h_j}\right]f(x) \tag{19-100}$$

with
$$f(x) = \exp\left(-\frac{i\omega}{v_{sr}}\sin\alpha_r x\right) = \exp\left(-\frac{i\omega}{v_{sj}}\sin\alpha_j x\right)$$

where v_{sr} = shear wave velocity in the rock
α_r = angle of incidence in rock
v_{sj}, α_j = corresponding quantities for the jth sublayer
u_j, u_{j+1} = displacements of nodal points on z axis ($x = 0$)

Then

$$\gamma_{xy} = -\frac{i\omega}{v_{sr}}\sin\alpha_r\left[u_j\left(1 - \frac{z_j}{h_j}\right) + u_{j+1}\frac{z_j}{h_j}\right]f(x)$$

$$\gamma_{yz} = \frac{1}{h_j}(-u_j + u_{j+1})f(x) \tag{19-101}$$

The stiffness matrix for the segment $j, j+1$ (finite element) becomes

$$[k]_j = \begin{bmatrix} \dfrac{G_j}{h_j} & -\dfrac{G_j}{h_j} \\ -\dfrac{G_j}{h_j} & \dfrac{G_j}{h_j} \end{bmatrix} + \frac{G_j\omega^2}{v_{sr}^2}\sin^2\alpha_r \begin{bmatrix} \frac{1}{3}h_j & \frac{1}{6}h_j \\ \frac{1}{6}h_j & \frac{1}{3}h_j \end{bmatrix}$$

$$= [k_0]_j + \frac{\omega^2}{v_{sr}^2}\sin^2\alpha_r[k_2]_j \qquad (19\text{-}102)$$

The consistent mass matrix is still given by Eq. (19-98). A similar derivation can be made for SV and P waves.

19-7 NONLINEAR SOIL AMPLIFICATION

Most of the formulations and methods of analysis discussed in previous sections are applicable only for a linear problem. This implies linear geometry and, more importantly, linear material properties. It has always been recognized, however, that an important factor in the correct application of these theories to practical situations (and perhaps the most important) is the consideration, at least approximately, of nonlinear soil behavior.

From laboratory tests (usually cyclic loading but also normal loading tests) the secant modulus and damping ratio of the soil as a function of strain are obtained. The effect of the overburden pressure at various depths of the actual soil profile is taken care of either by appropriate consolidation of the samples or by adjustment from field shear wave velocity measurements. Figure 19-23 shows the general form of these curves.

For one-dimensional amplification studies, mass density (or unit weight), shear modulus, and damping ratio are all the parameters necessary to describe the soil. Three approaches are then possible in increasing order of complexity:

1. To select estimated values of shear modulus and damping corresponding to the expected level of strains and to perform a unique linear analysis by any of the procedures already described.
2. To compute with the linear analysis characteristic values of strain at different depths and to select new values of modulus and damping consistent with these results, iterating until the values from two sequential cycles of iteration differ by less than a specified tolerance. Each cycle involves a linear analysis and can thus be performed by any method, in the time domain, in the frequency domain, or through a modal analysis.
3. To use a discrete model of the soil (finite differences or one-dimensional finite elements) and to select for each equivalent "spring" a nonlinear stress-strain or force-deformation law consistent with the experimental data (as far as variations of modulus and damping with strain are concerned). The system must

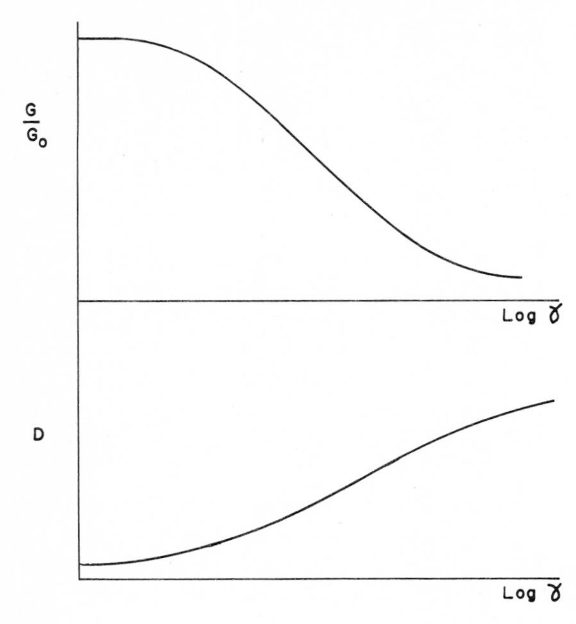

Figure 19-23 Secant modulus and damping vs. average strain.

then be solved in the time domain through direct integration of the nonlinear equations of motion.

The second procedure is probably the one most commonly used today. Several questions arise, however, about its validity:

1. The process seems to converge in a few cycles for most practical problems, but a proof of convergence for all cases is lacking. Some convergence difficulties have been found in fact when trying to solve the inverse problem, i.e., when trying to determine an input motion at bedrock which would produce a specified motion at the free surface (particularly for deep profiles).
2. The selection of modulus and damping ratio for each cycle requires the definition of a *characteristic strain*. For a harmonic, steady state condition this is obviously the maximum strain since its amplitude remains constant from one cycle of vibration to the next. It is not clear, however, what it should be for a transient motion. Several approaches have been followed in the past, e.g., taking the maximum strain, the average of the 10 largest values, or a fraction of the maximum strain. The last has become most popular and is normally used at present with a characteristic strain from 60 to 70 percent (typically two-thirds) of the maximum strain.
3. If the process converges and a representative value of strain is selected, the final question is how closely the motion calculated by the iterative scheme resembles the results of an exact nonlinear analysis.

A nonlinear analysis, on the other hand, requires the selection of a force-deformation characteristic representative of the soil. Several nonlinear models from elastoplastic or bilinear to multilinear, hyperbolic or Ramberg-Osgood relationships have been often used. In all these cases the goal is to obtain a relatively simple mathematical model which will produce a variation of modulus and damping with strain of the general form shown in Fig. 19-23. A perfect match to any experimental curve is not only difficult but also hard to justify considering the scatter in experimental data and the uncertainty involved in all phases of the analysis. Of all these models the Ramberg-Osgood relationship (see Chap. 2) seems to provide the most reasonable fit.[23] This stress-strain law, suggested in 1943 by Ramberg and Osgood,[24] has been extensively studied by Jennings.[25]

Figure 19-24 shows the general form of the load-deflection paths and the resulting secant modulus as a function of strain for $r = 2.25$ and different values of α. Values of r from 2 to 2.5 and values of α from 0.01 to 0.1 (typically 0.05) seem the most adequate.[26]

A comparison of the results of an iterative linear analysis with those of a nonlinear analysis for a bilinear soil model[7] indicated satisfactory agreement. A more comprehensive set of comparative studies, using a Ramberg-Osgood model with $r = 2$ and $\alpha = 0.05$ for the nonlinear analysis, and the modulus and damping curves corresponding to this stress-strain law for the iterative approach, so as to have consistent models,[26,27] shows that:

1. The iterative procedure reproduces well the basic qualitative features of nonlinear soil behavior. As the excitation and correspondingly the levels of strain increase, the shear modulus decreases and the effective internal damping increases. There is therefore a shift in the peaks of the amplification function and of the ratio of response spectra toward longer periods, accompanied by a

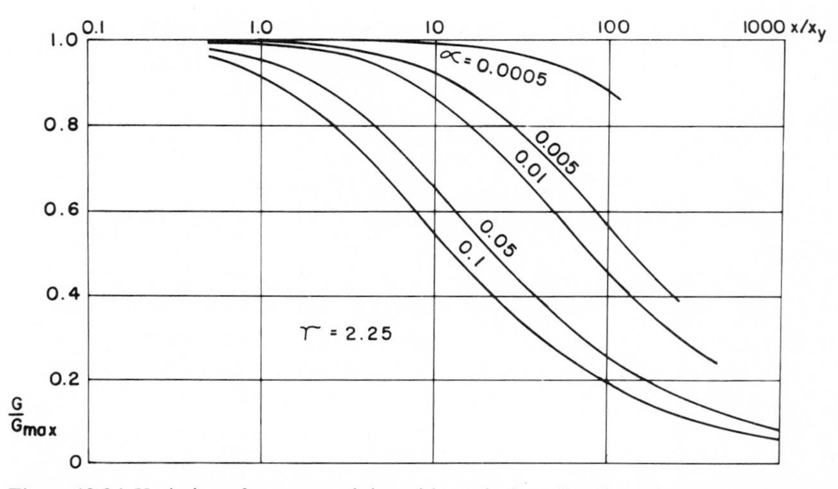

Figure 19-24 Variation of secant modulus with strain for a Ramberg-Osgood model with $r = 2.25$.

reduction in the peak amplification values due to the increased dissipation of energy.

2. From a quantitative point of view the iterative procedure with a characteristic strain equal to two-thirds of the maximum strain seems to overestimate slightly the maximum ground acceleration and the peak of the ratio of response spectra. On the other hand, it underestimates the maximum strains and ductility factors. This apparent discrepancy is the logical result of the $\frac{2}{3}$ factor (or any factor less than unity). Comparative studies selecting other factors (from 0.5 to unity) indicate, however, that it is in the range of 60 to 70 percent that the best overall agreement is obtained. The difference in results is small for the maximum acceleration or the peak ratio of response spectra (less than 20 percent) and on the conservative side. It may be somewhat larger (up to 50 percent) and unconservative for the maximum strains. The error seems to increase, in general, with increasing level of excitation.

More parametric studies are needed, using typical soil profiles, to assess the sensitivity of the results of a nonlinear analysis to the values of d_y, f_y, α, and r that define the Ramberg-Osgood model, within their logical ranges, or to the use of a different but equally sensible nonlinear model such as the hyperbolic stress-strain law. Some further attention should also be paid to the definition of the base rock boundary. When there is a clear and abrupt change in properties from the soil to the rock, the largest strains are likely to occur at the base of the soil deposit, decreasing the importance of the radiation damping with increasing level of excitation. In this case the distinction between rigid rock and elastic rock amplification is of little importance for high intensities of motion. If, on the other hand, the variation in elastic properties is gradual and a fictitious boundary is introduced at an arbitrary depth for mathematical convenience, one may end up with a model which does not reproduce the physical reality since nonlinear behavior is assumed for the soil but not for the underlying "rock."

Consideration of nonlinear soil behavior for two-dimensional amplification studies introduces another degree of complexity. For the iterative linear analysis it is necessary to define to what shear strain the experimental or assumed curves of shear modulus and damping correspond. The possible variation of a second elastic parameter (λ, E, ν, or the bulk modulus) must also be established. For a true nonlinear analysis several models are again available, as described in Chap. 2.

The practical treatment of nonlinear soil properties is thus based at present on engineering approximations rather than on a solid or rigorous mathematical theory. The main emphasis of the procedures most commonly used is to reproduce qualitatively the physical behavior that can be reasonably expected. The uncertainty in the soil properties, the scatter in experimental data, and the difficulty in finding a simple mathematical model that will match experimental results exactly make more elaborate procedures hard to justify. It seems, however, that additional research or more parametric studies in this area would improve present methods or at least increase the degree of confidence in the quantitative results they furnish.

19-8 FINAL CONSIDERATIONS

Soil amplification theories, even in their simplest one-dimensional form, have been used successfully to explain the different characteristics of earthquakes in soft soil deposits (such as the valley of Mexico), the difference in damage in various parts of a city under the same earthquake (valley of Caracas), and the difference in levels of motion at various depths. In recent years, however, they have been under attack due to a lack of distinct and noticeable soil effects in some earthquakes (particularly the San Fernando earthquake). It has also been argued that if amplification theories were true, all earthquakes recorded at a given place, or at least the two horizontal components of an earthquake, should have an almost identical frequency content. This argument ignores the effect of the type of waves, their angle of incidence and their relative amplitudes in the amplification function, and the variation of soil properties with level of excitation.

To assert that soil amplification studies can be carried out in practice with scientific rigor and accuracy would be presumptuous and even dangerous. On the other hand, to negate their validity and usefulness as an engineering tool for obtaining reasonable estimates of motions when applied with proper judgment is unwise. As in many other engineering problems, and particularly in the areas of geotechnology and earthquake engineering, there is a substantial amount of uncertainty in all phases of the analysis. Variation of the input parameters within logical ranges to bound the results is normally advisable. Special consideration must be given to the selection of an appropriate input motion with characteristics consistent with the location where it is specified (bedrock or the outcropping of rock), the establishment of boundaries and boundary conditions which correspond to this definition of the motion, and the estimation of soil properties and their variation with level of strains. Finally, the results, amplification functions, ratios of response spectra, or response spectra at the locations of interest should be properly smoothed.

REFERENCES

1. Thomson, W. T.: Transmission of Elastic Waves through a Stratified Solid Medium, *J. Appl. Phys.*, vol. 21, February 1950.
2. Kanai, K.: Semi-empirical Formula for the Seismic Characteristics of the Ground, *Bull. Earthquake Res. Inst.*, vol. 35, 1957.
3. Herrera, I., E. Rosenblueth, and O. A. Rascón: Earthquake Spectrum Prediction for the Valley of Mexico, *3d World Conf. Earthquake Eng., New Zealand, 1965.*
4. Arias, A., and L. Petit Laurent: Un Modelo Técnico para los Accelerogramas de Temblores Fuertes, *Rev. IDIEM*, vol. 4, no. 1, May 1965.
5. Donovan, N. C., and R. B. Matthiesen: Effects of Site Conditions on Ground Motions during Earthquakes, *State of the Art Symp., San Francisco, February 1968.*
6. Seed, H. B., and I. M. Idriss: The Influence of Soil Conditions on Ground Motions during Earthquakes, *J. Soil Mech. Found. Div. ASCE*, vol. 15, no. SM1, 1961.
7. Idriss, I. M., and H. B. Seed: Seismic Response of Horizontal Soil Layers, *J. Soil Mech. Found. Div. ASCE*, vol. 96, no. 3M4, 1968.

8. Roesset, J. M., and R. V. Whitman: Theoretical Background for Amplification Studies, *MIT Res. Rep.* R69-15, March 1969.
9. Roesset, J. M.: Effect of the Angle of Incidence on the Amplification of SH Waves, *Q. Progr. Rep. MIT Inter-Am. Prog., June 1–Aug. 31, 1969, Rep.* R69-73.
10. Jones, T. J., and J. M. Roesset: Soil Amplification of SV and P Waves, *MIT Res. Rep.* R70-3, January 1970.
11. Schnabel, P. B., J. Lysmer, and H. B. Seed: SHAKE: A Computer Program from Earthquake Response Analysis of Horizontally Layered Sites, *Univ. Calif. Rep.* EERC 72-12, Berkeley, December 1972.
12. Madera, G.: Fundamental Period and Amplification of Peak Acceleration in Layered Systems, *MIT Inter-Am. Prog.* R70-37 *Progr. Rep.* 10, Effect of Local Soil Conditions upon Earthquake Damage, June 1970.
13. Stern, S. H.: Characteristics of Filtered Earthquakes, M.S. thesis, Civil Engineering Department, Massachusetts Institute of Technology, Cambridge, Mass., September 1970.
14. Newmark, N. M.: Design Criteria for Nuclear Reactors Subjected to Earthquake Hazards, Interatomic Energy Association, Tokyo, June 1967.
15. Roesset, J. M., M. Sarrazin, and E. Vanmarcke: The Use of Amplification Functions to Derive Response Spectra Including the Effect of Local Soil Conditions, *MIT Civ. Eng. Dept. Res. Rep.* R69-48, July 1969.
16. Tsai, N. C.: Influence of Local Geology on Earthquake Ground Motions, Ph.D. thesis, California Institute of Technology, Pasadena, 1969.
17. Ang, A. H. S., and N. M. Newmark: Development of a Transmitting Boundary for Numerical Wave Motion Calculations, *Contr.* DASA-01-0090, *Rep. Defense Atomic Support Agency*, Washington, D.C., 1971.
18. Lysmer, J., and G. Waas: Shear Waves in Plane Infinite Structures, *J. Eng. Mech. Div. ASCE*, vol. 98, February 1972.
19. Waas, G.: Linear Two-dimensional Analysis of Soil Dynamics Problems in Semi-infinite Layered Media, Ph.D. thesis, University of California, Berkeley, 1973; also Analysis Method for Footing Vibrations through Layered Media, *U.S. Army Eng. Waterw. Expt. Stn. Tech. Rep.* S71-14, Vicksburg, Miss., September 1972.
20. Chang-Liang, V.: Dynamic Response of Structures in Layered Soils, *MIT Civ. Eng. Dept. Res. Rep.* R74-10, January 1974.
21. Kausel, E.: Forced Vibrations of Circular Foundations in Layered Media, *MIT Civ. Eng. Dept. Res. Rep.* R74-11, January 1974.
22. Kausel, E., and J. M. Roesset: Soil-Structure Interaction Problems for Nuclear Containment Structures, *ASCE Power Div. Spec. Conf., Denver, August 1974.*
23. Dobry, R.: Damping in Soils: Its Hysteretic Nature and the Linear Approximation, *MIT Civ. Eng. Dept. Res. Rep.* R70-14, 1970.
24. Ramberg, W., and W. R. Osgood: Description of Stress-Strain Curves by Three Parameters, *NACA Tech. Note* 902, July 1943.
25. Jennings, P. C.: Response of a General Yielding Structure, *J. Eng. Mech. Div. ASCE*, vol. 90, no. EM2, April 1964.
26. Constantopoulos, I. V.: Amplification Studies for a Nonlinear Hysteretic Soil Model, *MIT Civ. Eng. Dept. Res. Rep.* R73-46, September 1973.
27. Constantopoulos, I. V., J. M. Roesset, and J. T. Christian: A Comparison of Linear and Exact Nonlinear Analyses of Soil Amplification, *5th World Conf. Earthquake Eng., Rome, 1973*, sess. 5B, pap. 225.

CHAPTER

TWENTY

TWO- AND THREE-DIMENSIONAL DYNAMIC ANALYSES

John T. Christian, José M. Roësset, and Chandrakant S. Desai

The previous chapter described some of the fundamental concepts in dynamic analysis and showed how they are used when the geometry can be represented by one spatial dimension. Often structures and soil deposits cannot be characterized adequately in this way, and it becomes necessary to consider two or three dimensions. Then the analysis becomes more complicated from the point of view of both formulation and numerical programming. However, the physical principles set out in Chap. 19 remain valid.

Because a great deal of the work now being done in geotechnical engineering uses the finite element method, this chapter concentrates on this approach and does not emphasize finite difference techniques. The latter have advantages in many cases, e.g., those involving direct solutions for wave propagation in nonlinear media. There is no clearly superior approach for all problems, and many of the analytical concerns and difficulties are similar for all methods.

This chapter discusses the choices and options among, and distinctions between, the many techniques falling under the rubric of finite element methods rather than presenting extensive results obtained with one computer program. The selection of what specific approach to use for a particular analysis depends on the problem to be solved, the resources and time available, and the form of solution required.

20-1 BASIC CONSIDERATIONS, ELASTIC MATERIAL

Dynamic analyses in geotechnical engineering fall into three categories, illustrated in Fig. 20-1. The first is the vibration of a machine foundation on soil or rock. The input is prescribed at the level of the foundation and usually consists of steady oscillation at known frequencies. The second is the earthquake analysis of a structure founded on soil. The input motion is prescribed simultaneously along the bedrock, although in some cases this is based on desired motions at the surface or elsewhere. It consists of a relatively long history of accelerations with a rich frequency content. The third involves the propagation of shock waves through a complex of soil and rock with an embankment, tunnel, or other structures. The input usually consists of a relatively short waveform, but the boundary points move separately, and the nonlinearity of the soil behavior is important. It should be obvious that the same general purpose computer program will not handle all three cases efficiently. The different approaches can best be understood by con-

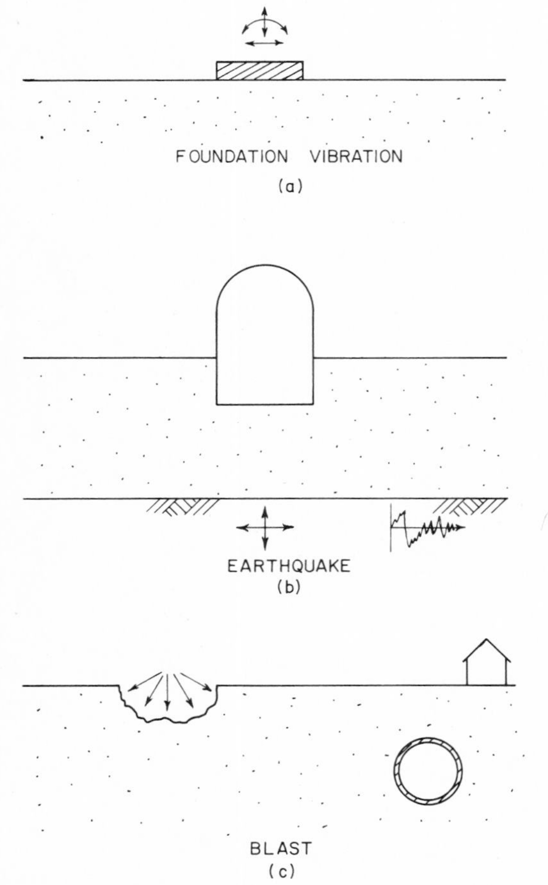

Figure 20-1 Types of dynamic problems.

sidering first the simplest formulation for the material—isotropic linear elasticity with no internal dissipation of energy.

Linearly Elastic, Undamped Formulation

The equations of dynamic equilibrium for a linearly elastic, undamped continuous body are

$$\begin{aligned}
\frac{\partial \sigma_{xx}}{\partial x} + \frac{\partial \sigma_{xy}}{\partial y} + \frac{\partial \sigma_{xz}}{\partial z} &= \rho \frac{\partial^2 u_x}{\partial t^2} + P_x(t) \\
\frac{\partial \sigma_{xy}}{\partial x} + \frac{\partial \sigma_{yy}}{\partial y} + \frac{\partial \sigma_{yz}}{\partial z} &= \rho \frac{\partial^2 u_y}{\partial t^2} + P_y(t) \\
\frac{\partial \sigma_{xz}}{\partial x} + \frac{\partial \sigma_{yz}}{\partial y} + \frac{\partial \sigma_{zz}}{\partial z} &= \rho \frac{\partial^2 u_z}{\partial t^2} + P_z(t)
\end{aligned} \tag{20-1}$$

The right-hand sides of these equations include both the products of mass density ρ and the corresponding accelerations and arbitrary, time-dependent forcing functions (P_x, P_y, P_z).

Many techniques can be used to obtain finite element or finite difference formulations from Eqs. (20-1). The most straightforward is a simple extension of the finite element methods described in Chap. 1. First, the terms involving mass times acceleration are replaced by equivalent body forces of opposite sense, called *d'Alembert forces*:

$$f_x = -\rho \frac{\partial^2 u_x}{\partial t^2} \qquad f_y = -\rho \frac{\partial^2 u_y}{\partial t^2} \qquad f_z = -\rho \frac{\partial^2 u_z}{\partial t^2} \tag{20-2}$$

where u_x, u_y, and u_z are the components of displacements in the x, y, and z directions, respectively.

Then the equilibrium at a given instant of time must require satisfaction of the principle of virtual work:

$$\int_V \sigma_{ij}\, \delta\epsilon_{ij}\, dV = \int_V f_i\, \delta u_i\, dV + \int_V P_i\, \delta u_i\, dV + \int_S \bar{T}_i\, \delta u_i\, dS \tag{20-3}$$

where $\bar{T}_i$ is the prescribed traction over the surface S_1. If the interpolation functions are the same for displacement, velocity, and acceleration, we have

$$\{u\} = [N]\{q\} \tag{20-4}$$

$$\{\dot{u}\} = [N]\{\dot{q}\} \tag{20-5}$$

$$\{\ddot{u}\} = [N]\{\ddot{q}\} \tag{20-6}$$

where $\{u\}$ = vector of displacements at point
$\{q\}$ = vector of nodal displacements
$[N]$ = matrix of interpolation functions

and dots indicate differentiation with respect to time.

Substitution of these expressions into Eq. (20-3) together with the usual algebraic and variational operations leads to[1]

$$\delta\{q\}^T[k]\{q\} = -\delta\{q\}^T\left(\int_V \rho[N]^T[N]\,dV\right)\{\ddot{q}\}$$

$$+\,\delta\{q\}^T\left(\int_V [N]^T\{P\}\,dV + \int_{S_1} [N]^T\{\bar{T}\}\,dS\right) \qquad (20\text{-}7)$$

The first expression on the right-hand side describes the inertial part of the system. It can be written in terms of the mass matrix $[m]$ as $-\delta\{q\}^T[m]\{\ddot{q}\}$. The mass matrix for one element is

$$[m] = \int_V \rho[N]^T[N]\,dV \qquad (20\text{-}8)$$

This is called the *consistent mass matrix* because its derivation is consistent with that of the stiffness matrix. For example, a constant-strain triangle with constant distribution of mass and with area A has a consistent mass matrix

$$[m] = \frac{\rho A}{3}\begin{bmatrix}[m_1] & [0] \\ [0] & [m_1]\end{bmatrix} \quad \text{where} \quad [m_1] = \begin{bmatrix}\frac{1}{2} & \frac{1}{4} & \frac{1}{4} \\ & \frac{1}{2} & \frac{1}{4} \\ \text{sym} & & \frac{1}{2}\end{bmatrix} \qquad (20\text{-}9)$$

This form of $[m]$ assumes that the nodal displacements $\{q\}$ are arranged with all the x displacements first and all the y displacements following; there are no z displacements for this element.

An alternative approach is to assume that the mass is concentrated at the nodes, leading to a *lumped mass matrix*. The constant-strain triangle has a lumped mass matrix:

$$[m] = \frac{\rho A}{3}\begin{bmatrix}[m_1] & [0] \\ [0] & [m_1]\end{bmatrix} \quad \text{where} \quad [m_1] = \begin{bmatrix}1 & 0 & 0 \\ & 1 & 0 \\ \text{sym} & & 1\end{bmatrix} \qquad (20\text{-}10)$$

The lumped mass matrix is simpler to compute and assemble than the consistent mass matrix, especially for more complicated elements than the constant-strain triangle, and it often leads to much greater computational simplicity. The consistent mass matrix follows from a more coherent and rigorous derivation and is more pleasing to one's sense of mathematical elegance.

The previous chapter has demonstrated that for the one-dimensional cases, the lumped mass matrix underestimates the resonant frequencies of the system and the consistent mass matrix overestimates them by nearly the same amount. Therefore, the most satisfactory procedure from the point of view of numerical accuracy is to use a mass matrix combining the lumped and consistent formulations in equal or nearly equal proportions. This is the approach most favored today.

The remaining terms on the right-hand side of Eq. (20-7) represent the contributions of externally applied loads. These can be combined in a nodal load vector $\{Q\}$, giving a term $\delta\{q\}^T\{Q\}$.

Now the terms for all elements are added node by node in the usual manner of assembling finite element global equations. If $\{r\}$ is the global nodal displacement vector, $\{R\}$ the global nodal external load vector, $[K]$ the summation of the elemental $[k]$ matrices, and $[M]$ the summation of the elemental $[m]$ matrices, then

$$\delta\{r\}^T[K]\{r\} = -\delta\{r\}^T[M]\{\ddot{r}\} + \delta\{r\}^T\{R\} \tag{20-11}$$

The vectors $\{r\}$, $\{\ddot{r}\}$, and $\{R\}$ are functions of time, but at any instant the principles of virtual work and equilibrium require that the variation vanish. Thus

$$[K]\{r\} + [M]\{\ddot{r}\} = \{R\} \tag{20-12}$$

This is the basic finite element formulation of the dynamic problem for linearly elastic, undamped materials. Further details on formulation and other aspects of finite element equations for the dynamic problem are given by Desai and Abel.[1]

Modification for Earthquake Input

Equation (20-12) can be used directly for analysis of problems with general dynamic loadings. In most earthquake problems the loading does not consist of a set of prescribed body and surface forces $\{R\}$ but of a set of prescribed boundary accelerations. It is usually assumed that all the points with the prescribed accelerations move together and that this time history of acceleration $\ddot{u}_g$ has a component $\ddot{u}_{gx}$ in the x direction, $\ddot{u}_{gy}$ in the y direction, and $\ddot{u}_{gz}$ in the z direction. If the entire system moved as a rigid body, each point would have the same time history of motion, which would be described by vectors $\{u_g\}$, $\{\dot{u}_g\}$, and $\{\ddot{u}_g\}$. These are constructed so that each x component in $\{\ddot{u}_g\}$ is $\{\ddot{u}_{gx}\}$, and so on. $\{\dot{u}_g\}$ and $\{u_g\}$ are simply the results of integrating $\{\ddot{u}_g\}$ in time once and twice, respectively.

In the actual, flexible case the relative motions between the free nodes and the nodes on the boundary with prescribed motion are denoted by the vectors $\{v\}$, $\{\dot{v}\}$, and $\{\ddot{v}\}$, defined by

$$\{v\} = \{r\} - \{u_g\} \qquad \{\dot{v}\} = \{\dot{r}\} - \{\dot{u}_g\} \qquad \{\ddot{v}\} = \{\ddot{r}\} - \{\ddot{u}_g\} \tag{20-13}$$

Because $\{u_g\}$ describes rigid body motion,

$$[K]\{u_g\} = \{0\} \tag{20-14}$$

Also, $\{R\}$ must be $\{0\}$ because there are no directly imposed forces other than those already accounted for by $\{\ddot{u}_g\}$. Therefore, Eq. (20-12) becomes

$$[K]\{v\} + [M]\{\ddot{v}\} = -[M]\{\ddot{u}_g\} \tag{20-15}$$

This formulation in terms of relative motions is widely used in earthquake engineering. It should be noted that since only those degrees of freedom belonging to nodes without prescribed time histories of acceleration need be retained in $\{v\}$, the sizes of the matrices and vectors are significantly reduced.

Alternatively, the earthquake problem can be stated in terms of absolute motions as

$$[K]\{r\} + [M]\{\ddot{r}\} = \{0\} \tag{20-16}$$

Here it must be understood that $\{r\}$ and $\{\ddot{r}\}$ contain some terms that are prescribed and that the $[K]$ and $[M]$ contain more rows and columns than the corresponding matrices in Eq. (20-15).

Modal Decomposition

Modal decomposition and solution involve a coordinate transformation such that a set of coupled partial differential equations, for example, Eqs. (20-12), is converted into a set of uncoupled ordinary differential equations. In other words, the stiffness and mass matrices are diagonalized.

The solution is assumed to consist of a set of sinusoidal motions under conditions of free vibration. This means that in Eq. (20-15) the ground accelerations $\{\ddot{u}_g\}$ are zero and the response $\{v\}$ has a number of components, each of which can be described by

$$\{v\}_i = \{\phi_i\}e^{i(\omega_i t - \theta_i)} \tag{2-17}$$

where $i = \sqrt{-1}$
$\{\phi_i\}$ = mode-shape function independent of time
ω_i = resonant or modal frequency
θ_i = phase angle

Substitution into Eq. (20-15) with $\{\ddot{u}_g\}$ set to zero gives

$$([K] - \omega_i^2[M])\{\phi_i\} = \{0\} \tag{20-18}$$

This is a classic eigenvalue problem that can be solved by a number of numerical methods to find ω_i and the corresponding $\{\phi_i\}$. The response of the actual problem with nonzero $\{\ddot{u}_g\}$ is then assumed to consist of the sum of the contribution of M modes, each of which has N degrees of freedom that correspond to the N degrees of freedom in $\{v\}$. The mode shapes are assembled into a modal matrix $[\Phi]$:

$$[\Phi] = [\{\phi_1\}\{\phi_2\} \cdots \{\phi_i\} \cdots \{\phi_M\}] \tag{20-19}$$

This matrix has dimensions N by M. The solution is expressed as

$$\{v\} = [\Phi]\{p\} \tag{20-20}$$

where $\{p\}$ is a dimensionless vector with M terms, each of which is the weighting factor by which the corresponding $\{\phi_i\}$ is multiplied before summing to give $\{v\}$. The terms in $\{p\}$ are time dependent. It follows that

$$\{\dot{v}\} = [\Phi]\{\dot{p}\} \tag{20-21}$$

$$\{\ddot{v}\} = [\Phi]\{\ddot{p}\} \tag{20-22}$$

Substitution into Eq. (20-15) gives

$$[K][\Phi]\{p\} + [M][\Phi]\{\ddot{p}\} = -[M]\{\ddot{u}_g\} \tag{20-23}$$

Premultiplication by $[\Phi]^T$ converts this into

$$[\Phi]^T[K][\Phi]\{p\} + [\Phi]^T[M][\Phi]\{\ddot{p}\} = -[\Phi]^T[M]\{\ddot{u}_g\} \tag{20-24}$$

Because $[K]$ and $[M]$ are real and symmetric and have all ω_i^2 positive, the orthogonality relations hold:

$$\{\phi_i\}^T[M]\{\phi_i\} = \begin{cases} 1 & \text{if } i = j \\ 0 & \text{if } i \neq j \end{cases} \tag{20-25}$$

$$\{\phi_i\}^T[K]\{\phi_i\} = \begin{cases} \omega_i^2 & \text{if } i = j \\ 0 & \text{if } i \neq j \end{cases} \tag{20-26}$$

The mode shapes must be normalized to the proper dimensions if these equations are to be valid. Otherwise a multiplicative constant must be used, but the normalization is very simple.

Equations (20-24) become

$$\begin{bmatrix} \omega_1^2 & & & & & \\ & \omega_2^2 & & & 0 & \\ & & \ddots & & & \\ & & & \omega_i^2 & & \\ & 0 & & & \ddots & \\ & & & & & \omega_M^2 \end{bmatrix} \{p\} + [I]\{\ddot{p}\} = -[\Phi]^T[M]\{\ddot{u}_g\} \tag{20-27}$$

This is now a set of uncoupled equations, each of which is of the form

$$\omega_i^2 p_i + \ddot{p}_i = -\{\phi_i\}^T[M]\{\ddot{u}_g\} \tag{20-28}$$

They can be solved by many numerical or analytical schemes, some of which will be discussed below in conjunction with step-by-step integration.

Modal analysis can be attractive because it reduces the number of unknowns from N to M, for in many cases only a modest number of modes need be considered. On the other hand, the solution of the eigenvalue and eigenvector problem can be time consuming, and it is sensitive to numerical error. When a large number of modes is required, the advantages of solution by modal decomposition begin to disappear. Moreover, the modal decomposition is relevant only to linear problems.

Solution in the Frequency Domain

The Fourier transform of a function f is defined by

$$F(\omega) = \int_0^\infty f(t)e^{-i\omega t}\,dt \tag{20-29}$$

where ω is the frequency and t is time. In this and the following description of

Fourier transform methods ω is a continuous variable that can be evaluated at discrete points, and it must not be confused with the resonant frequencies ω_i. The inverse operation gives

$$f(t) = \frac{1}{2\pi} \int_{-\infty}^{\infty} F(\omega) e^{i\omega t} \, d\omega \tag{20-30}$$

These equations describe a continuous transformation. Similarly, a discrete transformation can be defined in which a finite number of frequencies and times are used. If enough terms are used, the discrete transform is an excellent approximation to the continuous one. The discrete transform of a digitized time history, such as an earthquake accelerogram, can be computed very rapidly by the Cooley-Tukey algorithm, or fast Fourier transform, described in Chap. 1.

Applying the Fourier transform to Eq. (20-15) gives

$$[K]\{V(\omega)\} + [M]\{\ddot{V}(\omega)\} = -[M]\{\ddot{U}_g(\omega)\} \tag{20-31}$$

where $\{V(\omega)\}$, $\{\ddot{V}(\omega)\}$, and $\{\ddot{U}_g(\omega)\}$ are the Fourier transforms of $\{v(t)\}$, $\{\ddot{v}(t)\}$, and $\{\ddot{U}_g(t)\}$, respectively. For a particular frequency, ω_0, the time history of displacement, is

$$\{v_0(t)\} = \{A_0\} e^{i(\omega_0 t - \theta_0)} \tag{20-32}$$

where $\{A_0\}$ is the vector of amplitudes and θ_0 is the phase angle. It follows that

$$\{\dot{v}_0(t)\} = i\omega_0 \{v_0(t)\} \tag{20-33}$$

$$\{\ddot{v}_0(t)\} = -\omega_0^2 \{v_0(t)\} \tag{20-34}$$

Hence

$$\{\dot{V}(\omega)\} = i\omega \{V(\omega)\} \tag{20-35}$$

$$\{\ddot{V}(\omega)\} = -\omega^2 \{V(\omega)\} \tag{20-36}$$

It must be remembered that ω is a continuous function in these two equations. Then

$$[K]\{V(\omega)\} - \omega^2 [M]\{V(\omega)\} = -[M]\{\ddot{U}_g(\omega)\} \tag{20-37}$$

The solution is carried out by assembling for each chosen frequency a system of simultaneous equations

$$([K] - \omega^2 [M])\{V\{\omega)\} = -[M]\{\ddot{U}_g(\omega)\} \tag{20-38}$$

solving for $\{V(\omega)\}$, and taking the inverse transform to find $\{v(t)\}$. Because each frequency chosen means the solution of a complete set of linear equations, and because many frequencies are needed to compute a reasonable inverse Fourier transform, it is customary to evaluate Eq. (20-38) at a limited number of ω's and interpolate the solutions to obtain the required number of points for computing the inverse. When response at or near a particular frequency is to be completed, it is important that there be several frequencies near that value used in Eq. (20-38) and in the interpolation. It should be obvious that it is necessary to compute the inverse Fourier transform only for those terms in $\{V(\omega)\}$ or $\{v(t)\}$ that are desired.

A similar approach can be used starting from Eq. (20-16). It is useful to partition $\{r\}$ and $\{\ddot{r}\}$ into the unknown quantities and the specified time histories for the boundary points:

$$\begin{bmatrix} K_{11} & K_{12} \\ K_{12}^T & K_{22} \end{bmatrix} \begin{Bmatrix} r_1 \\ r_2 \end{Bmatrix} + \begin{bmatrix} M_{11} & M_{12} \\ M_{12}^T & M_{22} \end{bmatrix} \begin{Bmatrix} \ddot{r}_1 \\ \ddot{r}_2 \end{Bmatrix} = \{0\} \tag{20-39}$$

In this equation $[K_{11}]$ and $[M_{11}]$ are identical to $[K]$ and $[M]$ in Eq. (20-15) or (20-38), $\{r_2\}$ and $\{\ddot{r}_2\}$ are the vectors of known displacements and accelerations corresponding to u_g and $\ddot{u}_g$, and $\{r_1\}$ and $\{\ddot{r}_1\}$ are equal to $\{v\} + \{u_g\}$ and $\{\ddot{v}\} + \{\ddot{u}_g\}$, respectively.

The Fourier transform of Eq. (20-16) is

$$([K] - \omega^2[M])\{U(\omega)\} = \{0\} \tag{20-40}$$

Here $\{U(\omega)\}$ is used to designate the Fourier transform of $\{r(t)\}$ because $\{R\}$ has already been used. The matrices $[K]$ and $[M]$ are from Eq. (20-16) and are therefore different from the $[K]$ and $[M]$ in Eq. (20-38).

Because some terms in $\{r(t)\}$ are known, the corresponding terms in $\{U(\omega)\}$ are too, and they can be eliminated from the solution to give

$$([K_{11}] - \omega^2[M_{11}])\{U_1(\omega)\} = -([K_{12}] - \omega^2[M_{12}])\{U_2(\omega)\} \tag{20-41}$$

This is similar to Eq. (20-38) except for being expressed in absolute displacements.

Alternatively, the Fourier transform of Eq. (20-16) can be written

$$\left(-\frac{1}{\omega^2}[K] + [M]\right)\{\ddot{U}(\omega)\} = \{0\} \tag{20-42}$$

Multiplication by $-\omega^2$ gives

$$([K] - \omega^2[M])\{\ddot{U}(\omega)\} = \{0\} \tag{20-43}$$

Thus, the formulation is identical for absolute accelerations or absolute displacements.

The solution of Eq. (20-41) is

$$\{U_1(\omega)\} = -([K_{11}] - \omega^2[M_{11}])^{-1}([K_{12}] - \omega^2[M_{12}])\{U_2(\omega)\} \tag{20-44}$$

The individual terms in $-([K_{11}] - \omega^2[M_{11}])^{-1}([K_{12}] - \omega^2[M_{12}])$ are called *transfer functions*. They describe how a component of motion in $\{U_1(\omega)\}$ is affected by prescribed motion in $\{U_2(\omega)\}$. They are independent of the values used in $\{U_2(\omega)\}$ and can therefore be stored for future use with different earthquakes.

The transfer functions are identical for Eqs. (20-40) and (20-43). This means that the same functions can be used for prescribed displacements or prescribed accelerations. When the formulation of Eq. (20-38) is used, the transfer functions are *not* the same for prescribed displacements or prescribed accelerations.

20-2 DAMPING

The formulations in the previous section ignore the dissipation of energy by damping. Two kinds of damping exist, viscous and hysteretic. Viscous damping occurs in a dashpot or shock absorber. The dissipation of energy depends on the velocity of motion or strain, and the effects on the entire system vary with the frequency of motion. Hysteretic damping involves frictional loss of energy that is largely independent of frequency but depends on the magnitude of displacement or strain. A brake is an example. The internal dissipation of energy in soils is generally believed to be of a hysteretic nature, but the loss of energy due to propagation of waves away from the region of interest, also known as *radiation damping*, depends on the frequency of the waves. Both types of damping may be present in a case involving the dynamic behavior of soils. The mathematical treatment of linearly viscous damping is much simpler and until recently occupied the major place in dynamic analysis.

Modal Damping

Linearly viscous damping can be treated by first introducing the terms

$$-\eta \frac{\partial v_x}{\partial t} \qquad -\eta \frac{\partial v_y}{\partial t} \qquad -\eta \frac{\partial v_z}{\partial t}$$

on the right-hand side of Eqs. (20-2). The parameter η describes the energy loss, and v_x, v_y, and v_z are the displacements relative to the boundaries. The energy loss for each element is thus described by $\eta[N]\{\dot{q}\}$ and is introduced into the equilibrium equations and the virtual-work relation; then the stiffness formulation becomes

$$\delta\{q\}^T[k]\{q\} + \delta\{q\}^T[c]\{\dot{q}\} + \delta\{q\}^T[m]\{\ddot{q}\} = \delta\{q\}^T\{Q\} \tag{20-45}$$

where

$$[c] = \eta \int_V [N]^T[N]\, dV \tag{20-46}$$

which means that $[c]$ is proportional to $[m]$.

Alternatively, and more reasonably, the damping could be related to the strain rates rather than the rates of displacement. This would give a loss of energy for each element of $[\eta][B]\{\dot{q}\}$, in which $[\eta]$ is now a matrix of damping terms that accounts for the different components of strain. When this is introduced into the virtual-work expression, it must be related to strains and not displacements. An equation similar to Eq. (20-45) results, but $[c]$ is now defined by

$$[c] = \int_V [B]^T[\eta][B]\, dV \tag{20-47}$$

This will be similar in form to the stiffness matrix. If $[\eta]$ is proportional to the elastic stress-strain relation, $[c]$ will be proportional to $[k]$.

Assembly of the entire system and vanishing of the variation gives

$$[K]\{r\} + [C]\{\dot{r}\} + [M]\{\ddot{r}\} = \{R\} \tag{20-48}$$

For the case of relative motion in an earthquake, similar to that of Eq. (20-15),

$$[K]\{v\} + [C]\{\dot{v}\} + [M]\{\ddot{v}\} = -[M]\{\ddot{u}_g\} \tag{20-49}$$

Again, the number of degrees of freedom is reduced, and the matrices are smaller than those of Eq. (20-48).

It is possible to find the modal frequencies ω_i and mode shapes $\{\phi_i\}$ for the undamped system. From Eqs. (20-25) and (20-26) the mode shapes are orthogonal to $[M]$ and $[K]$. Most modal analyses of damped systems proceed from the assumption that they are orthogonal to $[C]$ as well. This is a significant assumption that is often glossed over and that is probably not true in many cases. Nonetheless, it leads to several useful conclusions. The assumption can be written as

$$\{\phi_i\}^T[C]\{\phi_j\} = \begin{cases} 2\omega_i\beta_i & i = j \\ 0 & i \neq j \end{cases} \tag{20-50}$$

where β_i, the ratio of damping to critical damping in mode i, is called the *critical damping ratio.*

By procedures identical to those that led to Eqs. (20-27) and (20-28) one obtains

$$\begin{bmatrix} \omega_1^2 & & & & \\ & \omega_2^2 & & 0 & \\ & & \ddots & & \\ & & \omega_i^2 & & \\ & 0 & & \ddots & \\ & & & & \omega_M^2 \end{bmatrix}\{p\} + \begin{bmatrix} 2\beta_1\omega_1 & & & & \\ & 2\beta_2\omega_2 & & 0 & \\ & & \ddots & & \\ & & & 2\beta_i\omega_i & \\ & 0 & & & \ddots & \\ & & & & & 2\beta_M\omega_M \end{bmatrix}\{\dot{p}\}$$

$$+ [I]\{\ddot{p}\} = -[\Phi]^T[M]\{\ddot{u}_g\} \tag{20-51}$$

and the individual uncoupled equations are

$$\omega_i^2 p_i + 2\beta_i\omega_i\dot{p}_i + \ddot{p}_i = -\{\phi_i\}^T[M]\{\ddot{u}_g\} \tag{20-52}$$

When the full system is used, the right-hand side will contain $-\{\phi_i\}^T\{R\}$.

Again, the solution can be effected by a number of means and the modal contributions superimposed to give the complete solution. The weakest link lies in the assumption that Eq. (20-50) is valid. It is also very difficult to postulate how the $[C]$ matrix is to be determined and by what means the material constants can be measured. Indeed, why should the modes be orthogonal with respect to $[C]$? In many cases, one can estimate values of damping for individual components, but how are they to be assembled and how are the values of modal damping to be found?

Damping Dependent on [K] and [M]

Let [C] be proportional to [M], as in Eq. (20-46):

$$[C] = s[M] \tag{20-53}$$

Then the orthogonality relations give

$$\beta_i = \frac{s}{2\omega_i} \tag{20-54}$$

Similarly, if [C] is proportional to [K]:

$$[C] = s[K] \tag{20-55}$$

and

$$\beta_i = \frac{s}{2}\omega_i \tag{20-56}$$

A generalization of these observations[2] is that if [C] can be related to [M] and [K] by

$$[C] = \sum_r s[K]([M]^{-1}[K])^r \tag{20-57}$$

then

$$\beta_i = \sum_r \frac{s}{2}(\omega_i)^{2r+1} \tag{20-58}$$

This means that if one knows the variation of β_i with modal frequency, one can compute a corresponding [C].

A more general method was developed independently by the second author and by Wilson and Penzien.[3] Since

$$[\Phi]^T[C][\Phi] = \begin{bmatrix} 2\beta_1\omega_1 & & & & \\ & 2\beta_2\omega_2 & & 0 & \\ & & \ddots & & \\ & & & 2\beta_i\omega_i & \\ & 0 & & & \ddots \\ & & & & & 2\beta_M\omega_M \end{bmatrix} \tag{20-59}$$

which can be abbreviated as

$$[\Phi]^T[C][\Phi] = [2\beta_i\omega_i] \tag{20-60}$$

and since

$$[\Phi]^T[M][\Phi] = [I] \tag{20-61}$$

it follows that

$$[\Phi]^T[C][\Phi] = [\Phi]^T[M][\Phi][2\beta_i\omega_i][\Phi]^T[M][\Phi] \tag{20-62}$$

Therefore

$$[C] = [M][\Phi][2\beta_i\omega_i][\Phi]^T[M] \tag{20-63}$$

will satisfy all the necessary conditions. For any arbitrary set of modal frequencies and damping ratios a damping matrix $[C]$ can be computed.

These approaches are useful only if one needs $[C]$ for the computations. In most practical cases it is much simpler, cheaper, and faster to solve the modal equations directly using the desired values of modal damping ratios instead of trying to find the $[C]$ that would yield those values of damping ratio. In the method of Eq. (20-57) one has to invert $[K]$ and in the method of Eq. (20-63) one has to compute first the $[\Phi]$, which is the major part of the modal superposition calculation in any case.

The difficulty with any of the above methods is that one seldom knows the modal damping ratios either. Instead the damping ratios appropriate to particular materials or to different regions of soil or rock may be known. The damping is thus determined element by element rather than mode by mode. It is usually not known in the form of viscous constants that can be used in equations like (20-46) or (20-47).

Three methods have evolved to evaluate the effects of damping when the damping is known only as a damping ratio for each element: Rayleigh damping, weighted modal damping, and complex formulations. The first two are described below; the last is the subject of the next section.

Rayleigh damping[4] starts from Eqs. (20-53) to (20-58). If the two types of proportional damping are combined,

$$[C] = a[M] + b[K] \tag{20-64}$$

and

$$\beta_i = \frac{a}{2\omega_i} + \frac{b\omega_i}{2} \tag{20-65}$$

As Fig. 20-2 shows, the addition of these two gives a curve that is reasonably flat over a limited range of frequencies. If ω_0 is the fundamental frequency of the system and β_s is the desired damping ratio for the range of frequencies, it is reasonable to make $\beta_i = \beta_s$ at ω_0. This can be achieved by setting

$$a = \beta_s \omega_0 \tag{20-66}$$

$$b = \frac{\beta_s}{\omega_0} \tag{20-67}$$

The approach is applied not to the system as a whole but to the individual elements. In other words, for a particular element, if $[m]$ and $[k]$ are the mass and stiffness matrices, and $[c]$ is the damping matrix,

$$[c] = \beta_s \omega_0 [m] + \frac{\beta_s}{\omega_0}[k] \tag{20-68}$$

where ω_0 is the fundamental frequency of the entire undamped system.

The Rayleigh damping method is reasonably simple to use and is intuitively plausible. Its major drawback is that the damping becomes very large for higher frequencies because of the contribution of the b term. Since neither Eq. (20-59) nor

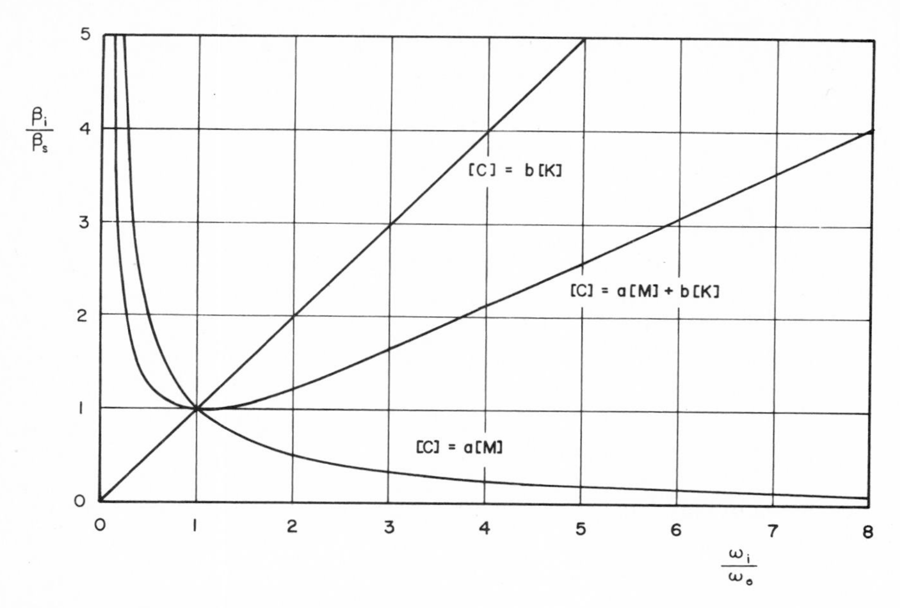

Figure 20-2 Frequency-dependent damping.

(20-64) is necessarily valid for the entire system, orthogonal modes may not exist and the solutions may not involve modal decomposition and superposition.

Weighted modal damping, described by Roesset et al.,[5] is a method of obtaining damping ratios β_i for different modes when the individual elements have different damping ratios. The maximum strain energy stored in the undamped system during an oscillation in mode i is $\frac{1}{2}\{\phi_i\}^T[K]\{\phi_i\}$. This must be the sum of the energy in the individual elements. If $[k_j]$ is the stiffness of element j and $\{\phi_{ij}\}$ is the portion of the modal displacement vector $\{\phi_j\}$ that involves nodes of element j,

$$\{\phi_i\}^T[K]\{\phi_i\} = \sum_{j=1}^{n} \{\phi_{ij}\}^T[k_j]\{\phi_{ij}\} \tag{20-69}$$

It is reasonable to evaluate the damping in mode i as the weighted average of the damping in all the elements and to use as a weighting function the proportion of the strain energy for mode i stored in each element. If β_i is the damping ratio in mode i, and β_j the damping ratio to be expected in element j,

$$\beta_i = \frac{\sum_{j=1}^{n} \{\phi_{ij}\}^T[k_j]\{\phi_{ij}\}\beta_j}{\{\phi_i\}^T[K]\{\phi_i\}} \tag{20-70}$$

The damping in each element β_j can be composed of portions that are independent of frequency to represent hysteretic effects and portions dependent on the frequency ω_j to represent viscous or radiation effects. For the viscous case the term β_j should be replaced by $\beta_j \omega_i/\omega_j$.

The calculation can be simplified considerably when a lumped mass matrix is used. Because

$$\{\phi_i\}^T[K]\{\phi_i\} = \omega_i^2\{\phi_i\}^T[M]\{\phi_i\} \tag{20-71}$$

the calculations can be done with the diagonal $[M]$ or $[m_j]$ for individual elements instead of the more fully populated $[K]$. In any case the denominator of Eq. (20-70) is ω_i^2.

This procedure allows one to use modal decomposition and superposition. However, when there is significant response at frequencies higher than the fundamental frequency, the method tends to underestimate the correct damping. For cases in which the response is primarily in the lower frequencies, it is an excellent method.

Complex Formulation with Damping

A complex formulation is especially useful when damping exists. Following the same procedures as in Sec. 20-1, one can find expressions for the Fourier transforms $\{U(\omega)\}$ of the displacements $\{r(t)\}$. For each frequency it must be true that

$$\{\dot{U}(\omega)\} = i\omega\{U(\omega)\} \tag{20-72}$$

$$\{\ddot{U}(\omega)\} = -\omega^2\{U(\omega)\} \tag{20-73}$$

For viscous damping, which permits a $[C]$ matrix to be computed in the time domain even if the specific values are not known, the transform of the basic equation (20-48) becomes

$$([K] + i\omega[C] - \omega^2[M])\{U(\omega)\} = \{S(\omega)\} \tag{20-74}$$

in which $\{S(\omega)\}$ is the transform of $\{R(t)\}$. Both $\{U(\omega)\}$ and $\{S(\omega)\}$ are complex. The combined effect of damping and stiffness can be expressed by a complex stiffness matrix $[K^*]$:

$$[K^*] = [K] + i\omega[C_V] \tag{20-75}$$

Here the subscript V indicates that the damping is viscous. The question is then how to obtain $[K^*]$ or $[C]$.

If for simplicity of exposition the system is assumed to have one degree of freedom, Eqs. (20-74) and (20-75) become

$$(k^* - \omega^2 m)U = S \tag{20-76}$$

$$k^* = k + i\omega c_V \tag{20-77}$$

c_V is a constant for linearly viscoelastic material. The resonant frequency of the undamped system is

$$\omega_0 = \sqrt{\frac{k}{m}} \tag{20-78}$$

and

$$c_V = 2\omega_0 \beta m \tag{20-79}$$

Then Eq. (20-76) becomes

$$k\left[\left(1 + 2i\beta\frac{\omega}{\omega_0}\right) - \left(\frac{\omega}{\omega_0}\right)^2\right]U = S \tag{20-80}$$

The complex stiffness thus shows that for a linearly viscoelastic system, the dissipation of energy into the viscous dashpot, which is represented by the imaginary terms in the complex stiffness, is proportional to the frequency of excitation. However, most experimental work on soils shows that the loss of energy or apparent damping is independent of the frequency. This is what would be expected of a frictional material; it would require a complex stiffness of the form

$$k^* = k(1 + 2i\beta) \tag{20-81}$$

or

$$k^* = k + ic_H \tag{20-82}$$

where the subscript H indicates hysteretic damping. This is called *structural* or *linear hysteretic damping.*

Linear hysteretic damping is a linearization of the nonlinear problem of energy loss dependent on displacement. Much confusion surrounds this procedure, and so it is well to emphasize its salient features:

1. It concerns the steady state response to a particular frequency of input. For more general input it can be used only in the frequency domain.
2. It does not apply for transient responses to arbitrary, time-dependent inputs unless the entire problem is first transformed into the frequency domain.
3. The *damping ratio*, β in Eq. (20-81), is chosen so that if the linear hysteretic system is excited at the resonant frequency of the undamped system ω_0 the response will be the same as a linearly viscoelastic one with a critical damping ratio β.
4. The imaginary part of the complex stiffness is independent of frequency.

Lysmer[6] has extended the formulation by considering the linearly viscoelastic system described by

$$m\ddot{r} + c\dot{r} + kr = e^{i\omega t} \tag{20-83}$$

and the linear hysteretic complex system described by

$$k^*r + m\ddot{r} = e^{i\omega t} \tag{20-84}$$

$$k^* = k_1 + k_2 i \tag{20-85}$$

and requiring that the amplitude of the response be the same for all values of ω. This leads to

$$k^* = k(1 - 2\beta^2 + 2i\beta\sqrt{1 - \beta^2}) \tag{20-86}$$

For small values of β Eqs. (20-81) and (20-86) are nearly equal. Figure 20-3 shows how the complex coefficient varies with increasing β. For $\beta < 0.30$ the two formulations are very close. For large β Eq. (20-81) describes simply an ever

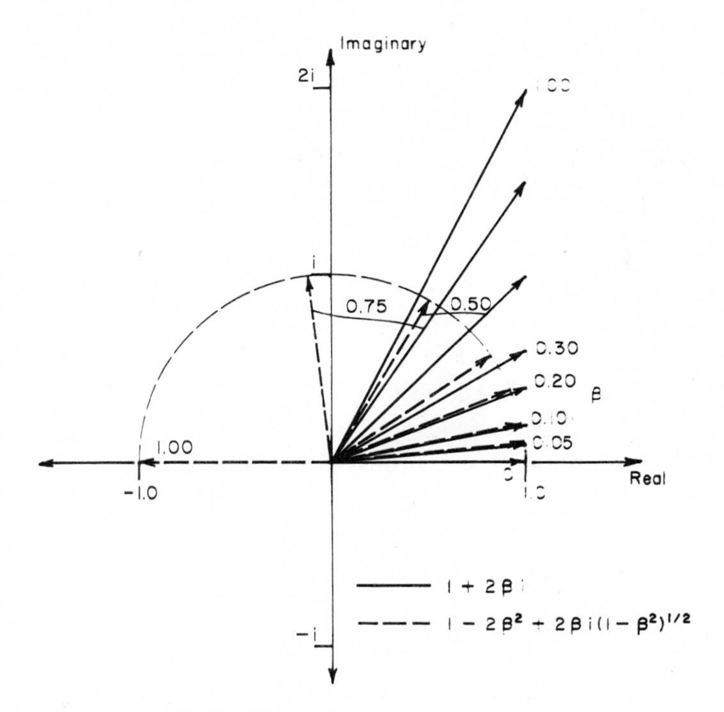

Figure 20-3 Complex damping factor.

increasing imaginary component, but Eq. (20-86) causes the real part to change as well and eventually leads to negative real values. At $\beta = 1$, Eq. (20-86) predicts a complex modulus equal to $-k$.

It is not clear which of the two formulations is to be preferred for $\beta > 0.30$. There are not adequate experimental data, and this is a range in which the linear approximation of a nonlinear system may break down anyway. Since the two formulations are nearly identical for $\beta < 0.30$, which is the expected range of β, the simpler form, $1 + 2i\beta$, would seem preferable.

For a system with both hysteretic and viscous damping, the complex stiffness would become

$$k^* = k + i(c_H + \omega c_V) \tag{20-87}$$

or

$$k^* = k(1 + 2i\beta_H + 2i\omega\beta_V) \tag{20-88}$$

When the system has more than one degree of freedom, the complex stiffness is computed by expressing the elastic coefficients in complex form. For example, the elasticity matrix for a linearly elastic material is $[C']$ as given in Eq. (2-49) or (2-50a). Then the complex elasticity matrix is

$$[C^*] = [C'](1 + 2i\beta_H + 2i\omega\beta_V) \tag{20-89}$$

For each element the complex stiffness is

$$[k^*] = \int_V [B]^T[C^*][B]\, dV \tag{20-90}$$

Here we have used $[C']$ for the $[C]$ in Chap. 2 in order to avoid confusion with the same symbol used for damping matrix.

The global stiffness is then assembled in exactly the same way as in a standard finite element problem, except that complex arithmetic must be used. The global equations are

$$[[K^*] - \omega^2[M]]\{U(\omega)\} = \{S(\omega)\} \tag{20-91}$$

which are identical in form to the results of Sec. 20-1, except that the stiffness is now complex. For the earthquake problem, in which $\{S(\omega)\} = \{0\}$, it is again true that the transfer functions for the acceleration and displacement formulation are the same when absolute motions are used and different when relative motions are used.

20-3 MARCHING SOLUTIONS

Marching solutions are numerical solutions in the time domain in which the solution is advanced by one discrete step at a time. Chapter 1 describes several such procedures. A great number of others have been developed for both special and general purposes, each having advantages and disadvantages. There is no one ideal method, and the engineer who wishes to use a marching procedure must choose one by weighing several factors including cost, efficiency, numerical accuracy, stability of the solution, treatment of boundary and initial conditions, compatibility of the method with the form of the input, desired completeness of the solution, and so on. For earthquake engineering, methods based on Galerkin's techniques, on Runge-Kutta techniques, and on various assumptions about the variation of acceleration or velocity during an increment of time have been most popular. To illustrate the approach, methods in the last category are described in this section.

Newmark[7] proposed that a Taylor series approximation to the solution of the dynamic problem in time together with various numerical integration schemes would lead to formulas of the form

$$r_{i+1} = r_i + \dot{r}_i\, \Delta t + [(\tfrac{1}{2} - \beta)\ddot{r}_i + \beta\ddot{r}_{i+1}]\, \Delta t^2 \tag{20-92}$$

$$\dot{r}_{i+1} = \dot{r}_i + [(1 - \gamma)\ddot{r}_i + \gamma\ddot{r}_{i+1}]\, \Delta t \tag{20-93}$$

In these equations r_i, $\dot{r}_i$, and $\ddot{r}_i$ refer to the displacement, velocity, and acceleration at time t_i, and r_{i+1}, $\dot{r}_{i+1}$, and $\ddot{r}_{i+1}$ refer to the same variables at time t_{i+1}. The difference between t_{i+1} and t_i is Δt. Different values of β and γ will give different specific formulas.

The basic equation for the linearly viscoelastic material is Eq. (20-48). It must be satisfied identically at t_{i+1}, or

$$[M]\{\ddot{r}\}_{i+1} + [C]\{\dot{r}\}_{i+1} + [K]\{r\}_{i+1} = \{R\}_{i+1} \tag{20-94}$$

Substituting Eqs. (20-92) and (20-93) and rearranging terms gives

$$\begin{aligned}([M] + [C]\gamma\,\Delta t + [K]\beta\,\Delta t^2)\{\ddot{r}\}_{i+1} &= \{R\}_{i+1} \\ &\quad - ([C](1-\gamma)\,\Delta t + [K](\tfrac{1}{2}-\beta)\,\Delta t^2)\{\ddot{r}\}_i \\ &\quad - ([C] + [K]\,\Delta t)\{\dot{r}\}_i - [K]\{r\}_i\end{aligned} \tag{20-95}$$

Thus, the accelerations at t_{i+1} can be computed from the known forces $\{R\}_{i+1}$ and the accelerations, velocities, and displacements at t_i. The following algorithm results:

1. The matrices $[K]$, $[C]$, and $[M]$ are computed and reduced to account for known boundary conditions. For earthquake problems with uniform motion on the boundaries, this results in the matrices of Eq. (20-49).
2. The constants β and γ are chosen, as well as Δt.
3. Since the initial condition is usually $\{\dot{r}\}_0 = 0$ and $\{r\}_0 = 0$, the initial accelerations can be found by solving

$$[M]\{\ddot{r}\}_0 = \{R\}_0 \tag{20-96}$$

4. The coefficient matrices in parentheses in Eq. (20-95) are computed. For linear materials they do not change during the solution, so that a single calculation suffices.
5. The matrix $([M] + [C]\gamma\,\Delta t + [K]\beta\,\Delta t^2)$ is inverted or triangularized to minimize future calculations.
6. From the initial conditions, $\{R\}_1$, and $\{\ddot{r}\}_0$, Eq. (20-95) is solved for $\{\ddot{r}\}_1$.
7. Equations (20-92) and (20-93) are used to find $\{\dot{r}\}_1$ and $\{r\}_1$.
8. Steps 6 and 7 are repeated for each subsequent time to find the solution from the conditions at the end of the previous time step and the load at the current time. The process continues as long as necessary to give the full time history of the solution.

Other algorithms are possible. One could, for example, solve for displacement first and then compute velocity and acceleration. Several values of β and γ have been used. If it is assumed that the acceleration is constant during the step from t_i to t_{i+1} and that it is equal to the acceleration at t_i, then $\gamma = 0$ and $\beta = 0$. This is Euler's method. If the acceleration varies linearly from t_i to t_{i+1}, $\gamma = \frac{1}{2}$ and $\beta = \frac{1}{6}$. If the acceleration is assumed constant in the time interval and equal to the average of the values computed at t_i and t_{i+1}, $\gamma = \frac{1}{2}$ and $\beta = \frac{1}{4}$. The last two assumptions are widely used (Chap. 1).

Two important questions in the selection of a method are those concerning accuracy and stability. Accuracy refers to how well the numerical solution matches the exact continuous solution. Stability refers to whether extraneous solutions are introduced in such a way that they increase rather than decay and thus come to dominate the results. Usually there is an upper limit to Δt that is necessary to guarantee stability, and the value of that limit depends on the type of element stiffness and mass matrices as well as on β and γ. For example, it is generally accepted that for $\gamma = \frac{1}{2}$ and $\beta = \frac{1}{6}$, Δt must be no larger than about one-eighth the shortest period contained in the system. For $\gamma = \frac{1}{2}$ and $\beta = \frac{1}{4}$, the linear problem is unconditionally stable.

It must be remembered that stability does not guarantee accuracy or vice versa. Many methods have been proposed that are stable for large values of Δt and are thus attractive for earthquake analyses in which long lasting inputs are used and short Δt's lead to very lengthy computations. These methods usually introduce so much extraneous damping that the accuracy of the solution is seriously impaired. Indeed, the fact that the solution does not oscillate wildly, producing overflows in the computer, gives an unfortunately misleading impression of the validity of the results. The subject of mathematical properties of numerical procedures is highly relevant for establishing ranges of applicability, limitations, and reliability of numerical schemes. This is particularly true of time-dependent problems, where the behavior of the numerical solution is influenced by spatial and temporal meshes, physical characteristics, and geometrical properties of the problem. Some of these aspects for time-dependent problems are discussed in Refs. 8 to 10.

If one wishes to reproduce the response at a frequency ω and corresponding period T, it is necessary to use a sufficiently small t to permit the oscillations to be described. Thus requirements that Δt be less than some fraction of T exist first so that the time history can be described. This is true regardless of considerations of stability.

One of the most attractive uses of marching methods is to solve for the individual modal responses described by Eq. (20-52). The coefficients in Eq. (20-95) become scalar constants rather than matrices, and the numerical work is correspondingly simplified. If one writes

$$f_{j,\,i+1} = -\{\phi_j\}^T\{R\}_{i+1} = -\{\phi_j\}^T[M]\{\ddot{u}_g\}_{i+1} \tag{20-97}$$

the basic recurrence relation for mode j is

$$\begin{aligned}(1 + 2\omega_j\beta_j\gamma\,\Delta t + \omega_j^2\beta\,\Delta t^2)\ddot{p}_{j,\,i+1}& \\ = f_{j,\,i+1} - [2\omega_j\beta_j(1-\gamma)\,\Delta t &+ \omega_j^2(\tfrac{1}{2} - \beta)\,\Delta t^2]\ddot{p}_{j,\,i} \\ &- (2\omega_j\beta_j + \omega_j^2\,\Delta t)\dot{p}_{j,\,i} - \omega_j^2 p_{j,\,i}\end{aligned} \tag{20-98}$$

This is easily solved step by step. For higher frequencies smaller Δt's can be used to maintain stability.

Nonlinear material properties present another condition in which numerical marching procedures are useful, and when the properties are specified directly rather than as iterative averages, such methods are virtually the only ones available. In Eq. (20-95) one could modify the stiffness terms at each step. This would be expensive. A better procedure is to keep the coefficient matrix constant and to assemble it only once but to compute the nonlinear terms and to use them as additional loads. This is in effect an initial-stress or initial-strain method. The details of the calculation will depend on the particular form of nonlinearity employed. Usually $\gamma = \frac{1}{2}$ and $\beta = \frac{1}{4}$ have been used, as well as fourth-order Runge-Kutta methods. Of course, accuracy and stability will be problems as in the linear case.

20-4 NONLINEAR MATERIALS

In many cases the magnitude of the strains is such that one must consider nonlinear material properties even though the displacements are still small enough to permit use of infinitesimal strain theory. One procedure is to use direct step-by-step integration in time, as described in the previous section. This has been used especially for analysis of blast effects. The second approach is to use an iterative linear solution to simulate the nonlinear effects. This has been described in Chap. 19 and is today most widely used in earthquake problems.

The iterative linear approach starts with a set of equations or curves that relate secant moduli and critical damping ratios to the level of strain. An initial assumption of modulus and damping is made for each element, and the problem is solved as a linear problem. The damping is usually taken as linear hysteretic rather than viscous. The strains are computed for each element, and updated values of modulus and damping are found for each element.[11] The new linear problem is solved, and new strains are evaluated. The process continues until the change in strains on modulus and damping is sufficiently small between two solutions.

For the one-dimensional wave propagation problem it is clear that the strain to be found is the shear strain in a vertical section. Usually an average value or representative value is computed as about 65 percent of the peak strain. For two- and three-dimensional problems the magnitude and orientation of all components of strain vary from time to time. It is not clear what should be used as the representative value. One procedure is to find the principal shear strain for each element at each step in time, then to find the maximum value of the principal shear strain over all times and to use 65 percent of this as the representative value for the element. This requires a great deal of computing effort. Kausel et al.[12] have shown that a more efficient procedure is to evaluate the root-mean-square values of the components of strain and find the principal shear strain from these components. The calculation can be done in the frequency domain so as to avoid many inverse Fourier transforms. It is much more efficient than the previous method. Other techniques could also be used, for it is not yet certain from experimental evidence what strain is the best determinant of the reduced modulus and damping.

Another problem in two- and three-dimensional analyses is that it is not certain whether the Young's modulus E should be determined by the strain level or the shear modulus G. In other words, is Poisson's ratio or the bulk modulus to be kept constant? Variation of E is attractive because it can be factored out of the stiffness matrix for the element. Then the element stiffness matrix can be computed once and simply multiplied by a different number as a function of strain level. Since the shear modulus cannot be factored out, the stiffness must be separated into two parts, one containing the bulk modulus and one containing the shear modulus. Then the shear portion is modified for strain level, and the two are added up to give the stiffness matrix for the element as a function of strain. This is more expensive than using E. However, the best physical understanding is that it is the shear stiffness that is affected by shear strains. Reducing E reduces the compressive stiffness as well. Therefore, modification of G, although more time consuming, is to be preferred.

The nonlinearities in two-dimensional earthquake problems can be divided into two categories, primary and secondary. The primary nonlinearities are those resulting from vertical propagation of the shear waves without the structure present. They can be studied by the one-dimensional techniques of the previous chapter. The secondary nonlinearities are the additional effects caused by the presence of the structure. Kausel et al.[12] have shown that if the modulus and damping are computed on the basis of primary nonlinearities only, the resulting behavior of the structure is very nearly the same as it is when both primary and secondary nonlinearities are considered. Thus, a great saving in time and money can be effected by evaluating the equivalent linear properties from one-dimensional analyses and ignoring further nonlinear effects in the two- or three-dimensional analyses. Since the secondary nonlinearities do have an effect on some local parts of the soil, this approach is not recommended if detailed behavior of the soil is to be studied. Usually it is the structural response that is of interest.

It must be remembered that the iterative methods are attempts to apply linear solutions to nonlinear problems. They may not converge, and they often do not converge uniquely. They may also not reproduce the complete nonlinear case accurately. The engineer should bear these limitations in mind when he uses iterative linear solutions.

20-5 BOUNDARY CONDITIONS

The discretization of a continuum through the use of finite elements or finite differences requires the existence of a finite domain with well defined boundaries, where conditions are specified for forces and displacements. If these boundaries do not exist naturally but are created artificially, it is necessary to determine appropriate boundary conditions that will simulate the physical behavior of the actual problem.

The effect of a bottom boundary on one-dimensional wave propagation was discussed in Chap. 19. It was shown that for a horizontally stratified soil deposit with shear waves propagating vertically, the existence of an underlying half-space could be reproduced in a discrete model by placing at the bottom a viscous dashpot and specifying an equivalent force rather than a displacement. This type of solution can easily be extended to the case of SH waves traveling at a specified angle. The same approach can be used to determine a bottom boundary matrix for a fixed train of SV and P waves.

For these simple cases, when the geometry is one-dimensional and only one train of plane waves is considered, appropriate conditions for the lateral boundaries can also be obtained easily. For example,

1. For SH waves with normal incidence the lateral boundary nodes should be left free.
2. For SV waves with normal incidence, the nodes on the lateral boundary should have free displacements in the horizontal direction and completely restrained displacements in the vertical direction.
3. For P waves with normal incidence the nodes on the lateral boundary should have free displacements in the vertical direction and completely restrained displacements in the horizontal direction.
4. For SH, SV, or P waves with a specified nonzero angle of incidence the nodes on the lateral boundary should have lumped dashpots or consistent damping matrices with properties that are functions of the angle.

The first three are usually referred to as *simple boundaries*. The last one involves viscous boundaries. Two possible alternatives, valid for all cases, are (1) to leave the boundary nodes free but to apply to them forces resulting from the stresses given by analytical solutions and (2) to fix the boundary nodes but to apply to them the displacements resulting from an analytical solution.

For more general two-dimensional situations, when the geometry can no longer be described properly by a horizontally stratified medium or when excitation does not consist of a single train of waves, these simple solutions are no longer rigorous. They are still used, however, as approximations. Their validity depends on the type of problem, the distance from the region of interest to the boundaries, and the amount of internal hysteretic damping in the material.

A case that is often encountered is one in which the soil can be divided into three regions: an irregular central core region, where forces are applied or disturbances can occur, and two horizontally layered regions extending to the left and to the right, respectively (Fig. 20-4). For this situation Waas[13] developed a boundary matrix that reproduces the effect of the lateral regions in a manner consistent with the finite element discretization of the core.

Figure 20-5 shows a two-dimensional plane strain problem. The core region is discretized with finite elements with a linear displacement expansion for simplicity, but other expansions could be used. The displacements in the horizontally stratified lateral regions can be expressed for the jth layer as

$$u_{xj} = \left[\left(1 - \frac{z_j}{h_j}\right)u_j + \frac{z_j}{h_j}u_{j+1}\right]f(x) \tag{20-99}$$

$$u_{zj} = \left[\left(1 - \frac{z_j}{h_j}\right)w_j + \frac{z_j}{h_j}w_{i+1}\right]f(x) \tag{20-100}$$

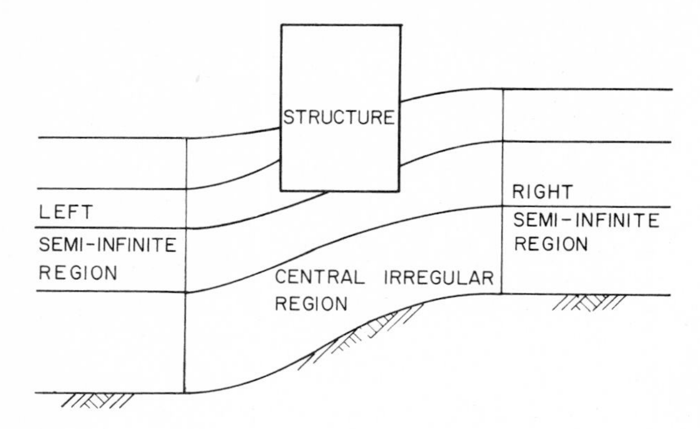

Figure 20-4 Three regions of dynamic problem.

The variation with z is consistent with that in the core region, and the variation with x is given by $f(x)$. If it is assumed that

$$f(x) = e^{-i\kappa x} \tag{20-101}$$

then

$$\epsilon_x = -i\kappa\left[\left(1 - \frac{z_j}{h_j}\right)u_j + \frac{z_j}{h_j}u_{j+1}\right]f(x)$$

$$\epsilon_z = \frac{1}{h_j}(-w_j + w_{j+1})f(x) \tag{20-102}$$

$$\gamma_{xz} = \left[\frac{1}{h_j}(-u_j + u_{j+1}) - i\kappa\right]\left(1 - \frac{z_j}{h_j}\right)w_j + \frac{z_j}{h_j}w_{j+1}\bigg]f(x)$$

A stiffness matrix can then be derived for the jth layer in the form

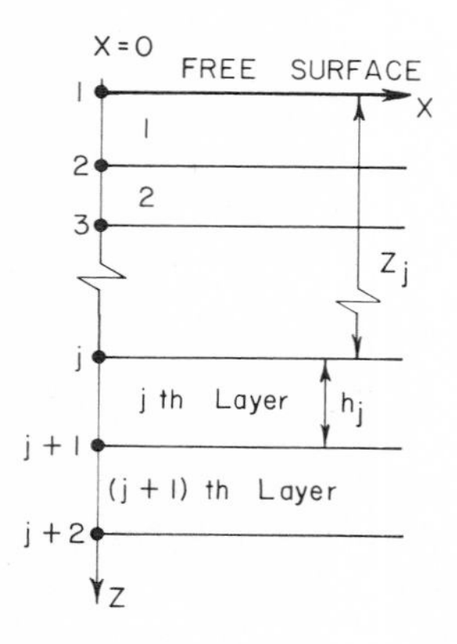

Figure 20-5 Layered boundary region (right side shown).

$$[k]_j = [k_0]_j + i\kappa[k_1]_j + \kappa^2[k_2]_j \tag{20-103}$$

where
$$[k_0] = \frac{1}{h_j}\begin{bmatrix} G & 0 & -G & 0 \\ 0 & \lambda + 2G & 0 & -(\lambda + 2G) \\ -G & 0 & G & 0 \\ 0 & -(\lambda + 2G) & 0 & \lambda + 2G \end{bmatrix} \tag{20-104}$$

$$[k_1] = [k_1]' - [k_1]'^T \tag{20-105a}$$

$$[k_1]' = \frac{1}{2}\begin{bmatrix} 0 & G & 0 & G \\ \lambda & 0 & \lambda & 0 \\ 0 & -G & 0 & -G \\ -\lambda & 0 & -\lambda & 0 \end{bmatrix}$$

$$[k_1] = \frac{1}{2}\begin{bmatrix} 0 & G - \lambda & 0 & G + \lambda \\ \lambda - G & 0 & \lambda + G & 0 \\ 0 & -\lambda - G & 0 & \lambda - G \\ -\lambda - G & 0 & G - \lambda & 0 \end{bmatrix} \tag{20-105b}$$

$$[k_2] = \frac{h}{6}\begin{bmatrix} 2(\lambda + 2G) & 0 & \lambda + 2G & 0 \\ 0 & 2G & 0 & G \\ \lambda + 2G & 0 & 2(\lambda + 2G) & 0 \\ 0 & G & 0 & 2G \end{bmatrix} \tag{20-106}$$

This stiffness matrix is a function of κ, the wave number, but it also has a complex term even for real soil properties, i.e., for no internal material damping. The wave number is $2\pi/L$, where L is the wavelength. One can interpret $[k_1]$ as a damping matrix and $[k_2]$ as part of the inertia matrix because κ is proportional to ω.

The consistent mass matrix is

$$[m]_j = \frac{\rho_j h_j}{6}\begin{bmatrix} 2 & 0 & 1 & 0 \\ 0 & 2 & 0 & 1 \\ 1 & 0 & 2 & 0 \\ 0 & 1 & 0 & 2 \end{bmatrix} \tag{20-107}$$

It is then possible to form, for the entire layered region, global stiffness matrices $[K_0]$, $[K_1]$, and $[K_2]$ and mass matrix $[M]$, all assembled in the same way from the matrices of the individual layers:

$$[K_0] = \begin{bmatrix} [k_0]_1 & -[k_0]_1 & & & \\ -[k_0]_1 & [k_0]_1 + [k_0]_2 & -[k_0]_2 & & \\ & -[k_0]_2 & [k_0]_2 + [k_0]_3 & -[k_0]_3 & \\ & & & & \ddots \end{bmatrix} \qquad (20\text{-}108)$$

If no external forces are applied at the top or at the bottom of the layered region, the dynamic equilibrium equations for a steady state periodic motion at frequency ω become

$$(\kappa^2[K_2] + i\kappa[K_1] + [K_0] - \omega^2[M])\{U\} = \{0\} \qquad (20\text{-}109)$$

For a given ω, this defines an eigenvalue problem. The eigenvalues κ are the possible wave numbers, and the eigenvectors $\{\phi\}$ are the corresponding mode shapes for the waves. If there are n layers, the matrices of Eq. (20-109) are of order $2n \times 2n$ and there will be $4n$ eigenvalues and eigenvectors. Of these, $2n$ correspond to waves whose amplitude decays with increasing x, and the other $2n$ correspond to waves decaying in the negative x direction. The former is of interest for the right lateral boundary, the latter for the left lateral one.

If one selects the $2n$ values of κ corresponding to the boundary of interest and defines $[\kappa]$ as a diagonal matrix with κ_i as the ith diagonal element, $[\Phi]$ as the modal matrix with $\{\phi_i\}$ as the ith column, and $\{\Gamma\}$ as a vector with Γ_i as the unknown participation factor of the ith mode, the displacements and forces at the nodes of the boundary can be expressed as

$$\{U\} = [\Phi]\{\Gamma\} \qquad (20\text{-}110)$$

$$\{P\} = -i[K_2][\Phi][\kappa]\{\Gamma\} - [K_1]'^T\{U\} \qquad (20\text{-}111)$$

From the first equation

$$\{\Gamma\} = [\Phi]^{-1}\{U\} \qquad (20\text{-}112a)$$

and therefore

$$\{P\} = -(i[K_2][\Phi][\kappa][\Phi]^{-1} + [K_1]'^T)\{U\} = -[R]\{U\} \qquad (20\text{-}112b)$$

Here $[R]$ is a dynamic matrix relating forces applied at the boundary of the layered region to the displacements of this boundary. The forces exerted on the boundary of the core region by the layered region would be the opposite. Therefore, it is sufficient to add the matrix $[R]$ to the part of the dynamic stiffness matrix of the core that corresponds to the boundary nodes. This accounts for the effect of the lateral soil in a consistent and accurate way.

The formulation of the consistent boundary matrix can be obtained in an alternate way that helps one to visualize the physical meaning of the terms. If it is assumed that the layered region is divided into columns of rectangular finite elements, all of the same width, the dynamic equilibrium equations of one column would be given by

$$[K_{11}]\{U_A\} + [K_{12}]\{U_B\} = \{S_A\} \qquad [K_{21}]\{U_A\} + [K_{22}]\{U_B\} = \{S_B\} \qquad (20\text{-}113)$$

where $\{U_A\}$ represents the displacements of the nodes on the left boundary of the column and $\{U_B\}$ the displacements on the right boundary. $\{S_A\}$ and $\{S_B\}$ are the corresponding nodal forces, and $[K_{11}]$, $[K_{12}]$, $[K_{21}]$, and $[K_{22}]$ are the dynamic submatrices of the column resulting from $[K] - \omega^2[M]$.

If one now considers several columns of finite elements and uses subscripts $n - 1$, n, and $n + 1$ to indicate forces or displacements along particular vertical lines, the dynamic equilibrium of the nodes along line n is

$$[K_{21}]\{U\}_{n-1} + ([K_{11}] + [K_{22}])\{U\}_k + [K_{12}]\{U\}_{n+1} = \{0\} \qquad (20\text{-}114)$$

if there are no external forces.

This is a system of recurrence equations that can be expanded by taking even wider systems, each twice the size of the previous one. The solution can be expressed in the form

$$\{U\}_n = \sum_i a_i(r_i)^n\{x\}_i \qquad (20\text{-}115a)$$

where r_i are the roots, or eigenvalues, of

$$(r^2[K_{12}] + r([K_{11}] + [K_{22}]) + [K_{21}])\{U\} = \{0\} \qquad (20\text{-}115b)$$

and $\{x_i\}$ are the eigenvectors.

It can be shown that this quadratic eigenvalue problem is the discrete equivalent of that represented in Eqs. (20-109). If the matrices are $2n \times 2n$, there will again be $4n$ eigenvalues, $2n$ for the left boundary and $2n$ for the right. (If r_i is an eigenvalue, so is $1/r_i$.) The coefficients a_i in Eq. (20-115a) are the equivalents of the participation factors Γ_i in the continuous formulation.

In particular, if $\{U\}_0$ represents the displacements of the boundary, and $\{S_0\}$ the corresponding forces,

$$\{S\}_0 = [K_{11}]\{U_0\} + [K_{12}]\{U\}_1 \qquad (20\text{-}116)$$

and

$$\{U\}_1 = [\Phi]\{r\}\{A\} \qquad (20\text{-}117)$$

$$\{U\}_0 = [\Phi]\{A\} \qquad (20\text{-}118)$$

where $[\Phi]$ = modal matrix with $\{X\}_i$ as ith column
$[r]$ = diagonal matrix with r_i as ith diagonal element
$\{A\}$ = vector with a_i as ith term

Then

$$\{A\} = [\Phi]^{-1}\{U\}_0 \qquad (20\text{-}119)$$

$$\{U\}_1 = [\Phi][r][\Phi]^{-1}\{U\}_0 \qquad (20\text{-}120)$$

and

$$\{S\}_0 = ([K_{11}] + [K_{12}][Q][r][Q]^{-1})\{U\}_0 = -[R]\{U_0\} \qquad (20\text{-}121)$$

The determination of the consistent boundary matrix $[R]$ by either of these two methods requires the solution of a quadratic eigenvalue problem. Because of the special form of the matrices involved, the solution of this problem is relatively simple and not more expensive than the solution of a linear eigenvalue problem with $4n$ degrees of freedom.

The original formulation of the consistent boundary by Waas[13] was applicable to the case where the bottom of the soil deposit is fixed and the excitation comes from the top. Chang-Liang[14] extended it to the case where the excitation is provided by a train of waves coming from the bottom. If then $\{U\}$ is the total displacement of the nodes on the lateral boundary, $\{U\}_{1D}$ the corresponding displacements from the one-dimensional solution, $\{U\}_2$ the increments due to two-dimensional effects, and $\{P\}$, $\{P\}_{1D}$, and $\{P\}_2$ the corresponding forces,

$$\{P\}_2 = -[R]\{U\}_2 = [R]\{U\}_{1D} - [R]\{U\} \tag{20-122}$$

$$\{P\} = \{P\}_{1D} + \{P\}_2 = \{P\}_{1D} + [R]\{U\}_{1D} - [R]\{U\} \tag{20-123}$$

For the equations of motion of the boundary points resulting from the finite element formulation of the core region, one would apply at these nodes fictitious forces equal to $\{P\}_{1D} + [R]\{U\}_{1D}$. The last terms in Eq. (20-123) are passed to the other side of the equations by adding the boundary matrix $[R]$ to the part of the total dynamic matrix that corresponds to the boundary nodes. Kausel[15] has extended the same formulation to the case of cylindrical geometries.

20-6 PRACTICAL CONSIDERATIONS

It was pointed out earlier that three types of lateral boundaries are normally used when solving two-dimensional dynamic problems with finite elements:

1. Elementary boundaries, where forces, displacements, or a combination of forces and displacements are specified
2. Viscous boundaries, where viscous dashpots with constant properties or with variable properties based on a specific type of waves are placed at the boundaries
3. Consistent boundaries, where a frequency-dependent boundary stiffness matrix is obtained by solving the wave propagation problem in an elastic layered system or by assuming that equal columns of finite elements extend to infinity

Two types of solutions are used for the bottom boundary:

1. A fixed boundary where displacements are specified either by setting them to zero or by imposing a base motion. This corresponds to the case of a soil stratum resting on a much stiffer rock that can be assumed rigid. When the excitation is in the form of waves coming from the bottom, the motion specified may be that computed from a one-dimensional solution allowing for an underlying elastic half-space. In this case the rigid bottom applies only to the two-dimensional disturbance.

2. A viscous boundary with constant dashpots corresponding to the one-dimensional case so as to reproduce a half-space or a very deep stratum arbitrarily chopped off at a finite depth. This approach will again furnish the exact boundary conditions for a one-dimensional case, but it will be only approximate for two-dimensional effects.

Of the three types of lateral boundaries, the consistent boundary is by far superior to the others with respect to accuracy. It provides results in excellent agreement with analytical solutions where these exist. While it requires the solution of a quadratic eigenvalue problem, this computational expense is compensated by the fact that the spatial domain needed is only the core region. There is a substantial reduction in the number of degrees of freedom. The major limitation of this boundary is that it is properly defined in the frequency domain alone and therefore the dynamic problem must be solved in the frequency domain. It is also strictly applicable to linear problems only. Nonlinear soil behavior must be simulated by an iterative linear procedure described earlier. The validity of such a procedure has not been demonstrated for two-dimensional cases.

Elementary and viscous boundaries must be placed some distance from the region of interest, and the number of elements and degrees of freedom are correspondingly increased. This is needed to avoid the so-called *box effect*. The distance is a function of the amount of internal damping in the soil, which is itself a function of the level of excitation since it arises from nonlinear hysteretic behavior, and of the type of problem and results desired. For a layer of soil resting on a rigid or much stiffer base there is no lateral radiation below the fundamental frequency of the layer. Therefore, for frequencies below this value it makes no difference what type of lateral boundary is used. An accurate determination of the stiffness of the soil for higher frequencies, e.g., for the design of the foundation of a rotating machine, may require that the distance from the region of interest to the lateral boundary be 10 to 20 times the radius of the foundation for an internal damping of 5 percent of critical and 5 to 10 times for 20 percent damping. On the other hand, the response of a soil and structure system to an earthquake excitation with a broad frequency spectrum is less sensitive to errors in the soil stiffness in the higher frequency range and is therefore less affected by lateral boundaries. Distances about half those mentioned may be sufficient for this case. It does not appear that viscous boundaries offer a significant reduction in these distances over elemental boundaries.

In some cases horizontally elongated elements have been used outside the core region to place the boundary at a sufficient distance while a relatively small number of degrees of freedom are retained. Using these elements near the foundation will distort the rocking stiffness considerably and must be avoided when rocking is important.

Either of the two approaches used for the bottom boundary will introduce errors in the disturbance from the one-dimensional solution caused by two-dimensional effects. Therefore the bottom boundary must be placed at a sufficient depth, which is again a function of the amount of internal damping as well as the characteristics of the soil and deposit and the variation of properties with depth. To determine the rocking stiffness of a foundation a depth of two to four radii is sufficient to reproduce the effects of a linearly elastic, homogeneous half-space on a deep soil deposit. To determine the horizontal or vertical stiffness under the same circumstances a depth of eight radii or more should be used. Of course, these limitations do not apply if there is a clear physical discontinuity at a shallow depth between the soil and an underlying stiffer material.

20-7 ADDITIONAL CONSIDERATIONS

The preceding descriptions have concentrated on finite element analyses with particular emphasis on linear methods appropriate for earthquake problems. When problems involving blast or foundation vibrations are to be solved, some modification of approach may be necessary. Christian[16] presents the chart of Fig. 20-6, illustrating the relations between the various options. The current state of the art of earthquake analysis emphasizes the use of the frequency domain, and this is especially limiting when nonlinearities are important.

Nonlinearities can involve material nonlinearity as well as large strains or displacements. Something has been said above about direct use of nonlinear stress-strain relations, but one additional special case involves *interface elements.*

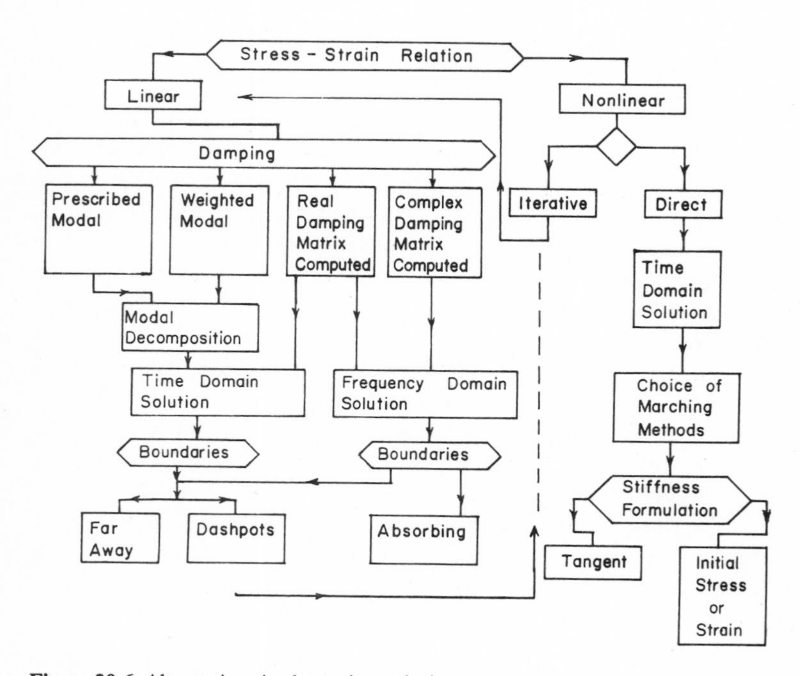

Figure 20-6 Alternatives in dynamic analysis.

When it is possible for slippage to occur along a discrete plane, it may be necessary to include this feature in the finite element model, and techniques for doing so in static problems have already been described in this book. The same approach can be used in dynamic problems, e.g., in the case of a structure slipping on its foundation. Although the idea is simple, there are many tricks and subtleties in application. In particular, the effects of the interface opening and closing with the resulting impact can be difficult to handle. The analysis must be done in the time domain.

Geometric nonlinearities can occur due to large strains or finite displacements. It is then necessary to reformulate the entire problem to incorporate geometric stiffness and the effects of changes in shape on the element stiffnesses. This is beyond the scope of the present chapter. The cases arises primarily in severe blast problems, which often have short times of loading. Of course, the calculations must be done in the time domain.

20-8 EXAMPLES

Figure 20-7 shows the finite element mesh used for an analysis of the dynamic behavior of a dam under the influence of an earthquake. The analysis was done in 1970 using techniques of modal analysis and superposition. The modulus and damping were assumed constant across the section, but several runs were made to obtain properties compatible with the average strain level. The earthquake was a small one with a peak acceleration of $0.12g$ at the level of the bottom of the cross section.

The analysis produced many results, and only a few are shown here. Figure 20-8 shows the first two mode shapes. It can be seen that the first mode involves primarily a shear deformation and the second mode primarily vertical compression and extension. The other modes show combinations and higher orders of these two basic phenomena. Seven modes were adequate to represent the behavior, even though there are a total of 112 modes.

Figure 20-9 is a plot of the maximum values of the principal shear stress. The maxima do not all occur at the same time, but they are within 0.10 s of each other.

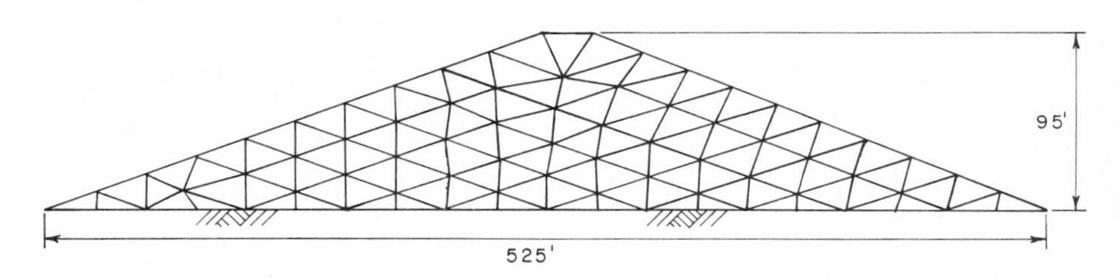

Figure 20-7 Finite element mesh, section on rock.

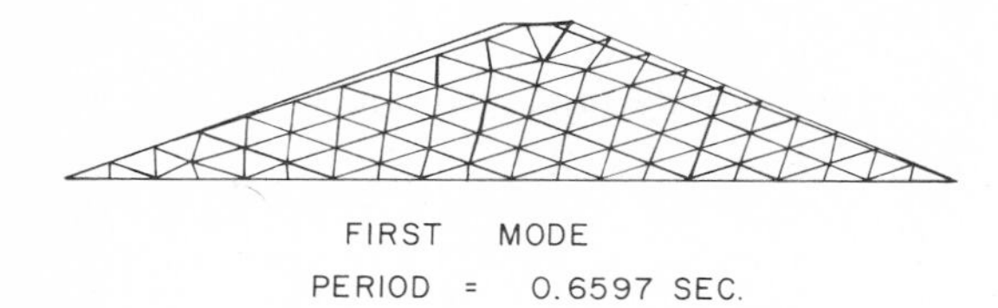

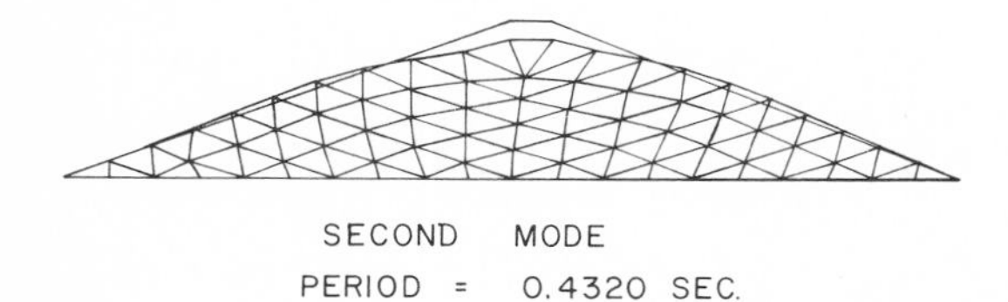

Figure 20-8 Mode shapes.

The excitation is the north-south accelerogram of the 1940 El Centro earthquake normalized to $0.12g$. The peak acceleration at the crest was more than twice the input peak of $0.12g$, but an average peak acceleration on a significant mass of the dam was about $0.18g$.

The above example did not require the special treatment of lateral boundaries because there were none. Figure 20-10 illustrates the mesh required for another cross section of the same dam, which had an underlying soil layer. The lateral boundaries were placed 12 depths away from the central region.

A more recent analysis by Kausel and Roesset[17] uses the mesh of Fig. 20-11 to represent an axially symmetric nuclear reactor containment. The lateral boundary is placed directly next to the structure with a great saving in elements and nodes. An artificial earthquake is specified at the free surface with a peak acceleration of $0.125g$ and a response spectrum for 1 percent damping, as shown in Fig. 20-12. Figure 20-13 shows the response spectra at the top of the dome

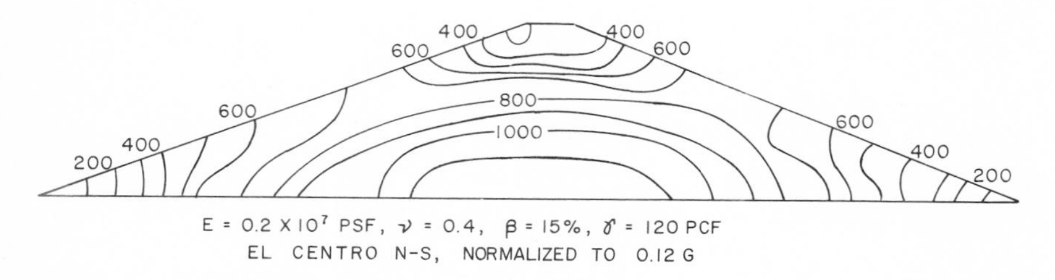

Figure 20-9 Maximum shear stresses (pounds per square foot).

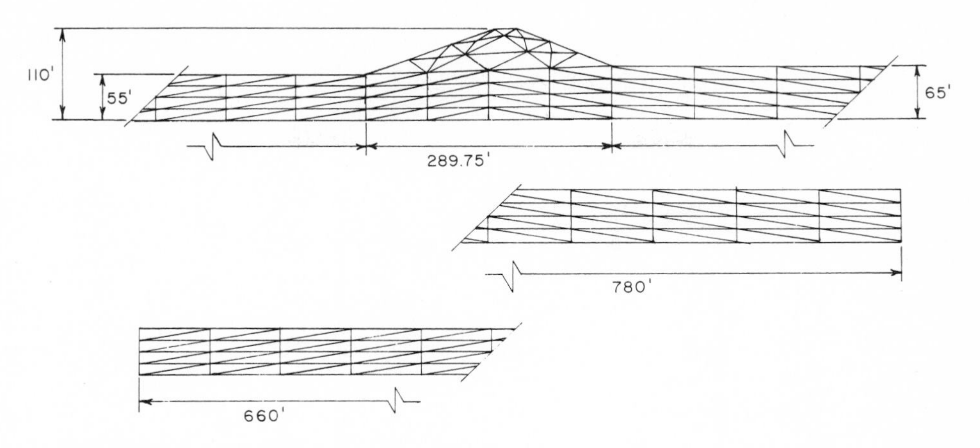

Figure 20-10 Finite element mesh, section on soil.

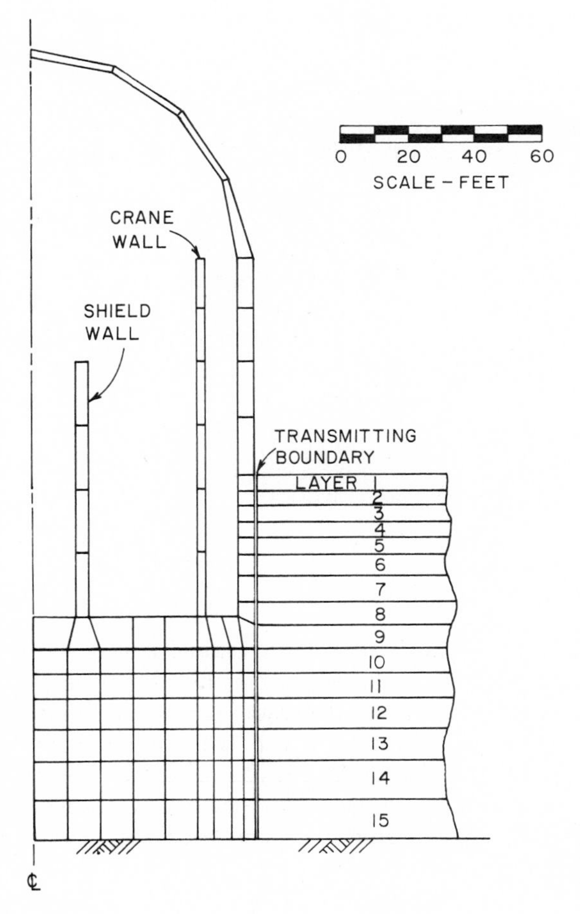

Figure 20-11 Three-dimensional finite element model.[17]

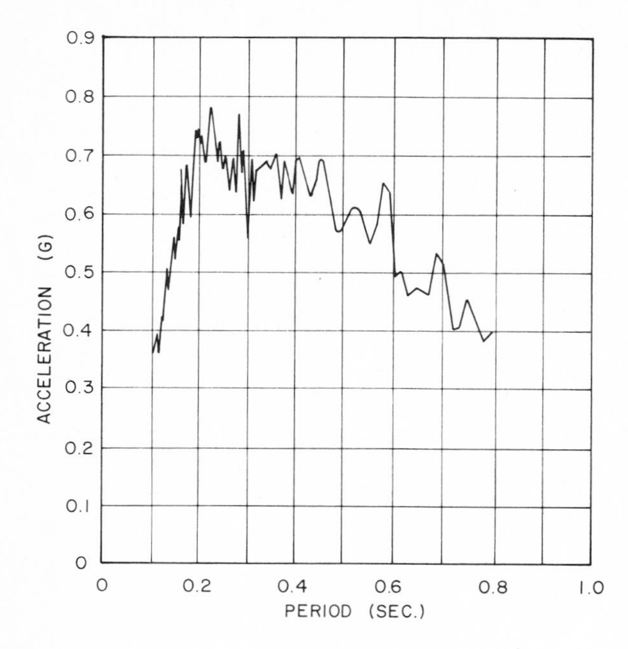

Figure 20-12 Response spectrum of input motion, 1 percent oscillator damping.[17]

computed with two finite element models. The three-dimensional model is an axially symmetric formulation with consistent boundaries and the capacity for arbitrary directions of dynamic loading. This is described in detail by Kausel.[15] The two-dimensional model is a plane strain analog of the three-dimensional case, described by Chang-Liang[14] and using the consistent boundaries developed by Waas.[13] It can be seen that the differences are not great.

These examples show some of the analyses and results obtained by currently used dynamic finite element programs. The technology is evolving rapidly with many developments reported frequently. At this time the major uncertainties remain in the description of the input earthquake or other dynamic motion and in determination of soil properties.[18] These are conditions common to all users of numerical methods in geotechnical engineering.

Additional Examples

As mentioned above, a large number of applications in dynamic analysis are frequently reported. In view of the space limitations, we have included here only a few of the applications. For additional applications, the reader can consult many references, including 19 and 20 for analysis of dams, and 21 to 23 for analysis of soil-structure interaction problems. Moreover, comprehensive reviews of theory and applications of various numerical methods such as finite element, finite difference, characteristics, integral-equation, and perturbation methods to dynamic analysis are reported in the proceedings of the two conferences.[24,25]

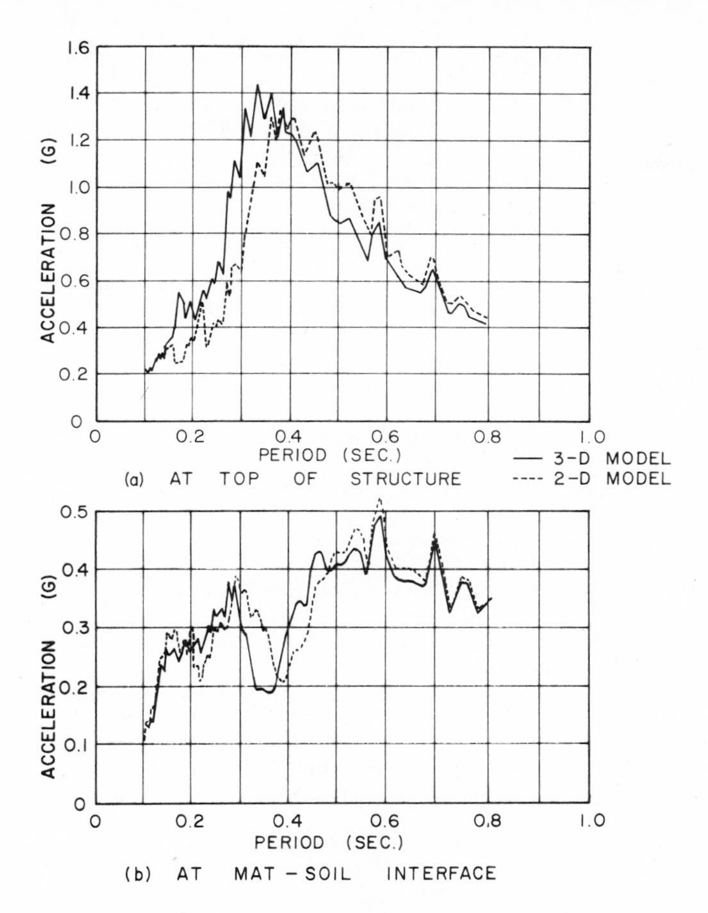

Figure 20-13 Amplified response spectra, 1 percent oscillator damping.[17]

REFERENCES

1. Desai, C. S., and J. F. Abel: "Introduction to the Finite Element Method," Van Nostrand Reinhold Company, New York, 1972.
2. Caughey, T. K.: Classical Normal Modes in Damped Linear Systems, *J. Appl. Mech.*, vol. 27, pp. 269–271, 1960.
3. Wilson, E. L., and J. Penzien: Evaluation of Orthogonal Damping Matrices, *Int. J. Numer. Methods Eng.*, vol. 4, no. 1, pp. 5–10, 1972.
4. Idriss, I. M., H. B. Seed, and N. Serff: Seismic Response by Variable Damping Finite Elements, *J. Geotech. Eng. Div. ASCE*, vol. 100, no. GT1, pp. 1–13, 1974.
5. Roesset, J. M., R. V. Whitman, and R. Dobry: Modal Analysis for Structures with Foundation Interaction, *J. Struct. Div. ASCE*, vol. 99, no. ST3, pp. 399–416, 1973.
6. Lysmer, J.: Modal Damping and Complex Stiffness, *Univ. California Note*, Berkeley, 1973.
7. Newmark, N. M.: A Method of Computation for Structural Dynamics, *J. Eng. Mech. Div. ASCE*, vol. 85, no. EM3, pp. 67–94, 1959.

8. Desai, C. S., and L. D. Johnson: Evaluation of Some Numerical Schemes for Consolidation, *Int. J. Numer. Methods Eng.*, vol. 7, pp. 243–254, 1973.
9. Desai, C. S., J. T. Oden, and L. D. Johnson: Evaluation and Analyses of Some Finite Element and Finite Difference Procedures for Time-dependent Problems, *U.S. Army Eng. Waterw. Expt. Stn. Misc. Pap.* S-75-7, Vicksburg, Miss., April 1975.
10. Oden, J. T.: "The Mathematical Theory of Finite Elements," Wiley-Interscience, New York, 1976.
11. Idriss, I. M., et al.: A Computer Program for Evaluating the Seismic Response of Soil Structures by Variable Damping Finite Element Procedures, *Univ. California Rep.* EERC 73-16, Berkeley, July 1973.
12. Kausel, E., J. M. Roesset, and J. T. Christian: Nonlinear Behavior of Soil-Structure Interaction, *J. Geotech. Eng. Div. ASCE*, vol. 102, no. GT11, pp. 1159–1170, November, 1976.
13. Waas, G.: Linear Two-dimensional Analysis of Soil Dynamics Problems in Semi-infinite Layered Media, Ph.D. thesis, Univ. of California, Berkeley, 1972.
14. Chang-Liang, V.: Dynamic Response of Structures on Layered Soils, *MIT Dept. Civ. Eng. Rep.* R74-10, Cambridge, Mass., 1974.
15. Kausel, E.: Forced Vibrations of Circular Foundations on Layered Media, *MIT Dept. Civ. Eng. Rep.* R74-11, Cambridge, Mass., 1974.
16. Christian, J. T.: Choices among Procedures for Dynamic Finite Element Analysis, *ASCE Nat. Conv., Denver, Colo.*, prepr. 2615, 1975.
17. Kausel, E., and J. M. Roesset: Soil-Structure Interaction Problems for Nuclear Containment Structures, *ASCE Power Div. Spec. Conf., Denver, Colo., 1974.*
18. Christian, J. T.: Uncertainties in Soil Structure Interaction, *2d Spec. Conf. Struct. Des. Nucl. Plant Facilities ASCE, New Orleans, 1975.*
19. Seed, H. B., K. L. Lee, and I. M. Idriss: Analysis of Sheffield Dam Failure, *J. Soil Mech. Found. Div. ASCE*, vol. 95, no. SM6, pp. 1453–1490, November 1969.
20. Seed, H. B., et al.: The Slides in the San Fernando Dams during the Earthquake of February 9, 1971, *J. Geotech. Eng. Div. ASCE*, vol. 101, no. GT7, pp. 651–688, July 1975.
21. Seed, H. B., J. Lysmer, and R. Hwang: Soil-Structure Interaction Analyses for Seismic Response, *J. Geotech. Eng. Div. ASCE*, vol. 101, no. GT5, pp. 439–457, May 1975.
22. Isenberg, J., L. C. Lee, and M. S. Agbabian: Response of Structures to Combined Blast Effects, *J. Transport. Eng. Div. ASCE*, vol. 99, no. TE4, November 1973.
23. Isenberg, J.: Interaction between Soil and Nuclear Reactor Foundations during Earthquakes, report, Agbabian-Jacobsen Associates, Los Angeles, June 1970.
24. Desai, C. S. (ed.): Applications of the Finite Element Methods in Geotechnical Engineering, *Proc. Symp. Finite Element Methods Geotech. Eng., U.S. Army Eng. Waterw. Expt. Stn., Vicksburg, Miss. September 1972.*
25. Desai, C. S. (ed.): Numerical Methods in Geomechanics, *Proc. 2d Int. Conf. Numer. Methods Geomech., Blacksburg, Va., spec. ASCE publ.*, June 1976.

CHAPTER
TWENTY-ONE

STABILITY ANALYSIS BY PLASTICITY THEORY

J. R. Booker and E. H. Davis

21-1 INTRODUCTION

For the last two decades there has been an increasing interest in the application of the theory of plasticity, based on relatively rigorous applied mechanics, to soil stability problems. This has been paralleled by increased sophistication in the development of more traditional engineering approaches to these problems. These approaches are approximate and inevitably involve intuitive assumptions concerning, for example, the shape of rupture surfaces for specific types of problems.

This chapter is concerned with the numerical solution of plane strain problems using the theory of plasticity. The theory is founded on a fully defined set of constitutive equations for a simple frictional plastic, i.e., a material which behaves elastically until it yields and then strains indefinitely at constant strength unless the stress state is changed. Definition of the plastic behavior is completed by specifying a simple flow rule, which need not be the associated flow rule (normality) of classical plasticity but, at least in this respect, can be specified to accord more closely with the observed behavior of real soil.

As in classical plasticity theory for metals, attention is focused on the collapse or ultimate load for the body or soil mass, and no information is sought regarding the behavior under a load less than that required for collapse. (The question of uniqueness of the collapse load is discussed elsewhere.[1]) Again as in classical plasticity, it is usually necessary to have some concept of the location and nature of the plastic zones in specific problems before solutions can be attempted;

frequently the initial solution can be regarded only as a formal solution and must be tested for kinematic admissibility and static extensibility before final acceptance, or alternatively the bracketing of the desired solution by upper and lower limit solutions may have to be accepted. In problems which defy ingenuity in postulating enough of the nature of the solution to enable the methods given in this chapter to be successfully started, the only recourse appears to be to follow a loading path approach, employing, say, finite element methods; however, this is outside the scope of this chapter.

Other matters which have been taken to be outside the scope of this chapter but which may be necessary for accuracy in some soil and rock applications are formulation of the theory to incorporate work hardening and softening and formulation to account for both initial and developing anisotropy.

21-2 FAILURE CRITERION

In this chapter we shall be concerned with a perfectly plastic material under conditions of plane strain. $0x$, $0y$, $0z$ are a set of cartesian reference axes chosen so that the stress state is independent of z. Stresses are denoted σ_x, σ_y, τ_{xy}, and compression is taken as positive.

Consider all possible combinations of stress which can be applied to an element of an elastoplastic material; these stresses can be divided into (1) elastic stress, for which all strains are recoverable, and (2) plastic or failure stresses for which there are irrecoverable strains and for which it is possible to have continued deformation under constant stress.

If the failure stresses are plotted as shown in Fig. 21-1*a*, it can be seen that they form a failure surface which encompasses the elastic stresses. It is not possible to attain states of stress lying outside the failure surface. The equation of the failure surface, or the failure criterion, can be written

$$f(\sigma_x, \sigma_y, \tau_{xy}) = 0 \tag{21-1}$$

If the material is isotropic, the orientation of the test element relative to the reference axes is immaterial and thus it is permissible to take an element whose axes are parallel to the principal stresses σ_1, σ_3. Thus for an isotropic material it is only necessary to plot the principal stresses, as shown in Fig. 21-1*b*. This means that for isotropic material the yield criterion [Eq. (21-1)] can be expressed solely in terms of the principal stresses σ_1, σ_3 or in fact in terms of any other pair of independent invariant quantities, for example,

$$p = \frac{\sigma_1 + \sigma_3}{2} \qquad R = \frac{\sigma_1 - \sigma_3}{2}$$

For an isotropic material it is often convenient to think of the failure criterion in terms of Mohr circles. To each failure stress state there will correspond a Mohr circle. The envelope of these Mohr circles (Fig. 21-1*c*) gives another representation of the failure criterion.

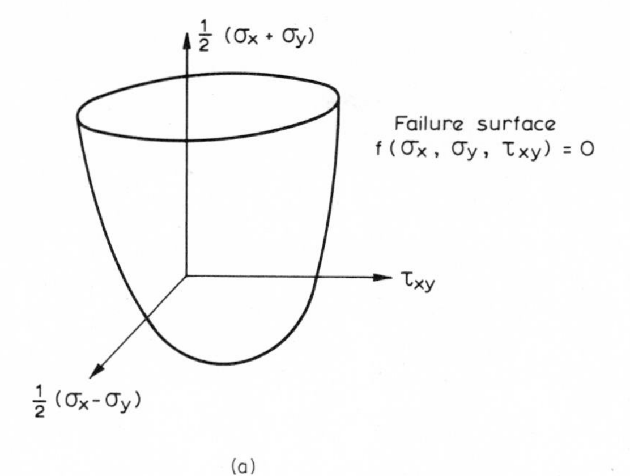

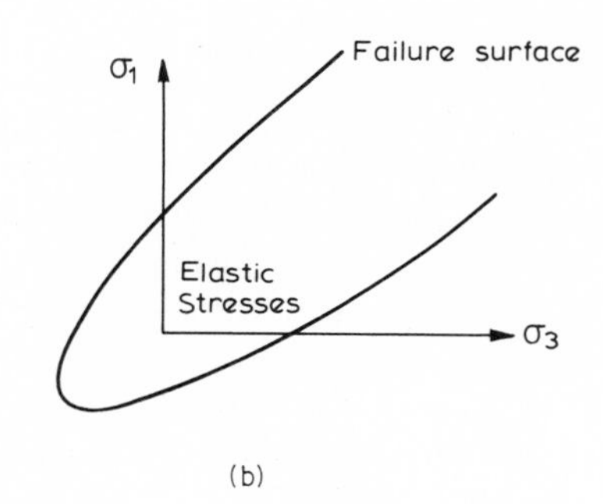

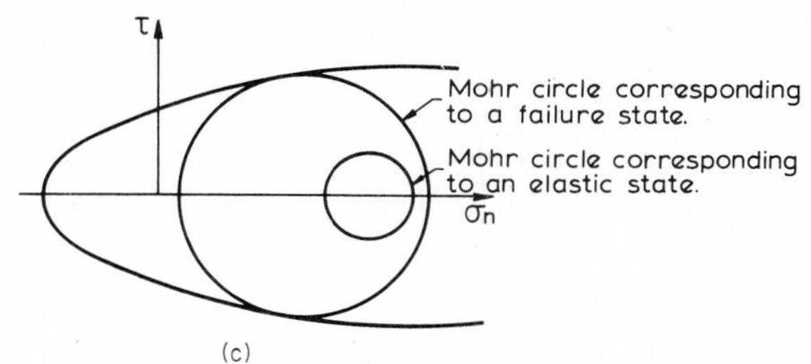

Figure 21-1 Representations of failure criteria.

For soils it is often found that failure can be described by the Mohr-Coulomb failure criterion shown in Fig. 21-2. In this case the failure criterion can be written in any of the alternative forms

$$R = p \sin \phi + c \cos \phi \tag{21-2a}$$

$$\frac{\sigma_1 + c \cot \phi}{\sigma_3 + c \cot \phi} = \tan^2 \left(\frac{\pi}{4} + \frac{\phi}{2} \right) = N_\phi \tag{21-2b}$$

$$(\sigma_x - \sigma_y)^2 + 4\tau_{xy}^2 = (\sigma_x + \sigma_y + 2c \cot \phi)^2 \sin^2 \phi \tag{21-2c}$$

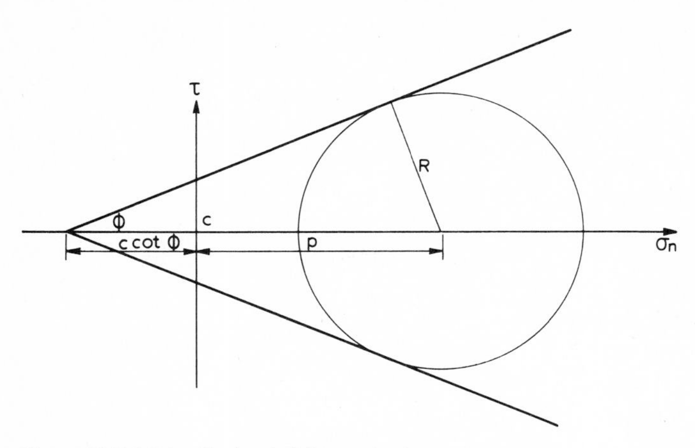

Figure 21-2 Mohr-Coulomb failure criterion.

In the previous discussion it has been assumed that the form of the failure surface is fixed; however, it may be that the form of loading or the previous stress path of an element will affect the position and form of the yield surface. Such a material will be called a *strain-hardening* or *strain-softening material* and will in the main lie outside the scope of this chapter.

21-3 FLOW RULES

When an elastoplastic material is subjected to a change in stress $\{d\sigma\}$,† it undergoes a change in strain $\{d\epsilon\}$; this strain increment can be broken into two components

$$\{d\epsilon\} = \{d\epsilon^E\} + \{d\epsilon^P\} \tag{21-3}$$

where $\{d\epsilon^E\}$ is the elastic component of the strain defined by equations of the type

$$d\epsilon_x = \frac{1}{E}[d\sigma_x - \nu(d\sigma_y + d\sigma_z)] \qquad d\epsilon_{xy} = \frac{1+\nu}{E}\, d\tau_{xy} \tag{21-4}$$

where E is Young's modulus and ν is Poisson's ratio.

Equation (21-3) then serves as a definition of $\{d\epsilon^P\}$, the plastic strain component. For a perfectly plastic material the plastic strain increments or, what is equivalent, the plastic strain rates are related to the current stress state by a flow rule of the form‡

$$\dot{\epsilon}_x^P = \lambda g_x(\sigma_x, \sigma_y, \ldots, \tau_{xy}) \tag{21-5}$$

† $\{d\sigma\}$ represents symbolically a change in the six stress components $d\sigma_x$, $d\sigma_y$, ..., $d\tau_{xy}$, with a similar definition holding for the increment in strain $\{d\epsilon\}$.

‡ In a theory involving strain hardening or softening the form of the flow rule (21-5) may depend on the previous history of the element, namely, on previous states of stress as well as the current stress state.

where λ is a one-signed multiplier reflecting the indeterminacy of the plastic strain rates in terms of stress alone. The sign of λ is determined by the condition that the local rate of plastic work is positive. If the stress state lies entirely within the yield surface (21-1), there will be no plastic strain component and $\lambda = 0$.

If the material is isotropic, the relation between stress and plastic strain rate will also be isotropic; this means that the principal directions of the stress tensor and the strain rate tensor will be coincident.

The behavior of an elastoplastic material is defined by its yield criterion (21-1) and its flow rule (21-5), both of which must be determined by experiment. The most convenient assumption mathematically and the one made by most investigators is that the flow rule is an associated one defined by

$$\dot{\epsilon}_x^p = \lambda \frac{\partial f}{\partial \sigma_x} \qquad \dot{\gamma}_{xy}^p = 2\dot{\epsilon}_{xy}^p = \lambda \frac{\partial f}{\partial \tau_{xy}} \tag{21-6}$$

Equation (21-6) has a useful geometric interpretation. If the strain rates are considered as components of a vector in stress space $\{\sigma_x, \sigma_y, \ldots, \tau_{xy}\}$, it can be seen that the strain rate vector is parallel to the outward normal to the yield surface. This geometric interpretation and the assumption of convexity of the yield surface enable limit theorems to be proved for such a material. These limit theorems[2,3] are extremely useful from a practical viewpoint, in that they allow the exact collapse load to be bracketed by upper and lower bounds. From the theoretical viewpoint they are useful in showing that the collapse load is independent of the loading path and providing a means of verifying whether a proposed solution is correct.

For a material with a yield criterion of the form (21-2) the flow rule takes the form

$$\dot{\epsilon}_1^P = \lambda \qquad \dot{\epsilon}_3^P = -\lambda N_\phi \tag{21-7}$$

where $\dot{\epsilon}_1^P$ and $\dot{\epsilon}_3^P$ are the principal plastic strain rates. Alternatively

$$\frac{\dot{\epsilon}_v^P}{\dot{\epsilon}_1^P} = \text{const} = 1 - N_\phi \tag{21-8}$$

where $\dot{\epsilon}_v^P$ denotes the rate of plastic volume strain.

Unfortunately the rate of volume strain predicted by Eq. (21-8) is far in excess of that exhibited by real frictional soils.† This led Davis[4] to propose the relationship

$$\frac{\dot{\epsilon}_v^P}{\dot{\epsilon}_1^P} = 1 - N_\psi \tag{21-9}$$

where $N_\psi = \tan^2(\pi/4 + \psi/2)$ is a constant which must be determined by experiment. Notice that if $\psi = 0$, the material deforms at constant volume, while if $0 < \psi \leqslant \phi$, the material dilates.

† An important exception is that of saturated clays failing sufficiently rapidly for no change in water content to occur. Under such circumstances $\phi = 0$, and there is no plastic volume change, so that the actual behavior satisfies (21-8).

In this chapter we shall primarily be concerned with a rigid plastic material ($E \to \infty$). In this case the elastic strains are regarded as small compared with plastic strains; using Eqs. (21-2) and (21-9) and noting that the principal directions of the stress and strain rate tensors coincide, we find that under conditions of plane strain the x, y velocity components u, v satisfy

$$\frac{\partial u}{\partial x} = -\lambda(\cos 2\theta - \cos 2\nu)$$

$$\frac{\partial v}{\partial y} = \lambda(\cos 2\theta + \cos 2\nu) \tag{21-10}$$

$$\frac{\partial u}{\partial y} + \frac{\partial v}{\partial x} = -2\lambda \sin 2\theta$$

where θ denotes the angle between the x axis and the principal stress direction

$$\nu = \frac{\pi}{4} - \frac{\psi}{2}$$

and λ is a positive multiplier.

21-4 STRESS EQUATIONS

In this section the equations governing the behavior of a plastic soil satisfying a Mohr-Coulomb failure law will be developed. It will be assumed that the unit weight of the soil is γ and that the direction of gravity is inclined at an angle ϵ to the x axis. The stresses σ_x, σ_y, τ_{xy} must satisfy the equilibrium equations

$$\frac{\partial \sigma_x}{\partial x} + \frac{\partial \tau_{xy}}{\partial y} = \gamma \cos \epsilon \tag{21-11}$$

$$\frac{\partial \tau_{xy}}{\partial x} + \frac{\partial \sigma_y}{\partial y} = -\gamma \sin \epsilon \tag{21-12}$$

and the Mohr-Coulomb failure criterion [Eq. (21-2c)]

$$(\sigma_x - \sigma_y)^2 + 4\tau_{xy}^2 = (\sigma_x + \sigma_y + 2c \cot \phi)^2 \sin^2 \phi$$

It proves more convenient to use the well known Mohr representation of the stresses

$$\sigma_x = p + R \cos 2\theta \qquad \sigma_y = p - R \cos 2\theta \qquad \tau_{xy} = R \sin 2\theta$$

where R is related to p via Eq. (21-2a) and θ is the angle between the major principal stress direction and the x axis. If these equations are substituted into the equilibrium equations, it is found that

$$\frac{\partial p}{\partial x}(1 + \sin \phi \cos 2\theta) + \frac{\partial p}{\partial y} \sin \phi \sin 2\theta + 2R\left(-\frac{\partial \theta}{\partial x} \sin 2\theta + \frac{\partial \theta}{\partial y} \cos 2\theta\right) = \gamma \cos \epsilon \tag{21-13}$$

$$\frac{\partial p}{\partial x}\sin\phi\sin 2\theta+\frac{\partial p}{\partial y}(1-\sin\phi\cos 2\theta)+2R\left(+\frac{\partial \theta}{\partial x}\cos 2\theta+\frac{\partial \theta}{\partial y}\sin 2\theta\right)$$
$$=-\gamma\sin\epsilon \quad (21\text{-}14)$$

Equations (21-13) and (21-14) are a pair of quasilinear hyperbolic equations; the properties of such equations are described by Courant and Hilbert;[5] an account relevant to the theory of plasticity is given by Hill.[6] Associated with a set of hyperbolic equations are two families of lines known as *characteristics;* the properties of hyperbolic equations are most readily understood when expressed in terms of the characteristics. The most important of these is that the field quantities [in this chapter (p, θ)] if continuous cannot have discontinuities in their derivatives except across a characteristic line. This has two important consequences: (1) characteristic lines are lines along which two stress fields which are continuous but analytically different may be joined; (2) if a solution is known on one side of a characteristic, it cannot be extended beyond that characteristic without additional information.

The characteristics of Eqs. (21-13) and (21-14) are

$$\frac{dy}{dx}=\tan(\theta-\mu) \qquad \frac{dy}{dx}=\tan(\theta+\mu) \quad (21\text{-}15)$$

where $\mu=\pi/4-\phi/2$. These are called α and β lines, respectively, and are shown schematically in Fig. 21-3. When Eqs. (21-13) and (21-14) are referred to these lines, it is found that

$$-\sin 2\mu\frac{\partial p}{\partial s_\alpha}+2R\frac{\partial \theta}{\partial s_\alpha}+\gamma\left[\sin(\epsilon+2\mu)\frac{\partial x}{\partial s_\alpha}+\cos(\epsilon+2\mu)\frac{\partial y}{\partial s_\alpha}\right]=0 \quad (21\text{-}16)$$

$$\sin 2\mu\frac{\partial p}{\partial s_\beta}+2R\frac{\partial \theta}{\partial s_\beta}+\gamma\left[\sin(\epsilon-2\mu)\frac{\partial x}{\partial s_\beta}+\cos(\epsilon-2\mu)\frac{\partial y}{\partial s_\beta}\right]=0 \quad (21\text{-}17)$$

where $\partial/\partial s_\alpha$ and $\partial/\partial s_\beta$ indicate differentiation with respect to arc length along the α and β lines, respectively. The more general case of an anisotropic material is dealt with in Ref. 7. Notice that Eq. (21-16) involves differentiation only along the α

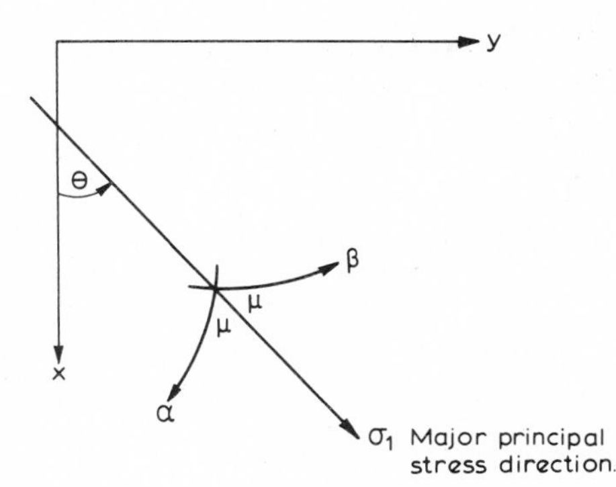

Figure 21-3 Definition of α and β lines.

line while Eq. (21-17) involves differentiation only along the β line. It is the fact that it is possible to represent Eqs. (21-13) and (21-14) in this form that leads to the previously stated properties of hyperbolic equations. This representation will prove particularly useful in developing numerical methods of solution for these equations.

21-5 NUMERICAL INTEGRATION OF THE STRESS EQUATIONS

Basic Calculation

The simplest method of solution of a pair of hyperbolic equations is due to Masseau.[8]† This method depends upon the following basic calculation, which will be called calculation 1.

Calculation 1 Suppose that p, θ are known at two adjacent points A, B shown in Fig. 21-4 and that it is desired to find their values at a point P which is the intersection of the α line through A and the β line through B. It must be emphasized that the position of P is not known at this stage and must be determined as part of the calculation. Obviously it is not possible to integrate along either of the characteristics AP, BP without knowing the exact variation of (p, θ); however, it is possible to integrate approximately, in which case, it is found that to sufficient accuracy, Eqs. (21-16) and (21-17) can be written

$$y_P - y_A = \tan\left(\frac{\theta_P + \theta_A}{2} - \mu\right)(x_P - x_A) \tag{21-18}$$

$$y_P - y_B = \tan\left(\frac{\theta_P + \theta_B}{2} + \mu\right)(x_P - x_B) \tag{21-19}$$

$$-\sin 2\mu(p_P - p_A) + (R_P + R_A)(\theta_P - \theta_A) = -\gamma \sin(\epsilon + 2\mu)(x_P - x_A) - \gamma \cos(\epsilon + 2\mu)(y_P - y_A) \tag{21-20}$$

† Finite difference techniques can also be used (see, for example, Ref. 9); however, since in plasticity problems the mapping of the characteristics is usually an important part of the problem, finite difference techniques find little application.

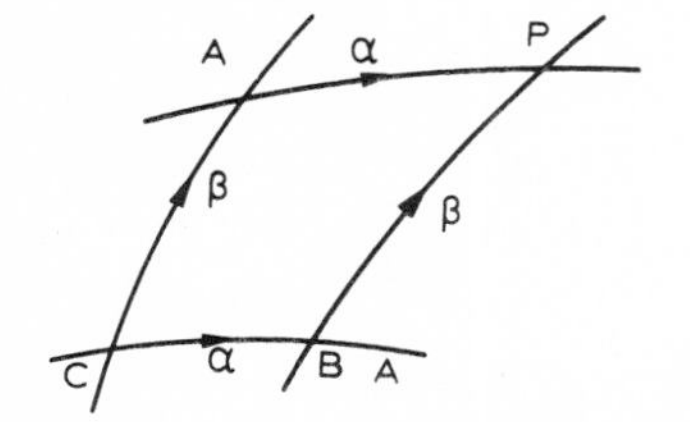

Figure 21-4 Characteristics in calculation 1.

$$\sin 2\mu(p_P - p_B) + (R_P + R_B)(\theta_P - \theta_B) = -\gamma \sin(\epsilon - 2\mu)(x_P - x_B) - \gamma \cos(\epsilon + 2\mu)(y_P - y_B) \quad (21\text{-}21)$$

where the subscript P, A, etc., denotes the value of that quantity at the points P, A, etc.

Equations (21-20) and (21-21) are a set of nonlinear equations. If an initial approximation to x_P, y_P, p_P, and θ_P is known, they can be solved iteratively according to the following scheme:

1. Solve Eqs. (21-20) and (21-21) for p_P, θ_P using the latest estimates of x_P, y_P, R_P.
2. Solve Eqs. (21-18) and (21-19) for x_P, y_P using the latest estimates of θ_P, R_P.
3. Check the new estimates of x_P, y_P, p_P, θ_P with their previous values to see whether the process has converged. If not, return to step 1.

It now remains to discuss how a first estimate of x_P, y_P, p_P, θ_P is obtained. Quite often the values of x, y, p, θ will be known at the point C shown in Fig. 21-4. In this case a first estimate is provided by the expressions

$$x_P \approx x_A + x_B - x_C \qquad y_P \approx y_A + y_B - y_C$$

$$\theta_P \approx \theta_A + \theta_B - \theta_C \qquad p_P \approx p_A + p_B - p_C$$

However, if the values at C are not known, it is usually quite adequate to use $x_P \approx \frac{1}{2}(x_A + x_B)$, etc.

The numerical process described above can be used to construct a subroutine CALCA, which obtains a numerical solution to calculation 1. A copy of this subroutine and the programs outlined in this chapter may be obtained on application to the authors.

As an example of this method consider the case of a cohesive frictional material with the material properties

$$c = 1 \qquad \phi = 30° \qquad \gamma = 1 \qquad \epsilon = 0°$$

such that

	x	y	p	$\theta°$
A	0.2887	0.5000	2.3095	90.0000
B	0.6947	0.7386	4.4295	78.2658
C	0.3491	0.3607	3.7036	76.8766

the successive estimates of x, y, p, θ at P are shown in Table 21-1. In this example the solution was assumed to converge when

$$|x' - x''| + |y' - y''| + |p' - p''| + |\theta' - \theta'| < \delta$$

Table 21-1

Iteration	x	y	p	$\theta°$
1	0.592189	1.022667	2.874661	89.716348
2	0.594103	1.020937	2.848294	89.237394
3	0.554000	1.021031	2.851673	89.263425
4	0.594010	1.021022	2.851374	89.260835
5	0.594009	1.021023	2.851401	89.261071
6	0.594009	1.021023	2.851399	89.261049

where $\delta = 10^{-5}$ and x', x'', y', y'', ... denote successive estimates of x, y, ... In constructing the subroutine CALCA it is usually advisable to provide a limit to the number of repetitions of steps 1 to 3 to guard against the possibility of an infinite loop in the event of nonconvergence.

Fundamental Problems

Three problems emerge as fundamental in the theory of hyperbolic equations:

1. The initial value, or Cauchy, problem
2. The characteristic initial value, or Goursat, problem
3. The mixed problem

These problems are considered to be fundamental because in practice it is found that the solution of many particular problems can be reduced to the solution of a sequence of fundamental problems. These problems together with their associated conditions for existence and uniqueness are described in Ref. 10, pp. 48–59. It will be assumed in this chapter that all such conditions have been met. We shall now discuss these problems together with their numerical solution for the stress equations of a perfectly plastic solid.

Numerical Solution of the Cauchy Problem

In Fig. 21-5, AB is a noncharacteristic arc. It is convenient to label the α, β lines 1, 2, 3, ..., as shown in Fig. 21-5, and denote the intersection of the α line i with the β line j by P_{ij}. The values of x, y, p, θ are known along the arc AB and in particular at $P_{11}, \ldots, P_{nn}$, and thus the basic calculation described previously can be used as follows:

1. The value of x, y, p, θ at P_{12} can be obtained from the known values at P_{11}, P_{22}.
2. The value of x, y, p, θ at P_{23} can be obtained from the known values at P_{22}, P_{33}.
3. The process is continued until the value of x, y, p, θ is known at $P_{12}, \ldots, P_{n-1,n}$.

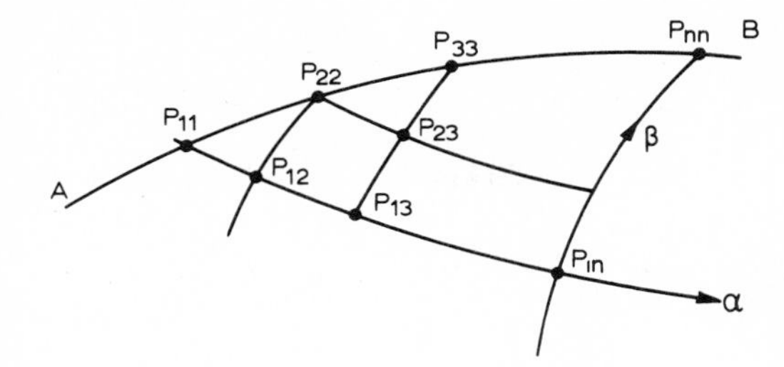

Figure 21-5 Characteristics for Cauchy problem.

Now the value of x, y, p, θ can be considered as known on the noncharacteristic arc P_{12}, $P_{n-1,\,n}$, and thus the above process can be repeated to find their values at P_{13}, P_{24}, ..., $P_{n-2,\,n}$ and similarly at P_{14}, P_{25}, ..., $P_{n-3,\,n}$. The process is continued until x, y, p, θ are known at all P_{ij} $(j \geqslant i)$.

It can thus be seen that the values of x, y, p, θ on the noncharacteristic arc $P_{11}P_{nn}$ are completely determined by their values within the region bounded by this arc and the α line 1 and the β line n. The determination of the values of x, y, p, θ throughout this region is known as a Cauchy problem. For an example of this calculation see Appendix 21A, Example 21A-1.

There are several points to note about the solution of a Cauchy problem.

1. The known values of x, y, p, θ on $P_{11}P_{mn}$ can also be used to calculate the values on the "other" side of the arc, i.e., in the region bounded by the arc, the β line 1, and the α line n. The solution is completely analogous to that already given. It will usually be evident which region is required. The region bounded by the α lines 1 and n and the β lines 1 and n is known as the *domain of influence* of the noncharacteristic arc AB.
2. The usual form of the Cauchy problem is that the surface tractions, namely the normal stress σ_n and the shear stress τ_{nt}, are given along some noncharacteristic arc. This information does not completely determine p and θ, as shown in Fig. 21-6, where it can be seen there are two possible values

$$p_1 + c \cot \phi = \frac{\bar{\sigma}_n - \sqrt{\sigma_n^2 \sin^2 \phi - \tau_{nt}^2 \cos^2 \phi}}{\cos^2 \phi}$$

$$p_2 + c \cot \phi = \frac{\bar{\sigma}_n + \sqrt{\sigma_n^2 \sin^2 \phi - \tau_{nt}^2 \cos^2 \phi}}{\cos^2 \phi}$$

where
$$\bar{\sigma}_n = \sigma_n + c \cot \phi$$

$$\theta_1 = \frac{1}{2} \tan^{-1} \frac{\tau_{nt}}{\sigma_n - p_1} \qquad \theta_2 = \frac{1}{2} \tan^{-1} \frac{\tau_{nt}}{\sigma_n - p_2}$$

For the solution of a particular problem one of these sets of values must be chosen. This can be done on kinematic grounds; that is, p_1 and θ_1 are selected because it is found that p_2 and θ_2 would not allow a correct velocity solution, namely the rate of plastic work would be found to be negative;

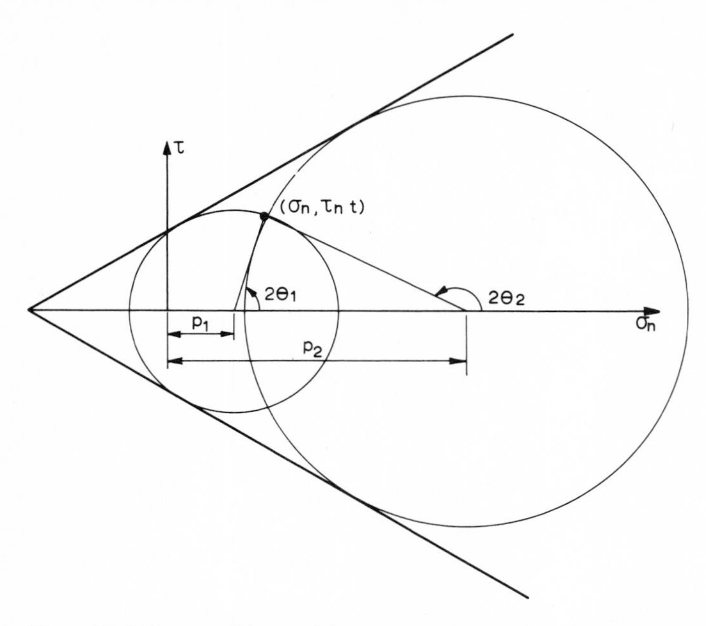

Figure 21-6 Two possible conditions for noncharacteristic boundary point.

but frequently the right choice is obvious from the general nature of the problem. It is interesting to note that the alternative vanishes only when $\tau_{nt} = \pm(\sigma_n + c \cot \phi) \tan \phi$, that is, when the line is itself a characteristic.

3. If the arc AB in Fig. 21-5 is allowed to tend toward a characteristic line, the domain of influence tends to vanish and thus it can be seen that a solution cannot be continued across a characteristic line without some additional information.

Numerical Solution of the Goursat Problem

In Fig. 21-7 OA is an α line and OB is a β line. Again it is convenient to number the α and β lines and adopt the notation of the previous section. The values of x, y, p, θ are assumed to be known along both OA and OB and in particular at the points P_{11}, P_{12}, ..., P_{1m} and P_{11}, P_{21}, ..., P_{n1}. The basic calculation described previously can now be used as follows:

1. The values of x, y, p, θ at P_{12}, P_{21} can be used to calculate their values at P_{22}.
2. The values of x, y, p, θ at P_{13}, P_{22} can be used to calculate their values at P_{23}.
3. The values of x, y, p, θ at P_{1m}, $P_{2,m-1}$ can be used to calculate their values at P_{2m}.

Now the values of x, y, p, θ can be considered as known on the α line $P_{21}P_{2m}$ and the β line $P_{21}P_{n1}$, and thus the above process can be repeated to calculate their

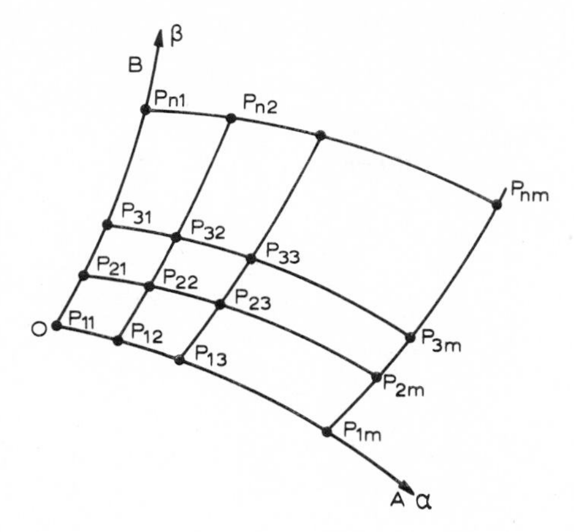

Figure 21-7 Characteristics for Goursat problem.

values at P_{31}, P_{32}, ..., P_{3m}, and similarly at P_{41}, P_{42}, ..., P_{4m}; the process continuing until x, y, p, θ are known at all points P_{ij}. For an example of this calculation see Example 21A-2.

Numerical Solution of the Mixed Problem

In Fig. 21-8 OA is a characteristic line which for definiteness will be assumed to be an α line. OD is a noncharacteristic line on which some condition is prescribed; e.g.,

1. OD may be a known straight line which is known to be shear-free.
2. OD may be a line whose position is unknown but which is prescribed by the condition that the normal and shear stress vanish on it.

The values of x, y, p, θ are assumed to be known along OA and in particular at the points P_{11}, ..., P_{1n} (the notation of the previous sections has been used again). In order to proceed with the computation it is necessary to assume that the following calculation can be performed. Suppose x, y, p, θ are known at some point P_{12} adjacent to the noncharacteristic line OD; then it is possible to use the values of x,

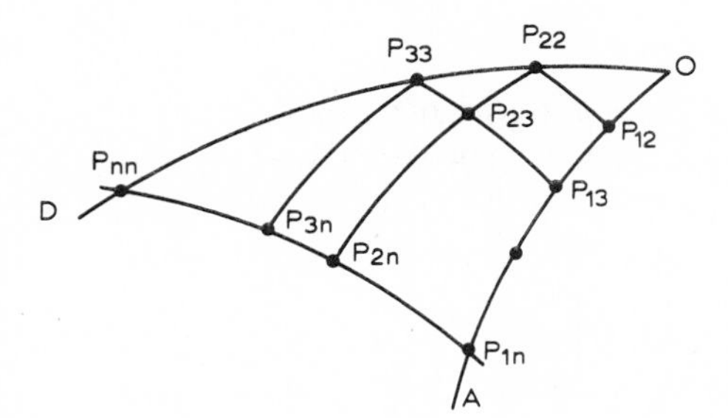

Figure 21-8 Characteristics for mixed problem.

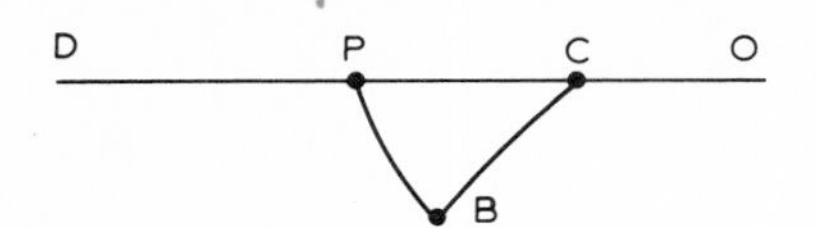

Figure 21-9 Typical mixed problem.

y, p, θ at P_{12} to obtain approximate values of x, y, p, θ at P_{22}, the intersection of the β line through P_{12} with OD. Because of the many types of mixed problem it is not possible to give a more precise definition. The ideas expressed above will become clearer with the following example. A mixed problem is defined as follows. Referring to Fig. 21-9, B is a point at which x, y, p, θ are known; OD is a prescribed straight line on which θ is known. It is desired to calculate the values of x, y, p, θ at P the point of intersection of OD and the β line through P. The performance of this calculation will be called calculation 2.†

Calculation 2 Since θ is prescribed along OD, the value of θ at P is known, and thus to sufficient accuracy the equation of the characteristic BP can be written

$$y_P - y_B = \tan\left(\frac{\theta_B + \theta_P}{2} + \mu\right)(x_P - x_B) \tag{21-22}$$

The equation of OD is also known and can be written

$$\frac{y_P - y_C}{y_D - y_C} = \frac{x_P - x_C}{x_D - x_C} \tag{21-23}$$

where C and D are two points on OD. Later it will be convenient to select C as the intersection of the α line through B with OD.

Equations (21-22) and (21-23) can be solved to obtain x_P and y_P. Now since x_P, y_P, and θ_P are all known, Eq. (21-17) can be integrated approximately to obtain p_P:

$$p_P = \frac{1}{\sin 2\mu + (\theta_P - \theta_B)\sin\phi}[(\sin 2\mu - (\theta_P - \theta_B)\sin\phi)p_B + 2c(\theta_P - \theta_B)\cos\phi - \gamma\sin(\epsilon - 2\mu)(x_P - x_B) - \gamma\cos(\epsilon - 2\mu)(y_P - y_B)] \tag{21-24}$$

Equations (21-23) and (21-24) can be used to construct a subroutine CALCB to perform calculation 2.

Once calculation 2 can be performed, the mixed problem of Fig. 21-8 can be solved as follows:

1. Calculate x, y, p, θ at P_{22} from the known values at P_{12} using calculation 2.
2. Calculate x, y, p, θ at P_{23} from the known values at P_{13}, P_{22} using calculation 1.
3. Calculate x, y, p, θ at P_{33} from the known values at P_{23} using calculation 2.

† Notice that calculation 2 includes the mixed boundary problem 1, described above.

The computation proceeds along successive β lines in the obvious way until the solution is known at all points P_{ij} $(i \leqslant j)$. For an example of this calculation see Appendix 21A, Example 21A-3.

21-6 SPECIAL CASES

Weightless Material

The treatment of the previous section can be substantially simplified if the plastic material can be considered weightless, i.e., if the dimensionless parameter $\gamma B/c$ is sufficiently small, B being some typical dimension of the problem. In this case Eqs. (21-16) and (21-17) reduce to

$$-\sin 2\mu \frac{\partial p}{\partial s_\alpha} + 2R\frac{\partial \theta}{\partial s_\alpha} = 0 \qquad \sin 2\mu \frac{\partial p}{\partial s_\alpha} + 2R\frac{\partial \theta}{\partial s_\alpha} = 0 \tag{21-25}$$

where $R = p \sin \phi + c \cos \phi$. These equations can be integrated immediately to give

$$\log \sigma - 2\theta \cot 2\mu = \text{const on } \alpha \text{ line} \tag{21-26}$$

$$\log \sigma + 2\theta \cot 2\mu = \text{const on } \beta \text{ line} \tag{21-27}$$

where $\sigma = p + c \cot \phi$.

If we now consider the calculation 1, described in Sec. 21-5, it is found that Eqs. (21-20) and (21-21) become

$$\log \sigma_P - 2\theta_P \cot 2\mu = \log \sigma_A - 2\theta_A \cot 2\mu$$

$$\log \sigma_P + 2\theta_P \cot 2\mu = \log \sigma_B + 2\theta_B \cot 2\mu$$

and thus

$$\begin{aligned} \theta_P &= \tfrac{1}{2}(\theta_B + \theta_A) + \tfrac{1}{4} \tan 2\mu \log \frac{\sigma_B}{\sigma_A} \\ \sigma_P &= (\sigma_B \sigma_A)^{1/2} \exp[(\theta_B - \theta_A) \cot 2\mu] \end{aligned} \tag{21-28}$$

Hence θ_P and σ_P are found immediately. Since θ_P is known, Eqs. (21-18) and (21-19) are merely linear equations and so x_P and y_P are easily found.† It should be noted that the process described above differs from that given in Sec. 21-5.1 in that no iteration is necessary.

Equations (21-28) can be simplified further if x, y, σ, θ are known at the point C, (Fig. 21-4). Then it can be shown that

$$\theta_P = \theta_A + \theta_B - \theta_C \qquad \sigma_P = \frac{\sigma_A \sigma_B}{\sigma_C} \tag{21-29}$$

† It is often convenient to perform this calculation graphically.[11]

If the first of Eqs. (21-29) is rewritten in the form

$$\theta_P - \theta_B = \theta_A - \theta_C$$

we have *Hencky's theorem*, which states that if we pass from one characteristic to another of, say, the β family along any characteristic, the angle turned through by this α characteristic is constant. An important corollary of Hencky's theorem is that if in a family of α characteristics one characteristic is straight, all α characteristics of that family are straight, and similarly for β characteristics.

Purely Cohesive Material

A simplified treatment is also possible for a purely cohesive material $\phi = 0$. In this case Eqs. (21-16) and (21-17) become

$$\begin{aligned} -\frac{\partial p}{\partial s_\alpha} + 2c\frac{\partial \theta}{\partial s_\alpha} + \gamma \cos \epsilon \frac{\partial x}{\partial s_\alpha} - \gamma \sin \epsilon \frac{\partial y}{\partial s_\alpha} = 0 \\ \frac{\partial p}{\partial s_\beta} + 2c\frac{\partial \theta}{\partial s_\beta} - \gamma \cos \epsilon \frac{\partial x}{\partial s_\beta} + \gamma \sin \epsilon \frac{\partial y}{\partial s_\beta} = 0 \end{aligned} \tag{21-30}$$

while Eqs. (21-15) become

$$\frac{\partial y}{\partial s_\alpha} = \tan\left(\theta - \frac{\pi}{4}\right)\frac{\partial x}{\partial s_\alpha} \qquad \frac{\partial y}{\partial s_\beta} = \tan\left(\theta + \frac{\pi}{4}\right)\frac{\partial x}{\partial s_\beta} \tag{21-31}$$

and so the α and β characteristics are orthogonal.

Equations (21-30) can be integrated exactly, so that

$$\bar{p} - 2c\theta = \text{const on } \alpha \text{ line} \qquad \bar{p} + 2c\theta = \text{const on } \beta \text{ line} \tag{21-32}$$

where $\bar{p} = p - \gamma \cos \epsilon \; x + \gamma \sin \epsilon \; y$.

As in the previous section this allows calculation 1 to be performed explicitly, and thus

$$\bar{p}_P = \frac{\bar{p}_B + \bar{p}_A}{2} + c(\theta_B - \theta_A) \qquad \theta_P = \frac{\bar{p}_B - \bar{p}_A}{4c} + \frac{\theta_B + \theta_A}{2} \tag{21-33}$$

Again Eqs. (21-33) can be simplified if x, y, p, θ are known at C (Fig. 21-4). Then

$$\theta_P = \theta_A + \theta_B - \theta_C \qquad p_P = p_A + p_B - p_C$$

It is evident from these equations that Hencky's theorem and its corollary remain valid.

Prandtl's Solution for the Bearing Capacity of a Smooth Strip Footing

Consider the bearing capacity of the smooth strip footing $A'A$ (in Fig. 21-10) indenting a half-space of a weightless cohesive frictional material. It seems reasonable to assume that the material immediately beneath and adjacent to the footing should be in a state of plastic failure. Suppose that the plastic region extends as far

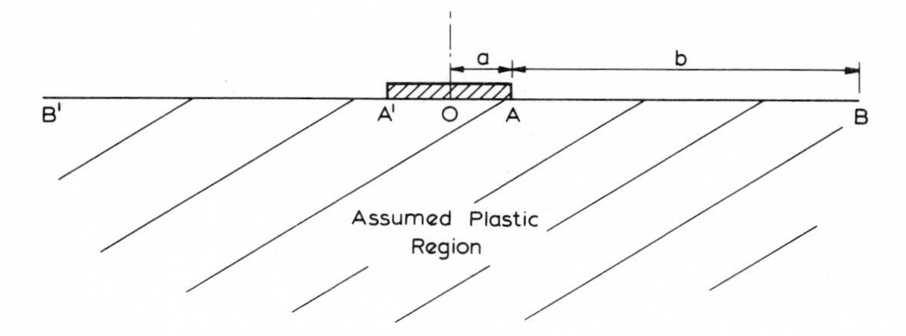

Figure 21-10 Smooth strip footing.

as B; then AB is a noncharacteristic line on which the tractions are specified, and it therefore defines a Cauchy problem. As stated before, there are two possibilities

$$\begin{aligned} \theta &= \frac{\pi}{2} \qquad \sigma = \frac{c \cot \phi}{1 - \sin \phi} \quad \text{on } AB \\ \theta &= 0 \qquad \sigma = \frac{c \cot \phi}{1 + \sin \phi} \quad \text{on } AB \end{aligned} \tag{21-34}$$

It is found on subsequent analysis that the second possibility would give rise to negative plastic work and must therefore be rejected.

When Eqs. (21-28) are applied to the first of Eqs. (21-34) it is readily seen that the stress state must be constant throughout the domain of influence of AB and that the characteristics must be straight lines, as shown in Fig. 21-11.

Now consider the determination of the remainder of the field. Clearly this field will be bounded by the α line AC, and so when the corollary of Hencky's theorem is used, all the α lines of this field must be straight. A little reflection shows that α lines cannot intersect the interior of the interval $A'A$, for the condition that $A'A$ is smooth would imply that these characteristics were parallel to AC, leading to the trivial solution that the half-space was in a state of passive failure and that the failure load was zero. Obviously it must then be inferred that the α lines all pass through the point A and that this is a singular point of the solution in that the stress state is multivalued there.†

† Such behavior would appear to be reasonable at the edge of a rigid footing.

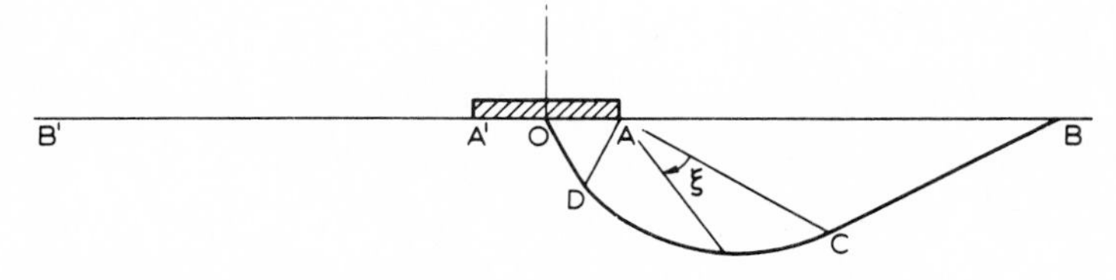

Figure 21-11 Prandtl's solution.

In the region ACD the stress state is defined by

$$\theta = \theta_0(\xi) = \frac{\pi}{2} - \xi$$

$$\sigma = \sigma_0(\xi) = \frac{c \cot \phi}{1 - \sin \phi} \exp (2\xi \cot 2\mu)$$

where ξ is as shown in Fig. 20-11. The characteristics in ACD are of course the family of straight α lines passing through A

$$\xi = \text{const}$$

and the family of β lines

$$r = (\text{const}) \exp (-\xi \cot 2\mu)$$

where r is the radial distance from A.

The extent of the region ACD is governed by the condition that $A'A$ is smooth; that is, $\theta = 0$ on $A'A$. This implies that the angle DAC is a right angle.

It now remains to determine the stress field beneath OA. This is a mixed problem, for we must find a solution such that

$$\theta = 0 \qquad \sigma = \frac{c \cot \phi}{1 + \sin \phi} \exp (\pi \cot 2\mu) \text{ on } AD$$

and $\theta = 0$ on OA.

The application of Eqs. (21-28) shows that this is a field of constant stress and that the characteristics are straight lines, as indicated in Fig. 21-11. The stress field is symmetric about a vertical line through O, and thus the solution is complete, giving an average failure pressure q of

$$q = c \cot \phi \left[\frac{1 + \sin \phi}{1 - \sin \phi} \exp (\pi \tan \phi) - 1 \right]$$

The solution given above is only formally complete, for in solving any plasticity problems it is necessary to divide the body into plastic and rigid regions and to show that:

1. It is possible to find an equilibrium stress field which satisfies the yield criterion and the stress boundary conditions in the plastic region.
2. It is possible to find a velocity field in the plastic regions which satisfies the flow rule of the material and the kinematic boundary conditions, involves no negative plastic work, and is compatible with the motion of the rigid regions.
3. It is possible to find an equilibrium stress field which does not violate the yield criterion in the rigid regions.

For a material with an associated flow rule satisfaction of conditions 1 and 2 shows that the solution provides an upper bound to the correct load, while satisfaction of conditions 1 and 3 shows that the solution provides a lower bound to the correct load,[2,3] indicating that the proposed solution provides the exact load. No such inference can be made for a material with a nonassociated flow rule, and indeed the solutions to such problems may be nonunique; i.e., the failure load may depend upon the loading path. In such cases it may be necessary to adopt an elastoplastic loading path technique although there is some evidence to suggest that the lack of uniqueness has little practical significance.[1]

21-7 VELOCITY EQUATIONS

It is not generally possible to determine the solution to a plasticity problem by merely considering equilibrium; the velocity field as well as the stress field must be determined. This will of course be obvious in problems which involve both kinematic and static boundary conditions; however, even when a problem contains no explicitly kinematic boundary condition, it is necessary to check that the stress field is compatible with some form of movement and that the internal rate of plastic work is everywhere positive. In passing it may be remarked that many finite element solutions to elastoplastic problems in which a loading path to general plastic collapse is followed fail to pay attention to this latter requirement.[1]

Analysis of the Velocity Field

The plastic flow of the material is governed by Eqs. (21-10), and if λ is eliminated from this equation, we obtain two quasilinear equations in u_x and v_y. These equations are hyperbolic and have the characteristics

$$\frac{dy}{dx} = \tan(\theta - \nu) \qquad \frac{dy}{dx} = \tan(\theta + \nu) \tag{21-35}$$

called α^* and β^* lines, respectively.

Along these characteristics (shown in Fig. 21-12) the velocity components vary as follows:

$$\frac{\partial u}{\partial s_{\alpha*}} + \tan(\theta - \nu)\frac{\partial v}{\partial s_{\alpha*}} = 0 \qquad \frac{\partial u}{\partial s_{\beta*}} + \tan(\theta + \nu)\frac{\partial v}{\partial s_{\beta*}} = 0 \tag{21-36}$$

where $\partial/\partial s_{\alpha*}$, $\partial/\partial s_{\beta*}$ denote differentiation along the α^* and β^* lines, respectively.

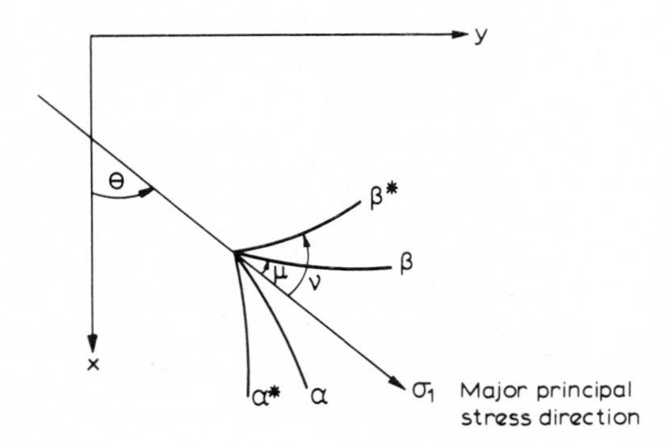

Figure 21-12 Definition of velocity characteristics.

Notice that if the material has an associated flow rule, that is, $\psi = \phi$, then $\nu = \mu$ and so it can be seen that the velocity characteristics and stress characteristic coincide.

The usual procedure in solving plasticity problems is to make sufficient assumptions to allow the determination of the stress field and then to verify the validity of these assumptions by showing that a velocity field involving no negative plastic work can be found and that the yield condition is not violated in the rigid portions of the body. If this procedure is adopted, when it comes to determining the velocity field, the stress field and the stress characteristics can be considered as known. From Eqs. (21-35) it can be seen that the α^* and β^* lines are inclined at an angle of $\pm\nu$ to the principal stress direction, respectively; similarly it can be seen from Eqs. (21-15) that the stress characteristics are inclined at angles $\pm\mu$ to the principal stress directions. Thus the α and α^* lines intersect at an angle of $\nu - \mu$, as do the β and β^* lines. The property is extremely convenient for determining the velocity characteristics.

In numerical work, a suitable procedure is to:

1. Determine the members of the family of β^* lines by using the result that they intersect at an angle of $\mu + \nu$ with the known family of α lines.†
2. Determine the members of the family of α^* lines by using the result that they intersect with the now known family of β^* lines at an angle of -2ν.

Of course, if the material has an associated flow rule, the stress and velocity characteristics coincide and no determination is necessary.

Once the velocity characteristics are known, the determination of the velocity field is quite straightforward. As mentioned before, the stress field may be considered as known, and thus θ is known throughout the plastic region. In particular, θ will be known along the velocity characteristics. Thus from Fig. 21-13 we see that if the velocity is known at points A and B and the α^* characteristic through A intersects the β^* characteristic through B at P, the velocity at P can be determined to sufficient accuracy by solving the two linear equations

$$u_P - u_A + \tan\left(\frac{\theta_P + \theta_A}{2} - \nu\right)(v_P - v_A) = 0 \tag{21-37}$$

$$u_P - u_B + \tan\left(\frac{\theta_P + \theta_B}{2} + \nu\right)(v_P - v_A) = 0 \tag{21-38}$$

It will be noticed that these equations are very similar to the corresponding equations (21-18) and (21-19) used in the determination of the stress characteristic, and the techniques employed to determine the stress characteristic may equally well be used to determine the velocity field.

† It would be possible to determine the β^* lines by using the result that they intersect the β lines at an angle of $\nu - \mu$; however, this angle may well be small and so result in an inaccurate numerical process.

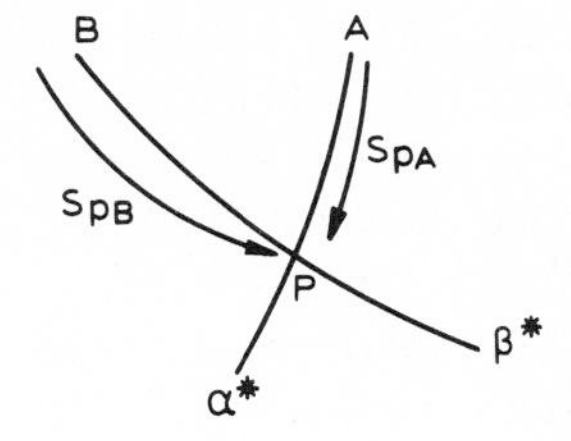

Figure 21-13 Solution along velocity characteristics.

Once the velocity field has been determined, it is necessary to show that the rate of plastic work is everywhere positive; that is, $\lambda > 0$.

An expression for λ convenient for numerical calculation is

$$-\lambda \sin^2 2\nu = \cos(\theta + \nu)\frac{\partial u}{\partial s_{\alpha*}} + \sin(\theta + \nu)\frac{\partial v}{\partial s_{\alpha*}} + \cos(\theta - \nu)\frac{\partial u}{\partial s_{\beta*}} + \sin(\theta - \nu)\frac{\partial v}{\partial s_{\beta*}} \tag{21-39}$$

From Fig. 21-13 it can be seen that an approximation sufficiently accurate for our purposes is

$$-\lambda \sin^2 2\nu = \cos\left(\frac{\theta_P + \theta_A}{2} + \nu\right)\left(\frac{u_P - u_A}{S_{PA}}\right) + \sin\left(\frac{\theta_P + \theta_A}{2} + \nu\right)\left(\frac{v_P - v_A}{S_{PA}}\right) + \cos\left(\frac{\theta_P + \theta_B}{2} - \nu\right)\left(\frac{u_P - u_B}{S_{PB}}\right) + \sin\left(\frac{\theta_P + \theta_B}{2} - \nu\right)\left(\frac{v_P - v_B}{S_{PB}}\right) \tag{21-40}$$

The velocity characteristics have several important properties:

1. A velocity discontinuity is possible only across a velocity characteristic.
2. If a velocity discontinuity exists across a velocity characteristic, the resultant velocity discontinuity must be inclined at an angle ψ to that characteristic.
3. A line separating a rigid from a deforming region must be a velocity characteristic.

The proofs of these results are very similar to those for a material with an associated flow rule[12] and will not be given here.

21-8 SAMPLE PROBLEM

As a sample problem let us consider the bearing capacity of a smooth strip footing indenting a half-space of c, ϕ, γ material. The case in which $\gamma = 0$ was dealt with in Sec. 21-6.

Again referring to Fig. 21-10, it seems reasonable to assume that the material beneath and adjacent to the footing is in a state of plastic failure. The line AB is again a noncharacteristic line on which the stress state is defined by the first of Eqs. (21-34). A numerical solution to this Cauchy problem is given in Example 21A-1, and the characteristics are shown in Fig. 21-14.

Consider the solution in the neighborhood of 0. An examination of Eqs. (21-15) to (21-17) shows that near this point the solution can be written

$$\sigma = \sigma_0(\xi) + \frac{\gamma r}{c}\sigma_1(\xi) + \cdots \qquad \theta = \theta_0(\xi) + \frac{\gamma r}{c}\theta_1(\xi) + \cdots$$

where r and ξ are polar coordinates defined in Sec. 21-6 and $\sigma_0(\xi)$ and $\theta_0(\xi)$ are the solutions corresponding to $\gamma = 0$, that is, those given in Sec. 21-6. Thus we see that the next stage of the solution reduces to the solution of the Goursat problem:

$$\sigma, \theta \text{ known on the } \alpha \text{ characteristic } OQ \tag{21-41}$$

$$\sigma = \frac{c \cot \phi}{1 - \sin \phi} \exp (2\xi \cot 2\mu) \qquad \theta = \frac{\pi}{2} - \xi \tag{21-42}$$

on the β characteristic localized at the point $r = 0$.† The extent of this characteristic is again determined by the condition that OS is shear-free and thus $0 < \xi < \pi/2$. The solution of the Goursat problem defined by Eqs. (21-41) and (21-42) is given in Example 21A-2. The characteristic field is shown in Fig. 21-14.

† A singularity involving point characteristics of the type (21-41) often arises in the theory of plasticity and is known as a *Prandtl singularity*.

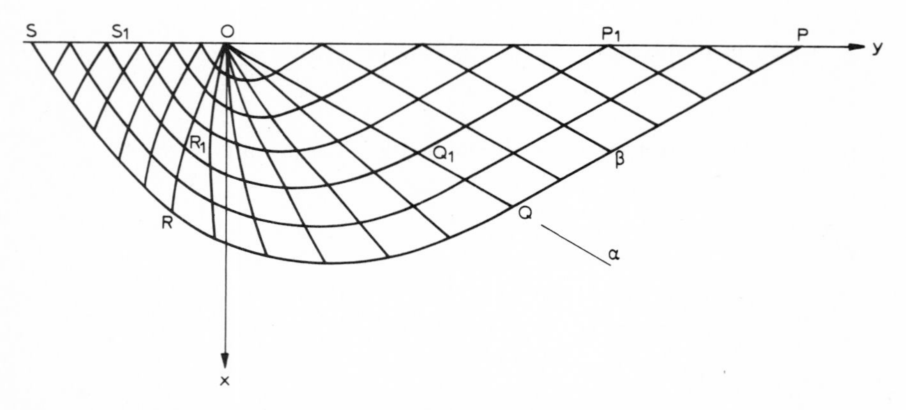

Figure 21-14 Characteristics for strip load on material with weight.

Table 21-2

$\frac{1}{2}\gamma B/c$	q_{av}/c
0	30.1396
0.2607	33.9463
0.5613	37.9392
0.8925	41.9498
1.2471	45.9958
1.6204	50.0699
2.0088	54.1629

The solution of the problem under consideration can now be completed by solving a mixed problem

$$\sigma, \theta \text{ known on the } \alpha \text{ line } OR$$

$$\theta = 0 \text{ on } OS$$

A numerical solution to this mixed problem is given in Example 21A-3. As before, the solution is symmetric about a vertical line through S, and thus the calculation of the stress field is complete.

Notice that in using these results the field $OP_1Q_1R_1S_1$ can be used to obtain the solution for a footing of semiwidth OS_1. Thus the calculations performed in this chapter enable us to find the relationship between the average pressure q_{av} and the dimensionless parameter $\frac{1}{2}\gamma B/c$ for a footing of breadth B. These results are given in Table 21-2. The solution for a particular problem can then be obtained by interpolation.

It must be emphasized that the proposed solution is still incomplete. To show that the solution is exact it must be demonstrated that

1. A satisfactory velocity field can be found.
2. The stress field can be extended into the rigid material without violating the yield criterion.

Details of these calculations are given in Ref. 13.

APPENDIX 21A

The Cauchy, Goursat, and mixed problems can be solved by the methods outlined in Sec. 21-5, and flow charts describing the computational process are shown in Fig. 21-15.

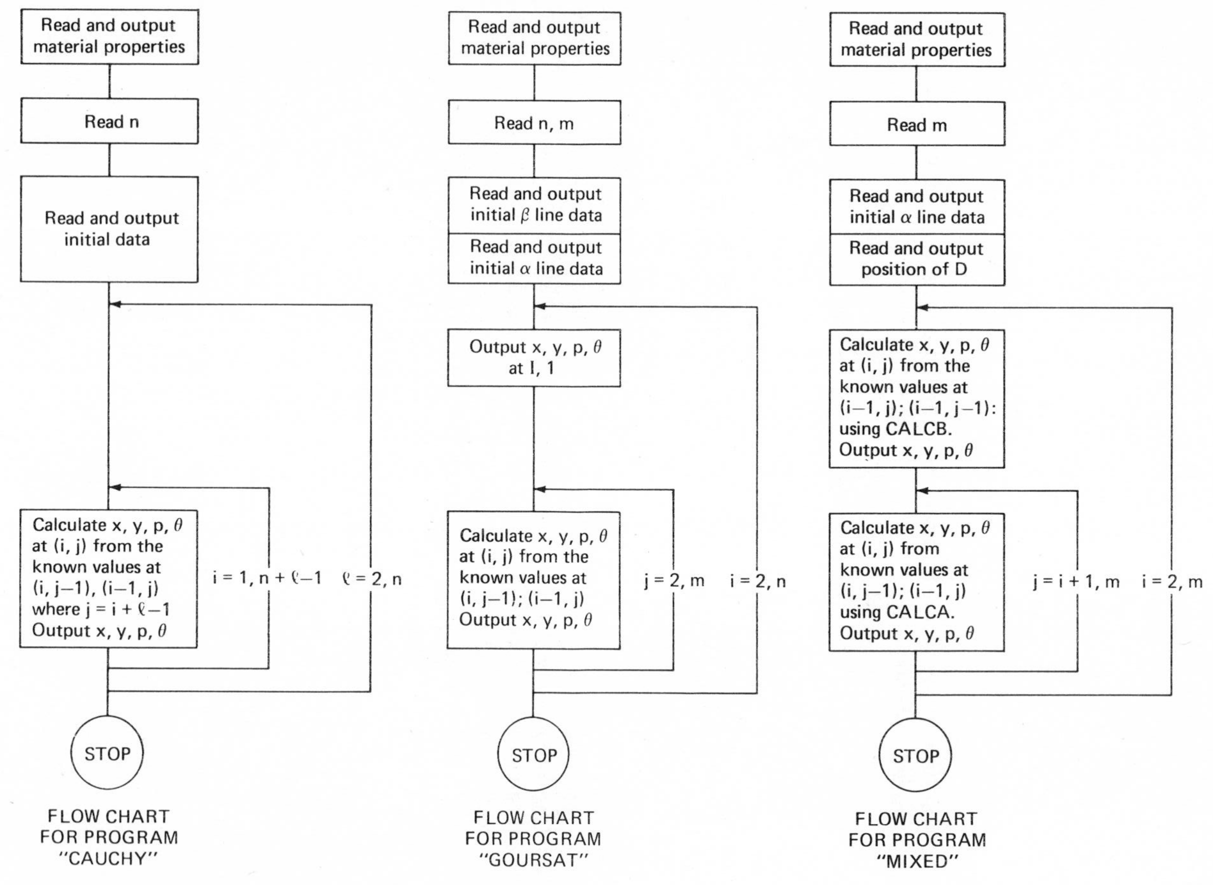

Figure 21-15 Flow chart for solution by characteristics.

Example 21A-1 Numerical solution of a Cauchy Problem As an example of the solution of a Cauchy problem consider the problem of determining the field defined by the noncharacteristic data $p = \sqrt{3}$, $\theta = 30°$ when $x = 0$, and $0 \leqslant y \leqslant 6$ for a cohesive frictional material with $c = 1$, $\phi = 30°$, $\gamma = 1$ and the direction of gravity parallel to the x axis. The network of characteristics OPQ is plotted in Fig. 21-14. In this simple case the characteristics are straight lines and OPQ is a Rankine passive zone.

```
CAUCHY   PROBLEM,

         C  =    1.0000

       PHI  =   30.0000

        FX  =    1.0000

        FY  =    0.0000


INITIAL  NON CHARACTERISTIC  DATA

I   J       X          Y          P          T

1   1    0.0000     0.0000     1.7321    90.0000
2   2    0.0000     1.0000     1.7321    90.0000
3   3    0.0000     2.0000     1.7321    90.0000
4   4    0.0000     3.0000     1.7321    90.0000
5   5    0.0000     4.0000     1.7321    90.0000
6   6    0.0000     5.0000     1.7321    90.0000
7   7    0.0000     6.0000     1.7321    90.0000
```

SOLUTION

I	J	X	Y	P	T
1	1	0.0000	0.0000	1.7321	90.0000
2	2	0.0000	1.0000	1.7321	90.0000
3	3	0.0000	2.0000	1.7321	90.0000
4	4	0.0000	3.0000	1.7321	90.0000
5	5	0.0000	4.0000	1.7321	90.0000
6	6	0.0000	5.0000	1.7321	90.0000
7	7	0.0000	6.0000	1.7321	90.0000
1	2	0.2887	0.5000	2.3095	90.0000
2	3	0.2887	1.5000	2.3095	90.0000
3	4	0.2887	2.5000	2.3095	90.0000
4	5	0.2887	3.5000	2.3095	90.0000
5	6	0.2887	4.5000	2.3095	90.0000
6	7	0.2887	5.5000	2.3095	90.0000
1	3	0.5774	1.0000	2.8868	90.0000
2	4	0.5774	2.0000	2.8868	90.0000
3	5	0.5774	3.0000	2.8868	90.0000
4	6	0.5774	4.0000	2.8868	90.0000
5	7	0.5774	5.0000	2.8868	90.0000
1	4	0.8660	1.5000	3.4642	90.0000
2	5	0.8660	2.5000	3.4642	90.0000
3	6	0.8660	3.5000	3.4642	90.0000
4	7	0.8660	4.5000	3.4642	90.0000
1	5	1.1547	2.0000	4.0415	90.0000
2	6	1.1547	3.0000	4.0415	90.0000
3	7	1.1547	4.0000	4.0415	90.0000
1	6	1.4434	2.5000	4.6189	90.0000
2	7	1.4434	3.5000	4.6189	90.0000
1	7	1.7321	3.0000	5.1962	90.0000

Example 21A-2 Numerical solution of a Goursat problem As an example consider the Goursat problem defined by the α line OQ shown in Fig. 21-14 and the Prandtl singularity (see Example 21-1) at O, for the material described in Example 21-A-1. The initial α line OQ was found in Example 21A-1, and the point β characteristic O was approximated by taking seven points with an angular spacing of 15°. The numerical solution is shown below, and the network of characteristics OQR is plotted in Fig. 21-14.

```
GOURSAT PROBLEM,

C = 1.0000

PHI = 30.0000

FX = 1.0000

FY = 0.0000
```

INITIAL BETA LINE DATA

I	J	X	Y	P	T
1	1	0.0000	0.0000	1.7321	90.0000
2	1	0.0000	0.0000	2.9548	75.0000
3	1	0.0000	0.0000	4.6091	60.0000
4	1	0.0000	0.0000	6.6473	45.0000
5	1	0.0000	0.0000	9.8755	30.0000
6	1	0.0000	0.0000	13.9725	15.0000
7	1	0.0000	0.0000	19.5157	0.0000

INITIAL ALPHA LINE DATA

I	J	X	Y	P	T
1	1	0.0000	0.0000	1.7321	90.0000
1	2	0.2887	0.5000	2.3095	90.0000
1	3	0.5774	1.0000	2.8868	90.0000
1	4	0.8660	1.5000	3.4642	90.0000
1	5	1.1547	2.0000	4.0415	90.0000
1	6	1.4434	2.5000	4.6189	90.0000
1	7	1.7321	3.0000	5.1962	90.0000

SOLUTION

I	J	X	Y	P	T
2	1	0.0000	0.0000	2.9548	75.0000
2	2	0.3491	0.3607	3.7036	76.8766

2	3	0.6945	0.7386	4.4295	78.2658
2	4	1.0364	1.1293	5.1417	79.3679
2	5	1.3750	1.5303	5.8424	80.2646
2	6	1.7108	1.9396	6.5337	81.0115
2	7	2.0438	2.3558	7.2167	81.6440
3	1	0.0000	0.0000	4.6091	60.0000
3	2	0.3724	0.2291	5.5547	63.2111
3	3	0.7479	0.4864	6.4557	65.6286
3	4	1.1235	0.7655	7.3319	67.5810
3	5	1.4981	1.0622	8.1868	69.1964
3	6	1.8708	1.3736	9.0236	70.5610
3	7	2.2412	1.6973	9.8444	71.7319
4	1	0.0000	0.0000	6.8473	45.0000
4	2	0.3645	0.1113	8.0372	49.1151
4	3	0.7436	0.2550	9.1541	52.2624
4	4	1.1308	0.4238	10.2344	54.8495
4	5	1.5224	0.6143	11.2831	57.0226
4	6	1.9162	0.8233	12.3044	58.8820
4	7	2.3108	1.0482	13.3015	60.4958
5	1	0.0000	0.0000	9.8755	30.0000
5	2	0.3324	0.0136	11.3867	34.6992
5	3	0.6910	0.0546	12.7816	38.3361
5	4	1.0668	0.1199	14.1243	41.3727
5	5	1.4542	0.2069	15.4218	43.9577
5	6	1.8495	0.3134	16.6804	46.1950
5	7	2.2501	0.4372	17.9045	48.1563
6	1	0.0000	0.0000	13.9725	15.0000
6	2	0.2836	-0.0627	15.9209	20.0607
6	3	0.6019	-0.1073	17.6865	24.0037
6	4	0.9453	-0.1333	19.3766	27.3381
6	5	1.3071	-0.1410	21.0020	30.2084
6	6	1.6828	-0.1314	22.5721	32.7167
6	7	2.0689	-0.1056	24.0936	34.9348
7	1	0.0000	0.0000	19.5157	0.0000
7	2	0.2252	-0.1165	22.0690	5.2770
7	3	0.4893	-0.2268	24.3391	9.3953
7	4	0.7833	-0.3272	26.4992	12.9115
7	5	1.1008	-0.4156	28.5661	15.9642
7	6	1.4371	-0.4913	30.5541	18.6527
7	7	1.7885	-0.5542	32.4732	21.0467

Example 21A-3 Numerical solution of a mixed problem As an example consider the mixed problem defined by the α line OR shown in Fig. 21-14 and the condition that $\theta = 0$ along the line $x = 0$ for the material described in Examples 21A-1 and 21A-2. The initial α line was found in Example 21A-2. The numerical solution is given below, and the network of characteristics are shown in Fig. 21-14.

MIXED PROBLEM,

C = 1.0000

PHI = 30.0000

FX = 1.0000

FY = 0.0000

INITIAL ALPHA LINE DATA

I	J	X	Y	P	T
1	1	0.0000	0.0000	19.5157	0.0000
1	2	0.2252	-0.1165	22.0690	5.2770
1	3	0.4893	-0.2268	24.3391	9.3953
1	4	0.7833	-0.3272	26.4992	12.9115
1	5	1.1008	-0.4156	28.5661	15.9642
1	6	1.4371	-0.4913	30.5541	18.6527
1	7	1.7885	-0.5542	32.4732	21.0467

T = 0.000000 ON OD

XO = 0.000000

YO = 0.000000

XD = 0.000000

YD =-10.000000

SOLUTION

I	J	X	Y	P	T
2	2	0.0000	-0.2607	24.5925	0.0000
2	3	0.2598	-0.3984	27.0948	0.0729
2	4	0.5481	-0.5269	29.4699	0.1353
2	5	0.8590	-0.6444	31.7376	0.1895
2	6	1.1884	-0.7500	33.9148	0.2373
2	7	1.5330	-0.8435	36.0130	0.2800

3	3	0.0000	-0.5613	29.4557	0.0000
3	4	0.2864	-0.7148	32.0215	0.0632
3	5	0.5944	-0.8572	34.4671	0.1181
3	6	0.9203	-0.9878	36.8117	0.1666
3	7	1.2609	-1.1065	39.0682	0.2100
4	4	0.0000	-0.8925	34.3853	0.0000
4	5	0.3068	-1.0584	36.9997	0.0556
4	6	0.6307	-1.2122	39.5030	0.1047
4	7	0.9688	-1.3540	41.9096	0.1486
5	5	0.0000	-1.2471	39.3652	0.0000
5	6	0.3230	-1.4231	42.0196	0.0497
5	7	0.6598	-1.5865	44.5692	0.0940
6	6	0.0000	-1.6204	44.3875	0.0000
6	7	0.3362	-1.8046	47.0740	0.0448
7	7	0.0000	-2.0088	49.4433	0.0000

REFERENCES

1. Davis, E. H., and J. R. Booker: Some Adaptions of Classical Plasticity for Soil Stability Problems, *Symp. Plasticity Soil Mech., Cambridge, 1973*, pp. 24–41.
2. Drucker, D. C., H. J. Greenberg, and W. Prager: The Safety Factor of an Elastic Plastic Body in Plane Strain, *J. Appl. Mech.*, vol. 10, no. 2, p. 371, 1952.
3. Drucker, D. C., W. Prager, and H. J. Greenberg: Extended Limit Design Theorems for Continuous Media, *Q. J. Mech. Appl. Math.*, no. 9, p. 381, 1952.
4. Davis, E. H.: Theories of Plasticity and the Failure of Soil Masses in I. K. Lee (ed.), "Soil Mechanics and Selected Topics," Butterworths & Co. (Publishers), Ltd., London, 1968.
5. Courant, R., and D. Hilbert: "Methods of Mathematical Physics," Interscience Publishers, New York, 1965.
6. Hill, R.: "The Mathematical Theory of Plasticity," Clarendon Press, Oxford, 1950.
7. Booker, J. R., and E. H. Davis: A General Treatment of Plastic Anisotropy under Conditions of Plane Strains, *J. Mech. Phys. Solids*, vol. 20, pp. 239–250, 1972.
8. Masseau, J.: Mémoire sur l'intégration graphique des équations aux dérives, *Ann. Ingen. Sortis Gand*, vol. 12, 1889.
9. Fox, P.: The Numerical Solution of Hyperbolic Equations, in A. Ralston and H. S. Wilf (eds.), "Mathematical Methods for Digital Computers," John Wiley & Sons, Inc., New York, 1967.
10. Courant, R., and K. O. Friedrichs: "Supersonic Flow and Shock Waves," Interscience Publishers, New York, 1948.
11. Symonds, P. S.: The Determination of Stresses in Plastic Regions in Problems in Plane Flow, *J. Appl. Phys.*, vol. 20, pp. 107–112, 1949.
12. Geiringer, H.: The Ideal Plastic Body in S. Flugge (ed.), "Elasticity and Plasticity," Springer-Verlag, Berlin, 1958.
13. Cox, A. D., G. Eason, and H. G. Hopkins: Axially Symmetric Plastic Deformation in Soils, *Phil. Trans. R. Soc., Ser.* A, vol. 254, pp. 9–46, 1962.

NAME INDEX

NAME INDEX

A

B

C

I

J

K

N

O

P

T

Y

Z

SUBJECT INDEX

SUBJECT INDEX

D

E

F

M

N

O

P

T

U

V